AF721018

ASTRONOMY AND ASTROPHYSICS LIBRARY

Series Editors: Martin A. Barstow, University of Leicester, Leicester, UK
Andreas Burkert, University Observatory Munich, Munich, Germany
Athena Coustenis, Paris-Meudon Observatory, Meudon, France
Michael A. Dopita, Mount Stromlo Observatory, Weston Creek, Australia
Roberto Gilmozzi, European Southern Observatory (ESO), Garching, Germany
Georges Meynet, Geneva Observatory, Versoix, Switzerland
Shin Mineshige, Department of Astronomy, Kyoto University, Sakyo-ku, Kyoto, Japan
Ian Robson, The UK Astronomy Technology Centre, Edinburgh, UK
Peter Schneider, Argelander-Institut für Astronomie, Bonn, Germany
Virginia Trimble, University of California, Irvine, USA
Derek Ward-Thompson, University of Central Lancashire, Preston, UK

More information about this series at http://www.springer.com/series/848

Maurizio Spurio

Probes of Multimessenger Astrophysics

Charged Cosmic Rays, Neutrinos, γ-Rays and Gravitational Waves

Second Edition

Springer

Maurizio Spurio
Department of Physics and Astronomy,
and INFN
University of Bologna
Bologna, Italy

Additional material to this book can be downloaded from http://extras.springer.com.

ISSN 0941-7834 ISSN 2196-9698 (electronic)
Astronomy and Astrophysics Library
ISBN 978-3-319-96853-7 ISBN 978-3-319-96854-4 (eBook)
https://doi.org/10.1007/978-3-319-96854-4

Library of Congress Control Number: 2018955970

© Springer Nature Switzerland AG 2015, 2018
This work is subject to copyright. All rights are reserved by the Publisher, whether the whole or part of the material is concerned, specifically the rights of translation, reprinting, reuse of illustrations, recitation, broadcasting, reproduction on microfilms or in any other physical way, and transmission or information storage and retrieval, electronic adaptation, computer software, or by similar or dissimilar methodology now known or hereafter developed.
The use of general descriptive names, registered names, trademarks, service marks, etc. in this publication does not imply, even in the absence of a specific statement, that such names are exempt from the relevant protective laws and regulations and therefore free for general use.
The publisher, the authors and the editors are safe to assume that the advice and information in this book are believed to be true and accurate at the date of publication. Neither the publisher nor the authors or the editors give a warranty, express or implied, with respect to the material contained herein or for any errors or omissions that may have been made. The publisher remains neutral with regard to jurisdictional claims in published maps and institutional affiliations.

This Springer imprint is published by the registered company Springer Nature Switzerland AG
The registered company address is: Gewerbestrasse 11, 6330 Cham, Switzerland

A Mariagrazia,
Eleonora e Federico

Du siehst, mein Sohn,
zum Raum wird hier die Zeit.

Non ho delle pretese,
il merito l'è tutto
della scuola bolognese!

Preface to the Second Edition

A real breakthrough in science occurred in 2016: the first observation of gravitational waves. High-energy astrophysics is now studied using different probes under at least three large experimental branches: the electromagnetic radiation, from radio to X-rays, with techniques developed by traditional astronomy; charged cosmic rays, γ-rays, and neutrinos, with experimental methods developed in high-energy laboratories; and gravitational waves (GWs), observed using laser interferometers that reached their maturity after a long experimental path.

The combined study of the Universe with all the aforementioned experimental probes offers incredible opportunities, as already demonstrated by the observation of the merging of two neutron stars on August 17, 2017. The prompt finding of the GW and of the γ-ray burst was followed by the most extensive worldwide observational campaign using 70 observatories on all continents and in space. The efforts produced results (extensively described in this book) that also led to a paper coauthored by almost 4000 physicists from more than 900 institutions.

The discovery of GWs pushed me to revise the first edition of this book, to include the impact of the discovery of GWs on the study of the Universe. The general layout of the book is unchanged, but I tried to offer a more global and *multimessenger* vision of the different probes used to study astrophysical phenomena.

The basics of GWs have been included in the completely new Chap. 13; here, the key features of the events observed with GWs are studied in terms of introductory physics and compared with the results presented in the discovery papers. I particularly stressed the detections' impact on astrophysics. Moreover, dedicated sections have been introduced in Chaps. 1, 6, 8, and 12 to include the consequences of GW observations on the understanding of the physics of compact objects (black holes and neutron stars), the short GRBs and the kilonovae, the jets in astrophysics, and the origin of the heavy elements in the Universe.

I also took the opportunity of a new edition to include:

- The new AMS-02 results (Chaps. 3 and 5) on direct measurement of cosmic rays (CRs);
- The ARGO-YBJ observations on the chemical composition of CRs at the knee region (Chap. 4);
- The evidence of extragalactic CRs above 8×10^{18} eV from the Pierre Auger Observatory and the updated results (together with the Telescope Array) on UHECRs (Chaps. 5 and 7);
- A completely revised description of the leptonic model for the production of γ-rays (Chap. 8);
- The updated catalogues from the Fermi-LAT satellite (Chap. 8);
- The description of the new HAWC telescope and the update of the TeV γ-ray catalogue (Chap. 9);
- A complete revision of the results on astrophysical neutrinos observed by IceCube, on cosmogenic neutrinos, and on new projects (Chap. 10);
- The new results on the measurement of solar neutrinos by the Borexino experiment (Chap. 12);
- An update on the searches for dark matter (Chap. 14).

The only negative remark to the first edition I received from a couple of colleagues is about the fact that SI units are not always used; in particular, in most electromagnetic formulas, the Gaussian unit system is used. Sorry, this is not a book for engineers; physicist not familiar with the (beautiful!) Gauss system can have the occasion to learn and use it. A dedicated section has been introduced at the end of Chap. 2.

Finally, I would really like to thank the many colleagues and students who gave such positive feedback on the first edition of the book. In some cases, they also alerted me to small imprecisions, mistakes, and misprints, which I have now taken the opportunity to correct or remove. In particular, I really appreciate the overall revision and comments by James Matthews (Louisiana State University) and Jim Linnemann (Michigan State University).

I hope readers will enjoy this new edition.

Bologna, Italy
June 2018

Maurizio Spurio

Preface to the First Edition

This book is designed as an introduction to the many different aspects that correlate particle physics with astrophysics and cosmology, as well as introducing astrophysics by means of the many experimental results recently obtained through the study of charged and neutral high-energy particles (including GeV and TeV photons).

The Standard Model of particle physics, which includes the theory of electroweak interaction and quantum chromodynamics for the strong interaction, explains all available experimental results quite well. The theory was recently crowned by the discovery of the last missing piece: the Higgs boson. A parallel Standard Model has been derived from observational cosmology, which describes the evolution of the Universe as a whole, as well as the objects within it.

"Multimessenger" astrophysics, connecting traditional astronomy with cosmic ray, γ-ray, and neutrino observations, has been made possible by the availability of experimental techniques and detectors used in particle physics. These developments have allowed for the construction and operation of experiments in space borrowing from the techniques used for accelerators. This has made possible, with space-based experiments, the study of cosmic matter and antimatter and the detection of high-energy γ-rays with good pointing capabilities. Underground laboratories, created to test particle physics beyond the Standard Model, offer an ideal low-background environment for detecting neutrinos produced by the nucleosynthesis in the Sun, or by the gravitational core collapse of a massive star. Deep underwater/ice neutrino telescopes have started to provide information on cosmic accelerators, while at the same time allowing tests of several physical properties of neutrinos.

This book is aimed primarily at those undertaking postgraduate courses, at PhD students, and at post doc researchers involved in high-energy physics or astrophysics research. It is also aimed at senior particle physicists usually interested in accelerator/collider physics, eager to understand and appreciate the mechanisms of the largest accelerators in the Universe. The reader is assumed to know, at an introductory level, particle and nuclear physics. Additional material (referred to as *"Extras"*) is freely available on the Springer website for this book.

The book is based on my lecture experiences at the University of Bologna on *Astroparticle Physics*. I have adopted here a systematic approach to cover the experimental aspects, as well as to introduce the theoretical background. In particular, I dedicated significant effort for first-order estimates of all the relevant processes described, referring to more advanced readings for deeper developments. The interpretations of experimental results of modern experiments rely, in most cases, on data comparison with model predictions obtained via Monte Carlo methods. These computational techniques are usually very detailed. It is very important that researchers acquire the habit of critically understanding the physical results of advanced simulation tools, even with simple back-of-the-envelope estimates. In this optic, I devoted great care to the measurement units of all relevant quantities, to develop toy-models for deriving observable quantities from intuitive arguments, and to compare the results of these simple predictions with published data.

Two words of warning about the literature: I usually quoted review papers. They are often easier to read and more helpful in introducing the reader to the inherent physics than specific research papers. Reviews, in turns, contain a detailed indication for further readings (on average, more than 200 citations for each review paper). For the permission to reproduce or adapt photographs and diagrams, I am very grateful to all of the authors, collaborations, institutions, and laboratories quoted in each figure caption. I made every effort to secure necessary permissions to reproduce the copyright material in this work. If any omissions are brought to my notice, I will be happy to include appropriate acknowledgments on reprinting.

I thank many colleagues for their cooperation and suggestions: the friends from the ANTARES and OPERA groups at Bologna University and from the former MACRO experiment, and in particular S. Cecchini, A. Margiotta, and Prof. G. Giacomelli (who recently passed away), and the colleagues from the Istituto Nazionale di Fisica Nucleare, especially the representatives of the National Committee 2—who are mainly involved in astroparticle physics—for the instructive discussions sometimes used in the content of this book. Particular mention must be made of the PhD students, those in our group and those with whom I interacted during the last 15 years in my research activity in regard to a neutrino telescope in the Mediterranean Sea. Special thanks to all members of the ANTARES collaboration. I am grateful to the many students who read the manuscript under the form of didactic material, for their suggestions and questions that have allowed me to form this work in a way that I hope will be useful for many.

Finally, I have to deeply and sincerely thank Prof. Vincenzo Flaminio, University of Pisa, who read the manuscript critically and in depth. He has been a severe, knowledgeable, affable reviewer, whose opinion served as an important means of support for the author and has contributed to the improvement of the final version.

I would also very much appreciate any corrections to mistakes or misprints (including the trivial ones), as well as comments or simple observations. Please address them to: maurizio.spurio@unibo.it.

Bologna, Italy
July 2014

Maurizio Spurio

Contents

Chapter 1
An Overview of Multimessenger Astrophysics

Abstract The Standard Model of the microcosm, which includes the theory of electroweak interaction and quantum chromodynamics for strong interaction, explains all available experimental results in particle physics quite well. On the other hand, few physicists believe that the Standard Model is the ultimate theory. Some considerations show that the model is incomplete and represents a sort of low energy limit of a more fundamental theory, which should reveal itself at higher energies. The threshold for this higher energy limit could be so high that no accelerator on Earth, or even in the far future, will be able to reach it. On the other hand, the study of the evolution and history of our Universe, has produced an equivalent Standard Model of the macrocosm. This chapter describes the connections among particle physics, astrophysics, and cosmology and introduces the content of the book: the use of different probes to test the status of our knowledge on the microcosm and macrocosm.

1.1 Introduction

The Standard Model (SM) of particle physics, which includes the theory of electroweak interaction and quantum chromodynamics for strong interaction, explains all available experimental results in particle physics quite well (Braibant et al. 2011, 2012). The SM predictions had precise confirmations from the measurements performed with the LEP and SLAC electron-positron colliders, and with the discovery of the *top* quark at the Tevatron $p\overline{p}$ collider. The theory was recently crowned with the detection by the CERN Large Hadron Collider (LHC) of the last missing piece of the theory: the Higgs boson.

On the other hand, few physicists believe that the SM is the ultimate theory. Some considerations show that the SM is incomplete and represents a sort of *low energy limit* of a more fundamental theory, which should reveal itself at higher energies. These considerations are based upon the following facts:

- the SM has many free parameters that need experimental input (the masses of leptons, quarks, and gauge bosons; the mass of the Higgs boson; the coupling constants ...);

© Springer Nature Switzerland AG 2018
M. Spurio, *Probes of Multimessenger Astrophysics*, Astronomy and Astrophysics Library, https://doi.org/10.1007/978-3-319-96854-4_1

- the three-family structure of leptons and quarks remains unexplained;
- the SM does not contain gravity;
- there are several unresolved "fine-tuning" problems;
- there are several unresolved "aesthetic" problems, such as the fact that the electric charge of the fundamental fermions and bosons is quantized in multiples of $\frac{1}{3}e$, without deeper justification.

The threshold for this *higher energy* limit could be so high that no accelerator on Earth, or even in the far future, will be able to reach it. For instance, Grand Unified Theories (GUTs) of electroweak and strong interactions predict that new physics would appear at extremely high energies, $>10^{14}$ GeV. It is in this context that *astroparticle physics* plays a fundamental role.

There are important connections between astrophysics, particle physics, and cosmology, in particular, in the early Universe, which is commonly described as a gas of very energetic particles. As time proceeded, the Universe expanded, the energy per particle decreased, phase transitions took place, the nature of particles changed, and there was a symmetry breaking from unified to nonunified interactions (Fig. 1.1). In recent years, some indication from the study of the early Universe has pointed out some features that are completely outside the SM, namely:

- a large fraction (about 70%) of the mass-energy is made of an unknown form of *dark energy*;
- a large fraction (about 25%) of the mass-energy is made of an unknown form of *dark matter*;
- the matter–antimatter asymmetry observed in the Universe is not fully justified by the charge-parity violation allowed within the SM.

Grand Unification Theories foresee the nonconservation of the baryon and lepton numbers. Some models from the 1980s predicted proton lifetimes on the order of 10^{30} years. This lifetime is much longer than the age of the Universe, but experimentally measurable. This motivated the construction of large apparatuses aimed at a search for proton decay. The detectors contained more than 1 ton ($\sim 10^{33}$ nucleons) of material, and were located underground to shield experiments from the radiation due to cosmic rays. Only the penetrating component, namely muons and neutrinos, is able to reach the detectors at such depths. Contrary to the optimistic expectation, no proton decays were observed (actually, the measured lifetime turned out to be longer than $\sim 10^{33}$ years), but the background events induced by the *atmospheric neutrinos* were particularly important. The events induced by atmospheric $\nu_e + \overline{\nu}_e$ were roughly in agreement with the expectation. However, the number of events induced by $\nu_\mu + \overline{\nu}_\mu$ was lower than expected. This was attributed to the phenomenon of *neutrino oscillations*, caused by a quantum-mechanical mixing between massive neutrinos. The definitive discovery of a nonzero neutrino mass in 1998 using atmospheric neutrinos was the result of long experimental investigation. This represented the first experimental hint for physics beyond the Standard Model.

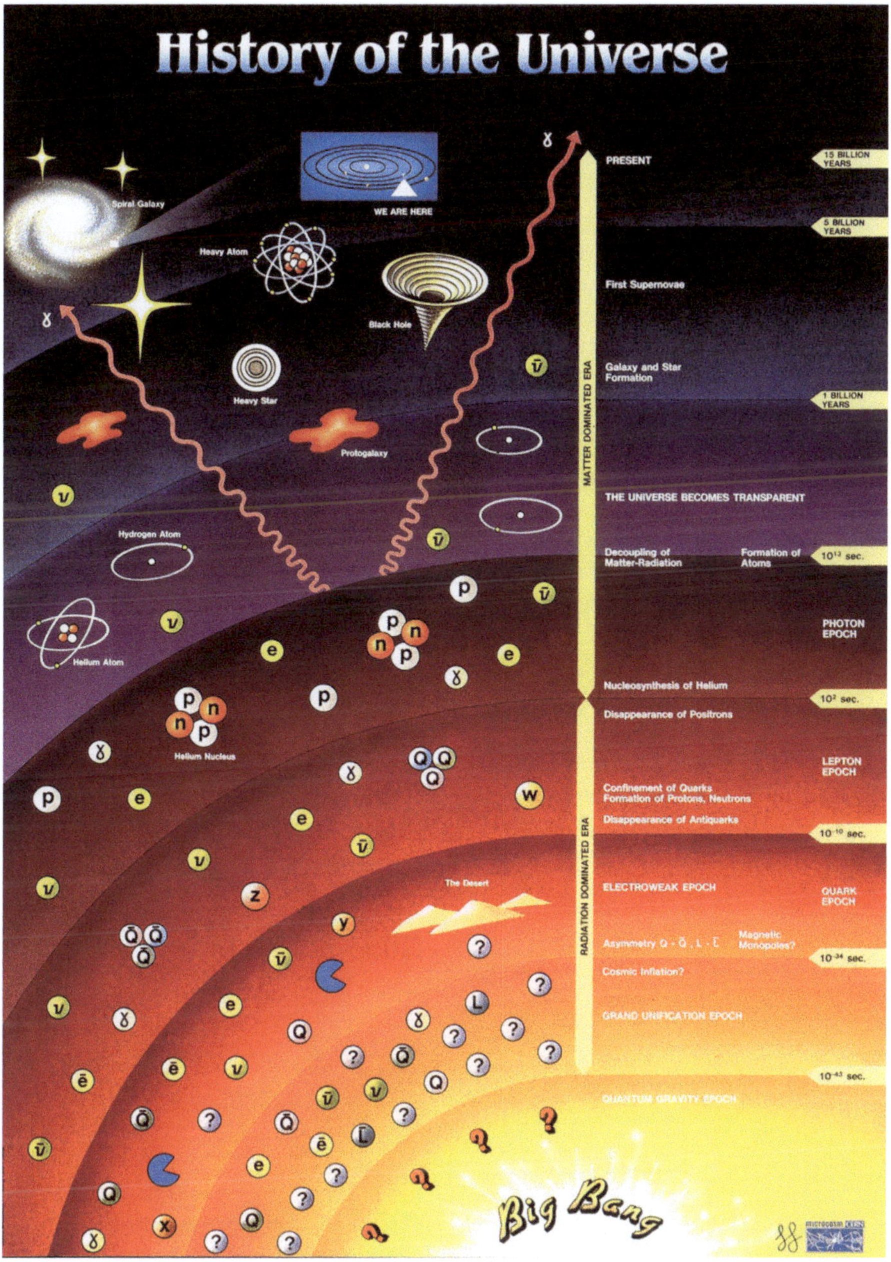

Fig. 1.1 An idealized sketch of the evolution of the Universe: from Big Bang to the present. The drawing was done by Giorgio Giacomelli

1.2 Astrophysics and Astroparticle Physics

Both *astrophysics* and *astroparticle physics* study the nature of cosmic objects, but using different experimental techniques. We currently use the term *astrophysics* when the studied astronomical messenger is electromagnetic radiation (see Fig. 1.2). We use *astroparticle physics* when the cosmic objects are studied with experimental techniques similar to those used in high-energy physics. Astronomical messenger probes observed by astroparticle detectors are charged cosmic rays, neutrinos, and high energy γ-rays. Astroparticle physics searches include those for unknown dark matter, for other relic primordial, or exotic particles, and the study of rare phenomena in a low-radioactivity environment.

The cultural difference between the communities of astronomers and particle physicists studying the Universe was significant up to a few years ago, after which it progressively decreased. In particular, the journals in which astroparticle and astrophysics researches are published were, in the past, disjoint, while at present, there is a tendency towards a larger exchange of information. This book shows many examples as to the way in which consciousness of the necessity of a stricter link has produced relevant results. From the experimental point of view, this is evident, for example, in the increased number of multimessenger observations (with the recent, impressive, example of GW170817, described in the next section) and with the protocols used for sharing real-time information between different experiments, e.g., the GCN (Sect. 8.11) or the AMON (Sect. 10.12) systems. Also, from the theoretical side, the interconnection between astrophysics and astroparticle physics has produced relevant results, such as, for instance, on computer codes for the description of core-collapse supernovae that include neutrino and gravitational waves emission; or on the diffusion of high-energy particles in the interstellar matter and galactic magnetic fields.

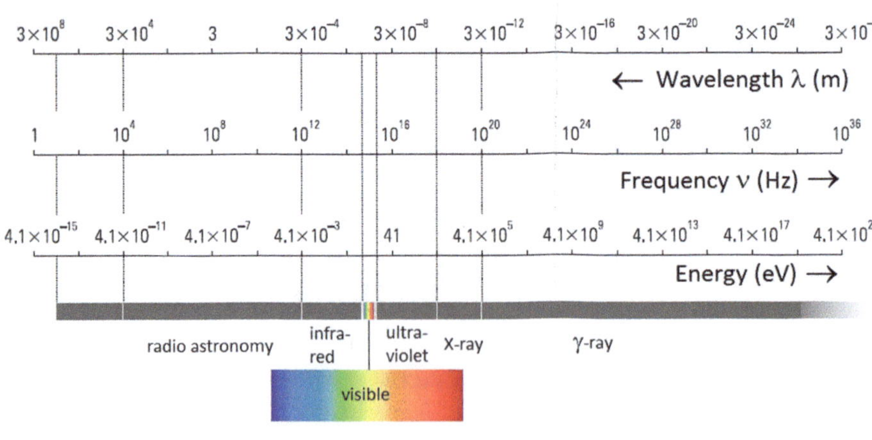

Fig. 1.2 Spectral range of electromagnetic radiation

To test whether your educational background is more similar to that of an *astrophysicist* or an *astroparticle physicist*, have a look at Fig. 1.2. If you are more familiar with the nomenclature of the electromagnetic spectrum through the electron Volt (eV) energy scale, you would appear to be an *astroparticle physicist*. On the other hand, if you prefer to classify electromagnetic radiation according to the wavelength or frequency, you are closer to an *astrophysicist*.

Astroparticle physics dates back to the discovery of cosmic radiation by Hess in 1912. One of the most important open questions is the nature of *cosmic accelerators*. After the discovery of radioactivity by Becquerel in 1896, it soon became evident that part of natural radioactivity does not originate on Earth, but has an extraterrestrial origin. Today, we know that cosmic objects exist that are able to accelerate protons and nuclei, the Cosmic Rays (CRs), up to 10^{20} eV. The bulk of the CRs is believed to be of galactic origin. Supernova remnants are the likely sites where galactic CR acceleration occurs. Models exist that predict that a fraction (up to $\sim$10%) of the kinetic energy released in supernova explosions is used to accelerate to high energy charged particles (protons, heavier nuclei, and electrons). The identification of these galactic engines (or *class* of galactic engines) able to accelerate protons to energy orders of magnitude larger that in the LHC is one of the major aims of astroparticle physics.

At the high-energy tail of the CR spectrum, the number of particles reaching the Earth's atmosphere is very low. At $E \sim 10^{20}$ eV, the flux is only about one particle per square kilometer per century. Experiments covering huge surfaces are needed to collect a reasonable number of events. One of the challenges of these experiments is the measurement, in addition to the primary CR energy, of its nuclear charge/mass. This information is necessary to understand which kind of galactic or extragalactic engine is at work in the acceleration process.

Astronomy (in the sense of the *localization* in the sky of peculiar cosmic objects) with charged particles is inhibited by the presence of irregular interstellar and intergalactic magnetic fields, which randomize the directions of charged CRs. Only particles at very high energies ($>10^{19}$ eV) travel along approximately straight lines through magnetic fields. The possibility of identifying galactic and extragalactic accelerators is open to neutral probes (γ-rays and neutrinos) of all energies and, possibly, to ultra-high-energy protons. From 2016, it has been demonstrated that the network of laser interferometers for the detection of gravitational waves is also able to identify particularly interesting astrophysical events.

The joint efforts of merging information from traditional electromagnetic measurements from radio to X-rays, observations of charged particles, γ-rays and neutrinos with instruments of high-energy physics and the detection of gravitational waves with laser interferometers correspond to the new field of multimessenger astrophysics.

1.3 Multimessenger Astronomy

Up to a couple of decades ago, our knowledge and understanding of the Universe was mainly based on observations of the electromagnetic radiation within a wide range of wavelengths. Historically, astrophysics was born when spectroscopy techniques were applied to optical observations of light from stars. As experimental techniques improved and new detectors were developed during the last century, wavelengths different than the visible radiation were used to improve our knowledge of astrophysical objects. For instance, the opening of the radio window after the Second World War made the 1960s a golden decade for astronomy, with the discovery of the cosmic microwave background, pulsars, and quasars.

Most electromagnetic radiation is blocked by the Earth's atmosphere, and therefore space observatories or some form of indirect detection is required (Fig. 1.3). Thus, another major advance came in the 1960s–1970s with the opening of the X-ray window, with the advent of rocket technology able to carry X-ray detectors above the atmosphere. The combined information from different instruments (usually, from radio to X-rays or soft γ-rays) covering a wide part of the electromagnetic spectrum is denoted as *multiwavelength astronomy*.

Starting in the 1980s, new kinds of detectors were developed, exploiting other forms of cosmic probes: individual photons with energy above the GeV, charged

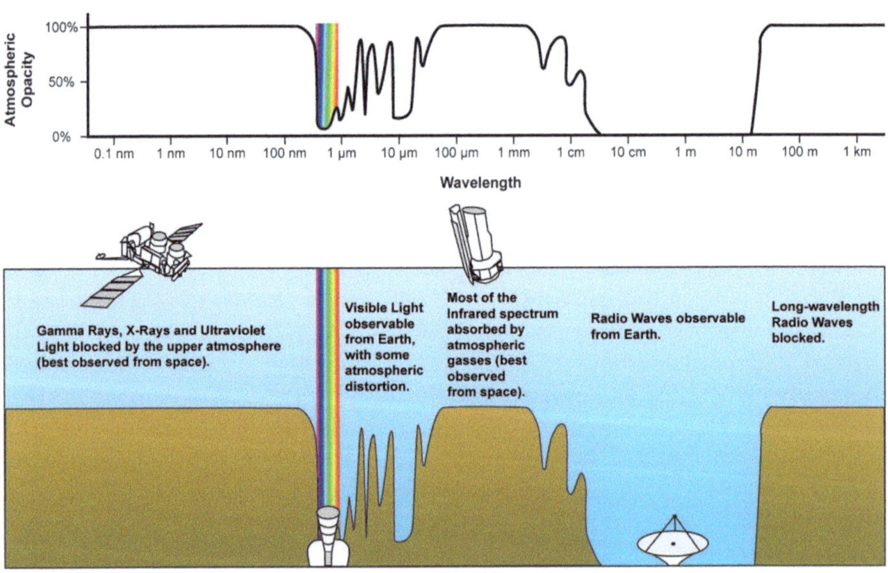

Fig. 1.3 The atmosphere opacity as a function of the wavelength is presented in the *upper part*. Opacity is represented by the percentage of electromagnetic radiation, which does not reach the ground. Space experiments are widely used to detect electromagnetic radiation that does not reach the Earth's surface. Note that the scale is in terms of the logarithm of the wavelength, so the energy scale decreases from left to right. Credit: NASA

cosmic rays, neutrinos and, finally, gravitational waves. These new "telescopes" led to unexpected breakthroughs. The measurement of neutrinos from the Sun and the detection in 1987 of a neutrino burst from a stellar gravitational collapse that occurred in the Large Magellanic Cloud opened the field of *neutrino astrophysics*, providing the first examples of multimessenger observations of astrophysical phenomena.

Multimessenger astronomy connects different kinds of observations of the same astrophysical event or system. The experimental opportunity to relate traditional astrophysics observations with the CR experiments (both in space and on the ground), γ-ray telescopes, neutrino detectors and gravitational wave observatories has been made possible by the availability of experimental techniques and detectors, mainly developed in high-energy particle physics.

As of this writing, the most impressive episode of the multimessenger astronomy saga started on August 17, 2017. At 12:41:04 Universal Time, the LIGO-Virgo detector network observed a gravitational wave (GW) signal from the inspiral of two low-mass compact objects consistent with a binary neutron star merger: the event was denoted as GW170817. Independently, a γ-ray burst (GRB) was observed less than 2 s later by the Gamma-ray Burst Monitor on board the Fermi satellite, and by the INTEGRAL satellite: the event was denoted as GRB170817A. This joint GW/GRB detection was followed by the most extensive worldwide observational campaign ever performed up to that point, with the use of space- and ground-based telescopes, to scan the region of the sky where the event was detected. Neutrino detectors also searched for a neutrino counterpart of the signal. Less than 12 h after the GW/GRB event (without the Sun on the signal region), a new point-like optical source was reported by different optical telescopes. The source was located in the galaxy NGC 4993 at a distance of 40 Mpc from Earth, consistent with the luminosity distance estimated by the Ligo/VIRGO detectors using the GW signal. The official designation of the optical counterpart in the International Astronomical Union (IAU) was AT 2017gfo. The source was intensively studied in the following weeks by all traditional astronomical instruments from radio to X-rays. These electromagnetic observations revealed signatures of recently synthesized material, including gold and platinum, solving a decades-long mystery as to where roughly half of all elements heavier than iron in the Periodic Table of elements are produced. As a summary, the interest and effort have been global: a large number of papers on different observations was published in the same issue of *The Astrophysical Journal Letters* (Vol. 848, n. 2) on October 20, 2017. This includes one paper describing the multimessenger observations (LIGO Scientific Collaboration et al. 2017), which was co-authored by almost 4000 physicists from more than 900 institutions, using 70 observatories on all continents and in space.

Figure 1.4 summarizes this multimessenger campaign. Two types of information are shown for each band/messenger. First, the shaded dashes represent the times when information was reported in a circular of the Gamma-ray Coordinates Network (GCN) https://gcn.gsfc.nasa.gov/. The GCN system distributes the locations of GRBs and other transients detected by spacecraft and ground experiments. The names of the relevant instruments, facilities, or observing teams are collected at

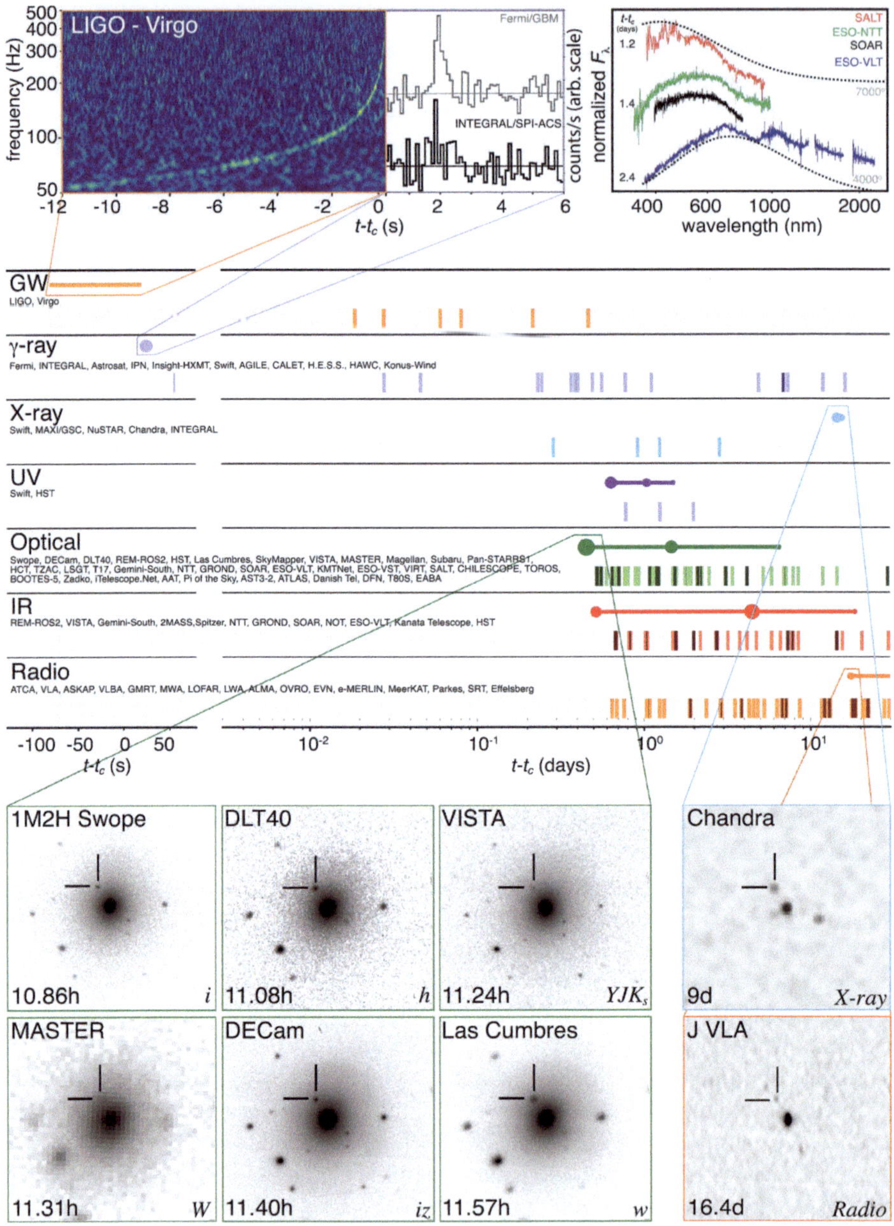

Fig. 1.4 One of the symbols of multimessenger astrophysics (LIGO Scientific Collaboration et al. 2017): the timeline of the discovery of GW170817, GRB 170817A, and AT 2017gfo. All observations are shown by messenger and wavelength relative to the time t_c of the gravitational-wave event. See text for details

the beginning of the row. Second, representative observations in each band are shown as solid circles; the solid lines indicate when the source was detectable by at least one telescope. Magnification insets give a picture of the first detections in the gravitational wave, γ-ray, optical, X-ray, and radio bands. More detail in Chap. 13.

The above example shows that observation in different electromagnetic bands and/or with different probes may not just be an advantage; it may be a necessity for solving some of the outstanding problems in particle physics and astrophysics:

- Can we definitively prove that the galactic CR acceleration mechanism is provided by supernovae shock waves?
- Are there galactic objects able to accelerate CRs up to $10^{18–19}$ eV?
- Are the Active Galactic Nuclei, the Gamma Ray Bursts, or some other still unknown mechanism the sources of CRs at $\sim 10^{20}$ eV?
- What is the mechanism(s) that triggers (trigger) the GRBs?
- Is the antimatter detected outside the Earth's atmosphere produced by secondary processes only? Is there antimatter of primary origin, or produced by non-Standard Model mechanisms?
- Are there fossil or primordial particles in cosmic radiation?
- Can we directly or indirectly detect dark matter and understand its nature?
- Is the proton really stable?
- What can we possibly learn from neutrino astronomy?
- Which is the astrophysical use of new information on the Universe provided by the observation of gravitational waves?

1.4 Experimental Results Not Covered in This Book

Since ancient ages, visible light was the only means to explore the Universe. Only in the twentieth century were we able to open new observational windows on other regions of electromagnetic radiation. All experiments covering the electromagnetic spectrum up to soft γ-rays (below a few tens of MeV), the relative discoveries and models are only marginally referred to in this book. Dedicated textbooks on these subjects are recommended (Courvoisier 2013; Lang 2013; Longair 2011; Rai Choudhuri 2012).

The launch of the first artificial satellite in 1957 opened new opportunities for astrophysics studies. The observation of X-ray sources and gamma radiation was made possible a few years later through dedicated instruments on board satellites. For X-rays, we should mention Uhuru (1970), the Einstein Observatory (1978), and Exosat (1983). The successful launch of the ROSAT, in 1990, paved the way for the discovery of numerous X-ray sources. More recent satellites include the XMM-Newton observatory (low to mid-energy X-rays 0.1–15 keV) and the INTEGRAL satellite (high energy X-rays 15–60 keV). Both were launched by the European Space Agency (ESA). The National Aeronautics and Space Administration (NASA) has launched the RXTE and the Chandra observatories.

Within the optical range, the Hubble space telescope (which was launched by NASA in 1990) provided optical images of stars and galaxies of unprecedented quality, the figure error of the primary mirror having been eliminated in 1993 by a spectacular repair in space.

Interest in the study of astrophysical sources through detection of their γ-ray emission started at the end of the 1950s. We should mention the following satellites: Vela (1967), OSO-3 (1969), SAS-2 (1972), COS-B (1975). At the beginning of the 1970s, the first firm detections of γ-rays from space came with the OSO-3 (Kraushaar et al. 1973) observation of radiation from the plane of our Galaxy. Real breakthrough event for γ-ray astronomy was the launch in 1991 of the Compton Gamma-Ray Observatory (CGRO), which was designed to incorporate the major advances in detector technology during the 1980s. The satellite carried four major instruments (BATSE, OSSE, COMPTEL, EGRET), which greatly improved the spatial and temporal resolution of γ-ray observations. EGRET was the first detector for tens of MeV γ-rays, opening the field of observation for very high energy γ-rays, which is a subject covered in this book.

All new observation windows provided a wealth of new data and new discoveries inaccessible within other spectral ranges, which widely improved our understanding of the high-energy processes in our Universe. See also Grupen (2005), Mészáros (2010), and Poggiani (2017).

1.5 Cosmic Rays

The nature of cosmic accelerators is one of the major unsolved astrophysical problems. It is an important subject, as the sites of cosmic ray acceleration in our Galaxy are strictly correlated with the dynamics of formation of massive stars, stellar evolution, and finally, supernova explosions. In some sense, the study of cosmic radiation helps to understand the formation of our own solar system.

The physics of cosmic rays is dealt with in some detail in this book. Chapter 2 introduces the physics of primary cosmic rays. *Primary* CRs are high-energy protons and nuclei (plus a minority electron component) produced in astrophysical environments, filling the galactic space and arriving on Earth. Primary CRs can be identified if measured before they interact with the atmosphere. *Secondary* CRs are those particles produced in interactions of the primaries with interstellar gas or with nuclei in the Earth's atmosphere. Nuclei such as lithium, beryllium, and boron (which are very rare end-products of stellar nucleosynthesis) are secondary particles, as well as antiprotons and positrons.

Chapter 3 deals with the experimental techniques used for the measurement of primary CRs below $\sim 10^{14}$ eV (*direct measurements*). Experiments mounted on balloons, satellites and on the International Space Station allow for a *direct* study of the CRs impinging on the top of the atmosphere, before their first interaction with Earth matter. This enables measurement of the so-called *chemical composition of cosmic rays*, which is the relative fraction of different nuclei present in cosmic radiation, and in some cases, of their isotopic composition.

When entering the Earth's atmosphere, CRs collide with nucleons of atmospheric nuclei (mainly oxygen and nitrogen) and produce a cascade of secondary particles, the so-called *air shower* (Gaisser 1991). The basic mechanism of air shower production of a CR proton on a nucleon N is the reaction

$$p + N \rightarrow \pi^{\pm}, \pi^{0}, K^{\pm}, K^{0}, p, n, \ldots \tag{1.1}$$

The decay of short-lived hadrons leads to a cascade of high-energy photons, electrons, and positrons (the electromagnetic component) and a penetrating component of muons and neutrinos. The measurement of the secondary particles with ground-based detectors (*indirect measurements*) allows for knowledge of the CR flux up to the highest energies, as described in Chap. 4.

Figure 1.5 shows the flux of primary CRs as measured by some *direct* and *indirect* experiments. Each experiment measures the CR flux within a given energy range. Different kinds of experiment cover more than 10 decades of energy. In the lower energy part of this spectrum, the Sun contributes with low energy cosmic rays (below the GeV scale), mainly associated with solar flares. In this book, we do not consider charged particles of solar origin. We focus on the origin, propagation, and detection of CRs with energies above several GeV, which are originated outside the solar system. On the opposite region of the CR spectrum, the highest-energy cosmic rays ($\sim 10^{20}$ eV) correspond to 16 J, i.e., to a mass of 1 kg moving at a speed of almost 6 ms^{-1}! Their origin is likely to be associated with the most energetic processes in the Universe.

The CR flux in Fig. 1.5 is shown in a double-logarithmic representation. We encounter this situation in a large number of cases: for the spectra of γ-rays, secondary particles in the atmosphere, atmospheric muons, and neutrinos.... In general, nonthermal particle spectra can usually be approximated by power laws of the type

$$\frac{d\Phi}{dE} = A \cdot E^{-\alpha}. \tag{1.2}$$

In a double-logarithmic representation, these power laws are represented by straight lines:

$$\log\left[\frac{d\Phi}{dE}\right] = \log[A \cdot E^{-\alpha}] = \log[A] - \alpha \cdot \log[E]. \tag{1.3}$$

The spectral index α corresponds to the slope of the line and the normalization A to the intercept with the y-axis. Since many spectra are steep ($d\Phi/dE \sim E^{-2} - E^{-4}$), it is useful to weight the y-axis with a power of the energy, E^{β}. In this way, the straight line has a different slope:

$$\log\left[E^{\beta} \cdot \frac{d\Phi}{dE}\right] = \log[A \cdot E^{-\alpha+\beta}] = \log[A] - (\alpha - \beta) \cdot \log[E]. \tag{1.4}$$

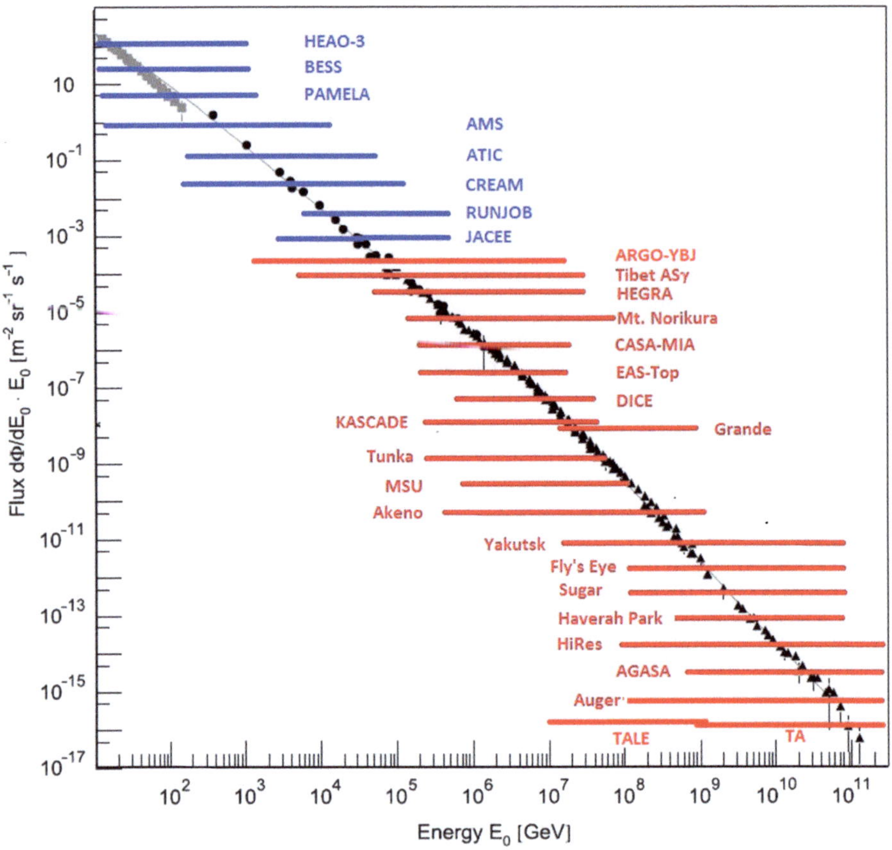

Fig. 1.5 Flux of cosmic rays as a function of energy. Below a few GeV, the contribution of particles coming from the Sun is not negligible. The energy range of the CR flux measured by some direct experiments is shown as a *blue line* and that measured by some indirect experiments as a *red line*

For instance, if $\alpha = 2.7$ (as for the differential spectrum of primary CRs), the choice of $E^\beta = E^{2.7}$ allows us to represent the flux with a flat line parallel to the x-axis. The double-logarithmic representation with a weight is chosen to simplify the reading off of fluxes characterized by a fast decrease in energy.

The propagation of primary CRs in interstellar space, continuously deflected by galactic magnetic fields, is an important piece of information necessary to understand the nature of cosmic accelerators, and it is studied in Chap. 5. During propagation of primary particles in the Galaxy, secondary CRs, as well as antiparticles, are produced by interactions with interstellar matter. Detailed modeling of this diffusion process with the production of secondary particles is of fundamental importance for understanding if anomalous signals eventually detected in the cosmic radiation could be a signal of physics beyond the Standard Model, indicating some evidence for Dark Matter. In the last two decades, excitement was induced by different results successively discarded as evidence of a dark matter signal, as a claimed excess of diffuse γ-rays within the GeV range. At present, the most intriguing deviation from the expectation from secondary CR production is an excess of positrons, detected with different space experiments.

The sources of CRs are still unidentified. Astronomy with charged CRs is prevented by the presence of galactic magnetic fields. Only neutral probes (such as photons and neutrinos) can unambiguously point to a potential source or class of sources. There are, however, different astrophysical candidate CR sources. Starting from the 1960s, models of cosmic acceleration at astrophysical shocks were developed. This led to a sort of standard model of cosmic ray acceleration known as the *diffusive shock acceleration model*, discussed in Chap. 6. In this model, about 5–10% of the nonrelativistic kinetic energy connected with shocks powered by supernovae explosions in our Galaxy is transferred to a few particles that become relativistic, those that we call cosmic rays. The key feature is that an energy power-law spectrum of the type $\sim E^{-2}$ is produced. The model is consistent with the balance between the energy transferred to the accelerated particles and the energy loss due to the escape of CRs out of the Galaxy.

It is plausible that the features in the CR spectrum, starting from the one clearly visible at $\sim 10^{15}$ eV and known as the *knee*, are due to transitions between different classes of galactic CR accelerator and/or propagation effects. Details of acceleration mechanisms and propagation of CRs at high energies are not completely understood. Some peculiar sources should transfer additional energy to already accelerated protons or nuclei through conventional electromagnetic or gravitational processes at the sites of acceleration to justify energies larger than that corresponding to the knee. However, the presence of galactic objects able to accelerate CRs to energies greater than $\sim 10^{19}$ eV seems unlikely.

Huge detector arrays are needed to measure the *Ultra High Energy Cosmic Rays (UHECRs)* with energies exceeding 10^{18} eV (Sokolsky 2004). At the highest energies, CRs are probably of extragalactic origin. The experimental techniques, results and hypotheses in support of an extragalactic origin of these UHECRs are described in Chap. 7. Above $\sim$50 EeV (1 EeV $= 10^{18}$ eV), the arrival directions of electrically charged cosmic rays are no longer significantly affected by galactic magnetic fields. It is still possible that the arrival directions of the highest energy cosmic rays provide information about the location of their sources. Proton astronomy at energies larger than the EeV range would probably be possible and it may also provide indirect information about extragalactic magnetic fields. The determination of the strength of these magnetic fields is a challenging issue for conventional astronomical methods.

1.6 Gamma-Rays of GeV and TeV Energies

High-energy processes, in general, manifest themselves with a *nonthermal spectrum* (i.e., their emission is not that of a black body with a given temperature). The measured diffuse photon flux indicates that the energy content in X- and γ-rays, produced by violent phenomena, is comparable to that associated with low-energy phenomena.

Gamma-ray astronomy at energies above some tens of MeV is an experimental discipline that reached full maturity during the last 10 years, as described in Chap. 8. The Earth's atmosphere is not transparent to high-energy photons (Fig. 1.3) and γ-rays must be detected either outside the atmosphere or exploiting the production of a cascade of secondary particles, following interactions in the atmosphere.

Within the GeV range, the Fermi-LAT experiment (launched in 2008) is made possible by the availability of new detectors coming from technologies typical of experimental particle physics. Fermi-LAT has opened a new window on the observation of the sky. Unlike that in the visible, the γ-ray sky is dominated by a diffuse galactic radiation. The majority of more than 3000 galactic and extragalactic sources present in the Fermi-LAT catalog appear as point-like, i.e., with angular dimensions smaller than the resolution of the detector. These discoveries have a strong impact on several topical areas of modern astrophysics and cosmology.

In addition to steady sources, there are occasionally (with a rate of 1–2 per day) very short bursts of γ-rays. Their duration ranges from a fraction of a second to a few hundred seconds. Spacecrafts equipped with gamma-ray burst detectors used to localize gamma-ray bursts (GRBs) and soft gamma repeaters are forming an Interplanetary Network (IPN) of satellites. One of the progenitors was the BeppoSAX satellite launched in 1996 and deorbited in 2003. It predominantly studied X-ray sources, but also increased significantly our understanding of γ-ray bursts, as well as the successive Swift. At energies above some hundreds of GeV, the γ-ray flux is too low to be detected by satellite detectors. As for charged CRs, the Earth's atmosphere itself can be used as a calorimeter for these very high-energy photons. After a long preparatory phase (Aharonian et al. 2008), the parallel branch of γ-ray astrophysics that uses ground-based detectors started to observe a large number of sources since mid-2000, as studied in Chap. 9.

When γ-rays collide with matter at the top of the atmosphere, they produce showers of particles emitting Cherenkov radiation, which can be detected with Imaging Atmospheric Cherenkov Telescopes (IACTs) on the ground. A milestone in this indirect way to conduct γ-ray astrophysics was reached in 1989 with high-confidence level detection of the Crab Nebula by the Whipple Observatory.

At present, the catalogue of objects emitting within the TeV energy range counts more than 200 galactic and extragalactic objects. Most of them are point-like, but extended emission originating from shells of supernova remnants (SNRs) has also been observed. Part of this emission could be interpreted as originating from hadronic interactions of CRs with energies up to 10^{14} eV in cosmic regions where CRs are accelerated.

Unlike charged CRs, once γ-rays emerge from the production regions, they are largely unaffected in the propagation. However, the Universe is not completely transparent to photons of TeV energy and above. For instance, PeV ($=10^3$ TeV; 1 TeV $= 10^{12}$ eV) energy photons produced at the edge of our own Galaxy cannot be detected, because they will convert into an electron/positron pair when interacting with a Cosmic Microwave Background (CMB) photon during their journey to Earth.

1.7 Neutrino Astrophysics

Neutrinos, due to the low interaction cross-section with matter, are *elusive particles*, but they are abundant in the Universe (Fig. 1.6). Neutrinos are special particles: they are several orders of magnitude lighter than all other fermions. Neutrinos are electrically neutral: they do not feel strong interactions, and interact only weakly. For example, the electron anti-neutrinos produced in nuclear reactors with $\sim$1 MeV energy, have a cross-section of $\sim 10^{-44}\,\mathrm{cm}^2$. This means that only one neutrino out of 10^{11} will interact when traveling along the Earth's diameter.

The fact that the neutrino interaction cross-section is so small is the reason why *neutrino astronomy* is a very recent discipline. In fact, massive and expensive detectors are required to achieve appreciable event rates. On the other hand, a small cross-section is also an extraordinary opportunity, because neutrinos can emerge from deep inside the core of astrophysical objects, directly revealing the physical processes in operation there. The appealing feature of neutrinos with respect to protons and γ-rays, is that they can travel through the Universe without being deflected or absorbed. In addition, different from the highest energy γ-rays, neutrinos of any energy can reach us practically without attenuation from the edge of the Universe.

The idea of a telescope for high-energy neutrinos based on the detection of secondary particles produced by neutrino interactions was first formulated in the 1960s by Markov. The basic principle of a neutrino telescope is a matrix of light detectors in water or ice, which offers a large volume of free targets for neutrino interactions, while at the same time providing shielding against secondary CRs.

The relativistic charged particles produced by neutrino interaction emit Cherenkov radiation in the transparent medium. A detector must measure with high precision the number and arrival time of these photons on a three-dimensional array of Photo Multiplier Tubes (PMTs), from which some of the properties of the neutrino (flavor, direction, energy) can be inferred.

At present, two large detectors exist in the world. In the Southern hemisphere, equipped with 86 detection strings embedded in the Antarctic ice, IceCube has been taking data in its full configuration since 2011. In the Northern hemisphere, in the Mediterranean Sea, ANTARES has proved the detection technique with a 12-string apparatus completed in 2008. The status of neutrino telescopes, their physical goals, and the recent first observations of cosmic neutrinos by IceCube are the subject of Chap. 10. This field is rapidly evolution and the construction of even larger detectors is foreseen: KM3NeT in the Mediterranean Sea, the GVD in Lake Baikal, and IceCube-Gen2 at the South Pole. These detectors will allow for studying the high-energy neutrino sky in much more detail than the present arrays permit. Neutrinos at energies exceeding 100 PeV could be detected by other experimental methods, for example, by large air shower arrays or radio-acoustic techniques.

Atmospheric neutrinos represent the irreducible background for neutrino telescopes. The study of atmospheric neutrinos started in the 1980s, with the advent of large underground laboratories for the study of phenomena predicted by Grand

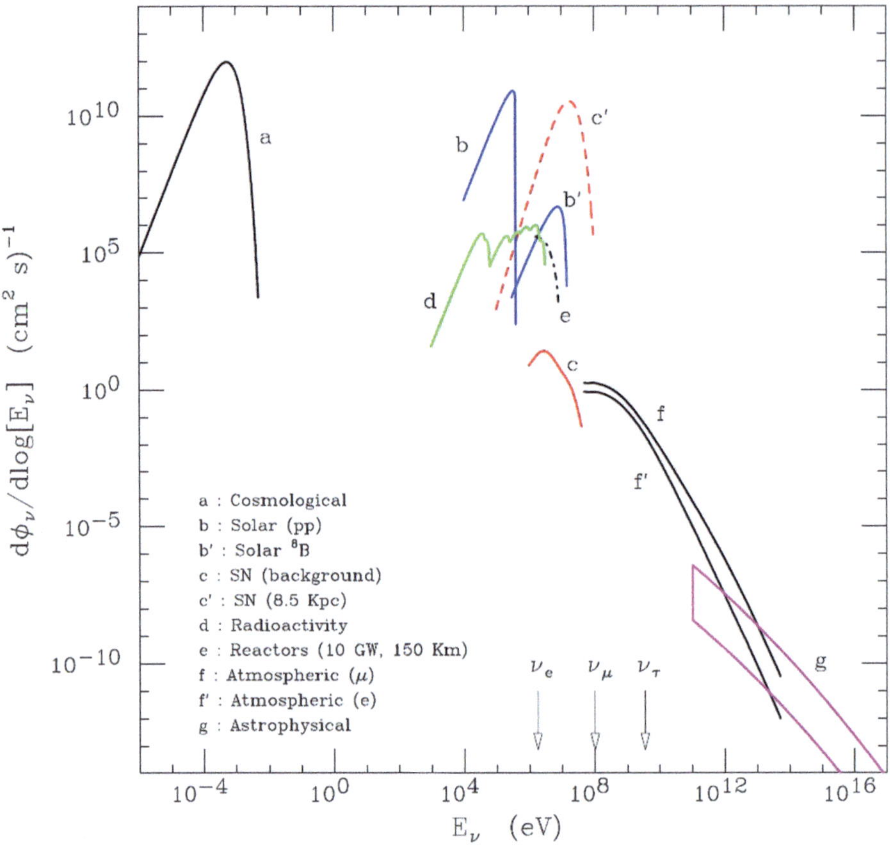

Fig. 1.6 Flux of neutrinos on the surface of the Earth. The three *arrows* near the x-axis indicate the energy thresholds for producing the corresponding charged lepton through charged current interactions on a free proton target. The line **a** that refers to cosmological neutrinos assumes that the neutrino mass is vanishing small. For solar neutrinos **b**, only ν_e from the *pp* and ^{8}B reactions are considered (see Sect. 12.2). The *lines* that refer to diffuse supernova neutrinos **c** and a supernova burst lasting 10 s (**c′**) correspond to $\overline{\nu}_e$ (Sect. 12.11). The other neutrino species have similar spectra, with differences that cannot be appreciated in the figure. The line **d** that describes geophysical neutrinos from radioactive decays includes the ^{238}U and ^{232}Th decay chains, whose flux depends weakly on the geographical location. The $\overline{\nu}_e$ produced by a 10 GW power reactor located at a distance of 150 km are considered **e**. The atmospheric muon and electron neutrino fluxes (**f**, **f′**, respectively) are calculated for the Kamioka (Japan) location. Only the lowest energy part depends on the location. A range of predictions for the flux of astrophysical neutrinos **g** is shown. Courtesy of Lipari (2006)

Unification Theories. This study led to the discovery of *neutrino oscillations*. All neutrino flavors undergo the mechanism of neutrino oscillations, and ν_μ disappearance was reported using atmospheric neutrinos in 1998 by the Super-Kamiokande and MACRO collaborations, using completely different experimental techniques (Chap. 11). The discovery that neutrinos are massive particles is the greatest contribution of astroparticle physics to particle physics since the discovery of the positron and of the muon in atmospheric showers.

In the cascade initiated by protons and nuclei, *atmospheric muons* represent a necessary counterpart to atmospheric muon neutrinos. Most long-lived charged mesons (such as pions and kaons) decay through weak interaction, preferentially into a μ and ν_μ pair. Relatively few high-energy muons are present in the cascade induced by a primary CR; these muons can penetrate more than 10 km of water and are measured by underground or underwater/ice detectors. Atmospheric muons represent a background for all measurements, which need a very low background, usually located in underground laboratories.

Among experiments that need a low-background rate, there are those devoted to neutrinos emitted by the Sun and by gravitational core-collapse bursts. The role of neutrinos for the comprehension of energy mechanisms that fuel the Sun, the main sequence stars and the core-collapse of massive stars is the subject of Chap. 12.

The Sun (like all main sequence stars) produces energy by fusion (Bachall 1989). The first reaction in the cycle occurs through weak interaction, that is, $p + p \rightarrow {}^2_1H + e^+ + \nu_e$; the Sun and the stars are sources of electron neutrinos. Approximately 6×10^{10} solar ν_e cross a surface of one square centimeter on Earth every second. The measurement of solar neutrinos represents the only experimental method for a detailed understanding of nuclear processes inside the Sun, a representative of the main sequence stars. Solar ν_e's come from nuclear fusion reactions in the inner core and directly tell us about the source of energy in the Sun. Photons in the ~MeV energy range are also emitted by the same nuclear reactions. It takes them at least 10^5 years to reach the solar photosphere, where they are emitted from the surface with a black body spectrum.

A major experimental milestone in astroparticle physics was the starting, in 1967, of the measurement of solar neutrinos by a radiochemical experiment in the Homestake Mine (the Davis experiment). This experiment, after a few years of data taking, indicated a deficit in the flux of solar neutrinos, which was confirmed by successive experiments, GALLEX/GNO, SAGE, Kamiokande, and later Super-Kamiokande. After a long experimental struggle, we learned that the neutrino deficit measured by the above experiments was not due to experimental problems or a deficit in the understanding of the astrophysical properties of stellar objects. It was due to oscillations of electron neutrinos during their travel from the Sun to the Earth.

The Sudbury Neutrino Observatory (SNO), which ended its data taking in 2006, was finally able to firmly confirm the oscillation scenario of solar neutrinos using heavy water. SNO showed that the total neutrino flux (the flavor-independent sum of the ν_e, ν_μ, ν_τ) from the Sun as measured on Earth was consistent with the total number of expected neutrinos predicted by the so-called *Standard Solar Model*. SNO demonstrated that 2/3 of the solar electron neutrinos had oscillated into neutrinos with different flavor (ν_μ or ν_τ), whose energy is below the threshold for charged current interactions (see the arrow positions in Fig. 1.6). Finally, the Borexino experiment running at the Gran Sasso Laboratory in Italy is the world's most radio-pure liquid scintillator calorimeter. Its primary aim is to make a precise measurement of the individual neutrino fluxes from the Sun and compare them to the standard solar model predictions. Specific goals of the experiment are to detect ^{7}Be, ^{8}B, *pp*, *pep* and CNO solar neutrinos. This will allow us to test and to further understand the functioning of the Sun (e.g., nuclear fusion processes taking place at its core, solar composition, opacities, matter distribution, etc.) when the properties of neutrino oscillations are taken into account.

An enormous number of neutrinos (some 10^{58}) are emitted in a few seconds following the gravitational collapse of a massive star that triggers a *supernova explosion*. The closest visible supernova over the last three centuries (the SN1987A) occurred in the Large Magellanic Cloud about 50 kpc from Earth. It was observed in 1987, when some sufficiently large neutrino detectors were in operation. They were able to observe about 25 neutrino interactions, which allowed for a first-order experimental confirmation of theoretical models on the supernova explosion mechanism. At present, new and larger detectors are in operation; the occurrence of a new galactic supernova (a few per century are expected) will open a new era for neutrino astrophysics and for the understanding of stellar gravitational collapse.

Finally, it should be mentioned that, similar to the cosmological microwave background radiation, we are immersed in a bath of cosmological neutrinos produced when the Universe was hot and young. The average number density ($\sim$340 cm^{-3}) of cosmological neutrinos is slightly smaller than that of cosmic microwave background photons. As they are of extremely low energy (meV, see Fig. 1.6), there is, so far, no reliable technique for their detection.

1.8 Gravitational Waves

Gravitation is not presently included in the Standard Model of particle physics, nor in advanced extensions, such as those foreseen by Grand Unification Theories. However, there is an aspect of gravitation that is strongly connected to particle physics, astrophysics and cosmology: gravitational waves. Gravitational waves are travelling ripples in space-time, generated when heavy cosmic objects accelerate. These distortions, described as waves, move outward from the source at the speed of light. They were predicted based on Einstein's general theory of relativity. Gravitational waves transport energy in a form of radiant energy similar to electromagnetic radiation. In contrast to the incoherent superposition of emissions from the acceleration of individual electric charges, gravitational waves result from coherent, bulk motions of matter. Because they transfer very small amounts of energy to matter, gravitational waves are able to penetrate the very densely concentrated matter that produces them.

The years 2016 and 2017 saw the dawn of astrophysics and cosmology with gravitational waves, awarded with the 2017 Nobel Prize in Physics. This first direct observation of a gravitational wave is a milestone, and not only for providing a means to investigate general relativity in a previously inaccessible regime. In fact, GWs allow for exploring the distant non-thermal Universe in a way completely independent of electromagnetic radiation. The events of September 14th, 2015 (denoted as GW150914, the first black hole-black hole merger) and August 17th, 2017 (GW170817, the coalescence of two neutron stars, producing a short gamma-ray burst and follow-up observed by 70 observatories on all continents and in space) represent true milestones in science. Chapter 13 shows the features of the observed events using concepts from Newtonian physics, dimensional analysis and analogies with electromagnetic radiation. The methods, the principal results and the perspectives from the multimessenger campaign are highlighted.

Gravitational waves are the last probes for multimessenger studies, after charged cosmic rays, γ-ray and neutrinos. The present network of laser interferometers used for the detection of gravitational waves will help researchers to determine the locations of sources in the sky and trigger "traditional" astronomical observations and neutrino telescopes for the study of high-energy processes in the Universe. Examples of the new insights in physics, astrophysics and cosmology that can be faced with the help of the new instruments for the detection of gravitational waves are:

- direct observation that gravitational waves emitted by different systems carry energy as expected from theoretical models;
- tests of general relativity, including under extreme strong-field conditions;
- direct observation of black holes in binary systems, including the measurement of their masses and a test of the fundamental *no-hair* theorem;
- information on the equation of state and other properties of neutron stars;
- measurement of the Hubble constant and the definition of a new "astronomy distance ladder" with a completely different technique;
- confirmation of the origin of short gamma-ray bursts by coalescence of neutron stars;
- study of the models for the accretion disks and jets;
- insights into the earliest stages of the evolution of the Universe through primordial gravitational waves;
- studies of galactic merging through the observation of coalescing massive black holes at their centers.

1.9 The Dark Universe

Another still unsolved question in astroparticle physics is the nature of *dark matter*. From the observation of orbital velocities of stars in nearby galaxies, and of the velocities of galaxies in galactic clusters, the conclusion emerges that the energy density of visible matter in the Universe is too low to correctly describe the dynamics. A common effort between astroparticle and particle physics concerns searches for *dark matter* candidates, as described in Chap. 14. Exotic, currently unknown particles or additional massive neutrino species may contribute to the missing matter of the Universe.

The framework became even more complicated after the discovery, starting in 1998, of the accelerating expansion of the Universe through observations of distant supernovae. The growing rate of expansion implies that the Universe is being pushed apart by an unknown form of energy embedded in empty space. This unknown *dark energy* (or vacuum energy density) makes up a large part (more than 70%) of the energy density of the Universe and probably plays a decisive function in its dynamics and evolution. The dark energy is an enigma, perhaps the greatest in cosmology today, which we do not consider in this book.

Dark matter candidates outside particle accelerators can be searched for in a *direct* or *indirect* way. Direct detection experiments typically operate in deep underground laboratories to reduce the background from cosmic rays.

Indirect experiments are looking for products of Weakly Interacting Massive Particle (WIMP) annihilations. These searches assume that WIMPs are Majorana particles (particles coinciding with their own antiparticles), or that WIMPs and anti-WIMPs exist. If two WIMPs (or a WIMP and an anti-WIMP) collide, they could annihilate each other, producing ordinary particle–antiparticle pairs or γ-rays. This could produce a significant number of γ-rays or particles with a characteristic energy distribution. Searches for γ-ray lines or an excess of e^+ or e^- over the known background are mainly in progress in space experiments.

WIMPs are expected to be gravitationally trapped inside massive objects, with a consequent possible increasing rate of WIMP-antiWIMP annihilation. Underground/water/ice experiments look for an excess of events induced by neutrinos from the direction of massive objects (such as the Sun or the Earth's core, and the galactic center).

Following the observations, the Universe has an age of about 13.7 billion years and consists of about 100 billion galaxies; each galaxy is made of about 100 billion stars. Taking into account the mass of each star, the observable Universe is made of about 10^{80} protons. However, this should account for less than 5% of matter and energy: most matter/energy is still invisible to us. The number of cosmic background radiation photons is about a billion times the number of protons.

1.10 Laboratories and Detectors for Astroparticle Physics

Astroparticle physics experiments, which are described in the following chapters of this book, are located underground, on the Earth's surface, in the atmosphere, and in space.

The direct detection of the CRs' flux, the measurement of GeV γ-rays, the search for antimatter in the Universe, and the indirect searches for WIMP-antiWIMP annihilation producing γ-rays or e^+e^- pairs takes place in space experiments.

As the cosmic ray flux strongly decreases with increasing energy, the study of CRs around and above the *knee* is done by measuring (with different techniques) the secondary particles produced in the atmosphere and reaching ground level.

The searches for rare phenomena benefit in the reduction of secondary CR flux in underground laboratories (Sect. 1.11). This is also true for very large and/or massive experiments needed to study neutrino physics and astrophysics, to search for proton decay and for relic particles from the Big Bang. A similar noise reduction is necessary for high-energy neutrino telescopes that use huge volumes of transparent medium, such as the Antarctic ice and the deep-seawater, to detect charged leptons produced by charged current neutrino interactions.

1.10.1 Space Experiments

The Earth's atmosphere corresponds to an absorption layer of about 25 radiation lengths. For this reason, the detection of particles interacting electromagnetically (including the high-energy photons) must be performed outside the atmosphere itself, on-board satellites or on-board the International Space Station (ISS).

The space environment has many unique advantages for research that cannot be realized on Earth. On-board satellites and on the ISS, it is characterized by micro-gravity, high vacuum, space radiation, and a wide field-of-view. Weightlessness (i.e., almost no gravity) means that buoyancy, sedimentation, static pressure, and thermal convection processes do not occur.

The ISS is a space station, corresponding to a habitable artificial satellite in a low Earth orbit. The ISS is the largest artificial body in orbit, and it can often be seen at the appropriate time with the naked eye from Earth. It is a modular structure, whose first component was launched in 1998, consisting of pressurized modules, external trusses, solar arrays, and other components. ISS components have been launched by USA Space Shuttles, as well as by Russian Proton and Soyuz rockets. The ISS serves as a microgravity and space environment research laboratory. Crew members can conduct experiments in physics, astronomy, meteorology, as well as biology, medicine, and other fields. The station is suited for the testing of spacecraft systems and equipment required for space missions.

Of particular relevance is the study of the possibility of long-term permanence of humans in space. The space environment is hostile to life. Unprotected presence in space is characterized by the intense radiation due to charged CRs and other charged particles from the solar wind, high vacuum, extreme temperatures, and microgravity.

The requirements for a space-borne high-energy physics experiment are extremely challenging. Several constraints are imposed on experiments that have to be launched using satellites or transported on the ISS. For instance, the Alpha Magnetic Spectrometer (AMS-02, Fig. 1.7) has flown with the Space Shuttle with

Fig. 1.7 The Alpha Magnetic Spectrometer (AMS) detector on board of the International Space Station (Photo: courtesy AMS-02 Collaboration)

the strict weight limit of 7 tons. For permanence on the ISS, the requirements are power consumption less than 2 kW (as a normal hairdryer) and data rate limited to 2 Mbits per second. In addition, the AMS-02 experiment must work properly in space without any external operation for many years, has survived accelerations up to 9 g and vibrations up to 150 dB during the shuttle launch, and withstand cyclic temperature variations between −80 and 50 °C in vacuum. Each subsystem and electronic component was produced in prototypes (engineering, qualifications, and flight models) tested to provide the expected physics performance and the mandatory space safety. In addition to AMS-02, other future detectors can be installed on the ISS.

1.10.2 Experiments in the Atmosphere

Scientific ballooning contributed significantly to the space science program, both directly, with physics results coming from measurements made by balloon-borne instruments, and indirectly, by serving as a test platform on which instruments have been tested that were subsequently flown on more constrained space missions.

One of the most widely recognized uses of ballooning was for the study of anisotropy in the 2.7 K cosmic microwave background in the millimeter wavelengths. Almost 20 flights have paved the way for the extremely successful measurements of the COBE, WMAP and Planck satellites.

The atmosphere itself has an important role in the development of showers induced by primary CRs. During propagation in the atmosphere, charged particles can emit Cherenkov light and/or excite the fluorescence[1] of nitrogen atoms. Experiments placed on the ground are able to measure TeV scale γ-rays and the ultra-high energy cosmic ray particles through the measurement of the light emission.

1.10.3 Ground-Based Experiments

The flux of primary CRs from $\sim 10^{15}$ eV up to more than 10^{20} eV is measured through the detection of secondary particles. The *Extensive Air-Showers Arrays* (EAS) are composed of a collection of detectors distributed over a large area. Scintillators or water-Cherenkov counters are typically used to detect charged particles reaching the ground. Other techniques include the *Cherenkov telescopes*, which detect the Cherenkov light emitted by the electrons in the atmosphere, and the *fluorescence detectors*, which observe the fluorescence light emitted by atmospheric nitrogen excited by the shower particles. These detectors can estimate many characteristics of the shower: the number of secondary particles, related to the primary CR energy; the shower lateral distribution with respect to the axis; the primary CR direction of incidence.

1.11 Underground Laboratories for Rare Events

In the Standard Model of particle physics, quarks and leptons are placed in separate multiplets and the baryon number conservation forbids proton decay. However, there is no known gauge symmetry that generates baryon number conservation, which must then be considered as an open experimental question. Starting from the 1980s, the search for proton decay was the main reason for constructing underground laboratories and large detectors. The simplest GUT model, SU(5), predicts a proton lifetime of $\tau_p \sim 10^{30}$ years for the process $p \rightarrow e^+\pi^0$ that corresponds to many proton decay events in a kiloton-scale detector, if the background from secondary CRs, in particular, from atmospheric neutrinos, is reduced by a large shield of surrounding matter.

Water Cherenkov (IMB, Kamiokande) and tracking calorimeters (KGF, NUSEX, Soudan) were the pioneering experiments in the search for proton decay. The

[1] The term "fluorescence" refers to the process by which atoms absorb photons of one wavelength and emit photons at a longer wavelength.

Cherenkov detection allows for larger masses, while tracking calorimeters provide better space resolution and good identification of the proton decay products, such as electrons, muons, and charged kaons. The limits provided by these experiments ruled the simplest GUT models out beyond a doubt.

Since these early experiments, it was easily realized that underground laboratories offer an ideal environmental situation for different experimental studies:

- the detection of low energy phenomena, such as solar neutrinos, neutrinos from stellar gravitational collapses, neutrinoless double-beta decay, and searches for rare phenomena. The main problem for all these experiments is the environmental purity of material and the radioactivity background; refined detectors, often of large mass, are needed. For detection of low energy neutrinos, the most important parameters are the detector mass and the energy threshold;
- the study of ∼1 GeV events, such as nucleon decays and neutrino oscillations. The main feature of a detector is its mass (1–50 kt) and the capability of identifying neutrino events;
- the detection of through-going particles, high-energy muons, magnetic monopole candidates, etc. The main feature of these detectors is the surface coverage.

The muon flux decreases with the thickness of the rock overburden, as shown in Fig. 1.8, which also reports the depth of some large national or international

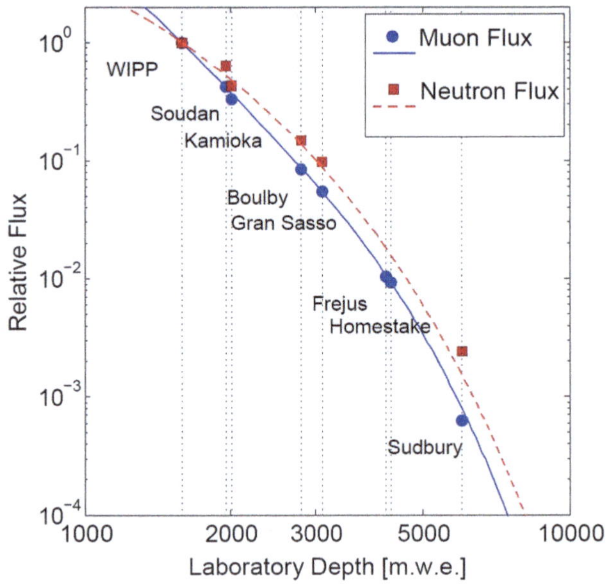

Fig. 1.8 Relative cosmic ray muons (*blue*) and energetic neutrons (>10 MeV, *red*) flux at some underground laboratories as a function of laboratory depth. The muon and neutron flux is assumed =1 at the depth of the Waste Isolation Pilot Plant (WIPP), a geological repository for radioactive waste in the USA. The *blue* and *red* points correspond to measured values, while the *lines* show a functional fit. Courtesy of Prof. T. Saab

underground facilities (Saab 2014). Operations at deep underground sites also reduce the flux of energetic neutrons, which are induced by the interaction of atmospheric muons within the materials and structures surrounding the experiment. This neutron flux represents a penetrating component that does not release a signal in most *veto* detectors, used for the above-mentioned rare event searches. Regardless of the experimental technique used in rare event searches, reducing the rate of background events plays an essential role. For instance, in the searches for WIMPs, energetic neutrons lead to interactions in a detector that are indistinguishable from a WIMP signal.

Underground laboratories differ in many other important aspects: horizontal or vertical access, interference with nearby activities (mine work, road traffic, etc.), quality of the support infrastructures (laboratories, office space, assembly halls, etc.) and personnel on the surface, degree of internationality of the user community, policy of space and time allocation, etc. Scientific sectors other than astroparticle physics, such as biology, geology, and engineering, can also profit from the very special underground environment provided by the laboratories and their facilities.

At present, there are a large number of underground experiments, with a dominant role being played by Europe, located in mines or in underground halls close to a highway tunnel. We give in the following a list of the major underground laboratories where some of the experiments described in this book are located.

- The Baksan Neutrino Observatory, Russia, at a depth of about 2100 m.w.e., where the SAGE and Baksan experiments were located. Started in 1966, it was one of the older underground laboratories. It is managed as an observatory, with very long-duration experiments.
- The Gran Sasso Laboratory (LNGS), L'Aquila (Italy), is the largest underground facility, the oldest in Europe after Baksan, located on the Roma-Teramo highway, 120 km east of Rome. The LNGS are operated by the Istituto Nazionale di Fisica Nucleare (INFN). The laboratory consists of three underground tunnels, each about 100 m long; it is at an altitude of 963 m above sea level, is well shielded from secondary CRs by an average amount of rock equivalent to 3700 m of water (or 3700 m.w.e.), and has a low activity environment. It has hosted and still hosts many important experiments (MACRO, Gallex/GNO, Icarus, Opera, DAMA, LVD, Borexino, Xenon…).
- The Kamioka Observatory in Japan is the largest and most important underground laboratory in Asia. The observatory was established in 1983 by M. Koshiba and is operated by the Institute for Cosmic Ray Research, University of Tokyo. The coverage corresponds to 2700 m.w.e. It has hosted or hosts the Kamiokande experiment, Super-Kamiokande, KamLAND, and XMASS. The KAGRA laser interferometer (in advanced construction) will be the first underground experiment for the detection of gravitational waves.
- The SNO-laboratory, Sudbury, Canada, hosted the SNO experiment at a depth of about 6000 m.w.e.; SNO will be replaced by SNO+, a new kiloton scale liquid scintillator detector that will study neutrinos. New structures are being excavated.

- The Soudan Underground Laboratory, Minnesota, USA, hosted the Soudan experiment, the CDMS dark matter experiment, and the MINOS long baseline experiment.
- The Homestake mine in South Dakota was a deep underground gold mine where the Clorine (or Davis) experiment took data until the mine's closure in 2002. A proposed Deep Underground Science and Engineering Laboratory (DUSEL) was a major project under consideration by founding agencies in the USA.
- The China Jinping Underground Laboratory is a laboratory in the Jinping Mountains of Sichuan, China, at a depth of 2400 m (6720 m.w.e.). It corresponds at present to the best-shielded underground laboratory in the world. It currently hosts two experiments for dark matter searches.
- Other infrastructures in Asia are: the OTO-Cosmo Observatory (Japan); Y2L, operated by the Dark Matter Research Centre (DMRC), Korea; INO, the India based Neutrino Observatory (India).

A complete review of the available facilities can be found in Bettini (2007).

References

F. Aharonian, J. Buckley, T. Kifune, G. Sinnis, High energy astrophysics with ground-based gamma ray detectors. Rep. Prog. Phys. **71**, 096901 (2008)

J.N. Bachall, *Neutrino Astrophysics* (Cambridge University Press, Cambridge, 1989)

A. Bettini, The world underground scientific facilities. A compendium (2007). arXiv:0712.1051

S. Braibant, G. Giacomelli, M. Spurio, *Particle and Fundamental Interactions* (Springer, Berlin, 2011). ISBN: 978-9400724631

S. Braibant, G. Giacomelli, M. Spurio, *Particles and Fundamental Interactions: Supplements, Problems and Solutions* (Springer, Berlin, 2012). ISBN: 978-9400741355

T. Courvoisier, *High Energy Astrophysics. An Introduction* (Springer, Berlin, 2013). ISBN: 978-3642436840

T.K. Gaisser, *Cosmic Rays and Particle Physics* (Cambridge University Press, Cambridge, 1991). ISBN: 978-0521339315

C. Grupen, *Astroparticle Physics* (Springer, Berlin, 2005). ISBN: 978-3540253129

W.L. Kraushaar, G.W. Clark, G.P. Garmire, R. Borken, P. Higbie, C. Leong, T. Thorsos, High-energy cosmic gamma-ray observations from the OSO-3 satellite. Astrophys. J. **186**, 401–402 (1973)

K.R. Lang, *Essential Astrophysics* (Springer, Berlin, 2013). ISBN: 978-3642359637

LIGO Scientific Collaboration et al., Multi-messenger observations of a binary neutron star merger. Astrophys. J. Lett. **848**, L12 (2017)

P. Lipari, *Introduction to Neutrino Physics* (2006). http://cds.cern.ch/record/677618/files/p115.pdf

M.S. Longair, *High Energy Astrophysics*, 3rd edn. (Cambridge University Press, Cambridge, 2011). ISBN: 978-0521756181

P. Mészáros, The high energy universe, in *Ultra-High Energy Events in Astrophysics and Cosmology* (Cambridge University Press, Cambridge, 2010)

R. Poggiani, *High Energy Astrophysical Techniques* (Springer International Publishing, Cham, 2017). ISBN: 978-3319447285

A. Rai Choudhuri, *Astrophysics for Physicists* (Cambridge University Press, Cambridge, 2012). ISBN: 978-1107024137

T. Saab, An introduction to dark matter direct detection searches and techniques (2014). arXiv:1203.2566

P. Sokolsky, *Introduction to Ultrahigh Energy Cosmic Ray Physics* (Westview Press, Colorado, 2004). ISBN: 978-0813342122

Chapter 2
Charged Cosmic Rays in Our Galaxy

Abstract This chapter introduces the primary Cosmic Rays (CRs). Primary CRs are high-energy and stable particles originating in astrophysical environments. They are protons, fully ionized atomic nuclei and electrons. Part of the primary radiation can interact in (or close to) the acceleration regions producing other high-energy charged and neutral secondary particles. Antiparticles (mainly positrons and antiprotons) are also present in the cosmic radiation, but in most cases, they are originated during the propagation of primary protons and electrons. All the particles produced far from the acceleration regions during the propagation of primary CRs, or produced on their arrival in the Earth's atmosphere, are denoted as secondary CRs. Particle detectors play a fundamental role in the history of CRs. As described, the intensity of this ionizing radiation is modulated by the solar activity. Below a few tens of GeV, the primary CR flux arriving on Earth also depends on the magnetic latitude, as it is affected by the geomagnetic field. The theoretical models on the origin of CRs rely on many astrophysical parameters and experimental observations about nature, energy density, confinement time, and chemical composition of CRs as a function of their energy.

Primary *Cosmic Rays* (CRs) are high-energy and stable particles originating in astrophysical environments. They are protons, fully ionized atomic nuclei and electrons. As discussed in the following, electrons are easily decelerated by their motion in local magnetic fields, and thus primary CRs are dominated (far from the sources) by protons and nuclei. Part of the primary radiation can interact in (or close to) the acceleration regions producing other high-energy charged and neutral secondary particles. Among them, γ-rays and neutrinos are stable as well and can arrive on Earth together with the primary component. The γ-ray and neutrino components will always be specified (for instance, their flux will be denoted as Φ_γ or Φ_ν, respectively). Due to their specificity, quantities referring to CR electrons will usually be distinguished. Antiparticles (mainly positrons and antiprotons) are also present in the cosmic radiation, but in most cases, they are originated during the propagation of primary CRs and electrons. All the particles produced far from the acceleration regions during the propagation of primary CRs, or produced on

© Springer Nature Switzerland AG 2018
M. Spurio, *Probes of Multimessenger Astrophysics*, Astronomy and Astrophysics Library, https://doi.org/10.1007/978-3-319-96854-4_2

their arrival in the Earth's atmosphere, are denoted as *secondary cosmic rays*. The discovery of CRs (Sects. 2.1 and 2.2) made it possible to verify both the Einstein relation between mass and energy and the Dirac theory about the existence of antimatter (Sect. 2.3). Particle detectors play (Sect. 2.4) a fundamental role in this history.

Some quantitative aspect of the CRs, such as the differential and the integral fluxes, are defined in Sect. 2.5, and the measured energy spectrum of the primary cosmic rays are presented in Sect. 2.6. Most CRs originate outside the solar system, and the physical properties of the Galaxy (dimension, matter density, magnetic fields, described in Sect. 2.7) have an effect on their journey from sources to the Earth. Only low-energy particles (below a few GeV) are of solar origin (Sect. 2.8). The intensity of this ionizing radiation is modulated by the solar activity. Below a few tens of GeV, the primary CR flux arriving on Earth also depends on the magnetic latitude, as it is affected by the geomagnetic field (Sect. 2.9).

The theoretical models on the origin of CRs rely on many astrophysical parameters and experimental observations about nature, energy density, confinement time, and chemical composition of CRs as a function of their energy. Supernova remnants are recognized as candidates for the acceleration of galactic CRs. The main motivation is the relation (described in Sect. 2.11) between the loss of kinetic energy due to CRs escaping from the galactic disk and the energy released by supernova shock waves. The details of the physical mechanisms through which kinetic energy is transferred to high-energy particles will be described in Chap. 6.

2.1 The Discovery of Cosmic Rays

The discovery of CRs is commonly attributed to Victor Hess (1912). After the discovery of radioactivity in 1896 by Henri Becquerel, while studying phosphorescent materials, it became known that some materials can produce *ionizations*. These substances are called *radioactive materials*, for their similarity to radium (^{88}Ra). In the presence of a radioactive material, a charged *electroscope* promptly discharges, while their golden leaves can stay apart from each other if the ionization level of the surrounding material is low. Radioactive elements emit charged particles, which ionize the gas therein, thus causing the discharge of electroscopes. The discharge rate was used at the beginning of the last century to measure the level of radioactivity.

Around 1900, Wilson and others developed a new technique for the insulation of electroscopes in a closed vessel, thus improving the sensitivity of the electroscope itself. As the discharge was also present when no radioactive elements were present inside the electroscope's shielding, the presence of ionizing agents coming from outside the vessel was assumed. The questions concerned the Earth or the extraterrestrial origin of such radiation. An experimental confirmation of one of the two hypotheses, however, seemed hard to achieve.

The original idea to perform a measurement far from any terrestrial material seems due to Theodor Wulf. In 1909, using an improved electroscope in which two metalized silicon glass wires replaced the two golden leaves, Wulf measured the rate of ionization at the top of the Eiffel Tower in Paris (300 m above ground). According to the hypothesis of the terrestrial origin of most of the radiation, a smaller ionization rate was expected at the top of the tower than on the ground. The measured ionization rate was, however, too similar to that on the ground to allow confirmation of the extraterrestrial origin of a part of the radiation.

The first measurement using a balloon (by K. Bergwitz) was made in the same year: the ionization at 1300 m altitude was found to be only about 24% of that at ground level. The final answer to the problem was given by the Austrian physicist Victor Hess (Fig. 2.1). After many balloon flights (starting from 1911) with different instruments on board, finally, in 1912, he reached an altitude of 5200 m. The results clearly showed that the ionization, after going through a minimum, increased considerably with height. Hess concluded that the increase of the ionization with height was originated by radiation coming from space. He also noticed the absence of day–night variations, and thus he excluded the Sun as the direct source of this hypothetical penetrating radiation.

The results by Hess were later confirmed by W. Kolhörster in a number of flights up to 9200 m. Hess was awarded the Nobel Prize in physics in 1936 for the discovery of the *cosmic rays*.

It should also be mentioned that several important contributions to the discovery of the origin of CRs have been forgotten. In particular, that of D. Pacini, who, in an earlier paper in Italian (Pacini 1912), wrote that *a sizable cause of ionization exists in the atmosphere, originating from penetrating radiation, independent of the direct action of radioactive substances in the soil.* A recommended historical overview of the early age of the history of cosmic rays is in Carlson and De Angelis (2010) and references therein.

After the discovery of CRs, a long scientific debate started about the nature of such an extraterrestrial radiation. Most believed at the beginning that cosmic rays were γ-rays, because of their huge penetrating power. In particular, R.A. Millikan hypothesized that those extraterrestrial γ-rays were produced during the formation of helium nuclei in interstellar space.

In 1927, J. Clay found evidence of a variation of the cosmic ray intensity with latitude. As this deflection was attributed to the presence of the geomagnetic field, the only plausible interpretation was that the primary cosmic rays were charged particles, not photons. This interpretation was principally due to Bruno Rossi (1930). He predicted a difference between the intensities of CRs arriving from the East and the West assuming positively charged particles. In 1932, Compton performed a worldwide survey, which verified the Rossi prediction of the so-called "East-West effect" and rejected the Millikan theory.

Fig. 2.1 Historical photograph of Hess preparing for a balloon flight (American Physical Society)

In 1938–1939 (in Paris first and then in the Alps mountains), P. Auger and collaborators showed that groups of particles could arrive coincident in time on detectors separated by distances as large as 200 m. This was the first indication that the observed particles in the atmosphere were secondary particles induced from a common primary CR (Auger et al. 1939). They concluded that there exist in the Universe mechanisms able to accelerate particles up to an energy of 10^{15} eV. Note that at that time, the largest energies from natural radioactivity or artificial acceleration were just a few MeV. Only in 1941 was it established (Schein et al. 1941) that CRs were mostly protons.

In 1962, John Linsley observed an event interpreted as being due to a cosmic ray of energy 10^{20} eV (Linsley 1962). This event was measured by an array of scintillation counters spread over 8 km^2 in a desert in New Mexico (USA).

2.2 Cosmic Rays and the Early Days of Particle Physics

CRs are very-high-energy particles entering the atmosphere, making possible, before the advent of particle accelerators, the creation of new, unstable particles through the Einstein relation between *energy* and *mass*. Around 1920, only the proton, the electron, and the photon were known as *elementary particles*. Before the advent of particle accelerators, up to the 1950s, the study and discovery of new particles and the study of the fundamental interactions between elementary particles was performed mainly using cosmic rays.

Starting in the 1930s, the experimental techniques for the detection and measurement of some physical quantities (e.g., electric charge, mass, lifetime) of the particles present in the secondary cosmic rays started to become more refined. Specifically, P. Blackett (Nobel laureate in 1948) used a cloud chamber inside a magnetic field that bent the trajectory of charged particles. Experimental techniques using CRs remained useful until well after the end of the Second World War, when particle accelerators started to be developed.

Using Blackett's experimental techniques, in 1932, C. Anderson (Nobel laureate in 1936) discovered a particle having the same mass as the electron, though with opposite electric charge (Fig. 2.2). It was the *antielectron* (or *positron*), that is, the antiparticle predicted by the relativistic quantum theory of the electron developed a few years earlier by P.A.M. Dirac (Nobel laureate in 1933). Immediately afterwards, in 1934, J. Chadwick (Nobel laureate in 1935) experimentally identified a particle with a mass similar to that of the proton, though without electric charge: the neutron.

Fig. 2.2 *Left* The first image of a positron obtained by Anderson in a cloud chamber (Anderson 1933). A charged particle interacting with the supersaturated vapor of the mixture inside the chamber produces ionization. The resulting ions act as condensation centers, around which a mist will form. Due to many ions being produced along the path of the charged particle, the tracks of particles having electric charge $Ze = 1$ or 2 have distinctive shapes (an alpha particle's track is broad, while an electron's is thinner and straight). The particle in this event has the same electric charge of the electron, but the opposite sign (Sect. 2.3). *Right* Carl D. Anderson, Nobel Prize for Physics in 1936, with his detector: a cloud chamber inside an electromagnet

In 1937, C. Anderson and S. Neddermeyer discovered a particle of intermediate mass between that of the proton and that of the electron: they called this new particle the *meson*. For some time, it was thought that this particle was the one necessary to glue protons and neutrons in nuclei. A theoretical model credited to H. Yukawa (Nobel laureate in 1949) predicted the existence of a particle with a mass very close to that of the just-discovered meson. Nevertheless, during World War II in Rome, M. Conversi, E. Pancini and O. Piccioni demonstrated, in a famous experiment using secondary cosmic rays, that the meson of Anderson and Neddermeyer (nowadays called the *muon*) was not the particle predicted by Yukawa. Even if the theory of Yukawa does not properly describe the physics of nuclei, the predicted particle (the *pion*) was discovered in 1947 by C. Lattes, G. Occhialini and G. Powell in secondary cosmic rays using nuclear emulsions (i.e., sophisticated photographic films) at high altitudes.

In 1947, in the interactions of cosmic rays in a cloud chamber with magnetic field, particles with a particularly strange behavior were discovered. They were thus named *strange* particles. Many years later [see Chap. 7 of Braibant et al. (2011)], it was realized that strange particles, as well as protons and neutrons, are composite objects. Protons and neutrons, and other short-lived particles with semi-integer spin, are made of three *u* and *d quarks*. The pions, like other mesons with null or integer spin, are made of a quark and an antiquark. Strange particles contain a new, heavier quark called *s* (for *strange*). With the advent of accelerators, the fields of particle physics and that of cosmic rays parted ways, though they have actually reconnected in recent years.

2.3 The Discovery of the Positron and Particle Identification

In this section, we will use Anderson's picture of the detected positron to introduce how particles are identified with the help of an external magnetic field.

2.3.1 The Motion in a Magnetic Field and the Particle Rigidity

The Lorentz force exerted by the magnetic field **B** on a particle with charge q, mass m, and velocity **v** is (in Gaussian unit system, see Sect. 2.12)

$$\Gamma m \frac{\mathrm{d}\mathbf{v}}{\mathrm{d}t} = \frac{q}{c}\mathbf{v} \times \mathbf{B} \,, \tag{2.1}$$

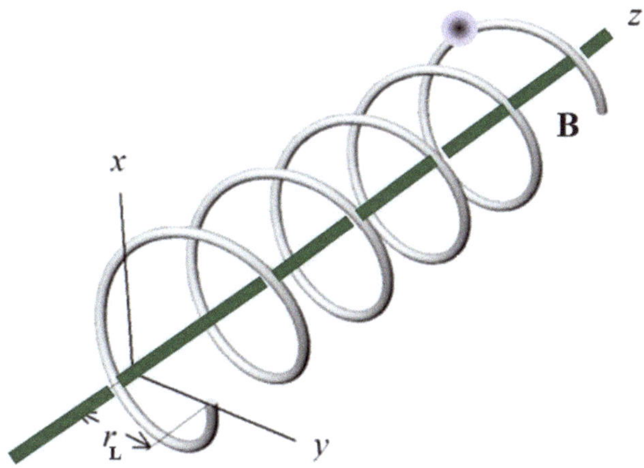

Fig. 2.3 The motion of a charged particle along the **B** field line

where Γ[1] is the Lorentz factor:

$$\Gamma \equiv \frac{1}{\sqrt{1 - v^2/c^2}} = \frac{E}{mc^2} \,. \tag{2.2}$$

Since the acceleration is perpendicular to both the magnetic field and the velocity vectors, in the case of a static uniform field, the orbit is a circle, plus a uniform motion along the direction of **B**. The resulting helicoidal motion is shown in Fig. 2.3. The integration of Eq. (2.1) yields, for a static and uniform magnetic field

$$\mathbf{v} = \mathbf{r} \times \frac{q\mathbf{B}}{\Gamma mc} \equiv \mathbf{r} \times \boldsymbol{\omega}_{\mathbf{L}} \,. \tag{2.3}$$

The angular frequency of the circular motion is

$$\omega_L = \frac{2\pi}{T_L} = \frac{qB}{\Gamma mc} \,. \tag{2.4}$$

The radius of the orbit is called the *cyclotron radius* or *Larmor radius*. For $q = Ze$, where e is the electric charge of the proton, the Larmor radius[2] is

$$r_L = \frac{v}{\omega_L} = \frac{\Gamma mvc}{ZeB} = \frac{\mathrm{pc}}{ZeB} \simeq \frac{E}{ZeB} \,. \tag{2.5}$$

[1] In the book, we use the Γ (uppercase) for the Lorentz factor to avoid confusion with the γ-ray.

[2] The formula appears different by a factor of c in the SI units, as pointed out in Sect. 2.12.

The last equality holds only for relativistic particles (as in most cases in this textbook). The cyclotron radius decreases with the increase of the charge Z of a nucleus, in units of the proton electric charge. As the effect of a magnetic field on a particle depends on the ratio between momentum and electric charge, the particle *rigidity* R is defined as

$$R \equiv \frac{\mathrm{pc}}{Ze} \simeq \frac{E}{Ze} = \frac{1.6 \times 10^{-12}(\mathrm{eV/erg})E(\mathrm{erg})}{4.8 \times 10^{-12}(\mathrm{statC})} = \frac{1}{300}E \quad [\mathrm{V}]\,, \tag{2.6}$$

when the energy is measured in eV. By definition, the eV is the product of an electric charge and a potential difference, and the rigidity is measured in Volts. If a multiple of the eV is used (for instance, the GeV), the rigidity is automatically expressed in the same multiple (GV). As discussed in the following chapters, propagation of CRs through cosmic magnetic fields and (probably) also their acceleration depends on rigidity.

2.3.2 *The Identification of the Positron*

Let us consider the picture in Fig. 2.2. The additional information that we have is:

- a 6 mm-thick lead plate is located in the middle of the chamber. A particle, while crossing the plate, loses energy;
- the intensity of the magnetic field inside the electromagnet in the region of the cloud chamber is 15 kG;
- the magnetic field is perpendicular to the plane of the paper;
- by comparing the number of ionization centers in the chamber with that released by radioactive nuclei emitting α (helium nuclei with charge$=2e$) or β radiation, the particle has an electric charge $+e$ or $-e$.

The simplest interpretation (an electron with charge $-e$ entering from above) is wrong. To demonstrate this we draw, in Fig. 2.4, two circles that interpolate the ionization points before and after the lead layer. The picture represents the projection on a plane of the 3-D particle trajectory[3] and our drawing is only approximate. In any case, we are lucky enough that the trajectory of the particle is mainly in the plane of the picture. Observing the circles, it is evident that the radius of curvature is larger in the bottom region. From (2.5), we conclude that the energy of the particle is larger in the region below the lead. Thus:

- the particle must have entered the chamber from below;

[3]For these reasons, in later *bubble chamber* techniques, two or more views of the event were used.

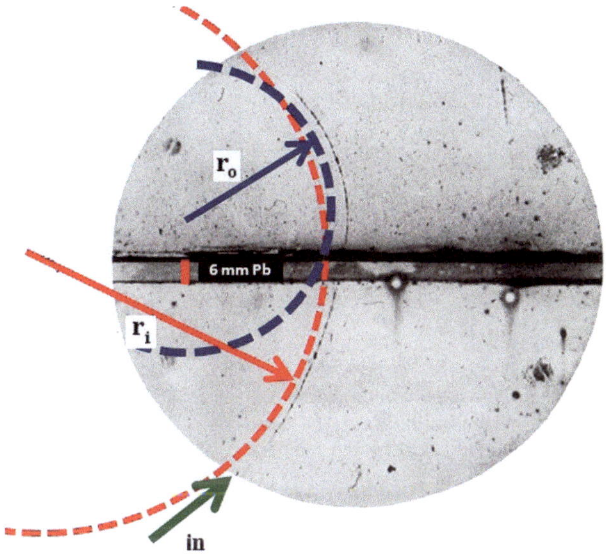

Fig. 2.4 Anderson's positron picture interpolated with two *arcs of circle* (slightly shifted to the *left*, to show the original track). The energy loss in the Pb plate is that characteristic of an electron-like particle (see text). The fact that $r_o < r_i$ indicates that the particle energy is $E_o < E_i$, and thus that it enters from below (*arrow*). However, from the knowledge of the magnetic field direction, the sign of the charge of this particle must be positive. It cannot be an electron

As we know the directions of the particle velocity and the magnetic field (entering the plane), using the relation $\mathbf{F} \propto \mathbf{v} \times \mathbf{B}$, we obtain that

- the particle has suffered a force toward the left. Thus, it is a positively charged particle.

It remains to show that

- the particle in the figure cannot be a proton and has almost the same mass as the electron. Thus, it is a *positron*, the particle foreseen by the quantum Dirac theory of the electron.

For this last step, we use the measurement of the radius of curvature of the two circles drawn in Fig. 2.4. The ruler in the figure is provided by the 6 mm Pb plate. The radius of the two circles can be scaled using this ruler in a way independent of the magnification of the picture. Our estimate is $r_i = 14$ cm, $r_o = 7$ cm for the particle when it enters and exits the cloud chamber, respectively. From (2.5), we obtain the respective particle momentum:

$$p_i c = r_i e B = 14 \times (4.8 \times 10^{-10}) \times (1.5 \times 10^4) = 1.0 \times 10^{-4} \text{ erg}$$
$$p_i = 63 \text{ MeV/c} , \quad (2.7a)$$

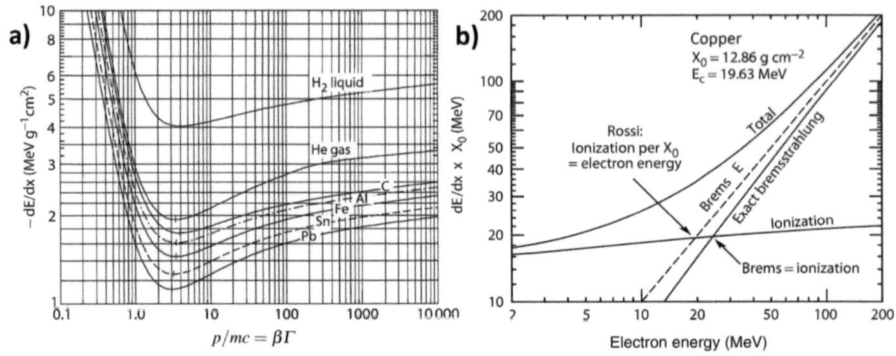

Fig. 2.5 (**a**) Mean energy loss in different materials of charged particles due to the electromagnetic interactions (excitation and ionization) with the electrons of the medium. The *horizontal scale* is in $\beta\Gamma$ units, which is independent of the incident particle type. (**b**) Electron energy loss in copper as a function of the electron energy. The excitation and ionization contribution remains roughly constant with increasing energy. Instead, the term due to the electron energy loss through interactions with nuclei (bremsstrahlung) increases. The intersecting point of the two *curves* defines the critical energy. In this figure, it corresponds to about 20 MeV. *Credit* the Particle Data Group (Beringer et al. 2012)

using the relation $1\,\text{eV} = 1.6 \times 10^{-12}\,\text{erg}$. Similarly, after the plate[4]:

$$p_o = r_o eB/c = 32\ \text{MeV/c}\,. \tag{2.7b}$$

The energy loss of charged particles through excitation and ionization of a material is given by the Bethe-Bloch formula as a function of the particle momentum p. It depends on the charge Ze and velocity $\beta = v/c$ of the incoming particle and only weakly on the properties of the crossed material. When p is expressed in terms of the adimensional Lorentz factors $p/mc = \beta\Gamma$, the energy loss assumes the behavior shown in Fig. 2.5a. The energy loss curves have a minimum for $\beta\Gamma \simeq 3$, that is, pc $= 3mc^2$. Neglecting the logarithmic rise when $E \sim$ pc $\gg mc^2$, the Bethe-Block formula can be approximated as

$$-\frac{dE}{dx} = \frac{Z^2}{\beta^2}\left(\frac{dE}{dx}\right)_{\min}; \quad \text{with} \quad \left(\frac{dE}{dx}\right)_{\min} \simeq 2\ \text{MeV g}^{-1}\text{cm}^2\,. \tag{2.8}$$

Let us consider the energy loss under the hypothesis of an incoming proton ($m_p c^2 = 938\,\text{MeV}$) with momentum p_i and

$$\beta\Gamma = \frac{p_i c}{m_p c^2} = \frac{63\ \text{MeV}}{938\ \text{MeV}} \simeq 0.07. \tag{2.9}$$

[4]In his work (Anderson 1933), Anderson reported *a 63 million volt positron passing through a 6 mm lead plate and emerging as a 23 million volt positron*. Our result is not so bad.

The energy loss for such low momentum is outside the scale of Fig. 2.5a. The energy loss can be estimated from (2.8), assuming $\beta\Gamma \sim \beta = 0.07$. The specific energy loss in lead (density $\rho_{Pb} = 11\,\text{g cm}^{-3}$) corresponds to

$$-\frac{dE}{dx} = (0.07)^{-2} \times 2\,(\text{MeV cm}^2/\text{g}) \times 11\,(\text{g cm}^{-3}) \simeq 5000\,\text{MeV/cm}.$$

To cross 6 mm of lead, a proton would lose ~3000 MeV. This is by far inconsistent with the observed variation of energy $\Delta E = E_i - E_o = 1.5\,\text{MeV}$, assuming an incoming proton having $p_i = 63$ MeV/c and $p_o = 32$ MeV/c. The positively charged particle in the picture cannot thus be a proton!

Let us consider the case of a *positive electron*. From the Dirac theory, its mass is identical to that of the electron ($m_e c^2 = 0.511\,\text{MeV}$), so that

$$\beta\Gamma = \frac{p_i c}{m_e c^2} = \frac{63\,\text{MeV}}{0.511\,\text{MeV}} \simeq 123\,. \tag{2.10}$$

In the case of the electron, in addition to the excitation-ionization energy loss, the *bremsstrahlung* (hard Coulomb scattering with atomic nuclei) process must also be considered. The electron energy loss in copper is given in Fig. 2.5b. Copper has a radiation length $X_0 \sim 12.9\,\text{g cm}^{-2}$. From the figure, an electron with $p \simeq 63$ MeV/c corresponds to an energy $E = 63$ MeV, and thus

$$\frac{dE}{dx} \times X_0 \simeq 70\,\text{MeV} \rightarrow \frac{dE}{dx} \simeq 70/12.9 = 5.4\,\text{MeV/g cm}^{-2}\,.$$

Thus, assuming the usual 6 mm of lead, i.e., $\Delta x = 6.6\,\text{g cm}^{-2}$

$$\Delta E \sim 5.4 \times 6.6 = 36\,\text{MeV}\,. \tag{2.11}$$

As the electron is relativistic, the energy loss from the picture is simply $E_i - E_o \simeq p_i c - p_o c = 31\,\text{MeV}$ in the 6 mm of lead, very close to the value obtained using (2.11). The mass and energy loss properties of the particle are compatible with that of the electron![5] Refer to Poggiani (2017) for more detail on the experimental methods used in astroparticle physics experiments.

[5]At the time of the Anderson picture, only protons and electrons were known as charged particles. Try to work out that the observed particle cannot be a charged π^+ meson, with $m_\pi = 139.6\,\text{MeV c}^2$.

2.4 A Toy Telescope for Primary Cosmic Rays

The aforementioned example of the positron discovery, although using very simple devices, is an example of how detectors for particle identification operate. Any experimental apparatus for CR detection should: (1) identify the particle, (2) measure its electric charge, and (3) measure its energy and momentum. We illustrate in the following how the *flux* of particles can be measured with a simple experimental device. The techniques used by modern experiments for the direct measurement of CRs are presented in the next chapter.

Our ideal experiment (a *toy telescope for charged CRs*) is made of two layers of counters of a given area A separated by the distance Δz (see Fig. 2.6). Each counter's layer is segmented with a characteristic spatial resolution. The signal in the counter is provided by the excitation/ionization energy loss of through-going charged particles (see Chapter 2 of Braibant et al. (2011)). The excitation/ionization energy loss (2.8) is proportional to the square of the electric charge Z of the particle. With *proportional counters*, the *amplitude* I of the signal thus depends on Z^2. This method for measuring the electric charge of the incoming particle through Eq. (2.8) will be referred to in the following as dE/dx *measurement*. Primary CRs can be measured if the telescope is carried to the top of the atmosphere by a balloon or outside the atmosphere by a satellite.

In Fig. 2.6, we assume that a charged particle crossing the lower counter induces a signal at the position labeled (x_7, y_4). Additional information are the z-position z_1

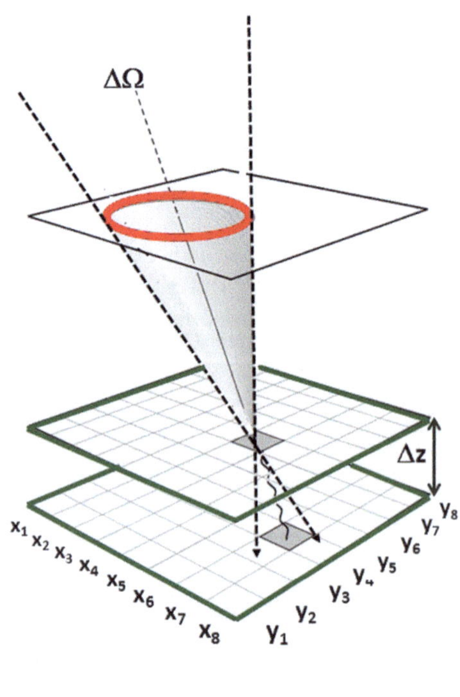

Fig. 2.6 Layout of a simple telescope for the measurement of CRs. The two counter's *layers* are assumed to be segmented, both in the x and y axes. A CR arriving within the *solid angle* $\Delta\Omega$ will produce one hit on each *layer* (see text)

of the layer, the crossing time (t_1) and the amplitude I_1 of the signal. The complete set of information (a *hit*) can be represented as $(x_7, y_4; z_1, t_1, I_1)$. The hits are stored as digital information on an *online* computer.

Each particle detector has some advantage/disadvantage with respect to others: some of them have good spatial resolution and poor timing resolution, and vice versa (see Sect. 2.7 of Braibant et al. (2011)). The design of a modern experiment is a compromise between the requirements necessary to have the best performances in terms of physics results and the cost, dimension, power dissipation, weight, etc., of each sub-detector.

A major requirement of any experiment is the *trigger logic*. This is a mandatory task, because the probability of a *fake signal* on a single counter is high. Due to the presence (for instance) of radioactive elements in the surrounding materials, or due to electronic noise, there are spurious signals in each detector plane with usually larger rates. By definition, these spurious hits are not correlated with a crossing particle and constitute the background.

In our simple example, a trigger is given by a *coincidence* between planes. This corresponds to having a signal both on the z_1 and z_2 layers within a given time interval T. A condition on amplitudes I_1, I_2 can also be added. The hits are permanently stored in the computer for further analysis if $|t_1 - t_2| \leq T$. The combination of signals in both planes, without constraints on time difference, will usually provide an overly large event rate with respect to the real CR rate.

Relativistic particles in vacuum cover 1 m in ~3.3 ns. Typical distances Δz between layers in CR telescopes (as in our example) are on the order of 1–2 m. The timing resolution of the detectors must be on the order of a ns (or better) to have the possibility of distinguishing between upward-(with $t_2 - t_1 > 0$) and downward-(with $t_2 - t_1 < 0$) going cosmic rays. In this case, a *time-of-flight* (ToF) measurement is performed. A very good timing resolution is characteristic, for instance, of most *scintillation counters*. Scintillation counters can be arranged in order to have sufficient spatial resolution to distinguish different directions (as in our telescope). In addition, their response depends on the ionization energy loss, and thus on the particle Z^2. Many ToF systems are also used to measure the Z of the detected particle.

In many real detectors a uniform *magnetic field* in the region between the counters performing a ToF is present. The magnetic field allows for the measurement of the particle momentum (if $|Ze|$ is known) and sign of the charge, as charged particles are deflected according to their *rigidity*. To measure the curved particle trajectory, additional detectors are needed inside the magnetic field region. If the magnetic field is along the y axis, $\mathbf{B} = B\hat{\mathbf{y}}$, the deflection is expected along the x-axis for particles entering with velocity along $\hat{\mathbf{z}}$. In this case, detectors with good spatial resolution inside the magnetic field (*tracking systems*) are used to accurately measure the x coordinate. The combination of the magnetic field and tracking detectors forms a *magnetic spectrometer*.

Finally, depending on some constraints (for instance, the total weight of the payload on a satellite or balloon), a *calorimeter* or other devices for the measurement of the particle energy can be added (see Chap. 3 for real experiments in space or in the upper atmosphere).

2.5 Differential and Integral Flux

The simple CR telescope in Fig. 2.6 is useful for the definition of the quantities related to the measurement of the number N of incident particles per unit time on the detector surface A at a given solid angle $d\Omega$. Usually, the area seen by particles depends upon their arrival direction (corresponding to given zenith and azimuth angles θ and ϕ, respectively) in a small angular region $d\Omega = \sin\theta d\phi d\theta$, and thus $A = A(\theta, \phi) = A(\Omega)$. The quantity

$$A\Omega \equiv \int A(\Omega) \cdot d\Omega \quad [\mathrm{cm^2 sr}] \tag{2.12}$$

is called *the geometrical factor*. The event rate in a detector (i.e., the number of events per second) is given by the particle flux (see below) times the geometrical factor.

The intensity vs. energy is determined using detectors able to measure the energy of the incoming particle. Thus, the number of CRs per unit of area A in a time T, $N/A \cdot T$, arriving in a given energy interval dE and solid angle $d\Omega$ represents the *differential intensity of particles of a given energy in the given solid angle*:

$$\Phi(E) \equiv \frac{d(N/A \cdot T)}{\mathrm{d}\Omega \cdot \mathrm{d}E} \quad \frac{\text{particles}}{\text{cm}^2\text{sr s GeV}} \,. \tag{2.13}$$

Sometimes, particles can be measured only if their energy is larger than a given energy threshold E_0. Equivalently, we could be interested in all particles measured with energy larger than E_0. In both cases,

$$\Phi(>E_0) \equiv \frac{\mathrm{d}(N/A \cdot T)}{\mathrm{d}\Omega}(E) = \int_{E_0}^{\infty} \frac{\mathrm{d}^2 N}{A \cdot T \cdot \mathrm{d}E\mathrm{d}\Omega}\mathrm{d}E \quad \frac{\text{particles}}{\text{cm}^2\text{sr s}} \tag{2.14}$$

represents the *integral intensity of particles with energy* $> E_0$, i.e., the measurement of the CR intensity for particles with energy larger than the given threshold.

According to the literature, the quantities $\Phi(E)$ (2.13) and $\Phi(> E_0)$ (2.14) are sometimes called the *differential flux* and the *integral flux* of particles, respectively. Note the different units in the two cases. In the following chapters, an index could appear: for instance, Φ_{CR}, Φ_e, Φ_γ, Φ_ν will indicate the flux of primary protons and nuclei, electrons, gamma-rays and neutrinos, respectively. Nuclei correspond to an important fraction of CRs. If the detector can measure the electric charge *Ze* of an incoming nucleus of species i, the flux Φ_i of that species can be determined, such as, for instance, in Fig. 3.10.

The arrival direction of CRs is largely isotropic. The particles' flux through a spherical surface is simply given by the integration over the solid angle of (2.14):

$$\frac{\mathrm{d}N}{A \cdot T \cdot \mathrm{d}E}(E) = 4\pi \frac{\mathrm{d}^2N}{A \cdot T \cdot \mathrm{d}E\mathrm{d}\Omega}(E) = 4\pi\Phi(E) \quad \frac{\text{particles}}{\text{cm}^2\text{s GeV}} . \tag{2.15}$$

In most cases (as for our ideal experiment in Fig. 2.6), we are interested in the flux through a planar surface. The differential particles' flux through a planar detector from one hemisphere is

$$\mathscr{F}(E) = \int \frac{\mathrm{d}^2N}{A \cdot T \cdot \mathrm{d}E\mathrm{d}\Omega} \cos\theta \mathrm{d}\Omega \quad \frac{\text{particles}}{\text{cm}^2\text{s GeV}} , \tag{2.16}$$

$d\Omega$ is, as usual, the elemental solid angle, θ the angle between the vector perpendicular to the area A and the direction of the incoming particle. For isotropic radiation (as in the case of the primary CRs), the flux on a plane is

$$\begin{aligned} \mathscr{F}(E) &= \Phi(E) \int_0^{2\pi} \mathrm{d}\phi \int_0^{\pi/2} \mathrm{d}\theta \sin\theta\cos\theta \\ &= \pi\Phi(E) \int_0^{\pi/2} \mathrm{d}\theta \sin 2\theta = \pi\Phi(E) \quad \frac{\text{particles}}{\text{cm}^2 \text{ s GeV}} . \end{aligned} \tag{2.17a}$$

Note that the form of the surface changes the numerical coefficient in front of Φ with respect to (2.15). Integrating (2.17a) over energies, we obtain the quantity:

$$\mathscr{F}(>E) = \pi\Phi(>E) \quad \frac{\text{particles}}{\text{cm}^2\text{s}} . \tag{2.17b}$$

An important quantity in astrophysics is the *number density* (units: cm^{-3}) of CRs moving with speed v. The number density corresponds to the number of particles present in a given volume at a given time. The setup in Fig. 2.6, for instance, is continuously crossed by CRs at a given rate. If we imagine taking a snapshot of the particles present in the detector volume at a given time and counting them, we could derive the number density of particles in the detector volume.

Using dimensional arguments, the number density can be easily obtained from the integral flux (2.14) integrated over the solid angle and divided by the speed (cm/s). It is easy to work out that, for an isotropic flux, the *particle number density* is

$$n = \frac{4\pi}{v}\Phi(>E) \quad \frac{\text{particles}}{\text{cm}^3} . \tag{2.18a}$$

If the flux is not isotropic, the integration over the solid angle gives a factor different from 4π. If the speed of particles is not constant, an integration over the velocity distribution spectrum is also needed.

Sometimes, we are interested in the *differential number density* of cosmic rays. Using (2.13), we obtain

$$\frac{\mathrm{d}n}{\mathrm{d}E} = \frac{4\pi}{v}\Phi(E) \quad \frac{\text{particles}}{\text{GeV cm}^3} . \tag{2.18b}$$

In this book, we are practically always dealing with relativistic particles, and $v \simeq c$.

2.6 The Energy Spectrum of Primary Cosmic Rays

After more than 100 years of research, we know that the solar system is permanently bombarded by a flux of highly energetic particles. Their energies extend from the MeV range to 10^{20} eV. The primary component arriving on the top of the atmosphere includes all stable charged particles and nuclei. Also, some unstable nuclei (with lifetimes larger than 10^6 years) are found. This (small) fraction of radioactive nuclei is important for estimating the *escape time* of CRs (Sect. 5.1).

Figure 1.5 shows the integral intensity of CRs as measured by a large number of different experiments, from small detectors on board balloons and satellites to huge air shower arrays on the ground, covering surfaces of more than 3000 km^2. Each experiment has measured the integral flux (2.14) in a given energy interval. The analytic interpolation of all available data shown in Fig. 1.5 is usually referred to as the *integral energy spectrum*.

The distribution of the differential flux (2.13) covering the whole energy interval of CRs is usually referred to as the *differential energy spectrum* (or simply the *energy spectrum*) of CRs, and it is shown in Fig. 2.7.

The energy spectrum falls steeply as a function of energy. The integral flux (2.17b) corresponding to different energy thresholds gives us

$$\mathscr{F}(> 10^9 \text{ eV}) \simeq 1000 \text{ particles/s m}^2 , \tag{2.19a}$$

$$\mathscr{F}(> 10^{15} \text{ eV}) \simeq 1 \text{ particle/year m}^2 , \tag{2.19b}$$

$$\mathscr{F}(> 10^{20} \text{ eV}) \simeq 1 \text{ particle/century km}^2 . \tag{2.19c}$$

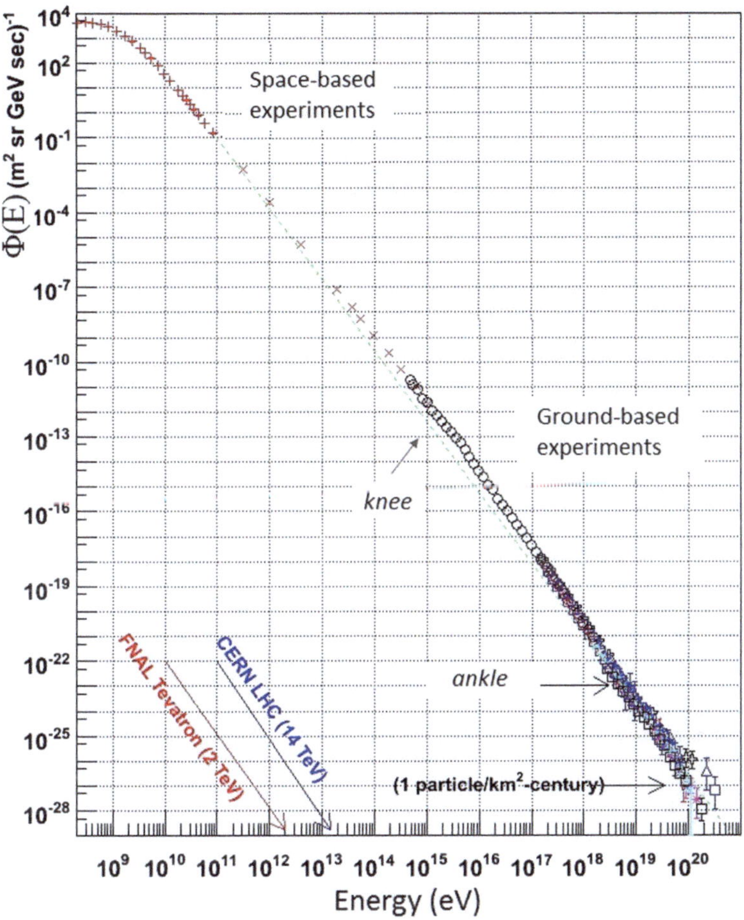

Fig. 2.7 The differential energy spectrum $\Phi(E)$ (units: particles/m^2 sr s GeV) of cosmic rays over eleven decades of energy. The *red/blue arrows* indicate the equivalent center of mass energy reached at the Tevatron collider at Fermilab and at the LHC collider at CERN. Note that the spectrum is remarkably continuous over the whole energy interval, and that the flux on the y-axis covers 33 decades. The *dashed line* shows a E^{-3} spectrum

Figure 2.7 seems almost featureless, but two transition points (where the slope of the spectrum changes) are clearly visible. This feature defines three energy intervals in the CR spectrum. The transition point at $\sim 3 \times 10^6$ GeV is called the *knee*. Below the knee, the integral CR flux decreases by a factor of ~50 when the energy increases by an order of magnitude.

At energies larger than a few GeV, where the contribution of particles coming from the Sun (Sect. 2.8) is negligible, the energy spectrum can be described by a power law where $E_0 = 1$ GeV:

$$\Phi(E) = K\left(\frac{E}{E_0}\right)^{-\alpha} \quad \frac{\text{particles}}{\text{cm}^2 \text{ s sr GeV}} \tag{2.20a}$$

$$\Phi(>E) = \frac{KE_0}{(\alpha-1)}\left(\frac{E}{E_0}\right)^{-\alpha+1} = \frac{KE_0}{(\alpha-1)}\left(\frac{E}{E_0}\right)^{-\gamma} \quad \frac{\text{particles}}{\text{cm}^2 \text{ s sr}} . \tag{2.20b}$$

The parameter α is the *differential spectral index of the cosmic ray flux* (or the *slope* of the CR spectrum) and K a normalization factor; $\gamma \equiv \alpha - 1$ is the *integral spectral index*. The numerical values of the parameters are determined through a fit to experimental data.

Different compilations of data exist that determine the parameters K, α using direct measurements of the CR flux (Chap. 3). These compilations give results in agreement within ~30%. In the energy range from several GeV to $\sim 10^{15}$ eV, we use that obtained in Wiebel-Sooth et al. (1998), with

$$K = 3.01; \quad \alpha = 2.68 , \tag{2.20c}$$

which includes the contribution of protons (~90%) and heavier nuclei. The compilation of Hörandel (2003) in the same energy region gives $K = 2.16, \alpha = 2.66$. The Beringer et al. (2012) gives the proton flux in terms of *energy-per-nucleon*, with $K = 1.8, \alpha = 2.7$.

The *energy-per-nucleon* is the *energy-per-nucleus*, divided by the number of nucleons. Conventionally, the two quantities can be distinguished by the indication that the energy-per-nucleus is measured in [GeV], and the energy-per-nucleon in [GeV/nucleon] or [GeV/A], where the "A" stands for "nucleons". The energy-per-nucleon can be assessed only through *direct* experiments, when both E and $Z \sim A/2$ are measured. When the atmosphere is used as a calorimeter in air shower experiments, generally, the energy-per-nucleus is measured.

Above the knee, the CR flux decreases by a factor of ~100 when the energy increases by a factor of 10. The spectral index of the CR spectrum becomes steeper, $\alpha \sim 3.1$. The measurements of the CR spectrum above the knee are presented in Chap. 4. At the energy of $\sim 10^{10}$ GeV, the spectrum becomes flatter again in correspondence to the second transition point, called the *ankle*. CRs above the ankle are thought to have an extragalactic origin (Chap. 7).

The *knee* and *ankle* structures are more evident in Fig. 2.8. It shows almost the same data set as in Fig. 2.7. The main difference is that the y-axis variable is multiplied by $E^{2.6}$. As explained in Sect. 1.5, this enhances the visibility for structures in the spectrum.

About 79% of the primary nucleons are free protons and about 70% of the rest are nucleons bound in helium nuclei (see Sect. 3.8). The flux of electrons corresponds to less than 1% of that of protons and nuclei at the same energy. This does not correspond to a charge-asymmetry in the CR sources. Electrons (as explained in

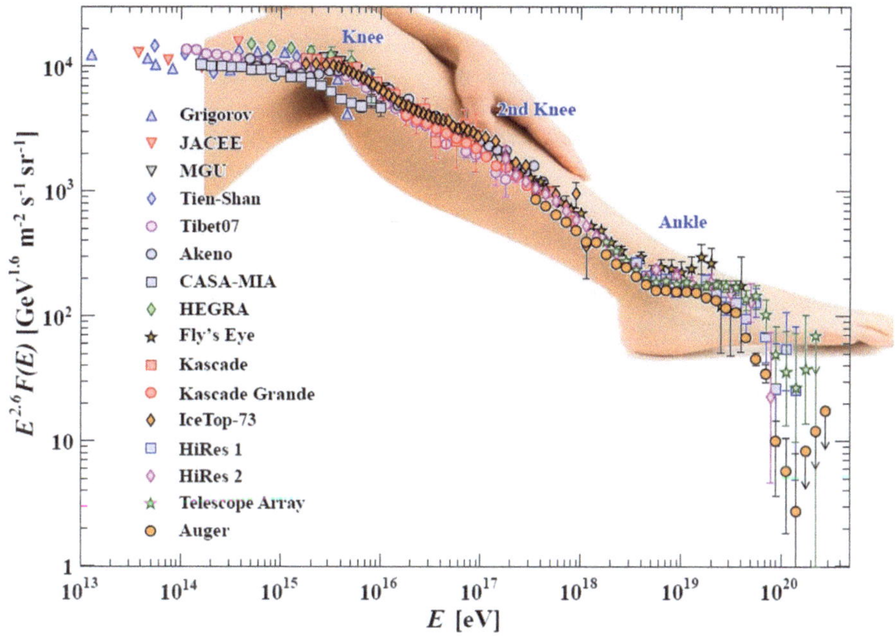

Fig. 2.8 The differential CR flux $\Phi(E)$ as measured by many *direct* and *indirect* CR experiments over eight decades of energy, almost the same as Fig. 2.7. Here, the flux is multiplied by a power of the energy: $\Phi(E) \times E^{2.6}$. The structures of the knee and ankle are more evident, as well as point-to-point differences between different experiments (most due to systematic uncertainties on the energy calibration). Adapted from a figure from Sect. 27: Cosmic Rays of Beringer et al. (2012)

the following chapters) suffer larger energy losses that reduce the number of those arriving with high energies on Earth. In general, fluxes of CR particles as measured on Earth are influenced by their travel through the galactic interstellar medium and magnetic field. In the next section, a description of the main properties of the Galaxy is given.

2.7 The Physical Properties of the Galaxy

Today, it is an established fact that the Sun is part of a system of stars, the *Galaxy* (or *Milky Way*), which is very similar to the spiral galaxies that we observe in the Universe. This conclusion was a sort of larger scale *Copernican revolution*, and was thus a nontrivial fact, also from the observational point of view.

The determination of the shape and size of the Galaxy with optical measurements has been difficult because of the position of the solar system within the Galaxy. In this context, the observation of distant galaxies was very important, as it revealed large-scale structures that were not visible in our Galaxy, due to light absorption by interstellar dust. Only around 1930 did it became unmistakably clear that the Galaxy is similar to objects that, at the time, were called *spiral nebulae*.

The recent images of the Galaxy using observations at different wavelengths show that it is basically a disk with a central bulge surrounded by a halo of globular clusters. It is convenient to distinguish two components: a *spheroidal* and a *disk* structure. Both contain stellar populations and other material with very different characteristics. These two components have different chemical compositions, kinematic and dynamic properties and a diverse evolutionary history.

Distances and sizes are usually expressed by astronomers in *parsecs* (symbol: pc). One parsec corresponds to about 3.26 light-years or to

$$1\ \text{pc} = 3.086 \times 10^{18}\ \text{cm}\ . \tag{2.21}$$

The spheroidal component has a very massive nucleus (smaller than 3 pc of radius) with a *black hole* at its center, with mass $2 \times 10^6\ M_\odot$,[6] a bulge with radius of $\sim$3 kpc and an extended halo of about 30 kpc. These three regions are approximately concentric. The disk is very thin ($\sim$200–300 pc thick) and has a radius of about 15 kpc. The Sun is about 8.5 kpc from the center.

The galactic volume, assuming a flat disk having a radius of $\sim$15 kpc and a thickness of $\sim$300 pc, corresponds to

$$\mathscr{V}_G = [\pi(15 \times 10^3)^2 \times 300] \times (3 \times 10^{18})^3 = 5 \times 10^{66}\ \text{cm}^3\ . \tag{2.22}$$

The volume of the galactic halo is more than an order of magnitude larger.

Stars and globular clusters are the characteristic components of the spherical region, where gas and dust are relatively scarce. Spectroscopy indicates that the stars in the spheroid component are *metal-poor*[7] (stars of population II). Metal-poor stars consist of material that has not undergone much recycling through previous generations of stars, are very old and represent the typical population of globular clusters.

The disk, contrastingly, is characterized by the presence of large amounts of dust and gas, which give rise to absorption of the interstellar radiation, and by young and metal-rich stars (Population I). These stars may be distributed more or less uniformly, or grouped in stellar associations along the spiral arms. These considerations suggest that the disk is made entirely of materials already processed in previous generations of stars. This situation is analogous to that observed in other spiral galaxies.

[6] The subscript $\odot$ conventionally represents the Sun and $\oplus$ the Earth.

[7] In the language of astronomers, all elements heavier than He are often called metals.

An estimate of the total number of stars in the Galaxy is about 10^{10}.

Important information on the Galaxy became available after the advent of radio astronomy. At low frequencies (150 MHz) the emission from the Galaxy shows a maximum of intensity along the galactic plane. The emission decreases steadily with increasing galactic latitude b,[8] covering the whole sky. The diffused radio emission is usually described in terms of two distinct structures: (1) a disk (which coincides with the optical disk) with an angular aperture of about $\pm 5°$ in galactic latitude b; (2) an ellipsoidal-shaped halo, which extends at high galactic latitudes and up to the poles. From the study of the emission spectrum, it can be inferred that the diffuse radiation is composed of a nonthermal component and a thermal one from the disk. The nonthermal component is due to the synchrotron radiation of electrons moving in the galactic magnetic field. The luminosity[9] of the component from the disk, integrated over the whole radio band, is equal to

$$\mathscr{L} \sim 10^{38}\ \text{erg/s}\ . \tag{2.23}$$

The luminosity from the emission of the halo is subject to greater uncertainty, but corresponds (within a factor of 2) to that of the disk.

Superimposed upon the thermal emission, discrete radio sources are observed. They are divided into two populations: (1) objects of large angular size, concentrated along the galactic plane; these are supernova remnants, surrounded by regions with a high concentration of ionized hydrogen; (2) radio sources of small angular diameter isotropically distributed and attributed to extragalactic objects. In both cases, the spectrum of the radiation is that typical of synchrotron emission.

2.7.1 *The Galactic Magnetic Field*

The presence of a magnetic field inside the Galaxy was discovered (1949) when it was realized that the observed light from the stars has a high degree of polarization. This polarization is correlated with the attenuation of starlight due to the presence of dust (Brown 2010).

Precise information on the galactic magnetic field comes from radio astronomy. Radio telescopes can measure the Faraday rotation angle and the extent of dispersion of the radiation emitted by pulsars and the Zeeman effect on the 21 cm line of neutral hydrogen.

---

[8] See *Extras # 2* for the astronomical coordinate systems.

[9] In astronomy, the luminosity measures the total amount of energy emitted by a star or other astronomical object per unit time over the whole electromagnetic spectrum or a defined part of it.

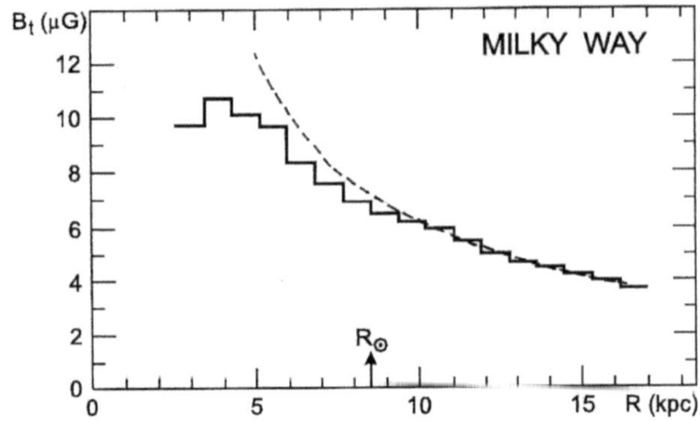

Fig. 2.9 Total magnetic field strength in the Galaxy as a function of the distance from the galactic center. The position of the Sun is indicated by the *arrow* (Battaner et al. 2007). Courtesy of Prof. E. Battaner

The Faraday rotation effect is based on the fact that the plane of polarization of linearly polarized electromagnetic waves rotates when they propagate in the presence of a magnetic field B in a medium with electron density n_e [cm^{-3}]. The rotation depends on the square of the wavelength λ and on the parallel component of the magnetic field $B_{\|}$ along the line of sight to the source:

$$\mathrm{RM} = \int_0^L B_{\|} n_e dr \,, \tag{2.24}$$

where L is the distance traveled by the radiation. By measuring the variation of the angle of polarization as a function of the wavelength λ from radio pulsars, RM can be estimated. From independent estimates of n_e and L, the value of B in the traversed region can be deduced.

Different estimates exist on the average intensity of the regular galactic magnetic field, which depends on the distance from the galactic center (Fig. 2.9). We assume approximately:

$$B \simeq 4\,\mu\mathrm{G} \,. \tag{2.25}$$

The galactic field is oriented mainly parallel to the plane, with a small vertical component along the z-axis ($B_z \sim 0.2$–$0.3\,\mu$G in the vicinity of the Sun).

The models of the large-scale structure of the galactic magnetic field provide a regular distribution of the **B** lines that follows the distribution of matter, i.e., a spiral shape (Stanev 2010). The spatial extension of regions in which the magnetic field is coherent is on the order of 1–10 pc. Figure 2.10 shows the direction and strength of the regular magnetic field in the galactic plane. The large-scale structure of the galactic magnetic field strongly influences the motion of charged particles. An example of this influence on charged particles with three different energies is visible in Fig. 7.4.

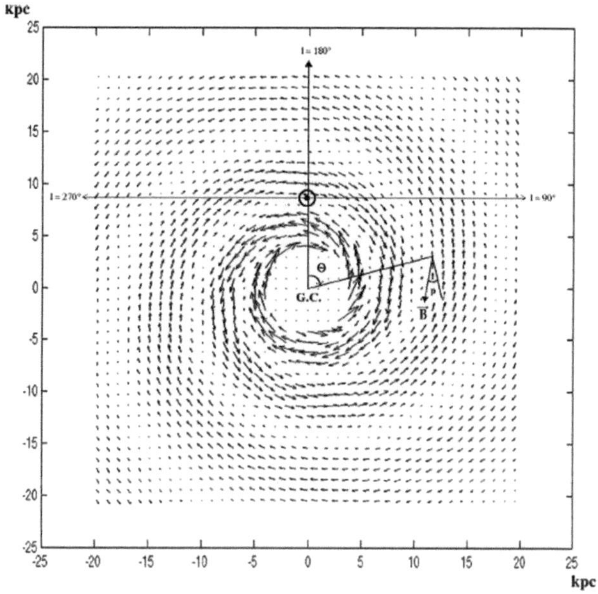

Fig. 2.10 The direction and strength of the regular magnetic field in the Galactic plane is represented by the length and direction of the arrows. The intensity of the field inside the circle of radius 4 kpc representing the bulge is assumed to be 6.4 μG (Prouza and Smída 2003). Courtesy of Dr. M. Prouza and Dr. R. Smída

An important problem, which is far from being solved, is the galactic magnetic halo, i.e., the extension of the magnetic field above and below the galactic plane. Recent measurements indicate an extended halo that can contribute significantly to the confinement of cosmic rays.

As the Galaxy is filled with a magnetic field whose average intensity is $B \sim 4\,\mu$G, we obtain, using (2.5), the following Larmor radii for protons at different energies:

$$
\begin{aligned}
r_L(E = 10^{12}\text{eV}) &\simeq 10^{15}\ \text{cm} = 3 \times 10^{-4}\ \text{pc}\ , \\
r_L(E = 10^{15}\text{eV}) &\simeq 10^{18}\ \text{cm} = 0.3\ \text{pc}\ , \\
r_L(E = 10^{18}\ \text{eV}) &\simeq 10^{21}\ \text{cm} = 300\ \text{pc}\ .
\end{aligned} \tag{2.26}
$$

These values of r_L should be compared with the dimensions of the Galaxy. Particles below 10^{18} eV are strongly constrained inside the galactic volume by the galactic magnetic field.

2.7.2 The Interstellar Matter Distribution

In outlying regions of the Galaxy, the ratio of the distances between stars and the star diameters is on the order of (∼1 pc)/ (10^6 km) $\simeq 3 \times 10^7$. Thus, only a small fraction of the space volume (∼4×10^{-23}) is occupied by matter in the form of stars. The rest is filled with large masses of gas (molecules, atoms, ions) and tiny solid particles, the interstellar dust. Dusts are made up of ice grains of various species, graphite, silicates and perhaps metals. The gas is revealed by the presence of absorption and emission lines, both in the optical and in radio. Dusts are observed only as large dark clouds obscuring the view of the stars behind them, or in reflection nebulae that shine owing to the presence of nearby stars. Another tracer of the presence of dust is the infrared emission in the vicinity of very hot stars.

As a whole, this gas and dust is called *InterStellar Matter* (*ISM*). It represents 5–10% of the total mass of the Galaxy. The average density of this medium is (Ferriere 2001):

$$n_{\rm ISM} \sim 1\ {\rm proton/cm^3} = 1.6 \times 10^{-24}\ {\rm g/cm^3}\ . \tag{2.27}$$

It is hard to detect the ISM in the visible range of the electromagnetic radiation, and it has been studied mostly using radio-astronomy techniques. Most of the ISM is made of neutral (HI) and molecular (H_2) hydrogen.

Neutral atomic hydrogen (HI) is the main component of the ISM, with an average density of approximately 0.4 atoms/cm^3. The presence of HI is revealed in the radio band through the 21 cm line. The line was mainly measured in emission along the galactic plane, and with lesser intensity at all galactic latitudes. The linear dimension of the regions in which neutral hydrogen is present is on the order of 100–150 pc.

The emission is due to the fact that the ground state of the hydrogen atom consists of two hyperfine levels. These correspond to configurations with spin of the proton and electron parallel (higher level) and antiparallel (lower level). The emission is due to the transition between levels, whose energy difference corresponds to an electromagnetic emission with frequency $\nu_H = 1420.40575$ MHz, or $\lambda_H = 21.1049$ cm. In thermodynamic equilibrium, the population of each level depends on its energy according to the Boltzmann law. The transition between the two levels has a lifetime $\tau \sim 1.1 \times 10^7$ years. Although this corresponds to a very low transition probability, there is such a large quantity of hydrogen in the Galaxy that the line is clearly detectable. In addition to the emission, the line can also be seen in *absorption* when a cloud of hydrogen is located on the line of sight between the observer and a radio source that emits a continuous spectrum.

About 1% of interstellar hydrogen is ionized (HII). It is generally found in the form of clouds with a density that exceeds 10 atoms cm^3. The ionization is due to the presence of very hot stars that emit photons of energy larger than the ionization energy of hydrogen (13.6 eV). The HII emission is due to *free-free* (or *thermal bremsstrahlung*) transitions, which produce a continuous spectrum.

Radio telescopes have revealed the presence in the ISM of the characteristic lines of many molecules. Molecules emit through their vibrational modes. For example, the simple OH molecule can rotate with respect to the axis that joins the two nuclei, or around an axis perpendicular to it. The vibrational modes of the molecules are quantized, and this allows for identification of the molecules themselves.

About 50% of the mass of the interstellar medium is in molecular form, and most of this seems to be H_2. Unfortunately, the H_2 molecule has no rotational energy levels in the radio band, and the estimates of its presence in the Galaxy are rather uncertain. A large fraction is gathered in clouds, both compact and diffuse, with dimensions that reach 50 pc and high densities (up to 10^{10} molecules/cm^3). The temperatures of these clouds can reach up to thousands of degrees. These clouds correspond to the star-forming regions.

In addition to the H_2 molecule, almost a hundred different molecules and molecular radicals were identified in the ISM. The more complex molecular systems have up to 13 atoms. Most of the complex molecules are organic. None of the inorganic molecules (except ammonia, NH_3) contain more than three atoms. Apparently, as on Earth, the bond with carbon is the key to the formation of complex molecules!

One of the most interesting molecules detectable in the radio is the CO, which is the most abundant molecule after H_2. It has three emission lines (with λ between 1 and 3 mm). CO is a tracer for molecular hydrogen, because its main excitation source is due precisely to collisions with the H_2 molecule.

2.8 Low-Energy Cosmic Rays from the Sun

The Sun is the main source of CRs of energy below ∼4 GeV. Episodic solar activities and the corresponding increase in the low-energy CR flux have a number of effects that are of practical interest. A radiation dose from energetic particles is an occasional hazard for astronauts and for electronics on satellites.[10] Such disturbances may damage power systems, disrupt communications, degrade high-tech navigation systems, or create spectacular aurora.

[10] I was always fascinated by "2001: A Space Odyssey", a science fiction film produced and directed in 1968 by Stanley Kubrick. A spaceship voyage to Jupiter tracing a signal emitted by an unknown object (a monolith) is organized. Most of the spaceship's operations are controlled by a computer on board, HAL 9000 (or simply "Hal", as Hal interacts and speaks with the human crew), and double-checked by a twin computer on Earth. Hal states that he is "foolproof and incapable of error". The main problem arises when Hal foresees an imminent failure on a device. The twin computer on Earth is of the contrary opinion. The humans on board the spaceship discover that Hal is, in fact, wrong. They decide to disconnect it and to assign the spaceship operations to the computer on Earth. This decision induces a fight between the humans and Hal. We are interested in the reason why there is a *discrepancy* between the prevision of the failure asserted by Hal and the conclusion of the twin computer on Earth. The only plausible reason is the fact that the processor units of the Hal on board the spaceship were damaged by cosmic rays. Although it is a science fiction movie, it fully grasped one of the main problems regarding the permanence of humans in space. Computer failures can be prevented by increasing the number of units. This solution cannot be adopted for humans.

Cosmic rays originating from the Sun were first observed in the early 1940s. They consist of protons, electrons, and heavy ions with energy ranging from a few tens of keV to a few GeV. They are originated mainly by *solar flares*.

A solar flare is a sudden brightening observed over the Sun's surface, which is interpreted as a large energy release. Flares occur in active regions around sunspots, where intense magnetic fields penetrate the photosphere to link the corona to the solar interior.

The Sun's activity also influences the probability that CRs with energy below a few GeV reach the Earth. When CRs enter our solar system, they must overcome the outward-flowing *solar wind*. This wind is a stream of charged particles continuously released from the upper atmosphere of the Sun and it consists mostly of electrons and protons with energies usually between 1.5 and 10 keV.

The flux of galactic CR nuclei with energies below ∼1 GeV/nucleon is strongly modulated by their interaction with the magnetic field carried by the expanding solar wind. The expanding magnetized plasma generated by the Sun decelerates and partially excludes the lower energy particles from the inner solar system. Consequently, the low-energy component of the CR flux undergoes a sizable variation over the solar cycle. This effect is known as *solar modulation*.

The magnetic activity and the solar modulations are manifested through *sunspots*, which have a *11-year cycle*. Most solar flares and coronal mass ejections originate in magnetically active regions around visible sunspot groupings. It should be noted that the earliest surviving record of sunspot observation dates from 364 BC, based on comments by Chinese astronomers (Early Astronomy and the Beginnings of a Mathematical Science, University of Cambridge, NRICH, 2007, http://nrich.maths.org/6843). From 28 BC, sunspot observations were regularly recorded by Chinese astronomers in official imperial records.

The intensity of low-energy CRs on Earth is measured through ground-based detectors called *neutron monitors*. Their measurements are anti-correlated with the level of solar activity, i.e., when solar activity is high and many sunspots are visible, the CR intensity on Earth is low, and vice versa. Neutron monitors are designed to measure neutrons produced by the interactions of CRs with the atmospheric nuclei. If the primary CR that started the cascade has energy over ∼500 MeV, some of its secondary by-products (including neutrons) will reach ground level. Figure. 2.11 shows the correlation between the number of observed sunspots (which measure the phase of the 11-years solar activity) and the number of neutrons detected at ground.

The correlation of the CR flux for energies below a few GeV with the solar phase is clearly visible in Fig. 2.12. Here, the flux of protons as measured by the PAMELA experiment (Sect. 3.4.2) in four different years is shown (Adriani et al. 2013). A higher proton flux at energies below a few GeV is evident during 2009, when the number of monthly sunspots in Fig. 2.11 was almost zero.

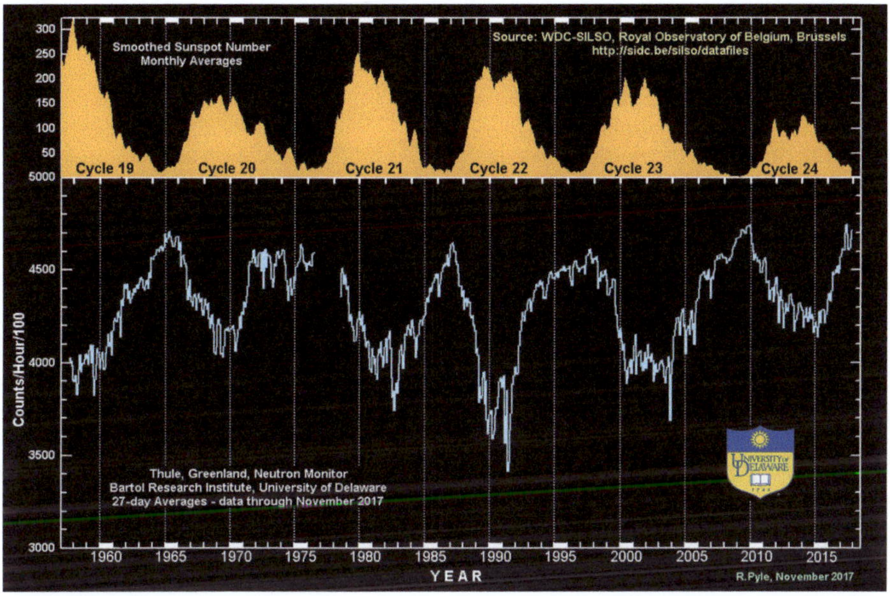

Fig. 2.11 The solar cycle (http://www.bartol.udel.edu/gp/neutronm/). The rate of the neutron monitor (updated monthly), which correlate the sunspot number (on the *top*) with the neutron rate measured at the McMurdo station (*bottom* of the figure). *Credit:* the Bartol Research Institute

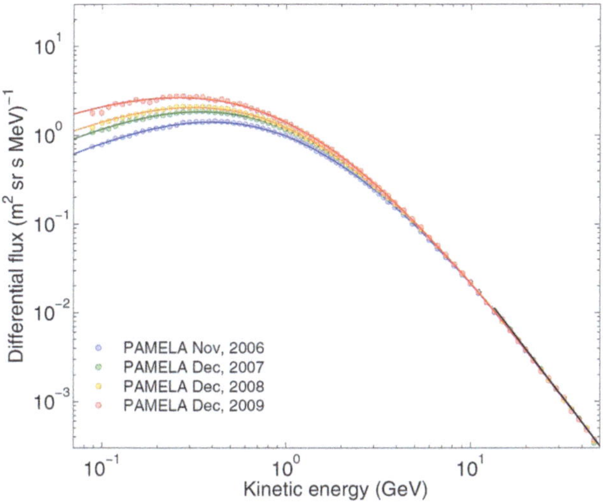

Fig. 2.12 The yearly proton energy spectrum measured by the PAMELA experiment (Sect. 3.4.2) from the beginning of the space mission in mid-2006 until the end of 2009. Compare the phase of the solar activity with Fig. 2.11. The variation in solar activity does not affect CRs with energy larger than a few GeV. Courtesy of the PAMELA collaboration

2.9 The Effect of the Geomagnetic Field

To reach the top of the atmosphere, particles and nuclei below the GeV range are guided by the Earth's dipolar magnetic field. Thus, the intensity of any low-energy component of the cosmic radiation depends both on the location and the time.

For certain magnetic field configurations, there exist regions of space for which the arrival of particles below a certain energy threshold is forbidden. These regions of space are said to be shielded from such particles. Using the concept of a magnetic potential barrier, Störmer first demonstrated (around the 1920s) the existence of a shielded region of the Earth's dipole magnetic field configuration. A simple condition necessary for a particle to reach the Earth's atmosphere is that all trajectories starting from the point considered on Earth (after reversing the charge of the particle) reach $r = \infty$. At low enough energies, this condition may be violated, because trajectories can be deflected back to the Earth or stay within a finite distance. In this case, the magnetic field does induce anisotropies in the observed flux.

Consider a particle of charge *Ze* with orbit in the equatorial plane of the dipole-like magnetic field of the Earth. Equating the centrifugal and Lorentz forces gives us (note that we express the following equations in SI units):

$$\frac{mv^2}{r} = Ze|\mathbf{v} \times \mathbf{B}| \, . \tag{2.28}$$

The Earth's magnetic field is induced by the Earth's magnetic moment M:

$$B = \frac{\mu_0}{4\pi}\frac{M}{r^3} \, . \tag{2.29}$$

At the surface ($r = R_\oplus = 6.38 \times 10^6$ m), the measured value is $B = 0.307 \times 10^{-4}$ T, therefore $M = 7.94 \times 10^{22}$ Am^2. It is easy, from (2.28) and (2.29), to work out the radius of the orbit:

$$r = \left(\frac{\mu_0}{4\pi}\frac{ZeM}{p}\right)^{1/2} , \tag{2.30}$$

where p is the particle momentum. Using the numerical values for $r = R_\oplus$,

$$\frac{p}{Ze} = \frac{\mu_0}{4\pi}\frac{M}{R_\oplus^2} \sim 59.6 \text{ GV}. \tag{2.31}$$

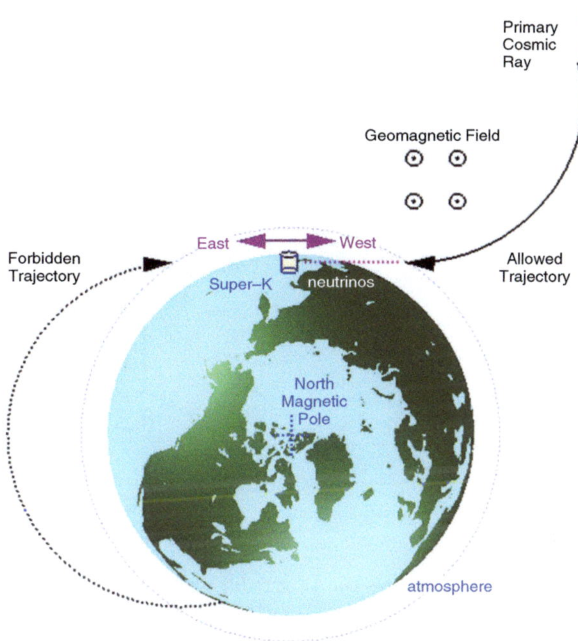

Fig. 2.13 Primary CRs with energy below ∼60 GeV are influenced by the Earth's magnetic field. In particular, low-energy cosmic rays from the East are suppressed compared to those from the West. The Earth itself effectively shadows certain trajectories, which are therefore forbidden (Futagami et al. 1999). This East-West effect was first detected in the 1930s and was used to infer that the charge sign of the primary cosmic rays is positive. Courtesy of Prof. E. Kearns

This value corresponds to the minimum rigidity for a particle to be able to reach the Earth from the East, if its orbit is exactly in the (magnetic) equatorial plane (see Fig. 2.13). In fact, the radius of curvature of the trajectory labeled as *forbidden trajectory* does not reach ∞ starting from the ground. Toward the poles, the influence of the dipole field becomes weaker (as the arriving particle velocity is almost parallel to **B**), and the cutoff rigidity (2.31) becomes smaller. The integrated CR intensity increases with the latitude for charged particles (*latitude effect*). This is exactly the property measured by Compton in 1932 that demonstrated that CRs are positively charged (as we mentioned in Sect. 2.1). The East-West effect also influences the production of low-energy atmospheric neutrinos detected in underground experiments (Chap. 11).

The Earth's dipole-like magnetic field also induces another relevant effect, known as the *Van Allen radiation belts* (named after its discoverer, J. Van Allen). These are two torus-shaped layers of energetic charged particles around the Earth, located in the inner region of the magnetosphere and held in place by the magnetic field (Fig. 2.14). The belt extends from an altitude of about 1000–60,000 km above the surface. The outer belt consists mainly of energetic electrons, while the inner belt is formed by a combination of protons and electrons. The belts pose a hazard to satellites, which must protect their sensitive components with adequate shielding if their orbit spends a significant amount of time inside the radiation belts.

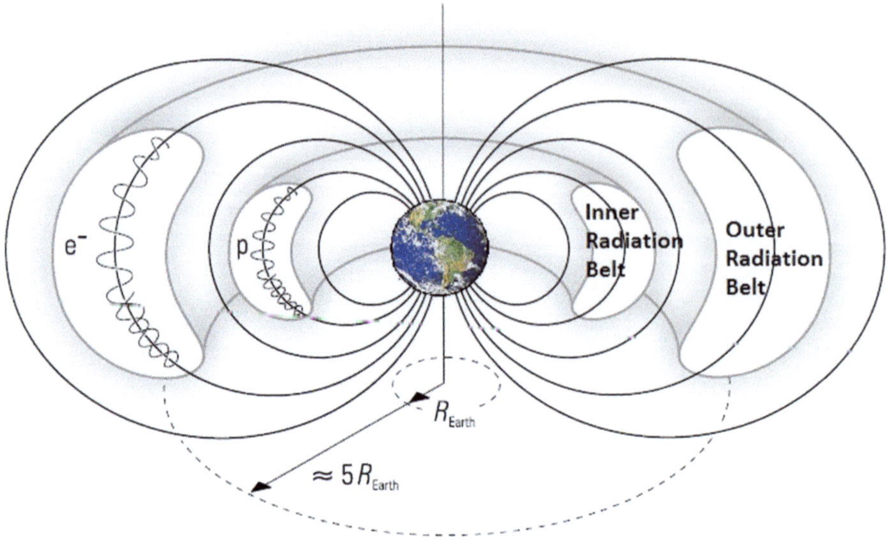

Fig. 2.14 Sketch of the van Allen radiation belts

2.10 Number and Energy Density of Cosmic Rays

After this long journey through the environment in which CRs propagate, let us now evaluate the number density of CRs, according to the definition (2.18a) for $v \simeq c$. We use the experimental flux as parameterized in (2.20b):

$$n_{\mathrm{CR}} = \frac{4\pi}{c}\Phi(>E_0) = \frac{4\pi \cdot K/(\alpha - 1)}{3 \times 10^{10}} \cdot E_0^{-\gamma} . \tag{2.32a}$$

The numerical value depends on the threshold energy E_0. We assume that the parameterization (2.20a) holds down to $E_0 \simeq 3$ GeV, where (somewhat arbitrarily) we put the threshold for the solar wind's contribution. This is also confirmed by Fig. 2.12, which shows no dependence on solar modulations for energies larger than this threshold. Using the values of (2.20c) and $E_0 = 3$ GeV,

$$n_{\mathrm{CR}} \simeq 1 \times 10^{-10}\ \mathrm{cm}^{-3} . \tag{2.32b}$$

A second important quantity is the *kinetic energy density* (or, simply, the *energy density*) of CRs. It can be obtained by the integration of (2.13) assuming the flux given by (2.20a):

$$\rho_{\mathrm{CR}} \equiv \frac{1}{c}\int_{E_0}^{\infty} E\frac{\mathrm{d}^2(N/A\cdot T)}{\mathrm{d}E\mathrm{d}\Omega}\mathrm{d}E\mathrm{d}\Omega \qquad [\mathrm{GeV\,cm^{-3}}]$$

$$= \frac{4\pi}{c}\int_{E_0}^{\infty} 3.01E^{-\alpha+1}\mathrm{d}E = \frac{4\pi}{c}\left[\frac{3.01}{\alpha-2}E^{-\alpha+2}\right]_{\infty}^{E_0}. \qquad (2.33a)$$

The numerical value depends on the threshold energy E_0. As above, using $E_0 = 3\,\mathrm{GeV}$, we obtain

$$\rho_{\mathrm{CR}} \simeq 1\ \mathrm{eV/cm^3}. \qquad (2.33b)$$

We need to compare (2.32b) and (2.33b) with some other astrophysical quantities, in order to understand whether they represent "small" or "large" quantities. The number density (2.32b) of CRs can be compared with the average number density of interstellar matter (2.27). Thus, only about one proton out of $\sim 10^{10}$ not bound in stars in the Galaxy is a relativistic particle, i.e., a cosmic ray.

Concerning the energy density (2.33a), let us consider the following astrophysical quantities.

The Energy Density of the Interstellar Magnetic Field Any magnetic field region is associated with a corresponding energy density that, in the Gauss system of units, corresponds to

$$\rho_B = \frac{1}{8\pi}B^2 \quad \mathrm{erg/cm^3}, \qquad (2.34)$$

(the relation in the SI is $1/(2\mu_0)B^2\ \mathrm{J/m^3}$). Using the numerical values for the average interstellar magnetic field ($B \sim 4\,\mu\mathrm{G}$), we obtain

$$\rho_B = \frac{(4\times 10^{-6})^2}{8\pi} = 6\times 10^{-13}\ \mathrm{erg/cm^3} \simeq 1\ \mathrm{eV/cm^3}. \qquad (2.35)$$

The coincidence within a small factor between (2.33b) and (2.35) suggests a connection between galactic magnetic field and CRs.

The Starlight Density From photometric measurements of the light coming from galactic stars, astronomers have evaluated the visible photon density:

$$n_{\gamma_{\mathrm{vis}}} \sim 2\times 10^{-2}\ \mathrm{cm^{-3}} \quad \rightarrow \quad \rho_{\gamma_{\mathrm{vis}}} \sim 4\times 10^{-2}\mathrm{eV/cm^{-3}}, \qquad (2.36)$$

assuming 2 eV/photon for the visible light. This is a much smaller value than (2.33b)

The Density of the Cosmic Microwave Background Radiation The CMB radiation is the thermal radiation that almost uniformly fills the observable Universe. A precise measurement of CMB radiation is extremely important to cosmology, since any proposed model of the Universe must explain this radiation. The CMB radiation has a thermal black body spectrum at a temperature of 2.725 K (Fixsen 2009), which corresponds to an energy of $E_{\text{CMB}} \sim 3\,\text{kT} = 7 \times 10^{-4}\,\text{eV}$, where $k = 8.61 \times 10^{-5}\,\text{eV K}^{-1}$ is the Boltzmann constant. Using the number density of the CMB radiation, derived from the measurement of the temperature T and Planck's law of black-body radiation,

$$n_{\gamma\text{CMB}} \sim 400\ \text{cm}^{-3} \quad \rightarrow \quad \rho_{\gamma\text{CMB}} \sim 0.3\ \text{eV/cm}^{-3}. \tag{2.37}$$

In this particular case, in spite of the similarity between (2.33b) and (2.37), there is no argued connection between the two phenomena.

2.11 Energy Considerations on Cosmic Ray Sources

Supernova remnants are energetically suitable candidates for the acceleration of CRs with energy below the knee. The main motivation is the equilibrium (first hypothesized by Baade and Zwicky in Baade and Zwicky (1934)) between the loss of CRs due to their escape from the galactic volume and the energy provided by supernova shock waves.

The Galaxy is uniformly filled with the relativistic radiation we detect on Earth. The CR sources are uniformly distributed throughout the Galaxy and the CRs are trapped by the galactic magnetic fields. According to the present observations, the total kinetic energy of CRs corresponds to:

$$\rho_{\text{CR}} \times \mathscr{V}_G = 8\ 10^{54}\ \text{erg}\,, \tag{2.38}$$

where the energy density ρ_{CR} is given in (2.33b) and the galactic volume ($\mathscr{V}_G \sim 5 \times 10^{66}\,\text{cm}^3$) in (2.22). If the particles are completely confined inside the galactic volume, this number should increase with time in the presence of new galactic core-collapse supernova explosions. This process, which represents the candidate injection mechanism for galactic CRs, started a very long time ago, as discussed in Chap. 6. Each supernova burst contributes to increasing the galactic CR density ρ_{CR}. A competitive effect that induces a decrease in ρ_{CR} is due to the escape of CRs out of the Galaxy with a characteristic *escape time* (or *confinement time*) τ_{esc}. This quantity corresponds to the average time needed for a CR, trapped by the galactic magnetic field, to reach the galactic boundary. From here, the particle can freely escape, because the magnetic field outside the galactic plane is negligible.

Anticipating the results that we derive in Sect. 5.2, the confinement time is $\tau_{esc} \simeq 10^7$ years$= 3 \times 10^{14}$ s. Assuming an almost steady value of the energy density ρ_{CR}, the energy loss rate due to the escape of CRs out of the galactic volume is

$$P_{CR} \simeq \frac{\rho_{CR} \times \mathscr{V}_G}{\tau_{esc}} = \frac{8\ 10^{54}}{3\ 10^{14}} = 3 \times 10^{40}\ \text{erg/s}\ . \tag{2.39}$$

Thus, the power required by cosmic accelerators to replenish the galactic volume corresponds to P_{CR}.

This number has a large uncertainty. The assumption of $\rho_{CR} \sim$ const for a time scale $\gg \tau_{esc}$ is completely reasonable from the astrophysical point of view. A large variation in the supernova rate in the last, say, billion years is not expected. Another uncertainty arises from the galactic volume, which is bigger if the *magnetic halo* is considered. As this magnetic field is poorly known, a galactic volume of about $10\mathscr{V}_G$ could be considered. In this case, the matter density in this extended volume is a factor of ~3 smaller. Compressively, the quantity estimated in (2.39) could be a factor of three larger, and $P_{CR} \simeq 10^{41}$ erg/s. Are these powers energetically compatible with the energy released by supernova explosions?

A supernova explosion of 10 solar masses ($10\, M_\odot$) releases about 10^{53} erg, 99% in the form of neutrinos (Sect. 12.11) and 1% in the form of kinetic energy of expanding particles (shock wave). The supernova rate f_{SN} in a galaxy like our own is about 3 per century ($f_{SN} \sim 10^{-9}\,\text{s}^{-1}$). If a physical process able to accelerate charged particles exists, it transfers energy from the kinetic energy of the shock wave to CRs with an efficiency η:

$$P_{SN} \simeq \eta \times f_{SN} \times 10^{51} = \eta \times 10^{42} \quad \text{erg/s}. \tag{2.40}$$

By requiring that $P_{CR} = P_{SN}$, the quantity η must be on the order of a few percent. In this case, the shock waves from supernova explosions are able to refurbish the Galaxy with new accelerated particles and maintain the stationary energy content of CRs. This condition makes the supernova model energetically compatible with the observations. A transfer mechanism with efficiency of a few % is known, and will be described in Sect. 6.2. With a rate of about three supernovae per century in a typical Galaxy, the energy required could be provided by a small fraction (~5–10%) of the kinetic energy released by supernova explosions.

2.12 A Note on Gaussian and SI Units in Electromagnetism

The Gaussian units constitute a metric system that is based on *c.g.s.* units (centimeter-gram-second) for the measurement of length, mass and time, respectively. Usually, classical physics books (and consequently, teachers) use International System (SI) units. The popularity of SI units is because the entire engineering world use volts, ohms, farads, etc. (all SI units). However, the

Table 2.1 Units of distance, mass and time in SI and c.g.s. and some derived quantities

Quantity	SI units	c.g.s. units	Conversion
Distance	Meter (m)	Centimeter (cm)	$1\,\text{cm} = 10^{-2}\,\text{m}$
Mass	Kilogram (kg)	Gram (g)	$1\,\text{g} = 10^{-3}\,\text{kg}$
Time	Second (s)	Second (s)	$1\,\text{s} = 1\,\text{s}$
Velocity	$\text{m}\,\text{s}^{-1}$	$\text{cm}\,\text{s}^{-1}$	$1\,\text{cm/s} = 10^{-2}\,\text{m/s}$
Force	$\text{kg}\,\text{m}\,\text{s}^{-2}$ = Newton	$\text{g}\,\text{cm}\,\text{s}^{-2}$ = dyne	$1\,\text{dyne} = 10^{-5}$ Newton
Energy	$\text{kg}\,\text{m}^2\,\text{s}^{-2}$ = Joule	$\text{g}\,\text{cm}^2\,\text{s}^{-2}$ = erg	$1\,\text{erg} = 10^{-7}$ Joule

Gaussian units make fundamental physical issues and theoretical relations involving electromagnetic phenomena clearer. For this reason, this system is commonly used by astrophysicists and theoretical physicists (the latter sometimes use some *natural units*, in which in addition to Gaussian units there is $\hbar = c = 1$). Because quantum electrodynamics and special relativity are simpler, more transparent, symmetric and elegant in Gaussian units than in SI units, I recommend that all physicist are able to use both SI and Gaussian units. As a bonus, generally the electromagnetism's formulas are simpler and easier to remember in Gaussian units than in SI units. Here, we present only some relevant differences between SI and Gaussian in formula that appears in the book. You can refer to the devoted appendix on Jackson (1999) for a complete discussion of units in electromagnetic formulas.

The differences in the Gauss and SI systems when using only mass, distance and time (and derived quantities) are trivial. In this case, the Gauss system corresponds to the c.g.s. and conversions only involve various powers of 10, see Table 2.1.

The non-trivial difference between SI and Gaussian units arises with the introduction of the unit of electric charge. In SI, a new unit (the Ampere) is introduced associated with electromagnetic phenomena. The electric charge is defined as 1 Coulomb = 1 Ampere × 1 s. By definition, the Ampere has a unique physical dimension, and it is not expressed in terms of the mechanical units (kilogram, meter, second). In Gaussian units, the electrical charge (the statcoulomb, statC) is a *derived* quantity. The starting point is that the intensity of the Coulomb force in Gaussian units is defined as:

$$F = \frac{q_1 q_2}{r^2} \,, \tag{2.41}$$

i.e., two charges of one statC each at one centimeter distance will feel an electrostatic force of one dyne. As a consequence, the electric charge, q, can be written entirely as a dimensional combination of the mechanical units (gram, centimeter, second) as: $1\,\text{statC} = 1\,\text{g}^{1/2}\,\text{cm}^{3/2}\,\text{s}^{-1}$. You may not like these fractional powers, but you are always free to treat the statC as an independent unit.

In SI units, because the electric charge is defined, one need a constant (the $\frac{1}{4\pi\epsilon_0}$ factor) in the Coulomb's law in order to have a force measured in Newton. In cascade, all electromagnetic formulas in SI units have factors such as the permittivity, permeability or impedance of *free space*. Notice that these are not

Table 2.2 Some electromagnetic formulas in Gauss and SI units

Name	Gauss units	SI units	Eq. #
Coulomb's law	$\mathbf{F} = \frac{q_1 q_2}{r^2}\,\hat{\mathbf{r}}$	$\mathbf{F} = \frac{1}{4\pi\varepsilon_0}\frac{q_1 q_2}{r^2}\,\hat{\mathbf{r}}$	I
Lorentz force	$\mathbf{F} = q\left(\mathbf{E} + \frac{1}{c}\mathbf{v}\times\mathbf{B}\right)$	$\mathbf{F} = q\,(\mathbf{E} + \mathbf{v}\times\mathbf{B})$	II
Gauss's law	$\nabla\cdot\mathbf{E} = 4\pi\rho$	$\nabla\cdot\mathbf{E} = \rho/\epsilon_0$	III
Gauss's law for magnetism	$\nabla\cdot\mathbf{B} = 0$	$\nabla\cdot\mathbf{B} = 0$	IV
Maxwell-Faraday equation	$\nabla\times\mathbf{E} = -\frac{1}{c}\frac{\partial\mathbf{B}}{\partial t}$	$\nabla\times\mathbf{E} = -\frac{\partial\mathbf{B}}{\partial t}$	V
Ampère-Maxwell equation	$\nabla\times\mathbf{B} = \frac{4\pi}{c}\mathbf{J} + \frac{1}{c}\frac{\partial\mathbf{E}}{\partial t}$	$\nabla\times\mathbf{B} = \mu_0\mathbf{J} + \frac{1}{c^2}\frac{\partial\mathbf{E}}{\partial t}$	VI
Poynting's vector	$\mathbf{S} = \frac{c}{4\pi}\,\mathbf{E}\times\mathbf{B}$	$\mathbf{S} = \frac{1}{\mu_0}\,\mathbf{E}\times\mathbf{B}$	VII
Vacuum energy density	$u = \frac{1}{8\pi}(E^2 + B^2)$	$u = \frac{\epsilon_0 E^2}{2} + \frac{B^2}{2\mu_0}$	VIII
Electric field	$\mathbf{E} = -\nabla\phi - \frac{1}{c}\frac{\partial\mathbf{A}}{\partial t}$	$\mathbf{E} = -\nabla\phi - \frac{\partial\mathbf{A}}{\partial t}$	IX
Magnetic B field	$\mathbf{B} = \nabla\times\mathbf{A}$	$\mathbf{B} = \nabla\times\mathbf{A}$	X

Refer to the appendix of Jackson (1999) for a more exhaustive discussion

fundamental physical properties of free space, but rather artifacts of the SI system of units, which disappear in Gaussian units.

Thus, the Coulomb's law in the two systems appears as in Eq. I of Table 2.2. As evident from the Maxwell's equations (Eqs. III to VI in Table 2.2), both the ϵ_0 and μ_0 disappear in Gaussian units, and since in the SI

$$\epsilon_0 \cdot \mu_0 = \frac{1}{c^2} \tag{2.42}$$

only factors c enter in Gaussian formulas. From Maxwell's equations, the vectors **B** and **E** have the same dimensions in Gaussian system. Thus, the magnetic field **B** differs by a factor of c in Gaussian units compared to SI units, see the Lorentz force (Eq. II of Table 2.2). Formulas defining quantities as the *cyclotron radius* or *Larmor radius* (2.5) or the angular frequency of the circular motion (2.4) appear different in SI and Gauss units; as guideline, notice that B/c (Gauss) $\rightarrow B$ (SI).

In Gaussian units all the fields **E**, **D**, **P**, **B**, **H** and **M** have the same dimensions, while in SI units the dimensions are all different. This is the reason why it is difficult to remember how to convert some electromagnetic formulas from SI units to Gaussian units (and vice versa). In addition, also the scalar, ϕ, and vector, **A**, potentials have the same dimensions in Gaussian units, but not in SI units. The uniform dimensions for fields in Gaussian units makes fundamental physical relations more transparent. For example, the presence of a dielectrics convert **E** into **D**, and relativity converts **E** into **B**, with coefficients that are dimensionless in Gaussian units. In SI units, it is necessary to insert ϵ_0, μ_0 factors into the conversion formulas, which mask the mentioned unified physical origin.

Finally, the beautiful symmetry of the Poynting's vector and of the vacuum energy density (Eqs. VII and VIII in Table 2.2) express the reason why the Gaussian system is superior for the description of the microscopic world.

References

O. Adriani et al., Time dependence of the proton flux measured by PAMELA during the 2006 July–2009 December solar minimum. Astrophys. J. **765**, 91 (2013)

C. Anderson, The positive electron. Phys. Rev. **43**, 491 (1933)

P. Auger et al., Extensive cosmic-ray showers. Rev. Mod. Phys. **11**, 288–291 (1939)

W. Baade, F. Zwicky, Remarks on Super-Novae and cosmic rays. Phys. Rev. **46**, 76 (1934)

E. Battaner et al., Magnetic fields in galaxies, in *Lecture Notes and Essays in Astrophysics. III Astrophysics Symposium* (2007). http://www.slac.stanford.edu/econf/C07091016/

J. Beringer et al. (Particle Data Group), The review of particle physics. Phys. Rev. D **86**, 010001 (2012). http://pdg.lbl.gov/

S. Braibant, G. Giacomelli, M. Spurio, *Particle and Fundamental Interactions* (Springer, Berlin, 2011). ISBN: 978-9400724631

J.C. Brown, The magnetic field of the milky way galaxy (2010). arXiv:1012.2932v1

P. Carlson, A. De Angelis, Nationalism and internationalism in science: the case of the discovery of cosmic rays. Eur. Phys. J. H **35**, 309–329 (2010)

K. Ferriere, The interstellar environment of our galaxy. Rev. Mod. Phys. **73**, 1031–1066 (2001)

D.J. Fixsen, The temperature of the cosmic microwave background. Astrophys. J. **707**(2), 916–920 (2009). https://doi.org/10.1088/0004-637X/707/2/916

T. Futagami et al. (The SK Collab.), Observation of the East-West anisotropy of the atmospheric, neutrino flux. Phys. Rev. Lett. **82**, 5194–5197 (1999). http://hep.bu.edu/superk/ew-effect.html

J.R. Hörandel, On the knee in the energy spectrum of cosmic rays. Astropart. Phys. **19**, 193–220 (2003)

J.D. Jackson, *Classical Electrodynamics* (Wiley, New York, 1999). ISBN: 978-0471309321

J. Linsley, Evidence for a primary cosmic-ray particle with energy 10^{20} eV. Phys. Rev. Lett. **10**, 146 (1962)

D. Pacini, La radiazione penetrante alla superficie ed in seno alle acque. Nuovo Cimento VI/3 (1912) 93. Translated and commented by Alessandro De Angelis: penetrating radiation at the surface of and in water. arXiv:1002.1810

R. Poggiani, *High Energy Astrophysical Techniques* (Springer International Publishing, Cham, 2017). ISBN: 978-3319447285

M. Prouza, R. Smída, The galactic magnetic field and propagation of ultra-high energy cosmic rays. Astron. Astrophys. **410**, 1–10 (2003)

M. Schein, W.P. Jesse, E.O. Wollan, The nature of the primary cosmic radiation and the origin of the mesotron. Phys. Rev. **59**, 615–633 (1941)

T. Stanev, *High Energy Cosmic Rays* (Springer Praxis Books, Berlin, 2010). ISBN: 9783540851486

B. Wiebel-Sooth, P.L. Biermann, H. Meyer, Cosmic rays. VII. Individual element spectra: prediction and data. Astron. Astrophys. **330**, 389–398 (1998)

Chapter 3
Direct Cosmic Ray Detection: Protons, Nuclei, Electrons and Antimatter

Abstract This chapter refers on the chemical composition of cosmic rays (CRs), i.e., the relative percentage of protons and heavier nuclei in cosmic radiation. Its detailed knowledge up to the highest energies is of crucial importance for the understanding of astrophysical sources of CRs and their propagation in the Galaxy. The chemical composition of CRs can be accurately measured through experiments carried out at a negligible residual atmospheric depth or outside the atmosphere. Here, we deal with the techniques and the experimental results of direct measurements performed with balloons and space missions. These accurately measured the flux and chemical composition of CRs up to about 100 TeV, allowing for the formulation of models around their galactic origin and propagation. One of the key feature derived by these observations is that the CR spectra are well-described by power laws, with similar spectral indices for protons and heavier nuclei, up to energies of $\sim 10^{15}$ eV. The CR sources up to these energies should be concentrated near the galactic disk, with a radial distribution similar to that of supernova remnants.

The relative percentage of protons and heavier nuclei in cosmic radiation is usually referred to as the *chemical composition of CRs*. A detailed knowledge of the chemical composition up to the highest energies is of crucial importance for the understanding of astrophysical sources of CRs and their propagation in the Galaxy. The chemical composition of CRs can be accurately measured through experiments carried out at a negligible residual atmospheric depth or outside the atmosphere. When arriving at the top of the atmosphere, primary CRs start to interact with nuclei of air molecules, producing a cascade of secondary particles. Primary nuclei undergo fragmentation processes, and the information about their mass cannot be easily derived from the *indirect measurements* that are the subject of the next chapter.

In this chapter, we deal with the techniques (Sects. 3.1 and 3.2) and the experimental results of *direct measurements* performed with balloons (Sect. 3.3) and space missions (Sect. 3.4). These accurately measured the flux and chemical composition of CRs up to $\sim$100 TeV (Sects. 3.6 and 3.8), allowing for the formulation of models around their galactic origin and propagation. One of the features predicted by the

© Springer Nature Switzerland AG 2018
M. Spurio, *Probes of Multimessenger Astrophysics*, Astronomy and Astrophysics Library, https://doi.org/10.1007/978-3-319-96854-4_3

standard model of CR acceleration is that the CR spectra are well-described by power laws, with similar spectral indices for protons and heavier nuclei, up to energies of $\sim 10^{15}$ eV. The CR sources are thought to be concentrated near the galactic disk, with a radial distribution similar to that of supernova remnants. The propagation of CRs in the Galaxy is usually studied with a *diffusion differential equation*. The theoretical models of CR acceleration and propagation in the interstellar medium presented in the following chapters are based on the data described here.

Measurements from early space-borne experiments refer mostly to energies lower than 1 GeV. They provided relevant information concerning the energy part of CRs affected by the dependence of the Sun's activity. Important information on the energy spectra of protons, helium, and heavier nuclei has arisen from the PAMELA satellite, launched in 2006. Even more important are the physical outputs of the AMS-02 experiment, launched in 2011 with the Space Shuttle and collecting data on the International Space Station (ISS). AMS-02 (Sect. 3.5) represents the most sophisticated particle detector ever sent into space, incorporating all the characteristics of the very large detectors used in large particle accelerators. AMS-02 is providing fundamental and detailed information concerning the chemical composition of cosmic radiation and the presence of primary antiparticles.

An important feature of the new experiments, including PAMELA and AMS-02, is the presence of magnetic spectrometers that enable the search for antiparticles and antimatter in space. Experimental evidence indicates that our Galaxy is made of matter. Antiparticles can be created as secondary particles by the interaction of CRs with the interstellar medium in our Galaxy. Whether or not there is a significant amount of primary antimatter is one of the fundamental questions of the origin and nature of the Universe. For instance, the observation of only one antihelium nucleus would provide evidence for the existence of antimatter in space. At present, searches for $\overline{p}$ and heavier antinuclei (Sect. 3.9) give no indication of primary sources of antimatter in our Galaxy. On the contrary, the measurements of electrons and positrons (Sect. 3.10) show unexpected features. In particular, an excess of positrons with respect to the expectation from secondary production reported with large statistical significance from PAMELA and AMS-02 has opened theoretical scenarios about their possible origin from dark matter (Chap. 14) annihilations.

3.1 Generalities on Direct Measurements

3.1.1 Generalities and "Data Mining"

Since the discovery of CRs, capabilities for their detection have steadily improved. A large variety of types of experiment (balloon- or satellite-borne, flown on a shuttle, installed on the international space station, or ground-based experiments) and technique (nuclear emulsions, drift chambers, Cerenkov counters, spectrometers...) have been used to refine our knowledge of the CR flux and chemical composition.

In the low energy region, up to about 1 GeV/n, detectors on satellites can identify individual CRs. In some cases (as described in Sect. 3.4.1), different isotopes of the same element can be separated, fully characterized by simultaneous measurements of their energy, charge and mass (E, Z, A). The charge and the time of flight (ToF) can be measured with the $\mathrm{d}E/\mathrm{d}x$ method described in Sect. 2.4. Usually, the ToF system also provides the trigger for other subdetectors.

Some experiments have redundant detectors for the measurement of the electric charge, in particular, if they are interested in searching for antiparticles. The redundancy is mandatory so as to distinguish, for instance, positrons from the more abundant CR protons when positively charged particles are detected. The measurement of the charge Z with different methods makes the nuclei selection very clean. However, the determination of the efficiency in particle selection is usually a difficult task. Nuclei can interact within the detector (or in the atmosphere above the detector, for balloon experiments) producing a hadronic shower and undergoing a fragmentation into lighter nuclei. The corresponding systematic uncertainties are thus important and sometimes difficult to assess.

One important task for the present generation of researchers is to find, have access to and use the huge amount of available data. The general solution to this problem in science is usually referred to as *data mining*. The overall goal of this process is to extract information from a data set and transform it into an understandable structure for further use.

An excellent example of how data can be made available to the community is provided by the database of charged cosmic-ray measurements (CRDB) (Maurin et al. 2014), available at http://lpsc.in2p3.fr/crdb. The database contains information on experimental setups (type, flight dates, techniques) from which the data originate, along with the references to all relevant publications. The database also has associated on-line tools for data selection, data export, producing plots, etc. At present, the information is restricted to cosmic rays with $Z \leq 30$ and a kinetic energy per nucleon up to a few tens of TeV/n. This corresponds to more than 200 different sub-experiments, i.e., different experiments, or data from the same experiment flying at different times, in as many publications.

Different recent review papers have used this database to produce figures and summarize results. Two examples in this book of plots obtained with the database are Figs. 3.13 and 5.4. I recommend that the reader enjoy using the CRDB to reproduce some of the data plots presented in this chapter, or to visualize the results of the experiments that are mentioned.

3.1.2 Energy and Momentum Measurements

Experimentally more challenging is the measurement of the CR energy (or momentum, or rigidity). In calorimeters (Sect. 3.2), the particles need to be (at least partly) absorbed. Within the range from the GeV to $\sim$1 TeV, the particle energy can also be derived using magnetic spectrometers or Cherenkov detectors. Individual elements are identified, characterized by their charge Z through the dE/dx method. At high energy, Transition Radiation Detectors (TRDs) are also used (Boezio and Mocchiutti 2012).

Calorimeters of limited dimension have to be used because of weight and size constraints of balloon and space experiments. The measurement of the kinetic energy up to energies of a few GeV can be obtained with a homogeneous calorimeter selecting noninteracting stopping particles. At higher energies, sampling calorimeters are used. The weight of a detector with a thickness of one hadronic interaction length (Sect. 3.2.1) and area of 1 m^2 amounts to $\sim$1 ton. In some cases, multiple energy measurements are needed in order to cover the largest possible energy range and to perform a cross-calibration of detectors with different systematic uncertainties.

Magnetic spectrometers consist of a tracking device system inside a region with a magnetic field generated by a solenoid (either permanent or superconductive). Magnetic spectrometers can measure the particle rigidity up to a maximum value that depends on the magnetic field and on the precision of the measurement of the curvature of particles traversing the magnet. The maximum detectable rigidity is reached when they are poorly described by arcs, and appear to be straight lines. The mass constraints on payloads limit these measurements to about $\sim$10–100 GV for experiments carried by balloon, 1 TV in PAMELA and 2 TV in AMS-02.

In some experiments, **transition radiation detectors (TRDs)** have been used to measure the Γ Lorentz factor of the incident particle, and thus the energy, replacing bulky and heavy hadronic calorimeters. The information from a TRD, together with data from other detectors, allows for separation of different nuclear species. Radiation (in the X-ray band, and usually denoted as *transition radiation*) could be produced when a particle with high Γ crosses several interfaces characterized by an abrupt change of the refraction index. Particles with large Γ induce X-rays with large probability. For a given energy, this allows for discrimination between light and heavy particles, such as $\Gamma = E/mc^2$. A TRD contains many layers of transparent materials with different indices of refraction n in order to increase the photon emission probability of a single layer.

3.2 The Calorimetric Technique

The *calorimetric* technique is another method used for determination of the energy of CRs. The calorimeters on board balloons or satellites are very similar to those widely used in high-energy physics experiments in accelerators, although with limits on payload weight.

In a *calorimeter*, the kinetic energy of the incident particle is converted inside an absorber into a cascade of many secondary particles (the *shower*). At the end, the primary energy of the incident particle is dissipated via excitation/ionization of the absorbing material.

A primary CR interacting with a nucleus produces a large number of secondary hadrons. They deposit energy through ionization/excitation of the medium and through successive interactions with nuclei (see Eq. (1.1)) yielding lower energy hadrons, the *hadronic cascade*. Neutral mesons produced in the cascade, mainly π^0, immediately decay into a $\gamma\gamma$ pair. In matter, each high-energy photon converts into an electron-positron pair; each e^- (e^+) is able to radiate energetic photons through bremsstrahlung. These radiated photons can convert into pairs that, in turn, radiate. In conclusion, one has an *electromagnetic shower* (*electromagnetic cascade*) with a large number of photons, electrons, and positrons.

3.2.1 Hadronic Interaction Length and Mean Free Path

The development of the hadronic cascade depends on the material medium through the *nuclear* (or *hadronic) interaction length* λ. This quantity corresponds to the mean path length after which the number of hadrons is reduced by the factor e. It is inversely proportional to the interaction cross-section σ and the medium number density n [cm^{-3}]:

$$\lambda = \frac{1}{n\sigma} \quad [\text{cm}] \,. \tag{3.1}$$

As the number density depends on the matter density ρ (g cm^{-3}), more frequently, the nuclear interaction length for a nucleus with mass number A is expressed in units [g cm^{-2}] as

$$\lambda_\text{I} = \frac{\rho}{n\sigma} = \frac{Am_\text{p}}{\sigma} \quad [\text{g cm}^{-2}] \,, \tag{3.2}$$

where m_p is close to the proton mass.

In the simplest model, the nuclear cross-sections are assumed energy-independent and proportional to the geometrical area of the interacting nuclei:

$$\sigma = \pi R^2 \quad \text{with} \qquad R \simeq R_T + R_P \,, \tag{3.3}$$

where R_P and R_T are the projectile (P) and target (T) radii. From nuclear models, the radius of a nucleus with mass A is

$$R_A = r_\circ A^{1/3} \quad \text{with} \quad r_\circ \simeq 1.2 \times 10^{-13}\text{cm} \,, \tag{3.4}$$

where $r_\circ$ is a parameter whose numerical value is close to the charge radius of the proton, $\sim 0.85 \times 10^{-13}$ cm. This model is confirmed by the experimental fact that the proton-proton cross-section (σ_pp) outside the region of the resonances is almost constant in a wide energy range [see Sects. 7.3 and 7.4 of Braibant et al. (2011)] at the value given by the geometrical cross-section, $\sigma_\text{pp} \simeq \pi r_\circ^2 \simeq 45$ mb.

Table 3.1 Main characteristics of detectors and absorber materials used in calorimeters

Material		X_0 (g cm^{-2})	λ_I (g cm^{-2})	E_c (MeV)
Active detectors	NaI	9.5	151	12.5
	BGO	8.0	157	7
Passive absorbers	Fe	13.8	132	28
	Pb	6.4	194	9.5
	U	6.0	199	9
Air [STP]	Mixture	36.7	90	86

For air, the Standard Temperature and Pressure (STP) conditions are considered. X_0 = radiation length, λ_I = interaction length for protons, E_c critical energy = energy where energy loss through radiation equals that due to excitation/ionization

The p-p cross-section increases very slowly with the center-of-mass energy $\sqrt{s}$ of the interaction. The σ_{pp} reaches ~70 mb at LHC energies ($\sqrt{s} \sim 10$ TeV), corresponding to primary CR protons with kinetic energy of ~10^5 TeV.

In most typical situations, the cross-section for a proton in a medium with mass number A is approximated by

$$\sigma \simeq \pi[r_\circ A^{1/3}]^2 = \sigma_{\text{pp}} A^{2/3} \,. \tag{3.5}$$

Some typical values of interaction lengths (Eq. (3.2)) using the above proton cross-sections are given in Table 3.1.

Exercise In the interaction of particles (or nuclei) with matter, the number of collisions depends on the number of scattering centers per unit volume. Often, the scattering centers are atomic nuclei. Consider, for example, the case of carbon, $A = 12$ and density $\rho \simeq 2.265\,\text{g cm}^{-3}$. Determine: the number of atoms per cm^3; the number of atoms per gram.

3.2.2 The Electromagnetic Radiation Length

In the electromagnetic component of the cascade, the electron and positron energy losses are mainly due to the excitation/ionization and bremsstrahlung processes (Fig. 2.5b) which can be parameterized as

$$-\frac{\mathrm{d}E}{\mathrm{d}X} \simeq \alpha(E) + \frac{E}{X_0} \,, \tag{3.6}$$

where the depth X is measured in units of (g cm^{-2}), the term $\alpha(E)$ due to the excitation/ionization energy loss is only slightly energy-dependent and the *radiation*

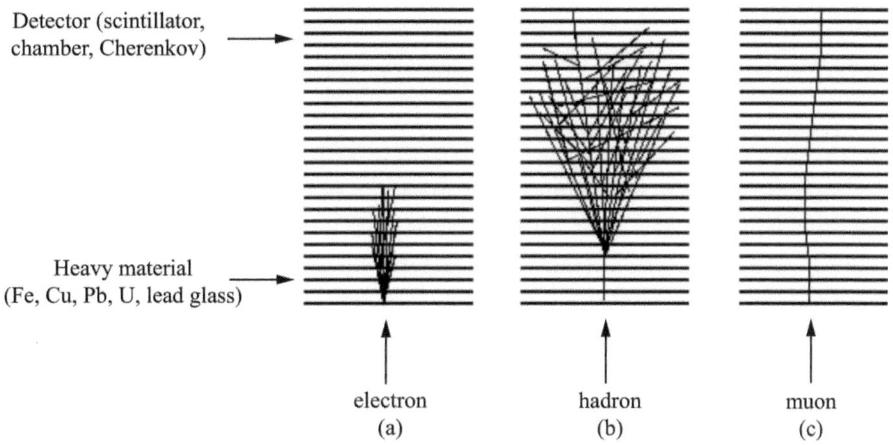

Fig. 3.1 Behavior **a** of an electromagnetic cascade, **b** of a hadronic cascade and **c** of a muon when traversing a sampling calorimeter constituted of an interleaved detector and absorber layers of heavy material. A muon behaves as a minimum ionizing particle and can be easily identified (Braibant et al. 2011)

length X_0 depends on the material. The quantity X_0 corresponds to the length after which the energy E_0 of the incident electron is reduced to E_0/e. Typical values of radiation lengths X_0 for some materials are listed in Table 3.1.

The development of the electromagnetic cascade continues until the energies of the electrons and positrons fall below the critical energy E_c. Below the critical energy, e^+, e^- lose energy mainly through the excitation/ionization process. In a medium with high Z, such a cascade has limited longitudinal and transverse dimension, as illustrated in Fig. 3.1a. An electromagnetic calorimeter measures the total energy produced through excitation/ionization due to e^+, e^-, γ.

Calorimeters for Space Experiments The most appropriate material for an electromagnetic calorimeter would be one with a short X_0. A hadron calorimeter should have a short interaction length λ_I to force hadronic interactions as the particle enters the detector and to completely absorb the cascade. The hadronic cascade is wider and longer than that of the electromagnetic (see Fig. 3.1b). Usually, a calorimeter for hadrons (or *hadronic calorimeter*) must have larger dimensions than the one for electromagnetic showers (or *EM calorimeter*) and in accelerator experiments, it usually has a thickness of six nuclear interaction lengths λ_I. Calorimeters carried by balloons at the top of the atmosphere (or outside the atmosphere by satellites) are limited in absorber thickness due to weight restrictions. The minimum depth depends on the energy resolution required for a particular experiment. Typically, the electromagnetic component is reliably measured up to the energy for which the maximum of the shower is contained within the calorimeter.

3.2.3 *Hadronic Interaction Length and Mean Free Path in the Atmosphere*

The Earth's atmosphere itself (vertical thickness $X_v^{\rm atm} = 1,030\,{\rm g\,cm^{-2}}$, Sect. 4.2) acts like a calorimeter of variable density with $\sim$11 interaction lengths and 28 radiation lengths (compare with Table 3.1). For this reason, direct CR measurements can occur only outside the atmosphere, or at a significant height.

If the considered medium is the Earth's atmosphere, with $A \simeq A_{atm} = 14.5$ with our approximated formula, Eq. (3.5), we obtain

$$\sigma_{p,air} = 270\,{\rm mb} = 0.27 \times 10^{-24}\,{\rm cm}^2\,, \tag{3.7}$$

which is in agreement with the measurement of the p-air cross-section for CR protons within the GeV-TeV energy range. At higher energies, the p-air cross-section is slightly higher, reaching $\sim$500 mb at CR proton energies of 10^6 TeV (Abreu et al. 2012).

Using (3.2), the mean free path for CR protons in the atmosphere is:

$$\lambda_I \simeq \frac{14.5 \cdot 1.6 \times 10^{-24}}{0.27 \times 10^{-24}} = 85\ {\rm g\ cm^{-2}}\,. \tag{3.8a}$$

An interesting consequence of the hadron quark model is that the cross-section for a secondary pion is $\sigma_\pi \simeq 2/3\sigma_p$ (see §7.14.3 of Braibant et al. (2011)). Thus, the mean free path of secondary pions in the atmosphere is

$$\lambda_I^\pi \simeq 120\ {\rm g\,cm^{-2}}\,. \tag{3.8b}$$

Finally, for a heavy CR nucleus (e.g., iron, with $A_{\rm Fe} = 56$) using Eq. (3.3), we obtain a ration between the Fe-air and p-air cross-sections of

$$\frac{\sigma_{Fe,air}}{\sigma_{p,air}} \simeq \frac{(A_{Fe}^{1/3} + A_{Atm}^{1/3})^2}{A_{Atm}^{2/3}} = 6.5 \quad \text{and thus} \quad \lambda_I^{Fe} \simeq \frac{85}{6.5} = 13\ {\rm g\ cm^{-2}}\,. \tag{3.8c}$$

This value is in agreement with dedicated measurements (Crane et al. 1983).

3.3 Balloon Experiments

The hypothesis of the existence of extraterrestrial radiation was confirmed with experiments using balloon ascensions. Balloon experiments have always been important throughout the history of CRs. As a curiosity, in the early 1930s, stratospheric flights were made with huge rubberized fabric balloons. Aeronauts in sealed, airtight capsules were able to survive up to a record altitude of 22 km

Fig. 3.2 (**a**) CREAM ballooncraft with the launch vehicle while a 10^6 m^3 balloon is being inflated at the launch site, Williams field near McMurdo, Antarctica; (**b**) the balloon trajectory of a 37-day flight of CREAM, which was launched on December 1, 2009, and terminated on January 8, 2010, during about three rounds of the South Pole. Courtesy of Prof. E. Seo (CREAM Collaboration) http://cosmicray.umd.edu/cream/

(in 1935). Scientific balloons used today (for instance, for NASA flights) are made of 20 μm thick polyethylene film; they are as large as a football stadium, with a diameter of about 140 m and a volume larger than 1.1 million cubic meters filled with helium gas. They can carry experiments (payloads) up to 3600 kg and fly at altitudes up to 42 km. The payloads are attached to a parachute. The flights are terminated by remotely firing an explosive squib that separates the payload from the balloon. The experiment descends slowly, suspended by the parachute, and it is recovered and refurbished for future flights.

In the early 1990s, remarkably successful long flights around Antarctica (1–2 weeks) began (Seo 2012). The duration of conventional 1–2 day flights was limited due to altitude excursions during day–night transitions. The continuous solar heating during local summer in the polar region ensures nearly constant altitudes with essentially no ballasting. From 2005 to 2010, the CREAM payload flew six times, for a cumulative exposure of 162 days and with a record duration of almost 40 days, while circumnavigating Antarctica three times (Fig. 3.2).

Apart from the measurement of the cosmic ray composition in the region below the knee (presented in this chapter), balloon experiments, as well as detectors on satellites, are devoted to searches for antimatter (Sect. 3.9) and dark matter in our Galaxy (Chap. 14). Here, we mention only some recent and large statistics experiments.

The improvement of flight duration and payload capability allows for the carrying of complex and heavy experiments. The major improvement has been the use of *superconducting magnet* spectrometers with a suite of particle detectors to identifying antiparticles. These experiments were the Balloon Experiment with a superconducting Solenoid Spectrometer (BESS), the Cosmic AntiParticle Ring-Imaging Cherenkov Experiment (CAPRICE) and the High-Energy Antimatter Telescope (HEAT). Usually, a date or an extension is used to distinguish different flights of the same detector when it has flown more than once, sometimes with a

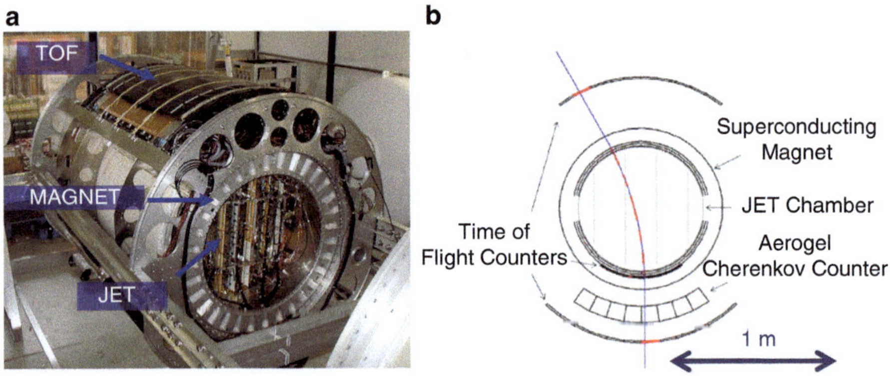

Fig. 3.3 (**a**) A picture of the original BESS apparatus. (**b**) A cross-sectional view of the BESS-Polar instrument also showing a particle's trajectory. Courtesy of Dr. J.W. Mitchell and Prof. A. Yamamoto (BESS Collaboration)

slightly different configuration. For instance, BESS has had nine successful flight campaigns since 1993, and the last one was denoted as BESS-polar.

In BESS, the particle rigidity (2.6) is measured in the superconducting spectrometer, where a uniform magnetic field of $\sim 10^4$ G acts for 1 m on the particles. The sagitta is measured with the central tracker made of *drift chambers*, having a spatial resolution of $\sim 150\,\mu$m. The ToF system provides the measurement of the particle direction, velocity $\beta = v/c$ and electric charge Ze. Figure 3.3a shows a picture of the BESS apparatus, in which the magnetic barrel is visible. The cross-sectional view of the instrument and a charged particle's trajectory is sketched on the right side of the figure.

The Advanced Thin Ionization Calorimeter (ATIC) was configured with a homogeneous Bismuth Germanium Oxide (BGO) calorimeter with about $20\,X_0$. The calorimeter is preceded by a silicon matrix for the measurement of the particle's electric charge. Since December 2000, ATIC had three successful flights from McMurdo, Antarctica (see Table 3.2). One of the main results of the ATIC flights was the reported excess of electrons and positrons around 600 GeV (see Sect. 3.10). Concerning the measurement of primary CR composition, the ATIC data filled the gap for elements from protons ($Z = 1$) to iron ($Z = 26$) of the measurements made by experiments using spectrometers and higher energy data from emulsion-based experiments.

Pioneering calorimeters using emulsions for measurements of CRs above 2 TeV were made by the Japanese-American Collaborative Emulsion Experiment (JACEE) and the RUssian-Nippon JOint Balloon Experiment (RUNJOB) collaborations. Both detectors had limited charge resolution and measured groups of nuclei with close electric charge. The experiments used the passive calorimetric techniques of emulsions and X-ray films. These passive techniques limit the exposures, because of the integrating effects of the background. Long exposures of experiments that use

Table 3.2 Comparison of balloon experiments for high-energy CR measurements

Instrument	Energy measurement	Charge range	Flight duration (days)	Atmospheric overburden (g/cm^2)	Exposure (m^2 sr days)	N_p $E > 6$ (TeV)
ATIC	Calorimeter	$1 \leq Z \leq 28$	48	4.3	5	~720
TRACER	TRD	$8 \leq Z \leq 28$	10	3.9	50	–
CREAM	Calorimeter	$1 \leq Z \leq 28$	160	3.9	48	~5000
	TRD	$3 \leq Z \leq 28$	42	3.9	55	–
JACEE	Emulsion	$1 \leq Z \leq 28$	60	5.3	10	~700
RUNJOB	Emulsion	$1 \leq Z \leq 28$	60	10	24	~700

The energy measurement techniques are identified in Column 2; the range of the electric charge measurement in Column 3. The charge resolution was $\sim \Delta Z = 0.3$ for all experiments, apart from JACEE and RUNJOBS

passive detectors would require frequent replacement of the emulsion plates and X-ray films.

The Transition Radiation Array for Cosmic Energetic Radiation (TRACER) experiment extended the flux measurements of heavy nuclei to higher energies. TRACER was configured with two layers of plastic scintillators ($2 \times 2\,\text{m}^2$), which measured the electric charge, and a TRD to determine the Γ of the incident particle. The TRD characteristics of this experiment preclude the measurement of light nuclei. In addition to the TRD, a Cherenkov counter (made of acrylic plastic) at the bottom of the detector was used to reject nonrelativistic particles. TRACER reported elemental spectra from oxygen ($Z = 8$) to iron ($Z = 26$) from a flight in Antarctica in 2003 (Table 3.2), and from boron ($Z = 5$) to iron in a second flight in 2006.

The Cosmic Ray Energetics and Mass Balloon Experiment (CREAM) used both a calorimeter (with 20 radiation lengths X_0 and half interaction length λ_I) and a TRD for the measurements of the CRs' energy. The two subdetectors had different systematic biases in determining the particle energy allowing an in-flight cross-calibration of the two techniques for particles with $Z \geq 4$. The CREAM calorimeter measured all elements, including nuclei with $Z = 1$ and 2, up to $\sim 10^{14}$ eV with energy resolution better than 45% for all energies. The highly segmented detectors comprising the instrument had about 10^4 electronic channels. The CREAM calorimeter was designed to be large enough to collect adequate statistics, within the weight limit for a balloon flight. It used a tungsten absorber (tungsten has high Z and a small radiation length, $X_0/\rho = 0.35$ cm) and thin scintillating fibers.

Table 3.2 compares the quoted balloon experiments for high-energy measurements: CREAM (Ahn et al. 2007), ATIC (Ganel et al. 2005), TRACER (Ave et al. 2008), JACEE (Asakimori et al. 1998) and RUNJOB (Derbina et al. 2005).

3.4 Satellite Experiments

Primary CR data from early space experiments refer mostly to energies lower than 1 GeV. Of particular importance was the first measurement of isotopic Li, Be, and B flux with experiments on satellites in the 1970s (Garcia-Munoz et al. 1975) with the Interplanetary Monitoring Platform-7 and -8 (IMP) spacecrafts, which measured CRs up to $\sim$100 MeV. The IMP was a series of 10 scientific satellites launched by NASA between 1963 and 1973. These experiments were sensible mainly in an energy region where the solar wind significantly affects CRs, and provided relevant information concerning solar activity. Other space experiments were on board the High-Energy Astrophysics Observatory (HEAO-3) satellite and the Cosmic Ray Nuclei (CRN) experiment, which flew for nine days on the Space Shuttle. The PAMELA experiment was a powerful particle identifier using a permanent magnet spectrometer with a variety of specialized detectors that accurately measures, with high sensitivity, the abundance and energy spectra of CR electrons, positrons, antiprotons and light nuclei over a very large range of energies from 50 MeV to hundreds of GeV.

In the following, we describe the Be isotopes flux measurement done by one of the first space experiments in the early 1970s, and the experiments representing the state of the art after 40 years of research: PAMELA and AMS-02.

3.4.1 *The IMP Experiments*

Among the different physics studies performed by the IMP-7 (launched in 1972) and IMP-8 (launched in 1973) experiments, one very important result was the detection of the ^{10}Be (Garcia-Munoz et al. 1977). This was the first measurement of this isotope (important for the determination of the "age" of CRs), which was repeated on successive satellite missions (Voyager, Ulysses, CRIS) (Sect. 5.2).

The general layout of the telescope is shown on the left side of Fig. 3.4. It is a small-scale version of our *toy* telescope, described in Sect. 2.4. Particles passing through the detector layers D1, D2, D3, and which have come to rest in D4, are considered. Counters D5 and D6 act as *veto*, to confirm that the particle is stopped in D4. The separation between different chemical species and isotopes is achieved using the $\mathrm{d}E/\mathrm{d}x$ technique (measured in D1, D2) as a function of the total energy released in D4. The events resulting from each isotope are located along distinct lines that are approximately equilateral hyperbolae, as shown on the right side of Fig. 3.4 (from IMP-8). Nuclei with different Z are well-separated by their energy loss, which depends on $(Z/\beta)^2$ of the particle. At a given total energy (proportional to the signal D4) of a nucleus with a given Z, the velocity $v = \beta c$ is smaller for the isotope with a larger A. Thus, ^{10}Be produces a larger signal in D1, D2 than the other isotopes ^{7}Be and ^{9}Be.

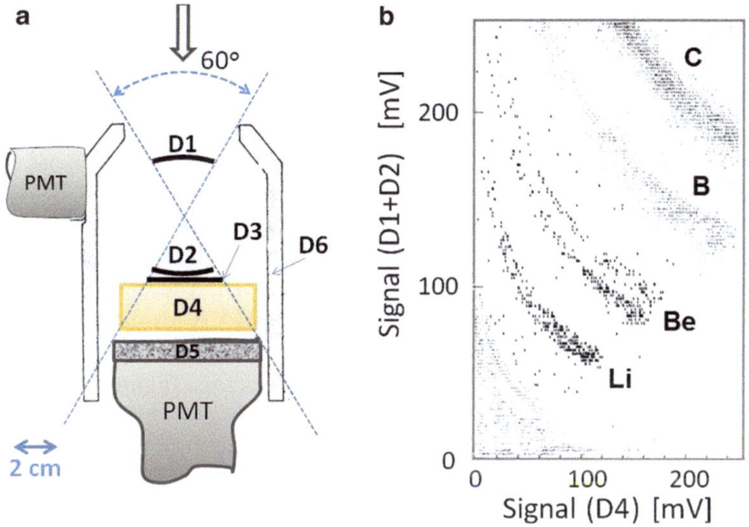

Fig. 3.4 (**a**) Layout of the telescope on board IMP-7, -8. The silicon detectors (denoted as D1, D2, D3) measure the particle energy loss. D4 is a CsI (Tl) scintillator viewed by four photodiodes. D5 and D6 are two additional scintillators. (**b**) Result from IMP-8 (data collected between 1974 and 1975). On the y-axis, the signal is proportional to the energy loss $\propto Z^2/\beta^2$ measured in (D1 + D2), on the x-axis, to the residual energy of the particle measured in D4. At a given energy, the value on the y-axis increases by Z^2. For a given Z and energy, nuclei with larger A have smaller β and undergo larger energy losses in D4

Very few ^{10}Be nuclei were collected by these small acceptance experiments. Assuming that all the Be isotopes are produced during propagation in the Galaxy of C, N, and O nuclei with the production cross-sections as estimated using accelerator data (see discussion in Sect. 5.1 and Table 5.1), we would expect a ratio between the three elements to be

$$^7\text{Be} : {}^9\text{Be} : {}^{10}\text{Be} = 9.7 : 4.3 : 2.3 \quad \text{(expected).} \tag{3.9}$$

The IMP measurements of the different Be isotopes give us

$$^7\text{Be} : {}^9\text{Be} : {}^{10}\text{Be} = 329 : 177 : 15 \quad \text{(measured).} \tag{3.10}$$

Through comparison of the expected and measured values, the ratio ^{7}Be:^{9}Be is almost equal to the ratio of production cross-sections. ^{10}Be is instead largely suppressed, and the only explanation is that a large fraction of this unstable isotope had time to decay after being originated, before detection. This allows for the measurement of the *CR escape time*, as discussed in Sect. 5.1.

3.4.2 The PAMELA Experiment

The Payload for Antimatter Matter Exploration and Light-nuclei Astrophysics (PAMELA) was an experiment mainly devoted to antimatter studies up to the hundreds of GeV region. Its size and flight durations were such as to allow an accurate measurement of the neutral and charged components of CRs up to the TeV region. PAMELA was launched on board the Russian Resurs-DK1 satellite by a Soyuz-U rocket from the Baikonur cosmodrome in Kazakhstan in June 2006. The operation of Resurs-DK1 was terminated in 2016. The main features of the instrument and a summary of the most relevant PAMELA results over a decade of observation are in Adriani et al. (2017).

The apparatus was composed of different subdetectors. Figure 3.5 shows, from top to bottom:

- a time of flight system, ToF (S1, S2, S3);
- an anticoincidence system (denoted in the figure as CARD, CAT, CAS);
- a magnetic spectrometer;
- an electromagnetic imaging calorimeter;
- a shower tail catcher scintillator (S4);
- a neutron detector.

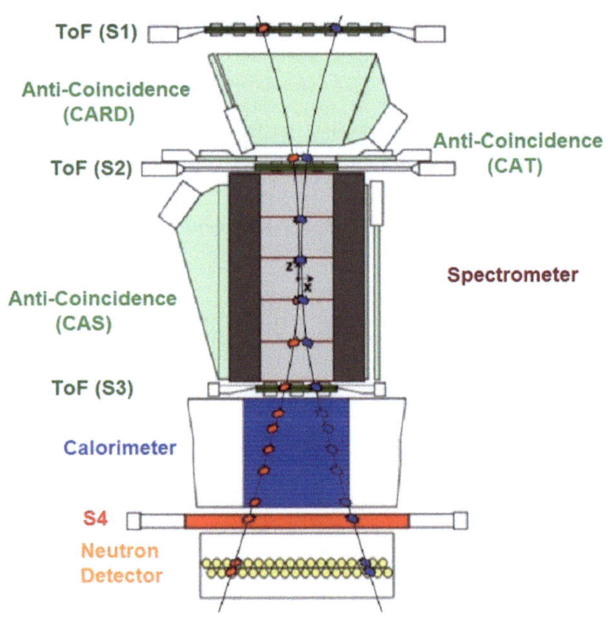

Fig. 3.5 Cross-section of the PAMELA detector with two opposite-sign charged particles. Combined measurements from the magnetic spectrometer, calorimeter, time of flight (ToF) system, and neutron detectors shown here distinguish the incident particles by their charge, momentum, and mass. Credit: PAMELA Collaboration http://pamela.roma2.infn.it

The ToF system comprised six layers of fast plastic scintillators arranged in three double planes (S1, S2 and S3). It provided a fast signal used to trigger the data acquisition and to measure the time-of-flight and $\mathrm{d}E/\mathrm{d}x$ of traversing particles. The ToF resolution of ∼0.3 ns allowed $e^-(e^+)$ to be separated from $\overline{p}(p)$ up to 1 GeV/c. Due to this very accurate time resolution, particles traversing the detector from the bottom (*albedo particles*) were rejected with a large statistical significance.

The central part of the PAMELA apparatus was the magnetic spectrometer, consisting of a 0.43 T permanent magnet and a microstrip silicon tracking system, with a spatial resolution of 3 μm. The maximum detectable rigidity was ∼1 TV. The dimensions of the magnet defined the geometrical factor (Eq. (2.12)) of 21.5 cm^2 sr. Ionization losses were measured in the ToF scintillator planes, in the silicon planes of the tracking system and in the first silicon plane of the calorimeter, allowing for a measurement of the electric charge up to $Z = 8$.

The sampling electromagnetic calorimeter comprised 44 silicon planes interleaved with 22 plates of tungsten absorber, for a total depth of the calorimeter of 16.3 radiation lengths and 0.6 interaction lengths. A plastic scintillator system mounted beneath the calorimeter helped in the identification of high-energy electrons; it was followed by a neutron detection system, which complemented the electromagnetic/hadronic discrimination capabilities of the calorimeter by detecting the increased neutron production associated with hadronic showers compared to electromagnetic ones in the calorimeter.

3.5 The AMS-02 Experiment on the International Space Station

The AMS-02 is the largest particle physics detector ever carried outside the atmosphere (Fig. 3.6). It was designed to operate as an external module on the International Space Station (ISS) (Fig. 1.7). It studies the composition and flux of cosmic rays with unprecedented precision. In addition, it could open new information on the Universe and its origin by searching for antimatter and dark matter candidates. Concisely, AMS-02 utilizes 15 among particle detectors and supporting subsystems in a volume of 64 m^3; its weight is 8500 kg and it dissipates 2.5 kW. It has a data downlink bandwidth of 9.6 Mbps. It was launched on May 16th, 2011, on board the shuttle Endeavour. The mission duration is expected to coincide with the lifetime of the ISS (2020 or longer): the detector will not come back to Earth. Its subdetector system is a sort of compendium of devices carried by preceding satellites and balloon experiments. For this reason, we will describe it in some detail.

The AMS-02 prototype was designated as AMS-01. It was a simplified version of the detector, which flew into space aboard the Space Shuttle Discovery in June 1998 for about 10 days. AMS-01 proved that the detector concept worked in space and provided some measurements quoted in the following sections.

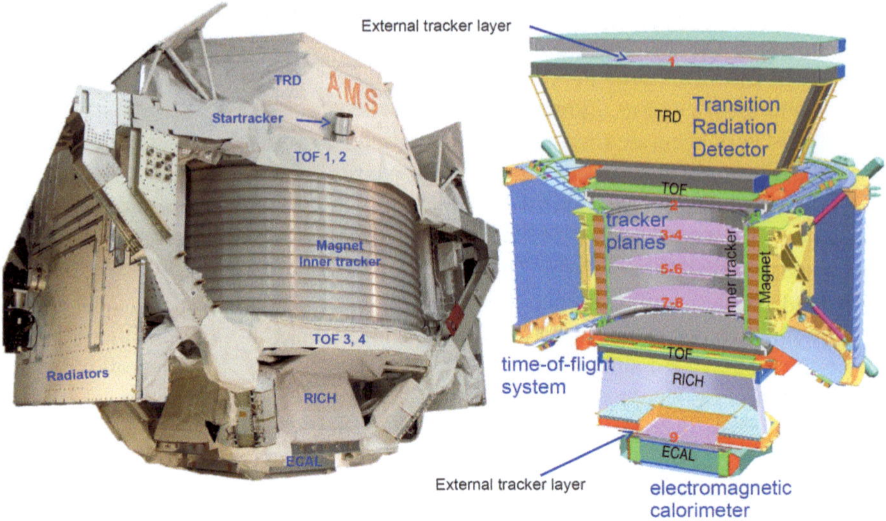

Fig. 3.6 The AMS-02 detector. See text for the details of different subdetectors. Credit: AMS-02 Collaboration (http://www.ams02.org/)

AMS-02 uses a very large permanent magnet (which is the heart of the detector) operating at ambient temperature, consisting of 6000 Nd–Fe–B pieces carefully magnetized and assembled. This magnet was successfully flown on the AMS-01 mission. The use in AMS-02 of a superconducting magnet was also tested, but this solution was discarded. The currently available technology does not allow for keeping a superconducting magnet operational on the ISS for more than 2–3 years. The configuration of the magnet has a negligible net dipole moment, to avoid coupling with the Earth's magnetic field, which would disturb the orbit of the ISS. The Nd–Fe–B magnets are the strongest permanent magnets, and that of AMS generates a magnetic field of 1.5 kG.

Inside the magnet, the *central Tracker* is able to precisely measure the curvature of the particles traversing the magnet and determine the particle rigidity. The tracker records the coordinate of a particle at eight different positions with a precision of $\sim$10 μm at each point. The radius of the best circular trajectory passing through the points defines the particle curvature, used to evaluate the rigidity through Eq. (2.6). The Tracker is made of 2264 double-sided microstrip sensors ($7.2 \times 4.1\,\text{cm}^2$, 300 μm thick) assembled in 192 read-out units, totaling 200,000 read-out channels. The large number of channels generates about 200 W of heat, which must be dissipated. The maximum detectable rigidity for the AMS-02 Tracker is about 2 TV, a very high value with respect to other space experiments.

The exact position of each module is of fundamental importance for determination of the trajectory. The procedure that determines the modules' misplacements is called *alignment*. The AMS-02 Tracker was aligned in 2010 using a straight proton beam at CERN. The proton beam was used as a reference. Translation and rotation constants for all Tracker modules have been determined with respect to it. Thermal conditions may change rapidly in space. This fact can introduce mechanical deformations and misalignments, which will affect the rigidity measurement. A Tracker Alignment System (TAS) made of laser beams that mimic straight tracks provide a fast and reliable monitoring of the Tracker's geometrical stability during the AMS-02 mission. Therefore, systematic effects due to misalignments can be monitored and corrected.

The time of flight (ToF) system is able to measure, with a precision of $\sim$0.15 ns, the particle transit time through the detector and provides triggers for the other subdetectors. Since the distance between the upper and lower ToF planes is approximately 1.2 m, the ToF is able to resolve particles velocity up to 0.98 c. It is composed of four planes of scintillation counters, two above and two below the magnet. Each ToF plane consists of *paddles* aligned along the x and y coordinates, respectively. A ToF paddle consists of a 1 cm thick plastic scintillator of dimension approximately $12 \times 120\,\mathrm{cm}^2$. The scintillators are coupled at both ends via light guides to photomultipliers.

The ToF system is also important for antimatter discrimination, as it is able to discriminate up-going/down-going particles with a rejection factor of 10^{-9}. In a magnetic field, the trajectory of an upward-going electron is equivalent to that of a downward-going positron and a wrong assignment of the direction of a particle would induce a wrong charge assignment.

The TRD of AMS-02 is made of many layers of plastic or felt and vacuum. The X-rays are measured by a gas detector (straw tubes) filled with a special $Xe:CO_2$ (80:20%) mixture regulated by a gas recirculation system. The TRD system is placed on top of the magnet vacuum case. This system provides the measurement of the Γ factor for protons and electrons of a given energy, improving the redundancy on the discrimination of positrons against CR protons.

The *Electromagnetic CALorimeter* (ECAL) helps in obtaining the highest rejection between protons and positrons at high energy. It is a pancake consisting of nine superlayers for an active area of $\sim 65 \times 65\,\mathrm{cm}^2$, and a total thickness of 16.65 cm. Each superlayer (1.85 cm thick) is made of 11 grooved lead foils interleaved with layers of scintillating fibers, glued together. The detector's imaging capability is obtained by stacking superlayers with fibers alternatively parallel to two orthogonal axes. The pancake has an average density of $6.9\,\mathrm{g\,cm}^{-3}$, 16 radiation lengths for a total weight of 496 kg.

The calorimeter is able to reconstruct a 3D shower profile at 18 different depths. These measurements will give an accurate description of the longitudinal and transverse shower shape that allows the positron/proton showers' separation with an identification power of one e^+ over $10^5 p$. From the shower's shape, it is also possible to reconstruct the direction of the incident particle. The ECAL can reach angular precisions of a few degrees. This is very important for the measurement of γ-rays converting inside the calorimeter. The γ-rays detected in the ECAL are complemented by those producing pair conversions in the Tracker. Both these measurements provide the γ-ray energy and direction with respect to the AMS-02 coordinate system.

The *Star Tracker* system uses a Global Positioning System receiver, and it is necessary to determine the orientation of the detector in the sky. This information is needed to measure the arrival direction of particles in the sidereal reference frame, i.e., with respect to fixed stars. The system provides continuous synchronization between the data acquisition and the Coordinated Universal Time (UTC).

AMS-02 has the maximum analyzing power for particles traversing the instrument from top to bottom. Particles with a high incidence angle cannot be well-measured, and an *Anticoincidence Counter* (ACC) is used to reject them. The ACC is a barrel of scintillation counters around the tracker and it is important for the rejection of events with bad topology. Indeed, high-energy particles incident on AMS-02 materials could interact inelastically. The result of such interaction is the production of a hadronic shower that will confuse the tracker pattern recognition. These events could be a significant background for the search of antimatter signals.

A particle from top to bottom of the detector will give a signal in the ToF and not on the ACC. Conversely, a horizontal particle may give a signal on the ACC and not in the ToF. Then, an event should be recorded following the logic: (ToF) AND NOT (ACC). This restrictive condition can be released under particular conditions. In general, the *Level 1 trigger* of AMS-02 is constructed using information from the ToF (for charged particles), the ACC (as veto for high inclination particles) and the ECAL (for neutral particles).

Because of the high levels of radiation in space, the electronics used for AMS-02 is also particularly challenging. The number of electronics channels for the correct operation is about 300,000 and it is equivalent to all remaining electronics channels on ISS. The data are supervised by more than 600 separate computers using special radiation-tolerant chips. Redundancy is systematically implemented: there are at least two of every card, cable, and connector.

Real-time transmission of data from AMS-02 to the NASA ground facilities occurs through a high-rate downlink system. The operations of the detector are monitored at a Science operation center at CERN. During the period May 19, 2011, to August, 2013, about 36.5×10^9 CRs were collected. First results on the measurement of the positron fraction in primary CRs were released in April 2013 (Aguilar et al. 2013a, b), followed by results on charged cosmic nuclei (protons, helium, boron, carbon) and on the fluxes, ratios, and anisotropies of leptons.

In August 2017, another experiment, the Cosmic Ray Energetics And Mass experiment for the International Space Station (ISS-CREAM) was installed on the Japanese Experiment Module of the ISS. The ISS-CREAM instrument (Seo et al. 2014) consists of complementary and redundant particle detectors for measuring elemental spectra of $Z = 1$–26 nuclei over the energy range from 10^{12} to $> 10^{15}$ eV. It comprises a silicon charge detector for identifying incident CRs, a carbon target, a sampling tungsten/scintillator calorimeter for the energy measurement of all nuclei, a segmented top and bottom counting detectors for the electron/proton separation, and a scintillator detector rich in boron for additional electron/proton separation and detecting neutron signals.

3.6 Abundances of Elements in the Solar System and in CRs

At low energies, the chemical composition of the CRs has been measured with the abundance of nuclei arriving above the Earth's atmosphere. Table 3.3 shows the relative abundances ($R.A.$) between the different components of the cosmic radiation above the energy threshold E_T of 2.5 GeV (column 3) and of $E_T >$ 10.6 GeV/n (column 4), where the solar contribution can be neglected. Nuclei heavier than helium only contribute about a few percent of the total flux on Earth. However, the relative abundance of these elements is an essential piece of information for understanding the origin and history of accelerated particles, in particular, when compared with the nuclear composition of the solar system (column 6).

A remarkable resemblance between the measured CR abundances of nuclear species and the abundances found in the solar system can be noticed. The latter are also representative of the *cosmic abundances* of elements, see Sect. 3.7. The relative abundances of elements in CRs as a function of the nuclear charge Z for all elements are shown in Fig. 3.7 and are compared with the solar system abundances. It is interesting to note that all the elements in the periodic table are present in the solar system, and have been found in cosmic rays as well. Elements up to iron are much more abundant than trans-iron elements. A peculiarity observed in Fig. 3.7 is the alternation between the relative abundance and scarcity of adjacent atomic numbers. This pattern is very similar to that expected from the energy levels of adjacent nuclei in the nuclear binding energy curve (see next section). Data shown in Fig. 3.7 and Table 3.3 represent what is probably one of the most important pieces of information about the nature of CR sources.

The two samples (CRs and solar system abundances) exhibit a striking similarity, the differences being within 20% in most cases. The first conclusion from the data shown in Fig. 3.7 is that the accelerated matter arriving on Earth is sampled from a region whose surrounding material has the same chemical composition as our Solar System. This material is plausibly originated by the same mechanism that originated the Sun and the planets.

However, some remarkable differences between the two exist. The most relevant corresponds to the overabundance of Li, Be, B elements in CRs with respect to the cosmic chemical composition. The observed abundance ratio (Li + Be + B)/(C + N + O) in CRs exceeds the value found in solar system material by a factor of $\sim 10^5$ (see Table 3.3). A similar excess occurs for the elements below the iron ($Z = 26$) and the lead ($Z = 82$) peaks. This difference is interpreted as being due to the effect of the propagation in the Galaxy (Sect. 5.1) and provide a measure of the material that CRs have encountered since they were accelerated. Carbon, nitrogen, and oxygen are considered primary cosmic rays, produced and accelerated by astrophysical sources. Lithium, beryllium, and boron are secondary components produced by fragmentation reactions of the heavier C, N and O elements during the journey of CRs through the interstellar medium. As the spallation cross-section of the relevant nuclei is known at GeV energies, the ratio

Table 3.3 Composition of the different CR components

Element (Z)	Group Z	$R.A.$ $E_T > 2.5$ GeV	$R.A.$ $E_T > 10.6$ GeV/n	Number of particles per 100 Si	
				CRs	SS
H (1)		26,000	540	1.9×10^5	2.93×10^6
He (2)		3600	26	2.63×10^4	2.47×10^5
Li (3	$\mathscr{L}$	18	0.121	130	5.6×10^{-3}
Be (4)		10.5	0.09	76	6.1×10^{-5}
B (5)		28	0.19	202	1.9×10^{-3}
C (6)	$\mathscr{M}$	100	0.99	720	7.2×10^2
N (7)		25	0.22	180	2.1×10^2
O (8)		91	1	655	1.6×10^3
F (9)	$\mathscr{H}$	1.7	0.015	12	8.0×10^{-2}
Ne (10)		16	0.152	115	330
Na (11)		2.7	0.026	20	5.8
Mg (12)		19	0.197	137	103
Al (13)		2.8	0.031	20	8.5
Si (14)		14	0.163	100	100
P (15)		0.6	0.005	4.3	0.8
S (16)		3	0.030	22	42
Cl (17)		0.5	0.005	3.6	0.5
Ar (18)		1.5	0.009	11	9.3
K (19)		0.8	0.006	5.7	0.4
Ca (20)		2.2	0.018	15.8	6.0
Sc (21)	$\mathscr{V}\mathscr{H}$	0.4	0.003	2.9	3.4×10^{-3}
Ti (22)		1.7	0.010	12.2	0.25
V (23)		0.7	0.005	5.0	2.9×10^{-2}
Cr (24)		1.5	0.011	10.8	1.3
Mn (25)		0.9	0.009	6.5	0.9
Fe (26)		10.8	0.110	78	84.8
Co (27)		< 0.2	4×10^{-4}	–	0.23
Ni (28)		0.4	0.007	2.9	4.9
(29-30)		–	–	–	0.2
(>30)	$\mathscr{U}\mathscr{H}$	5×10^{-3}	–	0.02	0.02
(-1)	e^-	260	5	3×10^4	2.93×10^6

The relative abundances (R.A.) measured with different methods and energy thresholds are reported in columns 3 (Grieder 2001) and 4 (Engelmann et al. 1990). The corresponding relative CR abundances (column 5) are compared with that of the solar system (SS, column 6) (Lodders et al. 2009). The last two columns are arbitrary normalized to 100 for Si

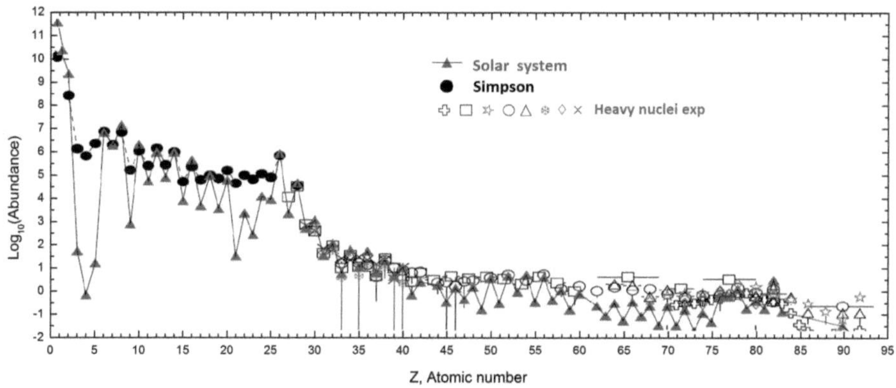

Fig. 3.7 Relative abundance of nuclei in cosmic rays as a function of their nuclear charge number Z at energies around 1 GeV/n, normalized to Si = 100. Abundances for nuclei with $Z \leq 28$ are drawn according to Simpson (1983). Heavier nuclei are measured by different experiments, as reported in Blümer et al. (2009). The abundance of elements (*triangles*) in the solar system according to Lodders (2003) is also shown

of secondary to primary cosmic rays will be used in Sect. 5.1 to infer the average escape time τ_{esc} of CRs in the Galaxy.

In addition to stable isotopes, CRs contain long-lived radioactive nuclides, mostly of secondary origin. The observed abundances of these isotopes can be used to establish various time scales related to the origin of CRs. In particular, secondary isotopes that decay through $\beta^{\pm}$ emission have been used (Sect. 5.2) for measuring τ_{esc}.

3.7 Cosmic Abundances and Origin of the Elements

The chemical elements existing in nature are a finite number: they are those that appear in the periodic table of elements. A stable nucleus does not spontaneously transform into a lighter nucleus though *radioactive decay*. The observed stable nuclei (or with extremely long half-lives, $>10^{16}$ years) are 264; the number of unstable ones is more than 1500. The latter number is increasing every year, as improved experimental techniques are developed, allowing for the measurement of shorter and shorter nuclear half-lives.

The nuclei can be classified in terms of the number of protons Z (the *atomic number*), the number of neutrons N and the number of nucleons A (the *mass number*): $A = Z + N = Z$ protons plus N neutrons. By sorting the nuclei on the basis of Z and N, the stable nuclei are distributed as shown in Table 3.4. The largest number of stable nuclei occurs when both Z and N are even. The number of nuclei with Z even and N odd is approximately equal to that with Z odd and N

Table 3.4 Number of stable nuclei, or with extremely long half-lives, according to the parity of Z and N

Z	$N = A - Z$	Number of stable nuclei
Even	Even	157
Even	Odd	53
Odd	Even	50
Odd	Odd	4
	Total	264

even. The content of Table 3.4 reflects the fact that the nuclear force is independent of whether the nucleons are protons or neutrons.

In the formation of a nucleus, binding energy is gained because a more stable system is obtained. The energy released should be compensated for by the decrease of the final mass, with respect to the sum of the initial masses of the constituents. For example, the bound state with the smallest nuclear mass is the deuterium nucleus (deuteron); the deuteron is a hydrogen isotope consisting of one proton and one neutron ($Z = 1$, $A = 2$). In this case, the mass deficit is 2.224 MeV, a small amount compared to $m_\mathrm{p} + m_\mathrm{n}$, but not quite negligible ($\sim 0.2\%\ m_\mathrm{p}$).

The binding energy (BE) is defined as the difference between the mass of the nucleus and the sum of the masses of the constituent nucleons:

$$M_A = \sum_{k=1}^{A} m_k - \mathrm{BE} = (Zm_p + Nm_n) - \mathrm{BE}\,. \tag{3.11}$$

The helium nucleus ${}^4_2\mathrm{He}$ (also called the α particle) is a particularly stable configuration whose binding energy is equal to 28.298 MeV. The binding energy of nuclei with small mass is not a regular function of A. For $A > 12$, the binding energy is approximately proportional to the number of nucleons (Fig. 3.8), with $BE/A \sim 8$ MeV/nucleon. The elemental abundances in the Universe as a function of the mass number A are determined by the *stellar nucleosynthesis* and by the nuclear binding energies. The curve of Fig. 3.8 increases up to $A \sim 60$ (the so-called *iron peak*). After the maximum, it decreases, ending at $A \sim 250$. The last stable nucleus has $A = 208$; nuclei with $A \gtrsim 240$ are so short-lived that their binding energy cannot be measured.

When the nuclear binding energy is plotted as a function of the nuclear charge Z, a characteristic alternation between low and high values is evident. This is caused by the higher relative binding energy of even atomic numbers with respect to odd atomic numbers, as explained by the Pauli exclusion principle. The *nuclear drop model* [Sect. 14.3.2 of Braibant et al. (2011)] provides a theoretical explanation for the above observations.

Cosmochemistry, or *chemical cosmology,* is the study of the chemical composition of matter in the Universe and the processes that led to the observed compositions. If the half-life of a nuclide is comparable to, or greater than, the age of the Solar System ($\sim 5 \times 10^9$ years), a significant amount will have survived

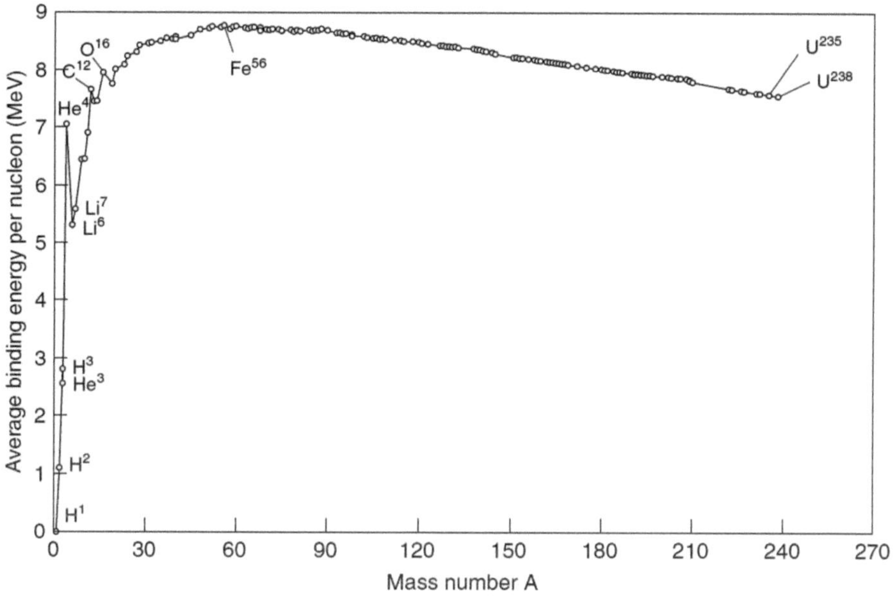

Fig. 3.8 The binding energy per nucleon, BE/A, of stable nuclei measured as a function of A. The binding energy is defined by Eq. (3.11). The peaks correspond to particularly stable nuclei. The curve has a maximum at $A \sim 60$

since the formation of the Solar System. Thus, the nuclide is said to be *primordial.* Meteorites are one of the most important tools for studying the chemical nature of the Solar System. Many meteorites come from material that is as old as the Solar System itself, and thus provides scientists with a record from the early solar nebula. Carbonaceous (C) chondrites[1] are especially primitive. C chondrites represent only a small proportion (4.6%) of meteorite falls. They have retained many of their chemical properties since their formation in the solar system about 5 billion years ago, and are therefore a major focus of cosmochemical investigations. In addition to meteoritic data, the abundances of elements in the solar system are derived from photospheric measurements of the light from the Sun. It has been known for some time that abundances determined from lines in the Sun's spectrum and abundances in CI-type carbonaceous meteorites agree quite well when normalized to the same scale. The CIs (named after the Ivuna meteorite) are a particular and rare type of chondrite. The elemental abundances determined from solar photospheric measurements and meteoritic CI chondrite are summarized in Fig. 3.7 and compared with the chemical composition of CRs.

[1]Chondrites are stony meteorites that have not been modified due to melting or differentiation of the parent body.

The abundance of elements in the solar system is believed to have been determined by the original matter composition after the Big Bang (roughly 76% hydrogen and 24% helium) and the *nucleosynthesis* in a number of progenitor supernova stars. For this reason, the chemical composition of the solar system is representative of the part of the Galaxy (the disk) with equal evolution history, and the term *cosmic abundances* is sometimes used as a synonym for solar system abundances. The production of elements heavier than iron and nickel is discussed in Sect. 12.16. For the relatively abundant elements (up to iron), energy spectra for individual elements have been measured: they are shown in the next section.

The fundamental properties of nuclear physics necessary to explain the binding energies as a function of A shown in Fig. 3.8 are more complicated than those necessary to understand atomic physics. However, our knowledge of nuclear interactions is advanced enough to conclude that the chemical elements everywhere in the Universe are the same as those found on Earth. There does not exist, therefore, some elusive stable element with physical properties unknown on Earth (e.g., the *unobtanium* on the Pandora planet of *Avatar*, a 2009 science fiction film by James Cameron). This is also proved by the fact that the chemical composition of CRs is not different from that found in our Solar System.

3.8 Energy Spectrum of CR Protons and Nuclei

The fluxes of nuclei in cosmic radiation follow a power law with a fast decrease with increasing energy. For all nuclear species, the dependence on energy is of the type ($E_\circ = 1\,\mathrm{GeV}$)

$$\Phi_i = K_i (E/E_\circ)^{\alpha_i}\ \mathrm{cm^{-2}s^{-1}sr^{-1}GeV^{-1}}\ , \tag{3.12}$$

as in Fig. 2.7. E can either be expressed in terms of energy-per-nucleon or energy-per-nucleus, as defined in Sect. 2.6. The parameters K_i, α_i are obtained from a fit to experimental data. Different compilations (based on experiments in which groups of nuclei are clumped together) exist, and the interested reader can refer to Wiebel-Sooth et al. (1998) and Hörandel (2003). When all the nuclear species are summed together, we have the so-called *all-particle spectrum*, whose parameters are given in (2.20c).

Figure 3.9 shows the proton and helium energy spectra (the flux is multiplied by $E^{2.7}$ here) above 1 GeV/n measured by recent balloon and space experiments. Below ~10 GeV/n, the difference among experiments is mainly due to solar modulations. At high energies, some differences well beyond the quoted statistical-only errors are present. Likely, the main sources of discrepancy arise from the evaluation of the detector and selection efficiencies and from the technique used in the determination of the energy. In experiments using magnetic spectrometers (such as AMS and PAMELA), the rigidity (and thus the energy) is determined by measuring the curvature of charged particles. Consequently, the energy resolution depends on the

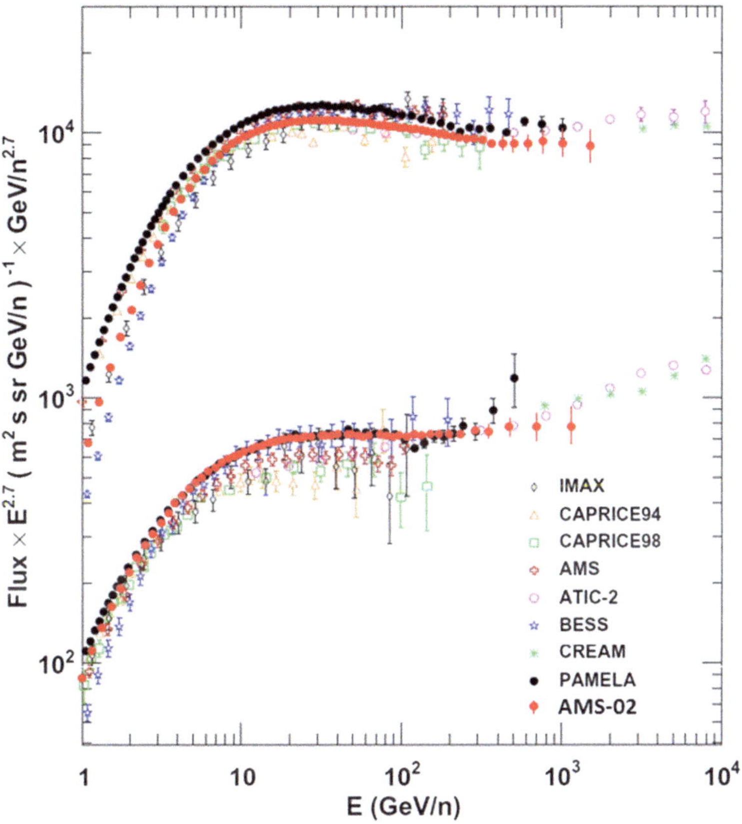

Fig. 3.9 Proton (upper set) and helium (lower set) energy spectra above 1 GeV/n obtained by balloon-borne [CAPRICE 94 (Boezio et al. 1999) and CAPRICE 98 (Boezio et al. 2003), IMAX (Menn et al. 2000), BESS (Haino et al. 2004), ATIC-2 (Wefel et al. 2007), CREAM (Ahn et al. 2010) and space-borne AMS-01 (Alcaraz et al. 2000), AMS-02 (Aguilar et al. 2015a,b), PAMELA (Adriani et al. 2011b)] experiments. Adapted from an original drawing by Dr. M. Boezio

spatial resolution of the tracking devices inserted into the magnetic field and on the topology of the event. The tracking alignment is a major ingredient for correct energy assignment. In fact, an incorrect assumption about the absolute position of the tracking sensor with respect to the magnetic field would result in a measurement affected by a systematic bias.

The rigidity dependence of the proton and helium fluxes are important for understanding the origin, acceleration, and propagation of CRs. In Fig. 3.9, the data from PAMELA and CREAM show a difference in the spectral index of proton and helium nuclei. In addition, a change of the slope above ~1 TeV/n is evident in the ATIC-2 and PAMELA data. The recent precise measurement of the proton flux from 1 GV to 1.8 TV of AMS-02 (Aguilar et al. 2015a), based on 3 10^8 events and detailed studies of the systematic errors, confirms that the flux deviates from

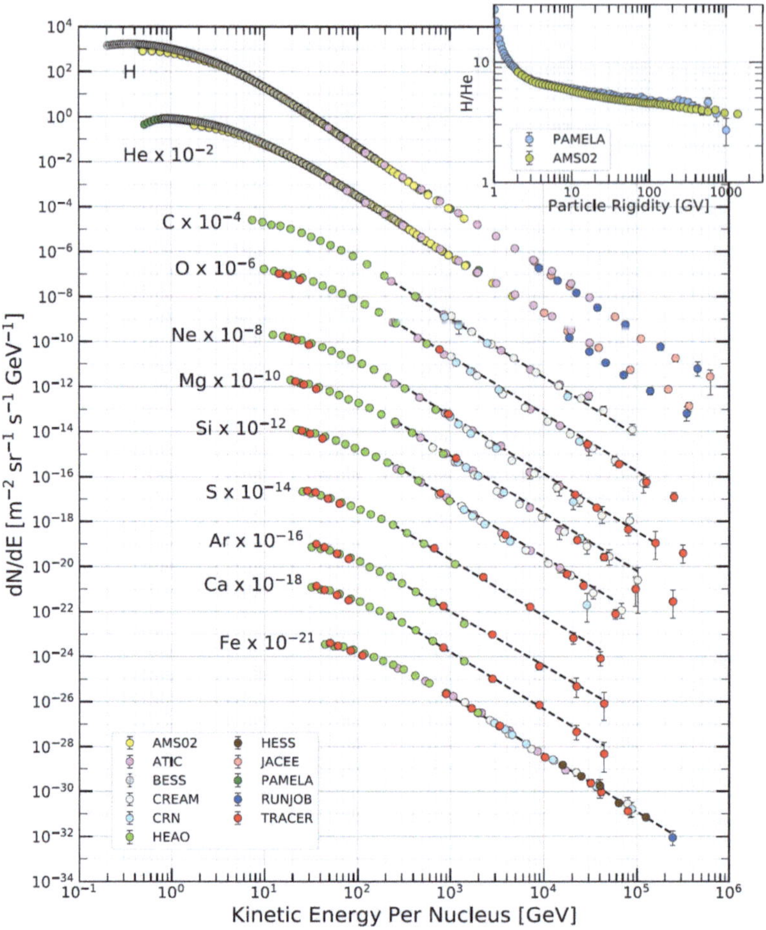

Fig. 3.10 Fluxes Φ_i of CR nuclei $i = Ze$ are plotted vs. energy-per-nucleus using data from direct experiments. For a better understanding of the figure, the flux of each nuclear species is multiplied for a scaling factor. The inset shows the H/He ratio at constant rigidity. From Sect. 30 of Patrignani et al. (2016)

a single power law and progressively hardens at high rigidities. Similarly, the He flux measured by AMS-02 from 1.9 GV to 3 TV with 5×10^7 of events deviates from a single power law and the spectral index progressively hardens at rigidities larger than 100 GV. The helium spectral index is different from that of the proton, but the rigidity dependence is similar for helium and protons, as shown in the inset on the upper right in Fig. 3.10. The astrophysical reasons for these deviations from a simple power law are still under debate.

Figure 3.10 shows the major nuclear components in CRs as a function of the energy-per-nucleus. The exponent α_i is almost the same for all nuclear species shown in Fig. 3.10, apart from protons. This fact is of primary importance for the theories studying the acceleration mechanisms of CRs. It should also be noticed that the data reported in Fig. 3.10 have been obtained mainly through balloon-borne experiments (CREAM, ATIC and TRACER) with a good agreement between the different measurements. At the highest energies, the measurement is usually limited by statistics. The results of AMS-02 for nuclei heavier than He have not been made public yet.

3.9 Antimatter in Our Galaxy

Equal amounts of matter and antimatter should have been produced at the beginning of the Universe as described by the Big Bang theory. The reason for the fact that only matter seems to surround us is one of the major unknowns in cosmology and particle physics. The possible presence of cosmological antimatter in the Universe is a fundamental issue of physics, which can be faced from the experimental point of view.

Antiprotons, as well as positrons, are a component of the cosmic radiation being produced in the interaction between CRs and interstellar matter. Positrons in CRs had already been observed in 1964 and antiprotons in 1979, with balloon-borne magnetic spectrometers. Secondary antiprotons are mainly produced by CR protons interacting with ISM protons

$$p + p \rightarrow p + p + (\overline{p} + p) \; . \tag{3.13}$$

This reaction was used in 1955 by E. Segrè and O. Chamberlain (Nobel Prize in 1959) at the Berkeley Bevatron, and occurs above the threshold of $E^{tr} \gtrsim 7$ GeV of the relativistic proton against the proton at rest. To give a first-order estimate of the antiproton flux in CRs, we should consider the $E^{\rm tr}$ and the steeply falling of the primary flux with energy, $\Phi(E) \propto E^{-2.7}$. Comparing the flux of secondary antiprotons at the threshold of ~0.1 GeV with that of protons at $E \sim 0.1$ GeV (where the maximum is, see Fig. 3.10), the expected under-abundance is of the factor

$$\frac{\Phi_{\overline{p}}}{\Phi_p} \sim \left(\frac{0.1}{7}\right)^{2.7} \sim 10^{-5} \; . \tag{3.14}$$

Being exactly the same as particles except for their opposite charge sign, antiparticles are readily distinguished, as they bend in opposite directions in the magnetic field. Magnetic spectrometers provide a clear and simple particle/antiparticle separation and probe the existence of antimatter in our Galaxy. Figure 3.11 shows the antiproton-to-proton flux ratio as a function of energy, as measured by the three largest experiments provided with a magnetic spectrometer: BESS, PAMELA (Boezio and Mocchiutti 2012) and AMS-02 (Aguilar et al. 2016). At threshold, this ratio is in agreement with the simple estimate (3.14). The antiproton spectrum

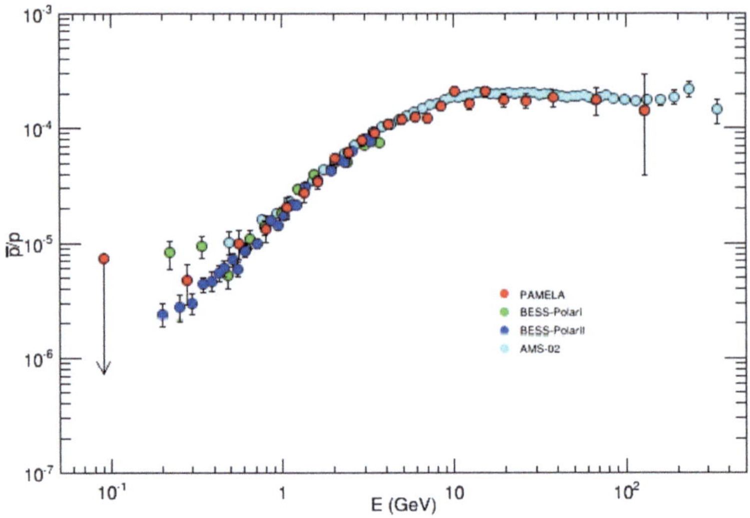

Fig. 3.11 Measurements of the antiproton-to-proton flux ratio provided by PAMELA, AMS-02 and BESS-Polar. See Adriani et al. (2017) and references therein. Courtesy Dr. M. Boezio and the PAMELA Collaboration

agrees with detailed theoretical simulations that assume pure secondary production of $\overline{p}$ during the propagation of primary CRs in the Galaxy and interactions with the interstellar medium (see Sect. 5.4).

No heavier antinuclei have been detected so far. The BESS experiment provides the lowest upper limit to date on the relative antihelium-to-helium ratio, 7×10^{-8}, within the rigidity range 1–14 GV. BESS also provides the lowest upper limit for the antideuteron flux of $2 \times 10^{-4}(\mathrm{m}^2\,\mathrm{s\,sr\ \ GeV/n})^{-1}$ at the 95% confidence level, between 0.17 and 1.15 GeV/n. The detection of a single antideuteron or antihelium nucleus would impact our understanding of the matter/antimatter asymmetry of the Universe.

3.10 Electrons and Positrons

Electrons and positrons constitute about 1% of the CRs, as shown in Table 3.3. This component provides additional information on the acceleration sites and CR propagation in the Galaxy. High-energy electrons are already subject to a number of electromagnetic energy loss processes in proximity to the sources, where the matter density and the magnetic fields are large. The accelerated electrons are the source of most of the nonthermal electromagnetic radiation measured by astronomers from radio to X-rays, as presented in Sect. 5.8 and widely discussed in Chap. 8. These processes cause distortions of e^- injection energy spectra as they propagate through the interstellar medium from their sources and they potentially provide information on the propagation, confinement, and production mechanisms of high-energy particles. Electrons directly produced at accelerator sites are called *primary electrons*.

In recent years, knowledge about the leptonic component in CRs has greatly benefited from new experimental results from the ATIC balloon-borne experiment (Chang et al. 2008), the Fermi-LAT (Ackermann et al. 2012) and PAMELA (Adriani et al. 2011a) satellite-based experiments. The Fermi-LAT experiment (Sect. 8.8.1) is mainly devoted to γ-ray astronomy and is also performing CR measurements (Thompson et al. 2012). It has not been instrumented with a magnetic spectrometer: electron and positron components are measured separately by exploiting Earth's shadow, which is offset in opposite directions for opposite charges due to the magnetic field of the Earth (Ackermann et al. 2012).

Finally, the AMS-02 (Aguilar et al. 2014) experiment has reported results on the largest sample of measured CR leptons. Their results on fluxes, ratios, and anisotropies of $e^{\pm}$ represent a fundamental step towards the comprehension of the lepton component of CRs. The measurement is based on data collected during the first 2 years of operation, with 6.8×10^6 $e^{\pm}$ events within the energy range 0.5–350 GeV.

Figure 3.12 shows the electron plus positron energy spectrum multiplied by E^3 as measured by different experiments. Within systematic errors, the entire electron spectrum from a few GeV to ~1 TeV can be fitted by a power law $\Phi_e(E) \propto E^{-3.1}$. In particular, the high-precision AMS-02 data above 5 GeV are compatible with a

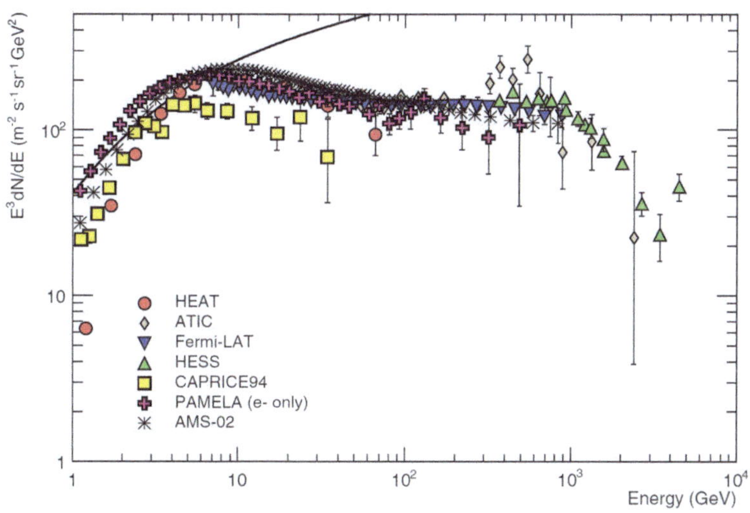

Fig. 3.12 The electron plus positron energy spectrum from different space-based, balloon, and ground-based experiments. The flux is multiplied by E^3. The *solid black line* shows, for reference, the proton spectrum multiplied by 0.01. Figure from Sect. 30 of Patrignani et al. (2016) and references therein

single power. If electrons are emitted by astrophysical sources with a spectral index α_e, a steepening of the spectrum by one power is expected because of the electron radiation losses, as we will show in Sect. 5.8. The measured spectral index ~3 indicates a spectral index of CR sources $\alpha_e \sim 2$. This expectation also corresponds to the results of calculations of the electron plus positron spectrum based on the propagation of CRs in the Galaxy. These computer-code calculations (Sect. 5.4) describe in detail the propagation of primary particles (protons, electrons, and nuclei) from the sources through the interstellar medium.

The presence of a structure in a smooth spectrum of the lepton component would represent an important signature for unexpected physics. In particular, from annihilation of dark matter candidates (Sect. 14.6) or from the presence of nearby and active accelerators of CR electrons in our Galaxy. An excess of electrons within the range 300–700 GeV with respect to that expected from conventional diffuse electron sources was reported by ATIC and PPB-BETS (a long duration balloon flight using the Polar Patrol Balloon in Antarctica). Fermi/LAT measured a not-entirely flat spectrum without confirming the peak of the ATIC excess at ~600 GeV. The HESS array (Sect. 9.1.2) also measured the electron flux above 400 GeV, finding indications of a cutoff above ~1 TeV, but no evidence for a pronounced peak below this. AMS-02 data has produced no evidence of structures in the electron energy spectrum, at present, up to 350 GeV.

The change in the spectral $(e^+ + e^-)$ distribution with increasing energies is compatible with an increase in the secondary positron component, discussed below. Secondary e^+ and e^- are produced by CR interactions with interstellar matter, as end products of the decay of short-lived secondary particles (mostly pions via the decay $\pi^\pm \rightarrow \mu^\pm \rightarrow e^\pm$). These secondary $e^\pm$ add to primary electrons.

3.10.1 The Positron Component

Experiments using magnetic spectrometers can distinguish the sign of the electric charge. This allows for the measurement of the positron fraction in the $e^\pm$ component of CRs as a function of the energy. The dominant background is represented by misidentified CR protons. Unlike electrons, which are present as a primary component in CR sources, the vast majority of positrons arise as secondary products of CR interactions in our Galaxy.

Figure 3.13 shows the positron energy spectrum $E^3\Phi_{e^+}(E)$ as a function of the energy E. The dashed line shows the prediction from one of the aforementioned computer codes of primary propagation in the interstellar medium (Sect. 5.4). Above a few GeV, a significant excess with respect to the secondary production by CR propagation is observed. First PAMELA (Adriani et al. 2009) and then Fermi-LAT observed that the $E^3\Phi_{e^+}(E)$ flattens up to ~30 GeV, before rising again above 30 GeV.

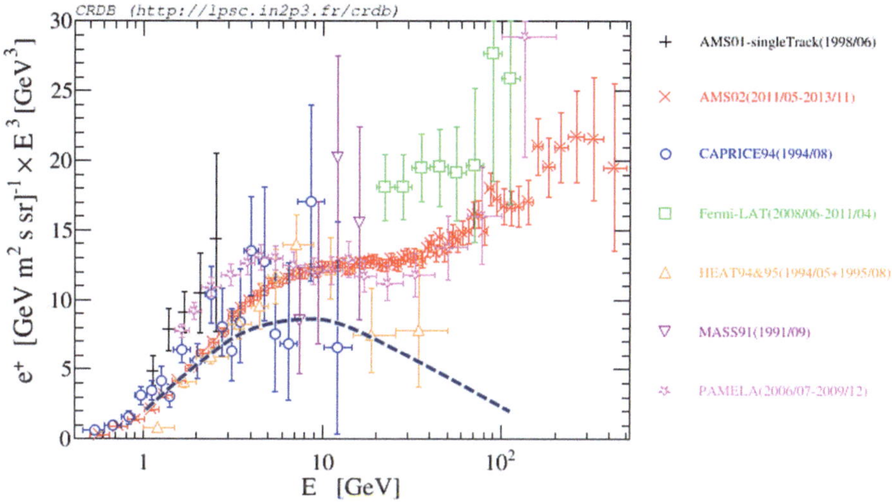

Fig. 3.13 The positron energy spectrum, Φ_{e^+} times E^3 measured by different balloon-borne (CAPRICE, HEAT, MASS) and space-borne (AMS-01, AMS-02, PAMELA, Fermi-LAT) experiments. The *dashed line* shows a calculation using the GALPROP program. Plot obtained from the CRDB (Maurin et al. 2014)

This behavior has been confirmed with high statistics and extended up to 350 GeV by AMS-02. In this experiment, the background is efficiently suppressed by requiring a minimal amount of material crossed in the TRD and ToF detectors. In addition, a good match between the particle momentum reconstructed in the nine tracker layers of the silicon spectrometer and the energy measured in the electromagnetic calorimeter is required. The performance of the TRD results in a proton rejection efficiency larger than 10^3 keeping 90% of positrons. The calorimeter provides a rejection factor better than 10^3 for protons with momentum up to 1 TeV/c. The combination of the two factors leads to an overall proton-to-positron rejection power of $\sim 10^6$ for most of the energy range under study.

The increase of the positron component is still more evident in Fig. 3.14, which shows the positron fraction, i.e., the ratio between $\Phi_{e^+}/(\Phi_{e^+} + \Phi_{e^-})$ measured by HEAT (Beatty et al. 2004), PAMELA (Adriani et al. 2013), and AMS-02 (Aguilar et al. 2013a, b) as a function of the energy E. The positron-fraction spectrum does not exhibit fine structures and steadily increases in the region between 10 and 250 GeV. In the high statistics AMS-02 sample, the ratio is on the order of $\sim$10 % above a few tens of GeV. As a consequence, since positrons are always created in pair with an electron,[2] *about 90% of the observed electrons must be of primary origin.*

[2]This is exact in the case of conversion of a γ-ray. Positrons can be produced as the end stage of hadronic interactions by the decay chain $\pi^+ \rightarrow \mu^+ \rightarrow e^+$. On average, isotopic spin invariance on $\pi^\pm$ production guarantees the presence of an electron through the decay $\pi^- \rightarrow \mu^- \rightarrow e^-$

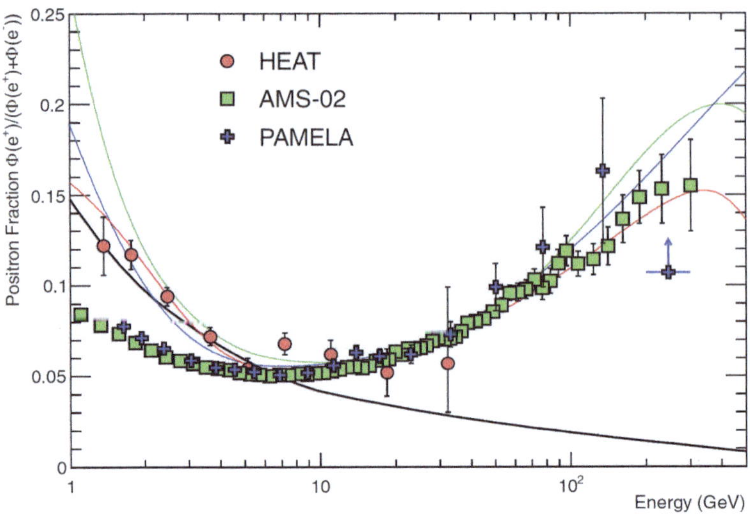

Fig. 3.14 The positron fraction (ratio of the flux of e^+ to the total flux of $e^+ + e^-$) as a function of the energy measured by HEAT, PAMELA, and AMS-02. The *solid black line* is a model of pure secondary production using a detailed propagation model of CRs (Sect. 5.4). The *three thin lines* show three representative attempts to model the positron excess with different phenomena discussed in Sect. 14.9.3: dark matter decay (*green*); propagation physics (*blue*); production in pulsars (*red*). The ratio below 10 GeV is dependent on the polarity of the solar magnetic field. Figure from the Sect. 30. of Patrignani et al. (2016)

The positron fraction at energies below ∼10 GeV is systematically lower than data collected during the 1990s through other experiments. This discrepancy is well-interpreted as being the result of solar modulation effects. At high energies (above 10 GeV), the positron fraction increases significantly with energy. This increase is well above that expected from a model in which all positrons are of secondary origin: the solid black line in Fig. 3.14 shows the result of a calculation based on such an assumption.

3.10.2 Considerations on the e^+, e^- *Components*

The $\overline{p}/p$ ratio ($\sim 10^{-5}$–10^{-4}) shown in Fig. 3.11 demonstrated that the antiproton flux is in overall agreement with a pure secondary component. The electron component declines faster with increasing energy than the baryonic one. At 1 TeV, the ratio between CR electrons and protons is $\sim 10^{-3}$. The $e^+/(e^- + e^+)$ ratio (about 10%) indicates that most of the detected electrons are also of primary origin,

---

with equal rate. However, due to the fact that CRs are positively charged, secondary positrons are in slight excess over electrons.

although the fraction of secondary leptons is much larger than that of secondary baryons. This is important information concerning the astrophysical origin of the leptonic component. Due to the presence of magnetic fields, the typical distance over which 1 TeV electrons lose half their total energy is estimated to be 300–400 pc when they propagate within $\sim$1 kpc of the Sun (Aharonian et al. 1995). Electrons are affected more readily by energy-dependent diffusive losses, convective processes in the interstellar medium, and perhaps reacceleration during propagation from their sources to us. For these reasons, at energies above a few hundred GeV, the majority of electrons is supposed to be originated by sources closer than a few hundred pc. We will return to this estimate in Sect. 5.8, after having presented the synchrotron energy loss. High-energy CR electrons really probe CR production and propagation in the nearby region of our Galaxy.

In conclusion, due to their large energy losses, it does not seem plausible (contrary to the case of protons and nuclei) that the observed electrons originate from a uniform distribution of sources in the Galaxy. If the source is too far, the probability that an electron reaches the Earth is extremely small. More likely, primary high-energy electrons observed on Earth originate from a small number of sources well-localized in space and relatively close (on a scale of galactic distances) to the Solar System.

Concerning the measured data on the positron fraction above 10 GeV (Fig. 3.14), it has stimulated significant scientific debate. The $e^+/(e^- + e^+)$ ratio cannot, in fact, be understood by models describing the production of secondary CRs during propagation in the Galaxy (see Chap. 5), without the introduction of *ad hoc* local sources. Several theoretical explanations have been proposed to explain the observed excess: an astrophysical origin, such as nearby pulsars or microquasars, or exotic sources, for instance, the annihilation of dark matter particles in the proximity of our Galaxy. We will return to the subject in Chap. 14. The agreement between PAMELA, Fermi-LAT and AMS-02 data reduces the possibility of a systematic bias and gives confidence that the increase of the positron flux is to be ascribed to a physical, albeit still unknown, effect. A detailed description of astrophysical models that can explain the origin of the positron excess can be found in Serpico (2012). It is possible that a final word will be given by the increased statistics from the AMS-02 experiment. The present data set covers only about 10% of the expected number of leptons. AMS-02 undoubtedly represents the leading experiment for the direct study of cosmic radiation in coming years.

References

P. Abreu et al. (AUGER Collaboration), Measurement of the proton-air cross-section at $\sqrt{s} = 57$ TeV with the pierre auger observatory. Phys. Rev. Lett. **109**, 062002 (2012)

M. Ackermann et al., Measurement of separate cosmic-ray electron and positron spectra with the Fermi-LAT. Phys. Rev. Lett. **108**, 011103 (2012)

O. Adriani et al. (PAMELA Collaboration), Observation of an anomalous positron abundance in the cosmic radiation. Nature **458**, 607–609 (2009)

O. Adriani et al. (PAMELA Collaboration), The cosmic-ray electron flux measured by the PAMELA experiment between 1 and 625 GeV. Phys. Rev. Lett. **106**, 201101 (2011a)
O. Adriani et al., PAMELA measurements of cosmic-ray proton and helium spectra. Science **332**, 69–72 (2011b). https://doi.org/10.1126/science.1199172
O. Adriani et al. (PAMELA Collaboration), The cosmic-ray positron energy spectrum measured by PAMELA. Phys. Rev. Lett. **111**, 081102 (2013). arXiv:1308.0133
O. Adriani et al., Ten years of PAMELA in space. Rivista Nuovo Cimento **40**, 473 (2017). https://doi.org/10.1393/ncr/i2017-10140-x
M. Aguilar et al. (AMS-02 Collaboration), AMS-02 provides a precise measure of cosmic rays. CERN Courier **53**(8), 23–26 (2013a). Also: B. Bertucci ICRC 2013 (ID 1267)
M. Aguilar et al. (AMS-02 Collaboration), First result from the alpha magnetic spectrometer on the ISS: precision measurement of the positron fraction in primary cosmic rays of 0.5–350 GeV. Phys. Rev. Lett. **110**, 14–141102 (2013b)
M. Aguilar et al. (AMS-02 Collaboration), Precision measurement of the $(e^+ + e^-)$ flux in primary cosmic rays from 0.5 GeV to 1 TeV with the AMS on the ISS. Phys. Rev. Lett. **113**, 221102 (2014)
M. Aguilar et al. (AMS-02 Collaboration), Precision measurement of the proton flux in primary cosmic rays from rigidity 1 GV to 1.8 TV with the AMS on the ISS. Phys. Rev. Lett. **114**, 171103 (2015)
M. Aguilar et al. (AMS-02 Collaboration), Precision measurement of the helium flux in primary cosmic rays of rigidities 1.9 GV to 3 TV with the AMS on the ISS. Phys. Rev. Lett. **115**, 211101 (2015)
M. Aguilar et al. (AMS-02 Collaboration), Antiproton flux, antiproton-to-proton flux ratio, and properties of elementary particle fluxes in primary CRs measured with the AMS on the ISS. Phys. Rev. Lett. **117**, 091103 (2016)
F.A. Aharonian, A.M. Atoyan, H.J. Voelk, High energy electrons and positrons in cosmic rays as an indicator of the existence of a nearby cosmic Tevatron. Astron. Astrophys. **294**, L41 (1995)
H.S. Ahn et al. (CREAM Collaboration), The cosmic ray energetics and mass (CREAM) instrument. T. Nucl. Instrum. Methods A **579**, 1034–1053 (2007)
H.S. Ahn et al., Measurements of the relative abundances of high-energy cosmic-ray nuclei in the Tev/nucleon region. Astrophys. J. Lett. **714**, L89 (2010)
J. Alcaraz et al., Cosmic protons. Phys. Lett. B **490**, 27 (2000)
K. Asakimori et al., Cosmic-ray proton and helium spectra: results from the JACEE experiment. Astrophys. J. **502**, 278–283 (1998)
M. Ave et al., Composition of primary cosmic-ray nuclei at high energies. Astrophys. J. **678**, 262–273 (2008)
J.J. Beatty et al., New measurement of the cosmic-ray positron fraction from 5 to 15 GeV. Phys. Rev. Lett. **93**, 24112 (2004)
J. Blümer, R. Engel, J. Hörandel, Cosmic rays from the knee to the highest energies. Prog. Part. Nucl. Phys. **63**, 293–338 (2009)
M. Boezio, E. Mocchiutti, Chemical composition of galactic cosmic rays with space experiments. Astropart. Phys. **39–40**, 95–108 (2012)
M. Boezio et al., The cosmic-ray proton and helium spectra between 0.4 and 200 GV. Astrophys. J. **518**, 457 (1999)
M. Boezio et al., The cosmic-ray proton and helium spectra measured with the CAPRICE98 balloon experiment. Astropart. Phys. **19**, 583 (2003)
S. Braibant, G. Giacomelli, M. Spurio, *Particle and Fundamental Interaction* (Springer, Berlin, 2011). ISBN: 978-9400724631
J. Chang et al., An excess of cosmic ray electrons at energies of 300–800 GeV. Nature **456**, 362 (2008)
J.H. Crane, D.D. Guo, M.H. Israel, J. Klarmann, Interaction mean-free-path of cosmic-ray Fe in air. Astrophys. Space Sci. **94**(1), 201–209 (1983)
V.A. Derbina et al., Cosmic-ray spectra and composition in the energy range of 10–1000 TeV per particle obtained by the RUNJOB experiment. Astrophys. J. **628**, L41–L44 (2005)

J.J. Engelmann et al., Charge composition and energy spectra of CR nuclei for elements from Be to Ni. results from HEAO-3-C2. Astron. Astrophys. **233**, 96 (1990)

O. Ganel et al., Beam tests of the balloon-borne ATIC experiment. Nucl. Instrum. Methods A **552**(3), 409–419 (2005)

M. Garcia-Munoz, G.M. Mason, J.A. Simpson, The cosmic-ray age deduced from the Be-10 abundance. Astrophys. J. **201**, L145 (1975)

M. Garcia-Munoz, G.M. Mason, J.A. Simpson, The age of galactic cosmic rays derived. Astrophys. J. **217**, 859–877 (1977)

P.K.F. Grieder, *Cosmic Rays at Earth* (Elsevier, New York, 2001). ISBN: 978-0444507105

S. Haino et al., Measurements of primary and atmospheric cosmic-ray spectra with the BESS-TeV spectrometer. Phys. Lett. B **594**, 35 (2004)

J.R. Hörandel, On the knee in the energy spectrum of cosmic rays. Astropart. Phys. **19**, 193–220 (2003)

K. Lodders, Solar system abundances and condensation temperatures of the elements. Astrophys. J. **591**, 1220 (2003)

K. Lodders, H. Palme, H.P. Gail. Abundances of the elements in the solar system, *Chapter 4 of Landolt-Börnstein*. New Series, Astronomy and Astrophysics (Springer, Berlin, 2009). Also: arXiv:0901.1149

D. Maurin, F. Melot, R. Taillet. A database of charged cosmic rays. Astron. Astrophys. **569**, A32 (2014). Arxiv:1302.5525

W. Menn et al., The absolute flux of protons and helium at the top of the atmosphere using IMAX. Astrophys. J. **533**, 281 (2000)

C. Patrignani et al. (Particle Data Group), Chin. Phys. C **40**, 100001 (2016, update). http://pdg.lbl.gov/2016/

E.S. Seo, Direct measurements of cosmic rays using balloon borne experiments. Astropart. Phys. **39–40**, 76–87 (2012)

E.S. Seo et al., Cosmic ray energetics and mass for the international space station (ISS-CREAM). Adv. Space Res. **53**, 1451–1455 (2014)

P.D. Serpico, Astrophysical models for the origin of the positron "excess". Astropart. Phys. **39–40**, 2–11 (2012)

J. Simpson, Elemental and isotopic composition of the galactic cosmic rays. Ann. Rev. Nucl. Part. Sci. **33**, 323–382 (1983)

D.J. Thompson, L. Baldini, Y. Uchiyama, Cosmic ray studies with the Fermi gamma-ray space telescope Large Area Telescope. Astropart. Phys. **39–40**, 22–32 (2012)

J.P. Wefel et al., *ICRC Conference*, vol. 2 (2007), p. 31

B. Wiebel-Sooth, P.L. Biermann, H. Meyer, Cosmic rays. VII. individual element spectra: prediction and data. Astron. Astrophys. **330**, 389–398 (1998)

Chapter 4
Indirect Cosmic Ray Detection: Particle Showers in the Atmosphere

Abstract Above 10^{15} eV, the cosmic ray (CR) flux drops below a few tens of particles per square meter per year. It is no longer possible to detect the incident particles above the atmosphere before they interact. Direct experiments are thus replaced with ground-based instruments that cover up to several thousands of km^2, the extensive air shower (EAS) arrays. A completely different experimental approach for CR measurements is used: EAS arrays are, in most cases, large area and long duration experiments studying, as accurately as possible, the nature, flux, mass, and direction of primary CRs up to the highest energies. This chapter describes: the developments of air showers initiated by primary protons and nuclei; the main shower features that characterize the electromagnetic and muonic components; some EAS array detectors using different experimental techniques; and the results obtained in knowledge of the CR flux in the energy region around the knee.

Above 10^{15} eV, the CR flux drops below a few tens of particles per m^2-year. It is no longer possible to detect the incident particles above the atmosphere before they interact. Direct experiments (characterized by a small geometrical factor $A \cdot \Delta\Omega$ [cm^2 sr]) must be replaced with ground-based instruments that cover up to several thousands of km^2, the extensive air shower (EAS) arrays. These use a completely different approach to CR measurements, starting with the pioneering experiments conducted soon after World War II by Auger, Kohlhörster and Rossi.

The EAS arrays are, in most cases, large area and long duration experiments studying, as accurately as possible, the nature, flux, mass, and direction of primary CRs up to the highest energies. Air showers are initiated by primary CRs, through interaction with a nucleus in the atmosphere. In addition to the hadronic component, the decays of short-lived hadrons lead to a shower of particles: photons, electrons, and positrons constitute the electromagnetic (EM) component; muons and neutrinos constitute the penetrating component. All these particles travel at the speed of light in the atmosphere, approximately along the direction of the primary CR.

High energy primary γ-rays induce an almost pure EM cascade. In Sect. 4.3, we present a simple model for the EM cascade initiated by a γ-ray, which can be

© Springer Nature Switzerland AG 2018
M. Spurio, *Probes of Multimessenger Astrophysics*, Astronomy and Astrophysics Library, https://doi.org/10.1007/978-3-319-96854-4_4

mathematically treated using differential transport equations of $e^{\pm}$ and γ in the atmosphere. Some simple features, as derived from approximate solutions of the cascade equations, are presented. The cascades initiated by primary CR protons or nuclei have additional features. They are also characterized by large event-to-event fluctuations. Their description is today achieved using full Monte Carlo simulations, which follow the details of the development of the EM and muonic components. It is interesting, however, to obtain (Sect. 4.4) a first-order estimate of the quantities, which can be measured by experiments, both for the muonic and the EM components. In the first sections of the chapter, we enumerate the main shower features (denoted as **SF1, SF2, … , SF12**), which characterize the EM and muonic components of the cascade initiated by γ-rays, protons and heavier nuclei. These characteristics are confirmed by detailed Monte Carlo simulations of air showers in the atmosphere (Sect. 4.5), which are used by the experiments to interpret their observations.

EAS arrays are installed on the ground and are sometimes referred to as *indirect detection* experiments. Modern shower arrays employ complementary techniques (Sect. 4.6), such as scintillators, air Cherenkov detectors, etc., to measure simultaneously as many shower parameters as possible, in order to reduce the model dependence in the energy and mass number A determination.

The features of the EM and muonic cascades will be used in Sects. 4.8–4.10 to illustrate how indirect experiments can derive the CR flux and properties in the energy region around the *knee* of CRs.

4.1 Introduction and Historical Information

Air showers are the particle cascades produced by the interaction of a cosmic ray with a nucleus in the atmosphere. The term “shower” is the English translation by P. Blackett of the Italian expression *sciame*, first used by B. Rossi. The atmosphere (Sect. 4.2) acts as a calorimeter and becomes part of the detection system. As this is not a fully controlled environment, in some cases, the atmospheric conditions are carefully monitored and recorded along with the air-shower array data.

The main techniques used to measure EAS can be classified as follows: detectors that measure the particle content of the shower at the ground; detectors that measure the light produced by the propagation of the EAS itself in the atmosphere. The main disadvantage of the experiments detecting light is that they can only collect data on dark, moonless nights, with a duty cycle of $\sim$5–10%. In EAS arrays, the CR direction is usually inferred by the relative arrival times of the signals on different counters, the energy from the integration of the measured density of secondary particles (the electromagnetic cascade, the muons or the visible light induced by the shower of charged particles in the atmosphere) at the detector level. The estimate of the mass of the incoming CR is a much more difficult task.

Direct measurements of the CR flux have provided a power-law dependence for the CR energy spectrum up to $\sim 10^{15}$ eV. Observations in 1959 of EAS indicated a steepening at around 3×10^{15} eV (the knee). In 1963, observations made with the first large shower array discovered another structure just above 10^{18} eV. For the early years of indirect measurements, see the review (Kampert and Watson 2012).

The main drawback of EAS experiments is that the interpretation of the observations depends on the model used to describe the hadronic interactions of CRs with air nuclei (Anchordoqui et al. 2004). Uncertainties in the development of cascades generated by CRs with energies above 10^{16}–10^{17} eV are unavoidable. In addition, relevant quantities for shower development (such as the number of secondary hadrons and their momentum distribution) occur in the very forward direction and are not accessible in collider experiments.[1]

One has, therefore, to rely on phenomenological interaction models, which differ in their predictions, making the task of retrieving information from air shower data even more difficult with increasing energies. Other uncertainties are related to the structure of the atmosphere, which is not a homogeneous calorimeter, and to its variations.

EAS arrays measure the *overall* cosmic ray spectrum, summing up the contribution of all nuclear species. The measurement of the total energy of a primary CR through observation of the induced cascades relies on a much firmer theoretical basis than the measurement of its mass. Uncertainties in the hadronic physics make it hard to separate the observable quantities produced by protons or by iron nuclei. No EAS experiment has up to now measured the primary composition of cosmic rays on an event-by-event basis. Instead, through the measurement of one or more of the observables, which are sensitive to the mass (Sect. 4.10), the flux of groups of nuclei as a function of the energy was estimated using statistical techniques.

4.2 The Structure of the Atmosphere

The main parameter concerning the development of a cascade of secondary CRs is the amount of matter above any atmospheric layer, in which the primary CR has interacted. This quantity is called the *(vertical) atmospheric depth*, X_v. This depth is the integral in altitude of the atmospheric density ρ above the observation level h (see Fig. 4.1):

$$X_v \equiv \int_h^\infty \rho(h')dh' \,. \tag{4.1}$$

[1]Before the LHC physics runs, someone expressed concerns over safety, and attempted to halt the beginning of the experiments through petitions to the US and European Courts. These opponents asserted that the LHC experiments had the potential to create micro black holes that could grow in mass or release dangerous radiation, leading to doomsday scenarios, such as the destruction of the Earth. Any doomsday scenario at the LHC was ruled out before starting of the physics runs, simply noting that the physical conditions and events created in the LHC experiments occur naturally and routinely in the Universe without hazardous consequences. In particular, ultra-high energy CRs that are impacting on Earth with energies considerably higher than those reached in any man-made collider have never destroyed the Earth!

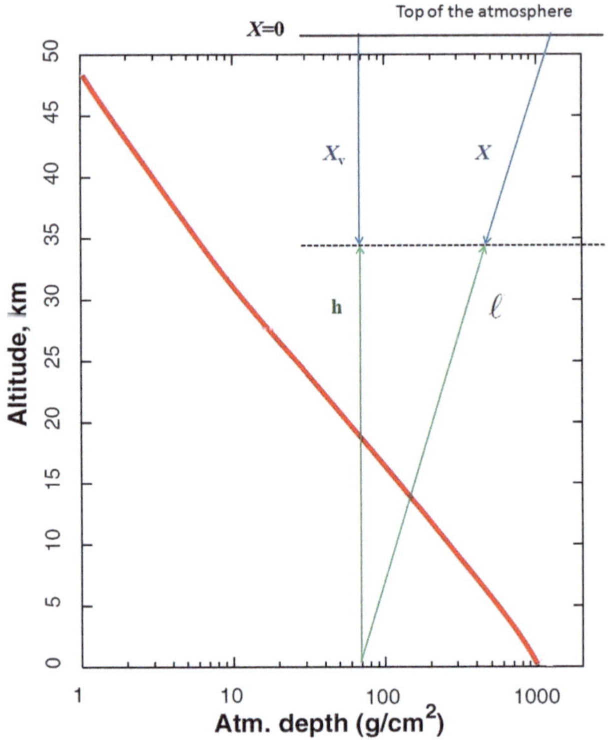

Fig. 4.1 The curve represents the altitude in km, which corresponds to a given residual atmospheric depth X_{υ}. For a proton, the interaction length is $\sim$85 g cm^{-2}: this is the amount of material crossed by a proton that reaches a height of about 20 km above ground. The atmospheric depth X for non-vertical directions is usually denoted as the *slant atmospheric depth*. The total depth of the atmosphere is $X_{\upsilon}^{atm} \simeq 1030\,\mathrm{g\,cm^{-2}}$. It is easy to verify that this corresponds to $\sim$10 m of a material with the density of water (meter of water equivalent = m.w.e.). Sometimes, it is useful to measure the atmospheric depth in terms of m.w.e. The geometrical definitions of the quantities given in the text (X_{υ}, X, h and ℓ) are also sketched in the figure

The variation of the atmospheric density with altitude is a very important piece of information for the modeling of shower cascades in the atmosphere, and for the competition between interaction and decay of secondary mesons.

The dependence of density on h, $\rho(h)$, can be determined using thermodynamics. The pressure p (atmospheric weight per unit of surface S) at the depth X_{υ} is $p = \frac{mg}{S} = \frac{g}{S}\int_h^{\infty} \rho(h')\cdot Sdh' = gX_{\upsilon}$ while, from (4.1), $\rho = -dX_{\upsilon}/dh$ (the $-$ represents the fact that ρ decreases as h increases). Assuming the atmosphere to be a perfect gas,

$$\rho = \frac{Mp}{kT}\,, \tag{4.2}$$

where k is the Boltzmann constant and M is the mean molecular mass of air. The average mass of the atmosphere, composed mainly of nitrogen (78%) and oxygen (21%), is $\langle A \rangle = A_{\rm atm} \simeq 14.5$. However, this average mass strongly depends on the altitude. As most molecules are biatomic, $M = 2A_{\rm atm} \times m_p = 4.8 \times 10^{-23}$ g and

$$T(h) = \frac{M}{k}\frac{p}{\rho} = -\frac{M}{k}\frac{gX_v}{dX_v/dh} , \tag{4.3}$$

where $T(h)$ is the temperature at the height h in the atmosphere. The temperature of the troposphere (extending from the ground up to 10–20 km of height) decreases approximately with increasing altitude at a rate $\sim$−6.5 K/km. Nevertheless, for a first approximation, the atmosphere is considered to be an isothermal gas with $T(h) = const$. Under this *isothermal approximation*, Eq. (4.3) can be easily integrated, and

$$X_v = X_v^{\rm atm} e^{-h/h_0}; \quad \text{with} \quad X_v^{\rm atm} = 1030\,{\rm g\,cm^{-2}}. \tag{4.4}$$

The *atmospheric scale height* h_0 is defined as

$$h_0 = \frac{kT}{Mg} . \tag{4.5}$$

Using the numerical value of M, at the surface of Earth ($T \simeq 290$ K), $h_0 \sim 8.4$ km. In the region where CRs interact, the temperature is between 210–240 K, and $h_0 \sim$ 6–7 km. The vertical thickness of the atmosphere $X_v^{\rm atm}$ corresponds, for incoming CRs, to a calorimeter of variable density with $\sim$11 interaction lengths and 28 radiation lengths (compare with Table 3.1).

If we consider that the temperature decreases with increasing altitude h, some analytical parameterizations are obtained (see, for instance, Gaisser (1991) and Stanev (2010)). These are useful as input for Monte Carlo simulations of the CR cascades. Figure 4.1 shows the dependence of X_v (here, along the x-axis) upon h, as derived with the parameterization reported in Stanev (2010).

Considering the curvature of the Earth ($R_\oplus$ is the Earth radius) and a non-vertical direction (zenith angle θ), the relation between h and path length ℓ in the atmosphere is

$$h = \ell \cos\theta + \frac{1}{2}\frac{\ell^2}{R_\oplus}\sin^2\theta . \tag{4.6}$$

The atmospheric depth for such an inclined direction is called the *slant depth* and corresponds to

$$X = \int_\ell^\infty \rho(h) d\ell . \tag{4.7}$$

For zenith angles $\theta < 60°$, one can scale the slant depth as

$$X \simeq \frac{X_v}{\cos\theta} \quad \text{and} \quad \rho = \frac{X_v}{h_0} \simeq \frac{X\cos\theta}{h_0} \tag{4.8}$$

(flat Earth approximation). For horizontal directions, the slant depth corresponds to about $X(\theta = 90°) = 36{,}000\,\mathrm{g\,cm^{-2}}$. This value does not arise from the flat Earth approximation.

4.3 The Electromagnetic (EM) Cascade

The showers containing only e^+, e^- and γ-rays (called *EM showers* henceforth) and initiated by photons or electrons are mathematically described using differential *transport equations*, which include the particle energy losses and particle production. The most popular description of the EM shower was developed in the 1930s by Rossi and Greisen.

EM showers are governed mainly by (1) bremsstrahlung of electrons (here, by *electrons*, we refer to both electrons and positrons) and (2) pair production by photons. In addition to bremsstrahlung, electrons are subject to excitation/ionization energy loss. The total energy loss $\mathrm{d}E/\mathrm{d}X$ of electrons is described by Eq. (3.6). The *radiation length*, X_0, is defined as: (a) the mean free path of a high-energy electron, corresponding to the distance over which it loses all but $1/e$ of its energy by bremsstrahlung, and (b) 7/9 of the mean free path for pair production by a high-energy photon. The factor 7/9 arises from QED arguments. For this reason, Eq. (3.6) can also formally be used to describe the propagation of high-energy photons in a medium.

The radiation length, X_0, and the critical energy (the energy at which radiation energy losses equal those derived from excitation/ionization) depend on the material. From Table 3.1, for electrons in air

$$X_0 \simeq 37\,\mathrm{g\,cm^{-2}}; \qquad E_c \simeq 86\,\mathrm{MeV}. \tag{4.9}$$

The bremsstrahlung (dominant for $E > E_c$) and the excitation/ionization energy loss (dominant for $E < E_c$) are competing processes in showers.

As a matter of nomenclature, in the following:

1. $N^i_e, N^i_\gamma, N^i_\pi, N^i_\mu$ represent, respectively, the number of electrons, hard photons ($= \gamma$-rays), pions and muons present in the cascade induced by a primary particle i. The index $i = \gamma, p, A$ corresponds to a primary γ-ray, proton or nucleus with mass A, respectively;
2. $N^i_{e_{\max}}$ corresponds to the number of $e^\pm$ at the maximum of the EM shower induced by a primary particle ($i = \gamma, p, A$);

3. X^i_{max} corresponds to the atmospheric depth (Sect. 4.2) in g cm^{-2}, where the maximum of the EM shower induced by a primary particle ($i = \gamma, p, A$) occurs in the atmosphere.

As shown in Sect. 3.10, the flux of electrons drops sharply at energies above 1 TeV. Contrastingly, pure electromagnetic showers initiated by γ-rays are increasingly interesting due to a sizeable flux above the TeV observed through Imaging Cherenkov telescopes (Chap. 9). The main difference between a shower initiated by a photon and that induced by a primary proton or nucleus is the presence in the latter of a hadronic component, which develops a significant muon cascade (Sect. 4.4.1).

4.3.1 *Heitler's Model of EM Showers*

Some properties of EM showers can already be understood within the very simple model credited to Heitler (Heitler 1944). In this model, the evolution of electromagnetic cascades is described as a perfect binary tree (Fig. 4.2). An incoming

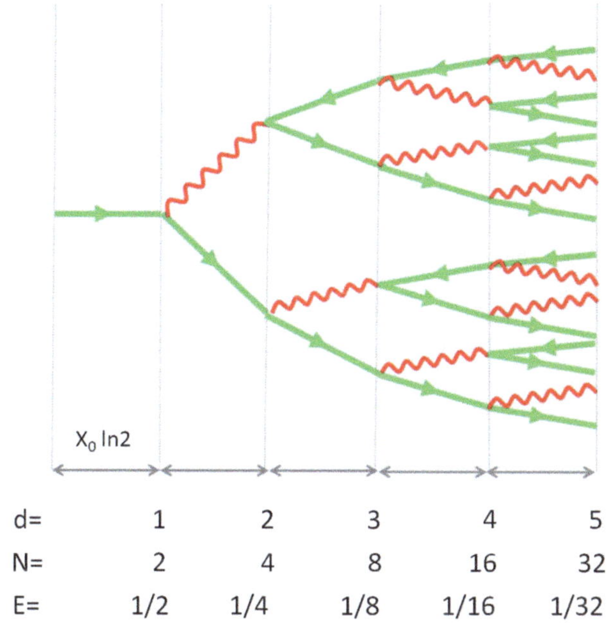

Fig. 4.2 Toy model evolution of an electromagnetic cascade. At each step of the cascade the number of particles is multiplied by two, through either pair creation or single photon bremsstrahlung. *Backward arrows* indicate a positron, as in Feynman diagrams. The evolution stops when individual particle energies fall below the critical energy E_c. The number N of particles at each step d and the average particle energy E in the Heitler's model are also indicated. Adapted from http://www.borborigmi.org/

electron interacts in the atmosphere after traveling a certain "*step*" and produces two new particles, each with half the energy of the initial particle. Similarly, a photon converts into an electron-positron pair if its energy $>2m_e$. In the following, the cross-sections of the two processes are assumed equal, independent of the energy, and additional energy loss mechanisms are ignored. The radiation length X_0 represents both the photon's and the electron's mean free paths in the matter, neglecting for the former the factor 7/9. In more detailed models for shower developments, the energy distribution of the bremsstrahlung photons should be considered (see Gaisser 1991).

Let us now quantify the length of the "step". The important quantity is the product $X = x\rho$ (g cm^{-2}) of the distance x (cm) traversed in a medium and the medium density ρ (g cm^{-3}). The energy loss of bremsstrahlung corresponds to the second term of (3.6). In the energy regime where excitation/ionization processes can be neglected ($\alpha = 0$), the solution to (3.6) is simply

$$E(X) = E_0 e^{-X/X_0} , \tag{4.10}$$

where E_0 is the energy of the particle initiating the shower. Let us define the "step" d as

$$d = \ln 2 \cdot X_0 . \tag{4.11}$$

By inserting (4.11) into (4.10), we have

$$\frac{E(X = d)}{E_0} = \exp\left(\frac{-X_0 \ln 2}{X_0}\right) = \frac{1}{2} . \tag{4.12}$$

Statistically, at each step, electrons halve their energy via bremsstrahlung emission of a single photon, while photons produce an electron-positron pair, each with half the energy of the photon. Therefore, at each step, the energy of each particle decreases by a factor of two and the total number of particles increases by a factor of two.

After k generations, the number of particles is $N_k = 2^k$ and their individual energy is $E_0/2^k$. This development continues until, at step k^*, the individual energy drops below the critical energy E_c. At such an energy, the multiplication process slows down, and soon after stops altogether. Therefore, at k^* the number of secondary particles reaches a maximum (the so-called *shower maximum*):

$$N_{\max} = 2^{k^*} \simeq \frac{E_0}{E_c} . \tag{4.13}$$

According to (4.13) and recalling (4.9), a 10 TeV photon reaching the top of the atmosphere produces, in Heitler's model, $\sim 10^5$ secondary particles at the shower maximum.

The depth in the atmosphere (in g cm^{-2}) where the maximum of the electromagnetic cascade occurs can be obtained from (4.13) and is given by

$$X^{\gamma}_{\max} = X_f + k^* \cdot d = X_f + k^* \ln 2 \cdot X_0 = X_f + X_0 \ln\left(\frac{E_0}{E_c}\right), \tag{4.14}$$

where X_f is the atmospheric depth at which the secondary photons or electrons start an electromagnetic cascade. More detailed computations, which consider the energy distribution of particles in the shower, agree with (4.14), with $X_f = X_0/2$ as an appropriate value.

The position of the shower maximum depends on the energy E_0 and the quantity

$$D_{10} \equiv \frac{\mathrm{d}X^{\gamma}_{\max}}{\mathrm{d}(\log_{10} E_0)} = 2.3 \frac{\mathrm{d}X^{\gamma}_{\max}}{\mathrm{d}(\ln E_0)} \tag{4.15}$$

is called the *elongation rate*. In this simple model, from (4.15), we obtain $D_{10} = 2.3X_0$, expressing the fact that D_{10} is proportional to the radiation length of the medium. This elongation rate corresponds to the slope of the curves representing the depth of the EM maximum as a function of the primary energy E_0, as shown in Figs. 4.7 and 4.20, together with the experimental data.

Heitler's model, although oversimplified, illustrates some important features of the electromagnetic component of the air showers. More accurate analytical modeling (Sect. 4.3.2) and Monte Carlo simulations (Sect. 4.5) confirm the properties of Heitler's model, although the particle number at maximum is overestimated by factors of ~2–3. Here (see Fig. 4.2), the ratio of electrons to photons is $N_e/N_\gamma = 2$, while direct measurements of showers in air report a value $N_e/N_\gamma \sim 1/6$, also confirmed by a more detailed computation (see below) and Monte Carlo simulations. This is explained by the fact that during the bremsstrahlung process, multiple photons are emitted and electrons lose energy through additional channels.

4.3.2 Analytic Solutions

Purely EM showers were historically studied by coupled differential equations, which describe the evolution of the number of photons, N_γ, and of the number of electrons, N_e, as a function of the atmospheric depth X (Kamata and Nishimura 1958; Rossi and Greisen 1941). They are sometimes expressed in terms of the adimensional variable $t \equiv X/X_0$. The equations governing $N_e(t)$, $N_\gamma(t)$ (which are called the electron and photon *sizes*, respectively) are coupled by the production processes. An electron can radiate a photon and a photon can create an $e^{\pm}$ pair with energy-dependent cross-sections. We do not present the mathematical details of these coupled differential equations (see for instance Gaisser (1991) and the review

Lipari (2009)). The resulting analytical solutions[2] are more complicated, but similar to those that we will present in Sect. 5.1 for the production of light Li, Be, and B nuclei during CR propagation. These solutions are combinations of exponential functions and contain an important parameter denoted as s, the *shower age*. The concept of shower age derives from the observation that all showers at the maximum of their development have similar characteristics (that is, they have the same "age").

The $N_e(t), N_\gamma(t)$ solutions from the analytical description increase as t increases, reach a maximum, and then decrease. As in Heitler's model (Eq. (4.13)) the number of particles at the maximum is directly correlated with the energy E_0 of the primary particle.

Based on the solutions to the cascade equations, Greisen developed a compact and still often used parameterization of the mean number of charged particles as a function of atmospheric depth X (Greisen 1960) for a γ-ray induced shower:

$$N_e^\gamma(X) = \frac{0.31}{\sqrt{\ln(E_0/E_c)}} \exp\left[\left(1 - \frac{3}{2}\ln s\right)\frac{X}{X_0}\right] . \tag{4.16}$$

The Greisen approximation for the shower size contains the shower age parameter s:

$$s \simeq \frac{3X}{X + 2X^\gamma_{max}} . \tag{4.17}$$

Figure 4.3 shows the electron size N_e as a function of t obtained from (4.16) and for different energies E_0 of the incoming particle. The shapes of the curves describing the electron size for showers initiated by primary photons with different energies E_0 as a function of t look very similar. Showers have age $s = 1$ at maximum and age $s < 1$ before the maximum ("young" showers). "Old" showers have $s > 1$.

At shower maximum, (4.16) corresponds, by definition, to $N^\gamma_{e\max}$, and thus

$$N^\gamma_{e\max} = \frac{0.31}{\sqrt{\ln(E_0/E_c)}} \exp\left[\frac{X^\gamma_{max}}{X_0}\right] . \tag{4.18}$$

Using (4.14) and neglecting the term X_f, we obtain:

$$N^\gamma_{e\max} = \frac{0.31}{\sqrt{\ln(E_0/E_c)}}\left(\frac{E_0}{E_c}\right) = \frac{1}{g}\left(\frac{E_0}{E_c}\right) , \tag{4.19}$$

where $g \equiv \sqrt{\ln(E_0/E_c)}/0.31$. In the Greisen approximation, g is weakly dependent on the primary energy. For a first approximation, within the energy range ($E_0 \simeq 10^{15}$–10^{18} eV) we deal with in this Chapter, the approximate value $g \sim 10$

[2]Two solutions exist, denoted as Approximation A when the electron excitation/ionization losses are neglected and Approximation B when they are included.

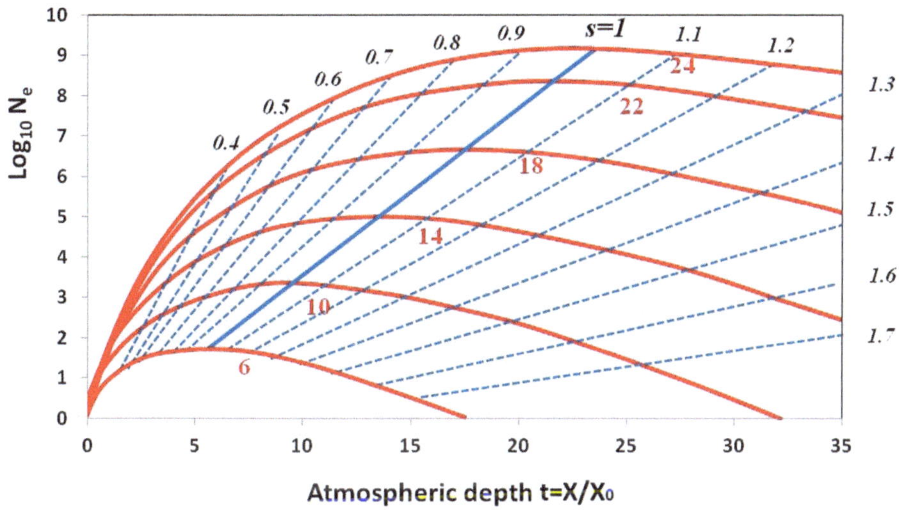

Fig. 4.3 Shower size $N_e(t)$ as a function of the atmospheric depth $t = X/X_0$ (the longitudinal variable in units of radiation lengths) for primary CR photons. The numbers close to the line $s = 1.1$ are the values of $\ln(E_0/E_c)$, where E_c is the critical energy. The corresponding six energies E_0 are in increasing order: 0.035, 2, 100 TeV; 5.7, 320, 2300 PeV. The diagonal line connecting the maxima of the different curves is labeled with $s = 1$. The different age values are shown by the small numbers. The dashed diagonal lines indicates the positions in the atmosphere with constant age s. The sea level corresponds to $t \simeq 27$

can be used. Thus, the EM size at maximum is about 10% of the total size (4.13) obtained from Heitler's model.

The Greisen profile accurately describes the average development of purely electromagnetic showers, and can be adapted for the description of proton and nucleus-induced showers, as presented in the following section. It cannot be easily adapted for the description of neutrino-induced showers (Lipari 2009).

From the analytical solution, it also follows that, for $E > E_c$, the energy spectrum of secondary particles in a shower is approximately described by a power law:

$$\frac{\mathrm{d}N_e}{\mathrm{d}E} \sim E^{-(s+1)}\,, \tag{4.20}$$

where s denotes the shower age parameter. Near maximum ($s = 1$), the energy spectrum is $\mathrm{d}N_e/\mathrm{d}E \propto E^{-2}$. As the shower becomes older, the energy spectrum of secondary electrons becomes *softer*: the fraction of high energy electrons decreases. Young showers have a *harder* spectral index. This nomenclature of *soft* and *hard* spectral indexes is quite common in astrophysics. A harder spectrum has a larger component of high-energy particles.

From the experimental point of view, detectors are able to measure charged particles above a given energy threshold E_t. Detectors are located at a fixed height in the atmosphere, and for each primary CR, they sample showers having different ages. This has an effect, because the energy distribution of electrons depends on the age s, and older showers have a smaller fraction of electrons contributing to the signal.

The detection of high-energy photons relies on the fact that they can produce charged particles (through pair creation or Compton Effect). For this reason, the effective EM size measured by EAS experiments is $N = N_e(> E_t) + \epsilon N_\gamma$, where $0 \leq \epsilon \leq 1$ is the probability of $\gamma \to e$ conversion in the detector. In some detectors (for instance, in thin layers of scintillator counters), $\epsilon \simeq 0$, while E_t is very low and the measured EM size corresponds to N_e. For this reason, in the following, we focus on the electronic component of the EM shower, and to its number $N_{e_{max}}$ at the position of the maximum.

The particle density as a function of the distance r to the shower core (the *lateral particle distribution*) is a measured quantity in most air shower experiments. The lateral particle distribution is mainly determined by multiple Coulomb scattering of electrons.[3] Results of detailed calculations of the lateral particle distribution by Kamata and Nishimura (1958) were parameterized by Greisen (1960) as

$$\frac{dN_e}{rdrd\phi} = N_e(X) \cdot \frac{C(s)}{2\pi r_1^2} \left(\frac{r}{r_1}\right)^{s-2} \left(1 + \frac{r}{r_1}\right)^{s-4.5} \frac{\text{particles}}{\text{m}^2} . \tag{4.21}$$

$C(s) = \Gamma(4.5 - s)/[\Gamma(s)\Gamma(4.5 - 2s)]$ is a normalization constant[4] obtained by imposing that $\frac{2\pi}{N_e(X)} \int_0^\infty r \frac{dN_e}{rdr} dr = 1$. The quantity r_1 is the Molière unit:

$$r_1 = X_0 \left(\frac{E_s}{E_c}\right) \simeq 9.2\,\text{g cm}^{-2} \tag{4.22}$$

in air, where E_c is the critical energy, $E_s = m_e c^2 \sqrt{4\pi/\alpha} \sim 21\,\text{MeV}$. m_e is the electron mass and $\alpha \simeq 1/137$ is the electromagnetic coupling constant. Note the units of this definition of r_1. When measured in units of length (dividing by the material density ρ in g cm^{-3}), the Molière radius

$$r_M = \frac{r_1}{\rho} \quad \text{cm} \tag{4.23}$$

is different for different materials. In air, it increases with decreasing air density in the atmosphere. At sea level, $r_M \sim 80\,\text{m}$, and at the position of the shower

[3]A charged particle traversing a medium is deflected by many small-angle scatters. This deflection is due to the superposition of many Coulomb scatterings from individual nuclei, and hence the effect is called *multiple Coulomb scattering*. The Coulomb scattering distribution is well-represented by a Gaussian distribution. At larger angles, the distribution shows larger tails and the behavior is more similar to that of Rutherford scattering.

[4]The Γ function is an extension for positive real numbers of the factorial.

maximum, $r_M \sim 200$ m. Showers developing at higher altitudes have larger lateral dimensions. Equation (4.21) is called the Nishimura-Kamata-Greisen (NKG) function.

The main properties of the EM component can be summarized as follow

(SF1) The number of particles at shower maximum is approximately proportional to the primary energy, Eq. (4.19).

(SF2) The depth in the atmosphere of the shower maximum increases logarithmically with energy, Eq. (4.14).

Cascades initiated by γ-rays are almost pure EM showers ($= e^{\pm}, \gamma$), without other particles. For instance, muon production depends on mechanisms such as the Drell-Yan process, characterized by a small cross-section (refer to Sect. 7.14 of Braibant et al. (2011)). Thus, an additional characteristic is that:

(SF3) In γ-ray induced showers very few muons are expected, and to a first approximation, $N_\mu^\gamma = 0$.

Monte Carlo simulations of the cascades confirm that the EM component of the showers exhibits a number of *universal features*. In particular, EM cascades induced by primary photons, electrons, protons and nuclei have properties that are independent of the primary type and rather insensitive to the primary energy. These features are:

(SF4) The longitudinal development of EM cascades depends on two parameters: the energy E_0 of the primary nucleus and the shower age s. They can also be described in terms of analytical formulas like that of Greisen, Eq. (4.16).

(SF5) Near the shower core, the electron energy distribution is a universal function of the age parameter, as in Eq. (4.20).

(SF6) The lateral distribution of the EM cascade, Eq. (4.21), at a given age s is a universal function if the lateral distance is measured in Molière units r_1, Eq. (4.22).

4.4 Showers Initiated by Protons and Nuclei

CRs are mainly protons and heavier nuclei, which initiate a *hadronic shower* by interacting with atmospheric nuclei after traversing, on average, one interaction length λ_I. The multiplicity of charged particles produced in the interaction increases with the center of mass energy $\sqrt{s}$. When expressed in terms of the laboratory energy E_0, the number of charged hadrons, n_{ch}, as found in $p\overline{p}$ and pp data (Eidelman et al. 2004), increases as

$$n_{ch} \propto E_0^{0.2} \,. \tag{4.24}$$

The same is true for $\pi^{\pm}$ collisions: the multiplicity in $\pi + ^{14}$N collisions increases with energy, and one finds: $n_{\rm ch} \simeq 5, 11$, and 27 at 10, 100, and 10^4 GeV, respectively (Hörandel 2007). After the first interaction, the $n_h = 3/2n_{\rm ch}$ produced hadrons (including the neutral ones, which are $n_0 = 1/2n_{\rm ch}$) carry a fraction κ of the primary CR energy E_0. They are mainly pions (see Sect. 10.9 of Braibant et al. (2011)). As an indicative value for the following first-order estimates, we fix[5] $n_{\rm ch} = 10$. The parameter $\kappa \sim 0.7$ is usually referred to as the *inelasticity*. It takes into account the fact that a significant fraction of the total energy is carried away by a single *leading particle* and is defined as

$$\kappa = \frac{E_0 - E'}{E_0 + M_T} , \tag{4.25}$$

where E' is the residual energy of the nucleon after the collision and M_T the mass of the target nucleon (which can be neglected in our energy regime). The residual energy E' is used by the leading particle for a successive interaction after traversing, on average, one more λ_I. After k iterations, the leading particle carries a fraction $(1 - \kappa)^k$ of the initial energy E_0. In the following, the inelasticity will not be considered. See Matthews (2005) for more details. The process continues (see Fig. 4.4) until the hadron energy falls below an *interaction threshold*. Neutral pions in the cascade have a lifetime $\tau_{\pi^0} = 8.4 \times 10^{-17}$ s characteristic of the electromagnetic interactions and decay:

$$\pi^0 \rightarrow \gamma\gamma \quad \text{with} \quad d'_{\pi^0} = \Gamma c \tau_{\pi^0} = \Gamma \cdot 2.5 \times 10^{-6} \text{cm} , \tag{4.26}$$

where d'_{π^0} is the neutral pion *decay length*, i.e., the distance in the laboratory frame traversed at the speed of light during one lifetime. The decay length in (4.26) is denoted as d' when measured in cm, and as $d = d'\rho$ when multiplied by the material density and measured in g cm^{-2}. As usual, we indicate with Γ the relativistic Lorentz factor, to avoid confusion with the symbol used for the γ-ray. The daughter γ-rays start an electromagnetic cascade practically at the first interaction point.

Charged pions decay through weak interactions with a longer lifetime ($\tau_{\pi^{\pm}} = 2.6 \times 10^{-8}$ s):

$$\pi^+ \rightarrow \mu^+ \nu_\mu \; ; \; \pi^- \rightarrow \mu^- \overline{\nu}_\mu \quad \text{with} \quad d'_{\pi^{\pm}} = \Gamma c \tau_{\pi^{\pm}} = \Gamma \cdot 780 \, \text{cm} , \tag{4.27}$$

and, when considering the density ρ of the medium,

$$d_{\pi^{\pm}} = d'_{\pi^{\pm}} \cdot \rho \;\; \text{g cm}^{-2} \tag{4.28}$$

[5] The quantity $n_{\rm ch}$ is more easily measured in accelerator experiments than n_h.

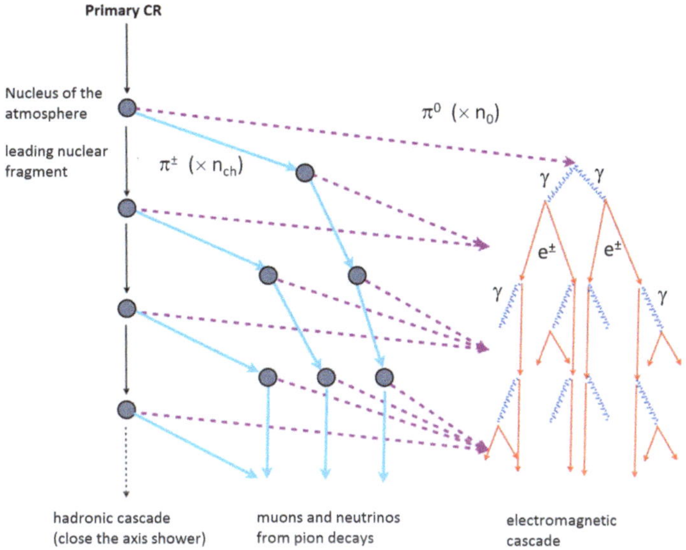

Fig. 4.4 Schematic evolution of cascades initiated by hadrons. At each step, roughly 1/3 of the energy is transferred from the hadronic cascade to the electromagnetic one (Letessier-Selvon and Stanev 2011)

is the charged pion decay length. The interaction length (3.2) of pions is $\lambda_I^\pi \simeq 120\,\mathrm{g\,cm^{-2}}$. For this reason, once produced, $\pi^\pm$ can either decay or interact with air nuclei. Whether decay or interaction dominates depends upon which of the two, $d_{\pi^\pm}$ or λ_I^π, is smaller. The role of the two processes is also complicated by the fact that the interaction probability depends on the material density, which in the atmosphere depends, in turn, on the height h. High energy $\pi^\pm$ prevalently interact because the relativistic Lorentz factor Γ is high, while low energy pions decay. At high altitudes, pions are more likely to decay than at low altitudes, where the atmosphere is denser. The stochastic interaction/decay competition of all charged mesons determines the details of the development of hadronic showers, Chap. 11.

In our first-order model (following Matthews 2005) we assume that $\pi^\pm$ always interact if their energy is above a certain threshold energy $E^\pi_{\rm dec}$, and decay if the energy is below $E^\pi_{\rm dec}$. In the case of interaction, additional $n_h = n_{\rm ch} + n_0$ hadrons of lower energy are produced. Here, we assume that all n_h are pions. We estimate $E^\pi_{\rm dec}$ as the energy at which the decay length of a charged pion, Eq. (4.28), becomes equal to their interaction length, Eq. (3.8b). The decay length depends, through the Lorentz factor, on the pion energy. The $E^\pi_{\rm dec}$ can be derived by imposing

$$\lambda_I^\pi = d'_{\pi^\pm} \cdot \rho = \left(\frac{E^\pi_{\rm dec}}{m_\pi c^2}\right) c\tau_{\pi^\pm} \cdot \rho \,, \tag{4.29}$$

and thus, using $\rho \simeq 10^{-3}$ g cm^{-2}

$$E^{\pi}_{\rm dec} = \frac{\lambda^{\pi}_{I}}{c\tau_{\pi^{\pm}} \cdot \rho} \cdot (m_{\pi}c^2) \simeq 160 \cdot (m_{\pi}c^2) \simeq 20\,{\rm GeV}\,. \tag{4.30}$$

The energy transferred to the EM component can be estimated within this simple model, neglecting the correction factor due to the inelasticity. In each hadronic interaction, 2/3 of the initial energy is transferred to the hadronic component. After k generations, the energies of the hadronic (E_h) and EM (E_{EM}) components are given by

$$E_h = \left(\frac{2}{3}\right)^k E_0 \quad ; \quad E_{EM} = E_0 - E_h\,. \tag{4.31}$$

Hence, after k interactions, the energy per pion is $E = E_0/(n_h)^k$. At a certain value k^*, E becomes smaller than $E^{\pi}_{\rm dec}$. The number of interactions needed to reach $E = E^{\pi}_{\rm dec}$ is:

$$k^* = \frac{\ln(E_0/E^{\pi}_{\rm dec})}{\ln n_h}\,. \tag{4.32}$$

Within the energy range we are considering (10^{15}–10^{18} eV), the value of k^* ranges from 4 to 6.

The basic properties of a cascade induced by a primary hadron with energy E_0 can be easily understood assuming the decay channels (4.26) and (4.27) for neutral and charged pions, respectively. Our simple cascade model consists of two interrelated processes: the development of a hadronic shower, whose observables are mostly the daughter muons, and an EM shower. The primary energy E_0 at the maximum of the shower is ultimately shared between N^p_{μ} muons and $N^p_{\rm max}$ electrons/photons. By analogy to Eq. (4.13), the total energy in this case is

$$E_0 = E_c N^p_{\rm max} + E^{\pi}_{\rm dec} N^p_{\mu}\,. \tag{4.33}$$

Scaling to the electron size using (4.19), $N^p_{e{\rm max}} = N^p_{\rm max}/g$:

$$E_0 = gE_c\left(N^p_{e{\rm max}} + \frac{E^{\pi}_{\rm dec}}{gE_c} N^p_{\mu}\right) \sim 0.85[{\rm GeV}](N^p_{e{\rm max}} + 24N^p_{\mu})\,. \tag{4.34}$$

The relative magnitude of the contributions from N_μ^p and $N_{e_{max}}^p$ depends on their respective critical energies, the energy scales at which electromagnetic and hadronic multiplication ceases. Different primaries produce different numbers of muons and $e^\pm$, also affected by shower-to-shower statistical fluctuations.

4.4.1 The Muon Component in a Proton-Initiated Cascade

Let us consider now the dependence of the number of muons in the cascade on E_0. Muons are produced in the decay of the k^* generation of charged pions, when $\pi^\pm$ reach an energy below the threshold E^π_{dec} and all decay into a muon-neutrino pair. Thus

$$N_\mu^p = N_\pi^p = (n_{ch})^{k^*} . \tag{4.35}$$

Using (4.32) and the properties of logarithms, we have:

$$\ln N_\mu^p = k^* \ln n_{ch} = \ln\left(\frac{E_0}{E^\pi_{dec}}\right) \cdot \frac{\ln n_{ch}}{\ln n_h} , \tag{4.36}$$

returning from logarithms to numbers:

$$N_\mu^p = \left(\frac{E_0}{E^\pi_{dec}}\right)^\beta \quad \text{where } \beta \equiv \ln n_{ch}/ \ln n_h . \tag{4.37}$$

Equation (4.37) represents a first order estimate, with many approximations hidden inside the two parameters β and E^π_{dec}. The value $\beta \sim 0.85$ is obtained for $n_{ch} = 10$. When the contribution of the inelasticity κ is included, $\beta \sim 0.9$. It should be noted that the number of secondary pions depends (slowly) on energy as the shower develops, and β can range between ~0.85–0.95. Only Monte Carlo computations can correctly reproduce the shower. The numerical values of β and E_{dec} for different particles and hadronic interaction models are given in Alvarez-Muniz et al. (2002). The hadronic energy in (4.31) can be completely accounted for in the muon component as

$$E_h = N_\mu^p E^\pi_{dec} . \tag{4.38}$$

The average energy of each muon in such a model is on the order of $E^\pi_{dec}/2 \sim 10$ GeV. At such an energy, the muon energy loss is $dE_\mu/dx \sim 2\times10^{-3}$ GeV g^{-1} cm^2. Thus, most muons have enough energy to cross the entire atmospheric depth (~1000 g cm^{-2}). Only muons with energy below a few GeV have a large decay probability, as the muon *decay length* is

$$d'_\mu = \Gamma \cdot c \cdot \tau_\mu = \Gamma \cdot 0.66 \text{ km} \tag{4.39}$$

for the process $\mu^- \rightarrow e^- \nu_\mu \overline{\nu}_e$ and $\mu^+ \rightarrow e^+ \overline{\nu}_\mu \nu_e$. For example, approximately 20% of 1 GeV muons ($\Gamma \sim 10$) produced at a height of 10 km will reach the sea level before decaying. Thus, the number of muons reaching the detection level (usually at depths between 800–1000 g cm^{-2}) is only slightly depleted with respect to N^p_μ at the position of the maximum.

4.4.2 The EM Component in a Proton-Initiated Cascade

The number of electrons is estimated using the relation $E_0 = E_{EM} + E_h$, where the hadronic energy is given by (4.38). The energy fraction carried by the electromagnetic component is

$$\frac{E_{EM}}{E_0} = \frac{E_0 - N^p_\mu E^\pi_{\text{dec}}}{E_0} = 1 - \left(\frac{E_0}{E^\pi_{\text{dec}}}\right)^{\beta-1} . \tag{4.40}$$

The number of electrons at maximum can be obtained with simple arguments. There are n_0 independent showers started by the EM decay of each neutral pion, each carrying E_0/n_h of the primary energy. Equation (4.19) can be traduced for a proton-induced shower to

$$N^p_{e\max} = n_0 \cdot \frac{1}{g}\left(\frac{E_0/n_h}{E_c}\right) = \left(\frac{E_0}{3gE_c}\right) = 4 \times 10^5 \left(\frac{E_0}{\text{PeV}}\right) \tag{4.41}$$

using the fact that $n_0/n_h = 1/3$. In the last equality, we inserted the numerical values $g = 10$, $E_c = 86$ MeV and 1 PeV= 10^9 MeV. This value underestimates by ~30% the electron size due to the additional contribution of successive interactions of the leading particle and of charged pions, producing additional neutral particles. A better estimate can be derived using Eq. (4.40), and it is left as an exercise; see Hörandel (2007) for a hint. This gives us

$$N^p_{e\max} = 6 \times 10^5 \left(\frac{E_0}{\text{PeV}}\right)^{1.046} . \tag{4.42}$$

The number of electrons at maximum grows as a function of energy slightly faster than exactly linear.

Monte Carlo simulations show that the fraction of energy transferred to the EM component at shower maximum increases from about 70% at 10^{15} eV to 90–95% at 10^{20} eV. For instance, a 10^{19} eV proton striking the top of the atmosphere vertically produces, at sea level, about 3×10^{10} particles with energy larger than 200 keV. 99% of these particles are photons and electrons. Their energy is mostly within the range of 1–10 MeV and they transport 85% of the total energy. The remaining particles are either muons with an average energy of about 4 GeV (carrying about 10% of the total energy), a few GeV pions (about 4% of the total energy), and, in smaller proportions, neutrinos and baryons. The shower footprint on the ground extends over a few km^2.

The average energy of electrons at the position of the maximum is below E_c, and thus on the order of tens of MeV. After the maximum, the energy of the EM cascade degrades faster than that of muons, due to the different energy loss of electrons and muons. The model yielding Eq. (4.42) does not take into account electron energy losses in the atmosphere and the size $N_e(X)$ is valid only at shower maximum. After the maximum, it reduces exponentially along the path to the detection level. Figure 4.5 shows the energy fraction (both in linear and logarithmic scales) of the electromagnetic, hadronic, muonic and neutrino components as functions of the atmospheric depth, as obtained with a full Monte Carlo simulation (CORSIKA, see Sect. 4.5) for a primary proton with $E_0 = 10^{19}$ eV. The *energy released into air* refers to the energy fraction transferred from high-energy particles to the excitation and ionization of the medium.

Figure 4.6 shows the number of muons N_μ^p and electrons/positrons $N_{e_{max}}^p$, using our simplified estimate for proton and iron primaries. The main features of the EM and muonic components produced by a proton are summarized as follow

(SF7) The number of muons (4.37) produced in an air shower increases almost linearly with the proton energy E_0, as $N_\mu^p \propto E_0^\beta$ with $\beta \sim 0.9$.

(SF8) The energy E_0 of a primary can be simply estimated if both N_e and N_μ are measured, Eq. (4.34). The relation is linear and almost insensitive to fluctuations in the EM and muonic size. If more primary energy than the average is on the muonic component, the EM one is depleted keeping their sum constant, and vice versa. It is also insensitive to primary particle type—see below. If only the EM size is measured, by inversion of (4.42), we obtain $E_0 \simeq (1.5\text{GeV})(N_{e_{max}})^{0.97}$. This is greatly dependent on the relative fluctuation of the EM/muonic size.

The results are in agreement with detailed Monte Carlo simulations (Sect. 4.5).

4.4.3 *Depth of the Shower Maximum for a Proton Shower*

The atmospheric depth at which the electromagnetic component of a proton-induced shower reaches its maximum is denoted as X_{max}^p. The cascade is provided by the *superposition* of many individual showers. The $n_0 = 1/3n_h$ neutral pions produced

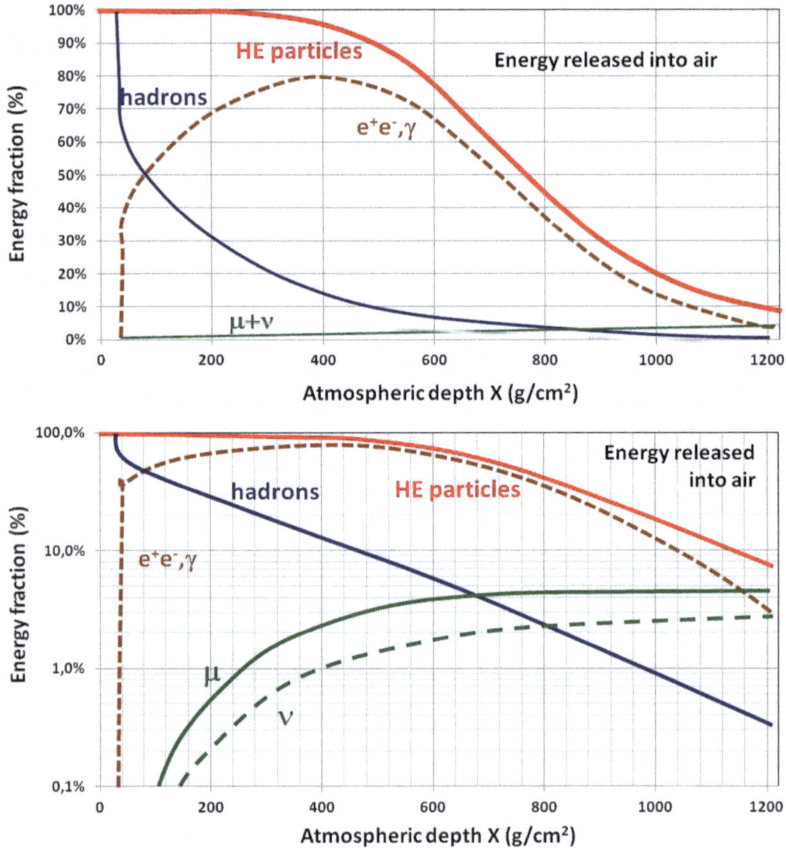

Fig. 4.5 Fraction of energy transferred to the different components of the cascade induced by a primary proton of 10^{19} eV. Part of the energy is released into air by excitation/ionization processes. The *top* graph uses a linear scale for the energy fraction; the *bottom* uses a log scale for a better visualization of the "older" part of the shower

in the first interaction generate (through $\pi^0 \rightarrow \gamma\gamma$ decay) $2n_0$ γ-rays, starting the EM cascade at the same position in the atmosphere. For a primary proton, the first interaction occurs, on average, at an atmospheric depth $X_f = \lambda_I$. Each γ-ray carries $E_0/2n_h$ of the primary energy. A simple estimate of $X^p_{\max}$ can be obtained using the result (4.14) for a shower initiated by a γ-ray with the substitutions $X_f \rightarrow \lambda_I$ and $E_0 \rightarrow E_0/2n_h$:

$$X^p_{\max} \simeq \lambda_I + X_0 \cdot \ln\left(\frac{E_0}{2n_h E_c}\right). \tag{4.43}$$

Let us estimate the difference between (4.43) and (4.14), i.e., between the position of the maximum in the atmosphere for a proton or a γ-ray initiated shower of the same

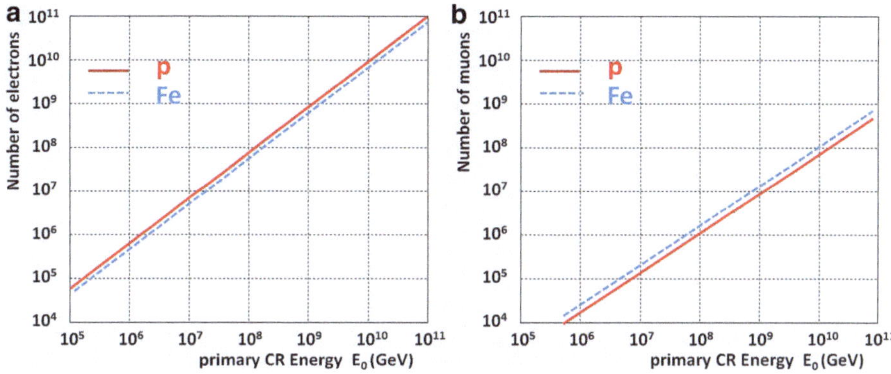

Fig. 4.6 Number of electrons **a** and number of muons **b** at shower maximum as a function of energy for primary protons (full lines) and iron nuclei (dashed lines). The proton lines are obtained using Eqs. (4.37) and (4.42)

energy E_0. The first interaction point for the proton is deeper in the atmosphere, $\lambda_I \simeq 2.5X_0$, using (3.8a). Then, (4.43) can be written as

$$X^p_{\max} = \lambda_I + X_0 \cdot \left[\ln\left(\frac{E_0}{E_c}\right) - \ln(2n_h) \right] \simeq 2.5X_0 + (X^\gamma_{\max} - X_0/2) - 3.4X_0 \,, \tag{4.44}$$

the last numerical factor ($\ln(2n_h) \simeq 3.4$) being obtained with $n_h = (3/2)n_{\rm ch} = 15$. Numerically, $X^\gamma_{\max} - X^p_{\max} \simeq 1.4X_0$, corresponding to about $50\,\mathrm{g\,cm^{-2}}$ in the atmosphere. The maximum of the EM shower induced by a proton occurs higher in the atmosphere than that induced by a photon of the same energy E_0. Also, the elongation rates, Eq. (4.15), are slightly different for protons and photons, as shown in Fig. 4.7. The difference $X^\gamma_{\max} - X^p_{\max}$ is thus dependent on energy. A correction arises owing to the fact that in (4.44), only the first-generation pions are accounted for, and the sub-showers generated in the following steps are neglected.

4.4.4 Showers Induced by Nuclei: The Superposition Model

To extend the simple approach from primary protons to nuclei, the *superposition model* is used. This assumes that a nucleus with atomic mass number A and energy E_0 is equivalent to A individual single nucleons, each having an energy E_0/A, and acting independently. The resulting EAS is treated as the sum of A individual proton-induced showers, all starting at the same point. The corresponding shower features are obtained by replacing $E_0 \rightarrow E_0/A$ in the expressions derived for proton showers and summing A such showers.

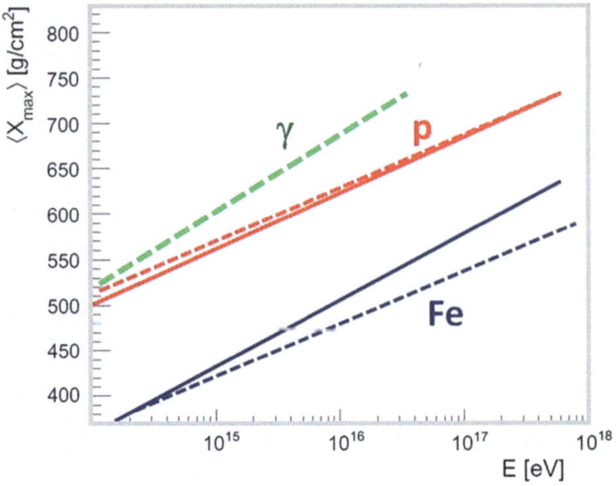

Fig. 4.7 Depth of maximum versus primary energy for the EM component of air showers. *Green dotted* showers are induced by a photon primary. *Dashed* showers are induced by proton (*red*) and iron (*blue*) primaries. Iron nuclei are uniformly shifted by ~150 g cm^{-2} w.r.t. protons according to Eq. (4.48). *Red* and *blue solid lines* are from full MC simulations of p and Fe showers

The number of the $e^{\pm}, \gamma$ (the EM component) at the maximum of the shower induced by a primary nucleus can be derived from (4.41) with the above assumptions:

$$N^A_{e_{\max}} = A \times \left(\frac{E_0/A}{3gE_c}\right) = N^p_{e_{\max}} \,. \tag{4.45}$$

This is an important result, which has equally important implications for experiments: from the measurement of the electromagnetic size, it is hard to distinguish a proton with energy E_0 from a nucleus A of the same energy E_0, as can also be deduced from the left panel of Fig. 4.6.

The corresponding number of muons in nucleus-induced showers can be obtained from (4.37):

$$N^A_\mu \simeq A \times \left(\frac{E_0/A}{E^\pi_{\rm dec}}\right)^\beta = A^{1-\beta} \cdot N^p_\mu \,. \tag{4.46}$$

The important feature is that N^A_μ increases slowly as a function of the mass A of the primary particle as $N_\mu \propto A^{0.1}$. The heavier the shower-initiating particle is, the more muons are expected for a given primary energy. For instance, using (4.46) with $\beta = 0.9$, the number of muons in a He ($A = 4$), O ($A = 16$) and Fe ($A = 56$) induced shower is 14, 32 and 50% larger than that induced by a primary proton of the same energy. Compare with the right panel of Fig. 4.6 for the case of Fe.

To evaluate the average depth of the shower maximum from a nucleus of mass A, we must use (4.43), assuming that the shower is originated from a nucleon in the nucleus with energy E_0/A. Thus, replacing $E_0 \rightarrow E_0/A$ and $\lambda_I \rightarrow \lambda_I^A$ in Eq. (4.43),

$$X_{\max}^A \simeq \lambda_I^A + X_0 \cdot \ln\left(\frac{E_0}{2\,A n_h E_c}\right), \tag{4.47}$$

from which we obtain

$$X_{\max}^A = \lambda_I^A + X_0 \cdot \ln\left(\frac{E_0}{2 n_h E_c}\right) - X_0 \ln A \simeq X_{\max}^p - X_0 \ln A \tag{4.48}$$

(neglecting in the last equality the difference between λ_I and λ_I^A). Using this relation, an air shower initiated by a He, O and Fe nucleus of the same total energy reaches its maximum ~50, 100 and 150 g/cm^2 earlier than that initiated by a proton with the same energy. The depth of the maximum as a function of the primary energy for proton and iron showers is presented in Fig. 4.7 as dashed lines. When a Monte Carlo simulation is used, the differences $X_{\max}^A - X_{\max}^p$ are smaller and with a mild dependence on the CR energy, as indicated by the solid lines in Fig. 4.7. A slight difference in the elongation rate is predicted from different interaction models, Sect. 4.5, used to estimate the hadron production in the simulation. This is visible in Fig. 4.20, which will be discussed in the following and compared with the experimental data.

The main qualitative results and properties of our first-order estimate for nucleus-induced showers are summarized below. They are confirmed by detailed simulations independent of the interaction models used in the Monte Carlo simulation (see next section).

- **(SF9)** The electromagnetic size is equal for a cascade initiated by a proton with energy E_0 and by a nucleus A of energy E_0, Eq. (4.45);
- **(SF10)** The number of muons in a nucleus-induced shower depends on A as $N_\mu^A \propto A^{1-\beta} \sim A^{0.1}$, Eq. (4.46);
- **(SF11)** The depth of the maximum of the EM component of a nucleus-induced shower differs from that of a proton-induced shower as $X_{\max}^A \simeq X_{\max}^p - X_0 \ln A$, Eq. (4.48).

This feature is evident in Fig. 4.8, which shows the longitudinal profile for two showers initiated by a proton (left) and an iron nucleus (right) with energy $E_0 = 10^{14}$ eV obtained using a CORSIKA-based Monte Carlo simulation (see next section). The different components of the cascade, the electromagnetic (e/γ), the muonic (μ) and the hadronic (h), are shown in different colors.

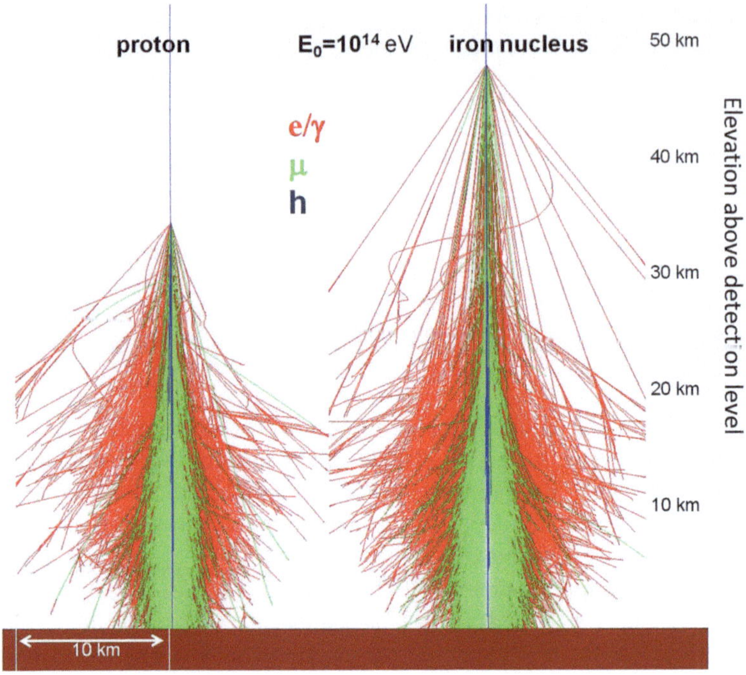

Fig. 4.8 Longitudinal profile for two showers with energy $E_0 = 10^{14}$ eV from a full MC simulation based on the CORSIKA code. *Left* shower initialized by a proton. *Right* shower initialized by an iron nucleus

(SF12) Due to the larger cross-section of a nucleus A, one has $\lambda_I^A \sim \lambda_I/A^{2/3}$ and the fluctuations on $X_{\max}^A$ are much smaller than those on $X_{\max}^p$. As a consequence, the root mean square fluctuation $\sigma(X_{\max}^A)$ on the position of the maximum of the EM component induced by a heavy nucleus is smaller than $\sigma(X_{\max}^p)$.

This peculiarity is noticeable in Fig. 4.9, which shows the number of charged EM particles from the simulation of 50 different showers induced by protons and iron nuclei of the same energy ($E_0 = 10^{19}$ eV) as a function of the atmospheric depth. Shower-to-shower fluctuations of the position of the maximum depth are present, although more evident in the case of protons. From this figure, the average atmospheric depth and its root mean square are $X_{\max}^p \simeq 770 \pm 60\,\mathrm{g\,cm^{-2}}$ for p-induced showers and $X_{\max}^{Fe} \simeq 700 \pm 20\,\mathrm{g\,cm^{-2}}$ for iron-induced showers.

Information from detailed Monte Carlo simulations are used by EAS experiments to estimate the parameters of the CR spectrum at energies around and above the knee. The properties of the primary CRs are deduced from the shape and the particle content of the EAS. The energy of the primary depends on the total number of secondaries produced, and information on the chemical composition can be deduced from the relative N_μ, N_e contributions or from $X_{\max}$.

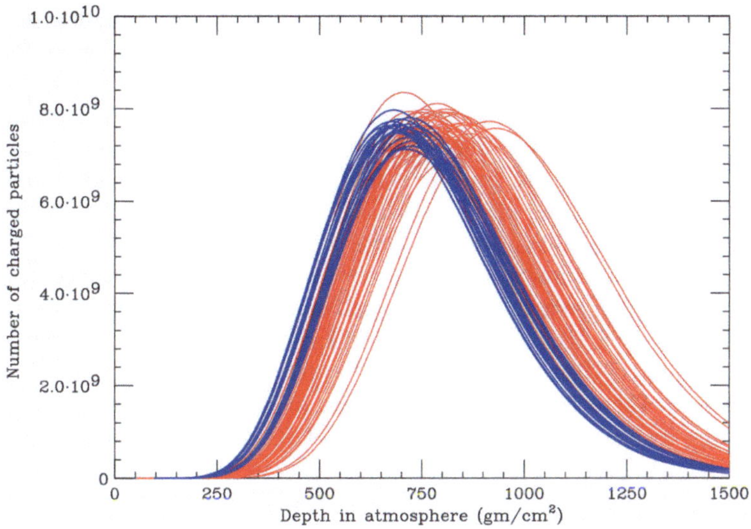

Fig. 4.9 Simulation of the longitudinal profile produced with the CORSIKA code for 50 proton-induced (*red*) and 50 iron-induced (*blue*) showers. The same total energy of 10^{19} eV is assumed. Shower-to-shower fluctuations on $N_{e_{\max}}$ and $X_{\max}$ are evident. Courtesy of Prof. J.W. Cronin

4.5 The Monte Carlo Simulations of Showers

EAS experiments cannot be exposed to a test beam for calibration. The interpretations of their measurements rely on comparison of the experimental data with model predictions of the shower development in the atmosphere obtained via Monte Carlo (MC) simulation of particle interactions and transport in the atmosphere.

While the electromagnetic and the weak interactions are rather well understood, the major uncertainties in MC simulations of EAS arise from the *hadronic interaction model*. The properties of an EAS for a given energy and primary are strongly dependent on the inelastic cross-sections $\sigma_{\rm inel}$ of primary and secondary particles with air nuclei.

At the present level of theoretical understanding, Quantum Chromo-Dynamics (QCD) cannot be applied to calculate the hadronic inelastic cross-section and the number of secondary particles from first principles. Therefore, hadronic interaction models are usually a mixture of basic theoretical ideas and empirical parameterizations tuned to describe the experimental data where measurements exist.

The data from the LHC at CERN allow for detailed tests of interaction models up to center-of-mass energies, which are equivalent to CR energies of $\sim 10^{17}$ eV, but an appreciable amount of extrapolation of the properties of hadronic interactions is still needed to interpret the air shower data. In fact, because of both the low transverse momentum (p_t) of secondary particles produced in hadronic interactions and momentum conservation, most secondaries are moving in the same direction as the primary CR (*forward production*). Due to the unavoidable presence of the beam pipe in accelerators, none of the large collider experiments registers particles emitted in the extreme forward direction (that of the beam), and the available information in this kinematic region is very poor. These low-p_t particles carry most of the hadronic energy and are of the greatest importance in EAS development, since they transport a large energy fraction deep into the atmosphere. Only one small-scale LHC experiment (LHC forward, LHCf) can detect neutral particles produced by hadron collisions in the very forward region. It consists of two small sampling calorimeters installed 140 m away from the ATLAS collision point. LHCf data are providing precise measurements of the particles produced in high-energy p-p and p-Pb collisions in order to tune hadronic interaction models used in CR experiments.

In present models for the σ_{inel}, the multiplicity distribution of secondary hadrons and their momenta distributions have to be extrapolated well beyond the range of knowledge. These model uncertainties are the dominant source of systematic uncertainties for the study of ultra-high energy CRs, as we discuss in Chap. 7.

One widely used Monte Carlo code, simulating in great detail the EAS initiated by photons, nucleons, or nuclei, is the CORSIKA program (Engel et al. 2011; Heck 1998; Knapp and Heck 1993). Each produced particle is propagated (*tracked*) along the atmosphere. All physics processes are considered: energy loss, deflection due to multiple scattering and to the Earth's magnetic field, decay modes (for unstable particles) and electromagnetic and hadronic interactions. Figure 4.10 shows the lateral and longitudinal profiles of all the components induced by a vertical CR proton of 10^{19} eV.

CORSIKA contains various computer codes for the hadronic interaction modeling of nucleons and nuclei, which allow for a comparison and an estimate of systematic errors. The available hadronic models are VENUS, QGSJET, DPMJET, SIBYLL, HDPM, and EPOS (see *Extras* # 3).

Detailed comparisons of the models available in CORSIKA have been performed, revealing differences on the 25–40% level. Some of the models are unable to describe all aspects of the experimental results. Most of the figures shown in this chapter contain predictions derived from the CORSIKA code with one or more of the quoted hadronic interaction models (usually, different versions and releases of the same code exist).

4.6 Detectors of Extensive Air Showers at the Energy of the Knee

Extensive air showers induced by CRs with energies above $\sim 10^{14}$ eV can be detected with arrays of sensors (scintillators, water Cherenkov tanks, muon detectors, Cherenkov telescopes, etc.) spread over a large area. When a CR falls within the array boundary, the subsample of detectors placed near enough to the shower core will observe the radiation reaching the detection level.

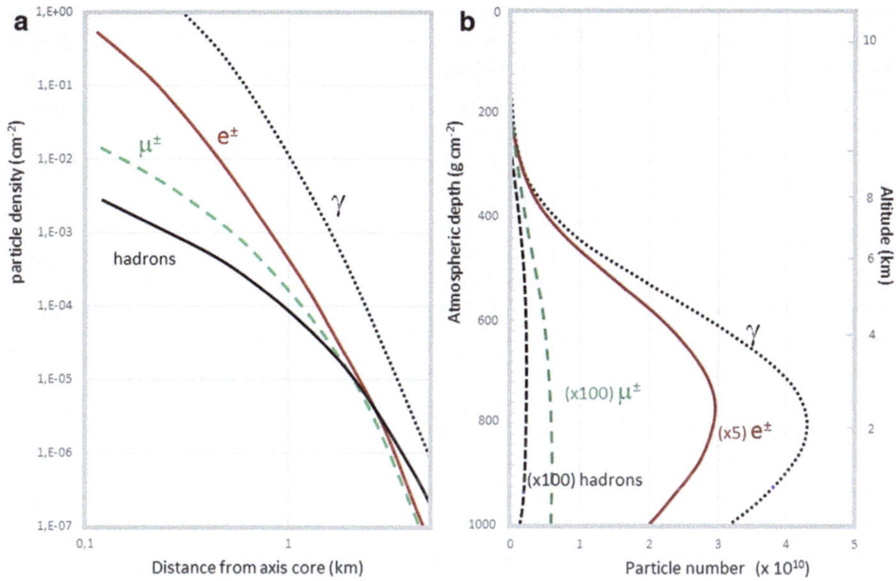

Fig. 4.10 Average **a** lateral and **b** longitudinal shower profiles of the hadronic, muonic and electromagnetic components generated with the CORSIKA code. The showers are induced by vertical protons of energy 10^{19} eV. The lateral distribution of the particles at ground level is calculated for 870 g cm^{-2}, the depth of the Pierre Auger Observatory (Sect. 7.8). Only photons and $e^{\pm}$ with energy larger than 0.25 MeV are followed in the simulation. For muons and hadrons, the energy threshold is 100 MeV

The accuracy of the description and reconstruction of EAS characteristics (energy, nature of the primary, direction) results from a compromise between the financial budget and the number of sensors distributed over the largest area A with the smallest detector spacing. The instrumented area A should meet the condition that $\Phi(>E) \cdot A \cdot \Delta\Omega \cdot T$ gives a reasonable number of events within a given observational time period T. The quantity $\Delta\Omega$ is the covered solid angle and the integral intensity $\Phi(>E)$ has units cm^{-2} s^{-1}sr^{-1} as usual. The number of secondary particles crossing each detector within a given time windows are counted. Detector spacing on the order of a few tens of meters appears optimal for 10^{15} eV CRs. At higher energies, the spacing must be increased to increase the instrumented area with sustainable economic budgets.

Most particles observed at ground level are close (about 200–300 m, corresponding to 2–3 Molière radii) to the shower axis. At the highest energies ($>10^{18}$ eV), particles can be observed up to some kilometers away. The incident CR direction and energy are measured assuming that the shower has an axial symmetry in the plane perpendicular to the shower axis. This assumption is valid for zenith angles $\theta < 60°$ and not too far from the axis.

In addition to the EM and muonic cascades, relativistic charged particles emit visible photons through the *Cherenkov effect* in the atmosphere. These visible photons are concentrated around the shower axis, and reach the ground essentially unaltered. Cherenkov light array detectors can measure the time and intensity profile of the emitted light, which carries memory of its production point along the shower axis. Another method used to detect CRs of energy larger than 10^{18} eV exploits the excitation of nitrogen molecules by the particles in the shower and the associated fluorescence emission of light. The light is detected by photomultipliers and the profile of the shower in the atmosphere can be inferred rather directly. This method will be discussed in Chap. 7.

Figure 1.5 shows the flux of primary CRs as measured by direct and indirect experiments. Each EAS experiment measures the CR flux within a given energy range. Different kinds of experiments cover more than 8 decades of energy.

Many arrays were built in different countries to study CRs in the region of the knee between 10^{14}–10^{17} eV, making important contributions. The early measurements have been replicated with superior statistics in the modern arrays built in Germany (KASCADE and KASCADE-Grande), Italy (EAS-TOP) and Tibet. EAS arrays are normally located at atmospheric depths between 800 g cm^{-2} and the sea level. Some exceptions exist, such as the Tibet air-shower array, built at Yangbajing (altitude 4300 m above sea level), with an atmospheric depth of 606 g cm^{-2}.

The first of the giant shower arrays operating at the beginning of the 1960s was built by J. Linsley, L. Scarsi and B. Rossi at Volcano Ranch, New Mexico. The array was constituted of 20 scintillators, 3 m^2 each and 900 m apart, covering a total surface area of about 8 km^2. Scintillator arrays are usually made of flat pieces of plastic scintillators displaced on the ground and connected by cables. As most EAS experiments are carried out in remote regions, serious problems regarding power supply, safety, data transmission, communication, etc. usually exist. Plastic scintillators are robust, cheap, easy to handle and simple to use and deploy; they are equally sensitive to all charged particles, thus they mostly measure the dominant EM component of the cascade. The sensitivity of an array of flat scintillators for inclined CRs drops quickly towards the horizontal direction, because of the decrease of their effective surface and because of the absorption of the EM component. Usually, data of scintillator arrays are restricted to zenith angles below 45°.

One alternative to scintillator counters is the use of water Cherenkov tanks. The largest pioneering EAS to use this technique was the Haverah Park array, operating in England from 1964 to 1987, made of water tanks of various sizes spread over about 12 km^2. In water, charged particles emit Cherenkov radiation. These detectors are heavy and require extra pure water with excellent protection against contamination. In addition, they are not as easy to deploy as scintillators. These disadvantages are compensated for by the fact that the Cherenkov light generated in water is proportional to the path length of the particle. For this reason, water tanks are sensitive to both the electromagnetic and the muon components. On average (the correct numbers depend on the exact detector geometry), a muon releases about 10 times more light than a single 20 MeV electron. Because of their height (typically on the order of 1 m), water tanks also offer a nonzero effective surface for horizontal showers.

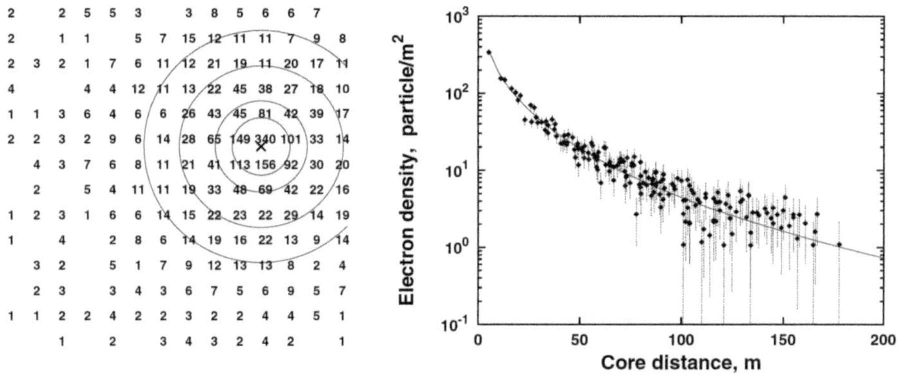

Fig. 4.11 *Left* A simulated event on an ideal detector with 196 counters (each having a surface of 1 m^2) on a 15 m grid. The geometry is similar to that of the KASCADE array. The lateral distribution was sampled from the NKG distribution, Eq. (4.21), with Poissonian fluctuations. At the position of each counter, we show the number of detected particles. Some empty counters are also present. The *circles* represent the regions within which the number of particles per counter exceeds 10, 20, 50 and 100 m^{-2}, respectively. *Right* The particle density as a function of the distance from the shower axis for the event shown on the *left*. The *line* represents the average particle lateral distribution. From Stanev (2010)

4.6.1 A Toy Example of an EAS Array

We use a toy model (Fig. 4.11) to illustrate some interesting features of the detection of the EM component using an EAS array. The figure on the left shows the footprint of a vertical shower generated by a primary iron nucleus with $E_* = 4 \times 10^{15}$ eV. We assume a detector at 2000 m above sea level, corresponding to about 820 g cm^{-2}. At this depth, the corresponding shower size is $N_e = 8 \times 10^5$ (as can be obtained from Fig. 4.3). The array consists of 14×14 counters (plastic scintillators, for instance, 1 m^2 each and spaced 15 m apart). We have superimposed on each counter the number of shower particles crossing it. The total instrumented surface is $A = 4 \times 10^8$ cm^2; the total surface of the detectors is $A_{\text{det}} = 196 \times 10^4$ cm^2, the ratio between $A_{\text{det}}/A = 5 \times 10^{-3}$.

The counters of this array detect mainly electrons and positrons; few photons convert in the detector. For a first approximation, each electron crossing a counter loses the same amount of energy ξ. A counter measures the total energy ΔE released by incoming particles within a given time window. Usually, the detection technique and the data acquisition electronics employed give a linear response in a wide interval of ΔE. In this case, the number of particles crossing the counter can be estimated as $\Delta E/\xi$. If the number of particles is too high, *saturation effects* can occur.

Each counter usually suffers from a continuous counting rate due to noise, mainly resulting from environmental radioactivity. Such noise has a random nature and occurs on different counters at different times. To get rid of such spurious signals, a time coincidence among several counters is normally required. The arrival of a shower is therefore identified by the array when many counters register, within a short time interval, a signal well above that registered in the absence of a shower. A *trigger* usually consists of a minimum number (*majority*) of counters fired within the same time window. In the presence of a trigger, the information (number of particles, arrival time, …) registered by each counter are stored on a computer.

The expected rate of CRs with energy at least equal to $E_* = 4 \times 10^{15}$ eV = 4 PeV producing events similar to that shown in Fig. 4.11 can be obtained from the particle flux (2.17b):

$$\dot{N}(> E_*) \equiv \mathscr{F} \cdot A = \Phi(> E_*) \times \frac{\pi}{2} \times A \times \epsilon \,. \tag{4.49}$$

The factor $\pi/2$ instead of π arises because only downward-going particles reach the detector. We need to anticipate the result on the CR flux above the knee provided in Sect. 4.8 by Eq. (4.53), i.e., $\Phi(> E) = 2.2 \times 10^{-10}/E^{2.06}$ when the energy is expressed in PeV. Thus

$$\dot{N}(> E_*) = \frac{2.2 \times 10^{-10}}{4^{2.06}} \times \frac{\pi}{2} \times 4 \cdot 10^8 \times \epsilon \simeq 10^{-2} \times \epsilon \text{ s}^{-1} \simeq 10 \text{ events/h} \,. \tag{4.50}$$

The quantity ϵ represents the overall array efficiency, which includes the fraction of solid angle covered, the fact that showers near the detector edges could not trigger the detector, etc. A factor $\epsilon \sim 0.3$ is assumed here.

Let us estimate the energy range in which this array can efficiently detect CRs. Let us start by considering a primary CR having the same core axis as the one just considered and a tenth of the energy E_*. The corresponding shower will have a particle density smaller by an order of magnitude. Each number in Fig. 4.11 will be a factor of 10 smaller and the outer ring will have a particle density of 1 m^{-2}. About 70 counters will have fired and the energy and direction of the CR originating the shower can be reconstructed. A shower produced by a CR with energy $E_*/100 = 4 \times 10^{13}$ eV will have a density one hundred times smaller than that of the event in Fig. 4.11. Now, the inner ring will have a particle density of 1 m^{-2}. The cascade will interest very few counters, below the majority needed to trigger the apparatus. We conclude that our EAS array is able to detect showers from primaries having energy $\gtrsim 10^{14}$ eV.

On the other side, a primary with energy $10E_*$ will have such a large number of particles to fire all counters, and some of them could saturate. As the integral flux decreases by $E^{-2.06}$ within this energy range, the event rate reduces to $\dot{N}(> 10E_*) = 10^{-2.06}\dot{N}(> E_*) = 3$/day. A shower produced by a CR with energy larger by another decade covers a much larger surface and saturates almost all counters. Only a rough estimate of the primary energy can then be given. However, due to the flux decrease, the event rate reduces by an additional $10^{-2.06}$ factor to a few events/year. In conclusion, the considered array could measure primary CRs efficiently within the energy range $10^{14} \lesssim E_0 \lesssim 10^{17}$ eV.

A viable solution for increasing the maximum value of measurable energy keeping the number of counters constant would be to increase the distance between counters. This solution is almost costless if the apparatus does not interfere with civil, engineering or environmental constraints, for instance, being installed in a desert region. It is intuitive that the same counters distributed over a larger area would increase the lowest measurable CR energy.

The right side of Fig. 4.11 shows the measured density distribution as a function of the distance from the shower axis. The position of the shower axis on the detection plane can be estimated in different ways, for instance, finding the center of gravity of the counting rate. The counter measuring the highest particle number (340) is very close to the shower core. Each point corresponds to the measurement of one counter at a given distance from the core. The error band corresponds to the statistical error: for Poissonian statistics, the error is the square root of the number of counted particles.

In the following, we present some selected EAS experiments covering the energy range up to $\sim 10^{18}$ eV and discuss the ways in which the direction, energy and mass of the primary can be derived from the data. For a review, see Nagano and Watson (2000), Blümer et al. (2009), Grieder (2010), Letessier-Selvon and Stanev (2011).

4.6.2 *Some EAS Experiments*

The Extensive Air Shower on Top (**EAS-TOP**) array was located at Campo Imperatore, in central Italy, at a height of 2005 m (above the underground Gran Sasso laboratory), at 820 g cm^{-2} atmospheric depth. It has been in operation, in different configurations, between 1989 and 2000. Its EM detector was made of 35 scintillator modules, 10 m^2 each, separated by 20 m in the central region and 80 m near the edges of the array. The detector was fully efficient for electromagnetic sizes $N_e > 10^5$. A central muon-hadron calorimetric detector, covering a surface of 140 m^2, was also present. It consisted of 9 layers of 13 cm thick iron absorbers and limited streamer tubes as active elements. Muons were counted if their energy was >1 GeV.

The KArlsruhe Shower Core and Array Detector (**KASCADE**) experiment started collecting data in 1996 and studied CRs within the range from 10^{14} to 10^{17} eV. It detected a combination of observables: electrons, muons at four energy thresholds, and energy and number of hadrons (Antoni et al. 2003). The main components were the field array, a central detector and the muon-tracking detector. The field array measured electrons and muons ($E_\mu > 230$ MeV) in the shower separately, using an array of 252 detector stations containing shielded and unshielded detectors, arranged on a square grid of $200 \times 200\,\text{m}^2$ with a spacing of 13 m. Figure 4.12 displays a sketch of a detector station. The EM size was measured by the e/γ detectors, which were positioned over a lead/iron plate inside a wooden

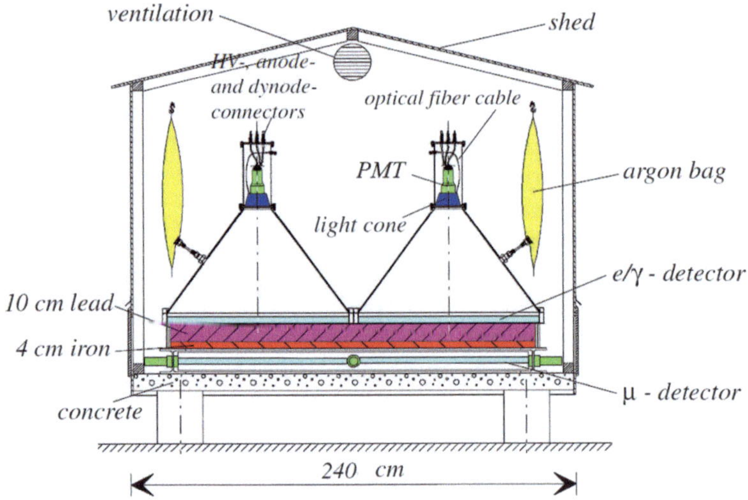

Fig. 4.12 Schematic view of one of the 256 sample array detector stations of the KASCADE experiment. Courtesy of the KASCADE collaboration (https://web.ikp.kit.edu/KASCADE/)

hut. Each e/γ detector consisted of a liquid scintillator covered by a light-collecting cone. On top, a light collector and a photomultiplier were mounted to read out the signals. The energy resolution was ~8% at 12 MeV, which corresponds to the mean energy deposit of a minimum ionizing particle (m.i.p.). Energy deposits up to ~2000 m.i.p. were measured. The time resolution was ~0.8 ns.

In the 192 outer stations, muon detectors were installed underneath the e/γ detectors, followed by a shielding of 10 cm lead and 4 cm iron, corresponding to 20 radiation lengths, to stop the EM component. Muons can cross the shielding, and the μ-detector consisted of four plastic scintillators of 3 cm thickness and 0.8 m^2 surface. The light was read out by 1.5-inch photomultipliers. The energy resolution was ~10% at 8 MeV, the mean energy deposited by a muon.

The surface covered with detectors was 1.3 and 1.5% of the total area for the electromagnetic and the muonic detectors, respectively. This large fraction of instrumented surface and the small distance between counters allowed for the reconstruction of the incident angles, the lateral distributions and the measurement of the size of N_μ and N_e for CRs down to 10^{14} eV. The layout of the KASCADE detector is shown in Fig. 4.13 (right).

As an extension of KASCADE, **KASCADE-Grande** was built by reassembling 37 stations of the electromagnetic detectors of the former EAS-TOP experiment. The goal of this enlargement was to extend the measurement of primary CRs up to 10^{18} eV. The area covered by the EM array was expanded to $700 \times 700\,m^2$ with the 37 stations, 10 m^2 each. The layout of the experiment is shown in Fig. 4.13 (left). KASCADE-Grande was in operation from 2003 until 2013, after which it was dismantled.

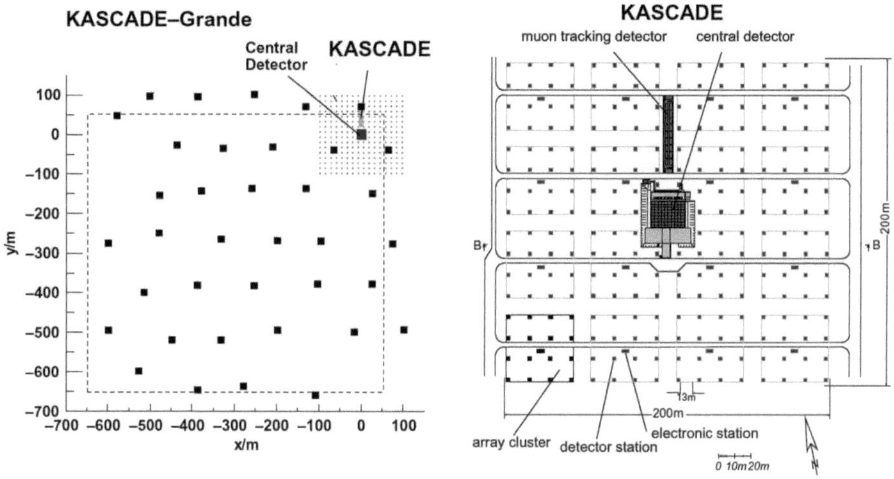

Fig. 4.13 *Left:* Schematic top view of the KASCADE-Grande experiment, with the KASCADE in the *top right hand corner* of the figure. *Right:* Layout of the KASCADE experiment and its subdetectors. The figure also shows the organization of the KASCADE array into 16 clusters of 16 or 15 stations. The highlighted *lower left* cluster is equipped with Flash-ADC. Courtesy of the KASCADE collaboration (https://web.ikp.kit.edu/KASCADE/)

The **Yakutsk** array represents the most complex, and most northerly, of the early giant arrays. It was operated by the Institute of Cosmophysical Research and Aeronomy at Yakutsk, Siberia (105 m above sea level) beginning in 1967 and was enlarged to cover an area of 18 km^2 in 1974. A particularly important feature was the presence of 35 Photo Multiplier Tube (PMT) systems of various areas to measure the air-Cherenkov radiation associated with the showers. These gave indirect information on the longitudinal development of showers and provided a calorimetric approach to the energy estimates for the primary particles. Measurements relating to CRs above 10^{17} eV have been reported.

The Chicago Air Shower Array (**CASA**) detector was an array of both surface and underground plastic scintillators, which measured the electromagnetic, and muon components of air showers. The array was located on the Dugway Proving Grounds, Utah, USA, and operated without interruption for 5 years, from 1992 to 1997, measuring CRs within the energy range 10^{14}–10^{16} eV. The 1089 CASA detectors were distributed on a regular grid with a 15 m spacing for the detection of the electromagnetic component. The area enclosed by CASA was 0.23 km^2 and included the Muon Array (**MIA**), which consisted of an array of 1024 scintillation counters distributed in 16 patches of 64 counters each. The MIA counters were buried beneath approximately 3 m of earth and had a typical muon energy threshold of $\sim$1 GeV.

4.6.3 Cherenkov Light Produced by EAS Showers

The majority of particles in EAS travel with relativistic velocities through the atmosphere. Cherenkov radiation is emitted when a particle has a speed exceeding that of light in the medium. The emitted radiation falls in the wavelength region of 300–350 nm (near ultraviolet) that matches the quantum efficiency of most photomultipliers very well. Due to their low mass, electrons (and positrons) have a low Cherenkov threshold $E_T = 21$ MeV at sea level, 35 MeV at 8 km of altitude. The reason for the change in E_T is the variation of the Cherenkov threshold velocity with the changing refraction index of the atmosphere $n = n(h)$. For Standard Temperature and Pressure (STP) conditions, $n = 1.00029$; as the pressure varies with altitude, n decreases with increasing h. The emission angle of Cherenkov light with respect to the particle direction is given by: $\theta_C = \cos^{-1}(1/n)$ with $\theta_C = 1.3°$ at STP. The angle increases when the EM shower propagates from the top of the atmosphere down to ground level (see Fig. 4.14).

For the detection of Cherenkov light, two techniques have been used: (1) *integrating detectors* (described below) consisting of arrays of photomultipliers looking upwards into the sky and distributed over a large area at ground level; (2) *imaging detectors or telescopes* (Chap. 9) composed of large area collection mirrors and a camera with segmented photomultiplier read-out. Optical devices such as

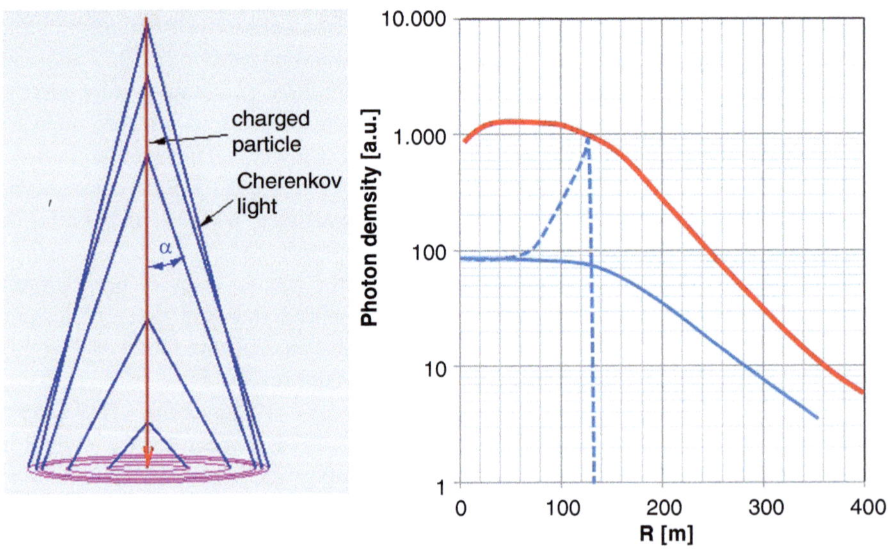

Fig. 4.14 *Left* Cherenkov light emission from EAS at five different atmospheric heights, showing the changing Cherenkov emission angle as a function of the depth, related to the variation of the atmospheric refraction index (courtesy of Dr. Konrad Bernloehr). *Right* Resultant photon density at ground level for two different primary energies differing by an order of magnitude. The *dashed line* represents the photon density for the case of no electron multiple scattering for a lower energy shower

Cherenkov detectors and fluorescence detectors can only be operated during clear moon-less nights. This restricts their duty cycle to about 10%.

Integrating detectors measure the lateral distribution of the Cherenkov light (see the right side of Fig. 4.14) with an array of PMTs. The longitudinal development in the atmosphere can also be measured, in particular, the position X_{max} of the shower maximum. These instruments are operating, in most cases, near array detectors measuring the EM and/or muonic component to increase the number of observed quantities of the same shower. This technique was pioneered by the **AIROBICC** experiment on the La Palma island.

The Broad LAteral Non-imaging Cherenkov Array (**BLANCA**) took advantage of the CASA-MIA particle array installation by augmenting it with 144 Cherenkov detectors. The Dugway atmospheric depth was 870 g cm^{-2}, and BLANCA used the CASA trigger (threshold at $\sim 10^{14}$ eV) to collect Cherenkov light from CR in the knee region. Each BLANCA detector contained a light-collecting cone (called a *Winston cone*) to concentrate the light striking an 880 cm^2 entrance aperture onto a PMT tube. The minimum detectable density of a typical BLANCA unit was approximately (one blue photon)/cm^2. BLANCA operated between 1997 and 1998, for a total of $\sim$460 h of Cherenkov observations.

The Dual Imaging Cherenkov Experiment (**DICE**) was an air shower detector consisting of two telescopes located at the CASA-MIA site with an effective geometrical factor of $\sim$3300 m^2 sr. Each telescope used a 2 m diameter spherical mirror with a focal plane detector consisting of 256 close-packed 40 mm hexagonal PMTs, which provide $\sim 1°$ pixels in an overall field-of-view of $16° \times 13.5°$ centered about the vertical. The telescopes were on fixed mounts separated by 100 m.

Another Cherenkov detector array is the **Tunka**, close to Lake Baikal in Siberia, where 133 PMTs with 8 inches diameter distributed over a $\sim$1 km^2 surface have been operational since 2009, replacing an older 25-PMTs array.

4.7 The Time Profile of Cascades

The different developments of the EM and muonic components of EAS induce a difference in the arrival times of electrons and muons at the observation level. Therefore, an alternative to the use of dedicated muon detectors is the separation of the muon and EM components by an adequate cut on the particle arrival time. Muons are, on average, produced higher in the atmosphere and scatter far less than electrons do; they are heavier than electrons and are less affected by the geomagnetic field. Their paths to the ground are nearly straight lines. Most $e^{\pm}$ are, on average, produced deeper in the atmosphere and reach the observation point after multiple scattering. This generates longer path lengths, and thus longer times of flight. The resulting effect of the different development is that muons arrive earlier at the observation level than the EM component.

These features, which can provide an interesting experimental tool for measuring the two components using the same counter array, are confirmed by detailed Monte

Carlo simulations. In fact, EAS detector arrays can be equipped with modern fast, time-resolving electronics to sample the time development of the detector responses. This offers an alternative method for determining the muon content of EAS.

A first measurement of such time delay was performed by the KASCADE-Grande array (Apel et al. 2008). The time structure of the EAS disk was analyzed as a function of the distance R from the shower axis up to $R = 600$ m. A Flash-ADC[6] based data acquisition system installed on a cluster (see Fig. 4.13) of the KASCADE array samples detector signals of the e/γ and the μ-detectors, respectively.

Figure 4.15 shows a sketch of the EM and muonic shower fronts. The quantity $\langle t\rangle(R)$ represents the delay of the EAS with respect to a flat front as a function of the radial distance R from the shower core. The delay (in units: ns) increases with increasing R (measured in m) as $\langle t\rangle[\text{ns}] \simeq 0.2R$ up to $R \sim 500$ m. For $R > 200$ m, the electronic system of the KASCADE-Grande detector was able to separate the contributions from the EM and muonic components. Close to the shower core ($R < 200$ m), the maximum difference in their average arrival time amounts to 4 ns.

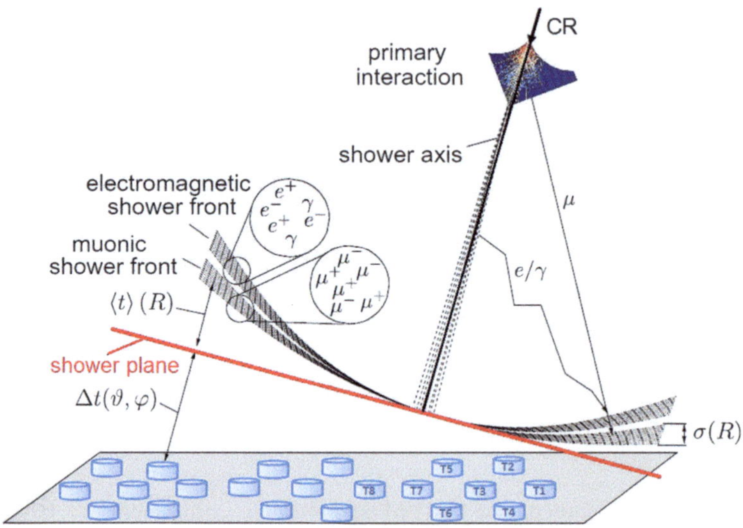

Fig. 4.15 Sketch of the development of EM and muonic showers in the atmosphere, produced by the interaction of a primary proton or nucleus (see text for details). Courtesy of the KASCADE collaboration (https://web.ikp.kit.edu/KASCADE/)

[6]Analog-to-Digital Converters (ADC) convert the height or the integral of an electronic signal into a digital number. For instance, the height of a signal between 0 and 5 V may be converted by a 10-bit ADC into a number between 0 and $2^{10} - 1 = 1023$. Flash-ADCs are very fast compared to other ADC types, so a single flash-ADC can be used to analyze various channels in sequence, or to analyze in a time-sequence the development of a pulse, functioning in this way as a Waveform Analyzer (WFA).

The measurements confirm that a shower is a thin pancake of particles propagating near the core axis, and the thickness (indicated with $\sigma(R)$ in Fig. 4.15) increases with increasing distance R from the shower axis. For $R < 400$ m, the thickness of the EM shower increases linearly according to the relation $\sigma(R)$ [ns] $\simeq 0.2R$ (R in [m]). For a particle propagating at the speed of light, 1 m thickness corresponds to ~3 ns. The muon shower disk is similar to that of the EM component.

The thickness of the high-energy components of the shower is expected to be very small. This has been measured using the arrival time of the penetrating component of the shower (namely, the muonic component). Their arrival times are compatible with that of a flat disk. We will return, in Chap. 11, to high-energy muons that are detected by underground and underwater experiments. These muons have high Lorentz factor ($\Gamma = E_\mu/m_\mu c^2 \gg 10^3$), and in their case, multiple Coulomb scattering as well as the deflection in the geomagnetic field, are negligible. The underground MACRO experiment (Ahlen et al. 1992) measured the arrival times of bundles of muons (each with $E_\mu > 1$ TeV at sea level) with a nanosecond resolution. The result was that muons in a bundle arrive at the same time, i.e., within the experimental time resolution of the detector.

4.8 The Arrival Direction of CRs as Measured with EAS Arrays

The direction of the shower axis, and hence of the primary CR, is obtained in an EAS array measuring the arrival time of the shower front in the intercepted counters. The position of the shower core (the cross on the left side of Fig. 4.11) is determined first. Various computational techniques have been adopted. Usually, a trial core position is determined and then the correct location is searched for using a fitting procedure between the densities observed and the densities expected from the lateral distribution. Usually, the NKG function (4.21) is used to describe the particle density of a shower with a given age s as a function of distance r to the core.

The shower direction (defined by its zenith θ and azimuth ϕ angles) is the normal to the propagating shower front, that is, approximately a plane in the proximity of the shower axis, but has a spherical-like shape away from the axis, see Fig. 4.15. The direction of the incoming CR can be deduced from geometrical considerations by measuring the time delay $\Delta t = T_i - T_j$ between different counters.

In principle, three counters are enough to identify a plane. However, due to statistical fluctuations, uncertainties on timing, and the finite thickness of the shower front, a larger number of counters is necessary for a sufficiently precise measurement. For these reasons, the CR direction is usually derived from a chi-squared minimization or a maximum-likelihood procedure using the measured arrival time of the shower front in a large number of counters, assuming the shower core to lie in the plane perpendicular to the shower axis. The quality of the reconstruction degrades when the measurement is performed far from the axis, because the thickness of the particle disk grows and the density decreases.

The uncertainties on the zenith θ and azimuth ϕ angles depend on the time resolution of the detection technique used by the EAS array and on the size of the cascade. Typical detectors have time resolutions within the range from 0.5 ns up to a few ns. For instance, the angular resolution of the KASCADE detector improves from 0.55° for small showers with size $N_e \sim 10^4$ to 0.1° for showers with $N_e \sim 10^6$.

After the shower core position and shower direction have been determined, the shower size can be obtained from a fit of the experimental lateral distribution using the NKG function. We recall that dN_e/rdr, at a given distance r, depends on the shower size N_e, on the Molière radius and on the shower age s. The difficulty is that s is not known and should be determined from the data themselves. Some experiments perform a global fitting procedure of the lateral distribution, which includes as free parameters the axis position, direction and the age s (the Molière radius can be determined from the atmospheric density at the position of the array). Another possible solution is to use an average value of s appropriate for the detector altitude and array layout in the fit. This average s is usually computed through a Monte Carlo simulation or empirically derived from the data. Once s is also fixed, the shower size can be derived from (4.21), assuming azimuthal symmetry in the plane perpendicular to the shower:

$$N_e = \int_0^\infty \frac{dN_e}{rdr} \cdot 2\pi r dr \ . \tag{4.51}$$

The size N_e is proportional to the primary CR energy E_0.

Figure 4.16 shows an example (from the EAS-TOP experiment) of the use of the NKG lateral distribution with a fixed value of s to fit the measured shower size as a function of the distance from the core. The agreement is good in every shower size interval; differences are, at most, 10% for large core distances. Other solutions are also adopted. The KASCADE experiment, for instance, found that the lateral distributions of the three shower components (electromagnetic, muonic, and hadronic) can all be parameterized using the NKG function. They simultaneously fitted the parameters N_e, r_1 (4.22) and s finding that the measured lateral distributions of the electromagnetic component can be reproduced with an accuracy of about 1%.

The arrival direction of charged CRs with primary energies between 10^{14}–10^{18} eV is remarkably isotropic. Different sky regions were observed by different EAS arrays, both in the Northern and Southern hemispheres. No excess of particles is measured from any particular direction in celestial coordinates. The trajectories of charged particles are largely tangled by regular and stochastic interstellar magnetic fields (Sect. 2.7.1). The EAS array measurements in the region between 10^{14}–10^{18} eV confirm that the CR flux on Earth is consistent with isotropy. No *astronomy of charged CRs* through the identification of an excess of events from a given direction is possible within this energy range.

A large-scale anisotropy that would reflect the general pattern of propagation of CRs in the Galactic environment is expected anyway, as presented in Sect. 5.7.

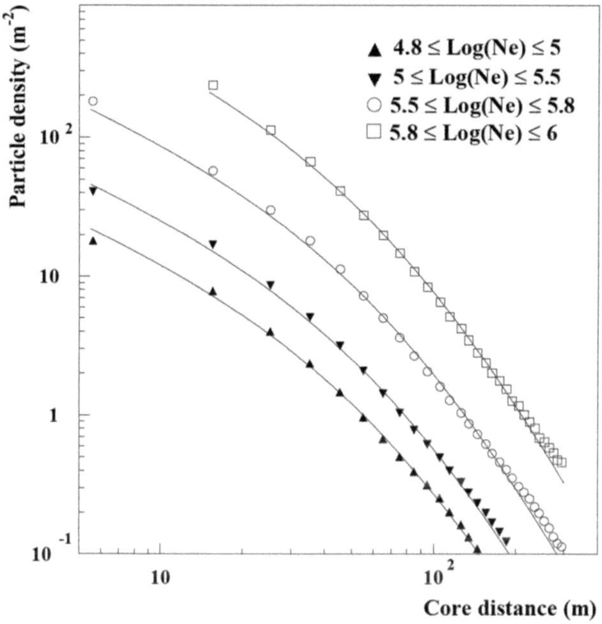

Fig. 4.16 Experimental lateral distributions in different intervals of measured shower size N_e. The average particle density as a function of the distance is fitted with the NKG function (4.21) (*solid lines*), with $r_1 = 100\,\mathrm{m}$ and a fixed age parameter $s = 1.21$. Courtesy of EAS-Top Collaboration

4.9 The CR Flux Measured with EAS Arrays

The CR flux in the region up to 10^{18} eV is still dominated by protons and nuclei originating in our Galaxy. Many experiments provided measurements of the number of events as a function of the energy E_0, the so-called *all-particle spectrum* (Hörandel 2007). This represents the sum over all nuclear species present in cosmic rays, without any separation. The energy E_0 can be derived through the measurement of the shower size. The experimental situation is complicated by the fact that surface detectors do not observe the number of electrons at shower maximum, but at a fixed depth $X_{\mathrm{det}}/\cos\theta$, where θ is the CR zenith angle. The number of particles N_e is calculated using (4.51) and the energy E_0 by inversion of (4.42), or similar relations. The CASA-MIA group used the measured N_e and N_μ sizes to derive the primary energy through (4.34). The KASCADE-Grande experiment found a more complex relation to estimate E_0 as function of the observed number of electrons (with $E_e > 3\,\mathrm{MeV}$) and muons (with $E_\mu > 300\,\mathrm{MeV}$) at sea level. More accurate size-to-energy conversions are obtained from detector-dependent Monte Carlo simulations.

As an example of an individual experiment, let us consider the EM size N_e measured by EAS-Top (Aglietta et al. 1999). Figure 4.17 shows the measured *size spectrum* for showers from almost the vertical direction (up to 17°). The statistical

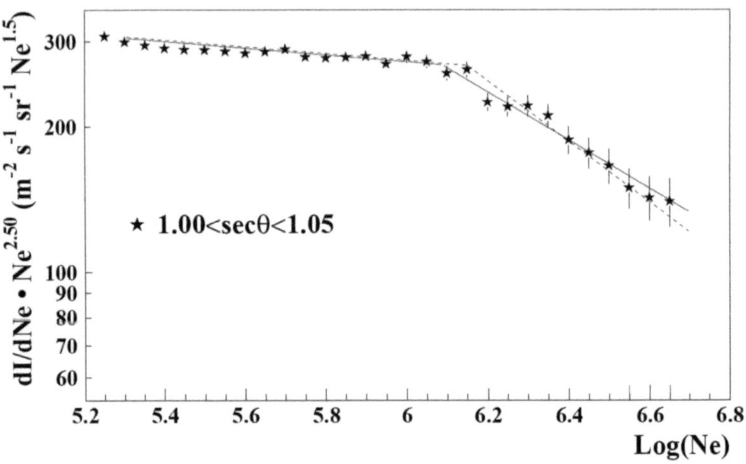

Fig. 4.17 Flux of CRs with zenith angle $\theta < 17°$ as a function of the shower size N_e measured at the atmospheric depth of $820\,\mathrm{g\,cm^{-2}}$ by the EAS-Top experiment. An almost linear dependence between measured size and energy exists, and the data on the x-axis, when converted into energy, range between 0.9×10^{15} and 1.0×10^{16} eV, covering the knee region. The flux has been multiplied by $N_e^{2.5}$ to make evident the change of slope occurring at $\log N_e \sim 6.1$. Courtesy of EAS-Top Collaboration

accuracy is better than a few percent in most bins. The size N_e is proportional to the primary energy E_0. From simulations (using CORSIKA with the HDPM hadron interaction model, Sect. 4.5), the energy interval of sensibility of the experiment was between $(0.9 < E_0 < 10) \times 10^{15}$ eV, exactly the region around the knee. The size spectrum was multiplied by $N_e^{2.5}$ (as N_e depends almost linearly on E_0, this is equivalent to the multiplication by $E_0^{2.5}$) to emphasize the change of slope at the knee.

The shower size N_e produced by primaries of a given energy will fluctuate from shower to shower because of differences in the stochastic development of the cascades. These considerations led Hillas to suggest that a more appropriate quantity for estimating the primary CR energy would be the measured particle density at a relatively large distance from the shower axis. Here, the fluctuations are small, and hence the density depends only on primary energy. Hillas proposed using the signal at an optimal distance, depending on the energy range and the array spacing. Monte Carlo simulations have confirmed that the density far from the shower axis depends only mildly on the hadronic interaction model used or on the primary composition, and may be used reliably as an estimator of the total energy. This method (Letessier-Selvon and Stanev 2011) was used, for instance, by the Haverah Park and AGASA experiments (Sect. 7.7).

A compilation of the results from a large number of different experiments is shown in Fig. 4.18. The y-scale variable has again been multiplied by $E_0^{2.5}$. At a fixed energy E_0, there is some dispersion of the points. A different compilation over an extended energy range is shown in Fig. 2.8.The red line in Fig. 4.18 shows

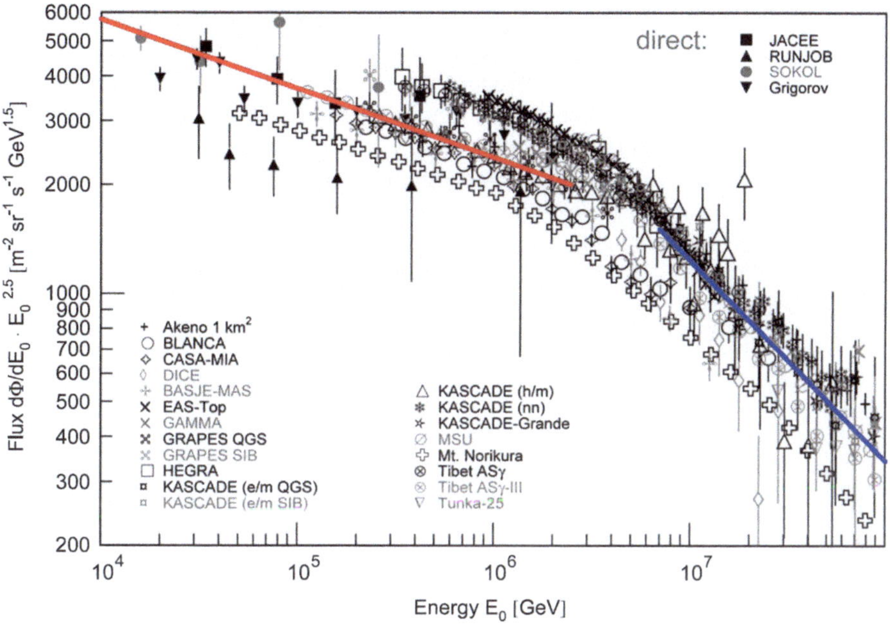

Fig. 4.18 All-particle energy spectra in the knee region from air shower experiments and a few direct experiments. The line below the knee refers to the power-law spectrum (2.20a), while the line above the knee to (4.52). See Blümer et al. (2009) for the references to the quoted experiments

the extrapolation up to the knee region of Eq. (2.20a) obtained from a fit to direct measurements. The position of the knee is at $\sim 4 \times 10^{15}$ eV. It is evident that the extrapolation of (2.20a) around and above the knee is no longer valid. After a smooth transition in the knee region, the new spectral index of the power-law function can be derived from EAS data in the energy region above 10^{16} eV. The differential energy spectrum between $10^{16} - 10^{18}$ eV can be expressed as

$$\Phi(E) \sim 1026 \times \left(\frac{E}{1\ \text{GeV}}\right)^{-3.06} \frac{\text{particles}}{\text{cm}^2\ \text{s sr GeV}} . \tag{4.52}$$

This function (when multiplied by $E^{2.5}$) corresponds to the blue line in Fig. 4.18 and is derived from the data compilation of Hörandel (2003). Here, an *energy-renormalized* scale of the individual experiments has been applied. We will return to that in Sect. 7.7. The compilation presented in Fig. 2.8 also indicates the possible presence of a second steepening of the spectrum near 8×10^{16} eV (*second knee*), with evidence from KASCADE-Grande (see following section) that this structure is accompanied by a transition to heavy primaries.

The dispersion of the points from different EAS experiments at a fixed value of the flux is mainly due to *systematic uncertainties* on the absolute energy calibration of each individual experiment. These systematic uncertainties are mainly due to the conversion from size N_e to energy E_0. Typical values of the absolute energy scale of each EAS experiment are about 15 to 25%. Most data sets agree within systematic uncertainties; this is a remarkable result, despite the fact that different experimental techniques are used, that the detectors are installed at different atmospheric depths and that different interaction models for interpreting the observed data are employed.

The integral spectrum above the knee (note that here, the energy is expressed in PeV, i.e., 10^{15} eV, which is a more natural unit for this energy range) corresponding to (4.52) is

$$\Phi(>E) \simeq 2.2 \times 10^{-10} \times \left(\frac{E}{1\ \mathrm{PeV}}\right)^{-2.06} \frac{\text{particles}}{\mathrm{cm}^2\ \mathrm{s\ sr}} . \tag{4.53}$$

The all-particle spectrum of Fig. 4.18 contains no information about the CR mass. The sensitivity of indirect measurements hardly allows for an estimate of the nuclear mass A for each individual shower. However, new techniques have been developed for the measurement of the relative abundances as a function of the primary energy, at least for groups of nuclear components.

4.10 Mass Composition of CRs Around the Knee

The mass A of the nucleus of energy E_0 originating the shower is another important quantity, because it is correlated with the nature of the accelerating astrophysical sources. At least two independent quantities have to be measured in order to estimate both the energy and mass of the primary CR that initiated the EAS. In addition to the shower size N_e, the position $X_{\max}$ of the maximum or the muon size N_μ are usually observed (Kampert and Unger 2012; Swordy et al. 2002). However, it is particularly difficult to determine E_0 and A on a shower-to-shower basis, because of the intrinsic fluctuation (at a fixed N_e) of the $X_{\max}$ and N_μ of each cascade. Those fluctuations arise from the stochastic nature of the interaction processes (in particular, the atmospheric depth at which the first interaction occurs) and from experimental limitations. The latter include the large spacing and limited sampling size of most EAS detectors.

The nuclear composition of CRs has been deduced on a statistical basis by some EAS experiments. These measurements rely to a large extent on the theoretical understanding of the shower development and on the modeling of the hadronic interactions generating the cascade.

4.10.1 The N_e Versus N_μ Method

In the N_e versus N_μ technique, the electron size N_e (or the weighted sum of electron and muon size) is a measure of the primary energy, while N_μ relates to the primary mass, Eq. (4.46). The electron-muon discrimination is achieved by employing a combination of unshielded and shielded scintillation detectors at ground level, as in the quoted examples of KASCADE, KASCADE-Grande and CASA-MIA.

The individual spectra of groups with different mass were first obtained by the KASCADE Collaboration (Antoni et al. 2005) with a two-dimensional *unfolding technique*. This approach yields a set of energy spectra of primary mass groups (three are considered in Fig. 4.19: [p+He], medium, heavy [Si+Fe]), such that their resulting *simulated* distribution of N_e and N_μ resembles the *observed* one.

To solve the underlying mathematical equations, the input needed is the so-called *response matrix* (or *kernel*) describing the correlation:

$$(E_0, A) \rightarrow (N_\mu, N_e) ,$$

including their shower-to-shower fluctuations. This relation means that for each simulated primary with given (E_0, A), an entry in the response matrix with a given number (N_μ, N_e) of muons and electrons corresponds.

The response matrix will depend on the chosen hadronic interaction model employed in the EAS simulations and on the detector response, which is included in the simulation. The kernel establishes the connection between the searched quantities and the measured ones by inversion of the matrix filled with the measured quantities. In the literature, it is known that a direct inversion of the matrix induces

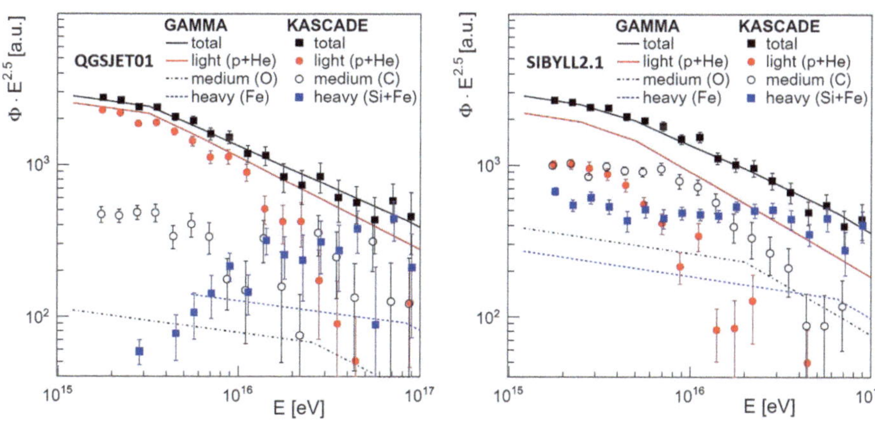

Fig. 4.19 Unfolded fluxes of different groups of nuclei as a function of the energy from the GAMMA and KASCADE experiments using two different interaction models (*left* QGSJET01; *right* SIBYLL2.1). The total represents the *all-particle spectrum*. Courtesy of Prof. Karl-Heinz Kampert

strong variations in the solution, mostly because of statistical fluctuations in the data. Small changes in the contents of nearby bins in the observed distribution can produce huge fluctuations when inverting the response matrix. Methods that produce regularizations of the solution and iterative methods are known in the literature (Cowan 1998).

The level of systematic uncertainties imposed by the interaction models was studied by the KASCADE Collaboration constructing kernel functions using different hadronic interaction models and comparing the corresponding unfolded results. Figure 4.19 shows the result obtained with QGSJET-01 and SIBYLL2.1. They concluded that neither model is able to describe the full range of energies consistently. However, they yield the same basic result. the CR composition at the knee is *light* (p-He dominated) with an evident change towards a heavy composition at higher energies. Moreover, the data are consistent with the assumption of a rigidity-dependent change in the knee energy. The figure also reports the result of the deconvolution from the data of the GAMMA Collaboration (Garyaka et al. 2007).

Data from the KASCADE-Grande experiment reached statistics sufficient to allow an extension of the unfolding analysis up to $\sim 10^{18}$ eV. The results (Apel et al. 2011) confirm the earlier findings of KASCADE and indicate a very heavy composition at about 10^{17} eV.

4.10.2 *Depth of the Shower Maximum*

The lateral distribution of Cherenkov light arriving on the ground is the result of a convolution of the longitudinal profile of charged particles in the shower (above the Cherenkov energy threshold E_T) and the Cherenkov emission angle θ_C. Both E_T,θ_C depend on the air density, and thus on the height. The electron energy distribution, and thus the number of electrons above E_T, is a universal function of the shower age s, Eq. (4.20).

The integrating Cherenkov technique provides a model-independent method for measuring both the calorimetric shower energy and $X_{\max}$ in each individual CR cascade with typical precision of a few tens of g cm^{-2}. The dependence of $X_{\max}$ on energy E_0 as measured by BLANCA, Tunka and Yakutsk is shown in Fig. 4.20. At low energies ($E < 10^{16}$ eV), the three measurements disagree by up to 40 g cm^{-2}. The three detectors agree on a change towards a heavier composition before $\sim 10^{16}$ eV. At $\sim 10^{17}$ eV, the values of $X_{\max}$ from Tunka and Yakutsk are approaching the results of the simulations that use heavy primaries. Beyond that energy, the average shower maximum increases again towards the air shower predictions for light primaries. At even higher energies, approaching the region where the contribution of extragalactic CRs starts, only data from Yakutsk exist. Measurements above 10^{18} eV will be discussed in Chap. 7.

The figure also shows the dependence of $X_{\max}$ upon the energy E_0 for different hadronic interaction models. It was assumed that all CRs are protons (red lines) or iron nuclei (blue lines). A variation of σ_{inel} and/or of the inelasticity parameter

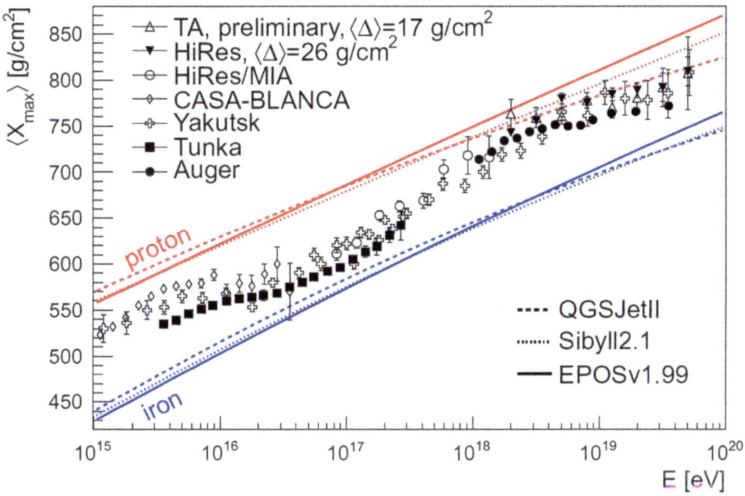

Fig. 4.20 Measurements of $X_{\max}$ with non-imaging Cherenkov detectors (Tunka, Yakutsk, CASA-BLANCA) and fluorescence (Sect. 7.6) detectors (HiRes/MIA, HiRes, Auger and TA). The results are compared to air shower simulations using three different hadronic interaction models (QGSJET, Sybyll, EPOS). Two extreme compositions are used: all showers induced by protons (*red lines*) or iron nuclei (*blue lines*). Courtesy of Prof. Karl-Heinz Kampert

κ in the model would directly translate into uncertainties about the primary mass estimates obtained from the height of the shower maximum. Larger cross sections and higher inelasticity produce short showers, while smaller cross-sections and lower inelasticity produce long showers penetrating deep into the atmosphere. The slopes of the curves represent the elongation rate defined in (4.15). The logarithmic increases of $X_{\max}$ as a function of energy, Eq. (4.43), as well as the difference between $X^p_{\max}$ and $X^{Fe}_{\max}$, Eq. (4.48), are evident.

4.11 Status and Future Experiments

The results of the KASCADE (see, in particular, Fig. 4.19) and KASCADE-Grande experiments provide the above standard picture of CR physics in the knee region, in which the knee of the all-particle spectrum at 3000 TeV is attributed to the steepening of the p and He spectra.

In contrast, recent results obtained by the ARGO-YBJ experiment (located at 4300 m asl) reported evidence, with different analyses, that the knee of the light component starts at ∼700 TeV, well below the knee of the all-particle spectrum (Bartoli et al. 2015). Other experiments located at high altitudes seem to confirm the ARGO-YBJ observation that the bending of the light (p+He) component is well below the PeV region. These results stimulated theoretical models that critically discussed the highest energies achievable in supernova remnants and the possibility that protons can be accelerated up to PeVs. We return to the problem of acceleration in Chap. 6.

The possibility of investigating the energy region around the knee thus demands a next generation of high altitude, high resolution and high statistics experiments using multi-component capability, able to continuously survey CRs and, possibly, the γ-ray sky. The determination of the chemical nature of excesses and deficits in the knee region is crucial for distinguishing between different models of CR acceleration and propagation in the Galaxy.

One advanced project is LHAASO (Large High Altitude Air Shower Observatory). This is a new generation instrument, to be built at 4410 meters of altitude (600 g/cm^2 of residual atmospheric depth) in the Sichuan province of China. It aims to study the energy spectrum, the elemental composition and the anisotropy of CRs in the energy range between 10^{12} eV to 10^{17} eV, as well as to act simultaneously as a wide aperture ($\sim$2 sr), continuously-operated γ-ray telescope in the >0.1 TeV energy range. LHAASO will consist of the following major components:

1. 1 km^2 array for measuring the electromagnetic component. A central part that includes 4931 scintillator detectors, 1 m^2 each in size and 15 m spacing, to cover a circular area with a radius of 575 m; and an outer guard-ring instrumented with 311 detectors with 30 m spacing up to a radius of 635 m.
2. An overlapping 1 km^2 array of 1146 water Cherenkov tanks, 36 m^2 each in size, with 30 m spacing. These will be located underground so as to be sensitive to the muonic component. The total sensitive area will be about 42,000 m^2.
3. A close-packed, surface water Cherenkov detector facility with a total area of about 78,000 m^2.
4. 12 wide field-of-view air Cherenkov telescopes.

The commissioning of one fourth of the detector is scheduled for 2018. The completion of the installation is expected by the end of 2021. Working at high altitude reduces the shower fluctuations, since the detector approaches the depth of the maximum longitudinal development of the shower. The key point is the possibility of separating, on an event-by-event basis, mass groups so as to measure their spectra and large-scale anisotropies. A hybrid array that uses the correlation between electromagnetic, muonic and Cherenkov components should allow for the selection, with high resolution, of the main primary mass groups on an event-by-event basis, without any unfolding procedure.

We will continue the description of the highest energy component of the CRs, likely produced by extragalactic sources, in Chap. 7. The detection methods used to catch CRs up to $\sim 10^{20}$ eV are similar to that described in this chapter.

References

M. Aglietta et al. (EAS-TOP Collaboration), The EAS size spectrum and the cosmic ray energy spectrum in the region $10^{15} - 10^{16}$ eV. Astropart. Phys. **10**, 119 (1999)

S.P. Ahlen et al. (MACRO Collaboration), Arrival time distributions of very high energy cosmic ray muons in MACRO. Nucl. Phys. **B370**, 432–444 (1992)

J. Alvarez-Muniz, R. Engel, T.K. Gaisser, J.A. Ortiz, T. Stanev, Hybrid simulations of extensive air showers. Phys. Rev. **D66**, 033011 (2002)

L. Anchordoqui et al., High energy physics in the atmosphere: phenomenology of cosmic ray air showers. Ann. Phys. **314**, 145–207 (2004)

T. Antoni et al. (KASCADE coll), The cosmic-ray experiment KASCADE. Nucl. Instr. Methods **A513**, 490–510 (2003)

T. Antoni et al., KASCADE measurements of energy spectra for elemental groups of cosmic rays: results and open problems. Astropart. Phys. **24**, 1–25 (2005)

W.D. Apel et al., Time structure of the EAS electron and muon components measured by the KASCADE-Grande experiment. Astropart. Phys. **29**, 317–330 (2008)

W.D. Apel et al., Kneelike structure in the spectrum of the heavy component of cosmic rays observed with KASCADE-Grande. Phys. Rev. Lett. **107**, 171104 (2011)

B. Bartoli et al. (ARGO-YBJ Collaboration), Knee of the cosmic hydrogen and helium spectrum below 1 PeV measured by ARGO-YBJ and a Cherenkov telescope of LHAASO. Phys. Rev. **D92**, 092005 (2015)

J. Blümer, R. Engel, J.R. Hörandel, Cosmic rays from the knee to the highest energies. Prog. Part. Nucl. Phys. **63**, 293–338 (2009)

S. Braibant, G. Giacomelli, M. Spurio, *Particle and Fundamental Interactions* (Springer, Berlin, 2011). ISBN: 978-9400724631

G. Cowan, *Statistical Data Analysis* (Oxford University Press, Oxford, 1998). ISBN: 978-0198501558

S. Eidelman et al., (Particle data group), Review of particle physics. Phys. Lett. B **592**, 1 (2004)

R. Engel, D. Heck, T. Pierog, Extensive air showers and hadronic interactions at high energy. Annu. Rev. Nucl. Part. Sci. **61**, 467–489 (2011)

T.K. Gaisser, *Cosmic Rays and Particle Physics* (Cambridge University Press, Cambridge, 1991)

A. Garyaka et al., Rigidity-dependent cosmic ray energy spectra in the knee region obtained with the GAMMA experiment. Astropart. Phys. **28**, 169–181 (2007)

K. Greisen, Cosmic ray showers. Ann. Rev. Nucl. Part. Sci. **10**, 63–108 (1960)

P.K.F. Grieder, *Extensive Air Showers* (Springer, Berlin, 2010). ISBN: 978-3-540-76940-8

D. Heck, CORSIKA: a monte carlo code to simulate extensive air showers. Forschungszentrum Karlsruhe FZKA 6019 (1998)

W. Heitler, *Quantum Theory of Radiation* (Oxford University Press, Oxford, 1944)

J.R. Hörandel, On the knee in the energy spectrum of cosmic rays. Astropart. Phys. **19**, 193–220 (2003)

J. Hörandel, Cosmic rays from the knee to the second knee: 10^{14}–10^{18} eV. Mod. Phys. Lett. A **22**, 1533–1552 (2007)

K. Kamata, J. Nishimura, The lateral and the angular structure functions of electron showers. Prog. Theor. Phys. **6**, 93 (1958)

K.H. Kampert, M. Unger, Measurements of the cosmic ray composition with air shower experiments. Astropart. Phys. **35**, 660 (2012)

K.-H. Kampert, A.A. Watson, Extensive air showers and ultra high-energy cosmic rays: a historical review. Eur. Phys. J. **H 37**, 359–412 (2012)

J. Knapp, D. Heck, Extensive air shower simulation with CORSIKA: a user's manual. Kernforschungszentrum Karlsruhe KfK 5196 B, 1993; for an up to date version see http://wwwik.fzk.de/CORSIKA/

A. Letessier-Selvon, T. Stanev, Ultrahigh energy cosmic rays. Rev. Mod. Phys. **83**, 907 (2011)

P. Lipari, The concepts of age and universality in cosmic ray showers. Phys. Rev. D **79**, 063001 (2009)

J. Matthews, A Heitler model of extensive air showers. Astropart. Phys. **22**, 387–397 (2005)

M. Nagano, A.A. Watson, Observations and implications of the ultrahigh-energy cosmic rays. Rev. Mod. Phys. **72**(3), 689–732 (2000)

B. Rossi, K. Greisen, Cosmic ray theory. Rev. Mod. Phys. **13**, 240–309 (1941)

T. Stanev, *High Energy Cosmic Rays* (Springer, Berlin, 2010). ISBN: 9783540851486

S.P. Swordy et al., The composition of cosmic rays at the knee. Astropart. Phys. **18**, 129–150 (2002)

Chapter 5
Diffusion of Cosmic Rays in the Galaxy

Abstract The observed spectra of Cosmic Rays (CRs) depend on two basic processes: the propagation in the interstellar medium of our Galaxy, described in this chapter, and the acceleration in the astrophysical sources. Upon leaving the source regions, high-energy charged particles diffuse in the random galactic magnetic field that accounts for their high isotropy and relatively long confinement time. The galactic diffusion model explains the observations on energy spectra, composition, and anisotropy of CRs. It also provides a basis for the interpretation of radio, X-ray, and γ-ray measurements, since a continuous radiation with a non-thermal spectrum is produced during propagation by the energetic electrons, protons, and nuclei. As discussed in this chapter, relevant information on CR propagation arise from the measurements of the abundances of some particular nuclei: the so-called light elements Li, Be, and B. Light elements are mainly of secondary origin, i.e., produced as the result of interactions of heavier primary nuclei with interstellar matter. We use the observed ratio between light and medium elements to assess an analytic description of the CR propagation and a first-order estimate of their escape time from our Galaxy. Electrons, as the lightest stable-charged particles, are subject to additional energy loss mechanisms with respect to protons and nuclei. The presence of magnetic fields induces synchrotron emission, which produces intense electromagnetic radiation in the proximity of the electron accelerators. In addition, a diffuse emission is produced during electron propagation in the galactic disk. Therefore, severe limits on the electron energy spectrum and on the distance of CR electron sources can be derived.

The observed spectra of CRs depend on two basic processes: the acceleration in the astrophysical sources and the propagation in the interstellar medium (ISM) of our Galaxy, described here. It is necessary first to study the latter (the propagation) in order to better understand the physics of acceleration mechanisms, the subject of Chap. 6.

Upon leaving the source regions, high-energy charged particles diffuse in the random galactic magnetic field that accounts for their high isotropy and relatively long confinement time. The galactic diffusion model explains the observations on

© Springer Nature Switzerland AG 2018
M. Spurio, *Probes of Multimessenger Astrophysics*, Astronomy and Astrophysics Library, https://doi.org/10.1007/978-3-319-96854-4_5

energy spectra, composition, and anisotropy of CRs. It also provides a basis for the interpretation of radioastronomical, X-ray, and γ-ray measurements (Chap. 8), since a continuous radiation with a nonthermal spectrum is produced during propagation by the energetic electrons, protons, and nuclei.

Most information on CR propagation arise from the measurements of the abundances of some particular nuclei: the so-called *light elements* Li, Be, and B, that is, elements with an atomic number just below the abundant C, N, and O elements (*medium elements*). Light elements are mainly of *secondary* origin, i.e., produced as the result of interactions of heavier primary nuclei with interstellar matter. The stellar nucleosynthesis processes explain the low cosmic abundance of Li, Be, and B. We use the observed ratio between light and medium elements to assess an analytic description of the problem (Sect. 5.1) and a first-order estimate of the *escape time* of CRs from our Galaxy. An independent method of "dating" the CR permanence in our Galaxy using radioactive nuclei is presented in Sect. 5.2.

The simple analytic description of CR propagation in the Galaxy will help us to build more general diffusion–convection equations for different cosmic ray species (Sect. 5.3). This *diffusion equation* also incorporates energy loss and gain processes in the interstellar medium, nuclear fragmentation, and radioactive decay of unstable nuclei. The galactic magnetic field randomizes the arrival direction of charged particles, making the flux *isotropic*.

Nowadays, the propagation process is well-described by solving the diffusion equation numerically or analytically, Sect. 5.4). The empirical transport coefficients of CRs, the properties of the sources (namely the total power, energy spectra of different components, elemental and isotopic composition), and the size of the confinement region of cosmic rays in the Galaxy are determined from fits to all available data on cosmic rays (Ptuskin 2012).

A knowledge of the effects of the propagation of CRs and the use of detailed diffusion models is very important not only for deducing the physical conditions at the sources, but also for estimating the dependence of the *escape time* upon particle energy. This is derived using the observed ratio of secondary to primary nuclei. For example, the boron-to-carbon ratio is decreasing with energy at $E > 1$ GeV/nucleon, as shown in Sect. 5.5. The dependence of the escape time on energy provides a very important constraint on the spectral index of CRs at sources, as obtained in Sect. 5.6.

Small anisotropies are expected for the escape process of CRs out the Galaxy, or to the possible contribution of sources near the solar system, or due to the motion of the solar system in the Galaxy. The expected amplitude and the status of experimental searches for large-scale anisotropies are in Sect. 5.7.

Electrons, as the lightest stable-charged particles, are subject to additional energy loss mechanisms with respect to protons and nuclei. The presence of magnetic fields induces synchrotron emission, which produces intense electromagnetic radiation in the proximity of the electron accelerators. In addition, a diffuse emission is produced during electron propagation in the galactic disk. As a consequence, severe limits on the electron energy spectrum and on the distance of CR electron sources can be derived (Sect. 5.8).

5.1 The Overabundance of Li, Be, and B in CRs

5.1.1 Why Li, Be, B Are Rare on Earth

The material formed in the early phase of the universe, at the time of the *primordial nucleosynthesis*, was, in mass, 3/4 protons and 1/4 helium nuclei. All heavier nuclei present in the periodic table of elements are produced by nucleosynthesis in stars, Chap. 12. Stellar nucleosynthesis is the process of nuclear fusions inside stars, producing energy to support their gravitational contraction.

Nuclear fusion in stars proceeds until the formation of nuclei with $A \leq 60$. The involved nuclear reactions, Sect. 12.1, do not increase the abundance of light nuclei (Lithium, Beryllium, and Boron) as these elements act as catalysts of nuclear reactions. The heavier elements up to iron are only synthesized in massive stars with $M > 8M_\odot$. Once Fe becomes the primary element in the core of a star, further compression no longe ignites nuclear fusion; the star is unable to thermodynamically support its outer envelope. This initiates the gravitational collapse (Sect. 12.13). All nuclei formed during stellar nucleosynthesis are released into the Galaxy and could be used for the formation of new stars.

Li, Be, and B act as catalysts of thermonuclear reactions in stars, and a low abundance after a stellar collapse is expected, as the collapse occurs when material for fusion is no longer available. The Li, Be, and B abundances in the solar system (and in the Universe) are generally very small for this reason.

The bulk of CRs is believed to be accelerated by galactic supernova remnants, Chap. 6. A similarity between the chemical composition of the solar system and that of nuclei present in the cosmic radiation is thus expected, as discussed in Sect. 3.6. A clear difference between the relative abundances of the Li, Be, and B elements in the CRs with respect to those in the solar system is evident in Fig. 3.7 and from Table 3.3:

$$R_{CR} \simeq 0.25; \quad R_{ss} \sim 10^{-5}. \tag{5.1}$$

Here, R_{CR} represents the ratio between the abundance of Li, Be, and B and that of C, N, and O nuclei found in CRs, and R_{ss} the corresponding ratio found in the solar system.

The discrepancy between the ratios in (5.1) is explained as being due to the CR propagation in the Galaxy before reaching the Earth. The interstellar medium is filled with matter (mainly hydrogen, Sect. 2.7), and Li, Be, and B elements are produced during the propagation as an effect of interactions of heavier nuclei with protons of the interstellar medium.

5.1.2 Production of Li, Be, and B During Propagation

To simplify the problem, let us globally identify the Li, Be, and B secondary elements with the symbol $\mathscr{L}$ (which stands for *light elements*) and the C, N, and

O primary elements with $\mathcal{M}$ (*medium elements*). The (relatively) abundant $\mathcal{M}$-type CR nuclei propagate in the Galaxy, constantly deflected by galactic magnetic fields until, in a random way, they reach the galactic border and exit the confinement volume. Along the way, $\mathcal{M}$ nuclei can interact with protons of the interstellar medium. This gives rise to the so-called *spallation process* (or *fragmentation process*)[1]; the result of this interaction process is the ejection of some nucleons from the nucleus that has been hit.

The production of secondaries by CRs distributed over the whole galactic volume depends on the nuclear cross-section, the average density of the interstellar material $\rho_{\rm ISM}$ (g cm^{-3}) and the distance x (cm) traveled between the production and the exit from the Galaxy. The relevant quantity for the production of secondaries is thus

$$\xi = \rho_{\rm ISM} \cdot x = c \cdot \rho_{\rm ISM} \cdot \tau \quad [\mathrm{g\,cm^{-2}}]\,, \tag{5.2}$$

where ξ represents the *path length* or the *grammage*, while τ represents a characteristic time of the phenomenon and c the speed of light, as the particles are assumed to be relativistic.

The problem consists in finding the value of $\xi = \xi_{\rm esc}$ that reproduces the observed ratio $R_{\rm CR}$ between the $\mathcal{L}$ and $\mathcal{M}$ abundances given in (5.1). This value corresponds to the mean amount of interstellar matter traversed by CRs before escaping from the confinement volume. In the simple model derived in this section, we obtain that $\xi_{\rm esc}$ = constant independent of the energy. We will work out in Sect. 5.4 that, more precisely, $\xi_{\rm esc}$ decreases with the increase of the particle energy.

The spallation of $\mathcal{M}$ nuclei by protons produces $\mathcal{L}$ nuclei and can be quantitatively studied with accelerator data using the reaction

$$p + X \rightarrow Y + \text{anything}\,, \tag{5.3}$$

with a high-energy proton interacting with a nucleus X at rest. In principle, the cross-section for the process (5.3) only depends on the center-of-mass energy of the initial system. It should, therefore, be equal to that for the process $X + p \rightarrow Y + anything$ (where the proton p is at rest, as in the case of the interstellar matter), provided that the center-of-mass energy is the same in the two cases. We are, moreover, assuming the cross-section to be approximately independent of the energy.

We are particularly interested in the experimental value of the spallation cross-section for processes (5.3) in which C, N, O nuclei correspond to the target X and the Li, Be, B isotopes are the fragments Y. The corresponding values for the spallation process are reported in Table 5.1 (Silberberg and Tsao 1990). From the table, we can deduce, for instance, that a proton interacting with a carbon nucleus produces a $^{11}_{5}$B with a probability of $31.5/252.4 = 12.5\%$.

[1] "Spallation" refers to inelastic nuclear reactions that occur when energetic particles interact with an atomic nucleus. Cosmic ray physicists usually refer to reactions induced by cosmic rays as "fragmentation". For our practical purposes, the two words are synonymous.

Table 5.1 Fragmentation cross-sections of C, N and O nuclei hit by protons

	Nuclear Fragment (Y)		Target (X) Fragmentation cross-section (mb)		
	Z	A	C	N	O
Li	3	6	12.6	12.6	12.6
Li	3	7	11.4	11.4	11.4
Be	4	7	9.7	9.7	9.7
Be	4	9	4.3	4.3	4.3
Be	4	10	2.9	1.9	1.9
B	5	10	17.3	16.0	8.3
B	5	11	31.5	15.0	13.9
Total cross-section (mb)			252.4	280.9	308.8

The considered fragments are the isotopes of Li, Be and B. We assume that the fragmentation cross-section does not depend on the energy

From Table 5.1, we obtain the average probability $P_{\mathscr{M}\mathscr{L}} = 0.28$ that a medium $\mathscr{M}$ element fragments into a lighter $\mathscr{L}$ element. This value is obtained by summing all the partial cross-sections in the table, divided by 252.4+280.9+308.8 mb. Thus, during propagation

$$N_{\mathscr{M}} + p \to N_{\mathscr{L}} + X \qquad \text{with } P_{\mathscr{M}\mathscr{L}} = 0.28. \tag{5.4}$$

The mean free path of hadrons and nuclei in matter was defined in Eq. (3.2). In the case of propagation of nuclei in the interstellar medium (dominated by the presence of protons), $A = 1$. The nucleus-proton cross-sections correspond to that of the spallation processes, proportional to the geometrical area of the nuclei, Eq. (3.3). The nuclear radius is given by (3.4). For $\mathscr{M}$ nuclei, the average value of the atomic mass is $A_{\mathscr{M}} \simeq 14$, while for $\mathscr{L}$ nuclei, we have $A_{\mathscr{L}} \simeq 8.5$. The target consists of protons, thus R_T = proton radius. Using these values, for the cross-section and mean free path, we obtain

$$\sigma_{\mathscr{M}} \simeq 280 \text{ mb} \quad \longrightarrow \quad \lambda_{\mathscr{M}} \simeq 6.0 \text{ g cm}^{-2} \tag{5.5}$$

$$\sigma_{\mathscr{L}} \simeq 200 \text{ mb} \quad \longrightarrow \quad \lambda_{\mathscr{L}} \simeq 8.4 \text{ g cm}^{-2}\,. \tag{5.6}$$

To determine, from (5.2), the escape length $\xi_{\rm esc}$ that reproduces the observed ratio between (Li, B, and Bo) and (C, N, and O), we need to set out a system of differential equations for the number of $\mathscr{M}$ and $\mathscr{L}$ nuclei as a function of ξ. The equation that describes the reduction of the number of $\mathscr{M}$ nuclei during their journey is

$$\frac{\mathrm{d}}{\mathrm{d}\xi} N_{\mathscr{M}}(\xi) = -\frac{N_{\mathscr{M}}(\xi)}{\lambda_{\mathscr{M}}}. \tag{5.7}$$

$\mathscr{L}$ nuclei in CRs are produced by spallation of heavier $\mathscr{M}$ elements. Their number increases with increasing path lengths of $\mathscr{M}$ nuclei. The differential equation that describes the number of secondary $\mathscr{L}$ nuclei as a function of path length ξ contains a positive *source* term and a negative *attenuation* term:

$$\frac{d}{d\xi}N_{\mathscr{L}}(\xi) = +\frac{P_{\mathscr{M}\mathscr{L}}}{\lambda_{\mathscr{M}}}N_{\mathscr{M}}(\xi) - \frac{N_{\mathscr{L}}(\xi)}{\lambda_{\mathscr{L}}} \,. \tag{5.8}$$

The source term increases with probability $P_{\mathscr{M}\mathscr{L}}$ as the spallation of $\mathscr{M}$ nuclei occurs during propagation. The attenuation term is similar to that affecting the $\mathscr{M}$ nuclei. The boundary conditions are $N_{\mathscr{L}}(\xi = 0) = 0$ and $N_{\mathscr{M}}(\xi = 0) = N^0_{\mathscr{M}}$. Here, $N^0_{\mathscr{M}}$ is a dummy parameter representing the production rate of $\mathscr{M}$ primary nuclei from astrophysical sources. It will disappear at the end when we compare the *ratio* between elements.

Equations (5.7) and (5.8) are coupled, since the number of $\mathscr{L}$ nuclei depends on $N_{\mathscr{M}}(\xi)$. Equation (5.7) can be immediately solved as

$$N_{\mathscr{M}}(\xi) = N^0_{\mathscr{M}} e^{-\xi/\lambda_{\mathscr{M}}} \,. \tag{5.9}$$

Some algebra is needed to solve (5.8). First, replace the $N_{\mathscr{M}}(\xi)$ in it with the value given by Eq. (5.9); then multiply both sides by $e^{\xi/\lambda_{\mathscr{L}}}$; the two terms containing $N_{\mathscr{L}}$ can be considered as the derivative of the product of two functions. As a result, (5.8) takes the form:

$$\frac{d}{d\xi}(N_{\mathscr{L}}(\xi)\cdot e^{\xi/\lambda_{\mathscr{L}}}) = \frac{P_{\mathscr{M}\mathscr{L}}}{\lambda_{\mathscr{M}}}N^0_{\mathscr{M}}\cdot e^{(\xi/\lambda_{\mathscr{L}}-\xi/\lambda_{\mathscr{M}})} . \tag{5.10}$$

As the equation contains exponential functions, the *ansatz* is of the form $N_{\mathscr{L}}(\xi) = c\cdot(e^{-\xi/\lambda_{\mathscr{L}}} - e^{-\xi/\lambda_{\mathscr{M}}})$, where c is a constant to be determined using the boundary conditions. By placing the test solution in Eq. (5.10), we obtain an identity if the constant c is $c = \frac{P_{\mathscr{M}\mathscr{L}}\cdot N^0_{\mathscr{M}}}{\lambda_{\mathscr{M}}}\cdot\frac{\lambda_{\mathscr{M}}\lambda_{\mathscr{L}}}{\lambda_{\mathscr{L}}-\lambda_{\mathscr{M}}}$. The solution to (5.8) is

$$N_{\mathscr{L}}(\xi) = \frac{P_{\mathscr{M}\mathscr{L}}}{\lambda_{\mathscr{M}}}\cdot N^0_{\mathscr{M}}\cdot\frac{\lambda_{\mathscr{M}}\lambda_{\mathscr{L}}}{\lambda_{\mathscr{L}}-\lambda_{\mathscr{M}}}\cdot(e^{-\xi/\lambda_{\mathscr{L}}} - e^{-\xi/\lambda_{\mathscr{M}}}). \tag{5.11}$$

The two functions (5.9) and (5.11) are shown in Fig. 5.1, where it is arbitrarily assumed that $N^0_{\mathscr{M}} = 1$.

The measured quantity is the ratio of $R_{CR} = N_{\mathscr{L}}/N_{\mathscr{M}} = 0.25$, which does not depend on $N^0_{\mathscr{M}}$. The value of $\xi = \xi_{\rm esc}$, which gives the measured value of R_{CR}, is determined using the ratio between (5.9) and (5.11), or through inspection of Fig. 5.1:

$$\xi_{\rm esc} = x_{\rm esc}\cdot\rho_{\rm ISM} = 5\ {\rm g\ cm^{-2}}. \tag{5.12}$$

This quantity is called the *average escape length* of CRs from our Galaxy. Since the value of the density of the interstellar material is $\rho_{\rm ISM} \sim 1\ {\rm cm^{-3}} = 1.6\times10^{-24}$g

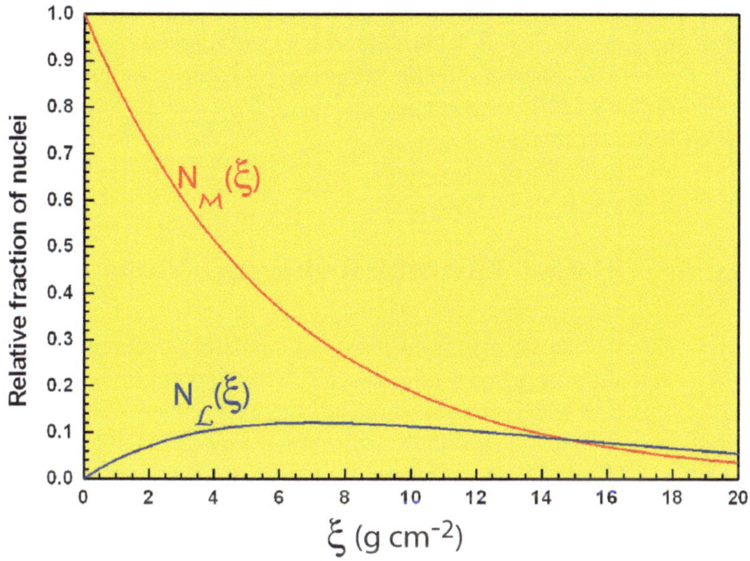

Fig. 5.1 Evolution of the number of $\mathscr{M}$ and $\mathscr{L}$ nuclei as a function of the galactic path length ξ. Near the astrophysical accelerators ($\xi = 0$), the $\mathscr{L}$ nuclei are absent. As ξ increases, $N_{\mathscr{L}}$ increases as light nuclei are produced by fragmentation of $\mathscr{M}$ nuclei. For instance, if the path length is equal to $\xi = 15$ g cm^{-2}, the ratio is $N_{\mathscr{L}}/N_{\mathscr{M}} = 1$. The measured ratio of $N_{\mathscr{L}}/N_{\mathscr{M}} \simeq 1/4$ yields $\xi_{esc} = x_{esc} \cdot \rho_{ISM} \simeq 5$ g cm^{-2}

cm^{-3}, the path x_{esc} corresponds to

$$x_{esc} = \frac{\xi_{esc}}{\rho_{ISM}} = \frac{5\,\text{g cm}^{-2}}{1.6 \times 10^{-24}\,\text{g cm}^{-3}} = 3 \times 10^{24}\,\text{cm} = 1\,\text{Mpc} \tag{5.13}$$

(1 parsec $= 3\times10^{18}$ cm). With a Galaxy having a radius of 15 kpc and a thickness of 300 pc, this result can be explained only if the propagation of cosmic rays resembles that of a *random walk*. Moreover, it suggests that the *propagation* and *acceleration* processes can be treated separately.

In Eq. (5.2), a characteristic time τ was introduced. When $\xi = \xi_{esc}$, this time corresponds to $\tau = \tau_{esc}$, the so-called *escape time*. It represents the average time of permanence of CRs inside the confinement volume before escaping our Galaxy. From (5.13)

$$\tau_{esc} = \frac{x_{esc}}{c} = \frac{3 \times 10^{24}\,\text{cm}}{3 \times 10^{10}\,\text{cm/s}} \simeq 10^{14}\,\text{s} = 3 \times 10^{6}\,\text{years}. \tag{5.14}$$

Since ξ_{esc} depends only on the ratio between the abundances of $\mathscr{L}$ and $\mathscr{M}$ nuclei, it does not depend on the observer's position: in any other position in the Galaxy, a hypothetical observer would measure the same $N_{\mathscr{L}}/N_{\mathscr{M}}$ ratio, obtaining the same value of τ_{esc}.

The fact that the value of x_{esc} (5.13) is orders of magnitude larger than the thickness of the galactic disk is a consequence of the tangled motion of charged particles in the galactic magnetic fields. As the gyromagnetic radius for a particle with charge Ze, energy E, in the magnetic field B is $R \simeq \frac{E}{eZB}$ (Eq. (2.5)) it is expected (Sect. 5.4) that the escape time τ_{esc} decreases as the particle energy increases.

5.2 Dating of Cosmic Rays with Radioactive Nuclei

The escape time τ_{esc} can be estimated using a completely different experimental method. The secondary-to-primary ratio described in the previous section provides a value of ξ_{esc} (or τ_{esc}) which critically depends on the assumed value of ρ_{ISM}. The decay of radioactive nuclei is used to derive τ_{esc} in a way that is independent of the estimate of ρ_{ISM}. The dating technique with radioactive isotopes relies on the fact that the half-life of the nucleus should not be too measured (Garcia-Munoz et al. 1977).

5.2.1 Dating "lived" Matter with ^{14}C

The radioactive isotope of carbon (^{14}C or *radiocarbon*) is the most commonly used element in the dating technique using radioactive isotopes. Radiocarbon is quite important, because it can be used to determine the age of matter that "lived" up to ~50,000 years ago (Bowman 1990).

The interactions of cosmic rays with atmospheric nuclei (as in a hadronic calorimeter) produce *showers* of hadrons. Among the hadrons, the neutrons bombard the nitrogen nuclei, $^{14}_{7}$N, which represent the major constituent of the atmosphere. This induces the reaction

$$n + ^{14}_{7}\mathrm{N} \rightarrow ^{14}_{6}\mathrm{C} + p. \tag{5.15}$$

Carbon has two stable isotopes: ^{12}C and ^{13}C. The $^{14}_{6}$C produced in the atmosphere has the (relatively) short half-life $t_{1/2} = 5730$ years. The amount of $^{14}_{6}$C in a sample is halved after 5730 years due to radioactive beta-decay ($^{14}_{6}\mathrm{C} \rightarrow ^{14}_{7}\mathrm{N} + e^- + \bar{\nu}_e$).

Due to the steady cosmic ray flux on Earth, the production of $^{14}_{6}$C in the atmosphere has been constantly occurring at a fixed rate for a very long time, so there is a fairly constant ratio of $^{14}_{6}$C to $^{12}_{6}$C atoms in the atmosphere. This ratio is approximately $(1.0-1.3) \times 10^{-12}$.

When plants fix atmospheric carbon dioxide (CO_2) into organic material during photosynthesis, they incorporate a given quantity of $^{14}_{6}$C corresponding to the level of its concentration in the atmosphere. After their "death," plants are used to make textiles or are consumed by other organisms (humans or other animals). Due to the metabolism of living organisms, humans and animals have also a $^{14}_{6}$C to $^{12}_{6}$C ratio at the level of the atmospheric concentration. From the instant of the vegetal or animal death, the concentration of ${}^{14}_{6}$C nuclei in the organic material decreases due to its radioactive decay according to the law $N(t) = N_0 e^{-t/\tau}$. The quantity τ is the lifetime that is related to the half-life, $t_{1/2}$, by the relation: $t_{1/2} = \tau \ln 2$.

After a time t, a measurement of the ratio ${}^{14}_{6}\mathrm{C}/{}^{12}_{6}\mathrm{C}$ allows us to derive the age of the sample. Its low activity limits the age determination by counting techniques to the order of 50,000 years.

As the level of atmospheric ${}^{14}_{6}$C is affected by variations in the cosmic ray intensity (which is, in turn, affected by variations in the Earth's magnetic field), high-accuracy measurements can only be achieved through a fine calibration of raw, i.e., uncalibrated, radiocarbon dates. The available standard calibration curves are based on the comparison of radiocarbon dates of samples that can be dated independently by other methods, such as the examination of tree growth rings, deep ocean or ice sediment cores, lake sediments, coral samples, and others. The most accurate curve extends back quite accurately up to 26,000 years. Any errors in the calibration curve do not contribute more than ± 16 years up to the last 6000 years and no more than ± 163 years over the entire 26,000 years (Reimer Paula 2004).

Radiometric dating was extended to many other elements. For instance, uranium–lead radiometric dating was used to date Earth rocks with a precision of less than 2 million years over a span of 4.5 billion years.

5.2.2 *Unstable Secondary-to-Primary Ratios*

The unstable secondary nuclei that live long enough to be useful probes of CR propagation are ^{10}Be (1.51×10^6 years), ^{26}Al (8.7×10^5 years), ^{36}Cl (3.1×10^5 years), and ^{54}Mn (6.3×10^5 years) (the figures in parentheses refer to their half-lives). The most frequently used one is the radioactive isotope ^{10}Be, which has a half-life similar to the escape time (5.14) and which is produced abundantly in the fragmentation of C, N, and O (see Table 5.1). The ^{10}Be undergoes β decay into ^{10}B. The relative abundances of the isotopes of Be and B provide a measure of whether or not all the ^{10}Be has decayed, and consequently an estimate of the time elapsed since production.

The experiments detecting ^{10}Be were carried out on satellites (IMP-7/8, ISEE-3, Voyager, Ulysses, CRIS). Let us try to derive an order-of-magnitude estimate t^* of the escape time using the data of the CR telescopes on board the IMP-7 (Explorer 47) and IMP-8 (Explorer-50) satellites, reported in Sect. 3.4.1. We use the ratio between the ^{10}Be and ^{7}Be. The latter is stable, while the number of ^{10}Be decreases with time:

$$N_{10}(t) = N^0_{10} e^{-t/\tau_{10}} \quad ; \quad N_7(t) = N^0_7 \, ,$$

with $\tau_{10} = 1.51 \times 10^6/\ln 2 \sim 2.2 \times 10^6$ years. At production, the ratio $N^0_7/N^0_{10} = 9.3/2.3 \sim 4$ depends only on the fragmentation cross-section, Eq. (3.9), obtained from the data in Table 5.1. Thus, at a time t^* after production, using the measured ratio $r_{mea} = 15/329 = 0.045$ between the two isotopes derived from (3.10, we obtain

$$r_{mea} = \frac{N_{10}^0}{N_7^0} e^{-t^*/\tau_{10}} \quad \rightarrow \quad \frac{t^*}{\tau_{10}} = -\ln\left(\frac{N_7^0}{N_{10}^0} r_{mea}\right) \sim 2\,.$$

The time spent during propagation in the Galaxy corresponds to about $t^* = 2\tau_{10} \sim 4 \times 10^6$ years. The error in this quantity is quite large, mainly due to the statistical error in r_{mea}. This rough estimate neglects some factors. The value obtained with a more detailed analysis in Garcia-Munoz et al. (1977) is $t^* = 17^{+24}_{-8} \times 10^6$ years.

The most precise estimate of the CRs' escape time using radioactive isotopes is due to the Cosmic Ray Isotope Spectrometer (CRIS) experiment, which was launched aboard NASA's Advanced Composition Explorer (ACE) satellite in 1997. The primary objective of CRIS was to measure the isotopic abundances of nuclei within the charge range $3 \leq Z \leq 30$ for energies below 500 MeV/n. The instrument consists of a scintillating fiber hodoscope, used as a tracking device, and four stacks of silicon wafers to measure the energy loss and the total energy. CRIS measured the abundances of the β-decay species ^{10}Be, ^{26}Al, ^{36}Cl, and ^{54}Mn. The determined values of τ_{esc} for different radioactive isotopes are shown in Fig. 5.2. Averaged over the different isotopes, CRIS obtained a confinement $\tau_{esc}^{CRIS} = 15.0 \pm 1.6$ My (Yanasak et al. 2001).

From the CRs' escape time, CRIS also estimated the hydrogen number density, $n_{ISM} = \rho_{ISM}/1.6 \times 10^{-24}$ g. The values corresponding to the different isotopes are shown in the upper plot of Fig. 5.2. The average value corresponds to

$$n_{ISM}^{CRIS} = 0.34 \pm 0.04 \quad \text{H atom cm}^{-3}\,. \tag{5.16}$$

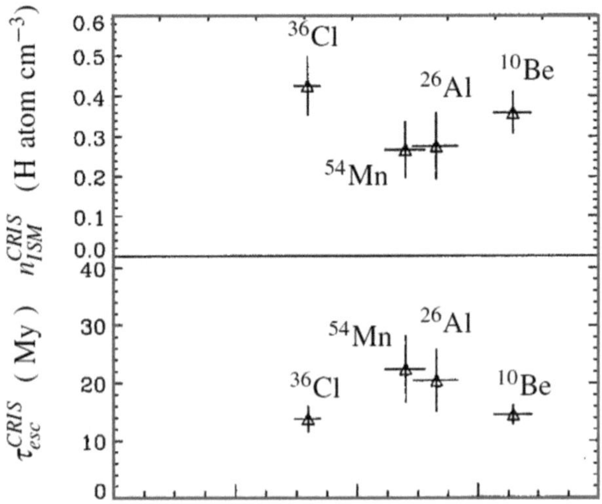

Fig. 5.2 CRIS result on the measurement of the different β-decay isotopes. *Upper plot* mean ISM Hydrogen number density n_{ISM}. *Lower plot* the galactic confinement time τ_{esc} derived from the CRIS observations. The different isotopes used for the measurements are shown in order of increasing half-live

The combination of the escape time and hydrogen number density measured by CRIS indicates an average escape length $\xi_{esc}^{CRIS} \simeq 7.6\,\text{g cm}^{-2}$, to be compared with the value (5.12) obtained with our simple estimate.

The value (5.16) represents evidence that galactic CRs spend time in the galactic halo, where the matter density is lower than the canonical value assumed for the number density in the disk ($n_{ISM} \sim 1\text{H atom cm}^{-3}$). A magnetic field confining CRs must therefore also be present in the galactic halo.

5.3 The Diffusion-Loss Equation

CRs undergo a diffusion process through the interstellar medium from their sources until they exit the Galaxy. Occasionally, some CRs can be intercepted by detectors near the Earth. In this section, we derive a *diffusion equation* that will be used to describe the CRs' journey through the Galaxy. This journey also modifies the CRs' energy spectrum from the sources to the observer. As the solar system has nothing peculiar with respect to any other point of our Galaxy, our observations are not influenced by the particular region where they are done. Particles having energies smaller than a few GeV, which are affected by the solar modulations, should not be considered. The galactic magnetic fields are the main factors that affect the CRs' motion, as the Larmor radius for particles below the knee ($E < 10^{15}\,\text{eV}$) is much smaller than the typical spatial dimension over which the magnetic fields are coherent (Sect. 2.7.1). A random component in the motion is induced by the presence of irregularities, associated either with fluctuations in the fields or with the induction of instabilities due to the streaming motions of the charged particles themselves. During their diffusion, CRs are subject to energy-loss mechanisms and absorption by dense media; nuclei may suffer spallation processes. Occasionally, CRs can gain energy by scattering with magnetized clouds.

To describe the changes resulting from this incoherent motion on the CR spectrum, a useful tool is to set a differential equation. This equation describes the energy spectrum at different points in the interstellar medium in the presence of diffusion (represented by a scalar diffusion coefficient D), energy losses, fragmentation, and other physics effects. To maintain secular equilibrium in the CRs' density, a term that describes the input from astrophysical sources is needed.

We give a derivation of the diffusion-loss equation, which closely follows the so-called *coordinate space approach* (Longair 2011).

Let us consider a *dummy state variable* ψ, which depends both on the spatial coordinates and on energy. *Dummy* means that it has no definite physical dimensions and it will disappear from the final result. The observable variable is $\mathcal{N}$, which has definite physical dimensions. For instance, it can represent the number of particles in a given volume and a given energy interval (units of $\text{cm}^{-3}\,\text{GeV}^{-1}$). The state of the system can be described in a Cartesian coordinate system, in which the x variable represents the spatial coordinates and the energy E is plotted along the y

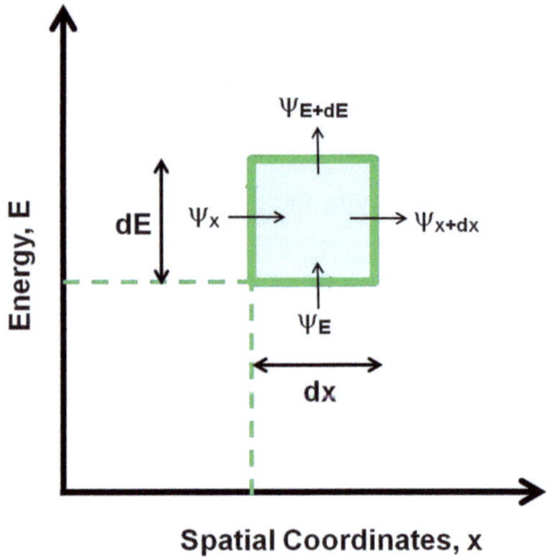

Fig. 5.3 The coordinate space diagram used to derive the diffusion-loss equation. The spatial coordinates are along the x-axis, the energy along the y-axis

axis, see Fig. 5.3. A change in the number of particles with a given energy E in a given region of space can be produced by two physical processes:

(1) The diffusion of particles out of the considered region of space. The process is visualized in Fig. 5.3 as a movement along the x-axis. As ψ_x is the particle flux at a fixed point x through the energy window $\mathrm{d}E$, the relation between ψ_x and $\mathcal{N}$ is

$$\psi_x \equiv -D\frac{\partial \mathcal{N}}{\partial x}. \tag{5.17}$$

D is a scalar quantity and it represents the *diffusion coefficient* (units of $\mathrm{cm}^2\,\mathrm{s}^{-1}$).

(2) The *loss* of energy within a given time interval $\mathrm{d}t$, due to processes such as excitation/ionization, bremsstrahlung, etc. A variation can also be induced by energy *gain*, due to acceleration mechanisms. The process is visualized in Fig. 5.3 as a flux along the y-axis. The relationship between the variable ψ_E, the number density $\mathcal{N}$, and the rate of energy change is

$$\psi_E \equiv \mathcal{N}(E)\frac{\mathrm{d}E}{\mathrm{d}t}. \tag{5.18}$$

We assume that energy is lost (or gained) at a rate:

$$-\frac{\mathrm{d}E}{\mathrm{d}t} = b(E)\,. \tag{5.19}$$

The function $b(E)$ is conventionally positive for energy losses, and negative for energy gains.

Considering the rectangle in Fig. 5.3, the number of particles in the interval dx and energy interval between E to $E + \mathrm{d}E$ is $\mathscr{N}(E, x, t)\mathrm{d}Edx$. Therefore, the rate of change of particle density in the defined phase space is:

$$\begin{aligned}\frac{\mathrm{d}}{\mathrm{d}t}\mathscr{N}(E, x, t)\mathrm{d}E\mathrm{d}x = &[\psi_x(E, x, t) - \psi_{x+\mathrm{d}x}(E, x + \mathrm{d}x, t)]\mathrm{d}E \\ &+ [\psi_E(E, x, t) - \psi_{E+\mathrm{d}E}(E + \mathrm{d}E, x, t)]\mathrm{d}x \\ &+ Q(E, x, t)\mathrm{d}E\mathrm{d}x \,. \end{aligned} \tag{5.20}$$

The quantity $Q(E, x, t)$ represents a *source term* (units cm^{-3} GeV^{-1} s^{-1} if $\mathscr{N}$ has units cm^{-3} GeV^{-1}) and it also represents the injection rate of particles per unit volume of coordinate space. After simplifying the notation, (5.20) becomes

$$\frac{\mathrm{d}\mathscr{N}}{\mathrm{d}t} = -\frac{\partial \psi_x}{\partial x} - \frac{\partial \psi_E}{\partial E} + Q \,. \tag{5.21}$$

Using (5.17) in (5.21), we get

$$\frac{\mathrm{d}\mathscr{N}}{\mathrm{d}t} = D\frac{\partial^2 \mathscr{N}}{\partial x^2} - \frac{\partial \psi_E}{\partial E} + Q \,, \tag{5.22}$$

which can be generalized in a three dimensional space:

$$\frac{\mathrm{d}\mathscr{N}}{\mathrm{d}t} = D\nabla^2 \mathscr{N} - \frac{\partial \psi_E}{\partial E} + Q \,. \tag{5.23}$$

The effect of energy losses is included by inserting (5.18) and (5.19) into (5.23), and thus

$$\frac{\mathrm{d}\mathscr{N}}{\mathrm{d}t} = D\nabla^2 \mathscr{N} + \frac{\partial}{\partial E}[b(E)\mathscr{N}(E)] + Q. \tag{5.24}$$

This represents the *diffusion-loss equation* for the time and spatial evolution of the energy spectrum of the particles. The solution of the most general case is, in principle, extremely difficult. Additional terms can be added to this equation to include other physical effects, such as the *escape probability*, the *radioactive decay*, and the *spallation* of nuclei during the propagation of cosmic ray nuclei from sources to Earth. We will adopt some approximations in what follows when we require a solution to this differential equation.

5.3.1 The Diffusion Equation with Nuclear Spallation

The production of a particular nucleus Z_i by the *spallation process* depends on the number of all nuclear species with $Z > Z_i$, on their cross-sections, and the matter number density of the crossed medium. In general, it is assumed that the spallation products have the same kinetic energy per nucleon as the progenitor and that the cross-section does not vary with energy. Mimicking what has been done for the case of Li, Be, B in (5.7) and (5.8), the diffusion equation (5.24) can include the spallation process with two additional terms:

$$\frac{\mathrm{d}\mathscr{N}_i}{\mathrm{d}t} = D\nabla^2\mathscr{N}_i + \frac{\partial}{\partial E}[b(E)\mathscr{N}_i(E)] + Q - \frac{\mathscr{N}_i}{\tau_i} + \sum_{j>i}\frac{P_{ji}}{\tau_j}\mathscr{N}_j . \tag{5.25}$$

τ_i and τ_j are the lifetimes of particles of species i and j. For the spallation process, they correspond to $\tau_i = \lambda_i/c$ and $\tau_j = \lambda_j/c$, where $\lambda_{i,j}$ are their interaction lengths (Sect. 3.2.3). P_{ji} is the probability that, in an inelastic collision involving the destruction of the nucleus j, the nucleus i is produced. The finite lifetime τ_{decay} of instable elements can also be accounted for by simply assuming that

$$\frac{1}{\tau_i} = \frac{1}{\tau_{\text{decay}}} + \frac{c}{\lambda_i}, \tag{5.26}$$

where the smaller between τ_{decay} and λ_i/c is the dominant term. In this way, both decay and interaction processes are taken into account.

The diffusion equation (5.25) is time-dependent. Normally, we are interested in the steady-state solution, corresponding to $\mathrm{d}\mathscr{N}_i/\mathrm{d}t = 0$. Electrons, positrons, and antiproton propagation can be described by the diffusion equation as well. They constitute special cases, differing principally due to the energy losses and production rates of these particles, and can be fully described with numerical simulations (see Sect. 5.4).

A high degree of isotropy is a distinctive quality of CRs observed on Earth. The motion of high-energy charged particles is influenced by magnetic fields that make the CR arrival distribution on Earth isotropic. The concept of CR diffusion explains why energetic charged particles are trapped in the Galaxy and have highly isotropic distributions. Before entering into the details of the techniques used to solve the problem, let us try to understand the meaning of the *diffusion coefficient* D in (5.25).

5.3.2 Numerical Estimate of the Diffusion Coefficient D

The long path length obtained in (5.13) indicates a sort of random walk of CRs in the Galaxy. In a random walk, after N steps of the same length $|\mathbf{l_i}| = l_0$, a particle moving from the origin of a reference frame arrives at the position $\mathbf{d} = \sum_{i=1}^{N}\mathbf{l_i}$. Assuming that the direction of each step is randomly chosen,

$$d^2 = \mathbf{d}\cdot\mathbf{d} = \sum_{i=1}^{N}\sum_{j=1}^{N}\mathbf{l_i}\cdot\mathbf{l_j} = Nl_0^2 + 2l_0^2\sum_{i=1}^{N}\sum_{j=1}^{N}\cos\theta_{ij} \simeq Nl_0^2 , \tag{5.27}$$

as the angles θ_{ij} between $\mathbf{l_i}$ and $\mathbf{l_j}$ are chosen randomly, and thus the off-diagonal terms cancel each other out.

Let us consider only the diffusion, the source and the time-dependent terms in (5.25):

$$\frac{d\mathscr{N}}{dt} - D\nabla^2\mathscr{N} = Q. \tag{5.28}$$

Note that in this way, the dependence on the energy variable is neglected and only the time variation of the spatial coordinates is taken into account. In addition, we assume a point-like source term Q, mathematically described as a Dirac delta function. Note that (5.28) can be formally transformed into the free Schrödinger equation with the substitutions $D \to \hbar^2/(2m)$ and $t \to -it$. As the same equations have the same solutions, we can borrow the free propagator for a nonrelativistic particle as Green's function $\mathscr{N} = G(r)$ for the diffusion equation, assuming D as a constant parameter:

$$G(r) = \frac{1}{(4\pi Dt)^{3/2}} e^{-r^2/(4Dt)} . \tag{5.29}$$

Thus, the mean distance travelled from the origin in a time t is $d \propto \sqrt{Dt}$. In the random walk, we get $d^2 \sim Nl_0^2$. Connecting the two pictures, we obtain

$$D \simeq \frac{Nl_0^2}{t} \simeq vl_0; \quad \text{with} \quad v = \frac{Nl_0}{t} . \tag{5.30}$$

Therefore, the diffusion coefficient D has the meaning of the product of the CRs' velocity $v \sim c$ times its mean free path l_0. An analysis in a 3-dimensional space more precisely gives us

$$D = \frac{vl_0}{3}. \tag{5.31}$$

An order of magnitude estimate for the diffusion coefficient D in a steady-state problem can be determined from (5.28) through dimensional arguments. We replace the spatial derivative with a division by the characteristic length scale L and the time derivative with a division for a characteristic time τ:

$$\frac{D\mathscr{N}}{L^2} \sim \frac{\mathscr{N}}{\tau} . \tag{5.32}$$

Then, we assume that the dynamics of CRs occurs within the galactic disc: the particles should diffuse to a distance L roughly equal to the thickness of the galactic disc, i.e., $L \simeq 300$ pc $= 9 \times 10^{20}$ cm. This occurs with a characteristic escape time given by Eq. (5.14), i.e., $\tau_{esc} = 10^7$ years $= 3 \times 10^{14}$ s. Thus

$$D \simeq \frac{L^2}{\tau_{esc}} = 3 \times 10^{27}\ \text{cm}^2\,\text{s}^{-1}\ . \tag{5.33}$$

By using the relation (5.30), we can evaluate the length l_0 of each step for charged particles moving in the Galaxy;

$$l_0 = \frac{3D}{c} = \frac{9 \times 10^{27}}{3 \times 10^{10}} = 3 \times 10^{17}\,\text{cm} = 0.1\ \text{pc}\ . \tag{5.34}$$

In the literature, estimated values of D up to 3×10^{28} cm^2 s^{-1} are found using more refined computations. Thus, the value of the step l_0 in Eq. (5.27) lies within the range

$$l_0 = 0.1{-}1\ \text{pc}. \tag{5.35}$$

This quantity can be interpreted as the typical scale of magnetic inhomogeneities in the interstellar medium. On the microscopic level, the diffusion of CRs results from particle scattering on random magnetohydrodynamic waves and discontinuities. There is evidence of galactic irregularities on the scale given in (5.35), associated with supernova shells, regions of ionized hydrogen, and so on.

5.4 The Leaky Box Model and Its Evolutions

The framework in which CRs propagate freely in a containment volume is called the *leaky box model*. In this model, the diffusion term in the diffusion equation is approximated by a leakage term:

$$D\nabla^2\mathcal{N} \rightarrow -\frac{\mathcal{N}}{\tau_{esc}}. \tag{5.36}$$

As the diffusion coefficient D should be energy-dependent, the characteristic escape time of CRs from the Galaxy $\tau_{esc} = \tau_{esc}(E)$ is also energy dependent. Consequently, (5.25) becomes

$$\frac{d\mathcal{N}_i}{dt} = -\frac{\mathcal{N}_i}{\tau_{esc}} + \frac{\partial}{\partial E}[b(E)\mathcal{N}_i(E)] + Q - \frac{\mathcal{N}_i}{\tau_i} + \sum_{j>i}\frac{P_{ji}}{\tau_j}\mathcal{N}_j. \tag{5.37}$$

The leaky box model provides the most common description of CR transport in the Galaxy at energies below $\sim 10^{17}$ eV. The model is based on particles injected by sources Q distributed uniformly over the galactic volume (the *box*) filled with a uniform distribution of matter and radiation fields. The particles get away from this volume with an escape time independent of their position in the box. The escape time $\tau_{esc}(E)$ depends on the particle energy, charge, and mass number, but it does not depend on the spatial coordinates. Secondary nuclei are produced during the propagation as a function of the path length (5.2).

The general problem translates into a system of coupled transport equations for all isotopes involved in the process of nuclear fragmentation, extending that shown in Sect. 5.1 for the $\mathscr{L}$ production from $\mathscr{M}$ nuclei. The solution of the problem today relies on powerful computer calculations. Before the modern computer epoch, the *weighted-slab technique* (Strong et al. 2007), which consists of splitting the problem into astrophysical and nuclear parts, was used. The fragmentation part is solved in the *slab model*, in which CRs traverse a thickness ξ of interstellar gas, and these solutions are integrated over all values of ξ, weighted with a distribution function $G(x)$ derived from an astrophysical propagation model. The solution of the leaky box model has an exponential distribution of path lengths ξ as $G(\xi) \propto \exp(-\xi/\xi_{esc})$, where ξ_{esc} represents the mean escape length.

The complexities of propagation of CRs, and in particular, of electrons and positrons, can be treated only through full computer simulations, because of their large energy and spatially-dependent energy losses. At present, propagation is described as diffusive, with a diffusion coefficient that is tailored to fit observations. Propagation codes (e.g., DRAGON (Evoli et al. 2008), PICARD (Kissmann 2014), Usine (Maurin et al. 2001) and GALPROP) provide the most advanced, explicit solution to date for the CR propagation problem.

For instance, the GALPROP code (Strong and Moskalenko 1998) enables simultaneous predictions of all relevant observations, including CR nuclei, electrons, antiparticles, γ-rays, and synchrotron radiation. The code [publicly available as a basis for further expansion (http://galprop.stanford.edu/)] incorporates current information on galactic structure and source distributions. Finally, the advances in computing power allow us to overcome the limitations of analytical and semi-analytical methods when CR, γ-ray, and other data become more accurate. The CR propagation (5.37) is solved numerically on a spatial grid, either in two dimensions with cylindrical symmetry in the Galaxy or in a full three dimensions. The boundaries of the model in radius and height, and the grid spacing, are user definable. In addition, there is a grid in the space of momentum, which is the natural quantity for propagation. The distribution of CR sources can be specified, typically to represent supernova remnants. Interstellar gas distributions are based on current HI (21-cm atomic hydrogen emission) and CO (molecular emission used to trace molecular hydrogen) surveys (see Sect. 2.7.2). Nuclear cross-sections are based on extensive compilations and parameterizations. The numerical solution proceeds in time until a steady-state is reached; a time-dependent solution is also a possible option. Starting with the heaviest primary nucleus considered (for example, ^{64}Ni), the propagation solution is used to compute the source term for its spallation products (which are then propagated in turn) down to protons, secondary electrons and positrons, and antiprotons. In this way, the production of secondaries, tertiaries, etc., is included.

Due to the above-mentioned properties of the lightest leptons, primary electrons are treated separately. The important features of diffuse γ-rays and synchrotron emission (Sect. 8.9) are computed using interstellar gas data (for pion decay and bremsstrahlung), the interstellar radiation field model (for inverse Compton), and the galactic magnetic field model.

5.5 Energy Dependence of the Escape Time τ_{esc}

An energy dependence of the CR escape time can be deduced from the measurement of the *secondary-to-primary ratios* of stable nuclei. The reference ratio is that between boron and carbon (B/C), because B is entirely secondary, i.e., produced by heavier primary CR nuclei. C, N, and O are the major progenitors of B, and the production cross-section is better known than those of Be and Li. A large number of different measurements exists up to ~1 TeV.

Figure 5.4 presents the measurement of the B/C ratio from different experiments as a function of energy. The figure shows a mild increase of the ratio starting from low energies and up to ~1 GeV/nucleon. This is due to the dependence, at low energies, of the nuclear cross-section on the relative velocity between nuclei. At

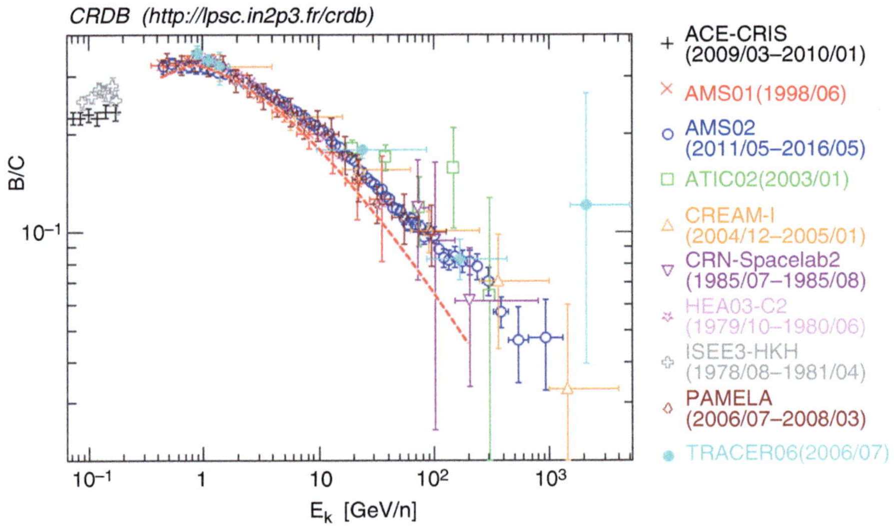

Fig. 5.4 Observed boron (B) to carbon (C) abundance ratio, measured as a function of the kinetic energy per nucleon, E_K, by different space and balloon experiments. The compilation includes the latest AMS-02 result, and the plot is obtained with the CRDB (Maurin et al. 2014) (see Sect. 3.1). The superimposed dashed line represents the result of a prediction with the *leaky box* model, assuming an energy-dependent escape path length $\xi_{esc} \propto E^{-0.6}$, and was derived from Obermeier et al. (2012)

higher energies, the fragmentation nuclear cross-section is almost constant. Above 1 GeV/nucleon, therefore, the decrease in the B/C ratio is only a consequence of propagation effects for nuclei with different energies. For this reason, the B/C ratio is used to constrain the spatial diffusion coefficient D, and consequently the escape time τ_{esc} in the leaky box model.

Empirically, we can assume a dependence of the path length on the particle rigidity R. Particles with low rigidity suffer a larger deflection during the motion in a magnetic field, because, according to (2.5), the Larmor radius is $r_L = R/B$. Thus, we assume that the path length decreases when the rigidity R increases as

$$\xi_{esc}(R) = \xi_0 \left(\frac{R}{R_0} \right)^{-\delta}, \tag{5.38}$$

where δ, ξ_0, R_0 are parameters that must be derived from theoretical models or fitting experimental data.

The above parameterization is consistent with the general diffusion equation. In fact, as

$$\tau_{esc} = \frac{\xi_{esc}}{c \cdot \rho_{ISM}} \qquad \longrightarrow \qquad \tau_{esc} = \tau_0 \left(\frac{R}{R_0} \right)^{-\delta} = \tau_0 \left(\frac{E}{E_0} \right)^{-\delta}, \tag{5.39}$$

the last equality assumes that at high energy, the rigidity is proportional to the energy for particles of given electric charge. In the leaky box equation (5.37), the energy dependence of the particles' loss is accounted for by the escape time τ_{esc} and, according to (5.39), the escape process dominates at high energy over fragmentation. In models using the diffusion coefficient D, this parameter becomes energy-dependent, $D = D(E)$. Appropriate parameterizations are introduced when the diffusion the form of (5.25) is used. This dependence of the τ_{esc} (or D) upon energy is of decisive importance for deriving the deriving the energy spectrum of CRs in the proximity of the sources.

Early predictions credited to Kolmogorov using a model of interstellar turbulence yielded $\delta = 1/3$ (Kolmogorov 1991); the value $\delta = 1/2$ was derived with a different model (Kraichnan 1965). A fit at high rigidities of the data before the recent AMS-02 measurement yields:

$$\delta = 0.6; \quad \xi_0 = 11.8 \text{ g cm}^{-2}; \quad R_0 = 5 \text{ GV/c}. \tag{5.40}$$

Notice that these values differ slightly from others obtained using different data sets [see, for instance, Engelmann et al. (1990), Strong et al. (2007) and Obermeier et al. (2012)]. One result based on the leaky box model is also presented in Fig. 5.4. Here, the time-dependent escape time due to the energy-dependent diffusion in the galactic disk is shown with a dashed line. Also, Iron produces secondary nuclei (Sc, Ti, and V) by a fragmentation process during propagation. The ratio between secondary nuclei (Sc, Ti, and V) and Fe can be described by a dependence on R similar to that reported in (5.38). In particular, the energy dependence of these ratios can be described using a value $\delta \simeq 0.6$. In contrast with these earlier results, the B/C ratio measured by AMS-02 (Aguilar et al. 2016) in the energy region above 10 GeV/n of Fig. 5.4 is well fit with $\delta = 0.33 \pm 0.02$, in good agreement with the Kolmogorov theory of turbulence.

The propagation of CRs in the Galaxy leaves an imprint not only on the B/C ratio and on heavier nuclei, but also in other measurable quantities. For instance, we will discuss in Sect. 8.9 that the emission of γ-rays from the direction of the Galactic plane reflects the density of hadronic and leptonic CRs and the density of gas and photons in the environment. In the disc of the Galaxy, most γ-ray emission is due to pp inelastic collisions, which result in meson production and decays: neutral mesons (mainly π^0) decay into two γ-rays. If the density of gas is reliably traced, the detection of gamma radiation from a given line of sight results in a measurement of the local density of CRs. The results of these observations (Acero et al. 2016) revealed that the density of CRs in the inner Galaxy (within a few kpc from the Galactic center) is rather peaked where the density of supernova remnants is also observed to be peaked, while the CR density decreases very slowly with the distance from the Galactic center in the outer Galaxy. Similar searches for neutrinos from the Galactic plane produced by $\pi^\pm$ decay have not yet given results (Albert et al. 2017).

As a summary, behavior of the B/C ratio and other observables is compatible with a power-law escape time of the type $\tau_{esc}(E) = \tau_0 E^{-\delta}$, with $\delta = 0.3$–0.6. In the following, in our simplified modeling of CR sources, to fix a value, we will assume $\delta = 0.6$.

5.6 Energy Spectrum of Cosmic Rays at the Sources

Let us derive a constraint for the spectral index of the CR energy spectrum near acceleration sites starting from the diffusion equation (5.37). We consider primary protons and stable nuclei ($\mathscr{N} = \mathscr{N}_P$) in a steady-state ($\mathrm{d}\mathscr{N}_P/\mathrm{d}t = 0$), neglecting the fragmentation processes ($P_{ij} = 0$). For primary protons and nuclei, the energy loss processes (excitation/ionization, bremsstrahlung, etc.) are negligible, and $b(E) \simeq 0$ (this is not true for electrons). The fragmentation processes can be neglected for the present purpose as well. Under these conditions from (5.26), we have $\tau_i = \lambda/c$, where λ is the interaction length of protons or nuclei in the ISM, and Eq. (5.37) becomes:

$$-\frac{\mathscr{N}_P(E)}{\tau_{esc}(E)} + Q_P(E) - \frac{\mathscr{N}_P \cdot c}{\lambda} = 0. \tag{5.41}$$

In Sect. 5.3, we adopted for $\mathscr{N}$ the units [GeV^{-1} cm^{-3}]. This quantity can reproduce the primary CR intensity when multiplied by $c/4\pi$. The quantity

$$\frac{c\mathscr{N}_P}{4\pi} \equiv \Phi(E) \quad \text{has units [GeV}^{-1}\text{ cm}^{-2}\text{ s}^{-1}\text{ sr}^{-1}]. \tag{5.42}$$

For CR protons, the mean free path is $\lambda_I \simeq 40\,\mathrm{g\,cm^{-2}}$, as obtained from (3.2) using $A = A_{\rm ISM} = 1$ and $\sigma = \sigma_{pp} \sim 45$ mb. The solution to (5.41) is

$$\mathscr{N}_P(E) = \frac{Q_P(E) \cdot \tau_{\rm esc}(E)}{1 + c\rho_{\rm ISM}\tau_{\rm esc}(E)/\lambda_I}\,, \tag{5.43}$$

as $c\rho_{\rm ISM}\tau_{\rm esc}(E) = \xi_{\rm esc}(E) \sim$ few g cm^{-2}, the term $c\rho_{\rm ISM}\tau_{\rm esc}(E)/\lambda_I < 0.1$, and it can be neglected. Finally, remembering (5.39) and (5.42) we get

$$\Phi(E) \propto Q_P(E) \cdot E^{-\delta}\,. \tag{5.44}$$

The left-hand side of (5.44) represents the primary spectrum of CRs observed on Earth (2.20a), $\Phi(E) \propto E^{-\alpha}$, and thus

$$Q_P(E) = \frac{\Phi(E)}{E^{-\delta}} \propto E^{-\alpha} \cdot E^{\delta}\,. \tag{5.45}$$

Thus, as $\alpha = 2.7$ below the knee ($\sim 10^{15}$ eV), and using the value $\delta \simeq 0.6$, we get the important prediction for the energy spectrum of CRs at the sources:

$$Q_P(E) \propto E^{-\alpha+\delta} = E^{-2.1} \tag{5.46}$$

Models of CR sources should reproduce this energy dependence, with a spectral index ~ 2.

We note that (5.43) can give information concerning the energy spectrum of heavier nuclei. In general, the acceleration processes provide (almost) the same energy dependence $Q(E)$ at the sources for protons and heavier nuclei. However, the nuclear cross-section increases by $A^{2/3}$, A being the mass number. Consequently, the interaction length decreases by $A^{-2/3}$, and for Fe nuclei, it becomes $\lambda_I^{Fe} \sim 2.5$ g cm^{-2}. In this case, the effect of interactions in (5.43) cannot be neglected until $c\tau_{\rm esc}^{\rm Fe}(E)/\lambda_I^{\rm Fe} \ll 1$. At sufficiently high energy, the escape time for heavier nuclei is small enough to reproduce the proton's behavior.

5.7 Anisotropies Due to the Diffusion

The CR flux on Earth is consistent with isotropy when the low-energy particles affected by the Sun are neglected. Small anisotropies are expected due to the global leakage of CRs from the Galaxy, to the possible contribution of individual sources, and due to the motion of the solar system in the Galaxy.

We can estimate how anisotropic the flux of cosmic rays would be by estimating their net streaming velocity V because of the presence of the diffusion. Let us use again dimensional arguments regarding (5.28) in a region without sources ($Q = 0$).

Assuming the same approximation used before (5.32), we have

$$\frac{\mathcal{N}}{T} = D\frac{\mathcal{N}}{L^2} \quad \rightarrow \quad \frac{L}{T} \equiv V = \frac{D}{L} \sim 10^{-4}c. \tag{5.47}$$

The numerical value is obtained assuming for L the galactic disk height (300 pc) and the diffusion coefficient D in (5.33). Therefore, if we were located at the edge of the galactic disc, we would observe a net streaming velocity of about 10^{-4} c, which would correspond to an anisotropy of the CR flux.

In general, the presence of a cosmic ray anisotropy is strictly correlated with the *streaming velocity* **V** of the CR particles. This streaming velocity plays the same role as the drift velocity attained by electrons in the presence of an electric field. There is an anisotropy only if there is a net streaming velocity, which can be expressed with a particular *amplitude* and *phase*. The amplitude[2] of the *CR anisotropy* is defined as:

$$\Delta \equiv \frac{I_{\max} - I_{\min}}{I_{\max} + I_{\min}}, \tag{5.48}$$

where $I_{\max}$, $I_{\min}$ represent the maximum and minimum intensity of cosmic rays from a given direction. Usually, the arrival direction of CRs is referred to as the *equatorial coordinate system*, see *Extras # 2*.

By simple dimensional arguments, the amplitude of the anisotropy depends on the streaming velocity amplitude as $V \sim \Delta \cdot c$. It can be demonstrated using Lorentz invariance (Kachelriess 2008) that, for a flux of particles (such as the cosmic rays) with differential energy spectrum $\Phi(E) \propto E^{-\alpha}$, the streaming speed and the anisotropy amplitude are correlated through the spectral index α by

$$V = \left[\frac{\Delta}{(\alpha + 2)}\right] c\,. \tag{5.49}$$

Because the CR escape probability increases with energy, Eq. (5.39), the diffusion coefficient D is energy-dependent as well. This corresponds to an increase in the streaming velocity V, according to Eq. (5.47). The expectation from our simple considerations is that the amplitude Δ of the anisotropy increases with energy. This simple prediction is confirmed by accurate calculations (Candia et al. 2003). Values as large as $\Delta \sim 10^{-1}$ can be obtained, depending on particle energy and on the strength and structure of the galactic magnetic field.

Different predictions for an anisotropy amplitude of a *dipole type* (when the CR directions of arrival are expressed in the equatorial coordinate system) as a function of the CR energies are shown in Fig. 5.5. A dipole anisotropy is a form of anisotropy resulting from the difference in the arrival intensity of particles from opposite

[2] In the literature, this amplitude is usually denoted with δ (lower case). We adopt the upper case Δ to avoid confusion with the diffusion parameter defined above and with the declination, also used in this chapter.

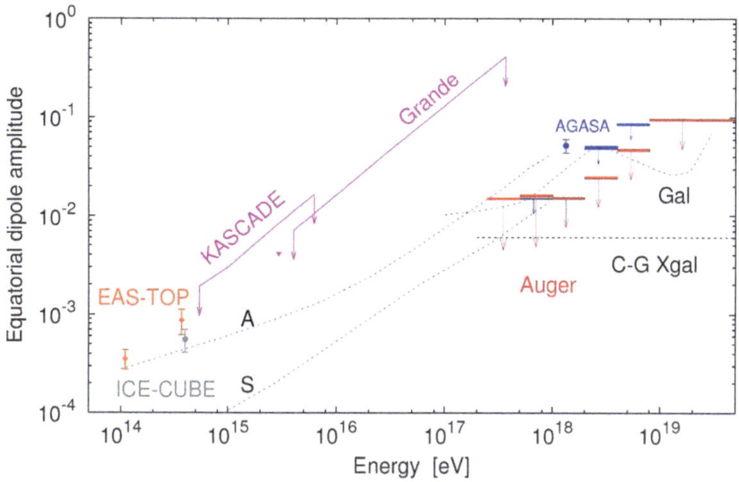

Fig. 5.5 The anisotropy amplitude as a function of energy. In red are the limits obtained by the Auger Observatory over the full energy range, as reported at the ICRC in 2013 [see Abreu et al. (2011) for the reference to the experiments]. The lines denoted as A and S up to 10^{18} eV refer to predictions for two different galactic magnetic field models. The predictions for a purely galactic origin of ultra-high energy cosmic rays (UHECRs) is denoted as "Gal", and the expectations from the Compton-Getting effect for an extragalactic component of CRs as "C-GXgal". In this case, the CRs are assumed to be isotropic in the cosmic microwave background rest frame. Courtesy of the Pierre Auger Observatory

directions, usually attributed to the motion of the observer relative to the source. As the CR diffusion depends on the modeling of the galactic magnetic fields, two predictions (labeled "A" and "S" in the figure) refer to two different configurations of the galactic magnetic fields. In the region of Ultra High Energy Cosmic Rays, above $\sim 10^{18}$ eV, a possible galactic origin of these particles (model labeled with "Gal") would produce an anisotropy of large amplitude, $\Delta \sim 0.1$. A much smaller anisotropy amplitude is expected if the sources of CR is of extragalactic origin, as indicated by the model labeled with "C-G Xgal." The level of different anisotropy amplitudes can be tested with experiments measuring the arrival direction of CRs in the region around and above the knee.

From the experimental point of view, because of the small amplitude of the expected anisotropy, a large data sample is necessary in order to have a statistically significant measurement of (5.48). Many experimental studies on dipole anisotropies (also called *large-scale anisotropies*) are based on the fact that the exposure of the detectors is uniform in the right ascension coordinate RA. The uniformity in the exposure guarantees that there is not bias from any particular sky region due to experimental effects.

As an example, let us consider the n events that a particular experimental apparatus has classified as CRs within the energy range between E and $E + \Delta E$. The n events are then analyzed in terms of the so-called *Rayleigh formalism of harmonic analysis*. The Rayleigh formalism gives the amplitude A (which can be identified with the Δ given in Eq. (5.48)) and phase ϕ of the first harmonic, and additionally, the probability P of detecting a spurious amplitude due to fluctuations from a sample of n events that are drawn from a uniform distribution (Mardia and Jupp 1999). Two quantities can be obtained from the dataset, with the sum including n right ascension values RA_i:

$$S = \frac{2}{n}\sum_{i=1}^{n}\sin(RA_i), \quad C = \frac{2}{n}\sum_{i=1}^{n}\cos(RA_i). \tag{5.50}$$

If the CR arrival directions are completely random, clearly, $S = C = 0$. In the presence of anisotropy, the amplitude A and the phase ϕ of the first harmonic are given by

$$A = \sqrt{C^2 + S^2}; \quad \phi = \arctan\frac{S}{C}. \tag{5.51}$$

The probability of obtaining an amplitude greater than A is given by

$$P(> A) = \exp\left(-\frac{nA^2}{4}\right). \tag{5.52}$$

The results of some experiments measuring CRs at different energies from the Galactic plane are reported in Fig. 5.5. Little evidences of large-scale anisotropies has been reported from extensive air shower experiments in the last two decades in the energy region 10^{14}–10^{15} eV, where the CRs are indubitably of Galactic origin. At high energy (when the asymmetry is larger), the CR flux is so low that it can be measured only by the largest shower array detectors. Note that some theoretical models for a purely galactic origin of UHECRs are strictly constrained by the established upper limits (lines with a downward arrow). The experiments require the anisotropy to be smaller than the values shown in the graph. In particular, an older large anisotropy value around 10^{18} eV reported by the AGASA experiment has been ruled out by the Pierre Auger Observatory.

5.7.1 *Evidence of Extragalactic CRs Above* 8×10^{18} *eV*

The non-observation of an anisotropy from the Galactic plane in the arrival direction of UHECRs is one of the main motivations for the hypothesis of their extragalactic origin, Chap. 7. Recently, the largest array for the measurement of UHECRs (the Pierre Auger Observatory, PAO in the following) reported the first evidence of a large-scale anisotropy in the arrival directions of CRs above 8×10^{18} eV (=8 EeV), suggesting an extragalactic origin (Aab et al. 2017). The PAO, designed to detect showers produced by primary CRs above 0.1 EeV, will be described in Sect. 7.8.

The paper presents a beautiful application of the method explained in the previous section, and its reading is recommended.

Extragalactic magnetic fields may be relevant for CRs propagating through intergalactic space, Sect. 7.5, and are poorly known. However, even if particles from individual sources are strongly deflected, it remains possible that anisotropies in the distribution of their arrival directions will be detectable on large angular scales, provided the sources have a non-uniform spatial distribution. The PAO detected, from 2004 to 2016, more than 110,000 CRs of energy >4 EeV, separated for the analysis into two energy bins, with a total exposure of 76,800 km^2 sr year.

To study the large-scale anisotropies in the arrival directions of CRs, the PAO collaboration applied the harmonic analysis in right ascension described in the previous section, Eq. (5.50). Table 5.2 shows the harmonic amplitudes, A, and phases, ϕ, within the two energy ranges, 4 EeV$< E <$8 EeV and $E > 8$ EeV. The quantity $P(>A)$ in the fifth column computed with (5.52) represents the probability that amplitudes larger than that observed could arise by chance from fluctuations in an isotropic distribution.

For the lower energy bin, the result is consistent with isotropy (the amplitude A is consistent with 0 within errors). For the events with E $\geq$ 8 EeV, the amplitude of the first harmonic is $\sim$(4.7 $\pm$ 0.8)%, which has a probability of arising by chance of $\sim 3 \times 10^{-8}$, equivalent to a two-sided Gaussian significance of 5.6σ. When trial factors are considered (the anisotropy is searched by PAO in 6 energy bins, including the additional four reported in Fig. 5.5), the significance is reduced to 5.2σ.

The table also reports the direction of the anisotropy in equatorial coordinates $(RA_d, \delta_d) = (100°, -24°)$. Notice that, because the Rayleigh formalism uses the RA value of detected events, the phase ϕ obtained with (5.51) corresponds to RA_d. The PAO also shows that the anisotropy is of *dipolar type* (the reason for the d subscript); a dipolar type means that the minimum is located approximately 180° away from the direction of the maximum. Finally, remember that the right ascension and declination of the Galactic center are (265°, −29°). Thus, the excess of events points to about 165° away from the direction of the center of the Milky Way, clearly excluding a galactic origin of the excess.

Table 5.2 Analysis in terms of the Rayleigh formalism of harmonic analysis of the two PAO UHECR samples

Energy (EeV)	Number of events	Amplitude A	Right Ascension (and phase) ϕ (°)	Probability $P(>A)$	Declination δ_d (°)
4 to 8	81,701	$0.005^{+0.006}_{-0.002}$	80 ± 60	0.60	-75^{+17}_{-8}
≥8	32,187	$0.047^{+0.008}_{-0.007}$	100 ± 10	3×10^{-8}	-24^{+12}_{-13}

The amplitude A, the phase ϕ and the probability $P(>A)$ are determined with the equations reported in the previous section. Notice that the phase ϕ coincides with the Right Ascension RA$_d$ of the anisotropy. The value of the anisotropy declination, δ_d has been derived with a similar Rayleigh analysis in the azimuthal angle

5.7.2 *The Compton–Getting Effect*

A small anisotropy is also expected as a result of the motion of the observer (on Earth) relative to the frame in which CRs have no bulk motion (the galactic disk). This effect has been known, since 1935, as the *Compton-Getting effect*. The Compton–Getting effect predicts that the intensity of CRs unaffected by the solar wind should be higher in the direction toward which Earth is moving. As the speed of the Sun (and consequently of the Earth) within the Galaxy is $V_\odot \sim 220$ km/s, the expected anisotropy amplitude due to the Compton–Getting effect can be estimated using (5.49):

$$\Delta_{CG} = \frac{V_\odot}{c}(\alpha + 2) \sim 10^{-3} , \tag{5.53}$$

thus a difference between the strongest and weakest cosmic ray intensities of about 0.1% would result. The amplitude of the anisotropy is on the same order of magnitude as the diffusion out of the galactic plane discussed above. We note in passing that the phase should be completely different.

Several ground-based extensive air shower detectors and underground experiments detecting cosmic-ray muons (Chap. 11) have observed that CRs within the energy range up to 100 TeV show a large-scale anisotropy with an amplitude $\Delta_{CG} \sim 0.1\%$. These observations seem to be related to the motion of the Earth in the Galaxy. The experimental values of Δ are very close to the predicted one. Figure 5.5 depicts the point relative to the EAS-TOP (Aglietta et al. 2009) and IceCube (Abbasi et al. 2012) experiments for energies below 10^{15} eV. For a compilation of results that includes lower energies, see Guillian et al. (2007).

5.8 The Electron Energy Spectrum at the Sources

In Sect. 5.6, using experimental observations, we derived that CR sources should have an energy spectrum of the type $Q(E) \propto E^{-2}$. The mechanism that provides accelerated particles with an E^{-2} spectrum will be described in Chap. 6, and it requires the existence of magnetic field regions in correspondence of sources.

Since ordinary matter in astrophysical environments is electrically neutral, we expect the presence of high-energy electrons, accelerated by the same mechanisms and with the same characteristics as protons. On the other hand, high-energy electrons are subject to additional energy loss processes, which cause distortions of their injection energy spectra. These distortions are mainly due to the presence of magnetic fields, which induce *synchrotron radiation*. In addition, electrons can also interact with matter and radiation. Their energy loss processes due to excitation/ionization and bremsstrahlung can be parameterized, as in Eq. (3.6).

Electrons accelerated in the presence of magnetic fields manifest the presence of acceleration regions through the production of electromagnetic radiation, which is detected with different experimental techniques (from radio-telescopes through X-rays, as described in a dedicated textbook, for instance, Ghisellini 2013). The highest tail of this radiation is detected as γ-rays by satellites or ground-based detectors, as described in Chaps. 8 and 9, respectively.However, the presence of the galactic magnetic field also has some consequences, as the radiation emission during propagation provide constraints on the possible distance of electron sources. For this reason, we start the presentation of the synchrotron radiation here, and then complete it later in Sect. 8.2.

5.8.1 Synchrotron Radiation

Synchrotron radiation is electromagnetic radiation generated by charged particles that are accelerated along a curved path or orbit. The radiation was named after its discovery in 1946 in a synchrotron accelerator in which charged particles were accelerated.

Classically, any accelerated charged particle emits electromagnetic radiation. The radiated power is given by the relativistic Larmor formula (Jackson 1999) (in Gaussian unit system, see Sect. 2.12):

$$P = -\frac{\mathrm{d}E}{\mathrm{d}t} = \frac{2e^2}{3c^3}a^2 , \tag{5.54}$$

where a is the particle acceleration and E its energy. For a non-relativistic circular orbit, the acceleration is just the centripetal acceleration, V^2/R. The relativistic acceleration is obtained from its definition, where $\tau = t/\Gamma$ is the proper time (Γ represents the Lorentz factor) and m_e the electron mass at rest:

$$a = \frac{1}{m_e}\frac{\mathrm{d}p}{\mathrm{d}\tau} = \frac{1}{m_e}\Gamma\frac{d(\Gamma m_e V)}{dt} = \Gamma^2\frac{dV}{dt} = \Gamma^2\frac{V^2}{R} . \tag{5.55}$$

The radiated power (neglecting the time dependence of Γ) is

$$-\frac{\mathrm{d}E}{\mathrm{d}t} = \frac{2e^2}{3c^3}a^2 = \frac{2e^2}{3c^3}\left[\Gamma^2\frac{V^2}{R}\right]^2 = \frac{2e^2\Gamma^4 V^4}{3c^3R^2} . \tag{5.56}$$

The velocity is $V \to c$ for relativistic electrons. In the case of a fixed radius R (that of an accelerator), the term $\Gamma^4 = [E/(m_e c^2)]^4$ depends on the fourth power of the particle energy. The radiated power is inversely proportional to the square of the accelerator radius R. Synchrotron radiation becomes the dominant factor in the energy loss rate (for instance, at the LEP e^+e^- collider at CERN) and is the limiting factor on the final beam energy that can be reached in such machines. The synchrotron radiation loss dependence suggests the construction of accelerators with a radius R as large as possible, or of linear accelerators.

Exercise The LEP collider was a storage ring of 27 km circumference. (a) Determine the energy loss due to the synchrotron radiation for each revolution of an electron of 50 GeV energy (LEP-Phase 1) and of 100 GeV energy (LEP-Phase 2). (b) Determine the intensity of the magnetic field needed to keep the electrons and positrons in orbit for both energies of 50 and 100 GeV; assume a uniform magnetic field along the ring. (c) Discuss what happens if the leptons are replaced by proton beams (as in the case of LHC).

In astrophysical environments, the radius of particles is dictated by the magnetic field itself. Using the Larmor radius (2.5) $r_L = R = E/eB = \Gamma m_e c^2/eB$ (for $Z = 1$) in (5.56), we obtain

$$-\frac{\mathrm{d}E}{\mathrm{d}t} = \frac{2e^2\Gamma^4 c^4}{3c^3 R^2} = \frac{2e^2\Gamma^4 e^2 B^2}{3c^3\Gamma^2 m_e^2} = \frac{2e^4}{3m_e^2 c^4} c\Gamma^2 B^2 \,. \tag{5.57}$$

Making use of the definition of the Thomson cross-section,

$$\sigma_T \equiv \frac{8\pi r_e^2}{3} = \frac{8\pi e^4}{3m_e^2 c^4} = 0.66 \times 10^{-24}\ \mathrm{cm}^2 \tag{5.58}$$

(where $r_e = e^2/m_e c^2$ is the classical electron radius), we can rewrite (5.57) as:

$$-\frac{\mathrm{d}E}{\mathrm{d}t} = \frac{\sigma_T}{4\pi} c\Gamma^2 B^2 = \frac{\sigma_T}{4\pi} c \frac{E^2}{m_e^2 c^4} B^2 \,. \tag{5.59}$$

The electron energy loss rate depends on the square of its energy and on the square of the magnetic field. The latter corresponds to the energy density of the magnetic field, $U_B = B^2/8\pi$.

More generally, if the electron has a velocity that is not perpendicular to the magnetic field direction, it moves in a spiral path at a constant pitch angle ψ and

$$-\frac{\mathrm{d}E}{\mathrm{d}t} = \frac{\sigma_T}{4\pi} c\Gamma^2 B^2 \sin^2\psi \,. \tag{5.60}$$

When considering a population of high-energy electrons, the distribution of pitch angles is expected to be isotropic because of the presence of irregularities in the magnetic field distribution and of random scattering processes during the motion of electrons. Averaging over an isotropic distribution of pitch angles and using the definition of the energy density of the magnetic field, $U_B = B^2/8\pi$, Eq. (5.60) becomes

$$-\frac{\mathrm{d}E}{\mathrm{d}t} = \frac{4}{3}\sigma_T c U_B \Gamma^2 \,. \tag{5.61}$$

The general feature of synchrotron emission is that the radiation is beamed along the particle direction. The derivation of the synchrotron frequency spectrum of the

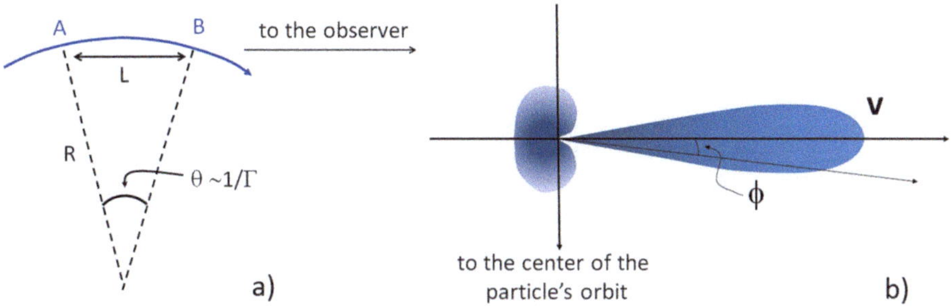

Fig. 5.6 (**a**) The geometry of the path of the electron during the time when the beamed radiation is observed by the distant observer. (**b**) The polar diagram of the radiation transformed into the laboratory frame of reference

radiation produced by a mono-energetic electron requires a considerable effort. We use here only a first-order approximation, remanding to Chapter 8 of Longair (2011) for a detailed derivation.

Relativistic beaming effects are associated with synchrotron radiation. A distant observer would receive the radiation only when the accelerated particle has a velocity toward the observer itself. This corresponds to a "pulse of radiation" that the observer notices every time the electron's velocity vector lies within a small angle. The duration of the pulse can be quantitatively derived with the help of Fig. 5.6a.

Consider an observer located at a distance D from the point A. The radiation from A, the leading edge, reaches the observer at time D/c. The radiation emitted at the point B occurs at a later time L/V and then travels a distance $(D - L)$ at the speed of light to reach the observer. The final part (the trailing edge) of the pulse therefore arrives at a time $L/V + (D - L)/c$. The duration of the pulse as measured by the observer is therefore

$$\Delta t = \left[\frac{L}{V} + \frac{(D-L)}{c}\right] - \frac{D}{c} = \frac{L}{V}\left[1 - \frac{V}{c}\right]. \tag{5.62}$$

The observed duration of the pulse is much less than the time interval L/V, which is true only if the light is propagated at an infinite velocity. The factor $1 - (V/c)$ in Eq. (5.62) takes into account the fact that the source of radiation is moving towards the observer, and it can be approximated with:

$$\left(1 - \frac{V}{c}\right) = \frac{[1-(V/c)][1+(V/c)]}{[1+(V/c)]} = \frac{[1-(V^2/c^2)]}{[1+(V/c)]} \simeq \frac{1}{2\Gamma^2}. \tag{5.63}$$

The electromagnetic field for a relativistic particle with Lorentz factor Γ is such that the angular distribution of the intensity of radiation with respect to the acceleration vector in the instantaneous rest frame is similar to that shown in Fig. 5.6b [see Jackson 1999]. In particular, the emitted radiation is beamed in the direction of motion of the electron within the angles $-1/\Gamma < \phi < 1/\Gamma$. The observer thus receives a pulse of radiation every time the electron's velocity vector lies within an angle of about $\pm 1/\Gamma$. This corresponds to the electron position in the trajectory characterized by the angle θ (Fig. 5.6a). Thus, we can assume that the radiation reaches the observer when emitted within the angle

$$\theta \simeq \frac{1}{\Gamma} . \tag{5.64}$$

Using the geometrical relations of Fig. 5.6a, we have

$$\frac{L}{V} = \frac{R\theta}{V} = \frac{1}{\Gamma \omega_r} \equiv \frac{1}{\omega_g} . \tag{5.65}$$

Classically, the angular gyrofrequency for an electron moving with velocity V on a circumference of radius R is $\omega_g = V/R$. The frequency of the emitted radiation would be expected to be $\nu_g = \omega_g/2\pi$. The relativistic gyrofrequency is defined as ω_r. Making use of (5.63) and (5.65), the relation (5.62) becomes

$$\Delta t = \frac{L}{V}\left[1 - \frac{V}{c}\right] \simeq \frac{1}{2\Gamma^3 \omega_r} . \tag{5.66}$$

This means that the duration of the pulse as observed by a distant observer in the laboratory frame of reference is roughly $1/\Gamma^2$ times shorter than the non-relativistic period $2\pi/\omega_g$.

The observed frequency (formally: the maximum Fourier component of the spectral decomposition of the observed pulse of radiation) corresponds to

$$\nu \sim (\Delta t)^{-1} \sim \Gamma^2 \nu_g = \Gamma^3 \nu_r = \frac{\Gamma^3 V}{2\pi R} \tag{5.67}$$

using the relation (5.65). This corresponds (for $V \sim c$) to the critical frequency

$$2\pi \nu_c = \omega_c \equiv \frac{3\Gamma^3 c}{2R} , \tag{5.68}$$

which includes the correction factor obtained from the average over all possible pitch angles between the velocity and the magnetic field. In general, we may interpret R as the instantaneous radius of curvature of the electron's trajectory and V/R as the angular frequency associated with it.

For order of magnitude calculations, it is sufficient to know that the total energy loss rate of the relativistic electron is given by (5.61) and that most of the radiation is emitted at the frequency given in (5.67). Figure 5.7 shows the shape of the spectrum of the synchrotron radiation (Longair 2011) emitted by a mono-energetic electron. The function is plotted in terms of the quantity $\omega/\omega_c = \nu/\nu_c$, where ω_c is the critical angular frequency (5.68). The spectrum presented in the figure has a broad

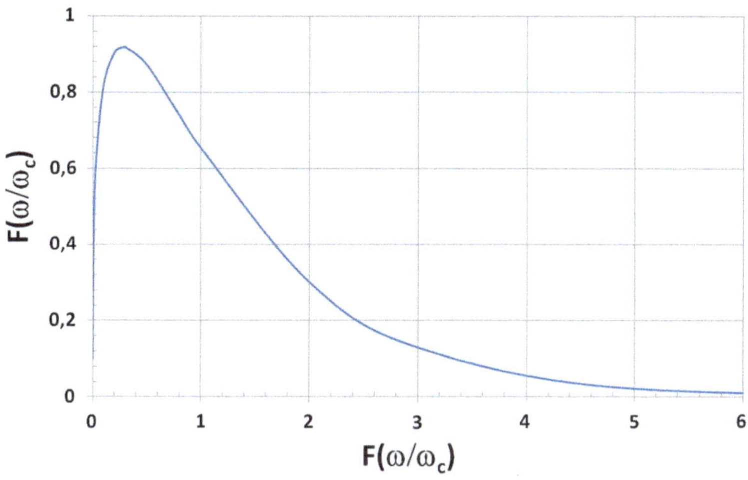

Fig. 5.7 Shape of the spectrum of the synchrotron radiation produced by a monoenergetic electron. The function is plotted in terms of the ratio $\omega/\omega_c = \nu/\nu_c$, where ω_c is the critical angular frequency

maximum, $\Delta\nu/\nu \sim 1$ centered roughly at the frequency $\nu \sim \nu_c$. The maximum of the emission spectrum corresponds to the value $0.29\nu_c$.

5.8.2 *Expected Spectral Index of Electrons*

The synchrotron emission of accelerated electrons near the sources generates radiation with an energy spectrum peaked in the infrared/X-ray range, as studied in Sect. 8.2. Here, we are interested in the way in which the energy loss rate (5.61) affects the energy spectrum of CR electrons, in a way independent of the peculiar characteristics of a source. Let us adapt the leaky box diffusion equation (5.37) for electrons:

$$\Phi_e(E) \equiv \frac{c}{4\pi}\mathcal{N}_e.$$

Most of the electron energy is lost near the sources because of the presence of magnetic fields. In this case, the dominant energy-loss process is due to the synchrotron radiation, and Eq. (5.59) can be written as $|dE_e/dt| = b_0 E^2$ with $b_0 = (\frac{\sigma_T}{4\pi})c\frac{B^2}{m_e^2 c^4}$.

In the diffusion equation (5.37), the energy-dependent term becomes $b(E) = b_0 E^2$. Because of this dependence on the squared energy, we can neglect the term due to diffusion and that due to other energy loss processes. Under these assumptions, all spatial dependences disappear from the diffusion-loss equation.

As usual, we assume that the spectrum has reached a steady-state ($d\mathcal{N}_e/dt = 0$) under the continuous injection of particles described by an infinite, uniform distribution of sources, each pumping in high-energy electrons with an injection spectrum

$$Q_e(E) = \kappa E^{-\alpha_e} \tag{5.69}$$

Equation (5.37) reduces to:

$$\frac{\mathrm{d}}{\mathrm{d}E}[b(E)\mathcal{N}_e(E)] = -Q_e(E) \quad \rightarrow \int d[b_0 E^2 \mathcal{N}_e(E)] = -\int Q_e(E) dE \,. \tag{5.70}$$

Assuming $\mathcal{N}_e(E) \rightarrow 0$ for energy $E \rightarrow \infty$, (5.70) can be integrated:

$$\mathcal{N}_e(E) = \frac{b_0}{(\alpha_e - 1)\kappa} \cdot E^{-(\alpha_e+1)} \,. \tag{5.71}$$

Thus, if synchrotron losses dominate,

$$\Phi_e(E) \propto E^{-(\alpha_e+1)}, \tag{5.72}$$

i.e., the observed spectrum is steeper by one power of E with respect to the spectrum at the sources.

We expect the energy distribution of electrons accelerated at the sources to follow the same behavior as that of CR protons and nuclei, thus $\alpha_e \simeq 2$. The measured energy spectrum of electrons follows, in a wide energy interval, a E^{-3} dependence (Fig. 3.12 in Sect. 3.10), demonstrating that most of the electron energy is released as electromagnetic radiation near the sources, where magnetic fields are large. The corresponding non-thermal electromagnetic radiation spectrum can be detected by means of different experiments.

5.8.3 Average Distance of Accelerators of Electrons

The presence of galactic magnetic fields induces synchrotron emission of electrons during the diffusion process of primary CR electrons in the Galaxy. This energy loss process constrains the maximum distance from which high-energy primary CR electrons can arrive on Earth.

An electron of energy E suffering energy loss processes $|dE/dt|$ lost its energy within a typical time-scale τ and typical path length-scale ℓ of

$$\tau = \frac{E}{|\mathrm{d}E/\mathrm{d}t|}; \quad \ell = c\tau. \tag{5.73}$$

If the energy loss is due to synchrotron radiation, using Eq. (5.59), we obtain the typical path length-scale ℓ_{syn}:

$$\ell_{syn} = \frac{4\pi m_e^2 c^4}{\sigma_T} \frac{1}{E} \cdot \frac{1}{B^2} = 1.25 \times 10^{13} \frac{1}{E\ [\text{erg}]} \cdot \frac{1}{B^2\ [\text{G}]^2}\ [\text{cm}]\ , \tag{5.74}$$

having inserted into the last equality the numerical constants to express energy in erg and the magnetic field in Gauss. For an electron of energy $E = 1$ TeV = 1.6 erg in the galactic magnetic field $B \sim 4\,\mu$G, we have $\ell_{syn} \simeq 5 \times 10^{23}$ cm $=1.6\times10^5$ pc.

This path length-scale ℓ_{syn} corresponds to a tangled trajectory in a diffusion process that can be considered as the sum of N_s steps of length l_0. If we consider for electrons the same value (5.35) for the dimension of magnetic inhomogeneities of the interstellar medium as for protons and nuclei, the number of steps corresponding to the length (5.74) is:

$$N_s(1\ \text{TeV}) = \frac{\ell_{syn}(1\ \text{TeV})}{l_0} \sim 1.6 \times 10^5 \quad \text{for} \quad l_0 = 1\ \text{pc}\ . \tag{5.75}$$

Using Eq. (5.27), we can estimate the average distance d traveled in this random way:

$$d = \sqrt{N_s} \cdot l_0 \simeq 400\ \text{pc} \quad \text{for}\ l_0 = 1\ \text{pc}. \tag{5.76}$$

Due to the presence of galactic magnetic fields, primary TeV electrons lose half their total energy within a distance smaller than a few hundreds pc from the source. This distance is even smaller if we assume a smaller value for l_0 or for higher energy electrons, due to the E^2 dependence of synchrotron energy loss. The bound (5.76) has important consequences in modelling the propagation of CR electrons in a computer program such as GALPROP and others or in the interpretation of experimental results concerning the leptonic component of CRs. We will return on this in Sect. 14.9.3 in relation to the excess of positrons found by PAMELA, AMS-02 and other experiments.

References

A. Aab et al. (The Pierre Auger Collaboration), Observation of a large-scale anisotropy in the arrival directions of cosmic rays above 8×10^{18} eV. Science **357**, 1266 (2017). Also arXiv:1709.07321

R. Abbasi et al., Observation of anisotropy in the galactic cosmic-ray arrival directions at 400 TeV with IceCube. Astrophys. J. **746**, 33 (2012)

P. Abreu et al., Search for first harmonic modulation in the right ascension distribution of cosmic rays detected at the Pierre Auger Observatory. Astropart. Phys. **34**, 627–639 (2011). The results were updated at the ICRC 2013 (see: arxiv:1310.4620)

F. Acero et al. (The Fermi-LAT Coll.). Development of the model of galactic interstellar emission for standard point-source analysis of Fermi large area telescope data. Astrophys. J. Suppl. **223**(2), 26 (2016)

M. Aglietta et al., Evolution of the cosmic-ray anisotropy above 10^{14} eV. Astrophys. J. Lett. **692**(2), L130–L133 (2009)

M. Aguilar et al. (AMS-02 Collaboration) Precision measurement of the boron to carbon flux ratio in CRs from 1.9 GV to 2.6 TV with the AMS on the ISS. Phys. Rev. Lett. **117**, 231102 (2016)

A. Albert et al. (ANTARES Collaboration). New constraints on all flavor Galactic diffuse neutrino emission with the ANTARES telescope. Phys. Rev. **D96**, 062001 (2017)

S. Bowman, *Radiocarbon Dating (Interpreting the Past)* (University of California Press, Berkeley, 1990). ISBN: 978-0520070370

J. Candia, S. Mollerach, E. Roulet, Cosmic ray spectrum and anisotropies from the knee to the second knee. J. Cosmol. Astropart. Phys. **05**, 003 (2003)

J.J. Engelmann et al., Charge composition and energy spectra of cosmic-ray nuclei for elements from Be to NI. Results from HEAO-3-C2. Astron. Astrophys. **233**, 96–111 (1990)

C. Evoli, D. Gaggero, D. Grasso, L. Maccione, Cosmic-ray nuclei, antiprotons and gamma-rays in the galaxy: a new diffusion model. J. Cosmol. Astropart. Phys. **10**, 018 (2008)

M. Garcia-Munoz, G.M. Mason, J.A. Simpson, The age of galactic cosmic rays derived. Astrophys. J. **217**, 859–877 (1977)

G. Ghisellini, *Radiative Processes in High Energy Astrophysics*. Springer Lecture Notes in Physics (Springer, Berlin, 2013). ISBN 978-3319006116

G. Guillian et al., Observation of the anisotropy of 10 TeV primary cosmic ray nuclei flux with the super-Kamiokande-I detector. Phys. Rev. D **75**, 062003 (2007)

J.D. Jackson, *Classical Electrodynamics* (Wiley, New York, 1999). ISBN 978-0471309321

M. Kachelriess, Lecture Notes on High Energy Cosmic Rays (2008). arXiv:0801.4376

R. Kissmann, PICARD: A novel code for the galactic cosmic ray propagation problem. Astropart. Phys. **55**, 37 (2014). https://doi.org/10.1016/j.astropartphys.2014.02.002

A.N. Kolmogorov, Dokl. Akad. Nauk SSSR **30**, 301 (1941); Reprinted on Proc. R. Soc. A 434, 9 (1991)

R.H. Kraichnan, Phys. Fluids **8**, 1385 (1965)

M.S. Longair, *High Energy Astrophys*, 3rd edn. (Cambridge University Press, Cambridge, 2011). ISBN 978-0521756181

V. Mardia, P. Jupp, *Directional Statistics* (Wiley, New York, 1999). ISBN: 978-0471953333

D. Maurin, F. Donato, R. Taillet, P. Salati, Cosmic rays below Z= 30 in a diffusion model: new constraints on propagation parameters. Astrophys. J. **555**, 585 (2001). https://doi.org/10.1086/321496

D. Maurin, F. Melot, R. Taillet. A database of charged cosmic rays. A&A **569**, A32 (2014). arxiv:1302.5525

A. Obermeier et al., The boron-to-carbon abundance ratio and galactic propagation of cosmic radiation. Astrophys. J. **752**, 69 (2012)

V. Ptuskin, Propagation of galactic cosmic rays. Astropart. Phys. **39–40**, 44–51 (2012)

J. Reimer Paula, INTCAL04 terrestrial radiocarbon age calibration, 0–26 Cal Kyr BP. Radiocarbon **46**(3), 1029–1058 (2004)

R. Silberberg, C.H. Tsao, Spallation processes and nuclear interaction products of cosmic rays. Phys. Rep. **191**, 351–408 (1990)

A.W. Strong, I.V. Moskalenko, Propagation of cosmic-ray nucleons in the Galaxy. Astrophys. J. **509**, 212–228 (1998)

A.W. Strong, I.V. Moskalenko, V.S. Ptuskin, Cosmic-ray propagation and interactions in the galaxy. Annu. Rev. Nucl. Part. Sci. **57**, 285–327 (2007)

N.E. Yanasak et al., Cosmic-ray time scales using radioactive clocks. Adv. Space Res. **27**(4), 727–736 (2001)

Chapter 6
Galactic Accelerators and Acceleration Mechanisms

Abstract In order to explain the features observed in experimental data and discussed in the previous chapters, Cosmic Rays (CRs) must be accelerated by non-thermal processes. As CRs with energies up to $\sim 10^{20}$ eV have been observed, theoretical model must consider sources and processes able to accelerate particles to these extraordinary energies. In most man-made accelerators, particles are accelerated by electric fields and deflected in circular orbits by magnetic fields. The magnetic fields also ensure that the particles remain confined in the acceleration regions. In most astrophysical environments, static electric fields cannot be maintained, because the matter is in the state of a plasma. In the standard model of CR production, the bulk of CRs is believed to be accelerated in recursive stochastic mechanisms in which low-energy particles, after a large number of interactions with a shock wave, will reach high energies. In this model, supernova remnants could accelerate protons up to 10^{15}–10^{16} eV, with a spectral energy index a ~ 2, as required by experimental data. However, the standard model of galactic CR acceleration has some limitations, and particular, it fails to describe the flux above the knee. Additional models have been put forward, such as the particle acceleration through electromagnetic mechanisms associated with time-varying magnetic fields. Some peculiar galactic objects can be involved in these processes. At present, no firm experimental proof is evident for any point-like source of CRs. In the chapter, we review the possible acceleration mechanisms and the involved astrophysical sources.

Cosmic Rays (CRs) are particles whose energies are typically much higher than the *thermal energies* found in astrophysical environments. By "nonthermal" emission, we mean continuum emission that cannot be originated by blackbody radiation or thermal bremsstrahlung. Their acceleration processes have to explain the features observed in experimental data and discussed in the previous chapters, namely that:

- The energy spectrum of cosmic rays (protons, heavier nuclei and electrons) has the form $\Phi(E) \propto E^{-\alpha}$ over a wide energy range (Sect. 2.6), indicating nonthermal acceleration processes;

© Springer Nature Switzerland AG 2018

M. Spurio, *Probes of Multimessenger Astrophysics*, Astronomy and Astrophysics Library, https://doi.org/10.1007/978-3-319-96854-4_6

- The measured exponent is $\alpha \sim 2.7$ for protons and nuclei up to the knee region (several PeV); protons dominate below the knee;
- The observed chemical abundances of CRs below the knee are similar to the abundances of elements found in our Solar System, Sect. 3.6. This indicates that CRs are accelerated out of a sample of well-mixed interstellar matter. In other words, CRs are "normal" matter accelerated to extremely high energies;
- The observed exponent α becomes $\sim$3.1 above the knee, Sect. 4.8. The chemical composition seems to become heavier, although at such energies, no measurement of the mass number A of individual CRs is possible;
- After corrections of the measured spectral index α for the effects due to the propagation in the Galaxy, the expected energy dependence near the sources should be $E^{-\alpha_S}$, with spectral index $\alpha_S \sim 2$, Sect. 5.6. This is true both for protons/nuclei and CR electrons;
- Above $\sim 5 \times 10^{18}$ eV, the energy spectrum flattens again to form the *ankle*;
- The Larmor radius of particles with energy above the ankle is larger than the galactic disk thickness;
- No preferred directions from the galactic plane are observed for the ultra-high-energy CRs, Sect. 5.7. The origin of UHECRs is thus probably extragalactic.

In addition, cosmic rays with energies up to $\sim 10^{20}$ eV have been observed (Chap. 7). Theoretical calculation must consider sources and processes able to accelerate particles to these extraordinary energies. In most man-made accelerators, particles are accelerated by electric fields and deflected in circular orbits by magnetic fields. The magnetic fields also ensure that the particles remain confined in the acceleration regions. This also occurs in an apparatus, like the *tokamak*, where high temperature plasma is confined. The *plasma* is the state of matter described as an electrically neutral and completely ionized medium of positive and negative particles.

In most astrophysical environments, static electric fields cannot be maintained, because the matter is in the state of a plasma. Ionized gases have a very high electrical conductivity, and any static electric field is rapidly short-circuited by the motion of free charges.

Cosmic rays below the knee are the dominant part of the energy density. In the standard model of CR production, they are accelerated by violent processes that produce shock waves and turbulences. The role of magnetized clouds is particularly important, Sect. 6.1. The bulk of CRs is believed to be accelerated in recursive stochastic mechanisms in which low-energy particles, after a large number of interactions with a shock wave, will reach high energies, Sect. 6.2. In this model, supernova remnants could accelerate protons up to $\sim 10^{15}$–10^{16} eV (Sect. 6.4), with a spectral energy index $\alpha \sim 2$ (Sect. 6.5), as required by experimental data.

The standard model of galactic CR acceleration has some limitations, Sect. 6.6. In particular, it fails to describe the flux above the knee. Additional models have been put forward, such as the particle acceleration through electromagnetic mechanisms associated with time-varying magnetic fields. Through Faraday's law, some special, very efficient galactic accelerators could explain the energy region of the CR spectrum above the knee. Some peculiar galactic objects can be involved in these processes, Sect. 6.9. At present, no firm experimental proof is evident for any point-like source of cosmic rays.

Due to the fact that the galactic magnetic field is not able to confine CRs with energy above $\sim 10^{18}$ eV, they are believed to be of extragalactic origin. They may be accelerated at Active Galactic Nuclei (AGN), at radio galaxies, in Gamma Ray Bursts (GRBs), or in other powerful astrophysical systems.

6.1 Second- and First-Order Fermi Acceleration Mechanisms

In this section, we study the acceleration mechanism of charged particles in regions where very strong inhomogeneous magnetic fields exist. The process can be seen as the scattering of the particle by magnetic field irregularities (*magnetic mirrors*). In the literature, this situation is sometimes called a *collisionless shock*. A collisionless shock can be defined as a shock wave in which the scattering occurs on a length scale much smaller than the mean free path $\lambda = (n\sigma)^{-1}$, Eq. (3.1), necessary for a particle collision. In such a structure, particles interact with each other not through collisions, but by the emission and absorption of collective excitations of the plasma (*plasma waves*). In nonrelativistic collisionless shocks, it seems that a pre-existing magnetic field is necessary to allow for the existence of such plasma waves. In astrophysical situations, the magnetic field is provided by a “frozen” cloud of interstellar matter, with a density much higher than that of the surrounding material. In addition, the magnetic field frozen inside the cloud has a higher intensity with respect to the average. The scattering between the particle (with mass m) and the magnetic cloud (with mass $M \gg m$) can be considered as elastic in the reference frame in which the cloud is at rest.

6.1.1 Magnetic Mirrors

If a charged particle moves in a nonuniform magnetic field, important differences arise in comparison to the case of the simple helicoidal motion that takes place in a uniform field, as discussed in Sect. 2.3.1. We adopt here a very simple approach to this problem; we refer to Chen (1984) for a detailed description used in plasma physics.

In the observer system frame, the particle moves in a nonuniform and static magnetic field. Because of its own motion, in the reference system integral to the instantaneous center of curvature, Fig. 6.1, the particle feels a time-varying magnetic field. In this reference system, the particle experiences an induced electric field $\mathscr{E}$, given by Faraday’s equation:

$$\nabla \times \mathscr{E} = -\frac{1}{c}\frac{\partial \mathbf{B}}{\partial t} \quad \text{(cgs units)} . \tag{6.1}$$

Fig. 6.1 Motion of a charged particle in a nonuniform magnetic field

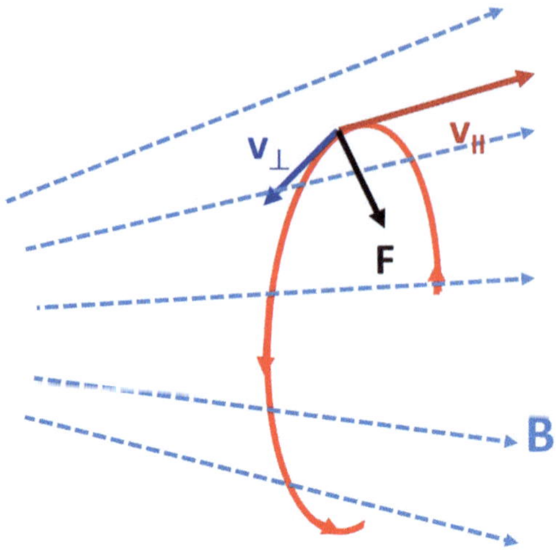

The induced electric field modifies the energy $E_\perp$ associated with the radial component of the velocity $v_\perp$ by an amount equal to the work done by $\mathscr{E}$. During one orbit, the change is given by

$$\Delta E_\perp = \Delta\left(\frac{1}{2}mv_\perp^2\right) = \oint q\mathscr{E}\cdot \mathrm{d}\ell = q\oint \nabla\times\mathscr{E}\cdot \mathrm{d}\mathbf{S} = -\frac{q}{c}\oint -\frac{\partial \mathbf{B}}{\partial t}\cdot \mathrm{d}\mathbf{S}\,, \tag{6.2}$$

where Stokes' theorem has been used. The surface integrals are extended to the circle delimited by the (approximately) circular orbit covered by the particle during one rotation.

Assuming that the field changes by only a small amount over a Larmor period $T_L = 2\pi/\omega_L$ (T_L is the time the particle takes to complete the circular orbit in the plane perpendicular to $\mathbf{B}$), the quantity $-\partial\mathbf{B}/\partial t$ can be approximated with $\Delta B/T_L$. From (6.2), we obtain

$$\Delta E_\perp = \frac{q}{c}\Delta B\frac{\omega_L}{2\pi}(\pi r_L^2) = \left(\frac{1}{2}mv_\perp^2\right)\frac{\Delta B}{B} = E_\perp\frac{\Delta B}{B}\,, \tag{6.3}$$

having used the relation[1] $\omega_L = qB/mc$ and the fact that $\omega_L r_L$ corresponds to the velocity $v_\perp$ along the instantaneous center of curvature. We can rewrite the above relation as

$$\Delta\left[\ln\left(\frac{E_\perp}{B}\right)\right] = 0 \quad \rightarrow \quad \frac{E_\perp}{B} = \text{const}\,. \tag{6.4}$$

[1]We assume here a nonrelativistic motion, and therefore $\Gamma = 1$ in Eq. (2.3).

Fig. 6.2 Mirror reflection of an electron beam in a magnetic field that converges to the *right*. Note that the guiding center (axis of spiral) of the reflected beam does not coincide with that of the incident. This is due to the gradient and curvature drift in a nonuniform field. Courtesy of professor Reiner Stenzel (http://www.physics.ucla.edu/plasma-exp/beam/)

Inhomogeneities in the magnetic field therefore produce variations of $E_\perp$. Note that, in the observer reference frame, the motion occurs in a static field, where the Lorentz force does not produce work. The kinetic energy must then remain constant in time, and for every change of $E_\perp$, there must correspond an equal and opposite variation of $E_\parallel$. For example, if the motion occurs in a region where the magnetic field B increases, then, according to (6.4), $E_\perp$ also increases, and $E_\parallel$ must therefore decrease. A force is established that opposes the motion of the particle along the lines of increasing magnetic field—see Fig. 6.2.

Because $v_\perp = v \sin\theta$, where θ is the angle between the velocity and magnetic field vectors, and taking into account the fact that v is constant, Eq. (6.4) can be rewritten as

$$\frac{\sin^2\theta}{B} = \text{const} \quad \rightarrow \quad \sin\theta = \sin\theta_0\sqrt{\frac{B}{B_0}}\,, \tag{6.5}$$

where B_0, θ_0 are the corresponding value at one reference, initial position. A relativistic particle in a magnetic field moves on a helicoidal trajectory, keeping constant the "adiabatic invariant" $\sin^2\theta/B$. If, during its journey, the particle enters a region of increasing magnetic field, it will be reflected. Equation (6.5) says, in fact, that the angle θ can increase up $\pi/2$. At this point, the particle cannot further penetrate the region in which B increases and it must reverse its longitudinal direction of motion and start moving along the direction of decreasing B. The regions where the magnetic field undergoes a strong increase behave as magnetic mirrors, for their property of reflecting the motion of charged particles. An electric charge between two magnetic mirrors will remain trapped in the region between the two mirrors. In the Earth's magnetic field, the trapped particles above the atmosphere constitute the van Allen radiation belts (Sect. 2.9).

6.1.2 The Second-Order Fermi Acceleration Mechanism

One of the first models for CR acceleration was created by Fermi (1949). He suggested a mechanism in which particles would be accelerated by collisions with moving clouds of gas.

Let us first consider the scattering between a particle with velocity $\mathbf{v}$ and a cloud moving with velocity $\mathbf{U}$, both along the x-axis. Let us assume, in addition, that $U \ll v$. Two situations are possible, as sketched in Fig. 6.3. By applying energy-momentum conservation in the elastic scattering, it is straightforward to derive that the velocity v' of the particle after the interaction with the cloud is

$$\mathbf{v}' = \frac{(m-M)\mathbf{v} + 2M\mathbf{U}}{m+M} \quad \rightarrow \quad v' \simeq -v \pm 2U \ , \tag{6.6}$$

assuming $M \gg m$. If $\mathbf{U}$ is opposite to $\mathbf{v}$ (type I collision, Fig. 6.3 left), then $v'_I = -v - 2U$. Otherwise, when $\mathbf{U}$ and $\mathbf{v}$ are parallel (type II collision, Fig. 6.3 right), $v'_{II} = -v + 2U$.

The kinetic energy (in the nonrelativistic case) of the particle before the collision is $E = mv^2/2$; after the collision, it is $E^* = m(-v \pm 2|U|)^2/2$. The sign $-$ occurs for type I, and the sign $+$ for type II collisions. The change in kinetic energy of the particle in first order in (U/v) during the collision is then:

$$\Delta E \simeq \pm 4\frac{U}{v} \cdot E \tag{6.7}$$

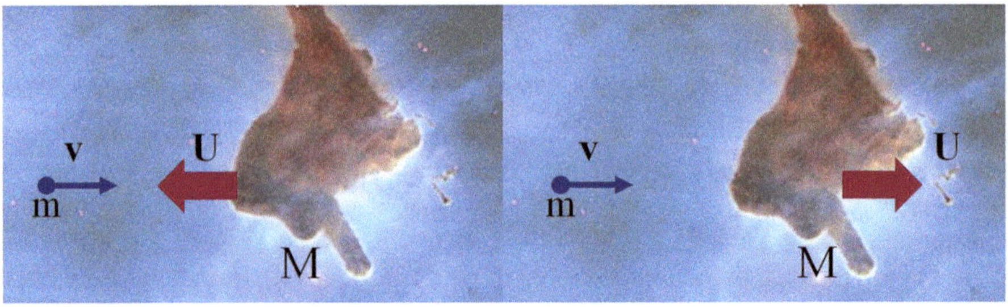

Fig. 6.3 *Left* Type I collision. The particle and the cloud velocities are opposite in direction. The particle gains energy in head-on elastic scattering. *Right* Type II collision. The particle and the cloud velocities are in the same direction. The particle loses energy in the elastic scattering

(here and hereafter, $U = |U|$). The variation can be either positive (ΔE_I) or negative (ΔE_{II}), according to whether the velocities of the particle and the cloud are opposite or in the same direction.

With a random distribution of clouds in both directions, the rates of type I and II interactions for a particle with positive velocity are, respectively,

$$f_I = \frac{v+U}{\ell}; \quad f_{II} = \frac{v-U}{\ell}. \tag{6.8}$$

Here, ℓ represents the particle mean free path. For this particular situation, the head-on collisions are thus more frequent than type-II collisions, and on average, the particle will gain energy. The average energy gained stochastically in unit time will be

$$\frac{\Delta E}{\Delta t} = f_I \Delta E_I + f_{II} \Delta E_{II} = 4 \cdot \frac{U}{v} \cdot E \cdot \frac{2U}{\ell} = \left(\frac{8U^2}{\ell v}\right) E. \tag{6.9}$$

We define here a constant τ_F'' with the dimension of a time:

$$\tau_F'' \equiv \frac{\ell v}{8U^2}, \tag{6.10}$$

which was first introduced by Fermi in the so-called *second-order Fermi process* (because (6.9) is quadratic in the cloud velocity U).[2]

This second-order process describes a situation similar to the mixing of two gases (in our case, clouds and particles) with different temperatures (different kinetic energies). Because of interactions between the two gas components, they reach equilibrium, i.e., energy equipartition between the two species. Here, the particles gain energy despite their large velocities, because the masses of the clouds are much larger.

In the 1949 paper, Fermi proposed that collisions with interstellar clouds were the main source of the CRs' energy. Fermi himself soon realized that such a mechanism is very inefficient for the acceleration of particles to very high energies. In fact, random velocities of interstellar clouds in the Galaxy are very small ($U/c \leq 10^{-4}, U \leq 3 \times 10^6$ cm/s). Furthermore, the mean free path ℓ for the scattering of cosmic rays corresponds to the typical scale of magnetic inhomogeneities in the interstellar medium (5.34), i.e., $\ell \sim 0.1$ pc. The constant

$$\tau_F'' = \frac{\ell v}{2U^2} \sim 10^7 \text{years} \tag{6.11}$$

[2]The coefficient 8 in Eqs. (6.9) and (6.10) is due to the uni-dimensional discussion. In the three-dimensional case, the correct coefficient is 2.

corresponds to the fact that a sizeable energy increase of a CR particle could only be reached on very large timescales.

6.1.3 The First-Order Fermi Acceleration Mechanism

A higher efficiency (as Fermi realized in a second work (Fermi 1954)), is obtained if the accelerated particles are moving between two clouds mutually approaching or stellar material separated by a shock front.

The basic idea of this mechanism is easily illustrated using the toy example of two approaching trains, shown in Fig. 6.4. If a ball is thrown from train B towards train A, at a speed v in the rest frame of train B, the speed of the ball in the laboratory frame will be $v_1 = v + V$ and $v_1^A = v + 2V$ in the rest frame of train A. Assuming the collision to be elastic, the ball will bounce back towards train B with speed $v_2^A = v + 2V$ in the rest frame of train A and $v_2 = v + 3V$ in the laboratory frame. Following the further collision of the ball with train B, it is easily seen that the ball will have a speed of $v_2^B = v + 4V$ in the rest frame of train B at the time of the new collision and will bounce back with a speed $v_3 = v + 5V$ in the laboratory frame. As the gain in each collision is proportional to the train velocity V, if the process could continue for a long time, we might thus see that the ball could acquire a very large amount of energy.

The "bouncing" back and forth of a proton or nucleus, between the material up-stream and down-stream of a shock-front, has a strong similarity to the toy example described. We shall, to start with, compute the energy gained by a particle of velocity v that undergoes a single collision with a magnetic cloud moving at velocity U in a direction opposite (*head-on* collision) or equal (*catching* collision) to that of the particle. Assuming that, as a result of many collisions, v will eventually become $\simeq c \gg U$, a relativistic approach is appropriate. Let us define S as the reference frame of the observer, with the cloud velocity directed along the x-axis, and S' as the frame in which the cloud is at rest. Only the p_x component of the particle momentum is relevant in the problem, as the y, z components are conserved in the interaction.

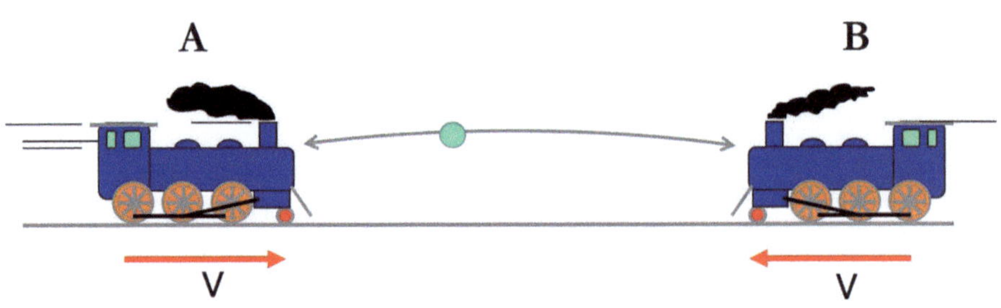

Fig. 6.4 Toy-model of a possible first-order Fermi acceleration mechanism

The four-momentum describing the particle is (E, p_x) in the frame S, and (E', p'_x) in S'. From the Lorentz transformations

$$E' = \Gamma(E + Up_x)\,, \tag{6.12a}$$

$$p' = \Gamma(p_x + \frac{U}{c^2}E)\,. \tag{6.12b}$$

The collision is *elastic* in the frame in which the cloud is at rest. This means that the energy and momentum after the collision are:

$$E' \overset{\text{collision}}{\longrightarrow} E'; \quad p'_x \overset{\text{collision}}{\longrightarrow} -p'_x\,. \tag{6.13}$$

The particle energy E^* after the scattering in the reference frame of the observer is obtained by inversion of (6.12a):

$$E = \Gamma(E' - Up'_x) \overset{\text{collision}}{\longrightarrow} \Gamma(E' - U(-p'_x)) \equiv E^*\,. \tag{6.14}$$

Inserting (6.12a), (6.12b) into (6.14), we obtain:

$$E^* = \Gamma\left[\Gamma(E + Up_x) + U\Gamma(p_x + \frac{U}{c^2}E)\right]. \tag{6.15}$$

Recalling that $p_x = mv\Gamma\cos\theta$ and $E = mc^2\Gamma$

$$\frac{p_x}{E} = \frac{mv\Gamma\cos\theta}{mc^2\Gamma} = \frac{v}{c^2}\cos\theta\,, \tag{6.16}$$

thus (6.15) becomes

$$\begin{aligned} E^* &= \Gamma^2\left[E + 2Up_x + \frac{U^2}{c^2}E\right] = \Gamma^2 E\left[1 + 2U\frac{p_x}{E} + \frac{U^2}{c^2}\right] \\ &= \Gamma^2 E\left[1 + 2U\frac{v\cos\theta}{c^2} + \frac{U^2}{c^2}\right]. \end{aligned} \tag{6.17}$$

Using now the definition of the Lorentz factor Γ and its Taylor approximation $\Gamma^2 = [1 - (U/c)^2]^{-1} \simeq [1 + (U/c)^2]$ and at the second order in U/c

$$E^* \simeq \left[1+\frac{U^2}{c^2}\right]E\left[1+2U\frac{v\cos\theta}{c^2}+\frac{U^2}{c^2}\right] \simeq E\left[1+2\frac{Uv\cos\theta}{c^2}+2\frac{U^2}{c^2}\right]. \tag{6.18}$$

The energy gained by the particle in the observer reference frame is

$$\Delta E = E^* - E = \left[2\frac{Uv\cos\theta}{c^2} + 2\left(\frac{U}{c}\right)^2 E\right]. \tag{6.19}$$

The first term in (6.19) is null when averaged over all directions. Energy is gained in head-on collisions ($\cos\theta > 0$) and lost in catching collisions when $\cos\theta < 0$. In this situation, the energy gain is $\Delta E \propto (U/c)^2$, as derived in the previous section (second-order Fermi model). The acceleration process is therefore rather inefficient.

Let us assume an astrophysical environment in which only head-on collisions occur (as in the toy situation in Fig. 6.4). Under such conditions, the quadratic term in (6.19) can be neglected, since $U/c \ll 1$. Then, also using the fact that the particle is relativistic and $v \sim c$, (6.19) becomes

$$\Delta E = E^* - E = \left(2\frac{U\cos\theta}{c}\right)\cdot E; \quad E^* = \left(1 + 2\frac{U}{c}\cos\theta\right)\cdot E\,. \tag{6.20}$$

The particle gains energy in each collision if $\cos\theta > 0$. We remind the reader that the average value $\langle x\rangle$ of a variable x distributed according to a function $f(x)$ is

$$\langle x\rangle = \frac{\int x\cdot f(x)\cdot \mathrm{d}\Omega}{\int f(x)\cdot \mathrm{d}\Omega}\,. \tag{6.21}$$

Here, $x = \cos\theta$, $\mathrm{d}\Omega = \mathrm{d}\phi\sin\theta\mathrm{d}\theta = \mathrm{d}\phi\mathrm{d}\cos\theta$ is the differential solid angle and $f(x)$ is given by (6.20). The average over all directions with $\cos\theta > 0$ gives us

$$\langle\cos\theta\rangle = \frac{\int_0^1 \cos\theta\cdot\cos\theta\cdot\mathrm{d}\cos\theta}{\int_0^1 \cos\theta\cdot\mathrm{d}\cos\theta} = \frac{2}{3}\,. \tag{6.22}$$

Finally, when averaged over directions, (6.20) becomes

$$\langle\Delta E\rangle = \left(\frac{4}{3}\frac{U}{c}\right)\cdot\langle E\rangle \equiv \eta\cdot\langle E\rangle\,, \tag{6.23a}$$

$$\langle E^*\rangle = \left(1 + \frac{4}{3}\frac{U}{c}\right)\cdot\langle E\rangle \equiv \mathcal{B}\cdot\langle E\rangle\,. \tag{6.23b}$$

The two quantities defined above, η and $\mathcal{B}$, are used in the following. A situation in which accelerated particles suffer only head-on collisions is assumed in the astrophysical diffusive shock acceleration model, which uses strong shock waves produced (for instance) by supernova explosions.

The above result can also trivially be obtained using (6.7) and (6.8), assuming only type I collisions. In this case, for $v \sim c \gg U$, (6.9) is simply

$$\frac{\Delta E}{\Delta t} = f_I \Delta E_I = 4 \cdot \frac{U}{v} \cdot E \cdot \frac{U+v}{\ell} \simeq 4\frac{U}{\ell}E = \frac{E}{\tau_F'} \,. \tag{6.24}$$

The characteristic time

$$\tau_F' \equiv \frac{\ell}{4U} = 2\tau_F''\frac{U}{c} \tag{6.25}$$

is now sensibly shorter than that expected in the second-order process (6.10).

6.1.4 The Power-Law Energy Spectrum from the Fermi Model

The important result of the Fermi mechanism is that it succeeded in deriving a power-law energy spectrum. To see this, let us use the diffusion-loss equation (5.37). As usual, $\mathscr{N}$ represents the (energy-dependent) number density of CRs. We are interested in a steady-state situation ($d\mathscr{N}/dt = 0$) and neglecting sources, $Q(E) = 0$. Then, (5.37) becomes

$$\frac{\partial}{\partial E}[b(E)\mathscr{N}(E)] - \frac{\mathscr{N}(E)}{\tau_{\rm esc}} = 0 \,. \tag{6.26}$$

The term $b(E) = -\mathrm{d}E/\mathrm{d}t$ represents now an energy gain term, obtained comparing Eq. (5.19) with (6.24) or (6.9):

$$\frac{\mathrm{d}E}{\mathrm{d}t} = \frac{E}{\tau_F} \quad \longrightarrow \quad b(E) = -E/\tau_F \,. \tag{6.27}$$

Therefore, (6.26) reduces to

$$-\frac{\mathrm{d}}{\mathrm{d}E}\left[\frac{E}{\tau_F}\mathscr{N}(E)\right] - \frac{\mathscr{N}(E)}{\tau_{\rm esc}} = 0 \,. \tag{6.28}$$

After differentiation and rearrangement of the terms, (6.28) becomes

$$-\frac{\mathrm{d}\mathscr{N}(E)}{\mathrm{d}E} = \frac{\tau_F}{E}\left[\frac{1}{\tau_F} + \frac{1}{\tau_{\rm esc}}\right]\mathscr{N}(E) = \frac{1}{E}\left[1 + \frac{\tau_F}{\tau_{\rm esc}}\right]\mathscr{N}(E) \,. \tag{6.29}$$

Therefore

$$\mathscr{N}(E) = \text{constant} \times E^{-\alpha_S} \qquad \text{with} \quad \alpha_S = 1 + \frac{\tau_F}{\tau_{\rm esc}} \,. \tag{6.30}$$

The Fermi acceleration mechanism succeeded in deriving a power-law energy spectrum with spectral index α_S. Note that in the above notation, we have used τ_F: it can represent both τ''_F and τ'_F. A power-law spectrum is thus obtained in both Fermi mechanisms. The efficiency of the two processes is completely different, and only the first-order one can account for galactic CRs' acceleration.

6.2 Diffusive Shock Acceleration in Strong Shock Waves

The first-order Fermi mechanism described in Sect. 6.1.3 is very efficient in accelerating particles if we consider regions where there are small-scale turbulences, for example, in the shells of young supernova remnants.

As this mechanism is associated with particle acceleration in strong shock waves, it is often referred to as *diffusive shock acceleration*. The key feature of this process is that the acceleration automatically results in a power-law energy spectrum (6.30) with spectral index $\alpha_S \simeq 2$.

Different modeling of the diffusive shock acceleration exists, according to different authors who have refined the model starting from the old first-order Fermi acceleration model (Axford et al. 1977; Bell 1978; Blandford and Ostriker 1978; Drury 1983; Jokipii 1987; Jones and Ellison 1991; Krymsky 1977). These works stimulated an enormous amount of interest, and different astrophysical environments are now considered as possible acceleration region candidates.

A shock wave carries energy and can propagate through a medium or, in some cases, in the absence of a material medium, through the electromagnetic field. Shock waves are characterized by an abrupt, nearly discontinuous change in the characteristics of the medium (Fig. 6.5a). Across a shock, there is always an extremely rapid rise in pressure, temperature and density of the flow. In supersonic flows (as in the case of supernova explosions), the shock wave travels through the medium at a higher speed than an ordinary wave. The speed v_s of the shock front is much larger than the thermal velocities of particles ($v_s \gg v_1$).

In the frame in which the shock front is at rest (Fig. 6.5b), the velocities of the material up-stream and down-stream are, respectively, $\mathbf{v_1} = -\mathbf{v_s}$ and $\mathbf{v_2} = \mathbf{v_s}/R$. The constant R depends only on the gas composition, and for a mono-atomic gas $R = 4$ (see the demonstration in Sect. 6.5.2).

Now, consider high energy particles *down-stream* of the shock. Scattering ensures that the particle distribution is isotropic in the frame of reference in which the gas is at rest (Fig. 6.5c). The shock front moves at velocity v_s but the gas *up-stream* (i.e., behind) the shock advances at a speed $U = (3/4)v_s$ relative to the material down-stream. When a particle crosses the shock front from right to left, an elastic collision occurs and the particle gains energy according to (6.23a). The velocity distribution of particles entering into the region behind the shock front will become isotropic with respect to that flow.

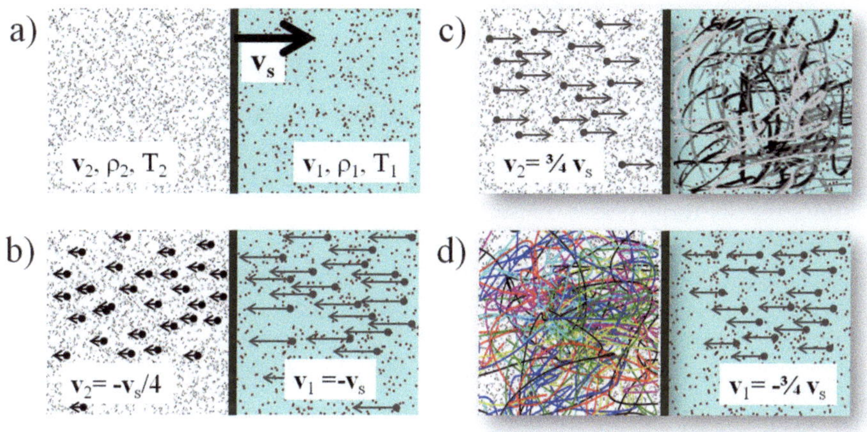

Fig. 6.5 (**a**) Conditions on the down-stream (*left*) and the up-stream (*right*) side of a shock wave in the laboratory system. The shock front propagates at supersonic velocity $\mathbf{v_s}$ through stationary interstellar gas with density ρ_1, temperature T_1 and particle velocities $v_1 \ll v_s$. Density, temperature and velocity behind the shock are ρ_2, T_2 and v_2, respectively. (**b**) The same situation seen in the reference frame in which the shock front is at rest. In this frame of reference, the ratio of the up-stream to the down-stream velocity is $v_1/v_2 = R$. For a fully ionized plasma, $R = 4$, as demonstrated in Sect. 6.5.2. (**c**) The flow of gas as observed in the frame of reference in which the up-stream gas is stationary and the velocity distribution of the high energy particles is isotropic. (**d**) The flow of gas as observed in the frame of reference in which the down-stream gas is stationary and the velocity distribution of high energy particles is isotropic

A symmetric situation occurs for a particle diffusing the shock from *up-stream*, moving towards the region ahead in front of the shock (Fig. 6.5d). The distribution of the particle velocities is isotropic behind the shock. When crossing the shock front, a particle encounters gas moving towards the shock front, again with speed $U = (3/4)v_s$. It receives the same increase in energy η on crossing the shock front from down-stream to up-stream as it did in traveling from up-stream to down-stream. The fact that, in every passage through the shock front, the particle receives an increase of energy is the clever aspect of the first-order Fermi acceleration mechanism. In addition, the energy increment $\eta = (4/3)U/c$ is the same in both directions. Thus, according to (6.23b), the particle energy after each collision is incremented by a factor $\mathscr{B}$:

$$\mathscr{B} = 1 + \eta = 1 + \frac{4}{3}\frac{3v_s}{4c} = 1 + \frac{v_s}{c} . \tag{6.31}$$

As v_s/c is small, many collisions are necessary to achieve a sizeable energy gain.

6.3 Supernova Remnants (SNRs) and the Standard Model of CRs Acceleration

There is a substantial consensus in the scientific community that galactic CRs are somehow related to one or more types of supernova explosion and that the acceleration process is mainly due to diffusive transport in the neighbourhood of strong shocks formed as a consequence of these explosions. In particular, the environment provided by the shock wave produced by Type II (or *core collapse* supernovae (SNe), described in Chap. 12) is the main candidate. Type II SNe occur at the end of the fusion process in very massive stars, $M \geq 8M_\odot$. These stars develop an onion-like structure (Fig. 12.10) with a degenerate Fe core. After the core is completely fused to iron, no further processes releasing energy are possible. Instead, photo-disintegration destroys the heavy nuclei, e.g., via $\gamma + ^{56}\text{Fe} \rightarrow 13^4\text{He} + 4n$, and removes the thermal energy necessary to provide pressure support. In the subsequent collapse of the star, the density increases and the free electrons and protons are forced to form neutrons via inverse beta decay. A neutron star forms in the core, with a density that equals that of the nuclear matter. The in-falling material of outer layers is reflected from the nuclear core and a shock wave propagates outwards heated by neutrino emission from the neutron star. The energy outfall from a supernova explosion (Sect. 12.11) is simply evaluated as the released gravitational binding energy, which is about (2–4) $\times\ 10^{53}$ erg per explosion. A large fraction (99%) of this energy is emitted as neutrinos, and only 1% (thus, about 10^{51} erg) is transferred into kinetic energy of the expulsed material forming the *shock wave*. The energy emitted as electromagnetic radiation is even less, about 1% of the kinetic energy.

6.3.1 SNRs as Galactic CR Accelerators

The reasons why SNRs are generally considered the candidate sites for galactic CR acceleration rely on the following facts:

- **The energy balance**. In a successful core-collapse SN around 10 $M_\odot$ are ejected with a velocity $U \sim 5 \times 10^8$ cm/s, see (6.33). Assuming three SNs per century in the Galaxy, the average output in kinetic energy from galactic supernovae is $P_{SN}^{kin} \sim 3 \times 10^{42}$ erg/s. This value is one or two orders of magnitude larger than the luminosity $P_{CR} \sim 5 \times 10^{40}$ erg/s requested to maintain a steady cosmic ray energy density $\rho_{CR} \sim 1$ eV/cm^3 (see Sect. 2.10).
- **The efficiency of the acceleration process**. To fulfill the energy balance requirement, the CR acceleration mechanism should convert part of the kinetic energy of material ejected by the SN to high energy particles: $P_{CR} = \eta P_{SN}^{kin}$, with efficiency $\eta = 0.01$–0.1. According to (6.23a), the particle gain in the first-order Fermi mechanism is $\sim 4/3(U/c)$. We will derive in Sect. 6.3.2 that $U/c \simeq$ a

few %. Thus, (U/c) is the order of the needed value for η, as first suggested by Ginzburg and Syrovatskii.

- **The chemical abundances**. The composition of cosmic rays as measured by direct experiments (Sect. 3.6), when the effects of propagation in the Galaxy are considered, is very similar to the abundances of the elements in the Solar System. The chemical composition of the Solar System is similar to that produced by core-collapse supernovae (Sect. 3.7). It follows that the CR acceleration sites are close (in this model, coincident) to regions where the abundance of the chemical elements is determined by a supernova explosion.
- **The maximum energy**. The supernova model provides CRs with energy up to the knee. The knee (in this model) is due to the different maximum energy given to nuclei with different charge, $E^{\max}(Z) \simeq 300 \cdot Z$ TeV, see Sect. 6.4. A prediction of this model is that the chemical composition of CRs becomes heavier as the energy increases from values below to above the knee.
- **The spectral index of the power-law energy spectrum**. A power-law energy spectrum for all types of charged particles is observed (protons, nuclei and electrons). The energy spectrum of cosmic rays and the electron energy spectrum from nonthermal sources have the form $\Phi(E) \propto E^{-\alpha_S}$, where the exponent α_S is typically $\sim$2. We derive this result in Sect. 6.5.

6.3.2 *Relevant Quantities in SNR*

The average energy emitted as kinetic energy K by a 10 $M_\odot$ supernova is roughly 1% of the total binding energy, i.e., for:

$$M = 10M_\odot = 2 \times 10^{34}\ \mathrm{g}\,, \tag{6.32a}$$

we obtain

$$\text{Gravitational Energy} \sim 2 \times 10^{53}\ \mathrm{erg} \tag{6.32b}$$

$$K \simeq 2 \times 10^{51}\ \mathrm{erg}\,. \tag{6.32c}$$

The velocity of the ejected mass (the *shock wave*) is on the order of

$$U \simeq \sqrt{\frac{2K}{M}} = \sqrt{\frac{4 \times 10^{51}}{2 \times 10^{34}}} \simeq 5 \times 10^{8}\ \mathrm{cm/s} \quad \rightarrow \quad \frac{U}{c} \simeq 2 \times 10^{-2}\,. \tag{6.33}$$

Density, velocity and temperature of the matter behind and ahead of the shock are determined by thermodynamical considerations in Sect. 6.5.2.

According to (6.33), a shock wave originated by a supernova explosion has typically $U/c \simeq 10^{-2}$, which corresponds to a nonrelativistic velocity but much larger than typical velocities of interstellar materials. More refined models (see Hillas (2005) for a review) assume that the velocity is higher for outer layers ($U/c \sim 10^{-1}$), while the inner layers expand more slowly. The range of values

$$\frac{4}{3}\frac{U}{c} \equiv \eta \simeq 10^{-2}\text{–}10^{-1} \tag{6.34}$$

corresponds to the needed *efficiency* η of the acceleration process defined in (6.23a) and required to explain the CRs' acceleration by supernovae explosions.

The shock front expands (we assume with constant velocity U and with spherical symmetry) across the interstellar matter (ISM), which has density $\rho_{ISM} \sim$ 1 proton cm$^{-3} \simeq 1.6 \times 10^{-24}$ g cm^{-3} (Sect. 2.7.2). During the expansion, the shock collects interstellar matter. When the mass of the swallowed material becomes comparable to the mass of the ejected shells of the supernova, the velocity of the shock decreases remarkably. As the radius R_{SN} of the shock front increases, the matter density $\rho_{\mathrm{SN}} \sim \text{mass}/R^3_{\mathrm{SN}}$ inside the shock volume decreases. We assume that the shock becomes inefficient when the $\rho_{\mathrm{SN}} \simeq \rho_{\mathrm{ISM}}$. The radius within which the shock wave is able to accelerate particles can be derived using the condition

$$\rho_{\mathrm{SN}} = \frac{10M_\odot}{(4/3)\pi R^3_{\mathrm{SN}}} = \rho_{\mathrm{ISM}}\ . \tag{6.35}$$

From this, we obtain

$$R_{\mathrm{SN}} = \left(\frac{3 \times 10M_\odot}{4\pi\rho_{ISM}}\right)^{1/3} = \left(\frac{6 \times 10^{34}}{4\pi \cdot 1.6 \times 10^{-24}}\right)^{1/3} = 1.4 \times 10^{19}\,\mathrm{cm} = 5\,\mathrm{pc}\ . \tag{6.36a}$$

The corresponding time interval T_{SN} during which particles are accelerated is:

$$T_{\mathrm{SN}} = \frac{R_{\mathrm{SN}}}{U} = \frac{1.4 \times 10^{19}\,\mathrm{cm}}{5 \times 10^8\,\mathrm{cm/s}} \simeq 3 \times 10^{10}\,\mathrm{s} \simeq \mathcal{O}(1000)\,\mathrm{y}\ . \tag{6.36b}$$

Figure 6.6 shows the SN 1572 (or Tycho's Supernova) as seen in different wavelengths by modern astronomers. It was a supernova of Type Ia in the Cassiopeia constellation, about $D = 3$ kpc from Earth. SN 1572 is one of about eight supernovae visible to the naked eye in historical records. It appeared in early November 1572 and was independently discovered by many individuals, including the famous astronomer Tycho Brahe. He described the new star in the book *De nova et nullius aevi memoria prius visa stella* ("Concerning the Star, new and never before seen in the life or memory of anyone"), published in 1573, which contains his own observations and the analysis of many other observers. From the scale of the figure, we obtain an angular radius of ~ 4 arcmin. Thus, the linear radius of the

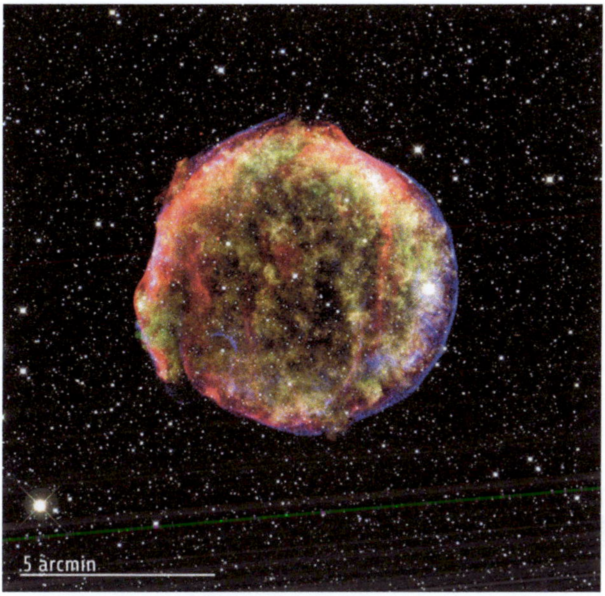

Fig. 6.6 This composite image of the Tycho Brahe SNR combines X-ray and infrared observations obtained with NASA's Chandra X-ray Observatory and Spitzer Space Telescope, respectively, and at the Calar Alto Observatory, Spain (http://chandra.harvard.edu/photo/2009/tycho/). It shows the scene more than four centuries after the brilliant star explosion witnessed by Tycho Brahe and other astronomers of that era. Credit: X-ray: NASA/ CXC/SAO, Infrared: NASA/JPL-Caltech; Optical: MPIA, Calar Alto, O.Krause et al.

object visible from the figure is (we need to convert the arcmin to radians)

$$R_{\text{Tycho}} = D\Delta\theta = 3 \text{ kpc} \times \frac{2\pi}{360}\frac{4}{60} = 3 \text{ kpc} \times 0.0011 \simeq 3 \text{ pc} , \tag{6.37}$$

in agreement with (6.36a) for a 450-year-old SN remnant.

A note of warning: the fact that protons or heavier nuclei are accelerated in the region shown in Fig. 6.6 is controversial (Sect. 9.7). Most of the detected nonthermal radiation, including that in the radio wavelength region, is almost certainly emitted by electrons.

Note that the timescale for particle acceleration is $T_{\text{SN}} \sim \mathcal{O}(10^3)$ year, while the timescale for CRs escaping from our galaxy is $\tau \sim \mathcal{O}(10^7)$ year, as derived in Sect. 5.2. A large number of supernova explosions, with a short acceleration time duration with respect to the CR escape time, contributes to filling the Galaxy with high energy particles. The CRs we detect could be accelerated by a number of supernovae as large as 10^4!

All the young galactic supernova remnants in historical records (Sect. 12.11) are in the evolution phase, when the swept-up mass is on the order of the mass of the ejected material and are entering into the so-called *Sedov phase*. This phase represents the situation when the shock has collected enough interstellar matter to initiate the slow-down period. An object such as the Crab Nebula follows a different evolution scenario, since a continuous energy flux is originating from a central pulsar (Pulsar Wind Nebula). We will return to the Crab in Sect. 9.5.

Equation (6.36a) tells us that supernova explosions compress inter-stellar material up to a few pc. This explains why the *average distance between stars* within galaxies (including our own) is on the order of the parsec. Also, planetary systems cannot exist on smaller scales. It is believed that planetary systems (like our own) were formed from a cloud of gas and dust. If the cloud is hit by a shock wave generated by the explosion of a pre-existing, nearby star with a larger mass, the shock wave causes the compression of the gas and dust cloud. These matters, because of the gravitational attraction, begin to pull other material inward, forming the star nebula. In the process of contraction, as a result of the gravitational attraction, pressure, magnetic fields and rotation, the nebula is flattened into a protoplanetary disk with a protostar at its center. In the protoplanetary disk, a certain number of planets are formed. If the protostar is too close to another massive star, the successive core-collapse restarts the process. The security distance within which long-lived stars (such as the Sun, which has been shining since 5 Gy, and will last for another 5 Gy) is distributed is thus on the order of a few pc.

6.4 Maximum Energy Attainable in the Supernova Model

With simple arguments, it is possible to derive the maximum energy that a charged particle can reach in the acceleration process due to the diffusive shock mechanism from a supernova explosion. Equation (6.23a) gives the energy gain for each iterative acceleration process. The rate of energy increase is given by the ratio between (6.23a) and the characteristic period T_{cycle} of the process, i.e., the time between two successive scattering processes:

$$\frac{\mathrm{d}E}{\mathrm{d}t} \simeq \frac{\eta E}{T_{\text{cycle}}} . \tag{6.38}$$

T_{cycle} can be evaluated with the following considerations. When the accelerated particles pass though the shock front in either direction, their velocity distribution rapidly becomes isotropic within the reference frame of the moving fluid on either side of the shock. This occurs because they are scattered by streaming instabilities and turbulent motions on either side of the shock wave and because they are trapped by the magnetic fields. The typical extension of the confinement region is given by the Larmor radius in the magnetic field B (2.5):

$$\lambda_{\text{cycle}} = r_L = \frac{E}{ZeB} . \tag{6.39}$$

Following the discussion of Drury (1983) and referring to Fig. 6.5b, the typical time between successive encounters in the rest frame of a shock front moving at velocity $v_S \simeq U$ is

$$T_{\text{cycle}} = \frac{\lambda_{\text{cycle}}}{U} = \frac{E}{ZeBU} \ . \tag{6.40}$$

Inserting (6.40) into (6.38), we obtain

$$\frac{\mathrm{d}E}{\mathrm{d}t} \simeq \eta \cancel{E} \frac{ZeBU}{\cancel{E}} \tag{6.41}$$

where, using (6.31), $\eta \simeq U/c$. Thus, the rate of energy gain is independent of the particle energy E. This is relevant, because the model is not constrained from a particular mechanism of pre-acceleration of the charged particles.

The maximum energy that a charged particle could achieve is then simply the rate of energy gain, times the duration T_{SN} of the shock (6.36b):

$$E^{\max} \simeq \frac{\mathrm{d}E}{\mathrm{d}t} \times T_{\text{SN}} = \eta ZeBR_{\text{SN}} \simeq \frac{ZeBR_{\text{SN}}U}{c} \ . \tag{6.42}$$

Inserting the numerical values for the velocity U of the shock (6.33), the proton electric charge ($e = 4.8 \times 10^{-10}$ statC), the maximum radius of the expansion R_{SN} (6.36a), and the typical value of the galactic magnetic field $B \simeq 4 \times 10^{-6}$ Gauss, we obtain

$$E^{\max} = eB(U/c)R_{\text{SN}}Z \tag{6.43a}$$

$$= (4.8 \times 10^{-10}) \cdot (4 \times 10^{-6}) \cdot (2 \times 10^{-2}) \cdot (1.4 \times 10^{19}) \cdot Z$$

$$\simeq 500 \cdot Z \text{ erg} \simeq 300 \cdot Z \text{ TeV} \ . \tag{6.43b}$$

The diffusive shock acceleration mechanism based on supernova explosions explains the spectrum of cosmic-ray protons up to a few hundred TeV, i.e., $\sim$ a few 10^{14} eV, which corresponds to the energy region where the knee begins.

An important consequence of (6.43) is that $E^{\max}$ depends on the particle charge Z. This means that a fully ionized heavy nucleus of charge Z could achieve much higher total energy with respect to a single-charged proton. Thus, in this model, the knee is explained as a structure resulting from the different maximum energy reached by nuclei with different charge Z (see Fig. 6.7).

As shown by (6.43), the diffusive shock acceleration model quite naturally explains the maximum *rigidity* (see Sect. 2.3.1) associated with acceleration effects. Models exist that also attribute this effect to propagation, or to both propagation and acceleration. More exotic effects (connected with a change in the behavior of nuclear cross-sections at very high energies) seem to have been ruled out. If cosmic rays are classified according to the energy per particle, as in the case of air shower measurements, then the spectrum should steepen first for protons, then for helium,

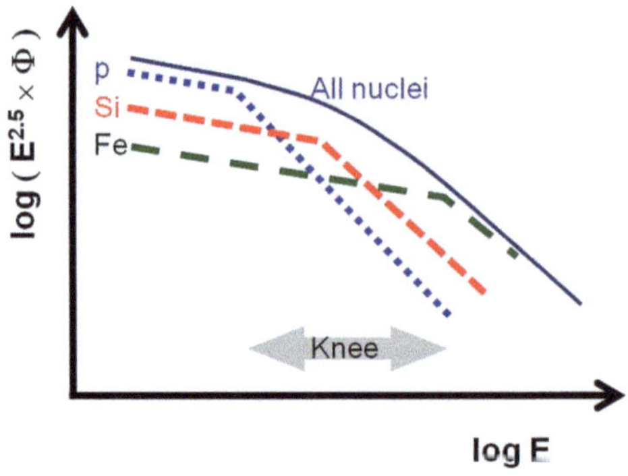

Fig. 6.7 The interpretation of the CRs' *knee* as being due to the correlation between the maximum energy and the nuclear charge Z. The flux of each nuclear species sharply decreases after a given cut-off, which depends on Z as $E_Z^{\max} = E_p^{\max} \cdot Z$, where $E_p^{\max}$ is the maximum energy reached by protons. The behavior of hydrogen, silicon ($Z = 14$) and iron ($Z = 26$) nuclei are depicted in figure

then for the C, N, O group, etc. The experimental evidence for the sequence of successive steepening for several groups of nuclei was discussed in Sect. 4.10.

6.5 The Spectral Index of the Energy Spectrum

The spectral index α_S of the energy spectrum of accelerated particles within the framework of the diffusive shock acceleration model can be estimated following the procedure used in Longair (2011).

In Sect. 6.4, we have shown that the maximal energy $E^{\max}$ for a particle accelerated at a specific source is determined by several factors. For a shock driven by a supernova explosion, a limited time $T_{\rm SN}$ is available for particle acceleration. This finite lifetime limits the maximum number $k^{\max}$ of possible iteration cycles. In each cycle, the particle energy increases by a finite amount $\mathscr{B}$ (6.31). In this simple model, all particles will have the same final energy if they undergo the same number of cycles. An *escape probability* $P_{\rm esc}$ per encounter can be defined. Let $P_{\rm esc}$ represent the possibility that, after a collision, the particle escapes the acceleration region and it is lost for successive iterations. In regions where the acceleration mechanism is efficient, $P_{\rm esc}$ is small. Thus, $P = 1 - P_{\rm esc}$ represents the probability that the particle remains inside the acceleration region after one collision process. We may further assume that $P_{\rm esc}$ is constant all over $T_{\rm SN}$.

The probability that a particle remains in the acceleration region after k encounters is $(1 - P_{\rm esc})^k$ and, if N_0 is the initial number of particles and E_0 their energy, after k cycles in the acceleration region, there will be

$$N = N_0 P^k \text{ particles with energy} \geq E = E_0 \mathscr{B}^k \,. \tag{6.44}$$

In fact, after the first encounter, there will be $N_1 = N_0 P$ particles with energy $E_1 = E_0\mathscr{B}$, with $\mathscr{B}$ given in Eq. (6.31); after two encounters, $N_2 = N_1 P = N_0 P^2$ particles with energy $E_2 = E_1\mathscr{B} = E_0\mathscr{B}^2$, and so on.

The unknown parameter k can be removed from both sides in (6.44) using the ratio

$$\frac{\ln(N/N_0)}{\ln(E/E_0)} = \frac{\ln P}{\ln \mathscr{B}} \, . \tag{6.45}$$

If k is large, the discrete process is seen as a continuous process, and hence

$$\frac{N(\geq E)}{N_0} = \left(\frac{E}{E_0}\right)^{\ln P/\ln\mathscr{B}} \, . \tag{6.46}$$

We explicitly write $N(\geq E)$ because (as shown above in the discussion of the discrete case) it represents the number of particles having energy at least equal to E:

$$N(\geq E) = \int_E^\infty N(E')\mathrm{d}E' \, .$$

The differential energy spectrum is thus given by the energy derivative of (6.46):

$$N(E) = \text{constant} \times E^{-1+\Lambda} \text{ where } \Lambda = \ln P/\ln\mathscr{B} \, . \tag{6.47}$$

Again, as resulted in (6.30), we obtain a power-law spectrum. The numerical value of Λ in (6.47) is determined by the values of P and $\mathscr{B}$. The latter was already derived in (6.31).

6.5.1 The Escape Probability

To estimate the escape probability P, we follow the argument from Bell (1978), also reported in Longair (2011). According to Sect. 2.4, if $\Phi(> E)$ is the flux of particles with energy larger than E, the flux $\mathscr{F}$ from one hemisphere through a planar surface is $\pi\Phi(> E)$ (2.17b). The number density of relativistic ($v \simeq c$) isotropic particles is given by (2.18a) as $n = 4\pi\Phi(> E)/c$. The continuity equation for mass (see Sect. 6.5.2) will ensure that the fluxes of particles from left to right and from right to left of the shock front are equal. Thus, the average flux of particles per square centimeter per second through the shock front is:

$$n = \frac{4\pi}{c}\Phi(> E) = \frac{4}{c}\mathscr{F} \quad \rightarrow \quad \mathscr{F} = \frac{nc}{4} \, . \tag{6.48}$$

The flux is the same in either direction with respect to the shock front. At each pass, the particle gains energy. In the down-stream region, since the particle velocities are randomized by the collisions, and are thus isotropic in that frame (Fig. 6.5c), there is a finite probability that the particle exits through the boundary on the left. In this way, particles are removed from the shock region and lost. Referring to Fig. 6.5b, these particles are moving with velocity $v_2 = v_s/4$ towards the left boundary and are removed from the region of the shock at a rate

$$\mathscr{F}_{\text{esc}} = nv_2 = n\frac{v_s}{4} \,. \tag{6.49}$$

Thus, the fraction of particles lost per unit time is the ratio between the two fluxes (6.49) and (6.48), and it represents the probability P_{esc} that a particle is removed from the shock region:

$$P_{\text{esc}} = \frac{\mathscr{F}_{\text{esc}}}{\mathscr{F}} = \frac{\frac{nv_s}{4}}{\frac{nc}{4}} = \frac{v_s}{c} \,. \tag{6.50}$$

This is an important result: if the shock is nonrelativistic ($v_s \ll c$), only a small fraction of the particles is lost per cycle. The probability of remaining in the acceleration region is thus:

$$P = 1 - P_{\text{esc}} = 1 - \frac{v_s}{c}. \tag{6.51}$$

We are able now to obtain the unknown spectral index Λ in (6.35) because:

$$\ln P = \ln\left(1 - \frac{v_s}{c}\right) \simeq -\frac{v_s}{c}; \quad \ln \mathscr{B} = \ln\left(1 + \frac{v_s}{c}\right) \simeq +\frac{v_s}{c} \tag{6.52}$$

and

$$\Lambda = \frac{\ln P}{\ln \mathscr{B}} = -1 \,. \tag{6.53}$$

Thus, the exponent of the differential energy spectrum in (6.47) is:

$$-\alpha_S = -1 + \Lambda = -2 \,. \tag{6.54}$$

This represents one of the principal results of the diffusive shock acceleration mechanism: it provides a power-law spectrum whose index is within the range of the experimental measurements. The power-law index of exactly 2 arises for the case of a strong nonrelativistic shock in an ideal gas and with constant escape probability P_{esc}. Inefficient situations in the shock conditions increase the spectral index between 2.0 and 2.4, as obtained using more refined computations (Hörandel 2007).

Harder spectral indexes (1.5–1.8) are obtained in models with shock-amplified magnetic fields, Sect. 6.6.

6.5.2 A Shock Front in a Mono-Atomic Gas[3]

We will use here thermodynamical arguments to derive the relations between the velocity and density of up- and down-stream materials, and in particular, the compression factor R introduced in Sect. 6.2. These considerations are valid within the reference frame in which the shock front is at rest (see Fig. 6.5b). Particles within the reference frame of an outside observer, having density ρ_1 up-stream of the shock, have velocities $v_1 \simeq 0$, much smaller than the shock velocity v_s (Fig. 6.5a). The material down-stream has density ρ_2. The gas behind the discontinuity moves at the velocity v_2, which depends on the shock front velocity.

The conditions of a material subjected to a shock are described by the ideal fluid equations. The three basic equations are the conservation of mass, the conservation of linear momentum, and the Poisson equation:

$$\frac{\partial \rho}{\partial t} + \nabla \cdot (\rho \mathbf{v}) = 0 , \tag{6.55a}$$

$$\rho \frac{\partial \mathbf{v}}{\partial t} + \rho \mathbf{v} \cdot (\nabla \mathbf{v}) = \mathbf{F} - \nabla P , \tag{6.55b}$$

$$\Delta \Phi = 4\pi G \rho . \tag{6.55c}$$

The left side of (6.55b) is Euler's equation, which measures the velocity change dv/dt of a fluid element, including both the time variation at a fixed coordinate $(\partial \mathbf{v}/\partial t)$ and the change due to the movement of the fluid element. The right side of the same equation includes the external force $\mathbf{F}$ and the force due to the pressure gradient ∇P. The Poisson equation (6.55c) connects the mass density ρ with the gravitational potential Φ. G is the gravitational constant.

In addition to the conservation laws for mass (6.55a) and momentum (6.55b), we can also consider the conservation law for energy:

$$\frac{\partial}{\partial t}\left(\frac{\rho v^2}{2} + \rho U + \rho \Phi\right) + \nabla \cdot \left[\rho \mathbf{v}\left(\frac{v^2}{2} + U + \frac{P}{\rho} + \Phi\right)\right] = 0 . \tag{6.56}$$

In (6.56), the term in the first brackets accounts for the change of kinetic, internal U and potential Φ energy with time; this variation has to be balanced by the energy flux through the boundary of the considered volume (the term in the second bracket).

[3] This section can be skipped in the early reading steps.

We will consider now shock waves in a perfect gas. Sound waves propagate, under most circumstances, adiabatically, i.e., without production of entropy, $dS = 0$. We tailor the situation to a mono-atomic gas, as the case of most interstellar matter. A mono-atomic gas is characterized by the adiabatic index[4] $\gamma = c_P/c_V = 5/3$. The sound velocity is

$$c_{\text{sound}} = \left(\frac{\gamma P}{\rho}\right)^{1/2}, \tag{6.57}$$

where P and ρ are the gas pressure and density, respectively.

We need to derive relations between the average gas velocity and that of the shock front, v_s. The shock front represents an abrupt discontinuity between two regions of gas. In the undisturbed region ahead of the shock wave, the gas is at rest with pressure P_1, density ρ_1 and temperature T_1, and the speed of sound is c_1. Behind the shock wave, the gas moves supersonically at a speed $> c_1$, and its pressure, density and temperature are P_2, ρ_2 and T_2, respectively (Fig. 6.5a). It is convenient to transform into a reference frame moving at velocity v_s, in which the shock wave is stationary (Fig. 6.5b). Within this reference frame, the undisturbed gas flows towards the discontinuity at velocity $v_1 = |v_s|$ and, when it passes through it, its velocity becomes v_2 away from the discontinuity.

We will consider, for simplicity, only the properties of a one-dimensional, steady shock *in its rest frame* and assume that magnetic and gravitational fields can be neglected. Then, the continuity equation for mass (6.55a) becomes simply

$$\frac{\mathrm{d}}{\mathrm{d}x}(\rho v) = 0. \tag{6.58}$$

The Euler equation simplifies, using the same assumptions and taking into account (6.58), to

$$\frac{\mathrm{d}}{\mathrm{d}x}(P + \rho v^2) = 0. \tag{6.59}$$

Equation (6.56) for a stationary flow (i.e., time independent) with no external potential ($\Phi = 0$) gives us

$$\frac{\mathrm{d}}{\mathrm{d}x}\left(\frac{\rho v^3}{2} + (U + P)v\right) = 0. \tag{6.60}$$

[4] In this section, as usual in thermodynamics, the symbol γ always refers to the adiabatic index of gases.

We can now integrate these equations over the discontinuity of the shock front:

$$\left[\rho v\right]_1^2 = \text{const.} \tag{6.61a}$$

This means that the product of density and speed in region 1 is equal to that in region 2. Analogously,

$$\left[P + \rho v^2\right]_1^2 = \text{const}\,, \tag{6.61b}$$

$$\left[\frac{\rho v^3}{2} + \frac{\gamma}{\gamma - 1} P v\right]_1^2 = \text{const}\,, \tag{6.61c}$$

where in (6.61c), we used the relation between internal energy and pressure: $U = P/(\gamma - 1)$. Since we assume a steady flow, these boundary conditions have to be evaluated in the shock rest frame, (cf. Fig. 6.5b). Otherwise, time-derivatives should be included.

Inserting first the condition obtained from (6.61a), $\rho_2 = (v_1/v_2)\rho_1$, into (6.61b), we obtain $P_2 = P_1 + \rho_1 v_1 (v_1 - v_2)$. The two relations above are used to eliminate ρ_2 and P_2 from (6.61c). Reordering the resulting equation according to powers of v_2

$$\left(\frac{\gamma + 1}{\gamma - 1}\right) v_2^2 - \frac{2\gamma}{\gamma - 1}\left(\frac{P_1 + \rho_1 v_1^2}{\rho_1 v_1}\right) v_2 + \left(v_1^2 + \frac{2\gamma}{\gamma - 1}\frac{P_1}{\rho_1}\right) = 0\,. \tag{6.62}$$

We can now use the fact that the pressure P_1 is related to the sound velocity c_1 by Eq. (6.57), and thus $P_1 = \rho_1 c_1^2/\gamma$. We use this relation to replace P_1 in (6.62) and we also divide by v_1^2 to obtain the quadratic relation

$$\left(\frac{\gamma + 1}{\gamma - 1}\right) x^2 - \frac{2\gamma}{\gamma - 1}\left(\frac{1}{\gamma}\frac{c_1^2}{v_1^2} + 1\right) x + \left(1 + \frac{2}{\gamma - 1}\frac{c_1^2}{v_1^2}\right) = 0\,, \tag{6.63}$$

where $x = v_2/v_1$. The ratio v_1/c_1 is defined as the Mach number $\mathscr{M}$. Since we are interested in fast flows, and $\mathbf{v_1} = -\mathbf{v_s}$, we also have that $v_1 \gg c_1$ and we can neglect the two quadratic terms in c_1/v_1. We obtain

$$\left(\frac{\gamma + 1}{\gamma - 1}\right) x^2 - \frac{2\gamma}{\gamma - 1} x + 1 = 0\,. \tag{6.64}$$

It is easily verified that the two solutions are

$$x = 1 \quad \longrightarrow \quad v_1 = v_2, \tag{6.65a}$$

$$x = \frac{\gamma - 1}{\gamma + 1} \equiv \frac{1}{R} \quad \longrightarrow \quad Rv_2 = v_1. \tag{6.65b}$$

The first solution is trivial (no shock), while the second one is the strong shock solution. The compression ratio R indicates how the density, the velocity and pressure in the up- and down-stream regions differ in the shock front reference frame. Remembering that $v_1 = |v_s|$, and that for a mono-atomic gas, $\gamma = 5/3$, and thus $R = 4$, we can write

$$v_2 = v_s/R = v_s/4, \tag{6.66a}$$

$$\rho_2 = R\rho_1 = 4\rho_1, \tag{6.66b}$$

$$P_2 = 3/4\rho_1 v_2^2. \tag{6.66c}$$

As a consequence of this model, no matter how strong a shock is, it can compress a mono-atomic gas only by a factor of four, and the velocity of the down-stream material after the transit of the shock front is $v_s/4$.

The ratio of the up- and down-stream velocities is independent of the particle nature (charge, mass,...) and valid for any strong shock in different astrophysical environments. This fact has the consequence that the same energy spectral index is obtained for any kind of accelerated particles.

6.6 Success and Limits of the Standard Model of Cosmic Ray Acceleration

The arguments presented in the previous sections on acceleration mechanisms have a number of key features that show why the diffusive shock acceleration is considered the most reliable among all those that try to explain the origin of galactic CRs (see Drury (2012) for a review). First, it is very natural and depends only on rather robust and simple physics arguments. Second, it produces power-law spectra without any unnatural fine-tuning; the exponent of the power-law is fixed entirely by the compression ratio $R = 4$ of the shock and the simple theory predicts a universal energy spectrum at relativistic energies of the form $\Phi(E) \sim E^{-2}$, close to that which is inferred for the cosmic ray source spectra (Sect. 5.6). Third, it does not require a separate pre-acceleration phase to produce seed particles for further acceleration; the process appears capable of accelerating particles directly from the thermal population up to the highest energies allowed by the "scale" of the shock.

The "scale" of the shock refers to a number of important constraints on the acceleration process that act to limit the maximum attainable energy. The first and simplest is that in any accelerator where particles are magnetically confined while being accelerated, the Larmor radius r_L of the particles has to be smaller than the size of the system. In the diffusive shock acceleration, this corresponds to the requirement that the diffusion length-scale of the particles is small compared to the shock radius. In addition, there are limitations from the finite age of the system, as worked out in Sect. 6.3.2. If we take fairly standard values for a SNR shock, the maximum particle rigidity reaches about 10^{14} V.

There exist alternative models to that presented in Sect. 6.4 concerning the origin of the knee. Some estimates (Hörandel 2007) give a maximum energy up to one order of magnitude larger than obtained in the previous section for some types of supernova (Berezhko 1996; Kobayakawa et al. 2002; Sveshnikova et al. 2003). According to these models, the effective magnetic field strength at the shock may be substantially larger than the standard values, and in this way, particles in galactic shocks can be accelerated up to substantially higher energies [models with a *shock-amplified magnetic field* (Drury 2012)). In this case, following Eq. (6.43), the maximum particle rigidity will be a factor of $\sim$10 higher than the above value. Also, the exponent of the differential energy spectrum could be significantly smaller than the value in (6.54), for instance, $\alpha_S = 1.5$ or 1.8. At an early stage of the SNR when the power of shocks is maximum and there is still significant field amplification, the maximum particle rigidity may extend up to 10^{17} eV. These shocks are, however, unable to tap the full power of the explosion, because the ejecta have only interacted with a small amount of matter surrounding the star. As the shocks continue to expand and slow down, the maximum rigidity drops as the field amplification becomes less and less effective. According to this picture, the steepening of the all-particle energy spectrum at the knee is due to the relative lack of power in the very fast early shocks responsible for the highest energy particles combined with the decrease in abundance as one moves to heavier elements.

In any case, it is worthwhile to remember that there are experimental uncertainties about the change of the spectral index in the knee region for different primary mass groups, as discussed in Sect. 4.11. Only further experimental efforts will help to clarify the details of CR acceleration models.

In the region of direct measurements, all nuclei up to uranium have been detected in CRs in proportions that are generally close to what one would obtain from a well-mixed sample of the local Galaxy (Sect. 3.6). This observation rules out some exotic models for CR origin, such as, for instance, the iron-nickel rich composition one would expect from a neutron star crust.

In general, direct production exclusively in one particular subclass of supernova seems to be ruled out. The CR composition, like the general composition of the Galaxy, requires the mixing of a variety of elements produced by stellar nucleosynthesis at different sites. There are, however, very interesting recent measurements (Sect. 3.8) suggesting that there are slight, but significant, differences between the proton and helium energy spectra, and that heavier nuclei behave like helium. Models that are more refined should consider the slight energy dependence of the composition. One of the suggested explanations for the difference between the spectral index of proton and helium nuclei is that they could be accelerated from different types of source or acceleration site. For example, most protons might come predominantly from the supernova explosion of low mass stars directly into the interstellar medium. Helium and heavier nuclei might come mainly from the explosion of more massive stars into the atmosphere swept out by the progenitor star, rather than directly into the general interstellar medium (Biermann et al. 2010). Another reason could be the randomness in the spatial and temporal distribution of SNRs (Blasi and Amato 2012).

In addition to those induced by the fragmentation during propagation, there are other significant differences in the chemical composition of CRs and of the cosmic abundances. Some significant differences from Solar System abundances have been reported for elements heavier than Fe by the CRIS experiment, as well as from the Trans-Iron Galactic Element Recorder (TIGER) balloon instrument (Rauch et al. 2009), see Table 3.3. These observations add support to the concept that the bulk of CR acceleration takes place in supernovae. Some of the parent stars of the supernova (see Sect. 12.14) are massive Wolf-Rayet stars, which form in loosely organized groups called *OB associations*. The abundances observed are consistent with a CR source mixture of about 20% ejecta of massive stars mixed with 80% material of solar system composition. According to some authors, the enhancements of these elements appear to correlate with their first ionization potential (FIP).[5] Since a similar effect is known to operate on high-energy particles originated in the Sun, this effect on CRs was long interpreted as a FIP-based bias. However, no satisfactory physical model for the FIP-effect exists for cosmic rays up to now, and thus the result should be seen as an empirical correlation (Webber 1997).

Cosmic rays of energies well above the knee have, in any case, been observed. The onset of extragalactic sources is expected at energies above the ankle. The continuously increasing quality of observations reveals new astrophysical aspects that force us to improve the quality of the effective models that we adopt to describe them: we invite the reader to refer to the literature for the latest evolutions, such as Aloisio et al. (2018).

In the next part of the chapter, we describe some possible galactic candidates that, in addition or as an alternative to the models characterized by a shock-amplified magnetic field, could provide CRs in the energy region between the knee and the ankle.

6.7 White Dwarfs, Neutron Stars and Pulsars

In order to overcome the rigidity-dependent limit (6.43), the possibility has been proposed that, under particular conditions, cosmic rays suffer additional acceleration by variable magnetic fields in the acceleration region. According to this mechanism, a neutron star (the remnant of the final stages of a massive star) or a powerful binary system can accelerate CRs up to a maximum energy of 10^{19} eV.

[5]The first ionization energy is the amount of energy it takes to detach one electron from a neutral atom.

We start with the description of *white dwarfs* (although they are not involved in cosmic ray acceleration), as most of the physics needed to describe neutron stars can more be easily understood. Stars with mass a few times that of our Sun will, at the end of their life, become white dwarfs. A small fraction of stars in our Galaxy (only $\sim$3%) are massive enough to become neutron stars.

6.7.1 White Dwarfs

A white dwarf is a small star made of electron-degenerate matter in which the gravitational pressure is counterbalanced by the pressure resulting from the electron degeneracy. The material in a white dwarf no longer undergoes fusion reactions, so the star has no source of energy, nor is it supported by the heat generated by fusion against gravitational collapse.

Let us evaluate the gravitational potential energy of a sphere having mass M and radius R. For our calculation, it is enough to remind ourselves that the gravitational potential energy of a system consisting of a sphere having radius $r < R$ and a thin spherical layer surrounding it having thickness dr, both of density ρ, is given by $dU_{\text{grav}} = GMdM/r = G_N\rho^2\frac{16}{3}\pi^2 r^4 dr$. Integrating this expression over r, between $r = 0$ and $r = R$, we obtain

$$|U_{\text{grav}}| = \frac{3}{5}\frac{G_N}{R}M^2 \,. \tag{6.67}$$

During the time a star is in the *main sequence*, thermonuclear reactions provide energy against the gravitational collapse. A white dwarf is supported only by electron degeneracy pressure, causing it to be extremely dense.

The Pauli exclusion principle disallows fermions from occupying the same quantum state. If one has a potential well (such as the one that, in first approximation, holds particles to form the star), fermions start filling up the quantum levels. The highest energy level filled by nuclei or electrons is called the *Fermi energy*. Degenerate states occur when the Fermi energy is larger than the typical thermal energy.

Let us consider first the case in which the degeneracy in a star is due to the atomic electrons. The degeneracy energy can be estimated using Heisenberg's uncertainty principle

$$\Delta x \cdot \Delta p \sim d \cdot p \sim \hbar \,.$$

If the number density is n [cm^{-3}], then each fermion is essentially confined to a cube of size $d^3 \sim 1/n$. This implies that the momentum of an electron is $p \sim \hbar/d \sim \hbar n^{1/3}$. The energy of an electron is therefore

$$E \sim \frac{p^2}{2m_e} \sim \frac{\hbar^2 n^{2/3}}{2m_e} \quad \text{in the non-relativistic case,} \tag{6.68a}$$

$$E \sim pc \sim \hbar n^{1/3} c \quad \text{in the relativistic case.} \tag{6.68b}$$

If the collapsing star has N_e electrons, the total degeneracy energy of the star is $E_{\text{deg}} = EN_e$.

Let us consider only the relativistic case (6.68b). This limit is reached when the relative distance $\Delta x = d$ between electrons decreases and the energy increases up to the value $pc = m_e c^2$. Thus, from Heisenberg's principle, one can write

$$d(pc) \sim \hbar c \quad \rightarrow d \sim \frac{\hbar}{m_e c}$$

$$\rho_{ec} = \frac{\mu}{d^3} = \frac{\mu}{(\hbar/m_e c)^3} \sim 3 \times 10^{13} \text{ g cm}^{-3} , \tag{6.69}$$

where ρ_{ec} is the *critical density of matter* for degenerate atomic electrons, and $\mu = m_p + m_e$. When the density ρ of the collapsing star is below ρ_{ec}, we are below the relativistic limit, and (6.68a) holds.

The total number of electrons in a star with mass M_* and radius R_* is

$$N_e = M_* Z/A\mu = M_* \zeta/\mu , \tag{6.70}$$

where A is the atomic mass, Z the number of electrons per atom and $\zeta = Z/A$. The total degeneracy energy of the star in the relativistic case (when the density reaches ρ_{ec}) is obtained using (6.68b):

$$E_{\text{deg}} = EN_e = \hbar n^{1/3} c N_e \sim \frac{\hbar c M_*^{4/3} \zeta^{4/3}}{R_* m_p^{4/3}} , \tag{6.71}$$

having used $n = N_e/V \sim N_e/R_*^3$.

Finally, we can use energy equipartition to estimate the equilibrium. This is done assuming that the gravitational binding energy $|U|$ (6.67) is on the order of the degeneracy energy. Hence

$$E_{\text{deg}} \sim |U_{\text{grav}}| \quad \longrightarrow \quad \frac{\hbar c M_*^{4/3} \zeta^{4/3}}{R_* \mu^{4/3}} \sim \frac{3}{5} \frac{G_N M_*^2}{R_*} . \tag{6.72}$$

Note that *the radius of the star cancels out!* As the mass M_* increases, the radius decreases, and once white dwarfs become compact enough for the electrons to be relativistic, there is a solution with only one mass (which we indicate as the Chandrasekhar mass limit), irrespective of the radius.[6]

---

[6]We leave for the student to work out the radius-mass relation for the nonrelativistic case. When the density in a white dwarf is below ρ_{ec}, as its mass increases, its radius becomes smaller and

The final (universal) value of the mass from (6.72) is (neglecting the 3/5 factor, as was done for the factor $4\pi/3$ in the expression of the volume)

$$M_* \sim \left(\frac{\hbar c \zeta^{4/3}}{G_N \mu^{4/3}}\right)^{3/2} = \left(\frac{\hbar c \zeta^{4/3}}{G_N \mu^2}\right)^{3/2} \mu . \tag{6.73}$$

In the last equality, as M_* and μ have the same dimension (mass), the quantity in the bracket is dimensionless. If we consider hydrogen, $\mu \sim m_p$ and $\zeta = 1$. The term $(\frac{\hbar c}{G_N m_p^2})$ (Braibant et al. 2011) corresponds to the inverse of the gravitational coupling constant, α_G. Thus, Eq. (6.73) can be written as

$$M_{\text{Ch}} \sim \frac{m_p}{\alpha_G^{3/2}} = \frac{1.6 \times 10^{-24}\ \text{g}}{(5.9 \times 10^{-39})^{3/2}} = 3.5 \times 10^{33}\ \text{g} = 1.4 M_\odot . \tag{6.74}$$

Equation (6.73) shows that $\hbar$ appears in the Chandrasekhar mass limit M_{Ch}: the Planck constant not only determines the interaction of elementary particles, but also the mass scale and the inner structure of stars. The evolution and structure of cosmic objects is determined by known laws of physics and by the values of fundamental constants.

The typical radius of a white dwarf can be worked out from the ratio between the mass M_{Ch} and its density ρ_{ec}. Note that in (6.69), the quantity $\hbar/m_e c$ corresponds to the electron Compton wavelength $\lambda\!\!\!^{-}_e = 3.8 \times 10^{-13}$ m, and $\rho_{ec} = m_p/\lambda\!\!\!^{-3}_e$. Thus, one can write

$$R_{\text{WD}} \sim \left(\frac{M_{\text{Ch}}}{\rho_{ec}}\right)^{1/3} = \left(\frac{m_p \lambda\!\!\!^{-3}_e}{\alpha_G^{3/2} m_p}\right)^{1/3} = \frac{\lambda\!\!\!^{-}_e}{\alpha_G^{1/2}} = \frac{3.8 \times 10^{-13}}{(5.9 \times 10^{-39})^{1/2}} \simeq 5 \times 10^6\ \text{m} . \tag{6.75}$$

The radius of a white dwarf is a few thousand kilometers, and depends on the electron Compton wavelength.

6.7.2 Neutron Stars

A neutron star is a type of stellar remnant that can result from the gravitational collapse of a massive star ($M > 8M_\odot$). As the core of a massive star is compressed during a supernova event, increases in the electron Fermi energy allow the reaction (with threshold 1.36 MeV):

$$e^- + p \rightarrow n + \nu_e . \tag{6.76}$$

---

smaller, scaling as $M_*^{-1/3}$. As the white dwarf approaches the mass limit M_{Ch}, the electrons become relativistic, and the dependence on mass becomes sharper than $-1/3$ as $M_* \rightarrow M_{\text{Ch}}$.

The weak interacting neutrinos escape, the matter cools down, the density increases and nuclei in the center of the star become neutron-enriched. At some point, the nuclei break into their components, and enough neutrons are created so that they become degenerate. The neutron degeneracy pressure immediately stops the collapse and an equilibrium state is established. The transition from collapse to equilibrium is very sudden, and the in-falling material experiences a *bounce* against the degenerate core, which creates an outward-propagating shock wave (the *supernova*). The shock wave is further boosted by the neutrino pressure from the core (Herant et al. 1997), as more extensively described in Sect. 12.13.

From the mathematical point of view, the description of the equilibrium state in a neutron star is similar to that provided above in the case of the white dwarf. The degenerated fermions are now the neutrons, and the critical density is obtained from (6.69) with the exchange $m_e \to m_n \sim \mu$, where m_n is the neutron mass. Thus, neutrons become relativistic when the density reaches the value

$$\rho_{nC} = \frac{m_n}{(\hbar/m_n c)^3} = \frac{m_n}{\lambda\!\!\!^{-}_n{}^3} \sim 10^{14} \text{ g cm}^{-3} . \tag{6.77}$$

The total degeneracy energy has the same value as that given in Eq. (6.71)! The only change is the replacement of the number of electrons $N_e = M_* \zeta/\mu$ with the number of neutrons: $N_n = M_*/m_n$, which corresponds to $\zeta = 1$. In this condition, the same equality (6.72) between the degeneracy energy and binding energy $|U_{\text{grav}}|$ holds, and the mass limit for a neutron star assumes the same value of the Chandrasekhar mass limit (6.74). The upper limit for a neutron star mass is the same as that of a white dwarf. As $\rho_{nC} \sim 10^3 \rho_{eC}$, the neutron star radius is much smaller than R_{WD}, and from (6.75),

$$R_{\text{NS}} \sim \left(\frac{M_{\text{Ch}}}{\rho_{nC}}\right)^{1/3} = \left(\frac{m_p \lambda\!\!\!^{-}_n{}^3}{\alpha_G^{3/2} m_p}\right)^{1/3} = \frac{\lambda\!\!\!^{-}_n}{\alpha_G^{1/2}} = \frac{2.1 \times 10^{-16}}{(5.9 \times 10^{-39})^{1/2}} = 3 \times 10^3 \text{ m} . \tag{6.78}$$

The radius of a neutron star, an object with a mass $M_{\text{NS}} \sim 1.4 M_\odot$, is $\sim$ a few km. If the remnant star has a mass greater than $\sim 3M_\odot$, it continues collapsing to form a black hole, Sect. 6.8. The maximum observed mass of neutron stars is $\sim 2.0 M_\odot$.

The estimated number of neutron stars in our Galaxy is $\mathcal{O}(10^8)$, a figure obtained from the number of stars that have undergone supernova explosions. Most of them are old and cold, and neutron stars can only be easily detected if they are young, rotating systems (in this case, they are usually referred to as *pulsars*, Sect. 6.7.3). Presently, eight binary neutron star systems are known in our Galaxy, including the Hulse-Taylor binary that provided the first indirect evidence for the existence of gravitational waves, Sect. 13.7.1. One extragalactic neutron star binary system was (before the merger) that which produced the gravitational wave and the gamma-ray burst observed in August 2017. The source was located in the galaxy NGC 4993 at a distance of 40 Mpc from Earth. This important event is discussed in detail in Sect. 13.7.1.

As a first approximation, neutron stars are composed entirely of neutrons; the electrons and protons present in normal matter have combined in the collapsing phase to produce neutrons through (6.76). However, current models indicate a possible onion-like structure. The surface of a neutron star should be composed of ordinary atomic nuclei crushed into a solid lattice, together with a plasma of electrons. Due to their high binding energy per nucleon, iron nuclei could be predominant at the surface, or immediately under the surface made of lighter nuclei. Proceeding inward, nuclei with ever-increasing numbers of neutrons would be present; such nuclei, if free, would decay quickly, but they are kept stable by enormous pressures. Then, the concentration of free neutrons increases rapidly up to the core. The equation of state for a neutron star is still not known; in particular, we do not know if the presence in the core of exotic forms of matter is allowed. These forms include degenerate strange matter (containing strange quarks in addition to up and down quarks), matter containing high-energy pions and kaons in addition to neutrons, or ultra-dense quark-degenerate matter.

6.7.3 Pulsars

A *pulsar* is a rotating neutron star that emits a beam of electromagnetic radiation, typically along its magnetic axis. They were discovered in the radio band by Hewish and Bell in 1967, and soon identified with isolated, rotating, magnetized neutron stars. The key observations were the very stable, short periods of the pulses and the observation of polarized radio emission. The radiation can only be observed when the axis is pointing towards the Earth. More than 2600 pulsars are known and present in the catalogue.[7] The rotation period, and thus the interval between observed pulses, is very regular, and the periods of their pulses range from 1.4 ms to 8.7 s. This rotation slows down over time as electromagnetic radiation is emitted.

The millisecond-rotating period for young pulsars can be estimated using basic physics arguments. A star like our Sun has a radius $R \sim 7 \times 10^5$ km and rotates at one revolution per 30 days, so that the angular velocity is $\omega \sim 2.5 \times 10^{-6}$ rad/s. After the collapse, the neutron star has a radius $R_{\rm NS} \sim 10$ km. From angular momentum conservation, one can write

$$MR^2\omega = MR_{\rm NS}^2\omega_{\rm NS} \,,$$

$$\omega_{\rm NS} = \left(\frac{R}{R_{\rm NS}}\right)^2 \times \omega = \left(\frac{7 \times 10^5}{10}\right)^2 \times 2.5 \times 10^{-6} = 12,500 \text{ rad/s} \,, \qquad (6.79)$$

so that $T_{\rm NS} = \frac{2\pi}{\omega_{\rm NS}} \sim 10^{-3}$ s.

[7] Available at http://www.atnf.csiro.au/people/pulsar/psrcat/.

The gravitational collapse amplifies the stellar magnetic field. As a result, the magnetic field B_{NS} near the NS surface is extremely high. To obtain an estimate of its magnitude, let us use the *conservation of the magnetic flux* during the contraction:

$$\oint \mathbf{B}_{star} \cdot d\mathbf{A}_{star} = \oint \mathbf{B}_{NS} \cdot d\mathbf{A}_{NS} , \tag{6.80}$$

and, assuming the magnetic field constant and the elements of surface $\mathbf{A}_{NS}, \mathbf{A}_{star}$ that of two spheres of radius R_{NS} and R, respectively, we obtain

$$4\pi B_{star} R^2 = 4\pi B_{NS} R_{NS}^2 \quad \rightarrow \quad B_{NS} = B_{star} \frac{R^2}{R_{NS}^2} . \tag{6.81}$$

For typical values $B_{star} = 1000\,\text{G}$, the magnetic fields on the NS surface become on the order of $\sim 10^{12}$ Gauss. This expectation has been experimentally confirmed by measuring quantized energy levels of free electrons in pulsar strong magnetic fields. Figure 6.8 shows a sketch of the rotating magnetosphere inside the *light cylinder* centered on the pulsar and aligned with the rotation axis. The radius of the cylinder corresponds to the distance at which the co-rotating speed equals the speed of light.

The pulsar loses energy by electromagnetic radiation, which is extracted from the rotational energy of the neutron star. The pulse periods of pulsars P, as well as the rate at which the pulse period changes with time, dP/dt, can be measured with very

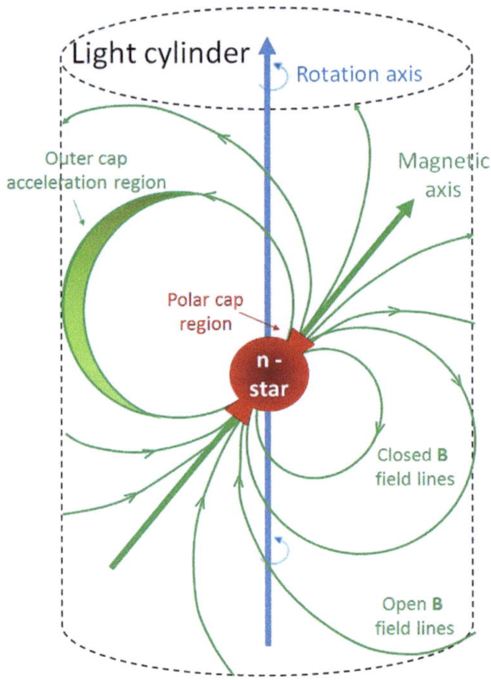

Fig. 6.8 A schematic model of a pulsar as a magnetized rotating neutron star in which the magnetic and rotation axes are misaligned. The radio pulses are assumed to be due to beams of radio emission from the poles of the magnetic field distribution

high accuracy. Normal radio pulsars are slowing down and $\mathrm{d}P/\mathrm{d}t$ is proportional to the rate of loss of rotational energy mainly due to *magnetic braking*. If the rotational energy of the pulsar is $E_{\mathrm{rot}} = I\omega_{\mathrm{NS}}^2$, where $I \propto MR_{\mathrm{NS}}^2$ is the inertial momentum, the rotational energy loss is

$$-\frac{\mathrm{d}E_{\mathrm{rot}}}{\mathrm{d}t} = 2I\omega_{\mathrm{NS}}\frac{\mathrm{d}\omega_{\mathrm{NS}}}{\mathrm{d}t} = 8\pi^2 I\frac{\mathrm{d}P/\mathrm{d}t}{P^3} . \tag{6.82}$$

The typical lifetime for normal pulsars can be estimated as

$$\tau_P = \frac{P}{2(\mathrm{d}P/\mathrm{d}t)} \tag{6.83}$$

and typical values are $\tau_P \sim 10^5$–10^8 years.

The pulsed radiation observed at a large distance is a result of the fact that the pulsar's magnetic dipole is oriented at an angle with respect to the rotation axis. The detected beamed radiation is associated with the beam sweeping the line of sight to the observer. Pulsars are known to emit radiation at all wavelengths (Lorimer and Kramer 2005). Because the neutron star is a spinning magnetic dipole, it acts as a *unipolar generator*, in which charged particles are subject to the total Lorentz force $\mathbf{F} = q(\mathbf{E} + \frac{1}{c}\,\mathbf{v} \times \mathbf{B})$. Electric charges in the magnetic equatorial region redistribute themselves by moving along closed field lines until they build up an electrostatic field large enough to cancel the magnetic force and give $\mathbf{F} = 0$ in the so-called *vacuum gaps* regions. This induces high voltages in which electrons and protons can be accelerated up to very high energies, Sect. 6.9. Vacuum regions occur (Fig. 6.8) at the polar cap, very close to the neutron star surface, in a thin layer along the boundary of the closed magnetosphere (slot gap) and in the outer region close to the light cylinder (outer gap).

If the co-rotating field lines emerging from the vacuum gaps cross the light cylinder, these field lines cannot close. Charged particles in the polar cap are magnetically accelerated to very high energies along the open but curved field lines (Aliu et al. 2008). The acceleration resulting from the curvature causes them to emit synchrotron radiation that is strongly polarized in the plane of curvature. Photons can reach energies up to the TeV region by the mechanisms described in Sect. 9.4.

6.8 Stellar Mass Black Holes

Another possible final state of the evolution of a massive star is a *stellar black hole* (or stellar-mass black hole). A black hole is a massive object exhibiting such strong gravitational effects that nothing (particles and electromagnetic radiation) can escape from inside its boundary, called the *event horizon*. In most cases, we can consider the event horizon equivalent to the *Schwarzschild radius*. This is correct for non-rotating massive objects that fit inside this radius.

The escape velocity, v_{esc}, from a body of mass m at a distance r from the center (assuming that $r \geq R_b$, with R_b being the radius of the spherical body) is $v_{esc} = \sqrt{2Gm/r}$. The Schwarzschild radius, $\mathscr{R}$, is defined as the dimension of an object of mass m such that $v_{esc} = c$. Using the above relation, we obtain:

$$\mathscr{R} = \frac{2Gm}{c^2} = 2.95\left(\frac{m}{M_\odot}\right) \text{ km} , \tag{6.84}$$

a quantity that scales linearly with the object mass. If the body is sufficiently dense and confined within $\mathscr{R}$, the Schwarzschild radius represents its event horizon and its inner region behaves like a black hole. Particles and light can escape the black hole only if they remain outside the event horizon.

By absorbing other stars and merging with other black holes, supermassive black holes of millions of solar masses may form. It is likely that supermassive black holes exist in the centers of most galaxies. Some particular galaxies show, in their nucleus, a particular activity, probably triggered by the presence of a hypermassive (billions of solar masses) black hole. These active galaxies, candidates to be at the origin of the highest energy CRs, will be discussed in Sect. 9.10.

Although Eq. (6.84) is obtained from Newtonian considerations, the same conclusion emerges from general relativity. Furthermore, in classical general relativity, a particle that is inside the event horizon can never emerge outside. More generally, black holes are particular solutions to the Einstein field equations. It has been demonstrated (by the so-called *no-hair theorem*) that a stable black hole is completely described at any time by the following quantities:

- its mass-energy, M;
- its angular momentum, or spin, $\mathbf{S}$ (three components);
- its total electric charge, Q.

In terms of these properties, four types of black holes can be defined: Uncharged non-rotating black holes (also called *Schwarzschild black holes*) and rotating black holes (called *Kerr black holes*). Then, there should also be charged non-rotating and rotating black holes. A rotating black hole is formed in the gravitational collapse of a massive spinning star or from the collapse of a collection of stars or gas with a total non-zero angular momentum. A rotating black hole can lose rotational energy through different mechanisms occurring just outside its event horizon. In that case, it gradually reduces to a Schwarzschild black hole, the minimum configuration from which no further energy can be extracted.

Although black holes do not emit radiation, under particular conditions, we can discover their presence. There is clear evidence of a supermassive black hole in our Galaxy, provided by the proper motions of stars near the center of our own Milky Way. The number of stellar black holes in our Galaxy is unknown, but there are several stellar-mass black hole candidates. Stellar black holes in close binary systems are observable when matter is transferred from a companion star to the black hole. The energy release by matter falling toward the compact object is so large that it radiates in X-rays. The black hole therefore is observable in X-rays, whereas the companion star can be observed with optical telescopes. Until 2016, there were $\sim$20 stellar black holes indirectly detected in such a way. The largest of them was $\sim 15 M_\odot$; the more likely mass was 5–10 $M_\odot$.

On September 14th, 2015, the LIGO gravitational wave observatory made the first-ever successful observation of gravitational waves, Chap. 13. The observed signal, denoted as GW150914, was consistent with theoretical predictions for the gravitational radiation produced by the merger of two black holes, the primary with $\sim 36 M_\odot$, and the secondary with $\sim 29 M_\odot$. This observation provides the most concrete evidence for the existence of stellar black holes to date.

GW150914 is not the only binary black hole merger observed by gravitational wave observatories. As of this writing, after the run ended in August 2017, five further merging black hole events (one of them with lower-significance) have been reported. The key parameters of the six observed events (and 18 black holes, including the ones formed in the final state after merging) are reported in Table 13.2; the consequences for the astrophysics of stellar black holes connected with the new gravitational wave observations are discussed in Sect. 13.6. This is one of the examples of the importance in astrophysics of the information arising from the new observational probes.

6.9 Possible Galactic Sources of Cosmic Rays Above the Knee

Let us use dimensional arguments to obtain the maximum attainable energy of a particle near an astrophysical object with a strong, rotating magnetic field. Small decreases in the energy associated with the magnetic field power the detected emission of high-energy electromagnetic radiation, particularly X-rays and γ-rays.

Let us start with some energy considerations for the cosmic ray spectrum, beginning with energies around the knee, $E_0 \sim 10^{14}$ eV = 100 TeV, and below the ankle, repeating those discussed in Sect. 2.11 for the energy region below the knee. Because of the steeply falling primary CR spectrum, the power needed to maintain a stationary presence of CRs above E_0 in our galaxy

$$P(> E_0) = \frac{\rho_{CR}(> E_0) \cdot V_G}{\tau_{esc}(> E_0)} \tag{6.85}$$

is much smaller than the total power requirement (2.39). Here, $\rho_{CR}(> E_0)$ is the energy density of CRs of energy above E_0.

The quantity $\rho_{CR}(> E_0)$ is determined from the indirect measurement of the CR spectrum, using the integral spectrum above 10^{16} eV given by Eq. (4.53). The number density is, as usual, given by $(4\pi/c)\Phi(> E)$ and the energy density as $(4\pi/c)\int \Phi(> E)\mathrm{d}E$. Note that at 100 TeV, the number density is about ten orders of magnitude smaller than (2.32a) for the whole range of energies above 3 GeV.

The value of the escape time $\tau_{esc}(> E_0)$ in (6.85) is much more difficult to estimate. We follow the arguments discussed in Gaisser (1991). Simple extrapolations of the escape time obtained using the direct measurements reported in Sect. 5.5 are not valid, as the direct measurements only reach $\sim$ 1 TeV. The simple $E^{-0.6}$ power-law does not fit well with the data in the high energy tail (Fig. 5.4). For instance, extrapolating this dependence of τ_{esc} up to 100 TeV, we would obtain $\tau_{esc}(> 100 \text{ TeV}) \sim 1000$ y. With such a small value, we expect a huge migration of $>$ 100 TeV cosmic rays out of the galactic disk. This will correspond to a large anisotropy of these CRs: the number of particles arriving from the galactic plane should be much larger than that of those arriving from outside of it. Such a strong anisotropy has not been observed by experiments, Sect. 5.7.

To obtain an order of magnitude estimate, from (5.47) and (5.49) we have

$$\tau \simeq T = \frac{L}{V} = \frac{L}{\Delta \cdot c/(\alpha + 2)} \simeq 10^6 \text{ year} , \tag{6.86}$$

having used $\Delta \simeq 10^{-3}$ for $E_0 = 10^{15}$ eV $= 10^3$ TeV $= 1$ PeV (below the upper limit in Fig. 5.5) and for the galactic half thickness $L \simeq 100\,\text{pc} = 0.3 \times 10^{21}$ cm. Estimates of the escape times as a function of the threshold energy E_0,

$$\tau \sim 2.5 \times 10^5 \cdot \left(\frac{E}{1 \text{ TeV}}\right)^{-0.13} \text{y} \qquad \text{for } 1 < E < 5 \times 10^3 \text{ TeV} , \tag{6.87a}$$

$$\sim 0.8 \times 10^5 \cdot \left(\frac{E}{5 \text{ PeV}}\right)^{-0.53} \text{y} \qquad \text{for } E > 5 \text{ PeV} , \tag{6.87b}$$

can be found in Gaisser (1991). Now, using (4.52) and (6.87) in (6.85), with the galactic volume ($V_G \sim 6 \times 10^{66}$ cm^3), we obtain

$$P(> E_0) \simeq 2 \times 10^{39} \text{erg/s for } E > 100 \text{ TeV} , \tag{6.88a}$$

$$\simeq 2 \times 10^{38} \text{erg/s for } E > 1 \text{ PeV} , \tag{6.88b}$$

$$\simeq 5 \times 10^{37} \text{erg/s for } E > 10 \text{ PeV} . \tag{6.88c}$$

The power needed to accelerate CRs above the knee is three orders of magnitude smaller than that required for the whole CR spectrum. Even one or two powerful galactic point sources could be important. Do sources exist in the Galaxy that can deliver such a quantity of energy per second?

6.9.1 A Simple Model Involving Pulsars

The rotation axis of pulsars usually does not coincide with the direction of the magnetic field. As an effect, the vector of these high magnetic fields spinning around the nonaligned axis of rotation will produce strong electric fields $\mathscr{E}$ through Faraday's law (6.1). This may, in turn, accelerate particles. From dimensional arguments, if $L \sim R_{NS}$ is the length of the region over which the magnetic field changes

$$\frac{\mathscr{E}}{L} = \frac{1}{c}\frac{dB}{dt} , \tag{6.89}$$

the maximum energy E_{max} gained from a particle over the length $L \sim R_{NS}$ is:

$$E^{max} = \int Ze\mathscr{E}dx = \int Ze\frac{L}{c}\frac{dB}{dt}dx = \int Ze\frac{L}{c}dB\frac{dx}{dt} = ZeR_{NS}B\frac{\omega_{NS}R_{NS}}{c} , \tag{6.90}$$

where ω_{NS} is the pulsar angular velocity.

As derived from the conservation of the flux of the magnetic field (6.81), $B_{NS} \sim 10^{11}$–10^{12} Gauss on the surface of a neutron star. By inserting the numerical values in (6.90), for protons near a neutron star ($R_{NS} = 10\,\text{km}$), one obtains (in Gaussian unit system, see Sect. 2.12):

$$\begin{aligned} E^{max} = ZeR_{NS}B\frac{\omega_{NS}R_{NS}}{c} &= 4.8 \times 10^{-10}\ [\text{statC}] \times 10^{6}\ [\text{cm}] \times 10^{11}\ [\text{Gauss}] \times 0.1 \\ &\simeq 5 \times 10^{6}\ \text{erg} \simeq 3 \times 10^{18}\ \text{eV} , \end{aligned}$$

where the conversion factor $1\,\text{e V} = 1.6 \times 10^{-12}$ erg has been used in the last equality. For the angular velocity, we used the value $\omega_{NS} \simeq 60\pi\ \text{s}^{-1}$, which holds for the Crab Nebula, Sect. 9.5. In this case, $\frac{\omega_{NS}R_{NS}}{c} \simeq 0.1$. Thus, some galactic accelerators could explain the presence of cosmic rays with energies up to a few 10^{18} eV. The validity of Eq. (6.90) will be made more general in Sect. 7.4.

Pulsars possess a rotational energy of about

$$E_{rot} = \frac{1}{2}I\omega_{NS}^2 = \frac{1}{5}M_{NS}R_{NS}^2\omega_{NS} \sim 10^{61\div 62}\,\text{eV} ,$$

using the values given above for $M_{NS}, R_{NS}, \omega_{NS}$. A very small fraction of this energy is sufficient to maintain the CR flux above E_0 from just one single object in the Galaxy (Berezinsky et al. 1990). However, the theoretical details of the mechanism that should work inside a pulsar, allowing it to transform a small fraction of its rotational and magnetic field energy into particle acceleration, is not known.

6.9.2 A Simple Model Involving Binary Systems

Binary systems are very often found in astrophysical environments. As candidate sites of CR acceleration, we are interested in particular binary systems consisting of a compact object (a pulsar, a neutron star or a black hole) and a normal star. Sources

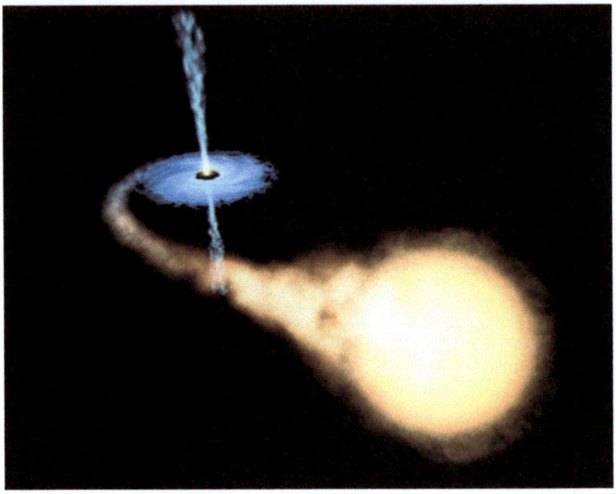

Fig. 6.9 Artist's impression of the microquasar GRO J1655-40. This microquasar (in *blue*) is the second discovered in our Galaxy. Microquasars are *black holes* of about the same mass as a star. They behave like scaled-down versions of much more massive *black holes* that are at the cores of extremely active galaxies, called *quasars*. Different microquasars have been found with masses ranging from 3.5 to approximately 15 times the mass of our Sun. The companion star had apparently survived the original supernova explosion that created the *black hole*. Credits: NASA/ESA

of electromagnetic radiation (electrons) and probably also high energy hadrons are produced by matter falling from one component, called the "donor" (usually a relatively normal star), to the other compact component, called the "accretor". The in-falling matter releases potential gravitational energy. Due to these enormous motions of ionized matter, very strong electromagnetic fields are produced in the vicinity of the compact object, and charged particles can be accelerated to high energies.

A particular class of binary system is the microquasars. Microquasars are galactic X-ray binary systems, which exhibit relativistic radio jets, observed in the radio band. The name is due to the fact that they turn out to be morphologically similar to the AGN, Sect. 9.10, since the presence of jets makes them similar to small quasars (Fig. 6.9). In quasars, the accretion object is a supermassive (millions of solar masses) black hole; in microquasars, the mass of the compact object is only a few solar masses. This resemblance could be more than morphological: the physical processes that govern the formation of the accretion disk and the plasma ejection in microquasars are probably the same ones as in large AGN. Microquasars have been proposed as galactic acceleration sites of charged particles up to $E \sim 10^{16\div18}$ eV. This hypothesis was strengthened by the discovery of the presence of relativistic nuclei in microquasars jets, like those of SS 433. This was inferred from the observation of iron X-ray lines. A part of the radio emission comes from relativistic jets, often showing apparent superluminal motion. Two microquasars, LS I +61 303 and LS 5039, have been detected as γ-ray sources above 100 MeV.

Also, in the case of binary systems, we can work out the maximum energy gain by using dimensional arguments. The rotating magnetic field of the neutron star, which is perpendicular to the accretion disk, will produce a strong electric field $\mathscr{E} = (\mathbf{v} \times \mathbf{B})/c$. A particle with charge e, moving with velocity $\mathbf{v}$ in the accretion disk plane, will gain energy during the in-fall towards the compact object. The energy E will be

$$E = \int e\mathscr{E} \cdot d\mathbf{s} \simeq evB\Delta s \,. \tag{6.91}$$

Under plausible assumptions ($v \simeq c$, $B = 10^{10}$ Gauss, $\Delta s = 10^7$ cm), particle energies up to 10^{19} eV are possible. Due to the similarity between microquasars and quasars, we expect larger energies from the accretion disks that form around black holes or the compact nuclei of active galaxies.

References

E. Aliu et al., Observation of pulsed gamma-rays above 25 GeV from the crab pulsar with MAGIC. Science **322**, 1221 (2008)

R. Aloisio, P. Blasi, I. De Mitri, S. Petrera, *Selected Topics in Cosmic Ray Physics* (Springer, Cham, 2018). arXiv:1707.06147

W.I. Axford, E. Leer, G. Skadron, The acceleration of cosmic rays by shock waves, in *Proceedings of the 15th International Cosmic Ray Conference*, vol. 11 (1977), pp. 132–135

A.R. Bell, The acceleration of cosmic rays in shock fronts. Mon. Not. R. Astron. Soc. **182**, 147–156 (1978)

E. Berezhko, Maximum energy of cosmic rays accelerated by supernova shocks. Astropart. Phys. **5**, 367–378 (1996)

V.S. Berezinsky et al., *Astrophysics of Cosmic Rays* (North Holland, Amsterdam, 1990)

P.L. Biermann et al., The origin of cosmic rays: explosions of massive stars with magnetic winds and their supernova mechanism. Astrophys. J. **725**, 184–187 (2010)

R.D. Blandford, J.P. Ostriker, Particle acceleration by astrophysical shocks. Astrophys J. **221**, L29–L32 (1978)

P. Blasi, E. Amato, Diffusive propagation of CRs from supernova remnants in the galaxy. I: spectrum and chemical composition. J. Cosmol. Astropart. Phys. **01**, 010 (2012)

S. Braibant, G. Giacomelli, M. Spurio, *Particle and Fundamental Interactions* (Springer, Dordrecht, 2011). ISBN: 978-9400724631

F.F. Chen, *Introduction to Plasma Physics and Controlled Fusion* (Springer, Basel, 1984). ISBN: 978-1-4419-3201-3

L.C. Drury, An introduction to the theory of diffusive shock acceleration of energetic particles in tenuous plasmas. Rep. Prog. Phys. **46**, 973–1027 (1983)

L.C. Drury, Origin of cosmic rays. Astropart. Phys. **39–40**, 52–60 (2012)

E. Fermi, On the origin of the cosmic radiation. Phys. Rev. **75**, 1169–1174 (1949)

E. Fermi, Galactic magnetic fields and the origin of cosmic radiation. Astrophys. J. **119**, 1–6 (1954)

T.K. Gaisser, *Cosmic Rays and Particle Physics* (Cambridge University Press, Cambridge, 1991)

M. Herant, S.A. Colgate, W. Benz, C. Fryer. Neutrinos and Supernovae. Los Alamos Science, No. 25 (1997). http://la-science.lanl.gov/lascience25.shtml

A.M. Hillas, Can diffusive shock acceleration in supernova remnants account for high-energy galactic cosmic rays? J. Phys. G: Nucl. Part. Phys. **31**, 39 (2005)

J. Hörandel, Cosmic rays from the knee to the second knee: 10^{14} to 10^{18} eV. Modern Phys. Lett. A **22**, 1533–1552 (2007)

J.R. Jokipii, Rate of energy gain and maximum energy in diffusive shock acceleration. Astrophys. J. **313**, 842–846 (1987)

F.C. Jones, D.C. Ellison, The plasma physics of shock acceleration. Space Sci. Rev. **58**, 259 (1991)

K. Kobayakawa et al., Acceleration by oblique shocks at supernova remnants and cosmic ray spectra around the knee region. Phys. Rev. D **66**, 083004 (2002)

G.F. Krymsky, A regular mechanism for the acceleration of charged particles on the front of a shock wave. Dokl. Akad. Nauk SSSR **234**, 1306–1308 (1977)

M.S. Longair, *High Energy Astrophysics*, 3rd edn. (Cambridge University Press, Cambridge, 2011). ISBN: 978-0521756181

D.R. Lorimer, M. Kramer, *Handbook of Pulsar Astronomy* (Cambridge University Press, Cambridge, 2005). ISBN: 978-0521828239

B.F. Rauch et al., Cosmic ray origin in OB associations and preferential acceleration of refractory elements: evidence from abundances of elements $_{26}$Fe through $_{34}$Se. Astrophys J. **697**, 2083–2088 (2009)

L. Sveshnikova et al., The knee in galactic cosmic ray spectrum and variety in supernovae. Astron. Astrophys. **409**, 799–808 (2003)

W.R. Webber, New experimental data and what it tells us about the sources and acceleration of cosmic rays. Space Sci. Rev. **81**, 107 (1997)

Chapter 7
The Extragalactic Sources and Ultra High Energy Cosmic Rays

Abstract Cosmic Rays with energies above 10^{18} eV are denoted as Ultra High Energy Cosmic Rays (UHECRs). Their Larmor radius is so large that their arrival direction could correlate with the position of the source, if their origin is Galactic. The non-observation of anisotropies from the Galactic plane, even at the highest energies, and the evidence for a large scale anisotropy above 8×10^{18} eV pointing away from the Galactic plane, strengthens the conjecture that UHECRs are likely produced in extragalactic objects. The method used to detect UHECRs is based on the use of the Earth's atmosphere as a target. Secondary charged particles are measured by arrays of particle detectors at ground level and the fluorescence light produced by the longitudinal development of the showers by mean of particular optical telescopes. The current generation of UHECR detectors, the Pierre Auger Observatory in Argentina and the Telescope Array experiment in Utah, are hybrids, with both surface detector arrays and fluorescence detectors observing at the same site. The goal of these very large experiments is to measure accurately the flux of UHECRs, to understand their nature and to provide evidence of correlations of arrival directions at the highest energies with the large-scale distribution of matter in the near-by Universe. After an introduction on the large-scale structure of the Universe, all these aspects are covered in this chapter.

In the previous chapters, we derived that, through the diffusive shock acceleration model, about 10% of the energy emitted by galactic supernova explosions can provide the power needed to account for the observed CRs up to $\sim 10^{15}$–10^{16} eV. Under particular conditions, already accelerated particles could gain additional energy through very high electric fields generated by rapidly rotating compact magnetized objects, such as young neutron stars. This represents a possible mechanism for the production of CRs up to the ankle.

CRs with energies above 10^{18} eV will be denoted as Ultra High Energy Cosmic Rays (UHECRs). Their Larmor radius is so large -in particular for proton primaries- that their arrival direction could correlate with the position of the source, if their origin is galactic. As UHECRs are likely produced in extragalactic objects, in Sects. 7.1 and 7.2, we extend our field-of-view outside our Galaxy. An additional reason for

© Springer Nature Switzerland AG 2018

M. Spurio, *Probes of Multimessenger Astrophysics*, Astronomy and Astrophysics Library, https://doi.org/10.1007/978-3-319-96854-4_7

this hypothesis is that no class of galactic source seems to be energetic enough for the production of particles at energies above a few 10^{19} eV. Moreover, the non-observation of anisotropies from the Galactic plane, even at the highest energies, and the evidence for a large scale anisotropy above 8×10^{18} eV pointing away from the Galactic plane, strengthens the extragalactic origin conjecture (Sect. 7.4). The interactions of UHECRs with the cosmic microwave background (CMB) radiation during their propagation and the effect of the extragalactic magnetic fields are discussed in Sect. 7.5.

The detection of UHECRs is based on the use of the Earth's atmosphere as a target. Arrays of particle detectors at ground level reconstruct the energy, direction, and some parameters related to their nuclear mass by observing the density and pattern of the showers. Arrays of optical telescopes detect the longitudinal development of the showers by observing the so-called fluorescence emission from atmospheric nitrogen (Sect. 7.6), deriving an essentially calorimetric measurement of the energy and the longitudinal profiles (Sect. 7.7).

The current generation of UHECR detectors, the Pierre Auger Observatory in Argentina and the Telescope Array experiment in Utah (Sect. 7.8) are hybrids, with both surface detector arrays and fluorescence detectors observing at the same site. Research and development efforts are under way for novel methods of air shower detection and measurement. The goal of these very large experiments is: (1) to measure accurately the flux of UHECRs (Sect. 7.9), in particular, above 3×10^{18} eV, where the hardening of the spectrum is suggestive of an extragalactic component; (2) to understand the nature (protons or heavier nuclei) of UHECRs (Sect. 7.9.2); (3) to provide evidence of correlations of arrival directions at the highest energies with the large-scale distribution of matter in the near-by Universe. The above information are crucial in order to elucidate the physical mechanisms acting at the acceleration sites.

7.1 Hubble's Law and the Cosmic Microwave Background Radiation

Observational cosmology consists of the study of the structure, the evolution and the origin of the Universe through experimental measurements. Up to a few years ago, cosmology was a speculative science based on a very limited number of observations. The dispute between the supporters of the *steady state* and the promoters of *Big Bang* cosmology is an example of the scientific debate before the advent of observational cosmology. A short summary of the Standard Cosmological Model is given in Sect. 14.1, while in this section, we describe two observations related to the cosmology of fundamental importance for the following.

Hubble's Law The beginning of modern cosmology starts with the Hubble observations, and thus from the correlation between the distance to galaxies and their *recessional velocities*, measured through the Doppler shift of the emission wavelengths. This shift can be measured because the emission and absorption spectra for atoms and molecules are distinctive and well known. Redshift (and blueshift) may be characterized by the relative difference between the observed (λ_{obs}) and emitted (λ_{emit}) wavelengths of the radiation emitted by an astrophysical object. In astronomy, it is customary to refer to this wavelength shift using the dimensionless quantity

$$z = \frac{\lambda_{\text{obs}} - \lambda_{\text{emit}}}{\lambda_{\text{emit}}} . \tag{7.1}$$

When the shift of various absorption and emission lines from a single astronomical object is measured, z is found to be remarkably constant.

The discovery of a linear relationship between the redshift and the *luminosity distance*, D_L (measured in Mpc: the quantity is defined below), of a Galaxy, coupled with the assumption of a linear relation between recessional velocity v (in km/s) and redshift, yields a straightforward mathematical expression for the so-called *Hubble's Law*:

$$v = cz = H_0 \cdot D_L \ , \tag{7.2}$$

where the Hubble constant, H_0, has physical dimensions of [time^{-1}]. The Hubble "constant" is a constant only in space, not in time, and the subscript "0" indicates the value of the constant today. H_0 is usually quoted in [km s^{-1} Mpc^{-1}], thus giving the speed in km/s of a galaxy 1 Mpc away. A plot of the luminosity distance D_L vs. redshift z is shown in the left-hand plot of Fig. 7.1. This plot is usually referred to as the *Hubble diagram*. The right-hand plot shows the deviation from the linear relation (7.2) at high redshift.

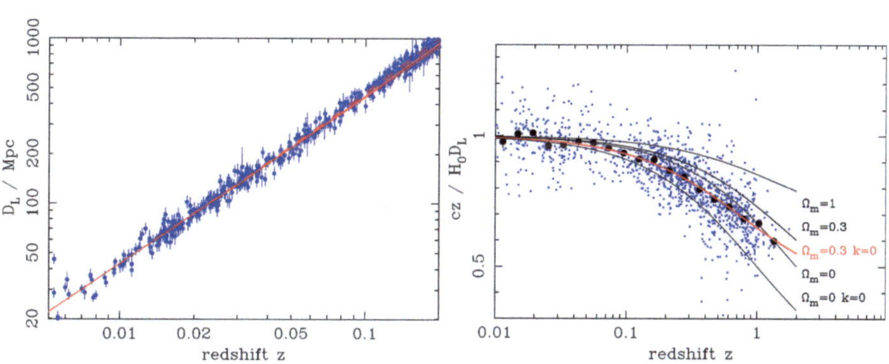

Fig. 7.1 Left: Hubble diagram, i.e., the distance estimate D_L of observed objects as a function of their measured redshift, z. This plot (from §22. of Patrignani et al. 2016) is based on over 1200 publicly available Type Ia supernova distance estimates. The panel shows that, for $z < 0.1$, we have $D_L \propto z$, yielding the Hubble relation (7.2). Right: Ratio between the velocity computed as $H_0 \cdot D_L$ with $c \cdot z$ shown on an expanded redshift scale. For large z, the Hubble law becomes nonlinear. Larger points with errors show median values in redshift bins. Comparison with the prediction of cosmological models (see Sect. 14.1) appears to favor a vacuum-dominated Universe

In astronomy, the luminosity distance D_L is defined in terms of the relationship between the absolute magnitude M and apparent magnitude m of an astronomical object. The effective luminosity of the object is determined using the inverse-square law and the proportions of its apparent distance and luminosity distance. Thus, the luminosity distance is expressed through the flux-luminosity relationship

$$\mathscr{F} = \frac{\mathscr{L}}{4\pi D_L^2} , \tag{7.3}$$

where $\mathscr{F}$ is the flux (measured in W m^{-2} or in erg cm^{-2} s^{-1}), and $\mathscr{L}$ is the luminosity (in W or erg/s). For cosmological distances, particular care must be taken, because the quantities are affected by relativistic effects such as spacetime curvature, redshift and time dilation.

As the flux $\mathscr{F}$ is measured, an estimate of D_L vs. z unbiased by cosmological effects can be performed using *standard candles*, i.e., objects with a known brightness. Although not perfect standard candles, Type Ia SNe (Sect. 12.12) luminosity at maximum brightness (after some corrections depending on the light curve shape and color) have a very small dispersion with respect to a central value. Thus, the peak luminosity of Type Ia SNe is used as an efficient distance indicator.

The first detection through gravitational waves of the merging of two neutron stars has sparked great interest in the possibility of having an independent measurement of the Hubble constant. With this method, the distance is inferred purely from the gravitational-wave signal, and the recession velocity z arises from the electromagnetic data. The gravitational signal from the merging of two massive objects depends only on the masses of the two objects, and the neutron star mass is constrained to be in a very narrow range of values. Because the frequency of the detected gravitational wave is within the range 100–1000 Hz (coincident with the acoustic region), a neutron star merging is called *standard siren*, in contraposition with the standard candles defined by purely electromagnetic quantities. We return to this in Sect. 13.9.

The physical dimension of the Hubble constant is [Time^{-1}]; the reciprocal of H_0 is called the *Hubble time*. The value of the Hubble time in the standard cosmological model corresponds to $\tau_{H_0} = 1/H_0 \sim 13.7$ Gy.

The Cosmic Microwave Background Radiation A second fundamental step in modern cosmology was the discovery of the cosmic microwave background (CMB) radiation, predicted in 1948 by G. Gamow and R. Alpher, as a consequence of a hot origin of the Universe. In 1965, A. Penzias and R. W. Wilson (Nobel Prize in 1978), with a radiometer that they intended to use for radio astronomy and satellite communication, discovered an excess of signal equivalent to a 3.5 K antenna temperature, which they could not account for. This antenna temperature was indeed due to the cosmic microwave background.

The birth of observational cosmology conventionally starts with the accurate measurement of the CMB radiation by the NASA Cosmic Background Explorer (COBE) satellite that orbited in the period 1989–1996. This experiment first

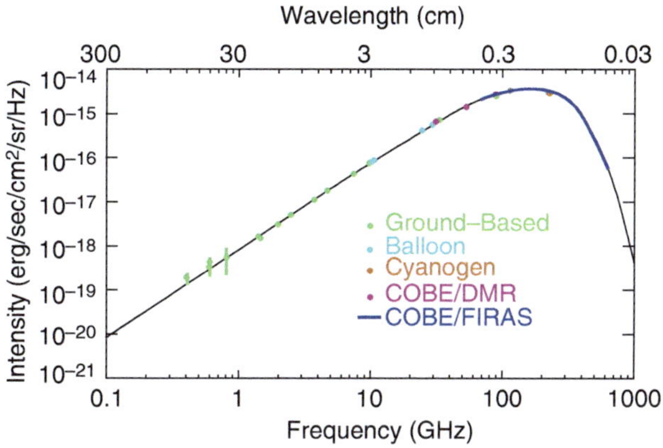

Fig. 7.2 Cosmic microwave background spectrum measured by the FIRAS instrument on the COBE (*blue points*) as a function of frequency (*bottom* x-axis) and wavelength (*upper* x-axis). This corresponds to the most-precisely measured black body spectrum in nature. The error bars are too small to be seen, even in an enlarged image, and it is impossible to distinguish the observed data from the theoretical curve. Other ground-based and balloon-based results at higher wavelengths than those measured by the FIRAS instrument are also shown (http://asd.gsfc.nasa.gov/archive/arcade/cmb_spectrum.html)

detected and quantified the large-scale anisotropies at the limit of its detection capabilities. Inspired by the COBE results of an extremely isotropic and homogeneous CMB, different ground- and balloon-based experiments quantified the anisotropies on smaller angular scales. COBE found that the CMB has a thermal black body spectrum (shown in Fig. 7.2) at a temperature of $2.7255 \pm 0.0006\,\mathrm{K}$. The spectral density in the Planck law $\mathrm{d}E_\lambda/\mathrm{d}\lambda$ peaks in the microwave range, at a wavelength of 1.06 mm corresponding to a frequency $\overline{\nu}$ of 283 GHz. Using this value, the average energy of the CMB is

$$E_{\mathrm{cmb}} = h\overline{\nu} \simeq 1.2 \times 10^{-3}\,\mathrm{eV}. \tag{7.4}$$

The average number density of CMB photons is given by the integral of the Planck spectrum and corresponds to

$$n_{\gamma_{\mathrm{cmb}}} \simeq 400\,\mathrm{cm}^{-3}. \tag{7.5}$$

In June 2001, NASA launched a second space mission dedicated to cosmological measurements, the Wilkinson Microwave Anisotropy Probe (WMAP) satellite, and a third space mission, Planck, was launched by the European Space Agency (ESA) in collaboration with NASA in May 2009. The measurements from these experiments played a key role in establishing the current Standard Model of Cosmology, Sect. 14.1.

7.2 The Large-Scale Structure of the Universe

The observable Universe consists of galaxies and other matter that can be observed from Earth. An interesting web site designed to give an idea of what our universe actually looks like, created by R. Powell, is: http://www.atlasoftheuniverse.com/. The interpretation of observations must take into account that light (or other particles) from those objects has had time to reach the Earth since the beginning of the cosmological expansion. Assuming the Universe to be isotropic, light can arrive in every direction from a distance $L = \tau_{H_0} c$ that corresponds to the observable Universe. The crucial point that removes every anthropomorphic hypothesis is that every place in the Universe is in the center of its own observable horizon, with the same extension L as the one centered on Earth.

Sky surveys and mappings in different wavelengths have yielded a great deal of information on the contents and structure of the Universe. The estimated number of stars is between 10^{22} and 10^{24}. Stars are organized into about 10^{11} galaxies, which, in turn, form clusters of galaxies and superclusters that are separated by immense voids, creating a vast foam-like structure sometimes called the "cosmic web". The organization of this structure appears to follow a hierarchical model, with classifications into groups, clusters, superclusters, sheets, filaments, and walls (see Fig. 7.3).

Our Galaxy (sometimes called the *Milky Way*) belongs to a galaxy supercluster, which also contains the Virgo Cluster near its center, and for this reason, it is called the Virgo Supercluster. It is thought to contain about 2500 large galaxies and over 47,000 dwarf galaxies, and its dimension is about 35 Mpc.

A galaxy is a gravitationally bound system consisting of stars, stellar remnants and an interstellar medium of gas and dust. These objects emit or adsorb and re-emit electromagnetic radiation. An additional component resulting from the unknown form of *dark matter* seems to largely contribute to the gravitational bound, Sect. 14.4. The dimensions range from dwarf galaxies with as few as 10^7 stars to giant galaxies with 10^{14} stars, each orbiting their galaxy's own center of mass. Observational data suggest that supermassive black holes may exist at the center of many, if not all, galaxies. Our own galaxy also appears to have a $(2\text{–}3) \times 10^6$ solar masses black hole in its center.

Galaxies are usually classified based upon morphological visual observations of three main types: elliptical, spiral, and irregular. Most elliptical galaxies are composed of older, low mass stars, with a sparse interstellar medium and minimal star formation activity. They are preferentially found close to the centers of galaxy clusters. Elliptical galaxies make up approximately 10–15% of the galaxies visible in the Virgo Supercluster, and they are not the dominant type of galaxy in the whole Universe. Spiral galaxies (such as our own Milky Way) consist of a rotating disk of stars and interstellar medium, along with a central bulge of generally older stars. See also Sect. 2.7. This classification may miss certain important characteristics of galaxies, such as star formation rate and activity in the core.

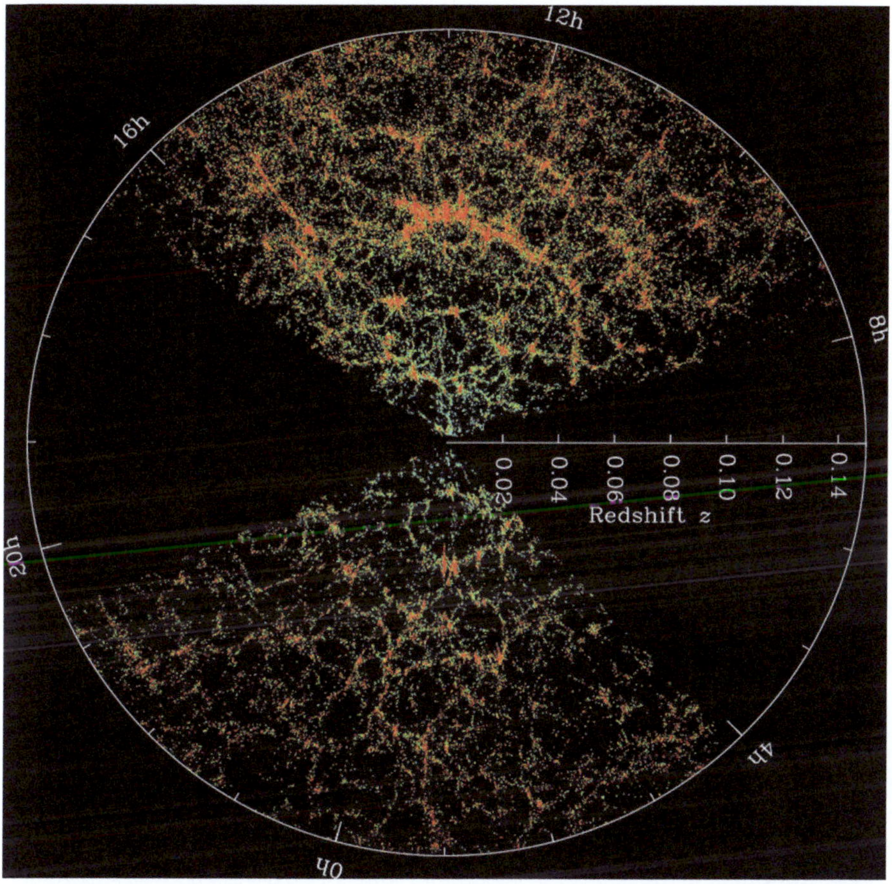

Fig. 7.3 Slices through the 3-dimensional map of the distribution of galaxies from the Sloan Digital Sky Survey (SDSS). The Earth is at the *center*, and each point represents a galaxy, typically containing about 10^{11} stars. The position of that point indicates its location in the sky, and the distance from the *center* of the image indicates its distance from the Earth. Galaxies are colored according to the ages of their stars, with the redder, more strongly clustered points showing galaxies that are made of older stars. The *outer circle* is at a distance of about 2×10^9 light years. The region between the wedges was not mapped by the SDSS, because dust in our own Galaxy obscures the view of the distant Universe in these directions. The *lower* part of the figure is thinner than the *upper*, thus it contains fewer galaxies. Another group completed a similar survey of the galaxies in the Universe called the 2 dF Redshift Survey. Credit: M. Blanton and the SDSS (https://www.sdss3.org/science/gallery_sdss_pie2.php)

Active galaxies are galaxies with an abnormal emission of electromagnetic radiation. The emission is mostly due to a small active core embedded in an otherwise typical galaxy, which may be highly variable and very bright compared to the rest of the galaxy. Galaxies with abnormal activity in their central region are called Active Galactic Nuclei (AGN). AGN (as described in Sect. 9.10) range from nearby galaxies emitting about 10^{40} erg s^{-1} to distant point-like objects (named *quasars*) emitting more than 10^{47} erg s^{-1} (Woo and Megan Urry 2002).

AGN are UHECRs source candidates and have been observed to emit also γ-rays up to energies of tens of TeV (Sect. 9.11) and probably neutrinos (Sect. 10.6.2).

7.3 Anisotropy of UHECRs: The Extragalactic Magnetic Fields

Magnetic fields affect the propagation of CRs, because charged particles are deflected, emitting synchrotron radiation. For protons, synchrotron losses are negligible, except in the strong magnetic fields present close to sources. Above 10^{19} eV, the galactic magnetic field would not even trap iron nuclei very effectively. The Larmor radius (2.5) of a particle with energy E and electric charge Ze in a magnetic field B can be expressed as

$$r_L = 110\,(\text{kpc})\, Z^{-1}\left(\frac{\mu\text{G}}{B}\right)\left(\frac{E}{10^{20}\,\text{eV}}\right), \tag{7.6}$$

which, for UHECRs, is much larger than the thickness of the galactic disk.

In intergalactic space, where the magnetic field intensities are expected to be much lower than in the Galaxy, the Larmor radius of UHECRs becomes extremely large. The possibility of *charged particle astronomy*, in the sense of the possibility of connecting the arrival direction of UHECRs to the coordinates of extragalactic objects, is still an open question. This possibility depends on the magnitude of the poorly known extragalactic magnetic fields.

The deflection angle of a particle of energy E moving in a direction perpendicular to a uniform magnetic field B after travelling the distance d is

$$\theta \sim \frac{d}{r_L} \sim 0.5^\circ Z\left(\frac{E}{10^{20}\,\text{eV}}\right)^{-1}\left(\frac{d}{\text{kpc}}\right)\left(\frac{B}{\mu\text{G}}\right). \tag{7.7}$$

A proton of energy 5×10^{19} eV will be deflected by 1°–5° in the galactic magnetic field, depending upon the direction and length of the trajectory. Each panel in Fig. 7.4 shows the simulated trajectories of 10 CR protons originating in a fixed position in the galactic plane. The galactic disk lies on the xy plane and the structure of the magnetic field is similar to that of Fig. 2.10. The energies are 10^{18}, 10^{19} and 10^{20} eV, as shown in the figure. The galactic magnetic field has little influence at

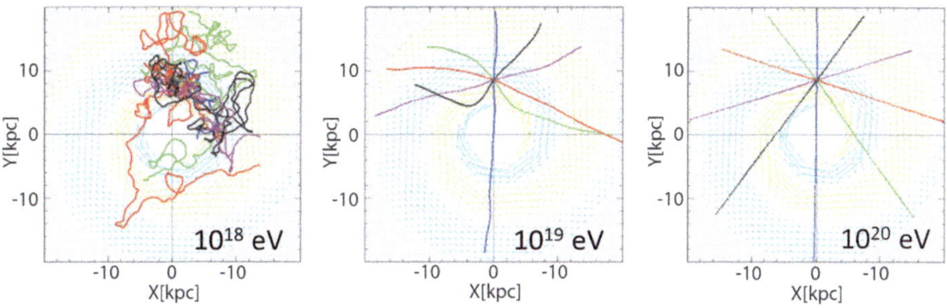

Fig. 7.4 Simulated trajectory of charged particles in the galactic magnetic field. Low-energy charged particles are bent and wound by magnetic fields, but those above 10^{20} eV travel along almost straight trajectories with little influence from magnetic fields, thereby retaining the original directional information. Credit: Prof. T. Ebisuzaki (http://www.asi.riken.jp/en/laboratories/chieflabs/astro/index.html)

energies $> 10^{19}$ eV, and a galactic source of CRs with that energy would produce a clear anisotropy on Earth measurements, which is not observed. The confinement mechanism in the Galaxy is not maintained at the highest energies, motivating the search for extragalactic sources of CRs.

Magnetic fields beyond the galactic disk are poorly known. In few clusters of galaxies, they have been estimated by observing the synchrotron radiation halos or performing Faraday rotation measurements. The two methods give somewhat different results for the field strength, with $B \sim 0.1$–$1\,\mu$G and $B \sim 1$–$10\,\mu$G, respectively. These regions enclose a small fraction of the Universe (less than 10^{-6}), and only upper limits on the extragalactic magnetic fields exist for regions outside galaxy clusters: $B < 10^{-9}$ G with coherent length of the field $l_c \sim 1$ Mpc. The value of l_c corresponds to the average distance between galaxies.

7.4 The Quest for Extragalactic Sources of UHECRs

Following the arguments used for galactic CR sources, the extragalactic acceleration mechanisms must satisfy the following criteria: they must provide enough energy to reach the largest observed energies, and the accelerated population should have an injection energy spectrum that would fit the observed UHECR spectrum after propagation.

The diffusive shock acceleration mechanism (Sect. 6.2) is based on the repeated scattering of charged particles on magnetic irregularities back and forth across a shock front. For non-relativistic shocks, the energy gain at each crossing is $\Delta E \propto \eta E \simeq \beta E$, see Sect. 6.1.3. As discussed in Chap. 6, the presence of plasmas in any real astrophysical condition destroys large-scale electric fields. Magnetic fields are instead almost omnipresent in astrophysical objects. Their space/time variations imply the existence of transient electric fields through the Faraday law that can supply a consequent amount of energy to charged particles (unipolar inductors, Sect. 6.9.1). The maximum energy (6.90) of a CR nucleus of charge Ze accelerated in a region where the magnetic field B changes in a spatial region of size L can be written as

$$\frac{E^{\max}}{Ze\beta} = LB. \tag{7.8}$$

To reach energies above $10^{18}\,\text{eV} = 1\,\text{EeV}$, the acceleration should occur with a necessary combination of scale L and magnetic field B: if the scale is small, the magnetic field must be huge, and vice versa.

Also, in man-made accelerators, the size L is related to the maximum energy obtainable. In the case of a collider, such as the Large Hadron Collider (LHC) at CERN, the maximum energy is a function of the radius of the machine and the strength of the dipole magnetic field. This field keeps particles in their orbits, see Fig. 7.5. The LHC uses the 27 km circumference tunnel that was built for the previous lepton accelerator, LEP. In an accelerator, particles circulate in a vacuum tube. The *accelerating cavities* are electromagnetic resonators that accelerate

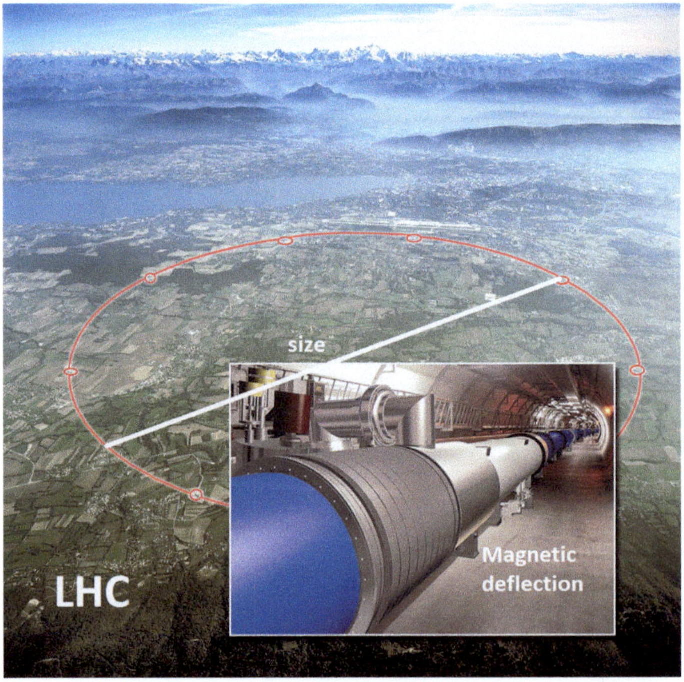

Fig. 7.5 In a man-made accelerator, the maximum energy is a function of the radius of the machine and the strength of the dipole magnetic field that keeps particles in their orbits. An accelerator of the size of the orbit of the planet Mercury would be needed to reach 10^{20} eV (=100 EeV) with the LHC technology. Adapted from CERN pictures

particles and then keep them at a constant energy by compensating for energy losses. Particles are constrained using electromagnetic devices: *dipole magnets* keep the particles in their nearly circular orbits, *quadrupole magnets* focus the beam, acting in a way analogous to converging lenses in optics. Heavy particles such as protons have a much lower energy loss per turn through synchrotron radiation than electrons. The LHC uses some of the most powerful existing dipoles and radio-frequency cavities. The size of the tunnel, magnets, cavities and other essential elements of the machine, represent the main constraints that determine the design energy of 7 TeV per proton beam.

The largest observed CR energy (about 2×10^{20} eV) corresponds to a macroscopic amount on the order of 20 J. The basic problem of the origin of the highest energy CRs is how an astrophysical source can efficiently transfer this enormous quantity of energy to a single particle. To work out some hypothesis, we rewrite (7.8) in more appropriate units as

$$E^{\max} = Z\beta(4.8 \times 10^{-10}) \cdot (10^{-6})\left(\frac{B}{1\,\mu\text{G}}\right) \cdot (3.1 \times 10^{21})\left(\frac{L}{1\text{kpc}}\right) [\text{erg}]. \quad (7.9)$$

Using the conversion $1\text{ eV} = 1.6 \times 10^{-12}$ erg and $1\text{ EeV} = 1.6 \times 10^{6}$ erg, we obtain

$$E^{\max} \simeq Z\beta \cdot \left(\frac{B}{\mu\text{G}}\right) \cdot \left(\frac{L}{\text{kpc}}\right) \quad [\text{EeV}]. \quad (7.10)$$

Equation (7.10) contains the factor $\beta = U/c$, where U is the characteristic velocity of magnetic scattering centers and the above relation is usually called the Hillas criterion.

In the case of one-shot acceleration scenarios (when the particle escapes the accelerating region after the first iteration) the maximum reachable energy has quite a similar expression to the shock acceleration case. We already discussed the case of a pulsar (Sect. 6.9.1), in which the quantity β in Eq. (7.10) is replaced by $\omega_{NS} R_{NS}/c$. In addition to the pulsar magnetospheres, other astrophysical candidate regions are gamma-ray bursts, active galactic nuclei and radio jets.

In both the shock acceleration and one-shot acceleration scenarios, a relationship analogous to (7.10) holds. The maximum attainable energy is, for a given nuclear charge Ze, approximately equal to the product of the magnetic field B and the size L of the acceleration/confinement region, Fig. 7.6. The straight lines shown correspond to values of $E^{\max} = 10^{20}$ eV for protons and iron nuclei. Diagrams of this type were first proposed by Hillas in 1984, and in the figure, several possible galactic and extragalactic acceleration sources are considered, see Torres and Anchordoqui (2004) for details. Let us consider the full line that corresponds to 100 EeV protons. If potential sources lie to the left of the line, protons cannot be accelerated to 10^{20} eV by these objects. The diagram indicates that there are potential sites of particle acceleration within a wide range of high-energy astrophysical objects (Ostrowski 2002). Clearly, this criterion is a necessary, but not sufficient, condition.

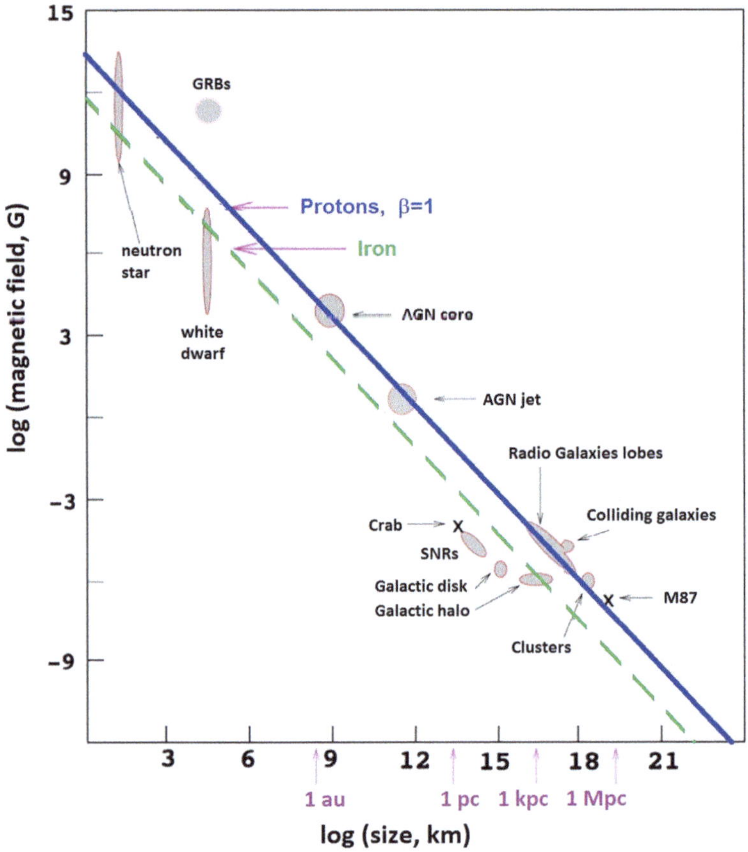

Fig. 7.6 Example of the diagram first produced by Hillas. Acceleration of cosmic rays up to a given energy requires magnetic fields and sizes above the respective line. The *full* (*dashed*) *line* corresponds to the condition for B, L to accelerate protons (iron) at 10^{20} eV. Some source candidates are still controversial

We will show in the next chapter that sources of CRs should also emit high-energy photons up to multi-TeV. The γ-rays emission mechanism is naturally connected to the acceleration of electrons and/or protons in astrophysical sources. The observed photon spectra also allow us to predict neutrino fluxes, since photons are the only messengers giving direct evidence of the properties of their sources (Becker 2008). The observation of neutrinos from astrophysical objects would be the key ingredient to identifying acceleration sources of CR protons and nuclei. Table 7.1 lists source classes with their intrinsic luminosity and possible contribution to the CR spectrum. The power measured in electromagnetic wavelengths should roughly correspond to the CR power at the source, since electromagnetic radiation originates from the emission processes of charged particles.

Table 7.1 The values for typical electromagnetic emissions and the characteristic lifetimes for different galactic and extragalactic objects

Source class	Electromagnetic output erg/s	Lifetime	Energy range
Galactic sources			
Supernova remnants (SNR)	10^{42}	10^3–10^4 y	$10^{10} \div 10^{15}$ eV
Wind-SNR	10^{44}	1000 y	$10^{10} \div 10^{18}$ eV
X-ray binaries	10^{38}	10^5 y	$10^{14} \div 10^{18}$ eV
Pulsars	10^{37}	10^6 y	$10^{14} \div 10^{18}$ eV
Extragalactic sources			
Galaxy clusters	10^{44}	10^7 y	$10^{18} \div 10^{21}$ eV
AGN	$10^{44} \div 10^{47}$	10^7 y	$10^{18} \div 10^{21}$ eV
GRBs	$10^{49} \div 10^{51}$	1–100 s	$10^{18} \div 10^{21}$ eV

The estimates are based on the multiwavelength observations of objects in the electromagnetic emission. It should be noted that luminosity and lifetime distributions can scatter, and there are objects within the given classes that have values deviating from the given typical ones. In the case of GRBs, a lower limit is given; the actual value depends on whether or not GRB afterglow emission contributes to UHECRs

AGN are potential sites where UHECR acceleration might take place. They are (almost) steady sources of electromagnetic emission, although variability in γ-ray emission is observed. The jets have transverse dimensions on the order of a fraction of a parsec and the magnetic field necessary to explain the synchrotron radiation from such objects of the order of a few Gauss. Using the Hillas condition (7.10), the maximum energy for protons in a region with magnetic field intensity on the order of 1 μG and $L = 0.1$ pc would be $\sim 10^{20}$ eV. The $E^{\max}$ attainable for protons in the AGN core, where $B \sim 10^3$ G in a size of $L \sim 10^{-5}$ pc, is almost the same.

One criticism of this model is that, under realistic conditions, the quoted $E^{\max}$ is unlikely to be achieved. In fact, the realistically attainable maximum energy should decrease because of energy loss mechanisms of charged particles, their synchrotron radiation in the magnetic field, and the interaction of high energy protons with the photons of the radiation field surrounding the central engine of AGN. The situation is worse for nuclei, which will photodisintegrate even faster in the photon field. This problem can be bypassed if we assume that energies up to the highest observed values can be reached when the final acceleration site is away from the active center. This should correspond to regions with a lower radiation density, as the terminal shock sites of jets.

Gamma Ray Bursts (GRBs) are the most energetic transient eruptions observed in the Universe, see Sect. 8.11. The transient nature of GRBs could possibly explain the lack of correlation between the arrival direction of the highest energy CRs and astrophysical objects. In this scenario, sources are not visible, since the detected CRs come from various bursts and reach the Earth long after the gamma-ray burst itself has occurred.

Based on γ-ray observations, the mechanism that produces bursts is likely to be due to the dissipation of the kinetic energy of a relativistic expanding fireball.

The value of the magnetic field at these shocks is estimated to be on the order of $B \sim 10^6$ G on a scale length $L \sim 10^{-6}$ pc from the center, fulfilling the Hillas conditions for an energy of $\sim 10^{20}$ eV. Consider that parameters can take different values at different times of the GRB explosion. The observation that the flux of γ-rays reaching the Earth from GRBs is generally comparable to the observed flux of UHECRs, implying a tight energetic connection, strengthens the hypothesis that GRBs are at least a part of UHECRs' acceleration sources.

Magnetars Like other neutron stars (Sect. 6.7.2), magnetars are around 20 km in diameter. Magnetars are different from other neutron stars because they have even stronger magnetic fields (up to 10^{15} G, three orders of magnitude larger than normal neutron stars). Their rotation period is comparatively slow, with most magnetars completing a rotation once every 1–10 s. The magnetic field gives rise to very strong and characteristic bursts of X-rays and γ-rays. The active life of a magnetar is short. According to some models, UHECRs could be accelerated through unipolar induction in the relativistic winds for rapidly rotating magnetars. The maximum energy attained by particles injected by these objects could reach 10^{20} eV, and only 5% of the extragalactic magnetar population needs to be fast-rotators to account for the observed UHECR energetics. Finally, magnetars, as well as GRBs, are transient sources and could not be associated with the detection of UHECRs.

The nature (protons or heavier nuclei - iron is sometimes considered as the extreme case) of the highest energy CRs plays an essential role in the understanding of acceleration mechanisms. If UHECRs are iron nuclei, energies of 10^{20} eV can be attained, even in shock regions with β significantly less than one, or in regions with smaller size/smaller magnetic fields, as shown by the dashed line in Fig. 7.6. If, instead, UHECRs are protons, an origin in AGN or in GRBs seems the most natural explanation, given their isotropy. The magnetar option is very promising if the UHECRs are heavy nuclei. In fact, the possibility of injecting large proportions of heavy nuclei into an acceleration region may be more easily fulfilled by young neutron stars than alternative sources, due to their iron-rich surface. See Kotera and Olinto (2011) for additional information on models.

The discrimination between different scenarios is the goal of present and future generations of experiments.

7.5 Propagation of UHECRs

There are three main energy loss processes for protons (or heavier nuclei) propagating over cosmological distances: adiabatic energy loss, pion-production on photons of the cosmic microwave background (CMB), and electron-positron pair-production, always on the CMB (Kotera and Olinto 2011). We can define the *energy loss length* ℓ as

$$\ell^{-1} \equiv \frac{1}{E}\frac{\mathrm{d}E}{\mathrm{d}x} \,. \tag{7.11}$$

7.5.1 The Adiabatic Energy Loss

The adiabatic loss of a cosmic ray with energy E is a general mechanism that affects particles and radiation and is due to the expansion of the Universe. The energy loss formula is similar to that valid for the electromagnetic radiation and depends on the Hubble constant

$$-\frac{1}{E}\frac{\mathrm{d}E}{\mathrm{d}t} = H_0. \tag{7.12}$$

The corresponding energy loss length is thus

$$\ell_{\text{adia}} = \frac{c}{H_0} \simeq 4 \text{ Gpc}, \tag{7.13}$$

where c is the light velocity and $1/H_0 \sim 13.7$ Gy. Noticeably, the energy loss length does not depend on the CR energy.

7.5.2 The Propagation in the CMB: The GZK Cut-Off

The propagation of UHECRs in the newly discovered CMB was independently studied in 1966 by K. Greisen, V. Kuzmin, and G. Zatsepin. They foresaw that the flux of CRs originating at cosmological distances would be greatly attenuated above a threshold energy $E_{\text{GZK}} \simeq 5 \times 10^{19}$ eV. This is the so-called Greisen-Zatsepin-Kuzmin (GZK) cut-off: the proton component of the CR flux from sources at cosmological distances drops sharply above the threshold energy E_{GZK}. As we can easily derive from the considerations below, a nucleus with mass A starts to be attenuated at an energy $A \cdot E_{\text{GZK}}$.

During propagation, protons would interact with the cosmic microwave background photons (γ_{cmb}) if the proton energy is large enough to achieve the resonant production of the Δ^+ hadron in the centre-of-mass system. The Δ^+ resonance immediately decays

$$p + \gamma_{\text{cmb}} \rightarrow \Delta^+ \rightarrow \pi^+ n \tag{7.14}$$

$$\rightarrow \pi^0 p\,. \tag{7.15}$$

Neutral pions decay in two photons, while the π^+ decays into $\mu^+\nu_\mu$. The produced neutrons also decay into $pe^-\overline{\nu}_e$. As a proton is always present in the final state, the final effect of the interaction is that the energy of the CR proton above threshold is reduced and high-energy photons and neutrinos are produced.

According to (3.1), the interaction length (in cm) is inversely proportional to the target number density (cm^{-3}) and the particle cross-section (cm^2). The *energy loss length* $\ell_{p\gamma}$ in (7.11) in this case corresponds to:

$$\ell_{p\gamma}^{-1} = \langle y\sigma_{p\gamma}n_\gamma\rangle \quad \longrightarrow \quad \ell_{p\gamma} \equiv \frac{1}{\langle y\sigma_{p\gamma}n_\gamma\rangle}, \tag{7.16}$$

where $y = (E - E')/E$ is the fraction of energy lost per interaction. Usually, the radiation field is due to the CMB radiation, where $n_\gamma = n_{\gamma_{\rm cmb}}$ is the corresponding photon number density (7.5). The brackets $\langle\ldots\rangle$ remind us that we should integrate the differential cross-section over the momentum distribution of the target photons. In our following estimates, we avoid this complication by considering only reactions well above the threshold energy when essentially all photons participate in the reaction, and using average values for the cross-section and photon momentum.

We derive now the threshold energy for protons to induce this reaction and their mean free path.

The Δ^+ resonance has mass $m_\Delta = 1232\,\text{MeV}$ (we use natural units with $c = 1$). The threshold for reaction (7.14) corresponds to the production of the Δ^+ resonance at rest. The condition is dictated by the center-of-mass invariant $\sqrt{s} = m_\Delta$ [see Sect. 3.1 of Braibant et al. (2011)]. If $p_p = (E_p, \mathbf{p_p})$, $p_\gamma = (E_\gamma, \mathbf{p}_\gamma)$ are the four-momenta of the proton and of the CMB photon, the resonance is produced when

$$s = (p_p + p_\gamma)^2 = m_\Delta^2 \quad \longrightarrow \quad m_p^2 + 2E_\gamma E_p - 2\mathbf{p}_p\mathbf{p}_\gamma = m_\Delta^2. \tag{7.17}$$

For a photon, $|\mathbf{p}_\gamma| = E_\gamma$, and at high energy, $E_p \simeq p_p$. In this case, (7.17) becomes

$$2E_p(E_\gamma - E_\gamma\cos\theta) = m_\Delta^2 - m_p^2,$$

and using the values $m_\Delta = 1.232\,\text{GeV}$, $m_p = 0.938\,\text{GeV}$, we obtain

$$E_p = \frac{m_\Delta^2 - m_p^2}{2E_\gamma(1-\cos\theta)} = \frac{0.32\,\text{GeV}^2}{E_\gamma(1-\cos\theta)}. \tag{7.18}$$

The minimum value for E_p occurs when $\theta = \pi$ (the directions of the photon and the proton are opposite). In this case, for collision with the CMB with average energy $E_\gamma = 1.2\;10^{-3}\,\text{eV}$, the threshold energy is

$$E_p = \frac{m_\Delta^2 - m_p^2}{4E_\gamma} = 1.2\times 10^{20}\,\text{eV}. \tag{7.19}$$

The threshold decreases when the interaction occurs with CMB photons in the high-energy tail of the spectrum. The effect starts to become significant for protons with $E_p \simeq 5\;10^{19}\,\text{eV}$.

The cross-section of the processes $\gamma p \to \pi^0 p$ and $\gamma p \to \pi^+ n$ were studied in laboratory (using high-energy photons and protons at rest), as a function of the photon energy in the laboratory frame. The cross-section for these two processes

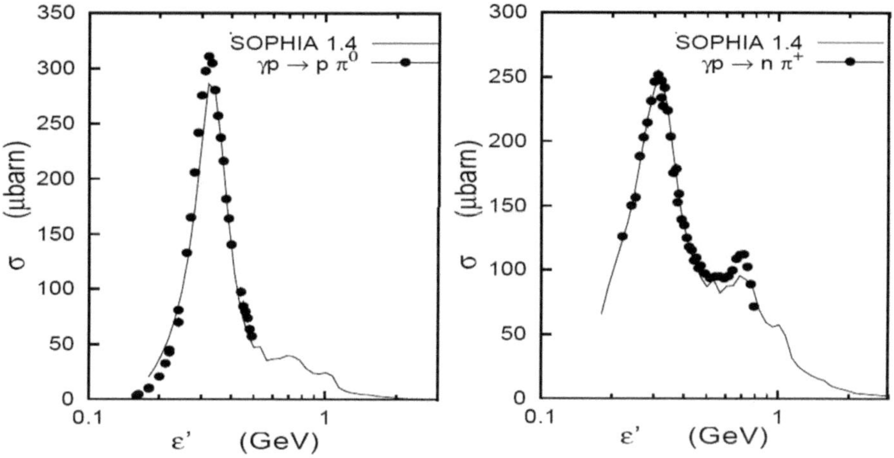

Fig. 7.7 Total cross-section of $\gamma p \rightarrow \pi^0 p$ (*left*) and $\gamma p \rightarrow \pi^+ n$ (*right*) as a function of the photon energy ϵ' in the laboratory frame. In this case, energetic photons interact with protons at rest and the resonance occurs at the energy $\epsilon' = m_\Delta - m_p$. The lines represent an analytical approximation of the data

at resonance is $\sigma_{p\gamma} \simeq 250\,\mu$b. The experimental results, as well as the prediction from an analytic parameterization used in CR propagation (Muecke et al. 2000), are presented in Fig. 7.7.

The energy loss per interaction is relatively small and can be qualitatively estimated considering that, in the final state of the process, a proton and a pion are present, and thus

$$y = \frac{\Delta E_p}{E_p} \sim \frac{m_\pi}{m_p} \simeq 0.1\,. \tag{7.20}$$

In each process, around 10% of the proton energy is lost. This means that within a few successive interactions, the proton energy decreases to a value below the threshold for the reaction. With the above quantities, from (7.16), the energy loss length $\ell_{p\gamma}$ of a proton in the CMB for this process is

$$\ell_{p\gamma} = \frac{1}{y \cdot \sigma_{p\gamma} \cdot n_{\gamma_{\rm cmb}}} = \frac{1}{0.1 \cdot 250 \times 10^{-30} \cdot 400} = 10^{26}\,\text{cm} = 30\,\text{Mpc}. \tag{7.21}$$

All protons originating at distances larger than $\sim$30 Mpc[1] from us arrive on Earth with energy below $\sim 10^{20}$ eV. This horizon is small in terms of the dimensions of the Universe. Thus, the energy loss length of a $\sim 10^{20}$ eV proton is comparable to

[1]Given the Hubble constant and the fact that $z \simeq v/c$, when $v \ll c$, it is straightforward to derive that 30 Mpc corresponds to $z \simeq 0.007$. Compare this value with the scale of Fig. 7.3.

the distance of the closest galaxy cluster (the Virgo). The mean free path of protons in the Universe above 5×10^{19} eV becomes small on a cosmological scale. The eventual detection of CR protons with energy exceeding that value corresponds to protons produced in "local sources".

For a heavier nucleus with mass A and energy E, the resonant reaction must occur through the interaction of one of the nucleons in the nucleus, which has energy E/A. In the nucleus center-of-mass, each nucleon can be considered as free with respect to a photon with energy ~300 MeV. The threshold energy $E_{\rm GZK}$ for heavier nuclei is consequently higher by a factor A.

7.5.3 $e^{\pm}$ Pair Production by Protons on the CMB

During the propagation in the CMB, electron-positron pairs can be produced in the process

$$p + \gamma \to p + e^+ e^- . \tag{7.22}$$

Also, this process has a threshold that can be determined, as in the case of the pion photoproduction, by the condition

$$s = (p_p + p_\gamma)^2 \geq (m_p + 2m_e)^2 \quad \longrightarrow \quad E_p \geq \frac{m_e m_p}{E_\gamma} \simeq 2 \times 10^{18}\,\text{eV} , \tag{7.23}$$

where E_γ is the average energy of the CMB photon.

The cross-section for this electromagnetic reaction is

$$\sigma_{p\gamma} = (\alpha/32)\sigma_T \cdot f(s) , \tag{7.24}$$

where σ_T is the Thomson cross-section (5.58), $\alpha \simeq 1/137$ is the fine structure constant and $f(s)$ is a function that depends on the center-of-mass energy of the proton-photon system. Below the production threshold, $f(s) = 0$; at the resonant energy, $f(s) = 1$. This cross-section is a factor $\sim 10^{-4}\sigma_T \sim 0.1$ mb, which is thus comparable with that of the production of the Δ^+ resonance, Fig. 7.7. For comparison, Compton scattering (which is relevant in many astrophysical processes) has a cross-section that is proportional to σ_T.

Because in the final state there is always a proton, Eq. (7.22) corresponds to an effective energy loss of the proton. The energy loss per interaction is relatively small, and the fraction of energy loss can be qualitatively estimated as

$$y = \frac{\Delta E_p}{E_p} \sim \frac{2m_e}{m_p} \simeq 10^{-3} . \tag{7.25}$$

The energy loss lengths for this process can be computed similarly to Eq. (7.21); the target is the same and the cross-section has a similar value. The quantity y has

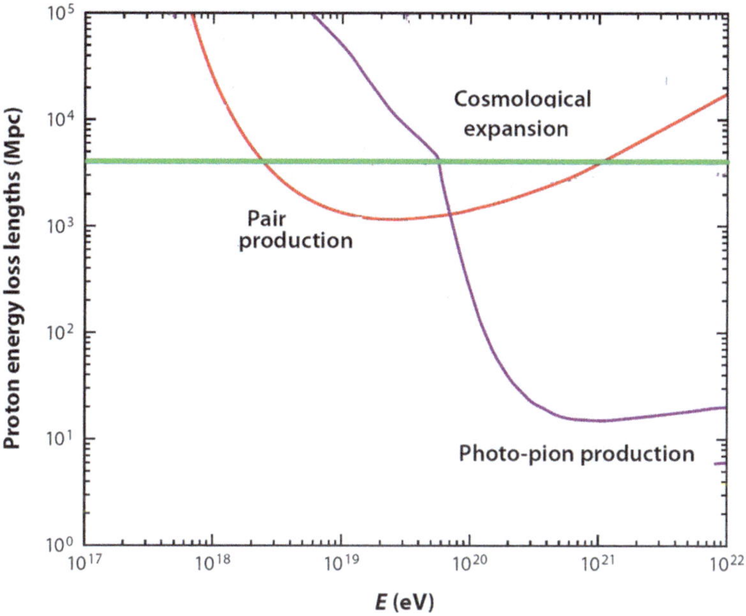

Fig. 7.8 The energy loss lengths (7.16) for a high-energy proton propagating through the CMB radiation field. Pair creation (7.22), pion production (7.14), and energy loss through cosmological expansion (7.13) are shown

a value about two orders of magnitude smaller. Figure 7.8 compares the relative energy loss lengths of a proton due to pion-production and with that of e^+e^- pair-production on CMB photons, as well as that due to the redshift. One clearly recognizes that the e^+e^- pair-production process has a lower energy threshold compared to pion photoproduction, and an average path length about two orders of magnitude larger.

7.5.4 Propagation in the Extragalactic Magnetic Fields

Upper limits on the extragalactic magnetic fields are about $B < 10^{-9}$ G over scale dimension l_c, Sect. 7.3. Even such small fields can affect the propagation of UHECRs. If we neglect energy loss processes, then the root mean square deflection angle $\theta_{\rm rms} = \langle\theta^2\rangle^{1/2}$ for charged particles travelling over distances $d \gg l_c$ is (Kachelriess 2008)

$$\theta_{\rm rms} \sim \frac{(2dl_c/9)^{1/2}}{r_L} \simeq 0.8^\circ Z\left(\frac{E}{10^{20}\,{\rm eV}}\right)^{-1}\left(\frac{d}{10\,{\rm Mpc}}\right)^{1/2}\left(\frac{l_c}{1\,{\rm Mpc}}\right)^{1/2}\left(\frac{B}{10^{-9}\,{\rm G}}\right). \tag{7.26}$$

The deflection $\theta_{\rm rms}$ can be estimated using observational data or reliable theoretical predictions, both for the magnitude and the structure of extragalactic magnetic fields. At present, no single theory for the generation of magnetic fields in extragalactic space has become widely accepted. The combination of poor observational data and the absence of a consistent theoretical picture prevents, at present, a reliable estimate of the influence of extragalactic magnetic fields on the propagation of UHECRs.

The random walk due to the extragalactic magnetic fields causes an increase in the CRs' propagation path length from their sources that causes a corresponding time delay. If UHECRs are generated in a gamma-ray burst, or in an active transient state of an AGN, we may not be able to correlate the observed X- or γ- rays with the resulting incoming direction of CRs. The observation of correlation between the arrival direction of UHECRs and the position of extragalactic sources would solve the mystery of the origin of UHECRs and place strong constraints on the particle acceleration mechanism in its extreme manifestations. On the other hand, if we could know the origin of (at least a set of) UHECRs, we could use the arriving particles as messengers probing the intervening unknown extragalactic magnetic fields, by studying the deflections they suffered along their journey to the Earth (Aharonian et al. 2012).

7.6 Fluorescent Light and Fluorescence Detectors[2]

Two detection techniques are principally employed to detect UHECRs. The first extends the use of extensive air shower arrays (Chap. 4) to energies above 10^{18} eV. The second method exploits the excitation of nitrogen molecules by the particles in the shower and the associated fluorescence emission of light. The light is detected by photomultipliers and the profile of the shower in the atmosphere can be inferred rather directly.

Fluorescence detectors are based on the effect that high-energy particles present in a cosmic ray air shower, colliding with nitrogen molecules or nitrogen ions in the atmosphere, exciting them to higher energy levels. The excited molecules undergo a very fast (10–50 ns) radiative decay, with emission of photons in the wavelength region between 300 and 440 nm, from the visible to the near ultraviolet. In dry air, the color of light produced by lightning is dominated by the emission lines of ionized nitrogen, yielding the primarily blue color observed. The fluorescent yield of an electron in the air shower is $\sim$4 photons per meter at ground level pressure (Arqueros et al. 2008).

As the shower advances, new nitrogen molecules, in rapid succession, are excited and decay, and new photons are emitted. It is worth noticing from the outset that the emission is *isotropic*, implying that, in fluorescence detectors, the air showers

[2]This section is largely due to V. Flaminio

can be observed from all directions, including from the sides. This is important, because it allows us to follow in detail the space development of the shower and to evaluate the variable $X_{\max}$. This makes fluorescence detectors conceptually and physically different from those of Cherenkov. If we could record, with a kind of movie camera, the photons emitted while the shower advances, we would have a perfect reconstruction of the space-time features and, if we could at the same time measure the number of photons emitted along the path of the shower, even obtain the energy of individual showers and an evaluation of $X_{\max}$.

A further advantage of this technique is related to the very good atmospheric transparency (at least in good weather conditions) to photons within this wavelength range. A drawback comes instead from the fact that the number of photons produced is relatively small, which makes the technique efficient mainly at the highest energies, where the number of particles in the shower is large.

Before going into a discussion of further advantages and disadvantages, it is important to clarify the way in which it is actually possible to follow, from the ground, the shower's development (the movie camera recording mentioned above). Assume that we have a single, small, light detector (a photomultiplier—PMT) pointing towards the sky, and that we have chosen a PMT of small angular acceptance (on the order of one squared degree). If the PMT happens to be pointing in the right direction, it will record fluorescent photons coming from the small region of the sky, falling within its angular acceptance. As the shower advances, new particles will pass through that small region and the PMT will record the progress of the shower through that region over time.

Of course, if we want to record the space development of the shower, we must have many PMTs of this type, pointing towards different but adjacent regions. This can be done using many PMTs, closely packed together, pointing in slightly varied directions, as shown in the toy example in Fig. 7.9.

Here, we have only sketched, in an oversimplified two-dimensional arrangement, five different PMTs. We see that when the shower passes through position #1, it will fall within the angular acceptance of PMT #1; it will fall within the angular

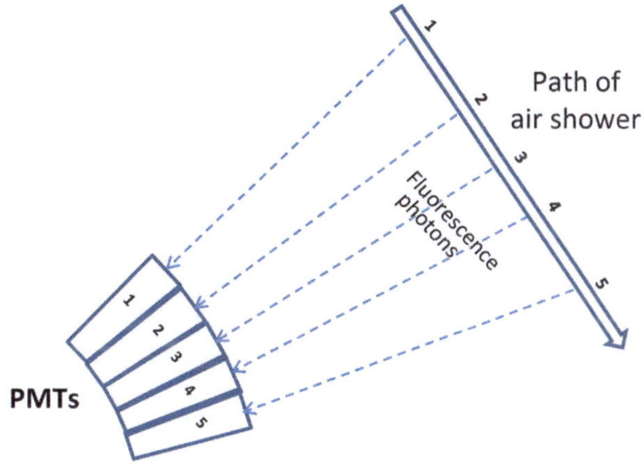

Fig. 7.9 The arrangement of the PMTs in the two-dimensional oversimplified arrangement described in the text

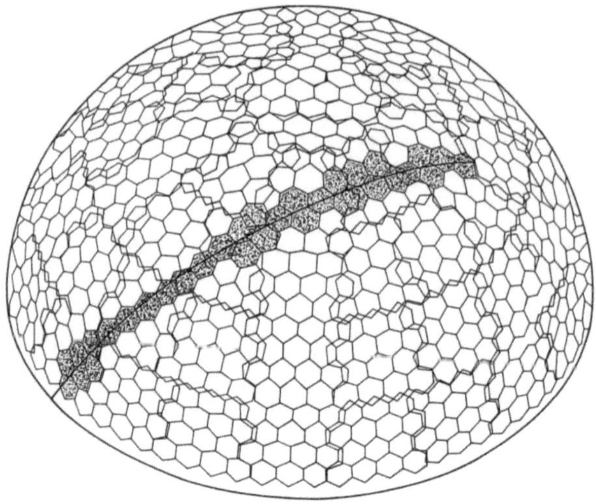

Fig. 7.10 The arrangement of the PMTs in the Fly's Eye experiment. Credit: Prof. P. Sokolsky

acceptance of PMT #2 when, a few microseconds later, it passes through position #2, and so on. Since the shower can occur anywhere in the sky and, moreover, will have a finite width in general, the oversimplified 2-D picture we have sketched must, in real life, be replaced by a 3-D arrangement like the one shown in Fig. 7.10. The type of detector shown in the figure reminds us of the eye of a fly, from which the earliest built detector of this type was given its name, the University of Utah Fly's Eye detector. This experiment was located in the western desert of Utah, USA, at the Dugway Proving ground. It collected data with different configurations from 1981 to 1992, paving the way for the High Resolution Fly's Eye detector (HiRes), Sect. 7.7, which collected data on the same site from 1997 until 2006. It is worth noting that in a detector of this type, the shower appears as a bright spot, moving across the sky at the speed of light.

Modern fluorescence detectors use large spherical mirrors to improve light collection. The light is then focused on an array of small PMTs, arranged in a very similar way to the one we have described. The latest (and largest) implementation of this technique, at the Pierre Auger Observatory, will be described later in Sect. 7.8. An illustrative example of such a telescope is shown in Fig. 7.11.

A fluorescence telescope is enclosed in its own building, to protect it from weather conditions, and in addition, a special large optical filter is placed in front of it, to remove undesired ambient light. The idea of an optical filter is to transmit most of the fluorescence signal in the near-UV while blocking other night sky background to which the PMTs are sensitive.

One limitation of this type of detectors is related to the fact that they may only operate in clear, moonless nights. Any spurious source of light would be a problem for the PMTs. Another limitation comes from the fact that a shower, travelling for a long distance in the sky, can move from regions of high air density to regions where the air density is low. These regions are characterized by different numbers of nitrogen molecules/ions per unit path length, and therefore different numbers of

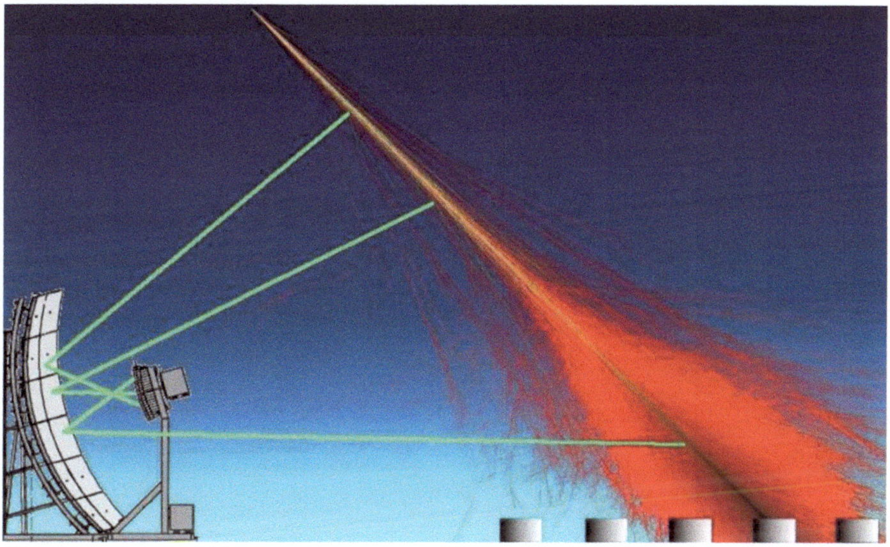

Fig. 7.11 An artist's impression of an air shower seen by a fluorescence detector. The fluorescent light is focused by the spherical mirror onto the matrix of PMTs analogous to that shown in Fig. 7.10. Courtesy of the Auger Observatory (http://www.auger.org/)

emitted photons. However, this effect finds a partial compensation in the fact that in regions of higher density, excited nitrogen molecules/ions undergo more frequent collisions, which cause de-excitation without emission of light.

Fluorescence detectors can observe showers developing even several km away, with a very large field-of-view. Because of the large distance between shower and detector and of the weak intensity of fluorescent light, aerosols and dust, which may cause absorption and scattering must be kept under careful control.

Cherenkov radiation has the property of emitting a much larger number of photons than fluorescence. However, the former radiation is strongly beamed in the very forward direction and does not constitute an important source of background noise.

It is easy to show (Kuempel et al. 2008) how the geometrical details of the shower (distance from telescope, direction in the sky) can be derived from observations made even with a single "eye", like the one shown in Fig. 7.10. Those pixels that have recorded a signal, with their space orientation, define a plane. This is known in the literature as the shower-detector-plane and clearly, the shower lies in such a plane, as shown in Fig. 7.12. In this figure, R_p is the distance of closest approach of the shower to the detector, t_0 is the time of transit of the shower by this point, t_i the time when pixel #i has received light emitted at the corresponding position, χ_i the angle that pixel #i forms with the horizontal in the shower-detector-plane, and χ_0 the angle between the shower and the horizontal plane (still measured in the shower-detector-plane). It can easily be seen that the following relationship between

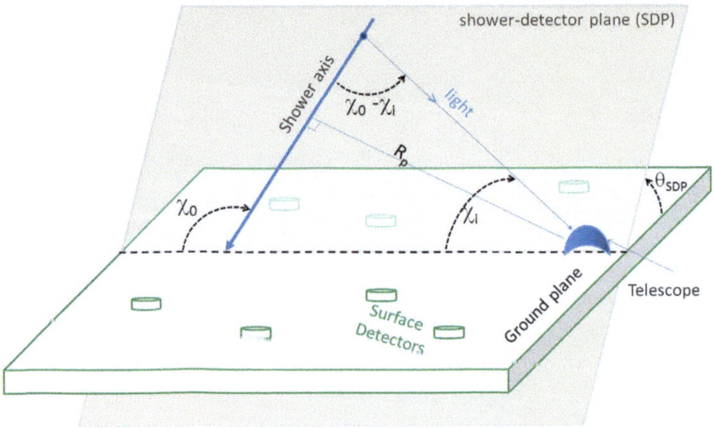

Fig. 7.12 Geometry of the detection of an air shower by a fluorescence telescope

the measured quantities t_i, χ_i and the needed variables χ_0, R_p, t_0 hold:

$$t_i = t_0 + \frac{R_p}{c} \tan[\frac{\chi_0 - \chi_i}{2}] \, . \tag{7.27}$$

Fitting all the data for a given shower to this expression provides the desired geometrical values of the shower parameters.

It may happen that the fit fails or provides ambiguous solutions. If more than one telescope is available, the solution may be found by using the fact that each telescope provides its own shower-detector plane, and the combined use of both planes easily provides the solution and constrains the geometry of the shower axis to within a fraction of a degree. This technique, first used by the HiRes experiment and later by the Pierre Auger Observatory, is known as *stereo reconstruction*. The other possibility is to complement the fluorescence detector with a surface shower detector. This independently provides the position of the impact point of the shower on the ground and of the t_0 parameter. This, known as the *hybrid technique*, was used in both the Pierre Auger and the Telescope Array experiments.

When the shower's geometry is determined, the fluorescence yield is proportional to the number of charged particles in the cascade at different depths X in the atmosphere. The total energy of the shower (dependent on the energy of the incident CR) is proportional to the sum of the light measured along the longitudinal development of the cascade. The depth in the atmosphere of the shower maximum, X_{max}, is directly determined. Despite this fact, the identification of the parent primary nucleus that originated the cascade cannot be done on a shower-by-shower basis, because of intrinsic fluctuations (see Fig. 4.9). Even at the highest energies, the measured fluorescence profile can give information on the CRs' chemical composition only through statistical methods.

7.7 UHECR Measurements with a Single Technique

Different experiments have measured the CR flux above 10^{18} eV. The pioneering **Volcano Ranch**, which took data from 1959 to 1963, and the **Yakutsk** array were already mentioned in Sect. 4.6. The Sydney University Giant Air shower Recorder (**SUGAR**) was the only one operational in the Southern hemisphere, from 1968 to 1979. The **Haverah Park** array was operational in the United Kingdom from 1968 to 1987. A detailed description of these array experiments and of early fluorescence detectors, as well as their results, can be found in Nagano and Watson (2000) and Sokolsky (2004).

The **Akeno Giant Air Shower Array (AGASA)** was the largest air-shower array before the advent of the Pierre Auger and Telescope Array detectors and it took data between 1984 and 2003. It consisted of 111 scintillation counters, each having an area of 2.2 m^2, placed at a distance of ∼1 km from each other and covering an effective surface of ∼100 km^2. In the Southeast corner of AGASA was the densely packed 1 km^2 array Akeno, which studied the CR spectrum from below 3×10^{18} eV. The signal in the AGASA scintillator counters was produced by the electromagnetic component of the cascade, with a small contribution from the shower muons. They used the methods described in Sect. 4.9 to estimate the primary CR energy, namely the particle density (ρ_{600}) as measured 600 m from the shower axis.

The **HiRes observatory** improved the pioneering Fly's Eye. Two air fluorescence detectors (HiRes-I working from 1997; HiRes-II completed at the end of 1999) were located on two hills separated by 12.6 km at an atmospheric depth of 870 g cm^{-2}. They operated on clear moonless nights with the typical fluorescence detector duty cycle of about 10%. HiRes-I consisted of 21 telescope units with a 360° view in azimuth. Each telescope was equipped with a 5 m^2 spherical mirror and 256 phototube pixels at its focal plane. At the focus of each mirror, a camera composed of 256 hexagonal photomultiplier tubes with 40 mm diameter was present, each tube viewing a ∼1° diameter section of the sky. Each telescope covered the elevation range between 3° and 17°. Detectors on the HiRes-II site were similar to those of HiRes-I, but with twice as many mirrors organized in two rings covering an elevation from 3° to 31°.

Although HiRes-I and HiRes-II could trigger and reconstruct events independently, an important fraction of events was measured stereoscopically, allowing for the reconstruction of the shower geometry with a precision of 0.4°. HiRes-I and HiRes-II took data until April 2006 for an accumulated exposure in stereoscopic mode of 3460 h. The "monocular" mode had better statistics and covered a much wider energy range. The precision in the evaluation of the primary CR energy in monocular mode was better than 20% at energies above 3×10^{19} eV.

Globally, the number of UHECR events before the advent of the Pierre Auger Observatory (PAO) was too small, and the low statistics and large systematic uncertainties plagued a number of analyses based on this data set.

The HiRes energy spectrum was based on monocular observations with HiRes-I and HiRes-II (Abbasi et al. 2008), indicated by the full bullets in Fig. 7.13. Most of the highest energy events were observed with HiRes-I. The CR spectrum measured by HiRes has two features: one at an energy of 4×10^{18} eV (the ankle) and another at 5.6×10^{19} eV. The second one corresponds to the expected feature from the GZK cut-off. Without the suppression, assuming a primary spectrum that continues as a power law with the same spectral index seen at lower energies, 43 events were expected above 6×10^{19} eV, while only 13 events were observed. Using Poisson statistics, the number of observed events is 5.3 standard deviations away from that expected in absence of the UHECR suppression.

During its live time, AGASA has observed 11 events above 10^{20} eV with zenith angle smaller than 45°. Differently from HiRes, the AGASA results (full squares in Fig. 7.13) seemed to indicate that the energy spectrum continues without a visible discontinuity beyond 5×10^{19} eV, in violation of the GZK effect (Takeda et al. 2003).

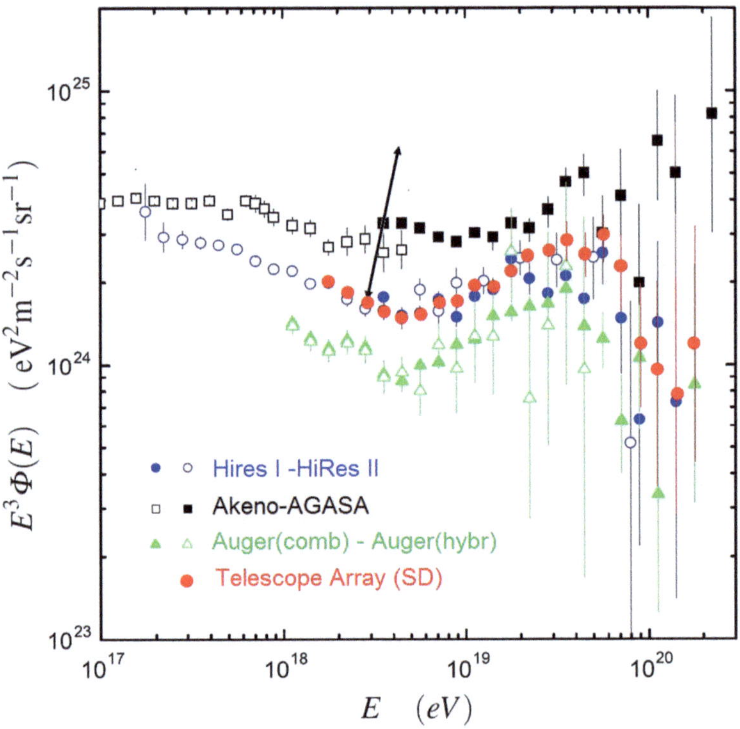

Fig. 7.13 Flux of UHECRs multiplied by E^3as measured by Akeno-AGASA, HiRes, Telescope Array and the Pierre Auger Observatory. The values as published by the Collaborations using the nominal calibration of the detectors are reported. The end of the arrow on the first point of AGASA indicates the position of the point with a ±25% shift in the energy scale

The discrepancy between the results of the two experiments was largely discussed before the advent of the PAO results. In fact, the HiRes spectrum was compatible with the existence of a UHECR suppression, while the AGASA spectrum was not. The ultrahigh energy events published by AGASA motivated the exotic top-down models of particle acceleration (see Sect. 7.11).

One possible explanation for the discrepancy between the two experiments is a possible bias regarding the energy assignment of the parent CR from one (or both) of the two methods. The energy in AGASA is estimated from the electromagnetic component of the shower with a small contribution from muons. In HiRes, it is determined from the observed development of the shower, with the critical parameter of the detector aperture.

One of the main objectives of a hybrid detector such as the PAO was to disentangle these contradictory results. If one assumed wrong the AGASA result, the simplest interpretation of the HiRes result was that UHECRs are protons generated outside the local Superclusters of Galaxies (35 Mpc), where no reasonable accelerators seem to be present. The flux of protons above 10^{20} eV is then attenuated by the GZK cut-off. On the other hand, if one assumed wrong the HiRes result, the non-evidence of a UHECR suppression could be due to heavier nuclei, or to their exotic production.

Beware of the Plots It is common to present the CR flux multiplied by E^{κ} to enhance deviations from a $dN/dE \propto E^{-\alpha}$ power law. The proper choice is $\kappa \simeq \alpha$, and for the UHECRs, usually $\kappa = 3$, as in Fig. 7.13. In this representation, characteristic features such as the ankle at $\sim 10^{18.5}$ eV are more evident. However, it should be emphasized that scaling the flux with energy to some power could induce a bias regarding the interpretation of the results from the visual inspection of the figure. All experiments, in fact, quote a systematic uncertainty ΔE about the estimated energy E. This means that points referring to a given experiment can be shifted along the x-axis by $\pm\Delta E$. Consequently, points along the y-axis must also be shifted by a quantity $(1\pm\Delta E/E)^3$. As an example, if we assume that the AGASA points suffer from a 25% systematic overestimation of the energy, all points should be shifted left by a factor of 0.75 E along x and pushed down by $(0.75)^3 = 42\%$ along y with respect to the nominal value. The arrow in Fig. 7.13 indicates such a shift for the first of the AGASA experimental points.

7.8 Large Hybrid Observatories of UHECRs

The Pierre Auger Observatory (PAO) and the Telescope Array (TA) Observatory are the two largest operating CR observatories ever built, operating, respectively, in the Northern and Southern hemispheres of the Earth. The two experiments are based on the hybrid concept in which both fluorescence and surface array detection techniques are used in order to combine and enhance the single detector capabilities and provide an accurate cross-check of systematic uncertainties of the two detection methods. A comparison of characteristics of the two experiments is given in Table 7.2.

The **Pierre Auger Observatory (PAO)** was completed in 2008, and has been collecting reliable data since 2004. The current layout of the PAO is shown in Fig. 7.14. **The surface detector (SD)** array consists of 1600 water Cherenkov detectors spaced by $\sim$1.5 km on a grid covering a total area of 3000 km^2. Sixty additional tanks were inserted inside the regular array with a smaller spacing (750 m) to extend the energy range to lower energies (down to 3×10^{17} eV). This denser array is coupled to the HEAT FD extension (see below).

Each tank of the SD array (Fig. 7.15) has a 10 m^2 surface and 1.2 m depth of purified water. Each station is equipped with three PMTs to measure the Cherenkov light produced in the water, Flash Analog to Digital Converters (FADC), a data acquisition and front-end electronic cards for control and trigger, a solar panel and two batteries for power, a Global Positioning System (GPS) receiver for the time tagging, and a custom radio emitter and receiver for trigger and data transfer. The station is connected to the Central Data Acquisition System in the Observatory building via a radio link to the nearest communication tower. The SD has a 100% duty cycle and the height of the individual SD tanks allows us to detect muons with also excellent sensitivity to horizontal showers.

The SD tanks are sensitive to muons, electrons, positrons, and photons. A vertical muon in the tank deposits about 240 MeV (the muon energy loss in water is $\sim$2 MeV/cm), while it is, on average, only a few tens of MeV for electrons. The

Table 7.2 Comparison of characteristics of the Pierre Auger Observatory (PAO) and the Telescope Array (TA)

	Pierre Auger Observatory	Telescope array
Average latitude	35.3° S	39.4° N
Location	Mendoza, Argentina	Utah, USA
Average altitude	1400 m	1400 m
Surface detectors		
Surface area	$\sim$3000 km^2	$\sim$760 km^2
Number of detectors	$\sim$1600	$\sim$500
Lattice distance, structure	1.5 km, hexagonal	1.2 km, square
Detector type	Water Cherenkov	Plastic Scintillator
Detector size	10 m^2 $\times$ 1.2 m	(2$\times$) 3 m^2 $\times$ 1.2 cm
Detector sampling	25 ns	20 ns
Fluorescence detectors		
Number of sites	4	3
Number of telescopes	24	36
Size of each telescope	13 m^2	6.8 m^2/3 m^2
Field of view	28.5° $\times$ 30°	16° $\times$ 14° /18° $\times$ 15°
Number of pixels	440	256

The low energy extensions for each observatory, HEAT and TALE (see text), are not included

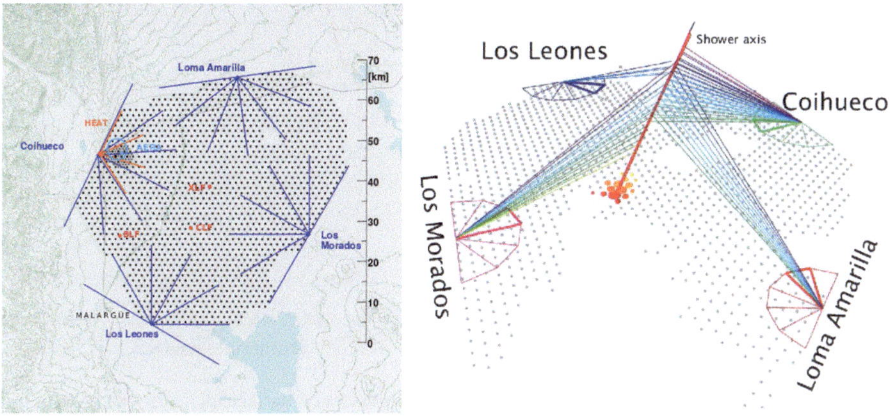

Fig. 7.14 The Pierre Auger Observatory. On the *left*, the position and field-of-view of the fluorescence detector (FD) eyes surrounding the array is displayed. The *dots* mark the positions of the 1600 surface detector (SD) tanks. On the *right* an event detected by the four FD eyes in coincidence with the SD array. Courtesy of the Auger Observatory (http://www.auger.org/)

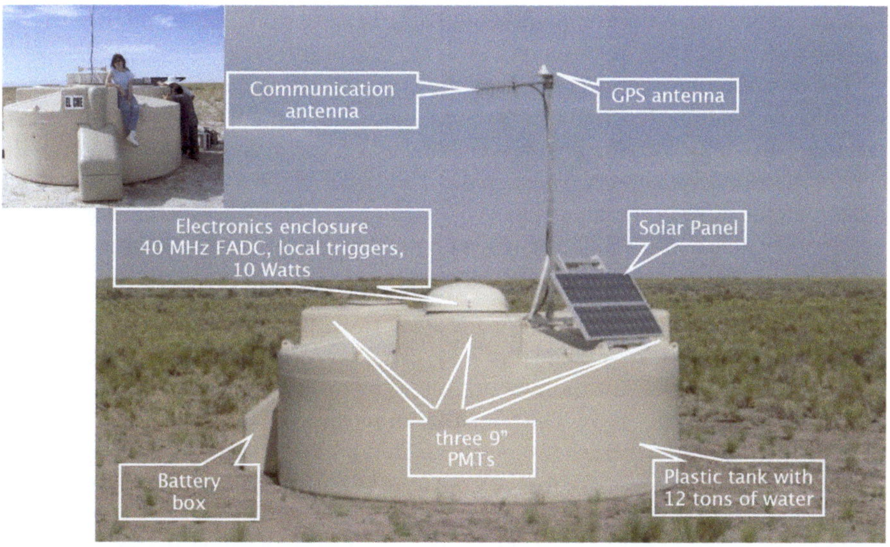

Fig. 7.15 One of the SD water Cherenkov tanks. The inset gives an idea of the size of a tank. Courtesy of the Auger Observatory and Dr. A. Castellina

recorded signal in each detector is expressed in vertical equivalent muon (VEM) units. One VEM is the average of the signals produced in the 3 PMTs by a vertical muon that passes centrally through it. The atmospheric muon rate provides an in situ calibration of the PMT gains. Local triggers are sent to the Central Data Acquisition

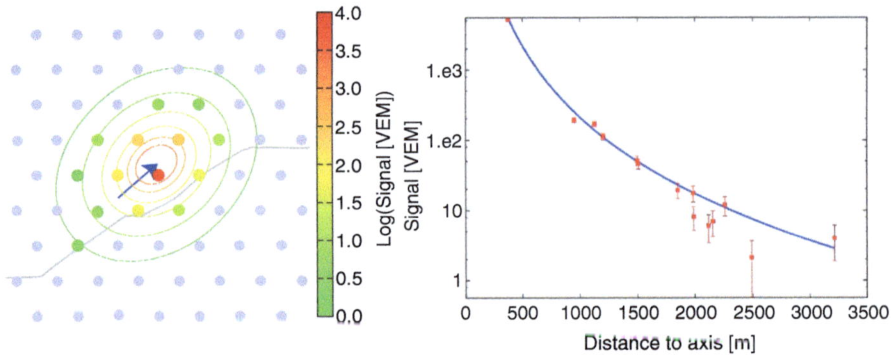

Fig. 7.16 *Left* zoomed view of the Pierre Auger Observatory surface detector interested by a cascade. The quantity is expressed in terms of a vertical equivalent muon (VEM) crossing a tank. The *concentric lines* represent the distance from the axis shower with an equal number of particles, obtained after reconstruction. *Right* fit of the event using the Auger Lateral Distribution Function. On the y-axis, the signal in units of VEM is proportional to the signal density ρ = particles/cm^2 as a function of the distance from the axis shower. Courtesy of the Auger Observatory

System, where space-time coincidences of at least three tanks are required to trigger the apparatus and allow for the permanent storage of the event.

The position of the shower core and the lateral distribution function are inferred with the techniques described in Sect. 4.8, see also Fig. 7.16. The energy of the primary particle is correlated with the signal at a fixed distance (1000 m) from the core of the extensive shower. The signal at 1000 m from the axis, S(1000), corrected for the attenuation in the atmosphere, is used as an energy estimator. For a given energy, the value of S(1000) decreases with the zenith angle of the incoming primary CR due to the attenuation of the shower particles and geometrical effects.

The fluorescence detector (FD) of the PAO uses the same detection method as HiRes and consists of four eyes (Fig. 7.14). Their geometrical arrangement ensures full detection efficiency for showers originated by primaries with $E > 10^{19}$ eV over the entire surface of the array. The High Elevation Auger Telescopes (HEAT) consist of three tiltable fluorescence telescopes, which represent a low energy enhancement (down to a primary energy of 10^{17} eV) of the FD system of the PAO. The HEAT has been collecting data since September 2009.

The presence of surface and fluorescence detectors permits the measurement of the development profile of the air shower with the hybrid technique. The first step is the geometrical determination of the shower axis using directions and timing information from the FD pixels, coupled with the arrival time of the shower at the SD station with the highest signal.

The primary CR energy is determined by the FD in a calorimetric way using the reconstructed energy deposited along the shower's profile. The sub-sample of extensive air showers that are recorded by both the FD and the SD, called golden hybrid events, is used to relate the energy reconstructed with the FD, E_{FD}, to

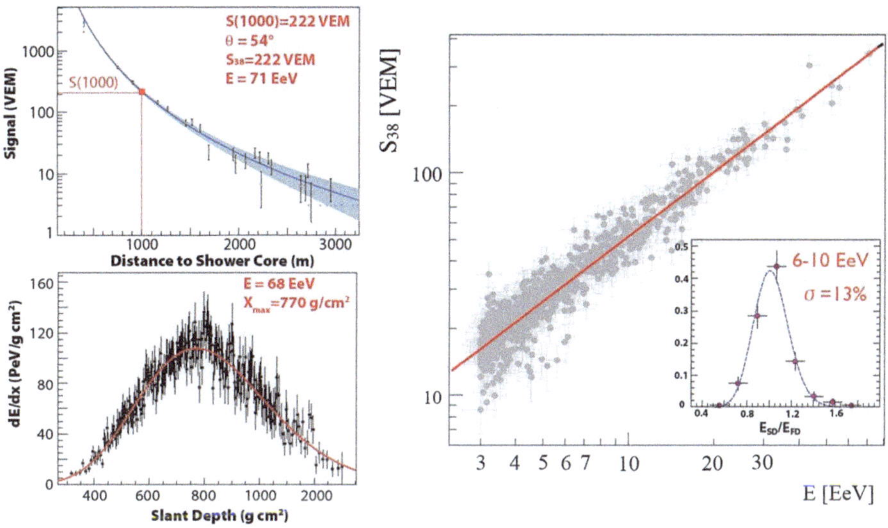

Fig. 7.17 The *left side* shows one golden hybrid event from PAO. *Upper panel*: the signal in the SD (in VEM units) as a function of the distance from the axis shower. The fit with the lateral distribution is also indicated with a line with error band. The SD information also provide the zenith angle of the incoming primary (in this case, 54°). The interpolated signal at 1000 m is used to estimate the energy (71 EeV). The *left bottom* panel shows the energy loss as a function of the atmospheric slant depth as measured by the FD. The maximum of the shower occurs at $\sim$770 g cm^{-2} and the estimated primary energy is 68 EeV. The *right panel* shows the correlation between the energy reconstructed in golden hybrid events, as presented at the ICRC conference in 2011. The energy from the FD is along the x-axis and that from the SD along the y-axis. The inset shows the distribution of the ratio between the two reconstructed energies and the estimated resolution. Courtesy of the Auger Observatory (http://www.auger.org/)

S(1000) (see Fig. 7.17). The energy scale inferred from this data sample is applied to all showers detected by the SD array. The angular and energy resolution of hybrid measurements at 1 EeV is better than 0.5° and 6%, respectively, compared with about 2.5° and 20% for the surface detector alone.

The **Telescope Array (TA) Observatory**, see Table 7.2, has been collecting data in the high desert in Millard County, Utah, USA, since 2007, observing CRs with energies above 10^{19} eV. It is a hybrid detector, mixing the information from both fluorescence detectors (FD) and surface detectors (SD), as in the case of PAO. The cosmic rays are observed at three fluorescence sites and with 507 SD consisting of 3 m^2 double layer scintillators powered by a solar panel. The array is divided into three parts that communicate with three control towers where information is digitized. A trigger is produced when the signals of three adjacent stations coincide within 8 μs. The SD reaches a full efficiency at 5×10^{18} eV for showers having a zenith angle smaller than 45°. This corresponds to a SD acceptance of 1600 km^2 sr. The fluorescence sites are about thirty kilometers apart from one another, forming an approximately equilateral triangle. Two of them are new and consist of 12 telescopes viewing elevations from 3° to 31° and 108° in the horizontal field-of-view. Each telescope has a camera consisting of 256 PMTs with a field-of-view of 1° × 1°. The signals are digitized by a 40 MHz sampling fast ADC converter and the waveforms are recorded when signals are found in five adjacent PMTs. The third station has 14 telescopes that use cameras and electronics from HiRes-I and mirrors from HiRes-II.

A low energy extension to the Telescope Array was also realized. TALE (the Telescope Array Low Energy extension) allows the observation of CRs with energies as low as 3×10^{16} eV. This is accomplished by means of 10 new high elevation angle telescopes, viewing up to 72°, at one of the telescope stations and with the addition of a graded infill array of scintillator surface detectors.

7.9 Recent Observations of UHECRs

7.9.1 *The Flux and Arrival Directions of UHECRs*

Almost at the same time as the final HiRes results, the first 3.5 years (2004–2007) of scientific data collected by PAO (Abraham et al. 2008) confirmed the existence of the UHECR suppression. The used data set had approximately four times the exposure of AGASA and 69 events above 4×10^{19} eV were measured, while 167 ± 3 were expected from an extrapolation of the power law measured at lower energies. Only 1 event above 10^{20} eV was found, instead of the 35 expected. Since then, the statistical sample has increased, confirming the early result (Aab et al. 2017).

Like HiRes, PAO observes the shower profile with its fluorescence detectors, which have a 13% duty cycle compared to that of the surface detector. In Fig. 7.13, the flux measured in the hybrid mode is represented by empty triangles. The result based on data from the surface detector array, with an energy calibration provided by the fluorescence detector, is represented by full triangles.

Later, the TA collaboration measured (Abu-Zayyad et al. 2013) the energy spectrum of UHECRs with energies above 1.6×10^{18} eV, using 4 years of observation by the surface detectors. The spectrum shows a depletion at an energy of 4.6×10^{18} eV and a steepening at 5.4×10^{19} eV, Fig. 7.13. To have an idea of the approximation involved in CR simulations, the energy scale of the SD determined from simulations can be reconciled with the calorimetric scale of fluorescence detectors by a renormalization of 27%. This result is in excellent agreement with that of the HiRes experiment.

Figure 7.18 shows a summary of all experimental results on the flux of UHECRs. In Fig. 7.13, the experimental points are plotted as they were presented in the original publications. Here, an (arbitrary) energy shift compatible with the quoted systematic error has been applied to all the experiments (Blümer 2009). The figure demonstrates the importance of the systematic uncertainty about the experimental

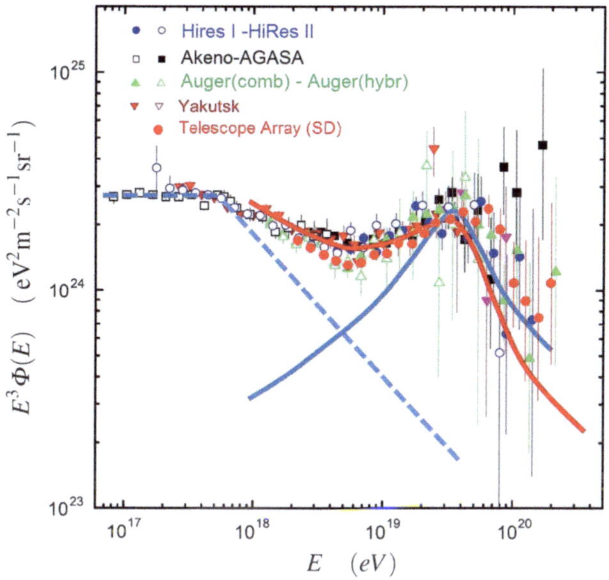

Fig. 7.18 The same data as in Fig. 7.13 (with the inclusion of the Yakutsk experiment) after re-scaling the energy of the experiments to obtain a common position of the depletion at ~5 10^{18} eV. The underlying theoretical assumption is that the depletion is due to the reaction $p + \gamma_{\text{cmb}} \rightarrow pe^+e^-$ (Aloisio et al. 2012). The nominal energy scales of the experiments have been multiplied by 1.2 (Auger), 1.0 (HiRes), 0.75 (AGASA), 0.95 (TA) and 0.625 (Yakutsk). Superimposed, the prediction from two theoretical models [adapted from Letessier-Selvon and Stanev (2011)]. The *red line* represents the *dip model* resulting from extragalactic protons, the *blue line* the *superposition model* of a galactic (*dashed line*) plus an extragalactic component (*full line*)

energy scale discussed above. Although the procedure is only motivated by a theoretical model and the value of the shift for each experiment is discretional, after the shift, a good overall agreement for almost all the different measurements is obtained.

The softening of the spectrum above $\sim 5 \times 10^{19}$ eV observed by HiRes, PAO and TA is consistent with the GZK effect, but it does not necessarily represent an unambiguous proof that the cut-off has been discovered. The GZK cut-off at $\sim 10^{20}$ eV necessarily implies that UHECRs are protons, and not heavier nuclei. The discussion about the chemical composition of UHECRs featured in the next subsection plays an important role. The softening of the spectrum can also be explained as an intrinsic feature of the source spectra themselves, indicating a maximum energy for the acceleration process.

7.9.2 The Chemical Composition of UHECRs

The determination of the charge (or mass) of the incoming primary UHECRs is difficult. Shower-to-shower fluctuations, which are larger for protons, prevent the individual measurement of the primary CR mass number. The general features of a cascade initiated by heavy nuclei are (Sect. 4.4.4): (1) the showers reach their maximum development, X_{max}, higher in the atmosphere, and (2) they generate more muons than showers induced by a proton primary of equivalent energy. The discrimination between the arrival of a proton or that of a heavier nucleus relies on the observation of either the longitudinal development of a shower (total number of photons and depth in atmosphere of the maximum) or by the simultaneous determination of the electromagnetic and muonic components of air showers at ground level. Unfortunately, the former technique can be used only on clear, moonless nights and suffers from the lack of statistics at the highest energies.

The first analysis of the X_{max} energy dependence with fluorescence detector data was done with the Fly's Eye. The HiRes Collaboration published, in 2005, an analysis of the UHECR composition above 10^{18} eV with the same method. The depth of the shower maximum as a function of the estimated energy of the incoming primary is shown in Fig. 7.19, which represents the high-energy section of Fig. 4.20. The HiRes result suggested a quick transition from heavy to light cosmic ray composition above a few 10^{18} eV. The Telescope Array measurement of the X_{max} position is consistent with that of HiRes.

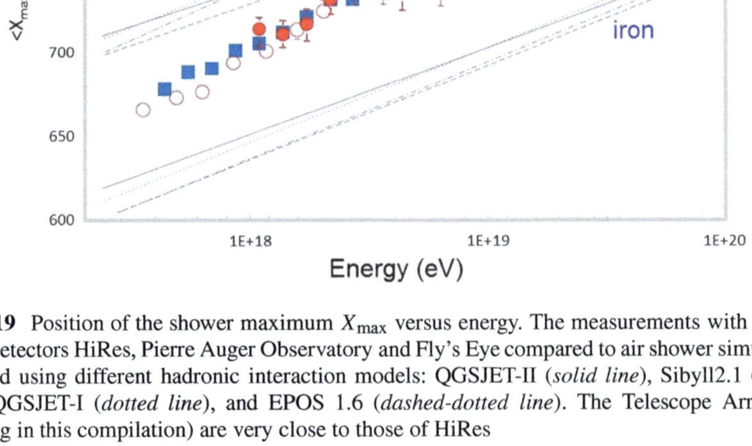

Fig. 7.19 Position of the shower maximum X_{max} versus energy. The measurements with fluorescence detectors HiRes, Pierre Auger Observatory and Fly's Eye compared to air shower simulations obtained using different hadronic interaction models: QGSJET-II (*solid line*), Sibyll2.1 (*dashed line*), QGSJET-I (*dotted line*), and EPOS 1.6 (*dashed-dotted line*). The Telescope Array data (missing in this compilation) are very close to those of HiRes

The Pierre Auger Observatory published results based on hybrid events in 2010. Their results are somewhat in disagreement with those of HiRes. In their data, the cosmic ray composition becomes lighter up to 2×10^{18} eV and then consistently heavier up to the highest energy measured. This interpretation is confirmed by the analysis of the width of the X_{max} distributions, $\sigma(X_{max})$. As shown in Sect. 4.4.4, a wider $\sigma(X_{max})$ distribution for light primaries with respect to heavier ones is expected. In the PAO data, while the $\sigma(X_{max})$ values from 0.7×10^{18} eV to 5×10^{18} eV look consistent with a light composition, at higher energies (up to the last point at 4×10^{19} eV), they are more consistent with a heavier composition. Note that the data in Fig. 7.19 terminate before the region where the GZK effect starts to be effective. The chemical composition of CR above 5×10^{19} eV is unknown.

The interpretation of the X_{max} and $\sigma(X_{max})$ data in terms of CR composition suffers from the fact that the experimental uncertainties are still too large. Additionally, hadronic interaction models also play an important role, as mentioned in Sect. 4.5. The HiRes and TA results are more consistent with the hypothesis that most of the CRs above 10^{19} eV are protons of extragalactic origin and that the strong decline of the cosmic ray flux above $\sim 5 \times 10^{19}$ eV is due to the interaction of primary protons with the CMB radiation background (the GZK cut-off). On the other hand, the PAO data seems to indicate that the overall proton contribution to the spectrum is mostly restricted to $E \leq 3 \times 10^{19}$ eV, well below the photo-pion production threshold (see Eq. (7.19) and following comment). According to PAO data, the ankle feature could be interpreted as the transition between two different classes of sources, the high-energy one with a large fraction of nuclei heavier than protons. The flux suppression at the highest energies could be generated by the photo-disintegration of nuclei and/or the maximum acceleration energy at the CR sources, instead of GZK cut-off.

A direct comparison of the X_{max} distributions measured by the PAO and TA experiments is not possible due to different detector acceptances and resolutions, as well as different analysis techniques used by the two collaborations. In order to assess a quantitative compatibility test, a joint PAO-TA working group compared the two measured X_{max} distributions (Abbasi et al. 2017). A set of showers compatible with the composition measured by the PAO was simulated; the events were reconstructed using the official TA software chain. This procedure simulates an energy-dependent composition mixture, which represents a good fit with the X_{max} distribution measured by Auger, exposed through the detector acceptance, resolution and analysis procedure of the TA experiment. Two compatibility tests were applied to the two resulting X_{max} distributions (the one from Auger in input, and the one reconstructed in TA obtained in output). Both tests indicate that the TA data are, within the systematic uncertainties, compatible with the mixed composition such as that measured by the PAO.

In the last decade, the PAO and TA experiments have accumulated UHECRs with an impressive amount of exposure, providing data with unprecedented quality. However, the detailed knowledge of the chemical nature of UHECRs is an open research field: the lack of statistics and the large systematic uncertainties require long and detailed future studies. As illustrated in the above example of comparing the X_{max} distributions, a coherent and final interpretation of the results concerning the nature of UHECRs has not yet been achieved. The mass composition of CRs in the energy region of the flux suppression is of key importance for testing astrophysical source models.

The improvement of the mass composition sensitivity of the surface detectors is the main goal of the PAO's major upgrade, called AugerPrime (Aab et al. 2017). The upgrade program will include: new plastic scintillator detectors on top of the SD; more powerful SD electronics; an extension of the dynamic range on the measurement of the light yields in the water-Cherenkov detectors with the installation of an additional PMT in the SD. The start of the commissioning of the upgraded stations is expected in 2018.

7.9.3 Correlation of UHECRs with Astrophysical Sources

The mass composition problem discussed above is strictly associated with the identification of UHECR sources. One related question is the origin of the depletion at around 5×10^{18} eV, evident in Fig. 7.18. According to the *superposition models*, the depletion is at the intersection of the rise of the extragalactic component and the decline of the galactic one. According to these models, the extragalactic CRs have a flat spectrum (from $\sim E^{-2}$ to $\sim E^{-2.5}$) and the galactic ones have a steep (about $\sim E^{-3.5}$) spectrum. Galactic cosmic rays, although in a minor way, also contribute above 10^{19} eV (the dashed line in Fig. 7.18).

A second model (the *dip model*) foresees that the depletion is caused by e^+e^- pair-production by the extragalactic protons with the CMB (Sect. 7.5.3). The transition from galactic to extragalactic cosmic rays would, in this model, take place at energies below 10^{18} eV. The extragalactic CRs have to be almost exclusively protons to allow $e^{\pm}$ pair-production, and their injection spectrum must be as steep as $E^{-2.7}$.

In both of the above scenarios, the cosmological evolution of the Universe would not much affect the predicted spectra, because the observed UHECR have to be produced in a relatively small region of the visible Universe. This property also seems to be common to other, more complex, models (Allard 2012; Letessier-Selvon and Stanev 2011).

The possible identification of extragalactic sources must exploit information both on the chemical composition and on the arrival direction distribution of the UHECRs. The latter possibility depends on the type of particle (protons or heavier nuclei) and on extragalactic magnetic fields. Attempts to identify the sources of cosmic rays can be divided into three general categories:

(1) searches for excesses from individual astrophysical objects or regions, for example, the Galactic center, using data from a single experiment;
(2) searches for statistically significant correlations of CR arrival directions with catalogs of astrophysical objects (multimessenger method);

(3) searches for anisotropies in the CR arrival direction distribution itself, independent of any catalogs. These include searches for large-scale anisotropies or small-scale clustering.

To date, no individual source of ultrahigh-energy CRs has been positively identified. Before the advent of HiRes, PAO and TA, some fine-tuning of the data set and the small number of events have led to a number of claims of anisotropies with marginal statistical significance. These early hypotheses have failed when tested with data from experiments with larger exposure and superior resolution, which ruled out several long-standing claims. See Beatty and Westerhoff (2009) and Letessier-Selvon and Stanev (2011) for a summary.

So far, the strongest positive result has been obtained with method (3) and corresponds to the evidence for a large scale anisotropy in the arrival directions of CRs above 8×10^{18} eV, presented in Sect. 5.7.1.

The field-of-view of Telescope Array is the Northern sky, with a small overlap with that of the PAO. The TA searched for intermediate-scale anisotropy in the arrival directions of UHECRs with energies exceeding 57 EeV using SD data collected over a 5-year period. A cluster of events (defined as a *hotspot*), found by oversampling using 20°-radius circles, has been reported. The hotspot is centered at RA=146.7°, δ=43.2°, about 19° off of the supergalactic plane,[3] and has significance at the 3.4σ level (Abbasi et al. 2014).

One of the intriguing results arising from the multimessenger method (2) is the recent analysis of PAO, using 5514 events above 20 EeV and with zenith angles up to 80°. The data provide evidence of anisotropy in their arrival directions on an intermediate angular scale, which is indicative of excess arrivals from strong, nearby sources (Aab et al. 2018). In order to ascribe the excess to possible classes of extragalactic sources, the UHECRs' information have been correlated with the observed γ-ray flux from the Fermi-LAT satellite (see the next chapter). *Sky models* for two distinct populations of extragalactic γ-ray emitters have been created: active galactic nuclei and starburst galaxies, with the starburst model having been found to fit the data better than the hypothesis of isotropy with a statistical significance of 4.0σ.

7.10 Measuring EeV Neutrinos with EAS Arrays

The two aforementioned theoretical scenarios (the *dip* and the *superposition models*) aimed at the interpretations of UHECR observations differ in their chemical composition. The dip model assumes a pure proton composition, while all the superposition models are based on a mixed composition, with the low energy part of

[3]The spherical *supergalactic coordinates* system has its equator aligned with the supergalactic plane, a major structure in the local Universe formed by the preferential distribution of nearby galaxy clusters (such as the Virgo cluster, the Great Attractor and the Pisces-Perseus supercluster).

the spectrum dominated by protons and the highest energy part dominated by nuclei. A powerful tool to distinguishing between the two scenarios is provided by the flux of secondary neutrinos, which is extremely sensitive to the chemical composition of UHECRs.

In fact, during their journey from the source, UHECRs cover cosmological distances and interact with the cosmic microwave background (CMB, Sect. 7.5) and the extragalactic background light (EBL, Sect. 9.14). Protons are affected almost exclusively by the CMB radiation field through both $e^{\pm}$ pair production and the resonant production of one pion. Nuclei are also affected by the EBL, and pair production and photo-disintegration are the relevant processes. As shown in Sect. 7.5.2, above $\sim 5 \times 10^{19}$ eV, the one-pion resonant production (7.14) with CMB photons produces neutrinos of energies typically 1/20 of the proton energy. Those ultrahigh energy neutrinos are called *cosmogenic* (Berezinsky and Zatsepin 1969). Cosmogenic neutrinos are thus expected at EeV ($= 10^{18}$ eV) energies. This mechanism is efficient only in the case of protons, while in the case of heavier nuclei, the neutrino production is significantly suppressed. Thus, a sizeable flux of cosmogenic EeV neutrinos is expected only if UHECRs around 10^{20} eV are protons.

The whole universe contributes to the flux of cosmogenic neutrinos, up to the maximum red-shift of astrophysical structures able to accelerate UHECRs. This maximum is $z_{max} \sim 10$, corresponding to the redshift of the first stars formed in the Universe.[4] Once produced at cosmological distances, neutrinos travel toward the observer almost freely, except for the adiabatic energy losses (Sect. 7.5) and flavor oscillations (Sect. 12.8).

While the cosmological evolution of the CMB is analytically known, the case of the EBL is model-dependent and several models have been proposed. As sources producing accelerated particles emit radiation as well, models of UHECRs sources are strictly related to models of EBL evolution. The simplest model corresponds to the no evolution, i.e., density and luminosity of UHECR sources are independent of their red shift. Other special evolution models assume that the UHECR sources have the same evolution as either the *star formation rate* (SFR), the gamma-ray burst (GRB) rate, or the active galactic nuclei (AGN) rate in the Universe. See Gelmini et al. (2012) and references therein for further details. Some constraints on these models will be discussed in Sect. 10.11.4. Hence, cosmogenic neutrinos are also a viable probe of the cosmological evolution of sources, while UHE protons and nuclei, given their energy losses, can be observed only if produced at $z < 3 - 4$.

Arrays of particle detectors at ground level able to measure almost horizontal showers (such as the surface detectors, SD, of the PAO) can be used to search for EeV neutrino candidates. In fact, the electromagnetic component of showers induced by CRs with highly inclined directions gets absorbed due to the large gram-

---

[4]The first stars formed in the Universe correspond to the hypothetical *population III stars*. Their existence is inferred from cosmology: they have not yet been observed directly and only indirect evidence exists. They are supposed to be extremely massive and hot, with virtually no metals, except possibly for intermixing ejecta from other nearby Pop III supernovas. Their existence may account for the fact that heavy elements are observed in quasar emission spectra.

mage of atmosphere from the first interaction point to the ground. Consequently, the shower front at ground level is dominated by the muonic component that induces sharp time signals in the water-Cherenkov stations. Downward-going neutrinos can start a shower deeper in the atmosphere, producing a young electromagnetic component that continuously spreads over time as the $e^{\pm}, \gamma$ pass through the detector. Neutrino identification uses the time structure of the shower signals in the SD stations; this selection is efficiently done for neutrinos of all flavors interacting in the atmosphere at large zenith angles.

In addition, tau neutrinos can produce *Earth-skimming* CC interactions. A horizontal ν_τ traveling below the Earth's crust, interacting near the exit point, induces a tau lepton that escapes the Earth and decays in flight in the atmosphere above the SD, as depicted in Fig. 7.20. In all cases, only neutrinos with energy above 10^{17-18} eV can produce a sufficient number of secondary particles to trigger the apparatus. Since the neutrino cross-section grows with energy, PeV ($= 10^{15}$ eV) and higher energy neutrinos cannot cross the Earth (see Sect. 10.7). The detection of these neutrinos is only possible if they interact in close proximity to the detector and produce a cascade.

Based on searches for near-horizontal or upward-going Earth-skimming showers, upper limits on the electron neutrino flux have been published by the HiRes collaboration, and limits on the diffuse flux of neutrinos of all flavor have been set by PAO (Aab et al. 2015). The present experimental limits on EeV neutrinos are used to constrain models on the cosmological evolution of sources, as discussed in Sect. 10.11.4.

The same method for the search of Earth-skimming neutrinos was used by PAO (Albert et al. 2017) to search for $E_\nu > 10^{17}$ eV eventually produced in the occurrence of the gravitational wave GW170817, Sect. 13.7. The position of the transient in NGC 4993 was just between 0°.3 and 3°.2 below the horizon during a window of $\pm$ 500 s centered on merging time. This region corresponds to the most efficient geometry for Earth-skimming ν_τ detection. No neutrino candidate was found.

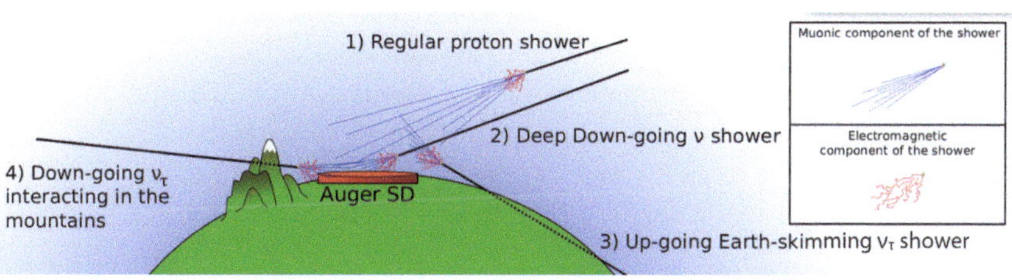

Fig. 7.20 Near-horizontal or upward-going Earth-skimming showers are distinguished by the PAO by measuring the electromagnetic and muonic components of the shower using the surface detector array. Near *horizontal* proton cascades are depleted in the EM component, which is dominant for the ν-induced cascades. Courtesy of the Auger Observatory (Guardincerri 2011)

7.11 Constraints on Top-Down Models

Before 2005, the results from AGASA about the non-evidence of a change in the CR energy spectrum above 10^{20} eV stimulated many theoretical ideas about non-standard acceleration mechanisms. One very attractive scenario (at least for particle physicists) was the mere suppression of the astrophysics accelerators themselves. In these *top-down models*, UHECRs are not accelerated as such, but rather directly produced, via the decay of some supermassive relic of the Big Bang, or by the collapse of topological defects. These models necessarily imply the existence of exotic particles with energies or masses largely exceeding 10^{20} eV. The disadvantage of top-down models is the replacement of the acceleration problem with the question of the nature and existence of such particles.

Direct implications of top-down models are both the non-existence of a CR suppression and the prediction of a large flux of UHE γ-rays and neutrinos. As mentioned above, around 2005, the UHECR suppression was observed by both the HiRes and PAO experiments, while the large flux of γ-rays and neutrinos was not. These observational facts have strongly reduced interest in top-down models.

We already mentioned that no candidates for an Earth-skimming ν_τ event have been observed so far, and the derived 90% confidence level upper limit excludes most top-down models. Let us describe now how it is possible to distinguish a photon-induced shower from those induced by primary hadrons. For a primary γ-ray of energy $\sim 10^{19}$ eV, the average depth of shower maximum is $\sim 1000\,\mathrm{g\,cm^{-2}}$, roughly $200\,\mathrm{g\,cm^{-2}}$ larger than for proton primaries. In addition, the corresponding particle density at a fixed distance is smaller. You can work out these results using the arguments reported in Chap. 4.

The PAO has set limits on the presence of primary photons in the CR flux, using the characteristics of both the surface and fluorescence detectors. A significant presence of photon primaries in the CR flux would, in fact, introduce a considerable bias in the energy determination based on the number of particles at a given distance from the impact point, i.e., the parameter S(1000) used by the surface detectors. In addition, the surface detectors measure the radius of curvature and the spread of the arrival time of the shower front at a fixed distance to the core, other quantities that behave differently for primary photons and hadrons. The fluorescence detector limit is based on the direct measurement of X_{max}. The result is that a significant presence of UHECR photons, as required by top-down models, is excluded.

7.12 Summary and Discussion of the Results

The detection of the highest energy CRs (above a few tens of EeV) requires large, expensive arrays and experiments of considerable duration. At present, the lack of large statistics on two important observables, the chemical composition and the arrival direction, is the main problem. Both the composition and the anisotropies will tell us about the UHECRs sources and their distribution, as well as about the involved acceleration mechanisms.

While the existence of the spectral softening above $\sim 5\times 10^{19}$ eV is no longer questioned, its interpretation depends on the nature of primary UHECRs. The HiRes and TA measurements of air shower maximum development are compatible with protons. The PAO measurements up to 3×10^{19} eV, however, suggest a gradual change to a heavy composition. If the UHECR flux is dominated by protons, the most likely explanation for the slope change is the GZK cut-off. If the composition is increasingly heavier to the highest energies, the structure could be the result of the maximum acceleration rigidity having been reached at the sources.

The other salient feature of the UHE energy spectrum is the so-called *ankle* near 4×10^{18} eV, whose concavity (Fig. 7.18) is that expected for a simple transition between two power-law spectra. As such, it may correspond to the energy at which extragalactic CRs become more abundant than those originated within the Galaxy. Despite their enormous Larmor radius, the arrival directions of CRs at the ankle do not favor the galactic center region or the galactic plane. An alternative explanation for the structure is that UHE protons suffer gradual energy loss to $e^{\pm}$ pair production by background microwave photons. The flux of extragalactic protons should exhibit a depletion due to this propagation effect in the energy region of $\sim 6\times 10^{18}$ eV, due to the creation of $e^{\pm}$ pairs. Even a modest admixture of other nuclei or protons of galactic origin would preclude the salient feature being due to this particular energy loss mechanism.

We also have to mention that there is tentative evidence of a second knee in the energy spectrum below 10^{18} eV, i.e., a softening of the spectrum before the hardening at the ankle, see Fig. 2.8. The PAO and TA are both attempting to make extended measurements of the energy spectrum across that energy range, and the results can have implications for the interpretation of the ankle.

In conclusion, the available data appear partially contradictory, and no model is able to explain all the observations in a coherent way (Sommers 2012). The experimental effort needs to be pursued along at least three lines: (1) increasing the statistics by instrumenting larger surface arrays; (2) improving the measurements by adding new detector components; (3) covering the whole sky. To make definite progress, the next generation of detectors should be able to measure independently, and possibly redundantly, all shower components.

The expansion of the PAO and Telescope Array techniques to larger exposures seems unreachable without a significant budget increase. New experiments such as the proposed JEM-EUSO in space should be a great step forward by providing detailed sky maps at energies for which the statistics are currently extremely low (http://jemeuso.riken.jp/).

References

A. Aab et al., Improved limit to the diffuse flux of ultrahigh energy neutrinos from the Pierre Auger Observatory. Phys. Rev. **D91**, 092008 (2015)

A. Aab et al., (The Pierre Auger Collaboration), *The Pierre Auger Observatory: Contributions to the 35th International Cosmic Ray Conference (ICRC 2017)*. arXiv:1708.06592

A. Aab et al., (PAO Collaboration), An indication of anisotropy in arrival directions of ultra-high-energy cosmic rays through comparison to the flux pattern of extragalactic gamma-ray sources. Astrophys. J. Lett. **853**, L29 (2018)

R. Abbasi et al., (HiRes Collaboration), First observation of the Greisen-Zatsepin-Kuzmin suppression. Phys. Rev. Lett. **100**, 101101 (2008)

R.U. Abbasi et al., (the Telescope Array Collaboration). Indications of intermediate-scale anisotropy of cosmic rays with energy greater than 57 EeV in the northern sky measured with the SD of the TA experiment. Astrophys. J. **790**, L21 (2014)

R.U. Abbasi et al., (The PAO and TA Collaborations), *Pierre Auger Observatory and Telescope Array: Joint Contributions to the 35th International Cosmic Ray Conference (ICRC 2017)*. arXiv:1801.01018

J. Abraham et al., (P. Auger Collaboration), Observation of the suppression of the flux of cosmic rays above 4×10^{19} eV. Phys. Rev. Lett. **101**, 061101 (2008)

T. Abu-Zayyad et al., The CR energy spectrum observed with the surface detector of the telescope array experiment. Astrophys. J. Lett. **768**, L1 (2013)

F. Aharonian, A. Bykov, E. Parizot, V. Ptuskin, A. Watson, Cosmic rays in galactic and extragalactic magnetic fields. Space Sci. Rev. **166**, 97–132 (2012)

A. Albert et al., (The ANTARES, IceCube, PAO, LIGO and Virgo Collaborations), Search for High-energy Neutrinos from Binary Neutron Star Merger GW170817 with ANTARES, IceCube, and the Pierre Auger Observatory. Astrophys. J. Lett. **850**, L35 (2017)

D. Allard, Extragalactic propagation of ultrahigh energy cosmic-rays. Astropart. Phys. **39–40**, 33–43 (2012)

R. Aloisio, V. Berezinsky, A. Gazizov, Transition from galactic to extragalactic cosmic rays. Astropart. Phys. **39–40**, 129–143 (2012)

F. Arqueros, J.R. Hoerandel, B. Keilhauer, Air fluorescence relevant for cosmic-ray detection-review of pioneering measurements. Nucl. Instrum. Methods Phys. Res. Sect. A **597**, 23 (2008)

J.J. Beatty, S. Westerhoff, The highest-energy cosmic rays. Annu. Rev. Nucl. Part. Sci. **59**, 319–345 (2009)

J. Becker, High-energy neutrinos in the context of multimessenger astrophysics. Phys. Rep. **458**, 173–246 (2008)

V.S. Berezinsky, G.T. Zatsepin, Cosmic rays at ultra high energies (neutrino?). Phys. Lett. B **28**, 423 (1969)

J. Blümer, R. Engel, J.R. Hörandel, Cosmic rays from the knee to the highest energies. Prog. Part. Nucl. Phys. **63**, 293338 (2009)

S. Braibant, G. Giacomelli, M. Spurio, *Particle and Fundamental Interactions* (Springer, Berlin, 2011). ISBN 978-9400724631

G.B. Gelmini, O. Kalashev, D.V. Semikoz, Gamma-ray constraints on maximum cosmogenic neutrino fluxes and UHECR source evolution models. J. Cosmol. Astropart. Phys. **1201**, 044 (2012)

Y. Guardincerri, The Pierre Auger observatory and ultra-high energy neutrinos: upper limits to the diffuse and point source fluxes. Contributions to the 32nd ICRC, Beijing, China (2011). arXiv:1107.4805

M. Kachelriess, Lecture Notes on High Energy Cosmic Rays (2008). arXiv:0801.4376

K. Kotera, A.V. Olinto, The astrophysics of ultrahigh-energy cosmic rays. Annu. Rev. Astron. Astrophys. **49**, 119–153 (2011)

D. Kuempel et al., Geometry reconstruction of fluorescence detectors revisited. Astropart. Phys. **30**, 167 (2008)

A. Letessier-Selvon, T. Stanev, Ultrahigh energy cosmic rays. Rev. Mod. Phys. **83**, 907 (2011)

A. Muecke et al., Monte-Carlo simulations of photohadronic processes in astrophysics. Comput. Phys. Commun. **124**, 290–314 (2000). Also arXiv:astro-ph/9903478

M. Nagano, A.A. Watson, Observations and implications of the ultrahigh-energy cosmic rays. Rev. Mod. Phys. **72**(3), 689–732 (2000)

M. Ostrowski, Mechanisms and sites of ultra high energy cosmic ray origin. Astropart. Phys. **18**, 229–236 (2002)

C. Patrignani et al., (Particle Data Group), Review of Particle Physics. Chin. Phys. C **40**, 100001 (2016); and 2017 update

P. Sokolsky, *Introduction to Ultrahigh Energy Cosmic Ray Physics* (Westview Press, Boulder, 2004). ISBN 978-0813342122

P. Sommers, Ultra-high energy cosmic rays: observational results. Astropart. Phys. **39–40**, 88–94 (2012)

M. Takeda et al., Energy determination in the Akeno Giant air shower array experiment. Astropart. Phys. **19**, 447 (2003)

D.F. Torres, L.A. Anchordoqui, Astrophysical origins of ultrahigh energy cosmic rays. Rep. Prog. Phys. **67**, 1663–1730 (2004)

J.-H. Woo, C. Megan Urry, AGN black hole masses and bolometric luminosities. Astrophys. J. **579**, 530–544 (2002)

Chapter 8
The Sky Seen in γ-Rays

Abstract The presence of galactic magnetic fields makes it impossible to localize Cosmic Ray (CR) sources using charged particles. The only way to gain information about their acceleration sites is by observing the neutral particles (γ-rays and neutrinos) generated by their interactions during acceleration. In recent years, a new window has been opened on the observation of the electromagnetic radiation up to the highest energies. The development has been made possible by the availability of new detectors coming from technologies typical of experimental particle physics. In most cases, electromagnetic radiation processes involving relativistic electrons could explain the photon flux up to the highest energies, which presents a non-thermal emission with a distribution with two distinct features. High-energy photons can be produced as well by accelerated protons though decay of secondary neutral mesons. In addition to physical mechanisms, in this chapter we describe the main experiments that allowed γ-ray astronomy: the Compton Gamma Ray Observatory, the Swift, AGILE and Fermi satellites. Unlike the sky at visible wavelengths, the γ-ray sky is dominated by a diffuse radiation originating in our Galaxy, due to the propagation of CRs in the interstellar medium. In most cases, galactic and extragalactic sources appear as point-like objects over the diffuse γ-ray background. In addition to these steady sources, flashes of gamma-rays were discovered serendipitously as early as the beginning of the 1970s. These Gamma Ray-Bursts (GRBs) are the brightest explosions in the Universe, observed at a rate of about 1/day. Their origin, classification, total energy output and the γ-ray differential flux have been experimentally investigated only recently. These objects are possible candidates as sources of ultra-high energy cosmic rays.

The presence of galactic magnetic fields makes it impossible to localize CR sources using charged particles. The only way to gain information about their acceleration sites is by observing the neutral particles (γ-rays and neutrinos) generated by their interactions during acceleration. For this reason, the present and the following two chapters are strictly related to the problem of the origin of cosmic radiation.

© Springer Nature Switzerland AG 2018

M. Spurio, *Probes of Multimessenger Astrophysics*, Astronomy and Astrophysics Library, https://doi.org/10.1007/978-3-319-96854-4_8

In recent years, a new window has been opened on the observation of the electromagnetic component (the γ-rays) of the cosmic radiation up to the highest energies, Sect. 8.1. The development has been made possible by the availability of new detectors coming from technologies typical of experimental particle physics.

In most cases, electromagnetic radiation processes involving relativistic electrons (in the so-called *leptonic model*, Sect. 8.2) could explain the photon flux up to the highest energies, which presents a nonthermal emission with a distribution with two distinct features. High-energy photons can be produced by accelerated protons though decay of secondary neutral mesons (mainly the π^0), as described in Sect. 8.4. Astrophysical candidates for γ-ray production (Sect. 8.6) include supernova remnants, pulsars, and quasars.

After a pioneering period (for a review of early experiments, see Thompson 2008; Thompson et al. 2012) a real breakthrough for γ-ray astronomy was the launch in 1991 of the Compton Gamma Ray Observatory, Sect. 8.7, with different instruments on board. Nowadays, the scenario is dominated by a few satellites, such as Swift (launched in 2004), AGILE (2007) and, up to the highest energies detectable in space, by the Fermi-LAT (launched in 2008), Sect. 8.8. Unlike the sky at visible wavelengths, the γ-ray sky is dominated by a diffuse radiation originating in our Galaxy, Sect. 8.9, due to the propagation of CRs in the interstellar medium. In most cases, galactic and extragalactic sources appear as point-like objects, i.e., with angular dimensions much smaller than the resolution of the detectors, over the diffuse γ-ray background, Sect. 8.10.

In addition to these steady sources (although some of them present time variability in their intensity), flashes of γ-rays were discovered serendipitously as early as the beginning of the 1970s. These Gamma Ray-Bursts (GRBs) are the brightest explosions in the Universe, observed at a rate of about 1/day. Their origin, classification, total energy output and the γ-ray differential flux have been experimentally investigated only recently, Sect. 8.11. These objects are possible candidates as sources of ultra-high energy cosmic rays, as mentioned in the previous chapter.

Space experiments cover a very broad γ-ray energy region, extending from a few MeV to hundreds of GeV. However, beyond a few hundreds GeV, the γ-ray fluxes are so small that the effective detection area of space-based experiments cannot provide adequate statistics, Sect. 8.13. The astrophysics studies of γ-rays at the highest energies rely on ground-based detectors, described in the next chapter.

Almost standardly, we use the definition "high energy" (HE) or "GeV" astronomy for the energy range from $\sim$30 MeV to $\sim$300 GeV, typically covered by space-based experiments. We will refer to "very high energy" (VHE), or "TeV" astronomy, for the range from $\sim$300 GeV to 100 TeV, covered by ground-based experiments.

8.1 The Spectral Energy Distribution (SED) and Multiwavelength Observations

A key role for the understanding of underlying physics processes in high-energy sources is played by the simultaneous observation of the same object using different experimental techniques, from radio to γ-rays. The electromagnetic emission is detected as incoming radiation over a wide range of photon energies (in the radio, microwave, infrared, visible,...). At the highest energies that we are dealing with (γ-rays), individual photons are detected, and their energy estimated. It is convenient that the quantity of energy emitted by the source in a given region of the electromagnetic spectrum be represented in a unified way. Multiwavelength (or multifrequency) astrophysics gathers and interprets the astronomical data collected using different instruments and detectors in many frequencies and/or energy bands. The *spectral flux density*

$$F(\nu) \quad \text{in units: erg cm}^{-2}\,\text{s}^{-1}\,\text{Hz}^{-1} \tag{8.1}$$

is the quantity that describes the rate at which energy is transferred by electromagnetic radiation through a real or virtual surface, per unit surface area and at a given frequency ν.

When the electromagnetic energy at a given frequency ν is measured as incoming radiation, the spectral flux density is determined using an appropriate detector of a given cross-sectional area A [cm^2], pointing directly towards the source. If the detector is sensitive to a wide range of frequencies, a narrow band-pass filter placed in front of the detector is necessary to select only that radiation whose frequency lies within a very narrow range $\Delta\nu$ centered on ν and to measure the rate (erg/s) of the incoming electromagnetic energy. The measured rate is then divided by $A\Delta\nu$ to obtain the spectral flux density (8.1).

When the electromagnetic emission from a source is measured through the detection of individual photons (the high-energy case), the energy E_γ of incoming photons must be determined. In most cases, the detector has an intrinsic efficiency $\epsilon(E_\gamma)$ to detect photons of different energies. In this case, a quantity called *effective area*, $A(E_\gamma)$, is defined. The effective area is the geometrical area multiplied by the detection efficiency $\epsilon(E_\gamma)$. As these detectors measure the number of incoming photons in a given energy interval $\frac{\Delta n_\gamma}{\Delta E_\gamma}$ and in a given time interval T, the quantity equivalent to the spectral flux density (8.1) corresponds to

$$E_\gamma \frac{1}{A(E_\gamma)\cdot T}\frac{\Delta n_\gamma}{\Delta E_\gamma} \to E_\gamma \frac{\mathrm{d}N_\gamma}{\mathrm{d}E_\gamma}, \quad \text{where} \quad N_\gamma = \frac{n_\gamma}{T\cdot A}. \tag{8.2}$$

A few comments concerning the measurement units. Traditional astronomy uses a particular units to measure the spectral flux density, the *Jansky (Jy)*, where 1 Jy $= 10^{-23}$ erg cm^{-2} s^{-1} Hz^{-1}. The Jansky is a suitable unit for radio astronomy. The brightest astronomical radio sources have spectral flux densities on the order of 1–100 Jy. For example, the Third Cambridge Catalogue of Radio Sources lists ~350 radio sources in the Northern Hemisphere brighter than 9 Jy at 159 MHz. When individual high-energy particles are measured, the energy units are, in most cases, GeV or TeV. Remember that 1 erg = 1.6 TeV. It is not infrequent to find in the literature the spectral flux density measured in mixed units: $\mathrm{erg\,cm^{-2}\,s^{-1}\,GeV^{-1}}$. This makes more evident the fact that the spectral flux density represents an energy flux per energy interval. Also, in (8.1), always keep in mind that frequency gives the corresponding energy, once multiplied by the Planck constant h. Gravitational waves also carry energy, and their spectral flux density is usually expressed in terms of $[\mathrm{W\,m^{-2}\,Hz^{-1}}]$. As we will see in Sect. 13.3, typical signals on Earth at 100 Hz are expected to be $10^{-6}\,\mathrm{W\,m^{-2}\,Hz^{-1}} = 10^{-3}\,\mathrm{erg\ cm^{-2}\,Hz^{-1}} = 10^{20}\,\mathrm{Jy}$. However, because of the poor coupling of gravitational waves to matter, such signals are difficult to detect.

The amount of power radiated through a given area in the given frequency interval $\Delta\nu = \nu_2 - \nu_1$ or energy $\Delta E_\gamma = E_2 - E_1$ in the form of electromagnetic radiation or individual photons, respectively, is the *flux density*

$$J(\nu) = \int_{\nu_1}^{\nu_2} F(\nu)\mathrm{d}\nu; \quad J(E_\gamma) = \int_{E_1}^{E_2} E_\gamma \cdot \frac{\mathrm{d}N_\gamma}{\mathrm{d}E_\gamma}\mathrm{d}E_\gamma \quad (\text{units: erg}\,\mathrm{cm}^{-2}\,\mathrm{s}^{-1}). \tag{8.3}$$

The flux density is the quantity that can be used to compare the electromagnetic energy emission in different regions of the electromagnetic spectrum as detected by different instruments. It is independent from the corpuscular or wavelike detection of the incoming radiation. Note that in most cases, the $F(\nu)$ and the $(\mathrm{d}N_\gamma/\mathrm{d}E_\gamma)$ distributions can be represented by power-law functions, at least within a limited range of frequency or energy. In this case, the flux density in the considered frequency/energy range is given by

$$J(\nu) \propto \nu F(\nu); \quad J(E_\gamma) \propto E_\gamma^2 \cdot \frac{\mathrm{d}N_\gamma}{\mathrm{d}E_\gamma} \quad (\text{units: erg}\,\mathrm{cm}^{-2}\,\mathrm{s}^{-1})\,. \tag{8.4}$$

The spectral energy distribution (SED) is a plot of the flux density $J(\nu)$ versus frequency and/or $J(E_\gamma)$ versus energy of the radiation. In some cases, the detected emission from a source starts in the radio frequency and ends in the TeV γ-rays. In this case, the SED for the source includes both measurement methods, and it can be plotted as a function of either frequency ν or energy $E_\gamma = h\nu$. The *radiative flux*, or *radiation flux*, is the amount of power radiated through a given area over the whole frequency/energy spectrum, also measured in $(\mathrm{erg\,cm^{-2}\,s^{-1}})$.

The SED is used in many branches of astrophysics to characterize sources, as shown in the following sections and chapters.

8.2 Astrophysical γ-Rays: The Leptonic Model

The electron is the lightest, stable charged particle. Thus, electrons are easily accelerated in the presence of astrophysical magnetic fields. This originates the *radiative processes in astrophysics* that are the subject of dedicated books (see, for instance, Ghisellini 2013; Longair 2011). Here, we present a synthesis of the energy emitted through these processes.

An astrophysical accelerator provides a continuous distribution of electron energies. As we know from Chap. 6, the energy spectra of cosmic rays and cosmic ray electrons can be approximated by power-law distributions. In the following, we assume the distribution of electron energies to be

$$\frac{\mathrm{d}n}{dE}\mathrm{d}E = \kappa E^{-\alpha_e}\mathrm{d}E, \tag{8.5}$$

where $(\mathrm{d}n/\mathrm{d}E)\mathrm{d}E$ is the number density of electrons in the energy interval E to $E+\mathrm{d}E$. This quantity is strictly related to the flux measured in $(\mathrm{GeV\,cm^2\,s})^{-1}$, as given in (2.18a). Usually, fully isotropic emission is assumed, and the dependence from the solid angle covered by the source is included in the normalization constant κ.

8.2.1 *The Synchrotron Radiation from a Power-Law Spectrum*

The power emitted as synchrotron radiation from a relativistic electron of energy E moving perpendicularly in a region with magnetic field of intensity B was obtained in Sect. 5.8.1. Let us now work out the synchrotron emission spectrum for a power-law distribution of electron energies given by (8.5). We use the fact that the spectrum of synchrotron radiation is sharply peaked near the critical frequency ν_c and that $\Delta\nu$ is much narrower than the breadth of the power-law electron energy spectrum. In our simple approximation, the electron of energy E produces electromagnetic radiation at the critical frequency ν_c, which can be approximated, using (5.67), by

$$\nu_c = \Gamma^2 \nu_g = \left(\frac{E}{m_e c^2}\right)^2 \nu_g; \quad \text{with} \quad \nu_g = \left(\frac{eB}{2\pi m_e c}\right). \tag{8.6}$$

Inserting the numerical values into the expression for the critical frequency ν_c, we obtain

$$\nu_c = \Gamma^2 \frac{eB}{2\pi m_e c} = 2.8 \times 10^6 \Gamma^2 B \,\mathrm{Hz}\,, \tag{8.7}$$

when B is measured in Gauss. It is easy to work out that for a magnetic field of $B \sim 100\,\mu\mathrm{G}$ (like that estimated for the Crab pulsar, Sect. 9.5), and for 1 TeV electrons

($\Gamma \sim 10^6$), the critical frequency corresponds to photon energies $E_\gamma = h\nu_c$ on the order of a fraction of eV.

The quantity usually used in astronomy is the *specific emissivity*, units [erg cm^{-3} s^{-1} sr^{-1} Hz^{-1}], that is, the power per unit solid angle, unit of volume and unit of frequency produced by the particles at sources. If the source is completely transparent (*optical thin*), the source's *monochromatic intensity* [erg cm^{-2} s^{-1} sr^{-1} Hz^{-1}] is simply given by the specific emissivity, times the scale dimension R of the source. Finally, assuming an isotropic source, the spectral flux density $F(\nu)$ defined in (8.1) is given by the monochromatic intensity times 4π. Thus, neglecting constant factors,

$$F(\nu)\mathrm{d}\nu \propto -\left(\frac{\mathrm{d}E}{\mathrm{d}t}\right)\left(\frac{\mathrm{d}n}{\mathrm{d}E}\mathrm{d}E\right) = -\left(\frac{\mathrm{d}E}{\mathrm{d}t}\right)\kappa E^{-\alpha_e}\mathrm{d}E\,. \tag{8.8}$$

The quantity $F(\nu)d\nu$ has units (erg cm^{-2} s^{-1}), as in (8.3).

To estimate $F(\nu)$, let us now consider the quantities on the right-hand side of (8.8), remembering that the power loss $-\mathrm{d}E/\mathrm{d}t$ was derived in Eq. (5.61):

$$E = \Gamma m_e c^2 = \left(\frac{\nu}{\nu_g}\right)^{1/2} m_e c^2, \tag{8.9a}$$

$$\mathrm{d}E = \frac{m_e c^2}{2\nu_g^{1/2}}\nu^{-1/2}\mathrm{d}\nu, \tag{8.9b}$$

$$-\frac{\mathrm{d}E}{\mathrm{d}t} = \frac{4}{3}\sigma_T c\left(\frac{E}{m_e c^2}\right)^2\frac{B^2}{8\pi}. \tag{8.9c}$$

The above quantities must be inserted into (8.8). Using the definition of ν_g given in Eq. (8.6), and after some algebra, the flux density as a function of the frequency ν is expressed in terms of the normalization constant κ, the magnetic field intensity B and of constant factors:

$$F(\nu) = (\text{constant})\cdot\kappa\cdot B^{(\alpha_e+1)/2}\cdot\nu^{-(\alpha_e-1)/2}. \tag{8.10}$$

In the case of the synchrotron flux received from a homogeneous and thin source of volume $V \propto R^3$ at a distance D_L, the (constant) in (8.10) is $\propto R^3/D_L^2 \sim R\theta_s^2$, where θ_s is the angular radius of the source.

It is very important to note that the power-law electron distribution (8.5) produces the power-law spectrum (8.10), and the two spectral indices are related. The spectral index of the radiation is denoted as α and is related to the electron spectrum through

$$\alpha = \frac{\alpha_e - 1}{2}\,. \tag{8.11}$$

Thus, (8.10) can be rewritten as

$$F(\nu) \propto \theta_s^2 \cdot R \cdot \kappa \cdot B^{1+\alpha} \cdot \nu^{-\alpha} . \tag{8.12}$$

This synchrotron flux is usually measured starting from radio telescopes to others astronomical instruments sensitive to higher frequencies. Observing the source at two (or more) frequencies allows for determining the spectral index α. Furthermore, if we know the source distance and extension R, the normalization depends on the particle density, κ, and the magnetic field B. This corresponds to two unknowns and only one equation; a second relation, presented in the following subsection, is necessary to define the system.

Before continuing, it is important to remind ourselves of the meaning of an optical *thin* and *thick* source for a photon of a given frequency ν. The free mean path, ℓ, of this photon, having interaction cross-section $\sigma(\nu)$ in a medium with number density n [cm^{-3}], corresponds to $\ell = 1/(n \cdot \sigma(\nu))$ [cm]. The *optical depth* at this frequency is defined as

$$\tau(\nu) = \int \frac{ds}{\ell} = \int n \cdot \sigma(\nu) \cdot ds , \tag{8.13}$$

where ds is a small path and the integral is over the source extension before the photon escape. Notice that τ is dimensionless. Thus,

$$\text{if } \tau(\nu) \ll 1 \text{ the source is optically } thin , \tag{8.14a}$$

$$\text{if } \tau(\nu) \gg 1 \text{ the source is optically } thick , \tag{8.14b}$$

for an homogeneous source of scale dimension R, $\tau \simeq n\sigma R$.

According to standard Fermi acceleration mechanisms, we expect $\alpha_e \sim 2$. Thus, from (8.11), $\alpha = 1/2$ for $\alpha_e = 2.0$, and thus $F(\nu) \propto \nu^{-1/2}$. This relation is divergent at low frequencies (an *infrared catastrophe*). This problematic behavior is solved by the fact that the photon cross-section increases at small frequencies (see, for instance, Fig. 2.16 of Braibant et al. 2011), and according to the definition (8.14) the source is no longer optically thin. Thus, the approximation used to derive (8.12) no longer holds. The absorption increases produce the self-absorbed flux derived below.

8.2.2 *Synchrotron Self-Absorption*

The above infrared divergence can be corrected by statistical thermodynamics considerations. The synchrotron emission is nonthermal, but the presence of matter in the region where the magnetic field is present modifies the emission spectrum.

Matter is able to absorb part of the radiation, reaching, at equilibrium, a given effective temperature, as explained below.

According to the principle of *detailed balance*, for every emission process, there is a corresponding absorption mechanism. In the case under consideration, synchrotron self-absorption corresponds to synchrotron radiation. Let us give a simple order-of-magnitude calculation of the basic physics of the process. From the *monochromatic intensity*, $I(\nu)$ in units: erg cm^{-2} s^{-1} Hz^{-1} sr^{-1}, we can define the *brightness temperature*, T_b. We recall that T_b is defined using the expression for the intensity of black-body radiation

$$I(\nu) = \frac{2h\nu^3}{c^2} \frac{1}{e^{(h\nu/kT_b - 1)}} \simeq \frac{2kT_b}{\lambda^2} , \tag{8.15}$$

where k is the Boltzmann constant; the last equality holds in the Rayleigh–Jeans limit. By inverting this relation, we obtain

$$T_b = \frac{\lambda^2}{2k} I(\nu) = \frac{(c/\nu)^2}{2k} I(\nu) . \tag{8.16}$$

The accelerated electrons follow a power-law energy distribution and are not in thermal equilibrium. However, we can associate a temperature T_e with electrons of a given energy through the relativistic formula relating electron energy to temperature

$$\Gamma m_e c^2 = 3kT_e . \tag{8.17}$$

As a result, the effective temperature T_e of the electrons now becomes a function of their energy. Since $\Gamma \propto (\nu/\nu_g)^{1/2}$, we can conclude that

$$T_e \sim \frac{m_e c^2}{3k} \left(\frac{\nu}{\nu_g} \right)^{1/2} . \tag{8.18}$$

For a source with self-absorption processes, the brightness temperature of the radiation must be equal to the effective kinetic temperature of the emitting electrons, $T_b = T_e$, and therefore, in the Rayleigh–Jeans limit,

$$I(\nu) = \frac{2kT_e}{\lambda^2} = \frac{2m_e}{3\nu_g^{1/2}} \nu^{5/2} \propto \nu^{5/2} \cdot B^{-1/2} . \tag{8.19}$$

This computation shows the physical origin of the suppression at the low-frequency of the expected spectrum in sources in which synchrotron self-absorption is important. Notice that the dependence on the magnetic field arises from the dependence of ν_g on B given in Eq. (8.6).

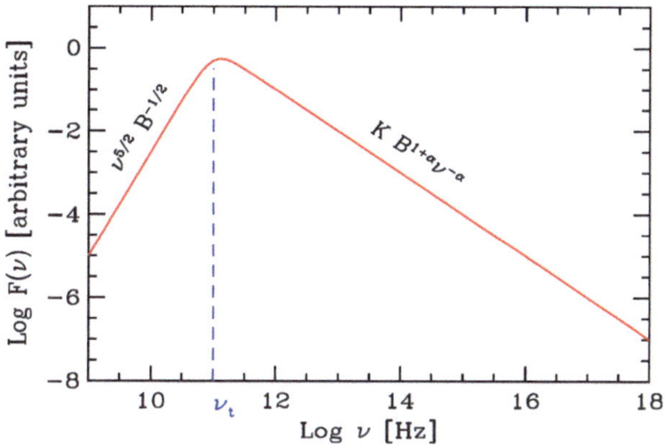

Fig. 8.1 The synchrotron spectrum from a partially self-absorbed source. The self-absorption frequency, ν_t, marks the transition between the frequency regions where the source is optically thick (left part) or thin (right part). Adapted from Ghisellini (2013)

Now, to obtain the flux $F(\nu)$ from the intensity, we must integrate I_ν over the angular dimension of the source, obtaining

$$F(\nu) \propto \theta_s^2 \frac{\nu^{5/2}}{B^{1/2}}. \tag{8.20}$$

Note that the spectrum depends, at low frequency, on $\nu^{5/2}$, if the magnetic field is uniform. In any case, the flux is no longer divergent. Note also that the particle density κ disappears: the more electrons present, the larger the emission, but also the larger the absorption.

Figure 8.1 shows the synchrotron spectrum as a function of the frequency; at low frequencies, the source is optically thick and the emitted spectrum is ruled by (8.20). At higher frequencies, the source is optically thin and the emitted spectrum is ruled by (8.12). The ratio between the two is

$$\frac{F(\nu)_{thin}}{F(\nu)_{thick}} \propto \frac{R \cdot \kappa \cdot B^{1+\alpha} \cdot \nu^{-\alpha}}{\nu^{5/2} B^{-1/2}} = R \cdot \kappa \cdot B^{3/2+\alpha} \cdot \nu^{-\alpha-5/2} . \tag{8.21}$$

Note the strong dependence upon frequency: at high frequencies, absorption is small. The separation between the two regimes (thick and thin) occurs when the ratio (8.21) is unit. If we have fixed the source dimension R and the normalization factor for the electron spectrum, κ, the condition $F(\nu)_{thin} = F(\nu)_{thick}$ depends only on a particular value of the frequency, called the *self-absorption frequency*, ν_t.

The self-absorption frequency is a crucial quantity when studying synchrotron sources, as the synchrotron spectrum peaks very close to ν_t, as shown in Fig. 8.1. Notice that ν_t is conditioned by the normalization factor κ and magnetic field B. According to (8.21), if, for any reason, the magnetic field increases, the ν_t value must also increase to keep the ratio equal to 1. The same applies if the number of electrons, and thus κ, increases.

The synchrotron spectra are observed in different astrophysical environments, and in our Galaxy, in nuclei of active galaxies and quasars. An important aspect of these observations (from radio to X-rays) is that they provide unambiguous evidence for the presence of relativistic electrons in the source regions.

8.2.3 Inverse Compton Scattering

Inverse Compton (IC) scattering is the process in which ultra-relativistic electrons scatter low energy photons so that the photons gain energy at the expense of the kinetic energy of the electrons. The Feynman diagram of this process is similar to that of the Compton effect and the corresponding computation of the transition probabilities are similar. The interaction occurs with photons of a radiation field u_{rad} (in the rest of frame of the Galaxy, and thus of the observer). The energy radiated by an electron is proportional to the total intensity of target radiation field and

$$-\left(\frac{\mathrm{d}E}{\mathrm{d}t}\right) = \sigma_T c u'_{\text{rad}} \ [\text{cm}^2][\text{cm/s}][\text{erg cm}^{-3}] \tag{8.22}$$

(the units in square brackets) where u'_{rad} is the energy density of radiation in the rest frame of the moving electron, with Lorentz boost Γ. The energy ϵ' of the interacting photon is much less than $m_e c^2$, and consequently the Thomson scattering cross-section can be used to describe the photon-electron interaction. We must consider the relativistic effects between u'_{rad} and u_{rad}. The energy density of target photons within the laboratory frame is simply $u_{\text{rad}} = n(\omega) \cdot \hbar\omega = n(E) \cdot E$. The quantity $n(E)$ is the number density of particles with energy between E and $E + \mathrm{d}E$. Thus, the number of photons in the differential three-volume $\mathrm{d}N(E) = n(E) \cdot \mathrm{d}x\mathrm{d}y\mathrm{d}z$ is a relativistic invariant. Because the four-volume element $\mathrm{d}t\mathrm{d}x\mathrm{d}y\mathrm{d}z$ is also a relativistic invariant, the quantity $n(E)$ behaves like the time-like component of a four-vector (it has the same Lorentz transformation as the time variable). Also, the energy E is the time-like component of a four-vector. For this reason, $E' = \Gamma E$ and $n'(E') = \Gamma n(E)$. Thus,

$$u'_{\text{rad}} = n'(E') \cdot E' = \Gamma n(E) \cdot \Gamma E = \Gamma^2 u_{\text{rad}}. \tag{8.23}$$

By inserting this result into (8.22), we obtain

$$-\left(\frac{\mathrm{d}E}{\mathrm{d}t}\right) \simeq \frac{4}{3}\sigma_T c \Gamma^2 u_{\text{rad}}, \tag{8.24}$$

where the factor 4/3 arises from averaging over the possible directions between the electrons and the photons. This result holds until $\Gamma\epsilon \ll m_e c^2$.

Notice the remarkable similarity of the result (8.24) to the expression (5.61) for the energy loss rate of the ultra-relativistic electrons by synchrotron radiation. The reason for this is that the energy loss rate depends on the electron acceleration in its rest frame. It does not matter if the acceleration is produced by the electric field **v x B** due to motion of the electron through the magnetic field (as in the synchrotron radiation) or by the electric fields of the electromagnetic waves incident upon the electron (as in the inverse Compton scattering).

The maximum energy that a photon can acquire corresponds to a head-on collision, in which the photon is scattered back along its original path, and can be determined using the kinematics of the Compton effect. Assuming an initial energy ϵ, after the collision, the maximum energy is

$$(E_\gamma)_{\max} = \epsilon \Gamma^2 (1 + v/c)^2 \simeq 4\Gamma^2 \epsilon \,. \tag{8.25}$$

It is easy to work out that, averaging over all relative photon-electron directions, the average energy of the scattered photons is

$$\overline{E_\gamma} = \frac{4}{3}\epsilon \Gamma^2 \quad \text{for} \quad \Gamma\epsilon \ll m_e c^2 \,. \tag{8.26}$$

This result is valid when the electron-photon interaction can be described using the Thomson cross section, i.e., until the condition $\epsilon \cdot \Gamma \ll m_e c^2$ holds. For highly relativistic electrons, when this condition is no longer satisfied, the Klein–Nishina cross-section for photon-electron scattering must be used. In the ultra-relativistic limit, the Klein–Nishina cross-section is

$$\sigma_{KN} = \frac{\pi^2 r_e^2}{h\nu}\left(\ln 2h\nu + \frac{1}{2}\right) . \tag{8.27}$$

The Klein–Nishina cross-section decreases as $\propto (h\nu)^{-1}$ at high energies. Consequently, high energy scatterings result in significantly reduced luminosities, as compared with the nonrelativistic calculation. In this high-energy limit, the average energy of the scattered photons replacing (8.26) is

$$\overline{E_\gamma} = \frac{1}{2}\overline{E_e} \quad \text{for} \quad \Gamma\epsilon \gg m_e c^2 \,. \tag{8.28}$$

The general result (8.26) that the frequency/energy of photons scattered by ultra-relativistic electrons with Lorentz factor $\Gamma = E/(m_e c^2)$ is increased by Γ^2 (in the Thompson regime) is of profound importance in high energy astrophysics. Different classes of astrophysical sources can provide accelerated electrons observed in the cosmic radiation, at least up to the TeV scale (see Sect. 3.10). This corresponds to $\Gamma = 10^6$. More frequent are situations in which $\Gamma = 10^2 - 10^3$.

Radio, infrared and optical photons scattered by electrons with $\Gamma = 1000$ (i.e., 0.5 GeV) then have an average frequency (or energy) roughly 10^6 times that of the incoming photons. In all situations, the condition $\Gamma\epsilon \ll m_e c^2$ holds. Radio photons with $\nu_0 = 10^9$ Hz become ultraviolet photons with $\nu = 10^{15}$ Hz (eV-scale photons);

scattering on far-infrared photons with $\nu_0 \simeq 10^{12}$ Hz, typical of the photons seen in galaxies that are powerful far-infrared emitters, produce X-rays with keV energies; optical photons with $\nu_0 = 4 \times 10^{14}$ Hz become γ-rays of the MeV energy-scale. The inverse Compton scattering process is an effective means of creating very high energy photons. On the other hand, the process is highly efficient for reducing the energy of high energy electrons whenever large fluxes of photons and relativistic electrons occupy the same volume.

The similarity of the synchrotron and inverse Compton scattering processes embodied by the similarity between (8.24) and (5.61) for the mean energy loss rate means that we can use the results of Sect. 8.2.1 to work out the spectrum of radiation produced by a power-law distribution of electron energies. We assume that α_e is the spectral index of the electron energy spectrum incident on the radiation field u_{rad} [erg cm^{-3}]. This radiation field is isotropic and made of photons of fixed energy (frequency) $\epsilon = h\nu_0$. The starting point for the determination of the spectral flux density is (8.8). To repeat the computation in the case of the Inverse Compton scattering, two changes must be considered:

- in synchrotron, Eq. (8.6) expresses the fact that the produced radiation ν_c is enhanced by a Γ^2 factor with respect to ν_g; in the IC, the produced photons have the energy (frequency) given by Eq. (8.26) in the Thomson regime. It has the same Γ^2 factor with respect to the frequency ν_0 of seed photons;
- the power loss (8.24) replaces (8.9b), thus u_{rad} appears instead of $B^2/8\pi$.

Those two changes lead to

$$F(\nu) \propto \cdot \kappa \cdot \frac{u_{rad}}{\nu_0} \cdot \left(\frac{\nu_c}{\nu_0}\right)^{-\frac{\alpha_e-1}{2}} , \tag{8.29}$$

which is analogous to (8.10) for IC. Note that, as in the case of synchrotron emission, (8.29) is a power-law energy distribution. The index $\alpha = (\alpha_e - 1)/2$ it is related to the electron spectral index by the same relation as (8.11). This is because both the IC and the synchrotron single electron spectra are peaked at a typical frequency that is a factor Γ^2 greater than the starting energy seed.

As the last step, consider the more realistic case of seed photons distributed in frequency; thus, (8.29) must be integrated over the photon frequencies (removing the subscript 0 for clarity):

$$F(\nu) \propto \nu_c^{-\alpha} \int_{\nu_{min}}^{\nu_{max}} \frac{u_{rad}(\nu)}{\nu} \nu^{\alpha} d\nu , \tag{8.30}$$

where $u_{rad}(\nu)$ [erg cm^{-3} Hz^{-1}] is the specific radiation energy density at frequency ν. The limits of integration in general, depend on photon scattered frequency ν_c. Note the important thing: if the radiation field $u_{rad}(\nu)$, where the electron population produces Inverse Compton, is produced by the same electron population that produces the synchrotron emission, then $u_{rad}(\nu) \propto \nu^{-\alpha}$ and the integrand function in (8.30) becomes a constant. Thus, the result of the integral is $\propto$ $\ln \nu_{max}/\nu_{min} = \ln \Lambda$, and

$$F(\nu) \propto \tau_c \nu_c^{-\alpha} \ln \Lambda \ , \tag{8.31}$$

where $\tau_c = \sigma_T \cdot R \cdot \kappa$ depends on the Thomson cross-section, on the source dimension R and on electron number density κ.

8.3 The Synchrotron Self-Compton (SSC) Mechanism

In most cases, the basic interpretation for the production of high-energy photons in the *astrophysical leptonic model* is the so-called Self-Synchrotron Compton (SSC) mechanism. As described in Sect. 8.2.1, synchrotron emission from electrons moving in a magnetic field generates photons with an energy spectrum peaked in a region ranging from the infrared to X-rays. Such photons, in turn, constitute a target for their own parent electron population. The process in which low-energy photons gain energy through collisions with high-energy electrons is the inverse Compton (IC) scattering. The IC mechanism has the effect of increasing the photon energy, and is important in regions where accelerated electrons coexist with a high energy density of soft-photons. Since electrons in some astrophysical environments are ultra-relativistic, with a Lorentz factor up to $\Gamma = E/(m_e c^2) \sim 10^9$, the energy of the scattered photon (in the Klein–Nishina high-energy limit), (8.28) gets boosted by a factor $\lessapprox \Gamma$. The energy distribution of γ-rays produced through the inverse Compton mechanism starting from an infrared/X-ray photon population can peak at GeV-TeV energies.

In the SSC mechanism, the processes of synchrotron radiation and inverse Compton are strongly correlated. An increase in the first also increases the number density of target photons. At equilibrium, we can assume that the ratio between the rates of energy loss of an ultra-relativistic electron by synchrotron and inverse Compton radiation in the presence of a photon energy density $u_{\rm rad}$ and a magnetic field of intensity B is constant:

$$\frac{(\mathrm{d}E/\mathrm{d}t)_{IC}}{(\mathrm{d}E/\mathrm{d}t)_{\rm sync}} = \frac{u_{\rm rad}}{B^2/8\pi} \ , \tag{8.32}$$

having used the relations (8.24) and (5.61). As explained in Sect. 8.2.2, estimates of the intensity of the magnetic field can be obtained through observations of the synchrotron spectrum from the radio band up to the X-rays. With high magnetic fields, the inverse Compton component is largely suppressed, as most of the electron energy is lost in the synchrotron emission.

Figure 8.2 shows the spectral energy distribution (SED), $\nu F(\nu)$, expected from high-energy electrons moving in a magnetic field. The resulting energy spectrum has two structures. The first distribution is produced by the synchrotron emission

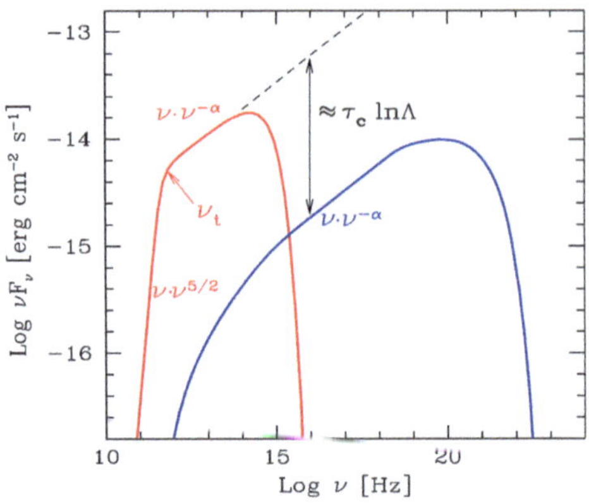

Fig. 8.2 Typical example of spectral energy distribution (SED) of photons produced in the leptonic model, shown in the $\nu F(\nu)$ vs ν representation. Synchrotron radiation is caused by relativistic electrons accelerated in a magnetic field. Photons from synchrotron emission also represent the target for inverse Compton scattering of the parent electrons. Adapted from Ghisellini (2013)

of decelerated electrons. The ν_t frequency indicates the self-absorption frequency, defining the transition from a source optically thick or thin for the radiation. The drop at large frequencies is due to some cut-off in the energy distribution of parent electrons. The second structure is due to the IC scattering of the same electrons with the produced radiation field. As derived in Sect. 8.2.3, it exhibits the same behavior as the SED induced by synchrotron radiation, damped by a factor $\tau_c \ln \Lambda$, where τ_c is the optical depth. At high energies, where no high-energy synchrotron photons are present, the spectrum is dumped.

Models of synchrotron self-Compton sources are worked out numerically and are strongly dependent upon the input assumptions. These computations assume that the source of radiation is stationary. Detailed predictions of the photon flux for both mechanisms (synchrotron and IC) can be derived, as we show for the case of the Crab nebula (Sect. 9.5). Many γ-ray sources are variable over short time-scales and display the features expected from synchrotron self-Compton radiation, but they must also involve relativistic bulk motions of the source regions to account for their extreme properties. Therefore, the predictions are somewhat model-dependent. In the following chapter, we will encounter different astrophysical objects that can be explained, totally or partially, in the context of the SSC mechanism.

In addition to synchrotron self-Compton, the *external synchrotron radiation* foresees electron scattering on radiation fields not produced by the primary electrons themselves. The cosmic microwave background radiation can represent a target for high energy electrons. In the case of external synchrotron radiation, the constraints derived from relation (8.32) do not hold.

Exercise Using the information contained in Fig. 8.2, estimate the maximum energy of the electrons accelerated by the source and producing the observed radiation. After reading the following section, explain the fact that no γ-rays from protons are expected.

8.4 Astrophysical γ-Rays: The Hadronic Model

High-energy photons can be produced by accelerated protons as well. A direct signature of the presence in astrophysical environments of accelerated protons is provided by the presence of neutrinos and γ-rays. They are mainly generated in the decay of charged and neutral pions, respectively; these mesons are produced in proton-proton collisions via

$$p + p \to \pi^{\pm}, \pi^{0}, K^{\pm}, K^{0}, p, n, \ldots \tag{8.33}$$

where … represents the presence of higher mass mesons and baryons. Because of the similarity with the process of production of secondary hadrons in a fixed-target accelerator experiment, process (8.33) is usually referred to as the *astrophysical beam dump* mechanism. The cross-section for (8.33) corresponds to about 40–50 mb (see Chap. 7 of Braibant et al. 2011), and it is dominated by inelastic processes with particles emitted with small transverse momentum (*soft* or low-p_t processes).

A second process that produces secondary mesons is due to high-energy protons interacting with low-energy photons in the surroundings of sources. The kinematics of the process (also called *photoproduction*) is similar to that discussed in Sect. 7.5.2 for CR protons interacting with CMB photons. Around astrophysical sources, there is usually a high density of radio, infrared, visible, and ultraviolet photons (the *ambient photons*, denoted in the following as γ_ϵ). As discussed in the previous sections, most ambient photons are produced by accelerated electrons in regions where high magnetic fields are present. The photoproduction occurs through the Δ^+ resonance:

$$p + \gamma_\epsilon \to \Delta^+ \to \pi^0 + p \,, \tag{8.34a}$$

$$\to \pi^+ + n \,, \tag{8.34b}$$

and cross-section (shown in Fig. 7.7) at the resonance of $\sim$0.250 mb for each individual channel. Although this cross-section $\sigma_{\gamma p}$ is two orders of magnitude smaller than the cross-section σ_{pp} of the beam-dump process (8.33), in some astrophysical environments, the probability that secondary mesons are produced by reaction (8.34) is much higher than the probability due to (8.33). This, because the number density of ambient photons, n_γ, could be much larger than that of environmental matter number density, n, and the event rate for the latter process is $\sigma_{pp} \cdot n \cdot c$, while for the former, it is $\sigma_{\gamma p} \cdot n_\gamma \cdot c$.

While protons may remain trapped because of high magnetic fields, neutrons and the decay products of neutral and charged pions escape. Neutrons are not confined by magnetic fields; they can escape the acceleration region and turn into protons through the weak interaction decay

$$n \rightarrow p e^- \overline{\nu}_e . \tag{8.35}$$

Neutral pions decay into γ-rays, via the process:

$$\pi^0 \rightarrow \gamma\gamma \,, \tag{8.36}$$

while the π^+ mesons undergo the decay chain:

$$\begin{aligned} \pi^+ &\rightarrow \nu_\mu + \mu^+ \\ &\hookrightarrow \mu^+ \rightarrow \overline{\nu}_\mu + \nu_e + e^+ \,. \end{aligned} \tag{8.37}$$

The energy escaping from the source is therefore shared among high-energy protons (=CRs), γ-rays and neutrinos produced by the decay of neutrons, π^0 and π^+ (π^-), respectively.

To be observed, this energy partition between CRs, γ-rays and neutrinos requires the source to be *transparent* enough. A transparent source is defined as a source of a much larger size than that of the proton mean free path, but smaller than the meson interaction length. For these sources, protons have a large probability of interacting once, and most secondary mesons can decay.

Under these conditions, the mechanisms that produce CRs also produce neutrinos, Eq. (8.37) and high-energy photons, Eq. (8.36), and the candidate neutrino sources are in general also γ-ray sources. In this hadronic model, there is a strong relationship between the spectral index of the CR energy spectrum $\Phi_{CR}(E) \propto E^{-\alpha_{CR}}$, and that of γ-rays and neutrinos. As derived in Sect. 11.2, the spectral index of secondary mesons (when their interaction processes can be neglected) is identical to that of the parent primary CRs. The same holds (in a transparent source, where the interaction probability of secondary mesons is low) for the pion daughters, γ-rays and neutrinos, and thus:

$$\alpha_{CR} \sim \alpha_\nu \sim \alpha_\gamma \,. \tag{8.38}$$

Hence, γ-ray measurements give crucial information on primary CRs, and they constrain the expected neutrino flux.

8.5 Energy Spectrum of γ-Rays from π^0 Decay

While neutrinos will be discussed in Chap. 10, here, we are interested in γ-rays produced by the decay of neutral pions, whose rest mass is m_π. Each photon has energy $E^*_\gamma = m_\pi c^2/2 = 67.5\,\text{MeV}$ and momentum opposite to the other within the rest frame of the neutral pion. In the following, masses, momenta and energies are measured in the same units (GeV) using natural units ($c = 1$). As the π^0 has spin 0, the angular distribution of the emitted photons in pion rest-frame (or c.m. system) is isotropic, since there is no preferential direction and $\frac{dN}{d\Omega^*} = \frac{1}{4\pi}$, where $d\Omega^*$ is the infinitesimal solid angle in the π^0 c.m. system. In the laboratory system, the pion has a relative Lorentz boost, with $\beta = \frac{|\mathbf{p}_\pi|}{E_\pi}$ and $\Gamma = \frac{E_\pi}{m_\pi}$.

For a given β, Γ, the photon energy can be determined in the laboratory system (remembering that this system is seen by the pion with a negative velocity) using the Lorentz transformation:

$$E_\gamma = \Gamma E^*_\gamma + \beta\Gamma p^*_\gamma \cdot \cos\theta^* \,, \tag{8.39}$$

from which one derives:

$$E_\gamma = \frac{E_\pi}{m_\pi}\frac{m_\pi}{2}(1+\beta\cos\theta^*) \quad\longrightarrow\quad E_\gamma = \frac{E_\pi}{2}(1+\beta\cos\theta^*) \,, \tag{8.40}$$

having used the relation $p^*_\gamma = E^*_\gamma$. Depending on the angle of emission in the c.m., the photon energy in the laboratory ranges between

$$E^{\min}_\gamma = \frac{E_\pi}{2}(1-\beta) \quad ; \quad E^{\max}_\gamma = \frac{E_\pi}{2}(1+\beta). \tag{8.41}$$

For very high energy pions ($\beta = 1$), these approach 0 and E_π, respectively. The shape of the energy spectrum of emitted photons within the laboratory frame, dN/dE_γ, is obtained by differentiating Eq. (8.39) with respect to the variable $\cos\theta^*$:

$$dE_\gamma = \beta\Gamma p^*_\gamma \cdot d\cos\theta^* \quad\longrightarrow\quad \frac{d\cos\theta^*}{dE_\gamma} = \frac{1}{\beta\Gamma E^*_\gamma} \,, \tag{8.42}$$

from which one finds

$$\frac{dN}{dE_\gamma} = \frac{dN}{d\cos\theta^*}\frac{d\cos\theta^*}{dE_\gamma} = \frac{1}{2}\frac{1}{\beta\Gamma p^*_\gamma}. \tag{8.43}$$

All quantities in (8.43) are constant for a given pion energy: that means that the probability of emitting a photon of energy E_γ is constant over the range $[E^{\min}_\gamma, E^{\max}_\gamma]$. For many decaying π^0's, the distribution of the number of photons is constant over the same energy range.

Exercise Show that the mean of the energies within the laboratory frame of secondary γ-rays produced in π^0 decays is equal to $m_\pi c^2 \simeq 70$ MeV, and is independent of the energy of the primary particle. Then, either in an analytic way or using numerical values in an electronic spreadsheet, show that, **in logarithmic plots** of the energy spectra of these γ-rays, the composite distribution is always symmetric around 70 MeV. Assuming a uniform number of π^0, construct the distribution of dN_γ/dE_γ from $E_\pi = 0.2$, 1, 10 and 100 GeV.

When a E^{-2} energy spectrum of accelerated protons at the source is considered, the π^0-decay spectrum for an observer within the laboratory frame in the $E_\gamma^2 \frac{dN}{dE_\gamma}$ representation rises steeply below ~200 MeV and approximately traces the energy distribution of parent protons at energies greater than a few GeV. This characteristic spectral feature is shown in different real cases in the next chapter (Figs. 9.13, 9.14, and 9.15). Gamma-rays from π^0 decay are also produced by the CR diffusion in our Galaxy. A nice example of the $E_\gamma^2 \frac{dN}{dE_\gamma}$ distribution of γ-rays from pion decay is in Fig. 8.6. The structure resulting from this mechanism is often referred to as the *pion-decay bump* and uniquely identifies the presence of γ-rays originated by π^0-decays. The discovery of the bump originated by the π^0 decay in the source would identify the presence of high-energy protons; this is one in the main challenges on γ-ray astronomy today.

8.6 Galactic Sources and γ-Rays: A Simple Estimate

As widely discussed in Chap. 6, Supernova Remnants (SNRs) have long been thought to be the main sources of galactic CRs through the diffusive shock acceleration process. The observation of γ-rays and neutrinos from SNRs is the most promising method of addressing the SNR paradigm for the origin of CRs. The detection of cosmic neutrinos is still problematic (see Chap. 10), and the objective of GeV–TeV γ-ray astronomy is to identify the sources where protons (and nuclei) are accelerated, through the measurement of the π^0-decay into γ's. The presence of π^0's is the signature of the presence of accelerated hadrons interacting with the surrounding material or radiation fields.

There are two epochs in the evolution of a supernova when γ-ray and neutrino emissions are expected. The first is shortly after the supernova explosion and lasts about 2–10 years. During this period, the density of the expanding supernova envelope is very high and the number density of the surrounding matter is large enough to allow hadronic interactions. The production of secondary particles will last for a long time, until the supernova reaches the Sedov phase, when most of accelerated particles start escaping the acceleration region and enter the Galaxy as CRs. Since this phase lasts for more than 1000 years, there should be some SuperNova Remnant (SNR) that acts as a γ-ray source.

We can produce a rough order-of magnitude estimate of the expected γ-ray flux from hadronic production using a very simple approach. Let us assume that the energy density of CRs energetic enough to produce secondary mesons, and thus γ-rays, corresponds to 10% of (2.33a), i.e., $\rho_{CR\rightarrow\gamma} = 0.1\,\mathrm{eV/cm^3}$.[1] Consequently,

[1] This is consistent with the argument given in Eq. (7.20) for photoproduction. In beam dump processes, above threshold, the simplest process for the π^0 production occurs through $pp \rightarrow pp\pi^0$. As an exercise, work out the threshold energy for a proton against a proton at rest to produce the above reaction.

the power of cosmic accelerators for this subsample of CRs is also one order of magnitude lower than (2.39), i.e., $P_{CR\to\gamma} \simeq 0.3 \times 10^{40}$ erg/s. This rate of energy is provided by the sum of all cosmic accelerators that occurred in our Galaxy in the last $\tau_{esc} \sim 10^7$ year, which corresponds to the escape time (Sect. 5.1) of galactic CRs. If SNRs represent the accelerators, as their occurrence is of the order of one every $t_{SN} = 100$ years then

$$N_{acc} = \tau_{esc}/t_{SN} \sim 10^5 \tag{8.44}$$

is the estimated number of explosions during the last ten million years. The power emitted as CRs by one individual explosion is

$$p_{CR\to\gamma} = \frac{P_{CR\to\gamma}}{N_{acc}} \simeq 3 \times 10^{34}\,\text{erg/s}. \tag{8.45}$$

If sufficient target material is present around the accelerating source, a fraction $\epsilon_{CR\to\gamma}$ of the CR energy is transferred to secondary particles by pp interactions (8.33). The appropriate estimate of $\epsilon_{CR\to\gamma}$ depends crucially on the number density, n, of matter surrounding the source. For the interstellar medium, n is on the order of $1\,\text{cm}^{-3}$ (Sect. 2.7.2). A reasonable guess for this unknown fraction could be $\epsilon_{CR\to\gamma} \simeq 0.1$. Under these assumptions, the power emitted by one source in the form of γ-rays is

$$L_\gamma = \frac{1}{3} \times \epsilon_{CR\to\gamma} \times p_{CR\to\gamma} \simeq 1 \times 10^{33}\,\text{erg/s}. \tag{8.46}$$

The factor (1/3) takes into account the fact that only ~1/3 of secondary particles in the hadronic shower are neutral mesons (mainly π^0). If we consider a SNR at distance $D = 1\,\text{kpc} = 3 \times 10^{21}$ cm, the flux of energy arriving on Earth as γ-rays is

$$\frac{L_\gamma}{4\pi D^2} \simeq 0.9 \times 10^{-11}\,\text{erg/(s cm}^2) . \tag{8.47}$$

For a Fermi-like mechanism at sources, we expect an energy spectrum of the E^{-2}-type. In the hadronic mechanism of γ-ray production, the energy spectrum of secondary particles follows that of the parent CRs. This is a consequence of the Feynman scaling discussed in Sect. 11.1. Therefore, we expect the number of arriving γ-rays $\mathrm{d}N_\gamma/\mathrm{d}E_\gamma$ to decrease as E^{-2} and the quantity (8.47) corresponds to the $E^2\mathrm{d}N_\gamma/\mathrm{d}E_\gamma$, for our hypothetic source. Hence, remembering that 1 erg = 1.6 TeV, our estimated spectral energy distribution for a galactic SNR is

$$E_\gamma^2 \frac{\mathrm{d}N_\gamma}{\mathrm{d}E_\gamma} \simeq 1.4 \times 10^{-11}\,\text{TeV/(s cm}^2) \quad \text{for} \quad \text{D}=1\,\text{kpc}. \tag{8.48}$$

Detailed computations exist of the flux of TeV γ-rays from a hadronic model of particle acceleration in SNRs. The derived flux depends on explosion and acceleration parameters, on the properties of the ambient medium, and on the distance of the SNR. A noticeable model is credited to Drury, Aharonian and Volk (Drury et al. 1994); they calculate that the expected integral flux of VHE γ-rays from SNRs generated from hadronic CR interactions above a given energy threshold E is

$$E_{\gamma}^{2.1}\frac{\mathrm{d}N_{\gamma}}{\mathrm{d}E_{\gamma}} \simeq 9\times 10^{-11}\cdot\eta\cdot\left(\frac{E_{\mathrm{SN}}}{10^{51}\,\mathrm{erg}}\right)\cdot\left(\frac{D}{1\,\mathrm{kpc}}\right)\cdot\left(\frac{n}{1\,\mathrm{cm}^{-3}}\right)\mathrm{TeV/(s\,cm^2)}, \tag{8.49}$$

where η is the efficiency of the particle acceleration (it has the role of $\epsilon_{CR\to\gamma}$ in our computation), E_{SN} the total kinetic energy released during the supernova explosion, D the distance to the SNR and n the density of the interstellar medium surrounding the SNR. Here, the authors obtain a power law with a spectral index of 2.1 instead of 2. For the typical values used in the paper ($\eta = 0.1$, $E_{\mathrm{SN}} = 10^{51}$ erg and $n = 1\,\mathrm{cm}^{-3}$), the result of Drury et al. (1994) is remarkably in agreement with (8.48).

The number of such accelerators in the Galaxy that can be detected is not large. As SNRs can efficiently accelerate CRs for 10^3 year at most, the γ-rays are produced within a limited time interval. Taking into account their frequency ($1/t_{\mathrm{SN}}$), there could be an order of tens of SNRs able to produce γ-rays in the whole Galaxy detectable in our epoch.

The discovery of a convincing case of a hadronic accelerator through the identification of γ-rays produced by π^0-decays is extremely difficult due to the additional γ-ray production mechanisms from relativistic electrons (Sect. 8.2). Gamma-rays originated from leptons can outnumber those produced by hadrons by a large amount. In particular, if the matter number density around the acceleration region is too low, no γ-rays from proton interactions are expected at all. A large effort is underway by the present generation of space- and ground-based experiments to identify the hadronic component of the γ-ray emission in SNRs, as we will see in the following chapter.

8.7 The Compton Gamma-Ray Observatory (CGRO) Legacy

Following the Hubble space telescope, the Compton Gamma-Ray Observatory (CGRO) was the second of NASA's great observatories to cover the widest interval of the electromagnetic radiation. It was launched using the Space Shuttle Atlantis on April 5, 1991, and operated successfully until it was de-orbited on June 4, 2000. A listing of the observations, along with other information about CGRO, can be found at the CGRO Science Support Center Web site http://cossc.gsfc.nasa.gov/docs/cgro/index.html. The Hubble and the CGRO observatories were followed (after some time) by the launches of NASA's Advanced X-ray Astrophysics Facility (Chandra) and the Spitzer Space Telescope (Infrared) when the CGRO mission was essentially complete.

The CGRO carried four instruments for γ-ray astronomy, each with its own energy range, detection technique, and scientific goals, covering energies from less than 15 keV to more than 30 GeV, over six orders of magnitude in the electromagnetic spectrum. The four instruments were:

- The Burst and Transient Source Experiment (BATSE). BATSE was the smallest of the CGRO instruments, consisting of eight modules located one on each corner of the spacecraft. Each unit included a large flat NaI(Tl) scintillator and a smaller thicker scintillator for spectral measurements, combined to cover an energy range from 15 keV to over 1 MeV.
- Oriented Scintillation Spectrometer Experiment (OSSE). This used four large, collimated scintillator detectors to study γ-rays within the range from 60 keV to 10 MeV. OSSE mapped the 0.5 MeV line from positron annihilation and provided detailed measurements of many hard X-ray/soft γ-ray sources.
- The Compton Telescope (COMPTEL), for the detection of medium energy γ-rays between 0.8 and 30 MeV, used a Compton scattering technique. Among its results, COMPTEL mapped the distribution of radioactive Aluminum-26 in the Galaxy, showing the locations of newly formed material.
- The Energetic Gamma Ray Experiment Telescope (EGRET) was the high-energy instrument on CGRO, covering the energy range from 20 MeV to 30 GeV.

8.7.1 The EGRET γ-Ray Sky

In the energy range above 10 MeV, the principal interaction process for γ-rays is pair production. Gamma rays cannot be reflected or refracted, and a high-energy γ-ray telescope detects $e^{\pm}$ with a precision converter-tracker section followed by a calorimeter.

The operational concept of EGRET, similar in most respects to the designs of other high-energy γ-ray telescopes, is shown in Fig. 8.3. The two key challenges for any such telescope are: (1) the identification of γ-ray interaction among the huge background of charged CRs; (2) the measurement of γ-ray energy, arrival time and arrival direction.

In EGRET, these objectives were obtained as follows:

1. The charged particles were vetoed through the *Anticoincidence System* (AS). The presence of a signal in the AS vetoed the tracking system electronics. A γ-ray candidate entered the detector without producing a signal in the anticoincidence. The AS rejected nearly all unwanted signals produced by charged CRs. The AS consisted of a single dome of plastic scintillator, read out by 24 PMTs mounted around the bottom.
2. The γ-rays interacted in one of 28 thin sheets of high-Z material (tantalum), converting via pair production into an electron/positron pair.

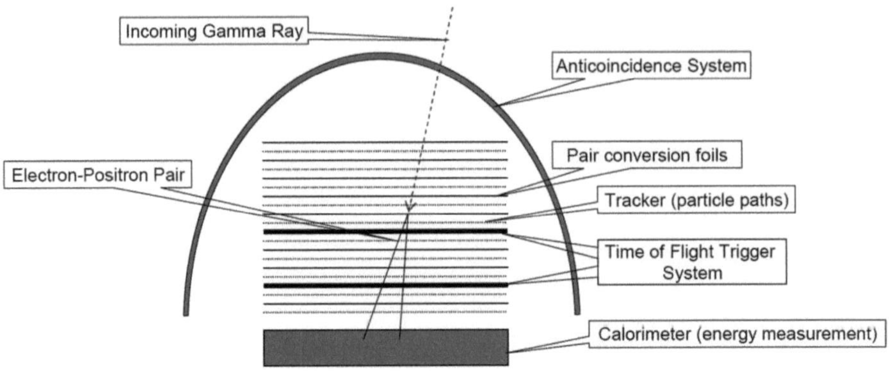

Fig. 8.3 Schematic diagram of a telescope which use the conversion of a γ-ray into a $e^{\pm}$ pair (pair-conversion telescope), reproducing the features of the EGRET experiment. The Fermi-LAT is similar, with an improved tracker device which avoids the use of a Time of Flight trigger system. Courtesy of the EGRET Collaboration

3. A *Tracker* device, consisting of 36 wire grid spark chambers and interleaved with converter plates, was used to record the paths of the electron and positron, to reconstruct the conversion point and the arrival direction of the γ-ray. The spark chambers were gas detectors and their performance slightly deteriorated with time due to gas aging.
4. The electron and positron passed through two scintillator detectors. These fast-devices triggered the readout of the spark chambers and provided the *time-of-flight* (TOF) measurement (used to confirm the direction of the particles).
5. The electron and positron entered the *Calorimeter*, producing an electromagnetic shower. Using standard methods derived from particle physics, the energies of the particles were measured and, therefore, the energy of the parent γ-ray determined. The calorimeter was made of 36 NaI crystals read out by 16 PMTs.

The angular resolution in the measurement of a single photon depends on the point spread function (PSF). The PSF is the quantity that characterizes the direction resolution of a detector. It is also used to obtain the angle containing 68% of the γ-rays emitted by an ideal point-like source. Usually, the PSF depends on the energy.

The EGRET angular resolution was $\simeq 6°$ at an energy of 100 MeV, and sources were localized with resolution of about 15 arcmin. The telescope has a field-of-view of about $\sim$0.5 sr. During its 9-years live time, EGRET detected over 1.5×10^6 γ-rays, allowing it to build up the first picture of the entire high-energy γ-ray sky. Figure 8.4 shows the summed photon map above 100 MeV, in galactic coordinates. The Milky Way runs horizontally across the figure, with the galactic center in the middle. One of the key features of this image, which provides a striking contrast to the sky at visible wavelengths, is the presence of a huge background of diffuse photons, particularly in the galactic plane. Over the diffuse background, some persistent sources are evident. In the galactic plane, the brightest sources were

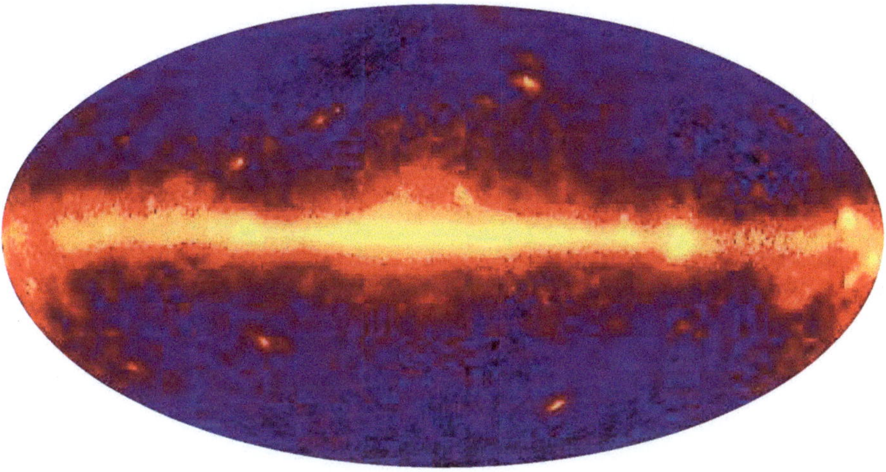

Fig. 8.4 The γ-ray sky seen by EGRET, shown in galactic coordinates. In this false color image, the galactic center lies in the middle of the image. Courtesy of the EGRET Collaboration

identified with pulsars. Many of the bright sources away from the galactic plane are blazars (a class of AGN, Sect. 9.10).

Gamma-Ray Sources: The Third EGRET Catalog A γ-ray source appears as an excess of photon counts above the diffuse emission, obtained from the map in Fig. 8.4. The last EGRET analysis of the sky produced the third catalog (Hartman et al. 1999), which contains 271 objects. A characteristic of the γ-ray sky is that it is highly variable, therefore not all sources were seen at all times.[2] As mentioned, the angular resolution of EGRET cannot be compared with that reached in other wavelengths of astronomical observations. In some cases, the 271 γ-ray EGRET sources have been associated with known astrophysical objects. The association is much easier far from the galactic plane. In particular, a part the Large Magellanic Cloud, which was detected as an extended γ-ray source and one sufficiently bright solar flare, 94 sources show a probable or possible association with blazars and five sources are known galactic pulsars. The remaining 170 sources, almost 60% of the total, had no identification with known astrophysical objects.

After EGRET, the situation regarding GeV γ-ray astronomy had a breakthrough in 2008 with the launch of the Fermi satellite.

[2]Aristotle's view of the eternal and immutable heavens depicted in *De Caelo* had an enormous influence on the medieval view of the Universe, despite having been modified at that time to correspond to Christian theology. A completely different history would have occurred if γ-ray sensors had been available to Aristotle.

8.8 Fermi-LAT and Other Experiments for γ-Ray Astronomy

8.8.1 The Fermi-LAT

"A revolution is underway in our understanding of the high energy sky". This is the introductory sentence of the Fermi-LAT technical paper (Atwood et al. 2009). In fact, immediately after its launch, on June 11 2008, an overwhelming amount of data has significantly improved our knowledge of high energy astrophysics. The Large Area Telescope (LAT) is a telescope for γ-rays within the energy range from 20 MeV to more than 300 GeV, detected over the large background of energetic charged particles at the 565 Km altitude orbit of the Fermi satellite. For each γ-ray, the LAT measures its arrival time, direction, and energy.

The key improvements of this experiment have been obtained because of the newer technologies, principally in particle detection and electronics, available since the construction of EGRET. These improvements provide a larger effective area over a much larger field-of-view, a better particle tracking that produces an improved angular resolution and background rejection, and a fast, flexible, multilevel trigger and data acquisition system. Fundamental is the rejection capability to discriminate between electromagnetic and hadronic showers, based on the different event topology in the three subsystems (the tracker, the calorimeter and the anticoincidence, as shown in Fig. 8.3).

The LAT therefore is a pair-conversion telescope with a precision converter-tracker section followed by a calorimeter. These two subsystems each consist of a 4×4 array of 16 modules (see Fig. 8.5). The large field-of-view results from the low aspect ratio (height/width) of the LAT made possible by the choice of particle tracking technology (i.e., silicon-strip detectors) that allowed for elimination of the time-of-flight triggering system used in EGRET.

For each γ-ray, the arrival time, direction, and energy are measured. The effective collecting area is $\sim$6500 cm^2 at 1 GeV with a wide field-of-view ($\sim$2 sr). The observing efficiency is very high, limited primarily by interruptions of data taking (13% over the live time) when Fermi passes through regions where charged particles are trapped by the Earth's magnetic field (Sect. 2.9) and the trigger dead time fraction (9%). As with EGRET, the angular resolution is strongly dependent on the photon energy; the 68% containment radius is about 0.8° at 1 GeV (averaged over the acceptance of the LAT) and varies with energy be approximately $E^{-0.8}$, asymptotically approaching 0.2° at the highest energies.

The tracking section of the LAT has 36 layers of silicon strip detectors to record the tracks of charged particles, interleaved with 16 layers of tungsten foil. There are 12 thin layers, 0.03 radiation lengths each, at the top (or Front) of the instrument, followed by 4 thick layers, 0.18 radiation lengths, in the Back section to promote the γ-rays' pair conversion. Unlike EGRET, the LAT basically triggers on all the charged particles crossing the active volume, with no built-in hardware trigger for photon selection that can induce inefficiencies. This drastic change of approach is

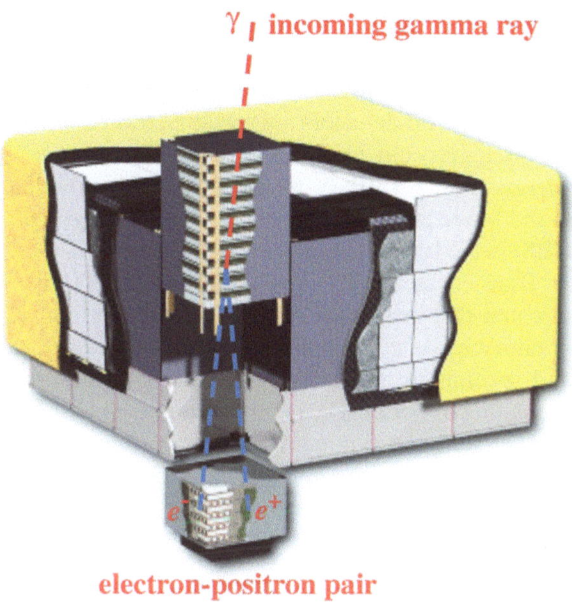

Fig. 8.5 Schematic diagram of the LAT. The telescope's dimensions are $1.8\times1.8\times0.72$ m. The power required and the mass are 650 W and 2789 kg, respectively. The upper part is the tracker, the lower part the calorimeter and the surrounding region (in *yellow*) the veto. Courtesy of the Fermi-LAT Collaboration

largely due to the use of silicon detectors, which allow for precise tracking with essentially no detector-induced dead-time.

Beneath the tracker is a calorimeter comprising an 8-layer array of CsI crystals (1.08 X_0 per layer) to determine the e^+, e^- energy. The calorimeter allows for imaging of the shower development, and thereby corrections of the energy estimate for the shower leakage fluctuations out of the calorimeter. The total thickness of the tracker and calorimeter is approximately 10 radiation lengths at normal incidence.

The tracker is surrounded by segmented charged-particle anticoincidence detectors (ACD) made of plastic scintillators with photomultiplier tubes, to reject CR background events. A programmable trigger and data acquisition system uses prompt signals available from the tracker, calorimeter, and ACD to form a trigger that initiates readout of these three subsystems. The onboard trigger is optimized for rejecting events triggered by CR background particles while maximizing the number of events triggered by γ-rays, which are transmitted to the ground for further processing. Additional information about Fermi-LAT can be found on http://fermi.gsfc.nasa.gov/.

8.8.2 *The Fermi-GBM*

The Fermi satellite also carries the *Gamma-ray Burst Monitor (GBM)*, which complements the LAT for observations of high-energy transients. The GBM is sensitive to X-rays and γ-rays with energies between 8 keV and 30 MeV. The GBM includes two sets of detectors: twelve sodium iodide (NaI) scintillators, and two cylindrical bismuth germanate (BGO) scintillators. Each NaI crystal is 12.7 cm in diameter by 1.27 cm thick, while each BGO is 12.7 cm in diameter and 12.7 cm in height. The NaI detectors are sensitive from a few keV to about 1 MeV and provide burst triggers and locations of a transient source. The BGO detectors cover the energy range $\sim$150 keV to about 30 MeV, providing a good overlap with the NaI at the lower end and with the LAT at the high end. The important feature of the GBM is the large field-of-view (FoV) of $\sim$9.5 sr ($\sim$75% of the whole solid angle).

Transient bursts of γ-rays (GRBs, Sect. 8.11) are detected by a significant change in count rate in at least two of the NaI scintillators; the trigger algorithm is programmable. After a trigger, the GBM processor calculate the preliminary position and spectral information for telemetry to the ground and possible autonomous repointing of Fermi. The combination of the GBM and the LAT provides a powerful tool for studying gamma-ray bursts, particularly for time-resolved spectral studies over a very large energy band. The GBM's large field-of-view and its capability for fast distribution of the parameters of detected GRBs had a fundamental role in the discovery of the gravitational wave due to the merging of two neutron stars, Sect. 13.7.

8.8.3 AGILE

The Fermi satellite was anticipated by the smaller-scale telescope *Astro-rivelatore Gamma a Immagini Leggero (AGILE)*. AGILE is a project of the Italian Space Agency (ASI) and was launched in April 2007. It is devoted to γ-ray observations within the 30 MeV–50 GeV energy range, with simultaneous hard X-ray imaging in the 18–60 keV band, and optimal timing capabilities for the study of transient phenomena. The AGILE instrument (Tavani 2008) consists of the Silicon Tracker, the X-ray detector SuperAGILE, the CsI(Tl) Mini-Calorimeter and an anti-coincidence system. The combination of these instruments forms the Gamma-Ray Imaging Detector (GRID). The very large field-of-view (2.5 sr) of the γ-ray imager coupled with the hard X-ray monitoring capability makes AGILE well suited to study galactic and extragalactic sources, as well as GRBs and other fast transients. AGILE reaches its optimal performance near 100 MeV with good imaging and sensitivity. Gamma-ray and hard X-ray sources can be monitored 14 times a day, and an extensive database has been obtained for a variety of sources.

8.8.4 Swift

The *Swift space observatory* (Gehrels et al. 2004) is a multi-frequency, rapid response observatory that was launched on November 20, 2004. To fulfill its purposes, Swift carries three instruments on board: the Burst Alert Telescope (BAT), sensitive in the 15–150 keV band, the X-Ray Telescope (XRT), sensitive in the 0.3–10.0 keV band, and the UV and Optical Telescope (170–600 nm) (UVOT). The primary objective of the Swift scientific program is the discovery and rapid follow-up of GRBs, Sect. 8.11. However, as these elusive sources explode at random times and their frequency of occurrence is subject to large statistical fluctuations, there are periods when Swift is not engaged with GRB observations and the observatory can be used for different scientific purposes. The sources observed through this secondary science program are usually called Swift fill-in targets. The very wide spectral range covered by these three instruments is of crucial importance for the study of AGN, as it covers the energy region where the transition between the synchrotron and inverse Compton emission usually occurs. Since the beginning of its activities, Swift has observed hundreds of blazars as part of the fill-in program.

8.9 Diffuse γ-Rays in the Galactic Plane

Unlike the sky at visible wavelengths, the γ-ray sky is dominated by a diffuse radiation originating in our Galaxy, Fig. 8.4, and diffuse radiation is seen at all galactic latitudes. This diffuse γ-ray radiation is largely produced by galactic CR hadrons and electrons interacting with the interstellar gas and photon fields. The spatial distribution of this high-energy radiation traces galactic structures as determined from radio and other measurements. These basic features had already been discovered by the SAS-2 and COS-B spacecrafts.

The physical processes that produce the observed γ-rays are due to:

- Inelastic collisions of CR protons and nuclei with the interstellar gas during their propagation. In the interaction, secondary particles are produced, particularly charged and neutral π mesons. The $\pi^0 \rightarrow \gamma\gamma$ decay has the features described in Sect. 8.4, and the expected γ-ray distribution has a broad energy range.
- CR electrons colliding with low-energy photons. The inverse Compton scattering boosts some low-energy photons into the γ-ray band. The principal targets are the optical and infrared photons found throughout the Galaxy.
- High-energy electrons interacting with the interstellar gas, producing γ-rays through bremsstrahlung.

The first detailed information on the galactic diffuse γ-ray radiation was provided by EGRET within the 30 MeV–40 GeV range. The spectrum was compared with a detailed modeling of the measured CR intensities combined with a three-dimensional model of the distribution of galactic photons and gas. The model reproduced most features of the observed gamma radiation, but also one unexpected feature: the flux above 1 GeV exceeded the model prediction by a significant amount. This discrepancy was known for some years as the "EGRET GeV excess".

Using the first 5 months of the science phase of the mission, Fermi-LAT was able to measure the diffuse flux for the galactic latitude range $10^\circ \leq |b| \leq 20^\circ$ (Abdo et al. 2009a). Figure 8.6 shows the spectral energy distribution (SED) of this

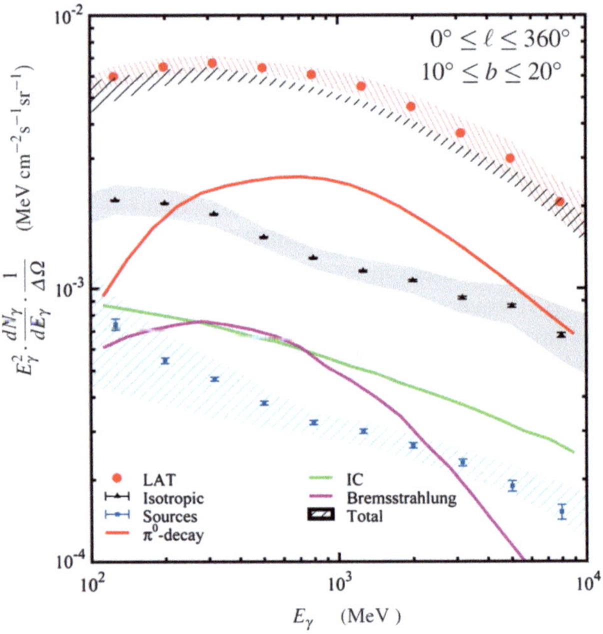

Fig. 8.6 Spectral energy distribution divided by the observed solid angle (SED/$\Delta\Omega$) of diffuse γ-rays as measured by Fermi-LAT. The SED is normalized to the solid angle region with galactic latitude $10° \le b \le 20°$ and longitude $0° \le \ell \le 360°$. Overlaid are the expected SEDs for the physical processes that produce the diffuse emission. The lines represent the contribution from: the dominant π^0 decay (*red*); bremsstrahlung (*magenta*); inverse Compton (IC, *green*). The unidentified isotropic background is represented by the *gray* shaded region. The *blue/hatched* distribution shows the unresolved contribution from sources. The *black/hatched* is the total theoretical prediction, in good agreement with data (*red* points with error band). Courtesy of th Fermi-LAT Collaboration

radiation in the considered region of the sky, along with the estimated contribution from the three afore-mentioned mechanisms due to the CR propagation (red line) and electron interactions (green and magenta lines).

The modeling necessary to produce the theoretical lines shown in Fig. 8.6 is not simple. It requires a good knowledge of the spatial distribution of CR protons and electrons and of all the radiation and matter distributions with which the CRs interact. The matter distribution was derived with the techniques described in Sect. 2.7.2. The most advanced models include the GALPROP code, which was discussed in Sect. 5.4 and which is used to compare the data in the figure.

The Fermi-LAT measurements are in agreement with the hypothesis that CRs fill the whole Galaxy. According to the modeling, most γ-rays with energies between 100 MeV and 50 GeV originate from the decay of π^0 produced in hadronic collisions when CR protons with energies from 0.5 to 10^3 GeV interact with InterStellar Matter (ISM) nuclei. The Fermi-LAT results do not confirm the unexpected excess in the energy range of a few GeV measured by EGRET. This excess could probably be explained by an incorrect energy assignment of the high-energy photons in EGRET.

As described in Sect. 9.13, extragalactic sources of γ-rays are probably objects characterized by jet activities, as in the case of *blazars*. It is interesting to note that some galaxies (for instance, the Large and the Small Magellanic Clouds and some *starburst galaxies*,[3] such as M82 and NGC 253) that do not exhibit apparent jet activities are also seen by Fermi-LAT as γ-ray sources. This observation provides information on the existence of CRs beyond our own Galaxy. In fact, GeV γ-rays in these galaxies come primarily from the interactions of hadrons and electrons produced by mechanisms similar to those providing our CRs, with interstellar matter and photon fields in that Galaxy. As a confirmation of this hypothesis, the luminosities of these normal and starburst galaxies show an approximately linear relationship with the product of the supernova rate and the total mass of gas in the galaxies (Thompson et al. 2012).

8.9.1 An Estimate of the Diffuse γ-Ray Flux

Let us work out an order of magnitude estimate of the diffuse γ-ray flux due to the propagation of CRs in the galactic ISM. While dedicated computer code are necessary for a complete description of the involved processes, we can obtain an order of magnitude estimate using the following ingredients.

1. Cross-section. We assume that both CRs and target materials are protons, with inelastic cross-section $\sigma_{pp} \simeq 40\,\text{mb} = 4 \times 10^{-26}\,\text{cm}^2$.
2. Number density of target material n. According to the discussion in Sect. 2.10, its average value can range from $n = n_{ISM} = 0.3\text{–}1\,\text{cm}^{-3}$, see also Eq. (5.16).
3. CRs' energy density. As in Sect. 8.6, we assume that the energy density of CRs producing the γ-rays observed by the Fermi-LAT corresponds to $\rho_{CR\to\gamma} \sim 0.1\,\text{eV/cm}^3$.
4. Energy transferred to γ-rays. In the process $pp \to hadrons$, the fraction ΔE of the energy E of the incoming proton converted into γ-rays depends on E. It is $\sim$10% of E at threshold for the process $pp \to pp\pi^0$, and rises to about $\sim$1/3 at very high energies, when 1/3 of the secondary particles are neutral pions. In the following, as a rule of thumb, we assume $\Delta E/E \sim 1/6$.

Using the above assumptions, the interaction rate of one relativistic CR (moving at the speed of light c) with target protons of the ISM corresponds to

$$R_{\text{coll}} = \sigma_{pp} \cdot n \cdot c = 4\ 10^{-26} \cdot n \cdot 3\ 10^{10} = 1.2\ 10^{-15} \cdot n \quad [\,\text{s}^{-1}]. \tag{8.50}$$

[3] In starburst galaxies, an exceptionally high rate of star formation, as compared to the star formation rate observed in most other galaxies, is observed.

This relation is important for verifying the connection between the total energy of a source and the photon luminosity.

As the σ_{pp} cross-section is assumed independent from the energy E of the CR, the energy emitted isotropically as γ-rays per unit of solid angle per cubic centimeter of the Galaxy per second corresponds to

$$L_\gamma^{\text{diff}} = \frac{\Delta E}{E} \cdot \frac{1}{4\pi} \cdot R_{\text{coll}} \cdot \rho_{CR\to\gamma} \quad \text{units: } [\text{sr}]^{-1}[\text{s}]^{-1}[\text{eV cm}^{-3}]. \tag{8.51}$$

Inserting the numerical values,

$$\begin{aligned} L_\gamma^{\text{diff}} &= \frac{1}{6} \cdot \frac{1}{4\pi} \cdot (1.2\ 10^{-15}) \cdot n \cdot (0.1) \simeq 10^{-18} \cdot n \quad \text{eV cm}^{-3}\,\text{s}^{-1}\,\text{sr}^{-1} \\ &= 10^{-24} \cdot n \ \ \text{MeV cm}^{-3}\,\text{s}^{-1}\,\text{sr}^{-1}. \end{aligned} \tag{8.52}$$

The spectral energy distribution per unit solid angle, $SED/\Delta\Omega$, (the quantity reported in ordinate in Fig. 8.6) depends on the linear distance D from which photons induced by pp interactions can arrive from the galactic plane as:

$$\frac{SED}{\Delta\Omega} = L_\gamma^{\text{diff}} \cdot D = 10^{-24} \cdot (nD)\,\text{MeV cm}^{-2}\,\text{s}^{-1}\,\text{sr}^{-1}. \tag{8.53}$$

The LAT observations reported in Fig. 8.6 are in the latitude range $10° < b < 20°$. To have a simple estimate of D, we consider the central value of the interval, $b = 15°$, and the position of the Sun (in a plane in the center of the Galactic disk of height of $h = 200$ pc). Thus, an estimate is $D = (h/2)/\sin 15° \simeq 400\ \text{pc} \simeq 1.2\ 10^{21}$ cm. The quantity (nD) in (8.53) represents the column density of material responsible for the photons seen by the experiment. Our estimate gives $(nD) \sim (0.4 \div 1.2) \times 10^{21}\ \text{cm}^{-2}$, for $n = n_{ISM} = 0.3 \div 1\ \text{cm}^{-3}$.

Column densities (nD) of atomic hydrogen gas were estimated by the Fermi-LAT collaboration from existing radio surveys of the 21 cm line of HI, and are in agreement with our simple evaluation. As an order of magnitude for Eq. (8.53), we have

$$\frac{SED}{\Delta\Omega} \simeq 10^{-24} \times 10^{21} = 10^{-3}\,\text{MeV cm}^{-2}\,\text{s}^{-1}\,\text{sr}^{-1}. \tag{8.54}$$

The value in (8.54) is in agreement with the Fermi-LAT estimate of the diffuse γ-ray flux within the 100 MeV–10 GeV range from π^0 decay. The red curve of Fig. 8.6 ranges between $(1 \div 2)\ 10^{-3}\,\text{MeV cm}^{-2}\,\text{s}^{-1}\,\text{sr}^{-1}$. Another prediction of our simple computation is the linear correlation between $\frac{SED}{\Delta\Omega}$ and the column density of matter, $n \cdot D$, which was effectively observed (Abdo et al. 2009b). This diffuse component of the γ-rays is the most direct evidence that CRs are filling our Galaxy with an energy density the same as that measured on Earth.

8.10 The Fermi-LAT Catalogs

The Fermi-LAT (LAT, in this section) collaboration is producing a large number of scientific results. Additional analyses of LAT data are also being carried out by researchers external to the collaboration due to the policy of data diffusion. During the first year, the release of LAT data was restricted to allow the LAT team to calibrate the instrument and carry out their proposed sky survey. After the first year, data are no longer proprietary, and are made freely available on the web. The LAT team has released data on transient sources and light curves for about 150 regularly monitored sources,[4] and continuously adds more sources to the list as they show significant brightening. Since the beginning of the second year of operations, all LAT science data is released as soon as possible.

At the time of this writing (early 2018), the LAT operations are still in progress. Different source catalogs are continuously produced and updated by the LAT collaboration. The following information are extracted from the official third catalog of high-energy γ-ray sources (3FGL) detected by LAT (Acero et al. 2015) and derived from data taken during the first 4 years of the science phase of the mission. The 3FGL catalog includes 3033 sources[5] above 4σ significance within the 100 MeV–300 GeV range, with source location regions, spectral properties, and monthly light curves for each. Among them, 238 sources are considered as having been identified based on angular extent or correlated variability (periodic or otherwise) observed at other wavelengths. For 1010 sources, no plausible counterparts at other wavelengths have been found. More than 1100 of the identified or associated sources are active galaxies of the blazar class. Pulsars represent the largest Galactic source class. The LAT catalogs and associated products are currently available as readable tables. Information are also available in file format to be used for data analysis within the Fermi Science Tools. Supporting tools and documentation have been provided and are available.[6]

As in the case of EGRET, the sources are obtained from the photon counting map. LAT collects about 150 million γ-rays per year (compared with the 1.5 million detected by EGRET in 9 years). Figure 8.7 shows the LAT photon map in galactic coordinates. This plot should be compared with that produced by EGRET, shown in Fig. 8.4. The diffuse galactic emission is a foreground for the identification of point sources, and hence affects the determination of their positions and fluxes. It is also a foreground for the much fainter extragalactic component, which is the sum of contributions from unresolved sources and truly diffuse extragalactic emission. The knowledge of the diffuse component, presented in Sect. 8.9, is a necessary first step, as the sources are found after subtraction of the diffuse component from the photon-counting map.

---

[4]https://fermi.gsfc.nasa.gov/ssc/data/access/lat/msl_lc/.

[5] The previous LAT catalog (2FGL) was released in 2012 (Nolan et al. 2012) and contained 1873 sources.

[6]http://fermi.gsfc.nasa.gov/ssc/data/access/.

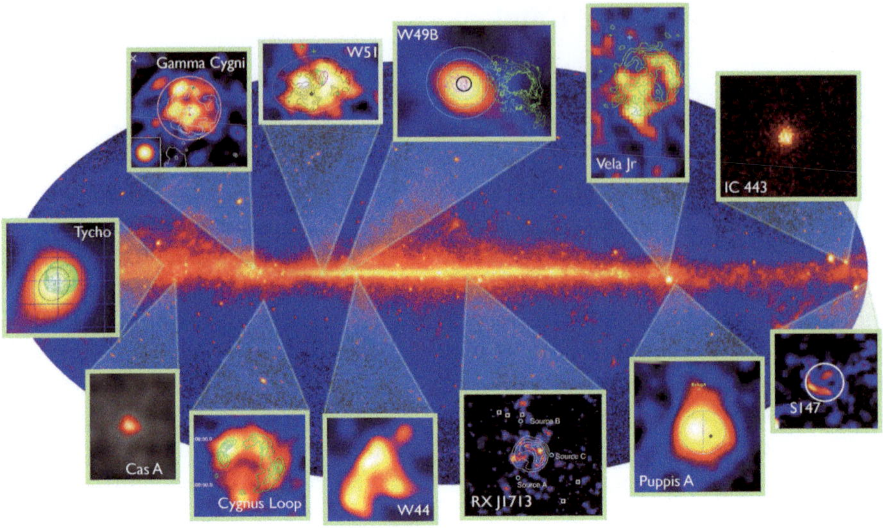

Fig. 8.7 Fermi-LAT sky map in galactic coordinates. Superimposed, the position of the principal supernova remnants in the LAT γ-ray sky. From Tibaldo (2013)

The basic analysis steps used to construct the 3FGL catalog (as in the 2FGL one) are source detection, localization, and significance estimation. Once the final source list is determined, the flux in 5 energy bands (corresponding to the SED between 100 and 300 MeV; 300 MeV–1 GeV; 1–3 GeV; 3–10 GeV; 10–100 GeV) and the flux history (light curve of the integrated flux) for each source is produced. The minimum flux from a source that LAT can discriminate from the background corresponds to $\sim 3 \times 10^{-12}\,\text{erg cm}^{-2}\,\text{s}^{-1}$ (Acero et al. 2015). Compare this value to that of Eq. (8.47).

A point-like object is spatially unresolved within the point spread function of the detector. The analysis and catalog also include a number of LAT sources (25, in the third catalog; there were 12 in 2FGL) that are known to be spatially extended. These extended sources are of particular importance and include twelve supernova remnants (SNRs), nine pulsar wind nebulae (PWNe) or candidates, the Cygnus X cocoon, the Large Magellanic Cloud (LMC) and the Small Magellanic Cloud (SMC), and the radio galaxy Centaurus A. Also, the Sun is a bright, extended source, due to CR interactions in its outer atmosphere and to IC scattering of CR electrons on the solar radiation field, which produces an extended γ-ray halo around our Star. The Moon is comparably bright to the Sun in γ-rays. Except for the diffuse emission and the sources explicitly considered as spatially extended, all remaining objects in the catalog are assumed to be point-like.

Source Association and Identification The designations of the classes that are used to categorize the 3FGL sources are listed in Table 8.1, along with the number of sources assigned to each class. Each source can either be *associated* or *identified*,

Table 8.1 Number of objects catalogued in the 3FGL

Description	Designator	Identified	Associated
Pulsar, identified by pulsations	PSR	143	–
Pulsar, no pulsations seen in LAT yet	–	–	24
Pulsar wind nebula	PWN	9	2
Supernova remnant	SNR	12	11
Supernova remnant/pulsar wind nebula	spp	–	49
Globular cluster	GLC	0	15
High-mass binary	HMB	3	0
Binary	BIN	1	0
Nova	NOV	1	0
Star-forming region	SFR	1	0
BL Lac type of blazar	BLL	18	642
FSRQ type of blazar	FSRQ	38	446
Blazar candidate of uncertain type	BCU	5	568
Non-blazar active galaxy	AGN	0	3
Radio galaxy	RDG	3	12
Seyfert galaxy	SEY	0	1
Normal galaxy (or part)	GAL	2	1
Starburst galaxy	SBG	0	4
Narrow line Seyfert 1	NLSY1	2	3
Soft spectrum radio quasar	SSRQ	0	3
Compact steep spectrum quasar	CSS	0	1
Total		238	1785
Unassociated			1010

The first section refers to galactic objects; the second to extragalactic sources. Identified (associated) objects are indicated with capital (lower case) designators. In the case of AGN, many of the associations have high confidence

with associations depending primarily on close positional correspondence. The identification, shown in the catalog by capitals in the designator column in Table 8.1, is based on one of three criteria:

1. Period variability. Temporal variability is relatively common in γ-ray sources and provides a powerful tool for associating them definitively with objects known at other wavelengths and studying the physical processes powering them. The method is mainly used to associate known pulsars.
2. Spatial morphology, which is used to identify spatially extended sources whose morphology can be related to that seen at other wavelengths. The method is used to identify SNR, PWNe, and galaxies.
3. Correlated variability. Variable sources, primarily AGN, whose γ-ray variations can be matched to variability seen at one or more other wavelengths, are considered to be firm identifications.

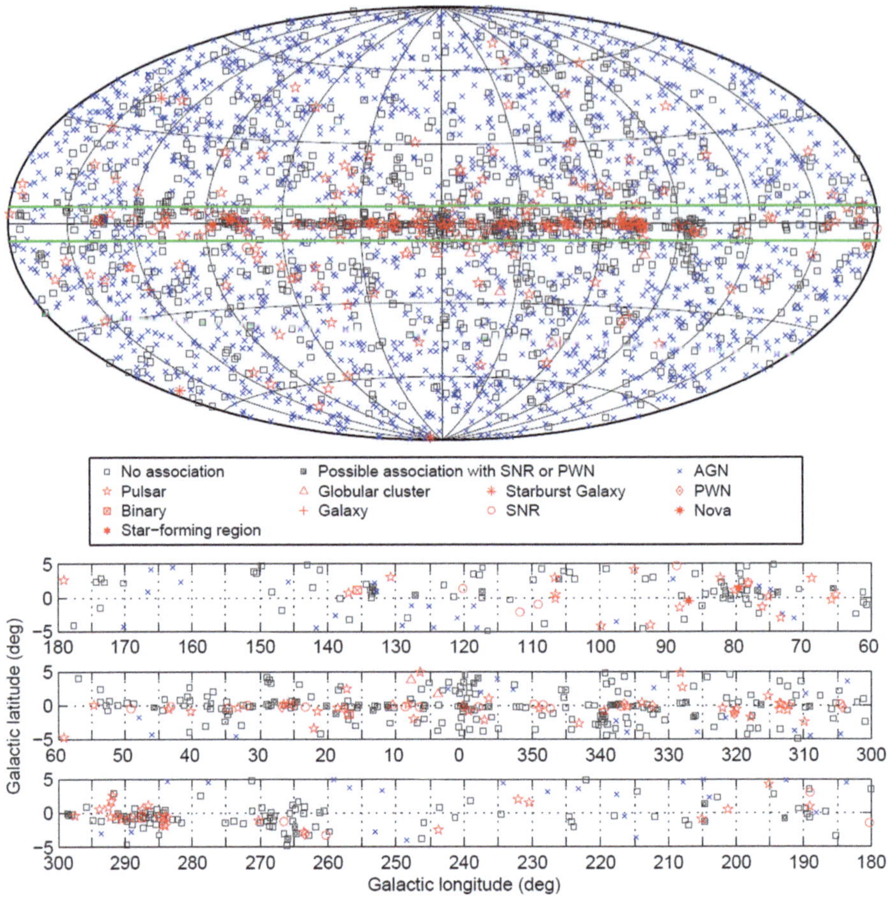

Fig. 8.8 Full sky map (*top*) and blow-up of the inner galactic region (*bottom*) showing sources by source class (refers to Table 8.1). Identified sources are shown with a *red* symbol, associated sources in *blue*. All AGN classes are plotted with the same symbol for simplicity. Courtesy of the Fermi-LAT Collaboration

In total, 238 out of the 3033 3FGL sources have been firmly identified (see table). The list of astronomical catalogs used for the identification/association is reported in Acero et al. (2015). Sources associated with SNRs are often also associated with PWNe and pulsars, and the SNRs themselves are often not point-like. Figure 8.8 illustrates where the different sources are located in the sky.

Source Spectral Shapes As the LAT measures the number of arriving photons as a function of their energy, the so-called *spectral shape* (photon flux as a function of the energy) can be constructed for each individual source. In most cases, a simple power law $dN/dE = K(E/E_0)^{-\alpha_\gamma}$ can be considered. Frequently, the flux shows a cut-off at high energy and the spectral shape is represented by exponentially cut-off power laws

$$\frac{\mathrm{d}N}{\mathrm{d}E} = K\left(\frac{E}{E_0}\right)^{-\alpha_\gamma} \exp\left(\frac{E-E_0}{E_c}\right)^b. \tag{8.55}$$

This is just the product of a power law and an exponential. The fit parameters are K, α_γ (as in the power law) and the cut-off energy E_c. A simple power-law form ($E_c \to \infty$) is used for all sources not significantly curved. The parameter b is fixed, for most sources, to one by the LAT collaboration. E_0 is a reference energy that is chosen freely for each source. This functional representation of the spectral shape is particularly suited to pulsars.

Other bright sources (mainly AGN) are also not very well-represented by power-law spectra. The LAT collaboration uses a functional form called *LogParabola* which adds only one parameter to the power law:

$$\frac{\mathrm{d}N}{\mathrm{d}E} = K\left(\frac{E}{E_0}\right)^{-\alpha-\beta\ln(E/E_0)}. \tag{8.56}$$

The left panel of Fig. 8.9 shows the distribution of the power-law index of all of the sources. Its average value is $\alpha_\gamma = 2.19 \pm 0.01$, which is very close to that expected for a Fermi mechanism for the acceleration of parent charged particles, $\alpha_{\rm CR} \sim 2$. The right panel shows the energy flux distribution of detected sources. In both cases, high galactic latitudes ($|b| > 10°$) are considered, so most Galactic sources are not included.

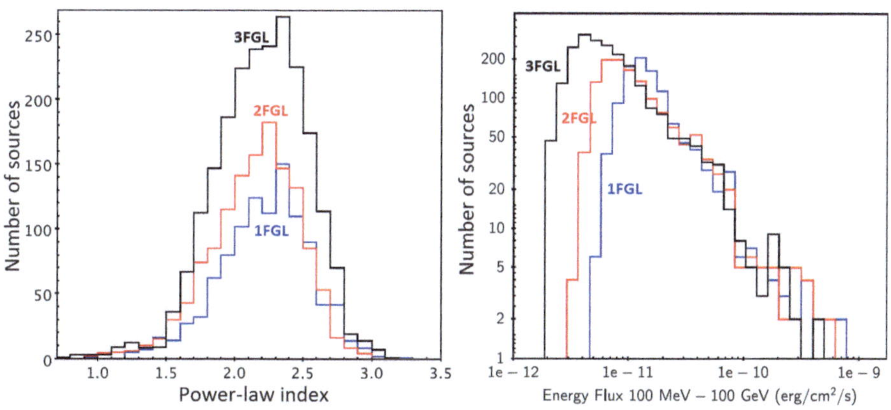

Fig. 8.9 Left: Distribution of the spectral index α_γ for the high-latitude sources ($|b| > 10°$) in the 3FGL (1960 sources). Superimposed, the corresponding distributions for the 1FGL (1043 sources) and 2FGL (1173 sources). Right: Distributions of the energy flux for the 1FGL, 2FGL, and 3FGL sources in the same Galactic latitude band. Courtesy of the Fermi-LAT Collaboration

8.11 Gamma Ray Bursts

Gamma-Ray Bursts (GRBs) are extremely intense and relatively short bursts of gamma radiation that occur a few times per day in the detectable Universe. Their emission exceeds the gamma emission of any other source. For instance, the GRB 080319B[7] was detected by the Swift (Gehrels 2004) satellite at 06:12 UTC on March 19, 2008, and was the brightest event ever observed in the known sky. The explosion set a new record for the most distant object that could be seen with the naked eye, it had a peak apparent magnitude[8] of 5.8 and it remained visible to the human eye for approximately 30 s (Racusin et al. 2008).

Present knowledge suggests that GRBs occur in random directions in the sky and at cosmological distances. The time integrated fluxes, or the *fluencies*, range from $\sim 10^{-7}$ to $\sim 10^{-4}$ erg cm^{-2}. The observed fluencies, combined with the distances determined from detections of the host galaxies for which optical redshift distances are obtained, show that GRBs are the brightest explosions in the Universe. If they were emitting isotropically, the γ-ray energy output would amount on average to a solar rest-mass energy, $M_{\odot}c^2 \sim 10^{54}$ erg, emitted in a few seconds.

In fact, there is evidence that the emission is anisotropic or *jet-like*, with a typical jet opening angle θ of a few degrees, corresponding to a solid angle $\Delta\Omega \sim \pi\theta^2$. This introduces, for a double jet configuration, an angular correction factor in the total energy emitted, which is, on average, $2\Delta\Omega/4\pi \sim 10^{-3}$. Thus, the actual average jet energy in γ-rays is $\sim 10^{51}$ erg emitted in a few seconds. This is to be compared with the isotropic kinetic energy content of a supernova explosion, $E_{SN} \sim 10^{51}$ erg, of which only $\sim 1\%$ emerges as visible photons over periods of months to years (Sect. 12.13).

In astronomy, many transient sources generally have rather simple time structures, which help to understand the underlying physics of the objects. GRBs are very peculiar from this point of view, as their light curves vary significantly one to another. There are no two identical GRBs: the duration, the number of peaks, the maximum brightness, in fact, every parameter can be different. See Fig. 8.10 as an example of the light curves for two GRBs.

History of the Discovery of GRBs GRBs were discovered accidentally. In the 1960s, both the United States and the former Soviet Union launched military satellites to monitor adherence to the nuclear test-ban treaty. These satellites were γ-ray detectors, as the signature of a nuclear detonation is a brief, but intense, pulse of γ-rays. While most satellites orbited at about 800 km above the Earth's surface, the U.S. Vela satellites orbited at an altitude of more than 100,000 km, above the Van Allen radiation belts, reducing the noise in the sensors. The extra height also

[7]GRBs are named according to the date (*yymmdd*) upon which they have been detected; if more than one burst per day is present, the extension A is used for the first, B for the second, and so on.

[8]The apparent magnitude of an object is a logarithmic measure of its brightness as seen from the Earth. For comparison, the value 0 is assigned to the star Vega.

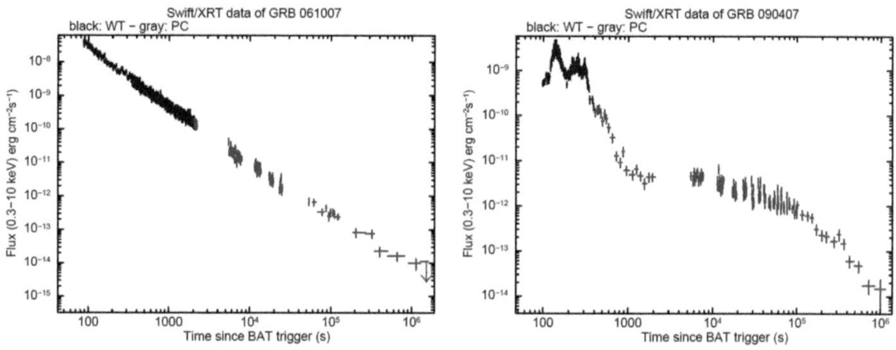

Fig. 8.10 Light curves for two GRBs measured by the X-Ray Telescope (XRT) in the Swift satellite. For details of how these light curves were produced, see Evans et al. (2007)

meant that the satellites could detect explosions behind the Moon, a location where the U.S. government suspected the U.S.S.R would try to conceal nuclear weapon tests.

On July 2, 1967, the U.S. Vela satellites indeed detected a γ-ray signal, but it had neither the intense initial flash nor the gradual fading that characterize nuclear weapon explosions. Instead, there were two distinct peaks in the light curve. It was only in 1973 that the observations were published, identifying a cosmic origin for the previously unexplained observations of γ-rays.

Another milestone in the history of GRBs was the launch of the BATSE experiment on the CGRO satellite, Sect. 8.7. The main result achieved by the BATSE detector is the conclusive proof that GRBs occur isotropically in the sky. In addition, it revealed that the GRBs can be split into two families: short and long duration GRBs, which have correspondingly different spectra. Thanks to the BATSE data, it has also been possible to measure the typical fluence (the flux integrated over time) of GRBs.

The emission spectra of the GRBs show a peculiar nonthermal behavior, peaking at around a few hundred keV and extending up to several GeV. The spectrum, that is, the number of photons per unit energy, is generally of the form of a broken power law (Fig. 8.11) called a Band Spectrum (Briggs et al. 1999), given by

$$\begin{aligned} N(E) \equiv \frac{\mathrm{d}N_\gamma}{\mathrm{d}E} &\propto E^{-\alpha} \quad \text{with} \quad \alpha \simeq 1 \quad \text{for} \quad E < E_b \\ &\propto E^{-\beta} \quad \text{with} \quad \beta \simeq 2 \quad \text{for} \quad E > E_b\,. \end{aligned} \tag{8.57}$$

The change of spectral slope occurs at a break energy E_b, which, for the majority of observed bursts, is within the range of 0.1–1 MeV. The bottom plot of Fig. 8.11 also shows the spectral energy distribution of the considered GRB.

Due to their short duration, GRBs were very difficult to localize precisely. In this respect, the real breakthrough was made by the Italian-Dutch satellite **BeppoSAX**,

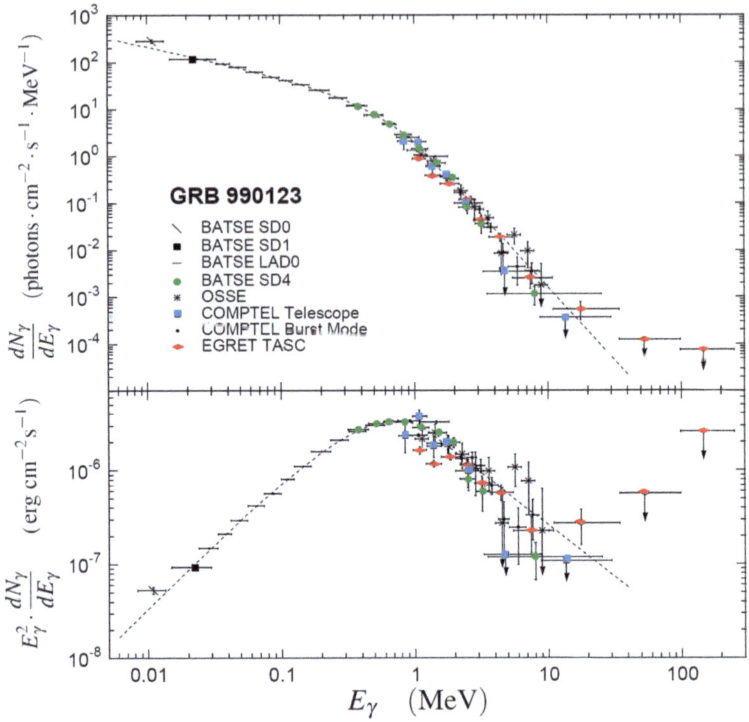

Fig. 8.11 Example of the spectrum for one GRB (GRB 990123) measured from all experiments on board the CGRO, showing a band-type broken power-law component. The *upper panel* shows the photon number spectrum $N_E = \mathrm{d}N/\mathrm{d}E$. The corresponding energy spectral energy distribution $E^2\mathrm{d}N/\mathrm{d}E$ is in the lower panel. Notice the change of energy units in the y-axes

launched in April 1996. Its Wide Field Camera, sensitive to the medium-hard X-ray energy range between 2–25 keV, was the first to allow for measuring the position of the GRBs with uncertainties of only a few arcminutes. The satellite could then observe the pinpointed region with the Narrow Field Instrument, covering the 0.1–10 keV range. It was then possible to detect a newly discovered feature: the X-ray *afterglow* of the GRB. In the meantime, a Gamma-ray burst Coordinates Network (the GCN system) that transmits a GRB alert to a network of selected instruments was set. The combination of BeppoSAX with the wide field camera trigger and the GCN system allowed for all ground-based telescopes to point in the direction of the detected GRB and detect optical, IR and radio afterglows.

The observation of these afterglows, in different wavelengths, allowed for the characterization of GRBs using additional information to that provided by the γ-ray data. For example, it was possible to measure the redshift (and thus the distance) of several GRBs. Knowing the redshift of a GRB, therefore, one has additional information extremely useful for the classification and study of each GRB. Determining the host galaxy of a GRB has been historically very challenging, as in most cases, there were several relatively bright galaxies contained within the position uncertainty of the GRB.

Optical observations after the GRB were extremely important, because they showed that the long GRBs came from not very bright galaxies, and that the location of the source is never very far from the center of its host galaxy. This information is useful for inferring that long GRBs are generally associated with massive and short-lived progenitors.

The Swift satellite succeeded BeppoSAX. The sensitivity of its Burst Alert Detector (BAT), is in the range 20–150 keV, was higher than the previous instruments. The speed with which the higher angular resolution X-ray and UV-optical detectors can turn toward the burst is less than 100 s from the occurrence of the trigger. These two advantages significantly improved the measurement capabilities for GRBs, especially those shorter than 2 s (short GRBs).

The latest milestone was the launch of the Fermi satellite, which allows for the investigation of the very high energy regime. Roughly one GRB per week is detected with the GBM between 8 keV–30 MeV, and roughly one a month is detected with the Fermi-LAT, 20 MeV–300 GeV. Several bursts have been detected by the Fermi-LAT at energies above 1 GeV (11 GRBs from August 2008 to January 2010), considerably improving our knowledge of high energy γ-ray emission. The most distant GRB is GRB 090423 (observed by Swift), with z $\sim$8.1, hence being produced only 600 million years after the Big Bang.

Today, the Gamma-ray Coordinates Network (GCN) provides the distribution of GRB and other transient locations detected by various spacecraft, and receives and automatically distributes to the GRB/transients community prose-style e-mail messages about follow-up observations on various GRBs/transient. In addition to the aforementioned LAT and GBM detectors on the Fermi satellite, and the instruments on the Swift, AGILE, and INTEGRAL satellites, there were others satellites in the past that were able to observe GRBs, and others are planned for the future.[9]

Most alerts are in real-time while the burst is still bursting, and others are delayed due to telemetry down-link delays. GRBs are reported at a rate of one or two per day. The *GCN circulars* allow the GRB community (extended to ground-based and space-based optical, radio, X-ray, TeV γ-rays, gravitational waves, cosmic-rays and neutrino observatories) to submit messages to a central queue, where they are automatically vetted and distributed (via email) to the entire GRB community. These are prose-style messages from follow-up observers reporting on their results (detections or upper limites) or for coordination with others. The GCN system had particular importance for the follow-up of the binary neutron star merger event observed in August 2017 as GRB170817A and a gravitational wave, as discussed in Sect. 13.7.

[9]The list is available at https://gcn.gsfc.nasa.gov/.

8.12 Classification of GRBs

Observations from BATSE led to the separation of γ-ray bursts into two families. The main classification is "long GRBs" and "short GRBs", in which the long population has an average duration of about 30 s, while the short lasts, on average, 0.3 s. As shown in Fig. 8.12, there is a clear separation between two families of GRBs. It has been noticed that the short bursts spectrum is significantly harder (with more high energy photons) than the long bursts. Long GRBs are the most frequently observed and, therefore, also the best understood. Each family of GRBs is associated with a different progenitor.

Progenitors of Long GRBs Observations show that host galaxies of long GRBs are active star-forming galaxies. In several cases, it has been proved that long GRBs happen in correlation with supernovae, linking them unambiguously to the death of a massive star. Core-collapse supernovae are the explosive deaths of massive stars that occur when their iron cores collapse into neutron stars or black holes (Sect. 12.13). The connection between supernovae and GRBs is given by the total emitted kinetic energy. The observation of GRB 980425 in conjunction with SN 1998bw showed this connection unambiguously. These two events were coincident both in time and space, and the energetic coincidence left few doubts about the connection. The connection of GRBs with the *hypernova* model is discussed in Sect. 12.13.3.

Progenitors of Short GRBs Models on the origin of short GRBs had to wait for the detection of a large sample of afterglows by Swift. Observation of afterglows allowed for the identification of host galaxies where short GRBs occurred. It was found that short GRBs are distributed uniformly among galaxies that contain a considerable quantity of old stars. In these galaxies, there is no evidence of

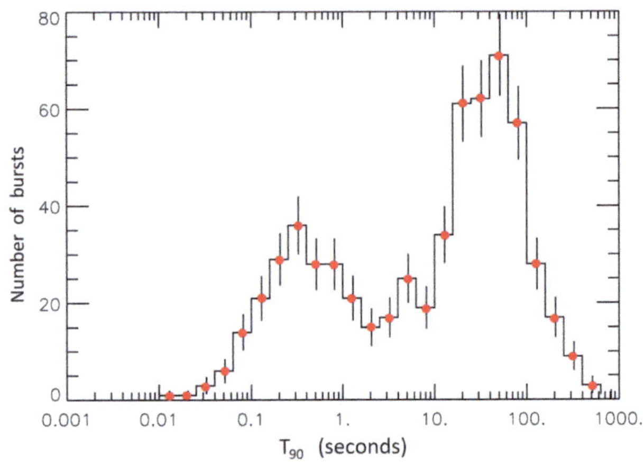

Fig. 8.12 Distribution of GRBs that occurred in 2008–2009 as a function of the log of the T90 in seconds. Usually, the duration of the GRB is expressed as the time during which 90% of the counts are detected, and denoted as T90. It is possible to see how the GRBs are split into two groups depending on their duration. "Long GRBs" last more than 2 s, the remaining ones are "short GRBs". Figure from http://f64.nsstc.nasa.gov/batse/

significant star formation, and such an old population is compatible with the presence of neutron star binary systems. The estimated total energy emitted by short GRBs exceeds that of long GRBs by many orders of magnitude. The current hypothesis attributes the origin of short GRBs to the merging of two compact objects. Possible candidates for such a process are mergers of neutron star binaries or neutron star-black hole binaries, which lose angular momentum and undergo a merger. Neutron star-neutron star binaries are observed in our galaxy, and the existence of neutron star-black hole binaries is plausible. Such structures lose energy due to gravitational radiation, as predicted by general relativity, and the two objects will spiral closer and closer. This process is thought to be extremely fast and to last no more than a few seconds, in agreement with the observation of short GRBs. This mechanism was spectacularly confirmed in the occurrence of GRB170817A, also detected in coincidence with a gravitational wave, as discussed in Sect. 13.7.

Mechanism of GRBs It is important to notice that, although the two families of GRBs are known to have different progenitors, the acceleration mechanism that gives rise to the γ-rays themselves (and probably to the production of neutrinos) is most likely independent of the progenitor of the event. The *fireball model* is the most widely used theoretical framework to describe the physics of GRBs. Independently of the details of the central engine, and based only on the release of the large amounts of energy (10^{51} erg on timescales of tens of seconds or less), the observed emission of γ-rays and the afterglow must arise from an emission region moving at relativistic velocities. The energy release in such short times in such compact regions produces a luminosity that exceeds the Eddington luminosity (see *Extras #4*), above which radiation pressure overwhelms gravity. The inner engine is attributed to a compact object, either the collapsing core of a massive star or the merger of two compact objects.

This inner engine causes an explosion that originates the relativistic blast waves moving through the star at relativistic speeds (the *fireball*). Two opposite jets form along the rotation axis of the accretion disk, and during the acceleration of matter in the jets, newly formed material accelerates faster and forms consecutive shells with different speeds. The fireball is expected to accelerate until it reaches a terminal bulk Lorentz factor Γ, which is estimated to be about $\sim$300. Interactions of shells with the external medium or collisions between shells reconvert the kinetic energy into internal energy, ready to be radiated in the form of γ-rays or transferred to baryons via baryon-photon coupling. Shocks between shells are responsible for the emission of γ-rays.

Behind the shock, the blast wave pushes the stellar material through the star surface and sweeps through space at nearly the speed of light, colliding with external gas and dust, producing additional emission of photons. These emissions are the afterglow. Figure 8.13 shows a schematic of this fireball shock model. Although this scenario is widely accepted, the discussion is still open regarding the form in which the energy is carried out. In some models, the energy outflow is carried out in the form of kinetic energy of baryons, while in other models, the flow is of an electromagnetic nature. Measurements of neutrinos from a GRB (Sect. 10.6.3) would then provide evidence for the presence of accelerated baryons.

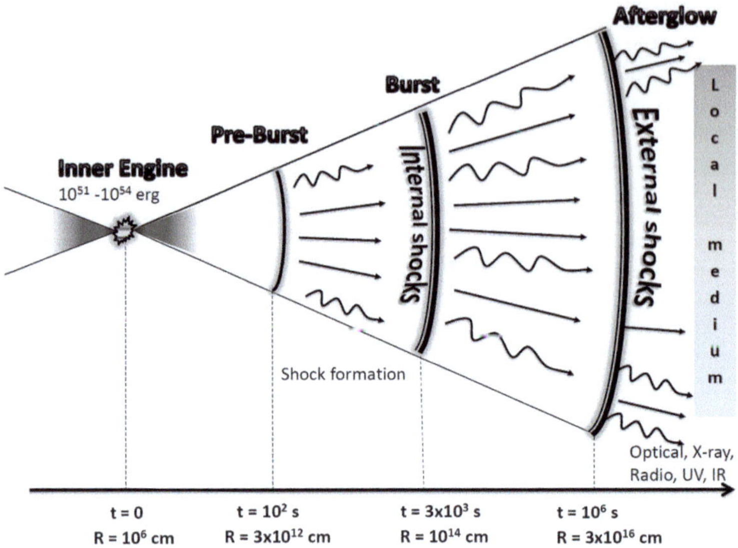

Fig. 8.13 Sketch of the internal-external fireball shocks model. A compact source produces a relativistic outflow. Internal shocks within the outflow produce the prompt γ-ray emission, while external shocks with the surrounding matter produce the lower energy and longer lasting afterglow. Courtesy of Dr. Presani (Presani 2011)

For detailed reviews on the subject, refer to Piran (2004) and Mészaros (2006).

8.13 Limits of γ-Ray Observations from Space

Despite the recent great achievements of high energy γ-ray astronomy, there are obvious limits to the performance of satellite-borne instruments. The photon flux threshold for which the Fermi-LAT can resolve a γ-ray source within the range between 1 and 100 GeV over the background corresponds to $N_{\mathrm{thr}} = 4 \times 10^{-10}$ photons cm^{-2} s^{-1} (Nolan et al. 2012). Sources with a lower photon flux cannot be resolved. Because the Fermi-LAT effective collecting area is $A \sim 6500\,\mathrm{cm}^2$ at 1 GeV in a large field-of-view of $\sim$2 sr, the corresponding number of events n_{thr} per year from a source with a flux equal to the threshold one, and assuming no dead-time, is

$$n_{\mathrm{thr}}(>1\,\mathrm{GeV}) = N_{\mathrm{thr}} \cdot A \cdot \left(\frac{2\ \mathrm{sr}}{4\pi\ \mathrm{sr}}\right) \cdot T = 12\,\mathrm{y}^{-1}. \tag{8.58}$$

$T = 3.15 \times 10^7$ corresponds to the number of seconds/y. Because many sources are simultaneously monitored within the large and homogeneous field-of-view, this is an adequate sensitivity for studying persistent γ-ray sources within the multi-GeV range. The details on the temporal and spectral characteristics of highly variable sources like blazars or GRBs is instead limited by the small effective area of the detector.

Due to the constraints on the maximum size and weight of an instrument that can be delivered into space, the effective detection area of any satellite experiments (using current launch vehicles) is limited to the order of a few m^2. Assuming a constant SED for the sources, also at higher energies, the photon flux decreases linearly with increasing energy. Thus, to detect the same number of events, the detector aperture $A \cdot T$ must increase by a factor of 100. Even with a mission live time of 10 years, this results in a practical limit for space-based observations of astrophysical sources at energies larger than a few hundreds GeV. Instruments with an effective area smaller than $\sim 100\,m^2$ cannot detect the expected astrophysical flux in the TeV energy region. Even with the advent of new, larger launch vehicles, the prospects for a space telescope with an area significantly larger than $\sim 10\,m^2$ are highly uncertain.

Apparently, after Fermi-LAT and AMS-02 (Sect. 3.5), space-based experiments for GeV γ-ray astronomy and for direct CR measurements, respectively, have achieved a point at which any further progress appears extremely difficult. However, in the near future, different projects have been envisaged to perform researches in space using particle identification, calorimetric and tracking devices. These are necessary to discriminate between charged particles (electrons/positrons, protons/antiprotons) and γ-rays. They will be devoted to searching for dark matter signals (Sect. 14.9), to extend the direct measurement of CRs in order to test theories of their origin, to investigate cosmic γ-ray emission, and to search for and study gamma-ray bursts.

CALET (http://calet.phys.lsu.edu/) is a Japanese-led international mission, with US and Italy. The detector is a calorimeter-based instrument with superior energy resolution and excellent separation between hadrons and electrons and between charged particles and γ-rays. It consists of an array of scintillation detectors used to determine the charge of the incoming CR particles; an imaging calorimeter of scintillating fibers used to determine the particle trajectory; and a deep lead tungstate calorimeter to measure energies up to 20 TeV. Unlike AMS-02, CALET has no magnetic spectrometer. CALET was launched on Aug. 19, 2015, and it is installed on the International Space Station. It is planned to detect electrons and γ-rays up to 10 TeV.

GAMMA-400 (http://gamma400.lebedev.ru/) is a Russian project for a satellite experiment that will extend the Fermi-LAT energy range up to 3 TeV with a better angular resolution. The launch is foreseen around 2020.

The HERD facility (http://herd.ihep.ac.cn/) is one of the Cosmic Lighthouse Program onboard China's Space Station, planned to be launched and assembled starting in 2020. The main constraints imposed on HERD are a total weight less than around 2 tons and total power consumption less than around 2 kw. HERD must have the capability for accurate electron and γ-ray energy and direction measurement (tens of GeV–10 TeV) and an adequate CR energy measurement with charge determination up to PeV energies.

Any further effort to improve significantly the sensitivities of space-based experiments will probably be difficult and expensive. One possibility is to use the Moon as a potential platform for the installation of a very large aperture γ-ray and CR telescope.

References

A.A. Abdo et al., Fermi large area telescope measurements of the diffuse gamma-ray emission at intermediate galactic latitudes. Phys. Rev. Lett. **103**, 251101 (2009a)

A.A. Abdo et al., FERMI LAT observation of diffuse gamma rays produced through interactions between local interstellar matter and high-energy cosmic rays. Astrophys. J. **703**, 1249–1256 (2009b)

F. Acero et al., Fermi large area telescope third source catalog. Astrophys. J. Suppl. Ser. **218**, 23 (2015)

W.B. Atwood et al., The large area telescope on the Fermi mission. Astrophys. J. **697**, 1071–1102 (2009)

S. Braibant, G. Giacomelli, M. Spurio, *Particle and Fundamental Interactions* (Springer, Heidelberg, 2011). ISBN: 978-9400724631

M.S. Briggs et al., Observations of GRB 990123 by the compton gamma-ray observatory. Astrophys. J. **524**, 82–91 (1999). ArXiv:astro-ph/9903247v1

L.O. Drury, F.A. Aharonian, H.J. Volk, The gamma-ray visibility of supernova remnants: a test of cosmic ray origin. Astron. Astrophys. **287**, 959–971 (1994)

P.A. Evans et al., An online repository of Swift/XRT light curves of γ-ray bursts. Astron. Astrophys. **469**, 379–385 (2007). This work made use of data supplied by the UK Swift Science Data Centre at the University of Leicester

N. Gehrels, The swift γ-ray burst mission. New Astron. Rev. **48**, 431–435 (2004)

N. Gehrels et al., The swift gamma-ray burst mission. Astrophys. J. **611**, 1005–1020 (2004)

G. Ghisellini, *Radiative Processes in High Energy Astrophysics*. Springer Lecture Notes in Physics (Springer, Cham, 2013). ISBN: 978-3319006116. https://www.springer.com/it/book/9783319006116

R.C. Hartman et al., The third EGRET catalog of high-energy gamma-ray sources. Astrophys. J. Supp. **123**, 79 (1999)

M.S. Longair, *High Energy Astrophysics*, 3rd edn. (Cambridge University Press, Cambridge, 2011). ISBN: 978-0521756181

P. Mészaros, Gamma-ray bursts. Rep. Prog. Phys. **69**, 2259–2321 (2006)

P.L. Nolan et al., FERMI large area telescope second source catalog. Astrophys. J. Suppl. Ser. **199**, 31 (2012)

T. Piran, The physics of GRBs. Rev. Mod. Phys. **76**, 1143–1210 (2004)

E. Presani, Neutrino induced showers from Gamma-ray Bursts, PhD thesis, University of Amsterdam, 2011. http://inspirehep.net/record/916768/files/

J.L. Racusin et al., Broadband observations of the naked-eye γ-ray burst GRB 080319B. Nature **455**, 183–188 (2008)

M. Tavani, G. Barbiellini, A. Argan et al., The AGILE space mission. Nucl. Inst. Methods Phys. Res. A **588**, 52–62 (2008)

D.J. Thompson, Gamma ray astrophysics: the EGRET results. Rep. Prog. Phys. **71**, 116901 (2008). ArXiv:0811.0738

D.J. Thompson, L. Baldini, Y. Uchiyama, Cosmic ray studies with the Fermi gamma-ray space telescope large area telescope. Astropart. Phys. **39–40**, 22–32 (2012)

L. Tibaldo, A tale of cosmic rays narrated in gamma rays by Fermi. Highlight Talk on the 33rd ICRC, Rio de Janeiro (2013)

Chapter 9
The TeV Sky and Multiwavelength Astrophysics

Abstract TeV γ-ray astronomy is outside the possibility of space-based experiments and can, at present, only be studied through ground-based experiments. The capability to suppress the high background induced by charged CRs and open the field of TeV γ-ray astronomy was made possible with the advent of the Imaging Air Cherenkov Technique and with some dedicated air shower particle arrays. The online source catalog now contains more than 200 galactic and extragalactic TeV sources. The remarkable achievements of these experiments now include the study of morphology, energy spectrum, and time variability of several galactic and extragalactic source populations. The largest class of galactic TeV emitting sources corresponds to that of pulsars with a wind nebula, among which the Crab is the most studied representative. Shell-type supernova remnants represent the major candidates as galactic sources of CRs, and thus largely studied through GeV-TeV γ-ray observations. Outside the galactic plane, it was discovered that the emission of radiation from jet-dominated AGN covers a large interval of the electromagnetic spectrum and is extremely variable. The coordinated efforts from the community are crucial for a detailed and unbiased study of AGN and other extragalactic objects: multiwavelength searches are becoming more and more important to obtain a complete picture of non-thermal processes in the Universe. Blazars, among the AGN, present some particular features, such as, for example, strong and rapid variability and a jet orientation toward the observer. As these objects are among the furthest observed objects in the Universe, their observation through TeV photons can constraint estimates of the presence of extragalactic background light, which is also relevant for cosmological models.

TeV γ-ray astronomy is outside the possibility of space-based experiments and can, at present, only be studied through ground-based experiments. However, because of the high background induced by CR showers, the large collection areas alone cannot provide adequate sensitivities for effective studies of cosmic γ-rays. The capability to suppress the events induced by charged CRs was made possible with the advent of the Imaging Air Cherenkov Technique (IACT), Sect. 9.1 and with some dedicated air shower particle arrays, Sect. 9.2.

© Springer Nature Switzerland AG 2018

M. Spurio, *Probes of Multimessenger Astrophysics*, Astronomy and Astrophysics Library, https://doi.org/10.1007/978-3-319-96854-4_9

Principally thanks to the successful realization of the stereoscopic technique adopted by IACT experiments, TeV astronomy rapidly evolved from an under-developed branch of CR studies to a true astronomical discipline. The online source catalog now contains more than 200 galactic and extragalactic TeV sources, Sect. 9.3. For an historical review and details on early experiments, see Aharonian et al. (2008), Holder (2012), and Hinton and Hofmann (2009). The high sensitivity and the relatively large field-of-view (about 4°) of IACT arrays not only allows for the study of targeted sources, but also all-sky surveys of at least of a part of the sky, such as the galactic plane.

The remarkable achievements of IACTs now include the study of morphology, energy spectrum, and time variability of several galactic and extragalactic source populations. The largest class of galactic TeV emitting sources corresponds to that of pulsars with a wind nebula (Sect. 9.4), among which the Crab is the most studied representative (Sect. 9.5). The open problem of the identification of galactic CR sources, Sect. 9.6, is connected to the identification of γ-rays originating from the hadronic mechanism. Shell-type supernova remnants represent the major candidates as galactic sources of CRs, and the status of the present GeV–TeV γ-ray observations is presented in Sect. 9.7.

Outside the galactic plane, it was discovered that the emission of radiation from jet-dominated AGN (Sect. 9.10) covers a large interval of the electromagnetic spectrum and is extremely variable. Space- and ground-based γ-ray experiments now also allow for a sensitive coverage of the MeV/GeV/TeV energy range, in addition to the band from radio to X-rays. The coordinated efforts from the community are crucial for a detailed and unbiased study of AGN and other extragalactic objects, Sect. 9.11. Multiwavelength searches are becoming more and more important in attempts to obtain a complete picture of nonthermal processes in the Universe, as demonstrated by the campaigns made to measure the spectral energy distributions of some blazars, Sect. 9.12. Blazars, among the AGN, present some particular features, such as, for example, strong and rapid variability and a jet orientation toward the observer (Sect. 9.13). As these objects are among the furthest observed objects in the Universe, their observation through TeV photons can constraint estimates of the presence of extragalactic background light, which is also relevant for cosmological models, Sect. 9.14.

Despite the substantial progress of GeV and TeV γ-ray observations, finding a convincing case for γ-rays due to π^0-decay remained extremely difficult, in particular, in the PeV region. Only the detection of neutrinos could unambiguously solve the problem of the origin of the highest energy CRs, as discussed in Chap. 10.

9.1 The Imaging Cherenkov Technique

The Earth's atmosphere is opaque to high-energy γ-rays. As discussed in Sect. 4.3, their mean free path for pair production is almost the same as the electron radiation length, $X_0 \simeq 37\,\mathrm{g\,cm^{-2}}$, in air. Gamma-rays interact electromagnetically, producing an electron/positron pair. These secondary particles yield a new generation of γ-rays through bremsstrahlung, starting the generation of an electromagnetic cascade. Any secondary charged particle in the shower produces Cherenkov light if its velocity exceeds the threshold $\beta = v/c > n$ (Sect. 4.6.3). The light is emitted at the Cherenkov angle θ, with $\cos\theta = 1/\beta n$. As the refraction index n of the atmosphere changes with atmospheric depth, the Cherenkov angle increases from $0.66°$ at a height of 10 km to $0.74°$ at 8 km. This results in a rough focusing of light onto the ground into a ring-like region with radius of $R \simeq 10\,\text{km} \times 0.012\,\text{rad} = 120\,\text{m}$ for a typical γ-ray shower. The number of Cherenkov photons emitted per unit length is $n_c \sim 0.1\,\text{photons}\,\text{cm}^{-1}$ at sea level. Multiplying n_c by the number of particles at maximum (N_{max}) and by the path length of shower particles, the total number of Cherenkov photons turns out to be $N_c \sim 10^6$ for 1 TeV γ-rays. N_c is proportional to E_γ.

Atmospheric Cherenkov detectors fall into two broad classes: sampling and imaging telescopes. Some sampling instruments used mainly for primary CR measurements have been described in Sect. 4.6. The imaging technique relies on the detection on the ground of the images of the Cherenkov light distribution from these electromagnetic cascades. From the measurement, it is possible to determine both the longitudinal and lateral development of the electromagnetic showers, and the arrival direction and energy of the primary γ-rays.

Imaging Cherenkov are essentially wide-field optical telescopes consisting of a large reflector of about 10 m radius, reflecting the light (the *image*) into a high-speed multi-PMT camera in the focal plane. Short exposures (less than 30 ns) are required to detect the faint flashes of Cherenkov light against the Poisson fluctuation in the night-sky background.

An IACT array must be operated (as with other telescopes) in almost total darkness, and thus must be installed far from human environments. These telescopes usually operate on moonless nights, thus limiting the duty cycle to 10–15%. High-speed detectors and electronics are required to minimize the integration time, the amount of time the chip spends "counting photons." Ideally, the integration time should be reduced down to the shortest intrinsic timescale of the Cherenkov light wavefront, which corresponds to a few nanoseconds (Fig. 9.1). Longer integration time reduces the signal-to-noise.

The field-of-view (FoV) of IACTs is $\sim 4°$, substantially larger than most optical telescopes. This FoV allows for to obtain images of showers whose impact parameters on the ground extend up to 120 m away. Thus, regions of the sky containing one or more source candidates are usually targeted for observation. Surveys can only be accomplished slowly, by tiling regions of the sky with overlapping FoVs. The energy interval from which the current generation of IACTs are sensitive ranges from 100 GeV to 100 TeV. Their angular resolution is on the order of $0.1°$, the energy resolution around 15% and an integral energy flux sensitivity of a few times $10^{-13}\,\text{erg}\,\text{cm}^{-2}\,\text{s}^{-1}$. Their sensitivity is sufficient to detect the Crab nebula

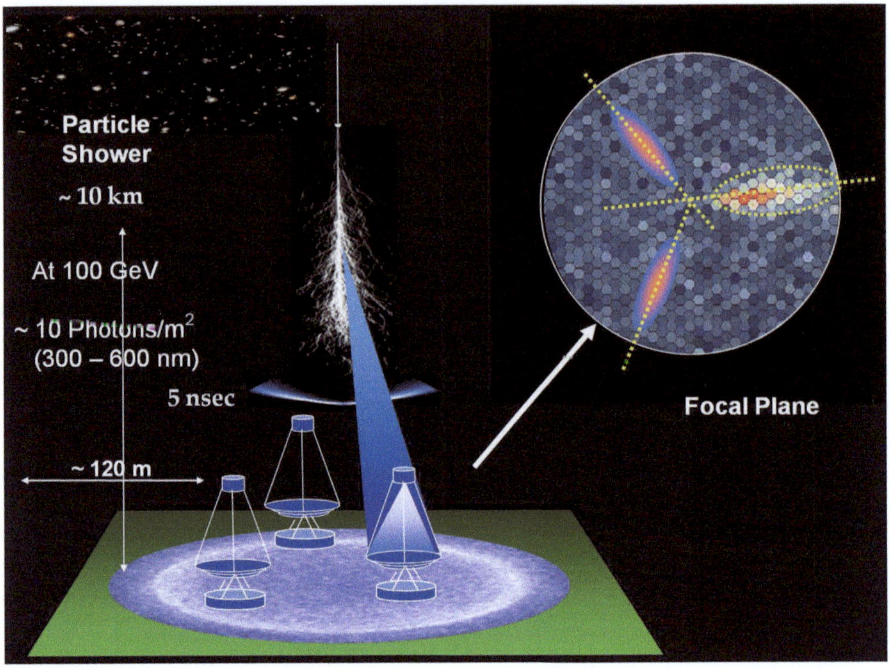

Fig. 9.1 Sketch of the imaging atmospheric Cherenkov telescopes technique. A shower initiated by a γ-ray of energy $\geq$100 GeV is observed through the Cherenkov radiation emitted by charged particles with $\sim$10 m diameter reflectors positioned about 100 m away from each other. The Cherenkov light arrives on the ground as a thin pancake, a few ns wide. The fraction arriving at the reflector is focused into a high-speed PMT camera in the focal plane. The images at the focal plane of different telescopes (*upper right inset*) allow for the determination of the shower's direction and energy. Modified from an original drawing by Prof. W. Hofmann

(Sect. 9.5) in about 1 min, and a source with 1% of the Crab flux in 25 h.[1] As with most very large optical telescopes, IACTs typically make use of an altitude-azimuth drive for tracking sources during large exposures. The angular resolution reached with the IACT technique allows for the resolution of important substructures of some sources.

Current telescopes are based on either simple parabolic reflectors (MAGIC) or many individual mirror segments having a radius of curvature equal to the focal length, placed on an optical support structure (HESS, VERITAS).

[1]Note that a source with a flux equal to 1% of the Crab is not detected in 100 min. The statistical significance of a signal excess depends on the background level, and this increases linearly with the observation time.

9.1.1 Gamma-Ray Versus Charged CR Discrimination

Pure electromagnetic showers, such as those initiated by γ-rays, have different characteristics from those initiated by protons and nuclei, Fig. 9.2. Images of EAS initiated by γ-rays have a compact elliptic shape, and the major axis of the ellipse indicates the shower axis projected onto the image plane (see the focal plane inset in Fig. 9.1). In contrast, the images of EAS produced by protons or nuclei show a complex structure due to electromagnetic sub-showers initiated by following generations of π^0 decay and to the presence of penetrating muons originated by decays of charged pions. For this reason, the images of Cherenkov light observed by an individual IACT are typically analyzed to obtain a set of quantities that characterize the shape of the images. IACTs achieve good γ-ray/hadron separation using the information on the image shape at the trigger level.

To give an idea of the background level and of the intrinsic rejection power of IACTs, we can use the all-particle CR spectrum (2.20b) for primary energies larger than 100 GeV

$$\Phi(>100\,\mathrm{GeV}) = 2 \times 10^{-3} \; \frac{\text{particles}}{\mathrm{cm}^2\,\mathrm{s}\,\mathrm{sr}} \,. \tag{9.1}$$

The background event rate R_{CR} is obtained by multiplying (9.1) by the solid angle $\Delta\Omega$ corresponding to the FoV of the telescope and by the pool area A covered by the telescope. The solid angle corresponding to a cone of aperture of 4° is $\Delta\Omega \sim 3\ 10^{-3}$ sr; the pool area A corresponds to a circle having a radius of about 100 m,

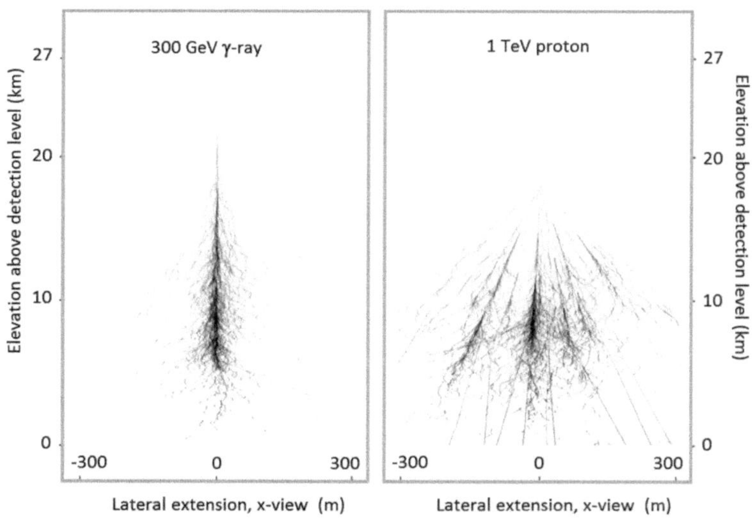

Fig. 9.2 Comparison of a pure electromagnetic shower (from a 300 GeV γ-ray) and a shower initiated by a 1 TeV proton. The plot shows the projection of secondary particle trajectories onto a *plane* in which the ordinate corresponds to the elevation (courtesy of Dr. Konrad Bernloehr)

and therefore $A \simeq 3\,10^8\,\text{cm}^2$. Hence, the event rate on a single IACT due to charged CRs is

$$R_{CR} = \Phi(>100\,\text{GeV}) \cdot A \cdot \Delta\Omega \sim 2000\,\text{s}^{-1} = 2\,\text{kHz}\,. \tag{9.2}$$

As the trigger rate of a single telescope of HESS or VERITAS is on the order of 100–200 Hz, a single IACT can reject charged CRs at the trigger level by a factor of ~10. The choice of a trigger logic is one of the most important decisions in setting an experiment. See, for instance, Moudden et al. (2011) for the case of HESS.

As shown in Fig. 9.1, the projection of shower images on the focal plane results in ellipses with their major axes pointing in the direction of the source. Observation of individual showers with a system of multiple IACTs allows for the reconstruction of the three-dimensional structure of γ-ray showers in the so-called *stereoscopic observation*. With stereo observations, background events are suppressed with higher efficiency (by a factor about 100) and the angular resolution reaches precisions of a few arcminutes. Such resolution enables the possibility of morphological studies of extended γ-ray sources, such as the SNRs.

A single telescope is triggered when several pixels in the PMTs of the camera exceed the threshold, within some time coincidence window. For the current generation of IACT instruments, this window is typically between 3 and 25 ns, and the pixel multiplicity ranges from 2 to 4. Stereo observations are enabled by an *array trigger*: the trigger signals from the individual telescopes are delayed and brought into coincidence within a coincidence window of ~40–100 ns.

9.1.2 HESS, VERITAS and MAGIC

There are currently three major imaging atmospheric Cherenkov telescope systems in operation, two in the Northern hemisphere, and one in the Southern.

The **High Energy Stereoscopic System** (HESS) Observatory is located in Namibia (−23°N, −16°W, altitude 1800 m), in the Southern Hemisphere. It is the IACT with the largest field-of-view and the only one in the Southern hemisphere with good observation conditions for the galactic plane. The initial four HESS telescopes (Phase I, completed in 2004) are arranged in the form of a square having a side length of 120 m, to provide multiple stereoscopic views of air showers. Each telescope of Phase I has a diameter of 13 m, with a total mirror area of 108 m^2 per telescope. The cameras that capture and record the Cherenkov images of air showers have a large field-of-view (~5°) to allow for observations and surveys of extended sources. It has 960 photon detector elements ("pixels") to resolve image details, and a triggering scheme which allows for identification of the brief and compact Cherenkov images and the rejection of backgrounds. The complete electronics for image digitization, readout and triggering is integrated into the camera body.

In Phase II of the project, a single huge dish with about 600 m^2 mirror area was added at the center of the array, increasing the energy coverage, sensitivity

Fig. 9.3 The HESS telescopes in Namibia. Credit: HESS Collaboration (http://www.mpi-hd.mpg.de/hfm/HESS/)

Fig. 9.4 The four IACT array VERITAS at Mt. Hopkins, Arizona. Credit: VERITAS Collaboration (http://veritas.sao.arizona.edu/)

and angular resolution of the instrument (see Fig. 9.3). This telescope started taking data in 2012. The camera of the new telescope follows the design of the others, but it is much larger—it contains 2048 pixels—and virtually every detail is improved.

The **Very Energetic Radiation Imaging Telescope Array System** (VERITAS) (Fig. 9.4) is in operation at the Fred Lawrence Whipple Observatory in southern Arizona, USA (32°N, 111°W, altitude 1275 m). It is an array of four 12 m optical reflectors with similar characteristics as HESS-Phase I. Each reflector uses 499-pixel cameras, with a field-of-view of 3.5°. The covered energy range is between 50 GeV and 50 TeV. The four-telescope array was completed in January 2007.

The **Major Atmospheric Gamma-ray Imaging Cherenkov** (MAGIC) (28°N, 17°W, altitude 2225 m) originally consisted of a single, very large reflector (236 m^2) installed on the Canary island of La Palma, with a 3.5° high-resolution camera composed of 576 ultra-sensitive PMTs. The first telescope has been fully operational since 2004. In 2009, a second telescope of essentially the same characteristics was added; MAGIC-II was installed at a distance of 85 m from MAGIC-I, Fig. 9.5. MAGIC is characterized by the largest collection surface of any existing γ-ray telescope worldwide, an assembly of nearly 1000 individual mirrors, together resulting in a parabolic dish with a 17 m diameter; the diamond-grinding and polishing of the individual aluminum mirrors and their mounting (in altitude/azimuth controlled position) on a lightweight carbon fiber structure was a real technological challenge.

Fig. 9.5 The two large telescopes of the MAGIC observatory. Credit: MAGIC Collaboration (https://magic.mpp.mpg.de/)

The large surface and the optimal light collection of the mirrors allow for the detection of γ-rays with an energy threshold of $\sim$25 GeV.

A very fast (average time 40 s) repositioning of the telescope axis is one of the major characteristics of MAGIC with respect to other IACTs. This is achieved by minimizing the device's weight and automating axis control. Repositioning in a matter of seconds is important when short-lived phenomena are signaled by other active devices, e.g., by satellite-based wide-angle detectors in the X-ray band, in particular, for GRBs.

In addition to the above detectors, **CANGAROO** was a Japanese and Australian observatory placed in Australia. In its final design (operating from 2004 to 2011), it consisted of four 10 m diameter telescopes.

The **Cherenkov Telescope Array (CTA)** will probably be the future of observational γ-ray astronomy, at least for the next 10–20 years. The research objectives of the next generation of IACT arrays, and especially of CTA, are devoted to (1) a significant improvement (by an order of magnitude) of the flux sensitivities in the standard 0.1–10 TeV energy interval, and (2) an expansion of the energy domain of IACT arrays in both directions—down to 10 GeV and well beyond 10 TeV. This ambitious research goal will be realized by increasing the number of telescopes with different sizes, from a few very large 20 m diameter class telescopes to a large number of modest area (10–30 m^2) reflectors. In this effort, all the three existing IACT communities are involved (Rieger et al. 2013).

9.2 EAS Arrays for γ-Astronomy

The EAS technique, designed for the detection of CRs at PeV and EeV energies, can also be adopted for γ-ray astronomy. The mandatory requirement is that the energy threshold be reduced by two or three orders of magnitude, using dense particle arrays located at very high altitudes. The feasibility of the measurement at ground level of showers initiated by a γ-ray has been successfully demonstrated by the Milagro and ARGO Collaborations.

The **Milagro** detector consisted of a large central water reservoir ($60 \times 80\,\text{m}^2$), which operated between 2000 and 2008 in New Mexico (36°N,107°W), at an altitude of 2630 m. The reservoir was covered with a light-tight barrier, and instrumented with PMTs. In 2004, an array of 175 small tanks was added, irregularly spread over an area of $200 \times 200\,\text{m}^2$ around the central reservoir. This configuration sampled with high-resolution the air shower over a relatively small area compared to the air shower footprint. Milagro developed analysis techniques for CR background discrimination that provided sufficient rejection for the first large-scale survey of the Northern γ-ray sky within the TeV range. The strong TeV sources Crab Nebula (Sect. 9.5) and Markarian 421 (Sect. 9.12.2) were observed, as well as three extended sources in the galactic plane, each with integrated fluxes comparable to the Crab nebula at 20 TeV.

HAWC These Milagro results, as well as the potential for continuous monitoring of a large fraction of the sky, have motivated proposals to construct larger EAS detectors like the High-Altitude Water Cherenkov Observatory (HAWC). HAWC is located at an altitude of 4100 m close to Sierra Negra, Mexico (19°N, 97°W), Fig. 9.6. The observatory was completed in 2015 and it consists of an array of 300 water Cherenkov detectors. Each Cherenkov detector has dimensions of

Fig. 9.6 The HAWC detector with the Pico de Orizaba in the background, August 2014. Credit: Jordanagoodman and HAWC Collaboration (http://www.hawc-observatory.org/)

7.3 m diameter and 4.5 m depth, each viewed by 4 upward-facing PMTs, with an overall 15-fold increase in sensitivity with respect to Milagro. HAWC monitors the Northern sky (instantaneous field-of-view of ~2 sr) and makes coincident observations with other wide field-of-view observatories.

HAWC observes a large ensemble of sources, measuring their spectra and variability to characterize TeV scale acceleration mechanisms: it reaches, in a 1-year survey, a 50 mCrab sensitivity at 5 standard deviations. Due to the large FoV, HAWC can observe diffuse γ-ray emission from the plane of the Galaxy over a broad range of Galactic longitudes reaching to the Galactic center. The observatory also started to produce physics results when in partial configuration. At present, 30 sources in the TeV catalog (next section) have been observed by, or also by, HAWC.

ARGO-YBJ A large plateau at a very high altitude (4300 m-30°N, 90°E) at Yangbajing in Tibet has hallowed for the installation of large surface array detectors. The Astrophysical Radiation with Ground-based Observatory (ARGO) experiment was in operation from 2001 until 2013 and consisted of a single layer of resistive plate chambers completely covering an area of 110×100 m^2. They observed the emission of γ-rays from the Crab nebula, Markarian 421 and two Milagro sources. The Tibet Air Shower Experiment (ASγ) air shower array, also at Yangbajing, consists of ~750 closely spaced scintillation detectors covering an area of 37×10^3 m^2.

At higher energies, a new project called LHAASO has been proposed, improving the features of ARGO, Sect. 4.11.

Figure 9.7 shows the sensitivity as a function of the γ-ray energy of current and future ground-based detectors. Here, the *sensitivity* represents the minimum

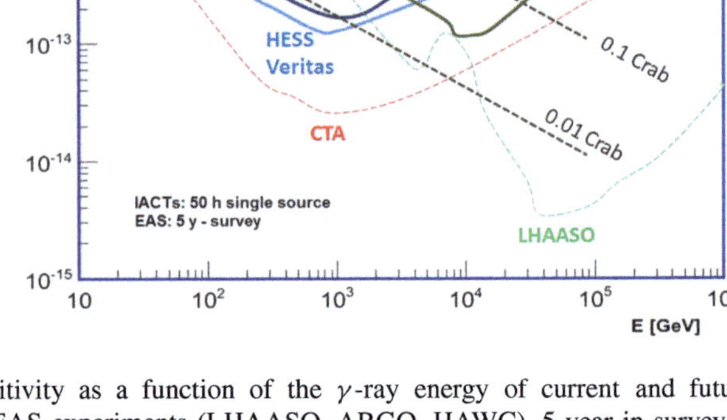

Fig. 9.7 Sensitivity as a function of the γ-ray energy of current and future ground-based detectors. For EAS experiments (LHAASO, ARGO, HAWC), 5 year in survey mode have been assumed, while 1 year has been assumed for Fermi-LAT and AGILE. For the remaining IACTs, 50 h of observation on a single source are considered. The thin *dashed lines* represent future experiments

value for the spectral energy distribution (SED) of a source that can produce an observable signal. Sources with a SED smaller than the sensitivity will be completely overwhelmed by the background. We will return to this concept in Sect. 10.10.1. The dashed line denoted as "Crab" represents the γ-ray flux from the Crab nebula. Also, fluxes 1/10 and 1/100 of the "Crab" are indicated. For comparison, the sensitivities of AGILE and Fermi-LAT are also reported. Satellite experiments, as well as Argo, Milagro, and HAWC, have a much larger field of view with respect to the IACTs and can simultaneously monitor a large number of sources. In the figure, 50 h of IACTs observations are considered, and 5 year of survey mode for large field-of-view detectors.

9.3 TeV Astronomy: The Catalog

Production of GeV (or HE) and TeV (or VHE) γ-rays are expected in astrophysical environments where acceleration of particles (protons, nuclei, and electrons) is accompanied by their intensive interactions with the surrounding gas and radiation fields. These interactions contribute significantly to high-energy emission from galactic objects such as young supernova remnants, star forming regions, pulsars, pulsar wind nebulae and compact binary systems. Gamma-ray emission is also expected from extragalactic objects such as active galactic nuclei and radio galaxies.

In 2003, ten TeV-emitting sources were known, detected with the so-called first generation of IACTs: 7 blazars, 2 supernova remnants and the Crab. The second generation of IACTs (Sect. 9.1.2) started scientific operation in $\sim$2004, and in only a few years, they brought data that transformed our view of the high-energy γ-ray sky. The sky positions of the 208 known TeV-emitting sources (early 2018) are shown in Fig. 9.8. The number of sources catalogued in different topologies is reported in Table 9.1. As discussed for the satellite observations within the GeV range, most of the sources are associated with objects already known through different wavelength observations. In this case, the object is denoted with the name used in other catalogues (for instance, Vela pulsar or 3C 279). However, new classes of emitting sources have been discovered. The fraction of unassociated TeV sources is $\sim$25%. In this case, the source maintains the sign of the first observatory that made the observation (MGRO for Milagro, HESS for H.E.S.S., HAWC, and so on) and the sky coordinates.

At TeV energies, the unidentified sources lie essentially on the galactic plane (only 4 out of 54 are far from the plane), as opposed to sources observed at the GeV energy range. This could be due to an observational bias. The wide field-of-view and the survey mode operation of the Fermi satellite provides LAT with a roughly (within a factor two) uniform exposure to the entire sky. The IACTs have very narrow effective fields-of-view and the galactic plane is the only large fraction of the sky that has been studied by HESS in detail. This dedicated survey of 2800 h has covered the range in galactic longitude between [$-85°$, $+60°$] and [$-3.5°$, $+3.5°$] in latitude. It has revealed more than fifty VHE γ-ray sources; the significance

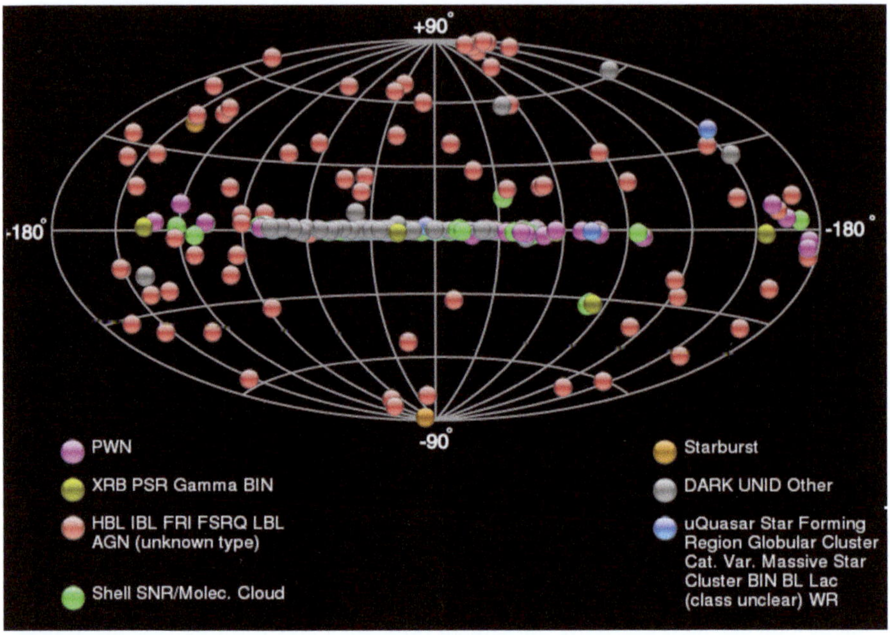

Fig. 9.8 Map of the 208 TeV sources (2018) retrieved from the online TeV catalog (http://tevcat.uchicago.edu/), which displays, with different color codes, the position in galactic coordinates of the various γ-ray sources detected from the ground. Most of the detections were performed by IACTs, which typically are sensitive to γ-rays well above 100 GeV

Table 9.1 Number of objects catalogued in the TeVCat in Spring 2018

Type	Designator	Objects	Representatives
Pulsar wind nebula	SNR/PWN	34	Crab, Geminga, Vela X
SNR with shell	SNR/Shell	14	See Table 9.2
SNR with molecular clouds	SNR/Mol. Cloud	10	W28, W51
SNR with pulsar	SNR/PSR	2	Vela psr, Crab psr
Binary systems	Binary	9	LS 5039, LSI +61 303
Massive star clusters, globular cl.	–	4	–
HBL Lac type of blazar	HBL	48	Mrk 421, Mrk 501
IBL Lac type of blazar	IBL	8	Bl Lac, W Comae
LBL Lac type of blazar	LBL	2	–
FSRQ type of blazar	FSRQ	7	3C 279
FRI type of blazar	FRI	4	Centaurus A, M87
Starburst galaxy	Starburst	3	
Unidentified	UNID	54	–
Total		208	

It comprises 78 Galactic sources, 75 extragalactic sources and 54 unidentified. From http://tevcat.uchicago.edu/

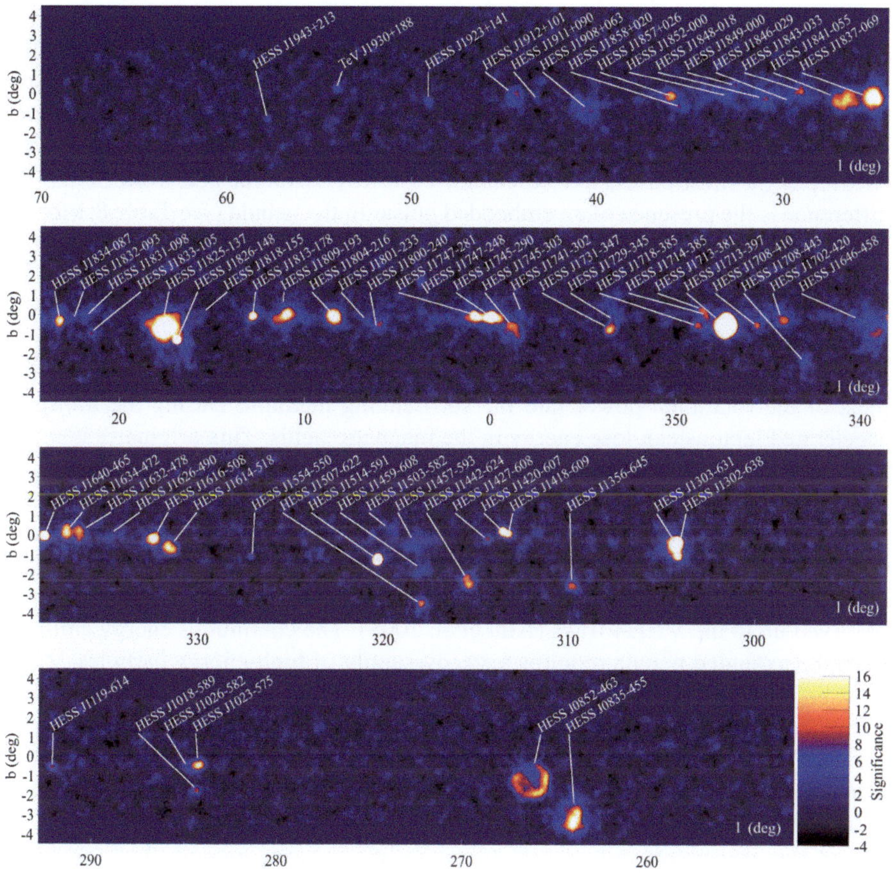

Fig. 9.9 Significance map of γ-ray objects in the galactic plane as observed by HESS. Credit: HESS Collaboration (Carrigan et al. 2013)

map of the discovered objects is shown in Fig. 9.9. A large fraction (more than half) corresponds to PWNe, located in close vicinity to young and energetic pulsars (Paneque 2012).

In the next section, we describe pulsar and pulsar wind nebulae, which are the dominant galactic populations of identified objects emitting γ-rays at GeV and TeV energies. The efficiency of converting spin-down power into γ-rays is typically within the range of 1–10%.

9.4 Gamma-Rays from Pulsars

Pulsars (PSR) are rotating neutron stars (Sect. 6.7.3) that have traditionally been a subject of radio astronomy. Some pulsars have an extended nebula emitting radiation. This constitutes the class of pulsar wind nebulae (PWNe). In some cases, PSR and PWNe also emit radiation at high energies. Since the launches of Fermi-LAT and AGILE, the number of γ-ray pulsars has increased from half a dozen to more than 150. Despite the high galactic background, the periodic γ-ray emission stands out due to the high fluxes, hard spectral index, and powerful timing identification. As a result of the relatively poor angular resolution of space γ-ray telescopes, in most cases, PWNe cannot be firmly distinguished from SNRs: the difference is the presence of an embedded pulsar in the nebula (see Table 8.1 for the Fermi-LAT survey). Concerning the TeV range, PWNe are the most abundant class of known emitters in the Galaxy, following the HESS survey of the galactic plane (see Table 9.1). Refer to Abdalla et al. (2018) for a detailed study of properties of TeV pulsar wind nebulae.

PWNe are objects in which a wind of energetic particles from a pulsar carries most of the rotational power into the surrounding medium. During the transport, accelerated leptons can lose energy in the magnetic fields. This originates nonthermal radiation, the "synchrotron nebula" around the pulsar, ranging from the radio to the X-ray and, in some cases, to the MeV band. In the steady-state, a second component is originated by the inverse Compton effect (IC) of electrons on low-energy radiations fields of synchrotron, thermal or microwave-background origins, as described in Sect. 8.2. The resulting extended nebula around the pulsar can be observed up to the VHE γ-rays (Aliu et al. 2008). The continuous energy emission from the extended nebula requires a steady supply of high-energy particles, which must be provided by the inner pulsar with a continuous injection of energy. In fact, a charged particle of energy E present in the nebula and with a characteristic energy loss (dE/dt) (for instance, due to synchrotron emission) produces radiation for a characteristic time $\tau \simeq E/(dE/dt)$. Most of the observed PWNe are associated with pulsars that are less than a few hundred kyr old.

For the considered production of GeV–TeV γ-rays, models that transfer a few percent of the rotational energy of pulsars to γ-rays exist. The regions of the magnetosphere where particle acceleration can occur are called vacuum gaps (Sect. 6.7.3). Also, the pulsed, periodic γ-rays are thought to be originated in gap regions. Models of particle acceleration differ, primarily, on the location of these gaps in the magnetosphere. The measured light curves and spectral energy distributions from space- and ground-based experiments can help to disentangle the various models.

Most of the γ-ray energy spectra measured by Fermi-LAT (Nolan et al. 2012) are well-fitted by a power-law function with an exponential cutoff, Eq. (8.55), with E_0 ranging between 1 and 10 GeV and $b \lesssim 1$. If the production of γ-rays occurs in the polar cup regions (refer to Fig. 6.8), strong absorption effects are expected due to interaction of high-energy photons with the magnetic field, as well as with the radiation field (production of $e^{\pm}$ pairs). In this case, a super-exponential cutoff, i.e., $b > 1$ in (8.55), is expected, along with a strong suppression of the flux at relatively low energies (few GeV). The value $b \lesssim 1$ is, in general, well-explained by the mechanism of synchrotron radiation emission and gives a preference to models of γ-ray production in the outer gap regions.

Measurements with ground-based observatories indicate that PWNe are the most effective galactic objects for the production of TeV gamma-rays, allowing for the detection of such systems, even in the neighboring LMC galaxy. Before 2004, only the Crab PWN was known as a source of steady TeV γ-rays. The development of the new sensitive IACTs has increased the number of likely PWNe detected, but the Crab still plays a fundamental role in the understanding of the physics involved in pulsars, as shown in the next section.

9.5 The CRAB Pulsar and Nebula

The Crab nebula is the strongest TeV γ-ray source, and it is used as a gauge for other sources (see Fig. 9.7). It is located at a distance of (2.0 ± 0.2) kpc from the Earth and belongs to the class of supernova remnant with a pulsar at its center and without any detected shell component. It is associated with the supernova explosion reported by Chinese astronomers in 1054 AD. The Crab Pulsar at the center of the nebula is also known as PSR J0534+2200 and is one of the most energetic of the known pulsars. The rotational energy loss (6.82) of the Crab corresponds to $dE_{\rm rot}/dt \sim 5 \times 10^{38}$ erg/s. Estimation of its characteristic age using its rotation period ($P = 33$ ms) and derivative ($dP/dt = 4.2 \times 10^{-13}$ s/s) yields an age of 1240 year, close to the elapsed time from the supernova of 1054.

The nebula and the inner pulsar have been well-studied in almost all wavelength bands of the electromagnetic spectrum from the radio (10^{-5} eV) to hundreds of TeV γ-rays. A simple magneto-hydrodynamical model for the interaction of a high-energy electron-positron wind with the interstellar medium satisfactorily describes the main features of the nonthermal emission. Figure 9.10 represents the observed $E^2 \frac{dN}{dE}$ distribution from the Crab nebula, which extends over 21 decades of energies/frequencies. This spectral distribution is well-explained by considering the SSC mechanism, Sect. 8.3 and details in Kennel and Coroniti (1984) and Aharonian et al. (2004). The contribution of hadronic interactions to the γ-ray production does not seem necessary.

The TeV emission from the Crab nebula was first established with very high significance by the Whipple group in 1989. This result demonstrated the power of the imaging atmospheric Cherenkov method. The Crab, well-sampled from radio to TeV γ-rays, is used for the cross-calibration of ground-based detectors. The integral flux above 1 TeV corresponds to

$$\mathscr{F}_\gamma(> 1\,\text{TeV}) = (2.1 \pm 0.1) \times 10^{-11} \quad \frac{\text{photons}}{\text{cm}^2\,\text{s}} \,. \tag{9.3}$$

The Crab nebula was considered as a steady "standard candle". Only recently, γ-ray flares have been detected by AGILE and Fermi-LAT and pulsed emission from the Crab pulsar up to beyond 100 GeV observed by MAGIC and VERITAS. The origins of these variations are still under investigation.

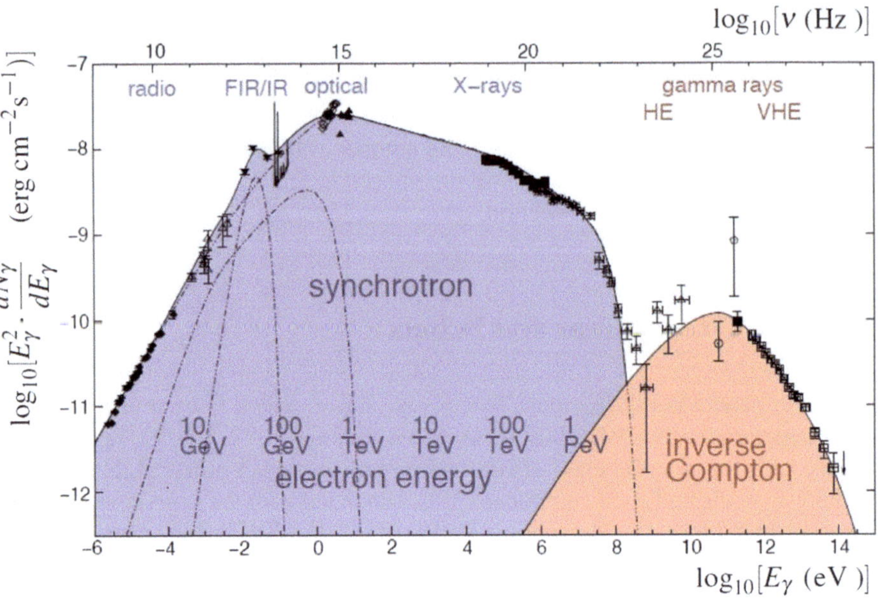

Fig. 9.10 Multiwavelength observations of the Crab nebula. The synchrotron emission is given by the superposition of the contributions of electrons with different energies. Mono-energetic electrons contribute with photons with peaked spectra, according to Eq. (8.7). For instance, 1 TeV electrons give photons in the visible wavelength. The synchrotron spectrum provides the target photons for the inverse Compton (IC) process (Funk 2011)

Although GeV–TeV γ-rays constitute only a small fraction of the luminosity of the nebula, they provide crucial information on the environmental conditions. The comparison of the X-ray and TeV γ-ray fluxes allows for, respectively, determination of the energy density of the magnetic field and that of the radiation energy density in (8.32). As the former exceeds the latter by more than two orders of magnitude, this has led to the estimate that the average nebular magnetic field is about 100 μG, as expected from magneto-hydrodynamical models (Rieger et al. 2013). Thus, as evident from the two regions labeled *synchrotron* and *inverse Compton* in Fig. 9.10, the Crab nebula is very inefficient at producing γ-rays through inverse Compton scattering. Only its extremely high spin-down power allows the production of the observed flux (9.3).

Particularly interesting is the transition region between the falling edge of the synchrotron component and the rising edge of the inverse Compton component. Figure 9.11 shows the results from high-energy experiments in detail. The data collected by COMPTEL and EGRET carry information about the fading synchrotron part of the spectrum. The Fermi-LAT data reveal a sharp transition from the synchrotron to the IC component at around 1 GeV (Abdo et al. 2010b). The measurements with ground-based observatories have almost approached 100 TeV. This is compatible with the fact that the IC component extends up to the maximum

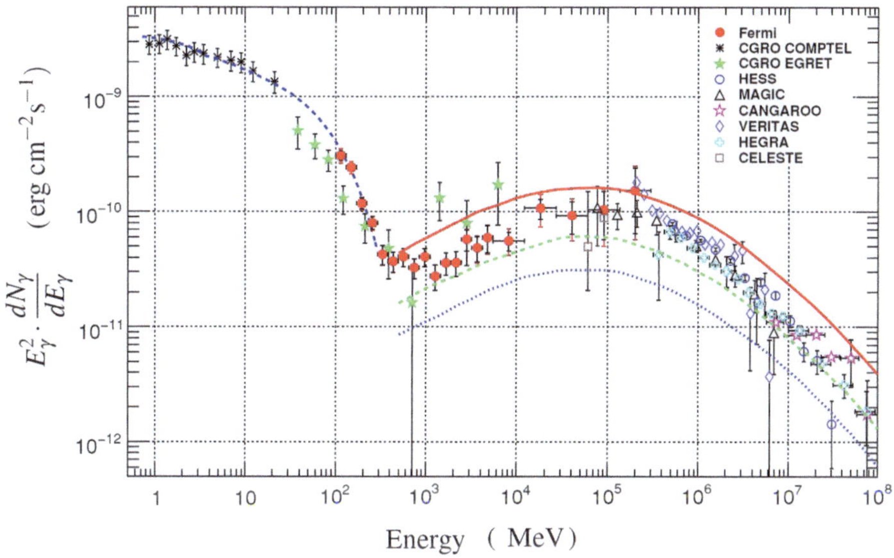

Fig. 9.11 Spectral energy distribution of the Crab nebula from soft to very high energy γ-rays. The fit of the synchrotron component, using COMPTEL and Fermi-LAT data (*dashed line* up to 300 MeV), is overlaid. The predicted inverse Compton spectra (from of 300 to 100 TeV) are overlaid for three different values of the mean magnetic field: 100 (*solid line*), 200 (*dashed line*), and 300 μG (*dotted line*). Credit: Fermi-LAT Collaboration

energy set by the maximum energy of the accelerated electrons, i.e., 1 PeV (see Fig. 9.10). At this energy, from Eq. (8.7), the critical energy for synchrotron emission corresponds to $E_c = h\nu_c \sim 1$ MeV. The maximum of the inverse Compton curve occurs in correspondence to $E_\gamma \simeq 60$ GeV. This is supported by both the Fermi-LAT and ground-based observatory measurements, which agree remarkably with each other. This is one of the first cases in which measurements made by ground- and space-based experiments overlap in energy. Note that the predicted flux of γ-rays from IC decreases as the magnetic field increases. This is due to the fact that the synchrotron energy loss of electrons increases with the magnetic field B, Eq. (8.12). As a result, the high region part of the electron spectrum decreases as B increases.

9.6 The Problem of the Identification of Galactic CR Sources

The diffusive shock acceleration model predicts the production of accelerated particles in SNRs that can interact with ambient magnetic fields, with ambient photon fields, or with matter. The amount of relativistic particles present in the acceleration region increases with time as the SNR passes through its free expansion phase, and reaches a maximum in the early stages of the Sedov phase (Sect. 6.3.2).

Correspondingly, the peak in γ-ray luminosity typically appears some $\sim 10^3$–10^4 years after the supernova explosion.

A straightforward test of the acceleration of CRs in SNRs up to PeV energies would be the detection of γ-rays produced through the hadronic mechanism directly from young remnants and/or from dense clouds overtaken by the expanding shells. The main challenge is to distinguish γ-rays emitted through hadronic processes (π^0-decay, Sect. 8.4) from those originating in leptonic processes, Sect. 8.2.

Multiwavelength observations of objects of different classes, as in the case of the Crab reported in Figs. 9.10 and 9.11, are fundamental in the quest to disentangle sources in which leptonic or hadronic mechanisms are involved. The identification of galactic γ-ray emitters with astrophysical objects known at other wave bands is thus an important prerequisite in the study of the origin of cosmic rays.

The IACTs have been able to image SNRs in TeV γ-rays, probing the SNR-shell acceleration of either electrons or hadrons up to at least 100 TeV (in the case of leptonic emission) or a few hundred TeV (for hadronic acceleration). Thanks, in particular, to the HESS survey of the galactic plane, we know that these acceleration sites are spatially superimposed with regions of nonthermal X-ray emission. This coincidence has strengthened the hypothesis that galactic CRs up to the knee are accelerated in SNRs. Figure 9.12 shows the morphological structure of one SNR, namely RX J1713.7-3946 (Aharonian et al. 2007b). This image has revealed a good correlation of the TeV emission sites with the nonthermal emission detected in X-rays.

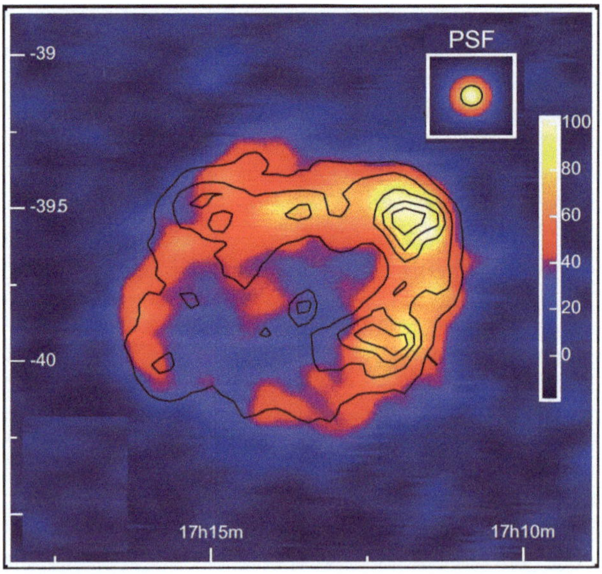

Fig. 9.12 HESS map of γ-ray excess events for RX J1713.7-3946—the first SNR shell to be resolved at TeV energies. The superimposed contours show the X-ray surface brightness as seen by ASCA within the 1–3 keV range. On the *upper right side*, the HESS point spread function (PSF). Credit: HESS Collaboration

However, even if radio and X-ray data suggest that SNRs are indeed the sources of CR electrons, no compelling evidence for the acceleration of protons in SNRs up to the PeV energies has been found up to now, and it is not clear whether proton and electron accelerators are of a different nature. We illustrate the reason in the following sections.

9.7 Extended Supernova Remnants

In radio, optical, and X-ray wavelengths, detailed observations of supernova remnants (SNRs) have been carried out. Several of them have been discovered as sources of TeV γ-rays, in particular, by the HESS survey of the galactic plane.

The Green catalog on radio observations of known galactic SNRs contains 295 objects,[2] and is based on results published in the literature up to June 2017. The basic summary data for each SNR include its designation, position, angular size, type, flux density at 1 GHz, spectral index, and any other names by which it is known.

The third Fermi-LAT catalog (3FGL, Sect. 8.10) lists, in addition to 12 identified SNRs, about 60 γ-ray sources associated with SNRs or PWNe, including such important objects as Cassiopeia A, Tycho's SNR, the Cygnus Loop, W51C, W44, IC 443, the TeV-bright SNRs RX J1713.7-3946 and RX J0852.0-4622 (Vela Junior), see Fig. 8.7.

The SNRs observed by Fermi-LAT allow for potential associations of these objects with the Green radio catalog. Two main classes have been identified: young SNRs and those that are interacting, often with molecular clouds. If radio and GeV emissions arise from the same particle population(s), e.g., leptons and/or hadrons accelerated at the SNR shock front, the indices of the energy spectra measured in GeV and radio bands should be correlated. The study of the correlation shows that several of the known, young SNRs are more consistent with a lepton-dominated emission via IC in the GeV regime. In other SNRs, the emission seems consistent with a production by a combination of leptonic and hadronic mechanisms.

Today, 11 shell-type SNRs have been firmly identified as γ-ray emitters at TeV energies (see Table 9.1). For these objects, their name, distance, size, age, luminosity (in units of the Crab), and spectral index in the TeV γ-ray emission are presented in Table 9.2. Three additional SNRs observed by HESS remain unidentified. Their γ-ray luminosities (derived from flux and distance) are about (0.1–10) $\times 10^{33}$ erg/s and have enabled the detection of these objects up to distances of $\sim$3.5 kpc with current instrument sensitivities. The source 0FGL J1954.4+2838 at larger distance was discovered by Milagro. The relatively large sizes of several of these shell-like SNRs ($>0.1°$) have allowed to resolve them for morphological studies. The

[2] https://www.mrao.cam.ac.uk/projects/surveys/snrs/.

Table 9.2 Shell-like SNRs firmly detected at TeV energies

Name	*D* (kpc)	Size *d* (pc)	Age (ky)	Flux (Crab u.)	α_γ
RX J1713.7-3946	1	12	1.6	0.66	2.2
IC 443	1.5	4	3–30	0.03	3.0
RXJ0852-4622 (Vela Jr)	0.2	3.5	$\sim$1	1	2.2
RCW 86	2.5	20	1.8	0.1	2.54
SN 1006	2.2	19	1	0.01	2.29
CTB 37B	2.2	14	$\sim$1	0.02	2.65
Cassiopeia A	3.4	2.5	0.35	0.03	2.3
Tycho	3.5	6	0.44	0.01	1.95
0FGL J1954.4+2838	9.2	–	–	0.23	–
G106.3+2.7 (Boomerang)	0.8	3.5	4	0.05	2.29
SNR G353.6-0.7	3.2	8	2.5–14	0.01	2.32

The table shows the distance D, the size d, the estimated age since the explosion, the flux in Crab units (9.3), and the spectral index of the power-law energy spectrum. From http://tevcat.uchicago.edu/

angular size (in radians) of each object can be obtained from the ratio $\theta \sim d/D$ (see Table 9.2).

This small number of detected shell SNRs is in agreement with our simple considerations in Sect. 8.6, related to the acceleration mechanism in SNRs. The diffusive shock acceleration, Sect. 6.4, foresees that shock waves in SNRs may be able to accelerate CRs up to PeV energies (*PeVatrons*) in the first $\sim$1000 year, while later, the high-energy hadrons escape from the system. The phase in which the maximum attainable energy is reached (from the model, up to some 10^{15} eV=1 PeV) can last less than several hundred years, largely reducing the number of SNRs emitting as PeVatrons that can be observed.

However, γ-rays produced by protons and nuclei could still be observed in the period when CRs diffuse away from the acceleration regions and start to get integrated in the galactic volume, if certain conditions in the surrounding environment are fulfilled. In particular, the presence of massive molecular clouds relatively nearby the acceleration regions is probably necessary. This seems the case with supernova remnants relatively near to molecular clouds; eight such objects are present in Table 9.1. These dense matter regions trap those running-away CRs, allowing them significant probability for hadronic interaction.

9.8 The SED of Some Peculiar SNRs

The spectral energy distribution of young SNRs observed in γ-rays by space- and ground-based experiments extends over almost five decades. These simultaneous observations have improved our understanding of the origin of the γ-rays, but have also evidenced a more complicated scenario in which different regions can contribute to the total emission.

One consideration, which seems to favor a hadronic scenario in some SNRs, is the presence of regions with high magnetic field amplification. These regions were discovered through the observation of synchrotron X-ray filaments. These high magnetic fields favor efficient confinement and acceleration of hadrons and prevent, in principle, a large inverse Compton contribution from leptons. In cases like RX J1713.7-3946 (see Fig. 9.12), Tycho or Cas A, the magnetic field has been estimated from multiwavelength observations to be $>100\,\mu$G, restricting the contribution of the IC emission and, in principle, favoring a hadronic origin of the TeV emission.

SNR RX J1713.7-3946 is one of the most powerful remnants and has long been the best candidate for γ-ray emission originated by hadronic interactions, although this claim has been extremely controversial. In fact, while the derived high magnetic field disfavors the leptonic acceleration mechanism, on the other hand the apparent low gas density ($n \sim 0.1\,\mathrm{cm}^{-3}$) in the shell of RX J1713.7-3946 poses troubles to standard hadronic scenarios.

The combined SED derived from observations by Fermi-LAT and HESS, Fig. 9.13, seems to be more in agreement with that expected from a leptonic scenario (Abdo et al. 2011b). This does not exclude the fact that CR protons and nuclei are accelerated in the source. As discussed in Sect. 8.9, the γ-ray luminosity is $L_\gamma \propto \rho_{CR} \cdot \sigma_{pp} \cdot n$, where ρ_{CR} is the density of accelerated CRs and n the target number density. CR protons (or heavier nuclei) might not have enough ambient target nuclei (i.e., low number density n) to yield a flux of γ-rays comparable with that produced by the leptonic mechanism.

The leptonic interpretation of γ-ray production by RX J1713.7-3946 is questioned using extensions of the standard diffusive shock acceleration models, Sect. 6.6. An increased acceleration efficiency could be reached if enhanced magnetic field regions were present in the expanding shells. These regions are

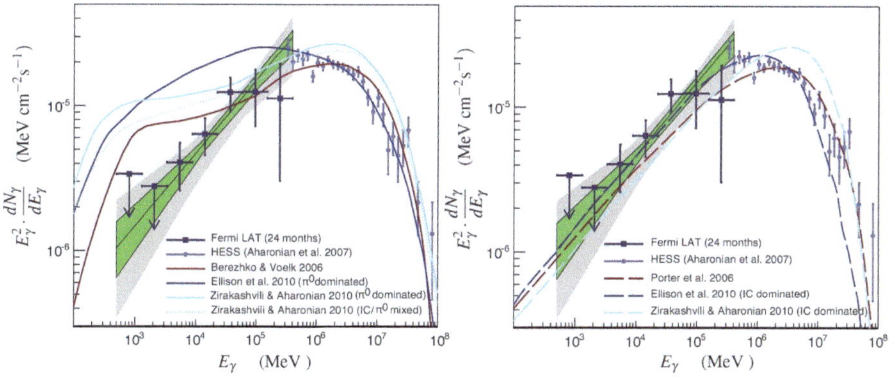

Fig. 9.13 Spectral energy distribution (SED) of γ-rays from RX J1713.7-3946 observed by Fermi-LAT and HESS. In the *left panel*, the lines represent the predictions derived before the Fermi-LAT measurements from different hadronic mechanisms. In the *right panel*, those derived assuming leptonic models. For details, see Abdo et al. (2011b). These observations suggest a leptonic origin of the emission, although hadronic models with a very hard proton spectrum cannot be ruled out. Credit: Fermi-LAT Collaboration

particularly efficient in the acceleration mechanisms, and they could produce an exponent of the differential energy spectrum $\alpha_{CR} \simeq 1.5-1.8$ instead of the standard value 2.0. Under these assumptions, the model can reproduce a γ-ray flux resembling that obtained in the leptonic model, and could therefore fit the Fermi-LAT and HESS data, as well as the leptonic model does.

Tycho SNR The composite image of the Tycho Brahe supernova remnant as seen by different instruments has been presented in Fig. 6.6. Within the γ-ray energy range, the Tycho SNR has been detected both by Fermi-LAT and VERITAS. As evident from Table 9.2, this source is much fainter than RX J1713.7-3946. The SED modeling, including GeV and TeV data shown in Fig. 9.14, seems to disfavor the leptonic model (Giordano et al. 2012). No cut-off is found in the VERITAS data. Taking into account the SNR's age, the expected maximum proton energy would be ~300 TeV, suggesting maximum acceleration below the CR knee. However, the hadronic interpretation is not completely compelling, given the large statistical errors in the measurements, the impact of various unknown parameters such as the SNR distance, and a possible enhancement of the γ-ray flux due to a nearby molecular cloud. Future measurements of the spectrum below 500 MeV, and better measurements at TeV energies, will further test the different interpretations.

SNRs Interacting with Molecular Clouds The brightest γ-ray sources associated with SNRs seen in the GeV region are middle-aged remnants that are interacting with molecular clouds. The SNRs detected by Fermi-LAT are generally also radio-bright objects. The luminosity in the 1–10 GeV band is typically $L_\gamma = (0.8-9) \times 10^{35}$ erg s^{-1}, larger than that of young SNRs like Cas A and RX J1713.7-3946. The corresponding luminosity within the TeV range, on the other hand, is much fainter.

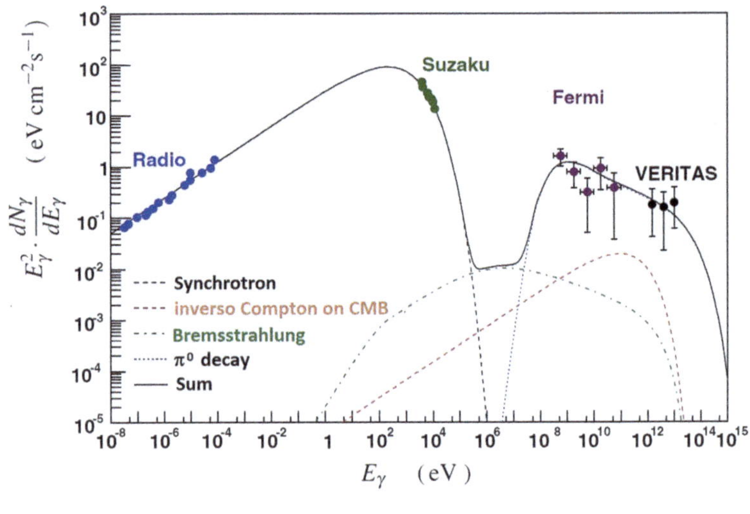

Fig. 9.14 Broadband SED data (in radio, X- and γ-rays) and the corresponding emission model of Tycho's SNR. Note that the shapes of the curves (synchrotron, Inverse Compton and π^0-decay) resemble that of Fig. 8.2, only with different relative weights. The hadronic contribution to γ-rays seems to be the dominant one. Credit: Fermi-LAT Collaboration

The predominance of this class in the SNRs detected by the Fermi-LAT and their high γ-ray luminosity indicate that the emission should be enhanced by the presence of matter. One model considers the γ-ray emission as being due to the interaction of runaway CRs, escaping from their acceleration sites, with nearby molecular clouds. Another scenario is the so-called *crushed cloud model* that invokes a shocked molecular cloud into which a radiative shock is driven by the impact of the SNR's blastwave. In all models, γ-ray emission from π^0-decay is enhanced because of more frequent *pp* interactions in the interstellar medium (Thompson et al. 2012).

Probably the best evidence of interactions between CRs accelerated by a SNR and dense clouds are the cases of the remnants IC 443 and W44. These objects are the two highest-significance SNRs in the Fermi-LAT catalog (3FGL). IC 443 and W44 are located at distances from the Earth of 1.5 and 2.9 kpc, respectively. The age of both remnants is estimated to be 10^4 years. The SED measured by Fermi-LAT, Agile, MAGIC and VERITAS for the SNR IC 443 is presented in Fig. 9.15, together with the theoretical predictions for hadronic and leptonic models of γ-ray production (Ackermann et al. 2013).

The γ-rays from π^0-decay are likely emitted through interactions between "crushed cloud" gas and relativistic protons. Filamentary structures of synchrotron radiation seen in a high-resolution radio map support this picture. The mass of gas is large enough to explain the observed γ-ray luminosity with an average matter density of $n = 20\,\text{cm}^{-3}$ in the case of IC 443.

The fact that the spectral measurements down to 60 MeV have enabled the identification of the π^0-decay feature in the case of IC 443 and W44 mid-aged

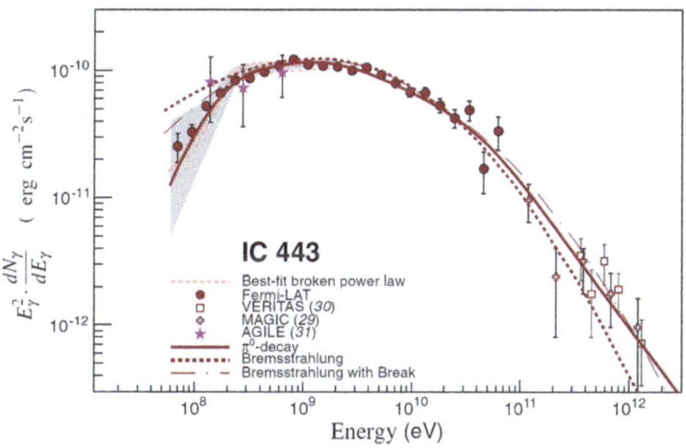

Fig. 9.15 SED in the γ-ray region of IC 443. Within the 0.1–60 GeV range, measurements from AGILE and Fermi-LAT are reported. The *gray-shaded* band shows the Fermi-LAT systematic errors below 2 GeV, due mainly to the uncertainties about background subtraction of the diffuse galactic emission. In the TeV domain, the points came from MAGIC and VERITAS. The *solid line* denotes the best-fit γ-ray spectrum assuming a hadronic mechanism, the *dashed line* the best-fit bremsstrahlung spectrum, and the *dash-dotted line* the best-fit bremsstrahlung spectrum when including an ad hoc low-energy break at 300 MeV/c in the electron spectrum. Credit: Fermi-LAT collaboration

SNRs, providing the first evidence for the acceleration of protons in SNRs, is largely stressed in Ackermann et al. (2013). However, these two objects are far from being able to accelerate CRs up to PeV energies, and the spectral index for the γ-ray energy spectrum is much greater than 2. The quest for PeVatron galactic accelerators is still open. In addition, the Fermi-LAT measurement is particularly difficult because of the presence of the background from diffuse emission (that must be subtracted to obtain the signal) and the uncertainties about the effective area of the detector at the lowest energies.

Other examples of observed enhancement of hadronic production due to the interactions of cosmic rays with dense gaseous complexes are W51, studied by MAGIC and HESS up to 5 TeV, and the 10^4 year-old SNR W28. In this last case, a clear correlation between the TeV emission and nearby massive molecular clouds emitting CO has been observed. More details and reference to the experiments can be found in Holder (2012) and Rieger et al. (2013).

9.9 Summary of the Study of Galactic Accelerators

The importance of multiwavelength astrophysics is compelling: the combination of γ-ray with radio and X-ray data observations suggests that SNRs are indeed the sources of CR electrons and/or hadrons. The first evidences of hadronic acceleration up to $\sim$10 TeV in some mid-aged SNRs interacting with molecular clouds are growing through the combination of Fermi-LAT data with observations from MAGIC, VERITAS, and HESS. However, the relative contributions of accelerated protons and electrons to γ-ray production in most objects still remain largely unknown, and it is not clear whether proton and electron accelerators are of a different nature.

As a general conclusion, if the magnetic field intensity near the acceleration region is low (i.e., smaller than $\sim$10 μG), the accelerated electrons lose a significant fraction of energy in IC γ-rays through interaction with the self-produced synchrotron radiation field. Thus, the observed VHE γ-ray spectrum is probably dominated by accelerated leptons. The contribution of the IC component will also dominate over the $\pi^0 \rightarrow \gamma\gamma$ if the fraction of accelerated leptons at the same energy as the protons is very small, for instance, $e/p = 10^{-3}$. The contribution of the hadronic component is expected to be dominant if the magnetic field in the shell significantly exceeds 10 μG and if the ambient matter density $n \gg 1\,\text{cm}^{-3}$.

The detailed morphological studies possible with IACTs at the level of $\sim$0.1° show that the acceleration sites are spatially coincident with the sites of nonthermal X-ray emission, strengthening the hypothesis that primary galactic CRs up to the knee are accelerated in SNRs. The identification of these objects is still an open field. Most likely, only neutrino telescopes (Chap. 10) can solve the problem.

The fact that a relatively small number of SNRs has been detected at TeV energies holds a possible explanation for the evolution of SNRs, which may only be able to accelerate CRs to PeV energies for a few hundred years, while later, the high-energy hadrons may escape from the system before undergoing further acceleration. The intensity of γ-ray emission from hadronic interactions depends upon the flux of high-energy nuclei and the density of the target material. The production of γ-rays via pp-interactions in dense gas condensations (e.g., molecular clouds with densities $>100\,\text{cm}^{-3}$) embedded in low-density shells represents an interesting scenario for the detection of hadron acceleration sites.

Future imaging experiments like CTA will allow for a precise measurement of the energy spectrum below 1 TeV (down to tens of GeV) and above 10 TeV (up to 100 TeV). In addition, the improved angular resolution could provide independent and complementary information about the radiation mechanisms through morphological studies.

In the absence of the detection of neutrinos, a proof of the origin of CRs up to the knee region in SNRs would be the detection of γ-rays of extremely high energy, in the PeV region. Synchrotron losses typically prevent the acceleration of electrons to energies beyond 100 TeV. In addition, at such high energies, IC emission is suppressed, because the Klein–Nishina cross-section replaces the Thomson one (Sect. 8.2.3). Because of these two effects, the contribution of inverse Compton γ-rays to the radiation above 10 TeV is expected to gradually fade out. Thus, the detection of γ-rays up to 100 TeV would establish a hadronic origin of the radiation.

9.10 Active Galaxies

The terminology of active galaxies is often confusing, since the distinction among different types of AGN sometimes reflects historical differences in the way in which the objects were discovered or initially classified, rather than real physical differences. Three main criteria used to classify active galaxies are as follows: (1) The emission of the source at radio wavelengths yields a division into radio loud and radio weak objects; (2) The optical luminosity of the object. Radio weak sources are subdivided into optically strong and optically weak sources, while radio loud sources are subdivided into low luminosity and high luminosity. (3) The orientation of the AGN toward the observer. The emission contribution from the jet(s) and the lobes that they inflate dominates the luminosity of the radio-loud AGN. The jet and jet-related emission can be neglected in the radio-quiet objects.

The radiation emitted by an AGN is usually attributed to one (or both) of the two following processes:

- *Thermal radiation* originating from in-falling matter strongly heated in the inner parts of an accretion disk close to the central black hole. Accelerated electrons can produce inverse Compton on the photons of the hot corona. The radiation produced in these processes ranges mostly in the optical, UV and X-ray bands. The AGN that are energetically dominated by thermal radiation can be classified as *thermal dominated* or *disk dominated AGN*.

- *Nonthermal emission* emitted in a magnetic field by highly energetic particles that have been accelerated in a jet of material ejected from the nucleus at relativistic speed. The radiation produced through this mechanism encompasses the entire electromagnetic spectrum, from the radio to γ-rays. These represent the *jet dominated AGN*.

The *disk dominated AGN* are objects usually called *Seyferts* galaxies (or quasi-stellar objects-QSOs).[3] They are radio weak objects that show continuum emission within the optical range from the central region. They also present narrow and occasionally broad emission lines, occasionally strong X-ray emission and sometimes a weak small-scale radio jet. The host galaxies of Seyferts are usually spiral or irregular galaxies. There is a correlation between the QSO's luminosity and the mass of its host galaxy, as the most luminous objects are located in the core of the most massive galaxies. Although thermal-dominated AGN are the large majority ($\sim$90%), none of the sources have been detected so far in the HE and VHE γ-rays.

Unified models exist supporting the hypothesis that different observational classes of AGN are really a single type of physical object observed under different conditions (Urry and Padovani 1995). The favored unified model predicts that the apparent differences arise simply because of their different orientations to the observer, see Fig. 9.16. If the jet is not pointing toward the observer, whose line of sight toward the core intercepts a significant amount of the dusty disk of material that lies in the plane of the galaxy, the typical emission from Seyferts galaxies is expected.

The class of *jet dominated AGN* corresponds mostly to radio loud AGN. These can be subdivided into *blazars* and *nonaligned blazars*, depending on the orientation of their jets with respect to the line of sight.

- **Blazars**. A strong and rapid variability (including superluminal motion, Sect. 9.13) is the observational indicator that these objects point their jet in a direction that is closely aligned to our line of sight. Because of this very special perspective, their emission is modified by relativistic effects (see Sect. 9.13.1). Blazars are divided into three main subclasses depending on their optical spectral properties.
  - **FSRQs** or Blazars of the QSO type, or BZQ. These are blazars showing broad emission lines in their optical spectrum, just like normal quasi-stellar objects. This category includes objects normally referred to as flat spectrum radio quasars (FSRQs) and broad-line radio galaxies.
  - **BL Lacs** or Blazars of the BL Lac type, or BZB. These are objects normally called BL Lacs or BL Lacertae objects. Their radio compactness and broad-band SED are very similar to that of strong lined blazars, but they have no strong and broad lines in their optical spectrum. In the literature, BL Lac

[3]Quasar is also the contraction of *quasi-stellar object*, because in optical images, they have optical luminosities greater than that of their host galaxy. Some astronomers use the term *quasi-stellar object* (QSO) to indicate radio-quiet quasars, reserving that of *quasars* for radio-loud objects.

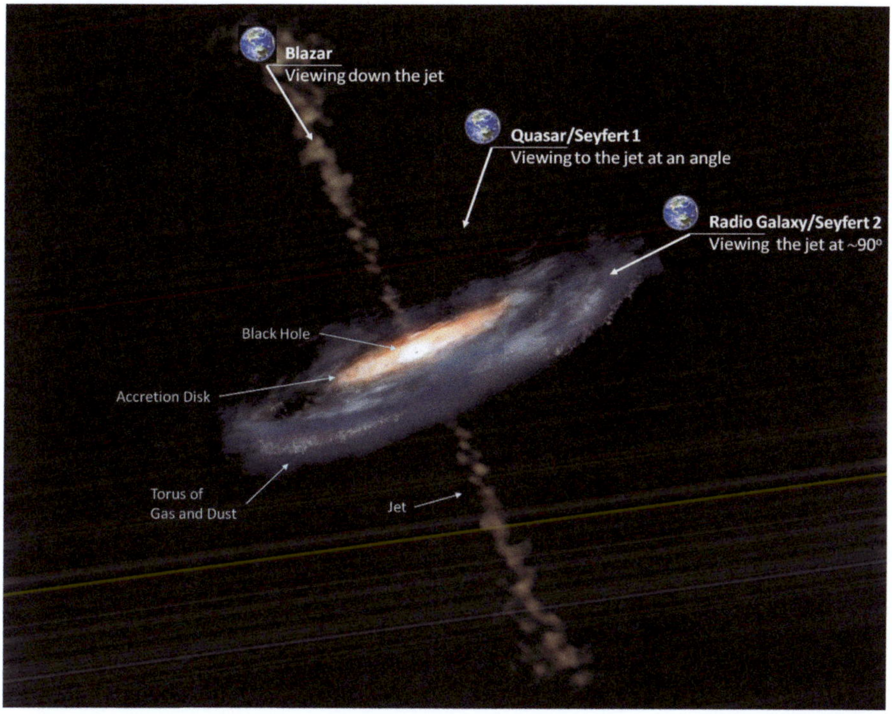

Fig. 9.16 This illustration shows the different features of an active galactic nucleus (AGN), and how our viewing angle determines what type of AGN we observe. The extreme luminosity of an AGN is powered by a supermassive *black hole* at the center

objects are often subdivided into three subclasses depending on their SEDs (see Sect. 9.12). If the frequency ν_s of the synchrotron emission is peaked in the far IR or IR ($\nu_s < 10^{14}$ Hz), they are classified as low-energy BL Lac (LBL); they are intermediate energy BL Lac (IBL) if the peak is between $10^{14} < \nu_s < 10^{15}$ Hz. Otherwise, if the synchrotron radiation peaks at higher energies ($\nu_s > 10^{15}$ Hz), they are called high-energy BL Lac (HBL).

- **Blazars of the unknown type** (BZU) are objects that show many of the hallmarks of blazars but do not have optical spectra of sufficient quality to safely determine the presence of broad emission lines or to accurately measure their equivalent width.

• **Nonaligned blazars**. These sources are radio loud AGN with jets pointing at large or intermediate (~15–40°) angles with respect to our line of sight. This category includes:

  - **Radio galaxies**. AGN with no broad emission lines. Often, they show extended, double-sided radio jets/lobes pointing in opposite directions with respect to the galactic plane and with a very large angle with respect to the line of sight. The nuclear emission is similar to that of blazars, but is usually fainter.

– **Steep Spectrum Radio Quasars (SSRQ)**. AGN with broad emission lines. The orientation of the jet in these sources is thought to be intermediate between that of blazars and radio galaxies.

Approximately 1% of all galaxies hosts an active nucleus. Around 10% of these AGN exhibit relativistic jets powered by accretion onto a supermassive black hole. Despite the fact that AGN have been studied for several decades, the knowledge of the emission characteristics up to the highest photon energies is mandatory for understanding these extreme particle accelerators.

One of the big achievements of the last decade is the large increase in the number of AGN that have been detected at γ-ray energies, allowing us to better understand some of their physical properties. Within the GeV energy range, the Fermi-LAT Collaboration reported, in the 3FGL catalog, more than 1700 AGN, while at TeV energies, the number of detected AGN is about 75, Table 9.1. There are, however, still many open questions regarding AGN: (a) the location and structure of their dominant emission zones; (b) the content of their jets; (c) the origin of their variability, observed on timescales from years down to minutes in different wavelengths; (d) the role of external photon fields in the mechanism that involves the inverse Compton effect to produce the observed γ-ray spectra; (e) the energy distribution and the dominant acceleration mechanism for the underlying radiating particles.

The other important experimental achievement is the organization and scientific interpretation of a large number of multi-instrument observing programs, which provide simultaneous observations from radio to γ-ray energies. Given the variability and the broadband nature of the jet emission, these coordinated efforts from the community are crucial for a detailed and unbiased study of AGN.

9.11 The Extragalactic γ-Ray Sky

The extragalactic sky is that observed at galactic latitudes $|b| > 10°$. The presence of γ-ray sources outside the galactic disk is evident both in Fig. 8.7 for the GeV sky and in Fig. 9.8 for the TeV sky. Most of the observed objects at HE and VHE energies are blazars. The presence in blazars of a jet oriented toward the observer is of particular interest, because the emission is dominated by relativistic beaming effects, which boost the observed photon energies and luminosity.

The GeV Range The large number of AGN in the Fermi-LAT catalog (Table 8.1) coupled with the all-sky monitoring of time-varying processes has given a tremendous boost to the study of the extragalactic γ-ray sky. Among the detected extragalactic objects, the collaboration has detailed 116 sources associated with high confidence at $|b| > 10°$ with AGN extracted from the 2FGL catalog, producing the *LAT Bright AGN Sample* (Abdo et al. 2010a). About 90% of the considered sources have been associated with AGN listed in radio catalogs, thus implying that the bright extragalactic γ-ray sky is dominated by the class of radio-loud AGN, namely FSRQs, BL Lacs, and radio galaxies. Only about one-third of the bright Fermi-LAT AGN was also detected by EGRET. This is a likely consequence of the larger Fermi-LAT effective area and of the strong variability and duty cycle of GeV blazars. In addition, the fraction of different AGN classes is different between the two experiments, and is probably due to a selection effect induced by the different energy response of the EGRET and Fermi-LAT instruments. Finally, the Fermi-LAT observations provide important criteria for the scheduling of observations with the rather narrow field-of-view TeV instruments at ground level.

At GeV energies, a significant number of AGN of uncertain type is present, and relatively few non-AGN objects have been discovered. Among non-AGN objects, there are several local-group galaxies (LMC, SMC, M31), as well as other star-forming galaxies (NGC 4945, NGC 1068, NGC 253, and M82). In each galaxy, as in our own, CRs should be accelerated by SNRs or other objects that are related to star-formation activity. These detections seem to confirm the relation between star-formation rate and γ-ray luminosity.

The TeV Range At TeV energies, the extragalactic γ-ray sky is completely dominated by AGN. At present, about 75 objects have been discovered and are listed in the online TeV Catalog, Table 9.1. The two most massive close-by starburst galaxies NGC 253 and M82 are the only non-AGN sources detected at TeV energies through the emission of γ-rays from their own CR propagation. The majority of them (90%) are sources with the jet pointing along the line-of-sight (high-frequency BL LACs, HBL). Only four radio galaxies (FRI) have been detected at TeV energies (Centaurus A, M87, NGC 1275 and PKS 0625-35).

The observed VHE spectra can be described, as usual, by a power law $dN/dE \propto E^{-\alpha_\gamma}$. For the HBL sources, inferred photon indices range in the interval from 2.3 to 4.5, with some indications of spectral hardening with increasing activity. Emission beyond 10 TeV has been established, for instance, in Mrk 501 and Mrk 421, where measured photon energies reach 20 TeV. Non-HBL sources are usually detected during high states only, with low states expected to fall below current sensitivities.

As γ-ray objects are, in most cases, associated with sources known in other wavelengths, most AGN distances have been estimated by redshift measurements. It is found that the objects in the Seyferts class (FRI) are much closer to us than quasars or blazars (FSRQ, HBL). Blazars of the QSO type up to redshift $z \sim 0.6$ (i.e., 3C 279 at $z = 0.536$) have been detected at VHE energies. Most of them have $z < 0.2$. Blazar population studies at radio to X-ray frequencies indicate a redshift distribution for BL Lacs objects that seems to peak at $z \sim 0.3$, with only a few sources beyond $z \sim 0.8$, while the FSRQ population is characterized by a rather broad maximum between $z \sim 0.6$–1.5.

Observed VHE flux levels for extragalactic objects typically range from 1% of the Crab nebula steady flux (for the average/steady-states) up to 10 times as much when the AGN are in high activity phases. Because TeV instruments are now able to detect sources at the level of 1% of the Crab, the variability down to the few-minute scale of the near and bright TeV-emitting blazars (Mrk 421 and Mrk 501) can be studied in detail. Another consequence of the sensitivity of IACTs instruments is that more than one extragalactic object could be visible in the same field-of-view. With respect to the early phase of the first decade of this century, the TeV instruments are now shifting their observation strategies to move toward higher-quality data sets of individual sources, rather than trying to simply increase the number of sources.

9.12 The Spectral Energy Distributions of Blazars

The study and classification of AGN and their acceleration mechanisms require observations from different instruments. The spectral energy distributions (SEDs) of blazars can span almost 20 orders of magnitude in energy, making simultaneous multiwavelength observations a particularly important diagnostic tool for disentangling the underlying nonthermal processes. Usually, SEDs of different objects were obtained using data not collected at the same time. Only very recently has the necessity of strictly contemporaneous (or at least as contemporaneous as possible) and broadband sampling of SEDs been strengthened. This effort is particularly relevant for time-varying sources in which changes in overall brightness are often accompanied by changes in the energy spectra.

As an example of these efforts, we describe in Sect. 9.12.1 the result of a quasi-simultaneous study of the SEDs of 48 blazars based on the Fermi-LAT Bright AGN Sample (Abdo et al. 2010a), presented by the Fermi-LAT Collaboration. Quasi-simultaneous means that the Fermi data have been collected continuously over a period of 3 months while all other data have been collected over much shorter periods (typically less than a few hours) and are not necessarily simultaneous among themselves. In Sect. 9.12.2, the results are shown from the 4.5-month-long multiwavelength observations of Mrk 421 (Abdo et al. 2011a). This campaign included the Fermi-LAT, VLBA, Swift, RXTE, MAGIC, and other collaborations and instruments that provided excellent temporal and energy coverage of the source throughout the year 2009. During this campaign, Mrk 421 showed a low activity at all wavebands. The extensive multi-instrument (radio to TeV) data set provides an unprecedented, complete look at the quiescent SED for this source.

9.12.1 Quasi-Simultaneous SEDs of Fermi-LAT Blazars

The SEDs of 48 blazars selected by Fermi-LAT were constructed using:

- The Swift (Sect. 8.8.4) database, based on information from the three instruments measuring at different frequencies (UVOT, XRT, and BAT);
- The AGILE data (Sect. 8.8.3) for energies greater than 100 MeV;
- The available TeV data from MAGIC, HESS, and VERITAS;

- Radio surveys from: the broadband monitoring program, covering the frequency range 2.6–42 GHz, of the Effelsberg 100 m radio telescope of the Max Planck Institute for Radio Astronomy; the 15 GHz observations made using the Owens Valley Radio Observatory (OVRO) 40 m telescope; the 1–22 GHz instantaneous radio spectra provided by the 600 m ring radio telescope RATAN-600 of the Special Astrophysical Observatory, Russian Academy of Sciences;
- The information from radio, mm, near infrared and optical frequency range from a dedicated program supporting the Fermi-LAT and AGILE scientific programs (http://www.oato.inaf.it/blazars/webt/);
- Mid-infrared observations carried out using VISIR, the ESO/VLT mid-infrared imager and spectrograph, composed of an imager and a long-slit spectrometer;
- Additional nonsimultaneous observations from the Spitzer Space Telescope. It provides images and spectra within the range between 3 and 180 μm through three instruments on board: the InfraRed Array Camera, the Multiband Imaging Photometer for Spitzer and the InfraRed Spectrograph;
- Nonsimultaneous multiwavelength archival measurements (included to illustrate the historical range of variability) at different frequencies from the NED (NASA/IPAC Extragalactic Database) and ASDC (ASI Science Data Center) online services.

Since HE γ-ray data have been accumulated over the relatively long period of 3 months, they likely represent the average of different intensity states. This is clearly a limitation, as flux and spectral variability in blazars often takes place within shorter timescales.

In all cases, the overall shape of the SEDs exhibit the typical broad double hump distribution, as shown in Fig. 9.17 for three AGN at different distances. The SEDs of all considered AGN show that there are considerable differences in the position of the peaks of the two components and in their relative intensities.

According to current models, the low-energy bump is interpreted as being due to synchrotron emission from highly relativistic electrons, and the high-energy bump is related to inverse Compton emission of various underlying radiation fields. UV photons generated by the accretion disk surrounding the black hole, or IR photons provided by the dusty torus, can also contribute as seed photons to the IC process. Large variability is also present, especially at optical/UV and X-ray frequencies. Within the GeV range, the γ-ray variability cannot be evaluated, as the Fermi data are averaged over the entire data taking period.

The studies on different blazar populations seem to suggest that there is a continuous spectral trend from FSRQ → LBL→ IBL→ HBL, often called the "blazar sequence". The sequence is characterized by a decreasing source luminosity, increasing synchrotron peak frequency and a decreasing ratio of high- to low-energy components. This sentence must be considered with some caveats due to selection effects and unknown redshifts of considered objects.

In Fig. 9.17, 3C 279 is a very far ($z = 0.536$) FSRQ low-frequency peaked. The BL Lacertae is a relatively near ($z = 0.069$) blazar that gives the name to the subclass of BL Lac objects, and is of LBL type. Finally, Mrk 421 is the closest AGN

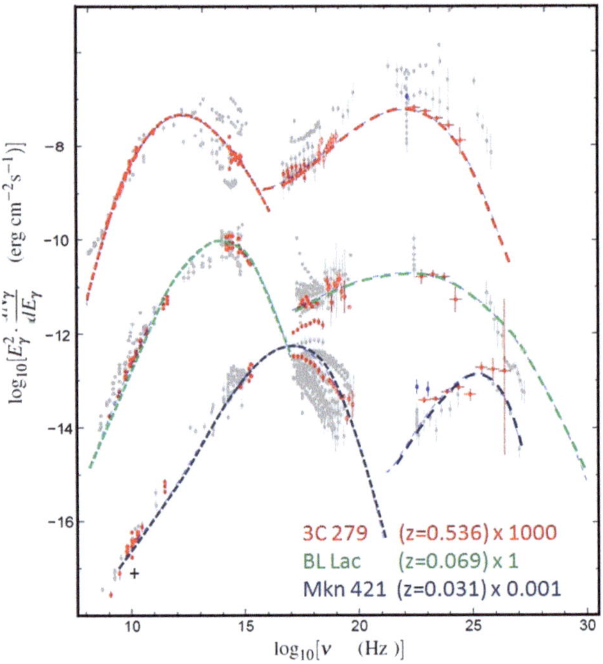

Fig. 9.17 The SED of three different AGN at different distances from the Earth and belonging to different subclasses. To improve the visibility of the spectra, the contents of the farthest (3C 279) have been multiplied by a factor of 10^3, while that of the nearest (Mrk 421) by a factor of 10^{-3}. The *dashed lines* represent the best fit to the data assuming leptonic production

($z = 0.031$) of the HBL type. Note that in the figure, the SEDs of the three objects are shifted for better visibility by different factors, as explained in the caption. The luminosities of the sources (L_γ, [erg/s]) scale almost as $1/z^2$ (for relatively small z, the redshift is proportional to the distance). When the effect of distances is considered, the luminosity in correspondence with the synchrotron peak decreases from $\sim 5 \times 10^{47}$ erg/s for FSRQ to $\sim 5 \times 10^{44}$ erg/s for HBL.

In the models of γ-ray emission, all high-frequency BL LACs (HBLs) seem to be well-described by the homogeneous (*one-zone*) leptonic Synchrotron Self-Compton (SSC) model. In the so-called one-zone SSC model, the emission, from radio through X-rays, is produced by synchrotron radiation from electrons in a homogeneous, randomly oriented magnetic field. The γ-rays are produced by inverse Compton scattering of the synchrotron photons by the same electrons that produce them. For IBLs, a combination of SSC plus inverse Compton on some additional external photons fields seems to be needed to fit the data. For FSRQs, the situation is rather complicated, and it is most likely that leptonic plus hadronic models are necessary to fully explain the data (Funk 2012; Rieger et al. 2013). The dashed lines in Fig. 9.17 represent the fits using a leptonic model, as given in Abdo et al. (2010a).

Exercise Estimate the peak luminosity for the three objects considered in Fig. 9.17 using the redshift-luminosity distance relation.

9.12.2 *Simultaneous SED Campaigns and Mrk 421*

All the TeV ground-based observatories are involved in massive, targeted, multi-wavelength and multimessenger campaigns. Of particular importance have been the large-scale multi-instrument efforts on Mrk 421 and Mrk 501. Figure 9.18 reports the most complete simultaneous SEDs collected for the BL Lac AGN Mrk 421 to date. The combination of the spectral observations from the γ-ray instruments Fermi-LAT and MAGIC have allowed us to characterize, for the first time, the entire bump due to the inverse Compton over five orders of magnitude in energy and without gaps. Similar broadband studies have been carried out for other AGN sources.

The SED of Mrk 421 looks similar to that of the Crab (Fig. 9.10): the same physical processes are providing the emission spectrum. The broadband SED has been modeled in Abdo et al. (2011a) with two different scenarios: a leptonic (one-zone SSC) model and a hadronic model. Both leptonic and hadronic frameworks are able to describe the average SED reasonably well, implying comparable powers for the jet emission, which constitutes only a small fraction (10^{-3}–10^{-2}) of the Eddington luminosity (see *Extras # 4*) of the source.

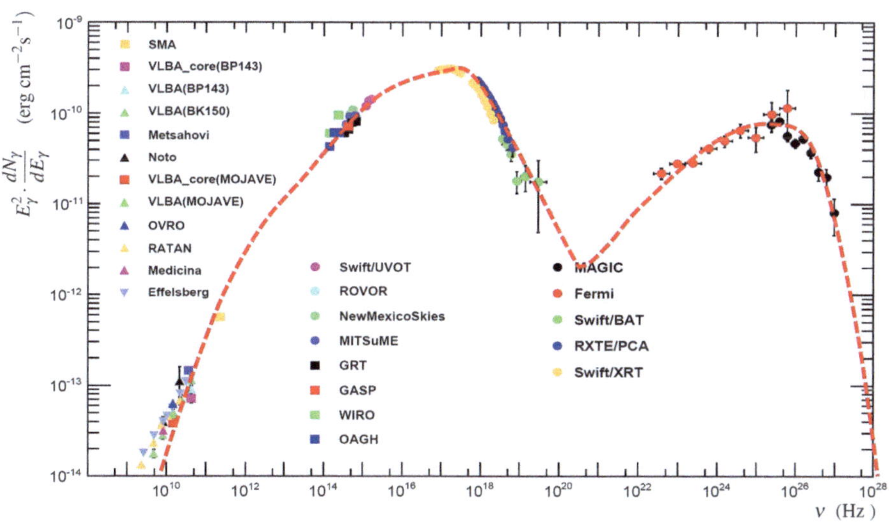

Fig. 9.18 Extensive multiwavelength measurements showing the spectral energy distribution of Markarian 421 from observations made in 2009. The legend reports the correspondence between the instruments and the measured fluxes. The *dashed line* is a fit of the data with a leptonic model. See Abdo et al. (2011a) for the references to the data

However, hadronic and leptonic models differ on the predicted environment for the blazar emission: the leptonic scenario constrains the size to being $R \leq 10^4\mathscr{R}$, the magnetic field to being $B \sim 0.05$ G and electrons to having energies up to $\sim 5 \times 10^{13}$ eV. $\mathscr{R}$ is the gravitational radius (or Schwarzschild radius) of the central massive black hole, defined in (6.84). For the hadronic model, and assuming the same power-law index of the injected electron and proton distributions $\alpha_e = \alpha_{CR}$, this implies a size of the emitting region of a few $\mathscr{R}$, a magnetic field $B \sim 50$ G and particles (protons) with energies up to a few 10^{18} eV.

The hadronic mechanism in AGN jets could explain the production of UHECRs in the Universe. In this case, neutrinos could also be produced by the decay of charged mesons, as the γ-rays arise from the decay of the neutral ones. Larger maximum energies can be reached by other classes of blazars. In the particular case of Mrk 421, the hadronic mechanism seems to require extreme conditions for particle acceleration and confinement.

On the other hand, the observed variability of the blazar seems to be better explained by a leptonic production model of γ-rays. We derive in Sect. 9.13.1 the way in which the source variability is connected to the size R of the emitting source. In the case of Mrk 421, it was observed that X-rays are significantly more variable than γ-rays, with timescales ranging from 1 h to 1 day. No significant correlations between variations in X- and γ-rays have been observed so far. In addition, within the X-ray and γ-ray energy bands, the variability increased with photon energy. The physical interpretation of this result within the context of the one-zone SSC model is that the flux variability increases with the energy of the electrons that produce them.

Another relevant multi-year campaign is that on M87 involving all the IACTs (VERITAS, MAGIC, HESS for 80 h in total), the Fermi-LAT, the X-ray satellite Chandra, the Hubble space telescope, and various other instruments all the way down to radio measurements at 1.7 GHz from VLBA. Some flaring episodes detected in these observations provide a rich data set for studying rise-and decay-times and spectral properties of flares in this close-by non-blazar AGN (Funk 2012).

9.13 Jets in Astrophysics

Powerful relativistic outflows and jets, with speeds approaching that of light, have been observed in different astrophysical sources, Fig. 9.19. This includes extragalactic objects (AGN) and galactic objects such as pulsars and microquasars. If these outflows point toward us, as in blazars, relativistic effects can significantly change their appearance. The dynamics of nonthermal processes in these outflows are important fields of research. The jet compositions and their production mechanisms are, in fact, not completely understood. Up to now, observations have not been able to provide enough evidence to support one of the various theoretical models among the many that exist. In the case of AGN, the leading model predicts that the matter in the disk falls into the black hole, converting a large fraction of the gravitational

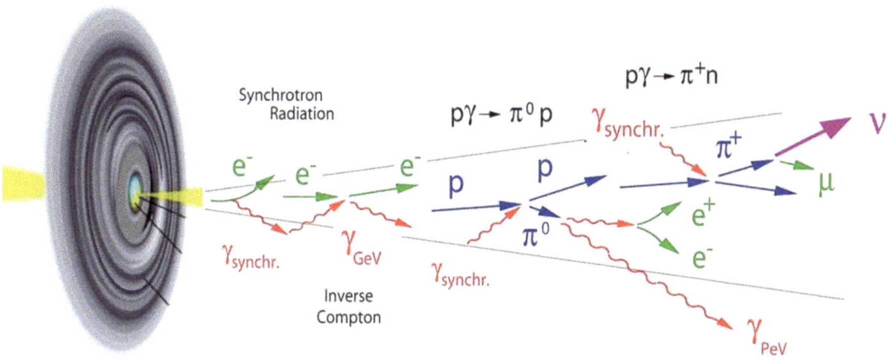

Fig. 9.19 A sketch of a jet. Charged particles are accelerated and could emit γ-rays and neutrinos through different processes

potential into kinetic energy. In some cases, the jet of particles moves outwards with relativistic bulk velocity (measured in the radio wavelength) up to Lorentz factors $\Gamma \simeq 50$ and with an extension that goes from fractions of kilo-parsecs up to hundreds of kpc.

The blazar emission zone is unresolved by all instruments (apart from interferometer radio telescopes using mm-waves), and the measurements of time variability are the only way of probing its structure. Additional information on the structure and dynamics of the jet arises from the study of the polarization at radio and optical frequencies.

Radio interferometric observations of blazar jets have revealed significant apparent superluminal motion ($\beta_a = v_a/c > 1$) of individual jet components on the parsec-scale propagating away from the core. This superluminal motion is not in contradiction with special relativity. The situation is similar to that of a laser beam illuminating high clouds. It can result in the appearance of a spot on the clouds moving faster than the speed of light if the laser is rotated fast enough. This apparent motion is the source's pattern speed. One of the first examples of such superluminal motions was observed as early as the beginning of 1980 in the quasar 3C 273, in which a source component moved away from the nucleus at a speed of roughly eight times the speed of light. The kinematics of this motion is known, and it is related to the bulk velocity of the jet material. We refer to Longair (2011) for additional information.

9.13.1 *Time Variability in Jets*

The leptonic or hadronic origin of the γ-ray emission in the AGN jets, as well as the location of the γ-ray emission along the jets, are still open important questions. This last problem can be faced on by studying the time variability in the photon flux.

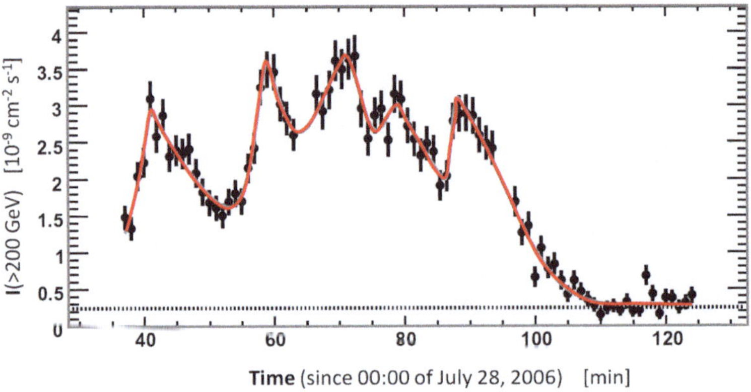

Fig. 9.20 Integrated flux $I(>200\,\text{GeV})$ versus time, as observed by HESS for PKS 2155-304 starting from midnight of July 28, 2006. The data are binned in 1 min intervals, and the horizontal *dashed line* shows, for comparison, the steady flux from the Crab nebula (Aharonian et al. 2007a). Courtesy of Prof. F. Aharonian

Despite the limited temporal coverage of the current IACTs to each single source, more than half of the AGN detected in the TeV domain have shown variability, albeit often weak. For the majority of them, variability timescales above 1 month have been found. In about a quarter of them, there is clear evidence for short-term TeV variability, Δt_{TeV}, on observed timescales of less than 1 h. In particular, for the HBL class, $\Delta t_{TeV} \simeq 3$ min have been measured for Mrk 501 and PKS 2155-304, $\simeq$10 min for PSK 1222+21 and Mrk 412. Also, the representative of the BL Lac class, the BL Lacertae, shows variability of $\simeq$15 min. Figure 9.20 shows the measured flux versus time for PSK1222+21.

The TeV variations observed in blazars with evidence of sub-hour variability are the fastest observed in any other waveband so far, and already imply a compact size of the γ-ray emitting region. The observation of time variability Δt_{TeV} in a jet of particles, moving outwards with a relativistic bulk velocity characterized by the Lorentz factor Γ, implies a restriction on the size R of the source given by

$$R < \frac{\Gamma \cdot c \cdot \Delta t_{TeV}}{(1+z)} \simeq 10^{15} \cdot \left(\frac{\Gamma}{10}\right) \cdot \left(\frac{\Delta t_{TeV}}{1\ \text{h}}\right) \text{cm}\,, \tag{9.4}$$

where the factor $(1+z)$ includes the general relativity effect of the expansion of the Universe. For the relatively nearby AGN observed in the TeV range, this factor can be neglected.

The sizes of massive compact objects M are ruled by general relativity, and the source size R could be compared with the Schwarzschild radius, $\mathscr{R} = 2\frac{GM}{c^2}$. Equation (9.4) for sources with $\Gamma \sim 10$ and time variability $\Delta t_{TeV} \sim 1$ h provides values of R comparable to the Schwarzschild radius of a very massive black hole in the AGN core: $\mathscr{R} = 3 \times 10^{14}$ cm for $M = 10^9\ M_\odot$. This led to the conclusion that the γ-rays are produced close to the central engine of the blazars.

However, some results (see Paneque (2012) for a review) have shown a clear correlation of some γ-ray outbursts with optical polarization changes and/or with the passage of radio knots through the core structure in the mm-wave radio-interferometer images. The radio knots are believed to be a standing shock situated several pc away from the central engine. Therefore, at least for some sources and detected outbursts, the blazar emission has been pinpointed as being far away (1–10 pc) from the supermassive black hole.

A variety of leptonic and hadronic emission models have been discussed in the literature (Böttcher 2010) to explain the SEDs of AGN. The short-term variability seems to favor a leptonic synchrotron-Compton interpretation. On the other hand, the detected "orphan TeV flares" (flares observed at TeV energies and not in the lower energy wavebands) are much more difficult to account for in the leptonic scenario.

As in the case of galactic sources, probably only the detection of neutrinos can demonstrate the presence of the hadronic mechanism at work. Vice versa, the nonobservation of neutrinos does not exclude hadronic acceleration mechanisms, because a necessary ingredient is the presence of enough matter or radiation fields with which accelerated hadrons can interact.

9.14 The Extragalactic Background Light

The detection of very far away γ-ray emitting blazars has started providing information useful for observational cosmology as well. In this section, we will derive the way in which the Universe is opaque to γ-rays whenever the energy-dependent photon mean free path is smaller than the distance of the source. The dominant process for the absorption of VHE photons of energy E produced by astrophysical sources is pair-creation

$$\gamma_E + \gamma_\epsilon \rightarrow e^+e^- \tag{9.5}$$

on low-energy extragalactic background photons of energy ϵ. These photons extend from the CMB to the near-ultraviolet (UV) wavelengths. The UV/optical/infrared background radiation is called the *extragalactic background light* (EBL). The cross-section for the pair-production process is described by the Bethe-Heitler formula

$$\sigma(E, \epsilon) \simeq 1.25 \times 10^{-25}(1 - \zeta^2)\,\text{cm}^2\,, \tag{9.6}$$

where $\zeta = \sqrt{1 - \frac{(m_e c^2)^2}{E\cdot\epsilon}}$. The cross-section in (9.6) is maximized when

$$\epsilon = 2\frac{(m_e c^2)^2}{E} \simeq \frac{520\,\text{GeV}}{E}\,\text{eV} \tag{9.7}$$

and the threshold energy is a factor of two lower. For energies above 100 TeV, the process (9.7) is dominant for the energies corresponding to photons of the cosmic microwave background.

Exercise Estimate the average energy E for which the interaction cross-section is maximal on the CMB photons, Eq. (7.4).Show that for photons in the TeV range, the dominant process is on infrared/optical photons.

The EBL is the totality of light emitted by stars, galaxies, and AGN over the lifetime of the Universe. Today, this pervasive photon background consists of light emitted at all epochs, modified by the redshift due to the expansion of the Universe. Therefore, in principle, the EBL contains information about the evolution of the baryonic components of galaxies and the structure of the Universe.

The bulk of the EBL falls within the range from $\sim 10^{-3}$ to 10 eV, corresponding to wavelengths from the far-infrared to the near-ultraviolet ($\sim$1000–0.1 μm). Figure 9.21 shows the estimated spectral energy distribution (SED) normalized for the considered solid angle of the background photons as a function of photon energy or wavelength [the curves are obtained for the model C in Finke et al. (2010)]. The quantity in the ordinate (left scale) is the same SED/$\Delta\Omega$ shown in Fig. 8.6 for the galactic γ-rays. To compare the two figures numerically, remember that 1 MeV $= 1.6 \times 10^{-6}$ erg. On the right-side scale, the energy density (in erg cm^{-3}) of the EBL has been obtained by multiplying the SED/$\Delta\Omega$ by $4\pi/c$. In the UV, optical and near-IR, most of the EBL is due to direct starlight, as well as a subdominant contribution from AGN. From the mid-IR to submillimetric wavelengths, the EBL consists of re-emitted light from dust particles, including both continuum thermal radiation and line emission from molecules. The background at longer wavelengths

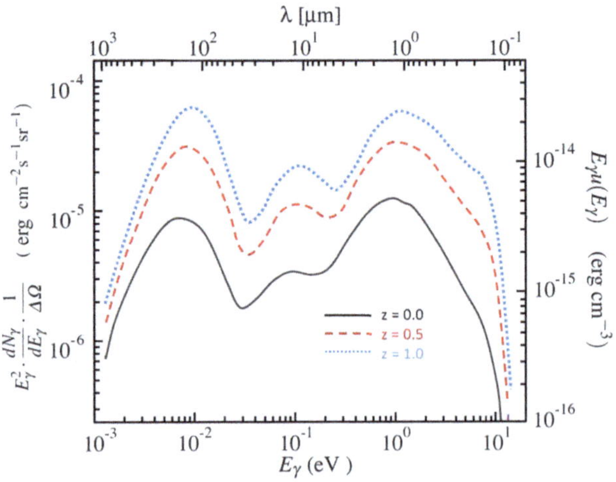

Fig. 9.21 *Left scale* SED per unit of solid angle for the extragalactic background light (EBL) energy density as a function of proper photon energy (*bottom scale*) of wavelength (*upper scale*) for three values of redshifts. *Right scale* the energy density of the EBL

(not shown in the figure) is dominated by the cosmic microwave background. The different lines in the figure represent the EBL background at three different epochs (redshifts) of the Universe.

Exercise Demonstrate that the CMB contribution in Fig. 9.21 would be represented by a parabola with negative concavity and vertex at about $5 \times 10^{-13}\,\mathrm{erg\,cm^{-3}}$ at $E_\gamma \sim 10^{-3}\,\mathrm{eV}$.

Direct measurement of the EBL is difficult due to contamination from nearby objects and galactic light. Due to different modeling approaches and uncertainties in underlying model parameters, the intensity and shape of the EBL spectrum remains controversial, and a wide range of EBL models has been developed (see, for instance, Finke et al. 2010 and Gilmore et al. 2012).

Once the energy density of the EBL has been assumed from a particular model, the *optical depth* $\tau_{\gamma\gamma}(E, z)$ for a photon of observed energy E produced in a source at redshift z can be calculated. The optical depth is the dimensionless quantity defined in Eq. (8.13). In this case, the target is provided by EBL photons and the cross-section σ of photons on matter in (8.13) must be replaced with the photon-photon cross-section $\sigma_{\gamma\gamma}$. The probability of the γ-ray of energy E to survive absorption along its path from the source at redshift z to the observer plays the role of an attenuation factor for the radiation flux, and it is usually expressed in the form

$$I(E, z) = I_0 e^{-\tau_{\gamma\gamma}(E,z)} , \tag{9.8}$$

where I_0 is the flux at the source. The computation accounts for the EBL photon number density and the cross-section (9.6), integrating over the distance and the energy of the background photons (De Angelis et al. 2008). Notice that QED, relativity and cosmology arguments are involved in the relation (9.8).

The energy dependence of $\tau_{\gamma\gamma}$ leads to appreciable modifications of the observed source spectrum with respect to the spectrum at emission, due to the exponential dependence in (9.8). As $\tau_{\gamma\gamma}$ increases with energy, the observed flux turns out to be more attenuated at higher energies, resulting in an effective mean free path of photons propagating in the Universe $\lambda_\gamma(E)$. Figure 9.22 shows the mean free path (in Mpc) of photons as a function of the photon energy. As a result, galactic sources of photons with energy of ~100 TeV also start being attenuated by the presence of the background photons. Photons from the nearby Universe (below some tens of Mpc) start to be attenuated above ~10 TeV by the presence of the CMB.

Absorption represents a drawback for γ-ray astronomy. On the other hand, the identification of absorption features in the spectra of GeV–TeV γ-rays should allow us to derive unique information about the poorly known EBL. This procedure requires precise measurements of the γ-ray flux from a large number of extragalactic objects located at different redshifts z, and, more importantly, a good understanding of the intrinsic γ-ray spectra at the sources, i.e., the quantity I_0 in (9.8).

At present, the HESS Collaboration has reported an upper limit on the EBL at near and mid-infrared wavelengths using measurements of γ-rays from blazars with

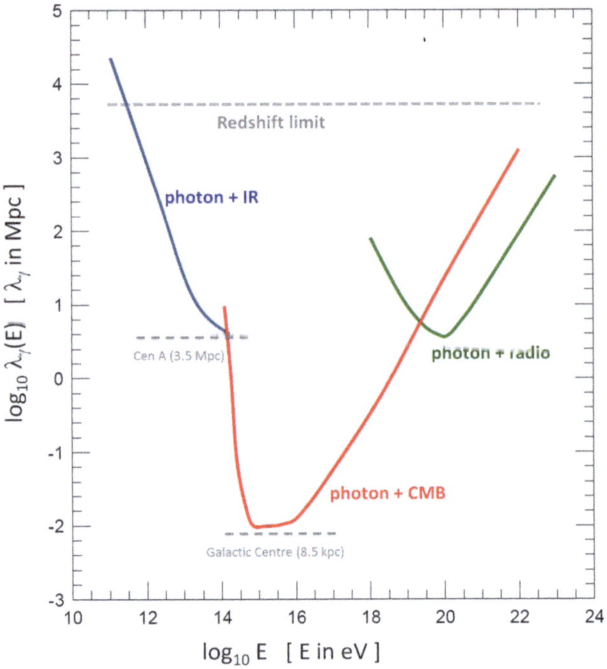

Fig. 9.22 Mean free path (Mpc) of photons as a function of the photon energy. The contribution of the scattering (9.5) on IR, CMB and radio background photons are considered separately. The positions of the galactic center and of the closest AGN are also shown

redshifts between $z \sim 0.1$–0.2. The inferred upper limit is very close to the lower limit given by the measured integrated light of resolved galaxies (galaxy counts). A similar result has been reported by the Fermi-LAT Collaboration for the EBL at optical and UV bands (Rieger et al. 2013). Although the derived upper limits agree with most of the theoretical/phenomenological predictions about the EBL, they are not free of model assumptions. It is believed that future measurements by next-generation detectors, in particular, by CTA, based on a much larger sample of AGN should significantly increase the source statistics and improve the quality of data, and consequently reveal details of the EBL.

References

H. Abdalla et al., The population of TeV pulsar wind nebulae in the H.E.S.S. Galactic Plane Survey. Astron. Astrophys. **612**, A2 (2018). ArXiv:1702.08280

A. Abdo et al., The spectral energy distribution of fermi bright blazars. Astrophys. J. **716**, 30–70 (2010a)

A.A. Abdo et al., FERMI Large area telescope observations of the Crab pulsar and nebula. Astrophys. J. **708**, 1254–1267 (2010b)

A.A. Abdo et al., Fermi large area telescope observations of Markarian 421: the missing piece of its spectral energy distribution. Astrophys. J. **736**, 131 (2011a)

A.A. Abdo et al., Observations of the young supernova remnant RX J1713.7-3946 with the Fermi Large Area Telescope. Astrophys. J. **734**, 28 (2011b)

M. Ackermann et al., Detection of the characteristic pion-decay signature in supernova remnants. Science **339**, 807 (2013)

F. Aharonian et al., The Crab nebula and Pulsar between 500 GeV and 80 TeV: observations with the HEGRA Stereoscopic air Cerenkov telescopes. Astrophys. J. **614**, 897–913 (2004)

F. Aharonian et al., An exceptional very high energy gamma-ray flare of PKS 2155–304. Astrophys. J. **664**, L71–L74 (2007a)

F. Aharonian et al., Primary particle acceleration above 100 TeV in the shell-type supernova remnant RX J1713.7-3946 with deep HESS observations. Astron. Astrophys. **464**, 235–243 (2007b)

F. Aharonian, J. Buckley, T. Kifune, G. Sinnis, High energy astrophysics with ground-based gamma ray detectors. Rep. Prog. Phys. **71**, 096901 (2008)

E. Aliu et al., Observation of pulsed gamma-rays above 25 GeV from the Crab pulsar with MAGIC. Science **322**, 1221 (2008)

M. Böttcher, Models for the spectral energy distributions and variability of blazars, in *Fermi Meets Jansky - AGN at Radio and Gamma-Rays*, ed. by T. Savolainen, E. Ros, R.W. Porcas, J.A. Zensus (2010). p. 41. Also arXiv:1006.5048

S. Carrigan et al., The H.E.S.S. Galactic Plane Survey—maps, source catalog and source population (2013). ArXiv:1307.4690v3

A. De Angelis, O. Mansutti, M. Persic, Very-high-energy gamma astrophysics. Riv. del Nuovo Cimento **31**, 187 (2008). https://doi.org/10.1393/ncr/i2008-10032-2

J.D. Finke, S. Razzaque, C.D. Dermer, Modeling the extragalactic background light from stars and dust. Astrophys. J. **712**, 238–249 (2010)

S. Funk, OG2 Rapporteurs talk at ICRC (2011)

S. Funk, The status of gamma-ray astronomy (2012). ArXiv:1204.4529v1 [astro-ph.HE]

R.C. Gilmore et al., Semi-analytic modelling of the extragalactic background light and consequences for extragalactic gamma-ray spectra. Mon. Not. R. Astron. Soc. **422**, 3189–3207 (2012)

F. Giordano et al., Fermi large area telescope detection of the young supernova remnant tycho. Astrophys. J. Lett. **744**, L2 (2012)

J.A. Hinton, W. Hofmann, Teraelectronvolt astronomy. Annu. Rev. Astron. Astrophys. **47**, 523 (2009)

J. Holder, TeV gamma-ray astronomy: a summary. Astropart. Phys. **39–40**, 61–75 (2012)

C.F. Kennel, F.V. Coroniti, Confinement of the Crab pulsar's wind by its supernova remnant. Astrophys. J. **283**, 694–709 (1984)

M.S. Longair, *High Energy Astrophysics*, 3rd edn. (Cambridge University Press, Cambridge, 2011)

Y. Moudden et al., The topological second-level trigger of the H.E.S.S. phase 2 telescope. Astropart. Phys. **34**, 568 (2011)

P.L. Nolan et al., FERMI Large area telescope second source catalog. Astrophys. J. Suppl. Ser. **199**, 31 (2012)

D. Paneque, Experimental gamma-ray astronomy. J. Phys. Conf. Ser. **375**, 052020 (2012)

F.M. Rieger, E. de Ona-Wilhelmi, F.A. Aharonian. TeV Astronomy (2013). ArXiv:1302.5603

D.J. Thompson, L. Baldini, Y. Uchiyama, Cosmic ray studies with the fermi gamma-ray space telescope large area telescope. Astropart. Phys. **39–40**, 22–32 (2012)

M. Urry, P. Padovani, Unified schemes for radio-loud active galactic nuclei. Publ. Astron. Soc. Pac. **107**, 803 (1995). Also arXiv:astro-ph/9506063

Chapter 10
High-Energy Neutrino Astrophysics

Abstract Neutrino astronomy shares with γ-ray astronomy the objective of understanding the sources and mechanisms of CR acceleration. Due to their much larger interaction cross-section, γ-rays are easier to detect than neutrinos, but neutrinos can only be produced through hadronic processes. No single source, either galactic or extragalactic, has been conclusively proven to accelerate CRs up to PeV energies. Neutrino astronomy is expected to be decisive in the quest for CR sources. The idea of a large volume experiment for cosmic neutrinos based on the detection of the secondary particles produced in neutrino interactions was first formulated in the 1960s by M. Markov. He proposed: "to install detectors deep in a lake or in the sea and to determine the direction of the charged particles with the help of Cherenkov radiation". Starting from the Markov idea, in this chapter we describe how the challenge of detecting galactic neutrinos is open for a multi kilometer-scale apparatus, deployed in the Antarctic ice or in deep seawater. At present a km^3 detector (IceCube) is operating in the ice of the South Pole and another smaller underwater telescope (ANTARES) is running in the Mediterranean Sea, waiting for the Mediterranean km3 telescope (KM3NeT) and another detector in Lake Baikal. All of them are made up of a grid of optical sensors (photomultipliers). The recent first measurement of an astrophysical high-energy neutrino flux, opening the field of neutrino astronomy for the next decade, is also reported.

Neutrino astronomy shares with γ-ray astronomy the objective of understanding the sources and mechanisms of CR acceleration. Due to their much larger interaction cross-section, γ-rays are easier to detect than neutrinos, and γ-ray astronomy is turning out to have fundamental importance in regard to several topical areas of modern astrophysics and cosmology. The existence of CR sources seems to guarantee the existence of high-energy neutrino sources, in addition to those of γ-rays. While γ-rays can be produced both by hadronic (through π^0 decay) and leptonic processes (inverse Compton, bremsstrahlung), neutrinos can only be produced by hadronic processes ($\pi^\pm$ decay), Sect. 10.1.

As shown in the previous chapter, the detailed modeling of the morphology and spectra of γ-ray sources seems to indicate a preference for their production through the leptonic mechanism in many source classes. At present, the hadronic mechanism

© Springer Nature Switzerland AG 2018
M. Spurio, *Probes of Multimessenger Astrophysics*, Astronomy and Astrophysics Library, https://doi.org/10.1007/978-3-319-96854-4_10

seems more plausible for certain supernova remnants surrounded by molecular clouds. In most cases, however, it is generally possible to fit the γ-ray data with either leptonic or hadronic production models by varying the model assumptions (for instance, the intensity of magnetic fields or the environmental matter number density). No single source, either galactic or extragalactic, has been conclusively proven to accelerate CRs up to PeV energies. Neutrino astronomy is expected to be decisive in the quest for CR sources.

The idea of a large volume experiment for cosmic neutrinos based on the detection of the secondary particles produced in neutrino interactions was first formulated in the 1960s by M. Markov. He proposed: "*to install detectors deep in a lake or in the sea and to determine the direction of the charged particles with the help of Cherenkov radiation*". A major challenge for these detectors is to separate the astrophysical signals from the large background of atmospheric neutrinos produced by CR interactions with atmospheric nuclei (Sects. 10.3 and 10.4).

The small neutrino interaction cross-section allows them to come from far away. Moreover, neutrinos, being neutral, are not deflected by magnetic fields. There are many theoretical and experimental reasons to believe that high-energy neutrinos are emitted in violent events taking place in many astrophysical objects, Sect. 10.6. Potential sources in our Galaxy are supernova remnants, pulsars and microquasars. Potential extragalactic sources are Gamma-Ray Bursts and AGN.

Starting from the Markov idea and from the present knowledge of TeV γ-rays sources, the challenge of detecting galactic neutrinos is open for a multi kilometer-scale apparatus, Sect. 10.7, deployed in the Antarctic ice or in deep seawater, Sect. 10.8.

At present (Sect. 10.9) a km^3 detector (IceCube) is operating in the ice of the South Pole and another smaller underwater telescope (ANTARES) is running in the Mediterranean Sea, waiting for the Mediterranean km^3 telescope (KM3NeT). All of them use the Markov idea and are made up of a grid of optical sensors (photomultipliers, PMTs) inside the so-called *instrumented volume*.

High-energy neutrino detectors are motivated by discovery and must be designed to detect neutrinos of all flavors over a wide energy range and with the best energy resolution. No astrophysical object has yet been identified as a high-energy neutrino source, and the status of the running experiments is discussed in Sect. 10.10. Lower energy neutrinos from the Sun and from core-collapse supernovae are discussed in Chap. 12. The recent first measurement of an astrophysical high-energy neutrino flux is reported in Sect. 10.11, opening the field of neutrino astronomy for the next decade.

10.1 The CR, γ-Ray and Neutrino Connection

The astrophysical production of high-energy neutrinos occurs via the decay of charged pions in the *beam dump* of energetic protons in dense matter or through photoproduction from CR protons interacting with ambient photons (Sect. 8.4).

In beam-dump processes, Eq. (8.33), almost the same number of π^0, π^+, π^- are produced. The π^0 decays immediately in two γ-rays; the charged pions decay as $\pi^- \rightarrow \mu^-\overline{\nu}_\mu$ and, in turn, there is the decay $\mu^- \rightarrow e^-\overline{\nu}_e\nu_\mu$ (and the charge-conjugate reaction for the π^+). Thus, there are three neutrinos for each pion, and six neutrinos for every two γ-rays.

In photoproduction processes, Eq. (8.34), the neutrino energy from the π^+ decay is related to the parent proton energy through the relation

$$E_\nu \simeq 0.05 E_p \,, \tag{10.1}$$

This arises because the average energy of the pion in (8.34) is $f_{p\pi} \sim 0.2$ that of the energy of the parent proton, and in the π^+ decay chain, four leptons are produced, each of which roughly carries 1/4 of the pion's energy. Note also that, because of the isospin conservation (Chapter 7 of Braibant et al. 2011), the branching ratio for the process (8.34a) is about twice as large as that for (8.34b), while the energy carried by the π^0 is slightly smaller. Taking into account the fact that in the final state of (8.37), there are three neutrinos and one positron, the energy transferred to neutrinos is about $(1/3) \times (3/4) = 1/4$. The fraction of energy transferred to photons is $(2/3)+(1/3)\times(1/4) = 3/4$ (because the positron annihilates, producing additional photons). Following our simple arguments, the ratio of neutrino to photon luminosity in photoproduction processes is

$$\frac{L_\nu}{L_\gamma} = \frac{1}{3} \,. \tag{10.2}$$

Here, and in the following, under the symbol ν, we always consider the sum of neutrinos and antineutrinos.

The existence of neutrino sources not observed in γ-rays is not excluded. If a source is occulted by the presence of thick clouds or material along the line of sight to the Earth, γ-rays are absorbed while neutrinos survive.

10.2 Neutrino Detection Principle

The basic structure of a detector for cosmic neutrinos is a matrix of light detectors inside a transparent medium. This medium, such as ice or water at great depths:

- offers a large volume of free target nucleons for neutrino interactions;
- provides shielding against secondary particles produced by CRs;
- allows the propagation of Cherenkov photons emitted by relativistic particles produced by the neutrino interaction.

High energy neutrinos interact with a nucleon N of the nucleus, via either charged current (CC) weak interactions ($l = e, \mu, \tau$)

$$\nu_l + N \rightarrow l + X \tag{10.3}$$

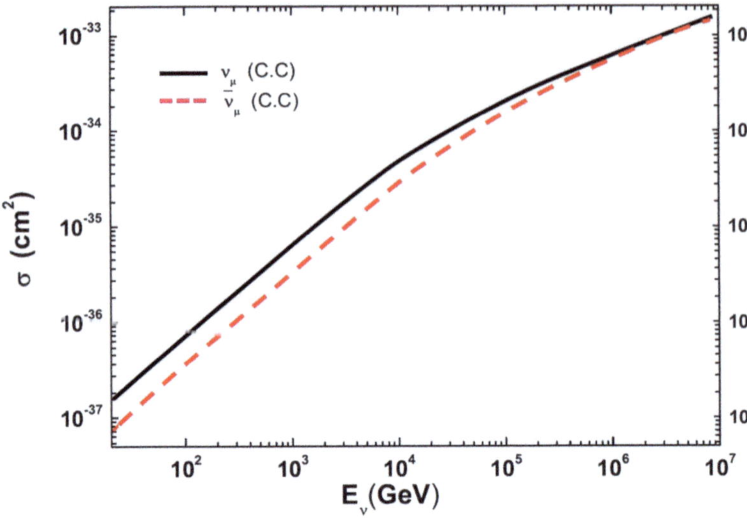

Fig. 10.1 Cross-section for ν_μ and $\overline{\nu}_\mu$ as a function of their energy according to one particular parton distribution function (CTEQ6-DIS)

or neutral current (NC) weak interactions

$$\nu_l + N \rightarrow \nu_l + X \; , \tag{10.4}$$

where the hadronic system X that is formed carries part of the incoming neutrino energy. Figure 10.1 shows the ν_μ and $\overline{\nu}_\mu$ cross-sections as a function of the neutrino energy. The neutrino cross-section is $\sigma \propto E_\nu$ up to $\sim 10^4$ GeV; at higher energies, the linear rise of the cross section starts flattening out. Outside the range measured with high precision at the HERA collider, no data constrain the quark and antiquark structure functions and the uncertainties about the total cross-section increase.

An undoubtable advantage of ν's over γ's as probes in astrophysical observations is related to their tiny cross-section. While a 1 TeV γ has an interaction length (in water) $\lambda \approx 42$ m, a ν of the same energy has $\lambda \approx 2 \times 10^9$ m. The increase of the ν cross-section with energy is such that at 1 PeV, its interaction length becomes a thousand times smaller, or 2×10^6 m. It may be seen that the ν interaction length becomes equal to the diameter of the Earth at energies on the order of 200 TeV.

Charged particles produced by the neutrino interaction travel through the medium until they either decay or interact. The mean distance traveled by secondary charged particles is called the *path length*. It depends on their energy and energy loss in the medium. Long *tracks* are produced by CC ν_μ interactions in or around the detector: the range of a 200 GeV muon corresponds to $\approx$1 km. In this case, the muon path length usually exceeds the spatial resolution of the detector, so that the trajectory of the particle can be resolved. *Showers* are induced by NC and by ν_e, ν_τ CC interactions inside the instrumented volume of the detector, Sect. 10.4.

The detection principle of operating detectors for neutrinos within the TeV–PeV range is based on the collection of the optical photons produced by the Cherenkov effect of relativistic particles (Chiarusi and Spurio 2010; Halzen 2006). The light is measured by a three-dimensional array of photomultiplier tubes (PMTs). The information provided by the number of photons detected and their arrival times are used to infer the neutrino flavor, direction and energy.

Cherenkov radiation is emitted by charged particles crossing an insulator medium with speed exceeding that of light in the medium. The coherent radiation is emitted along a cone with a characteristic angle θ_C given by $\cos\theta_C = \frac{1}{\beta n}$, where n is the refracting index of the medium and β is the particle speed in units of c. For relativistic particles ($\beta \simeq 1$) in seawater ($n \simeq 1.364$), the Cherenkov angle is $\theta_C \simeq 43°$.

The number of Cherenkov photons, N_C, emitted per unit wavelength interval, $\mathrm{d}\lambda$, and unit distance travelled, $\mathrm{d}x$, by a charged particle of charge e is given by the *Frank-Tamm formula*:

$$\frac{\mathrm{d}^2 N_C}{\mathrm{d}x\mathrm{d}\lambda} = \frac{2\pi}{137\lambda^2}\left(1 - \frac{1}{n^2\beta^2}\right), \tag{10.5}$$

where λ is the wavelength of the radiation. From this formula, it can be seen that shorter wavelengths contribute more significantly to Cherenkov radiation. The light absorption by water/ice will strongly suppress photons with wavelengths below 300 nm.

Figure 10.2 shows one typical optical module configuration (left side) used in ice and water experiments (Sect. 10.9). The PMT quantum efficiency (right side) is large within the wavelength range 300–600 nm, matching well the region in which ice and water are transparent to light.

Exercise Using the Frank-Tamm formula, evaluate whether the typical number of Cherenkov photons within the range of PMT sensitivity (300–600 nm) emitted by a relativistic ($\beta = 1$) muon per meter of water is $\sim 3.5 \times 10^4$.

10.3 Background in Large Volume Neutrino Detectors

Atmospheric muons and neutrinos are produced by CR interacting with atmospheric nuclei, Chap. 11. Here, we will show that up to $\approx$100 TeV, muons and neutrinos are produced mainly by decays of charged pions and kaons in the cascade, and their spectra are related by the kinematics of the $\pi \rightarrow \mu\nu$ and $K \rightarrow \mu\nu$ decays. Additional lower energy neutrinos are produced by muon decays. Before discussing the possible cosmic signal, we need to understand the main background sources.

The **atmospheric neutrino** flux from charged pion and kaon decays is dominated by ν_μ (Sect. 11.3). This flux is usually referred to as the *conventional atmospheric*

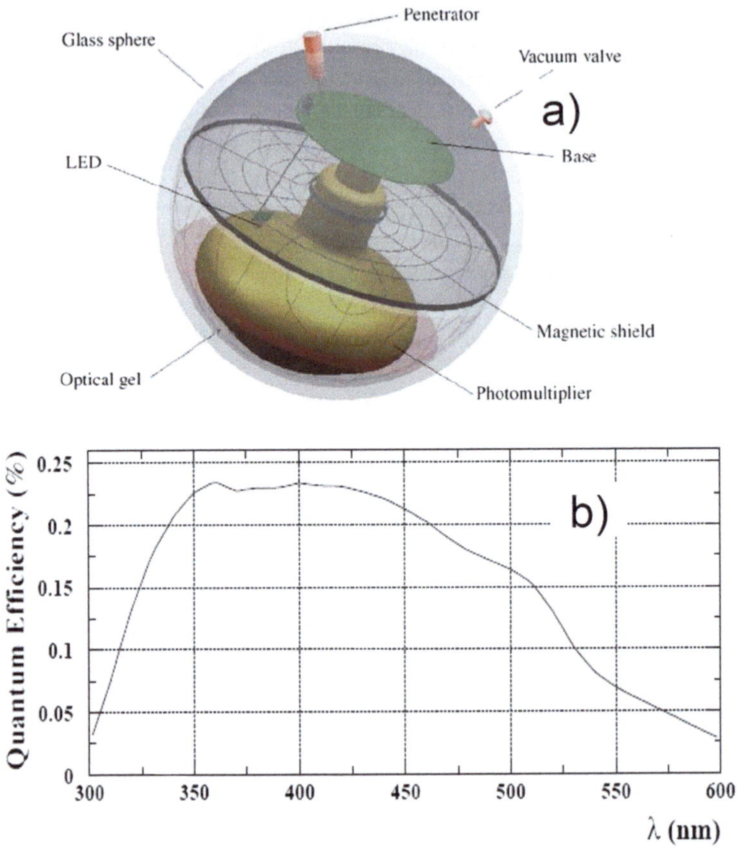

Fig. 10.2 (**a**) Sketch of an ANTARES optical module (OM). Those used by the IceCube experiment are similar. A large *hemispherical* (10 in. in diameter) *photomultiplier* (PMT) is protected by a pressure-resistant *glass sphere*. The outer diameter of the *sphere* is 43.2 cm. A mu-metal cage protects the PMT from the Earth's magnetic field. An internal *LED* is used for the calibration. (**b**) The quantum efficiency versus wavelength for PMTs commonly used in ice or water (from Hamamatsu)

neutrino flux and measured in $\mathrm{cm^{-2}\,s^{-1}\,sr^{-1}\,GeV^{-1}}$. At energies above 1 TeV and up to $\approx$100 TeV, the conventional atmospheric neutrino intensity can be expressed with a simple power-law spectrum:

$$\frac{d\Phi_\nu}{dE}(E) \propto E^{-\alpha_\nu^A} , \tag{10.6}$$

where $\alpha_\nu^A \simeq \alpha + 1$. The quantity $\alpha \simeq 2.7$ corresponds to the measured spectral index for CRs below the knee, Sect. 2.6. The ν_e flux at high energy is reduced by more than an order of magnitude with respect to the ν_μ.

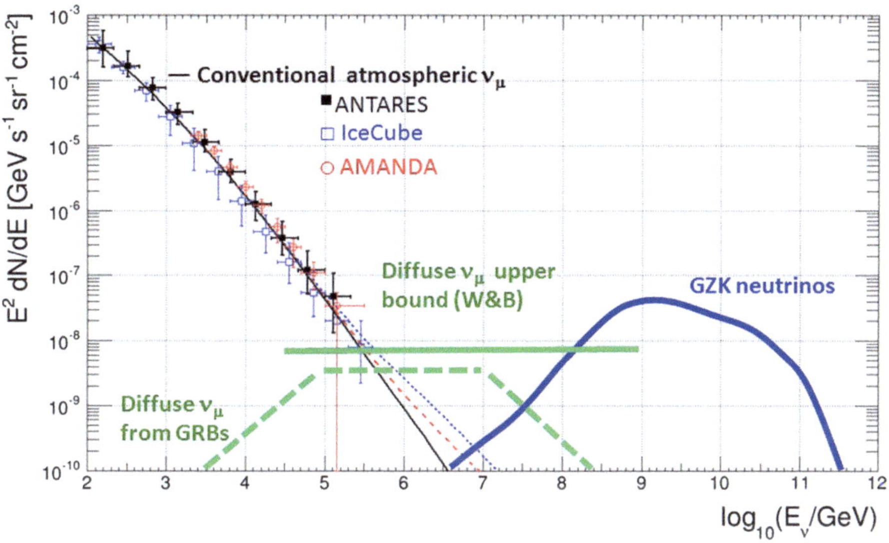

Fig. 10.3 Expected neutrino fluxes from different diffuse cosmic models and the atmospheric neutrino background. The points represent the measurements of atmospheric $\nu_\mu + \overline{\nu}_\mu$ flux as a function of the energy by three experiments. The *black line* is the expected conventional flux; the *dashed line* includes the contribution from two models of prompt neutrinos from charmed mesons decay. The *horizontal full green line* is the Waxman and Bahcall upper bound from a diffuse flux of neutrinos from extragalactic sources, Sect. 10.6.2. The *dashed green line* is the upper bound for ν_μ produced by GRBs, Sect. 10.6.3. The line with label *GZK neutrinos* indicates the possible contribution of cosmogenic neutrinos, Sect. 7.10

Interactions of primary CRs with atmospheric nuclei also produce charmed mesons. Their immediate decay yields a harder neutrino energy spectrum (*prompt neutrino flux*), Sect. 11.3.2, which is expected to exceed that of conventional neutrinos above ~100 TeV.

Figure 10.3 shows the expected flux of atmospheric ν_μ as a function of the energy. The conventional flux is shown by the black line, the contribution from two different prompt models as dashed lines. Since the spectrum is steeply falling, we have, for better visibility, multiplied the flux by E^2. The different symbols represent the measurements from experiments in ice (AMANDA, IceCube) and in seawater (ANTARES) (Sect. 10.9). Atmospheric neutrinos represent an irreducible background in experiments aimed at the detection of cosmic neutrinos.

The **atmospheric muons** (Sect. 11.3) can penetrate the atmosphere and up to several kilometers of ice/water, and they represent the bulk of reconstructed events in any large volume neutrino detector. Neutrino detectors must be located deep under a large amount of shielding in order to reduce the background. The flux of downgoing atmospheric muons exceeds the flux induced by atmospheric neutrino interactions by many orders of magnitude, decreasing with increasing detector depth, as shown in Fig. 10.4.

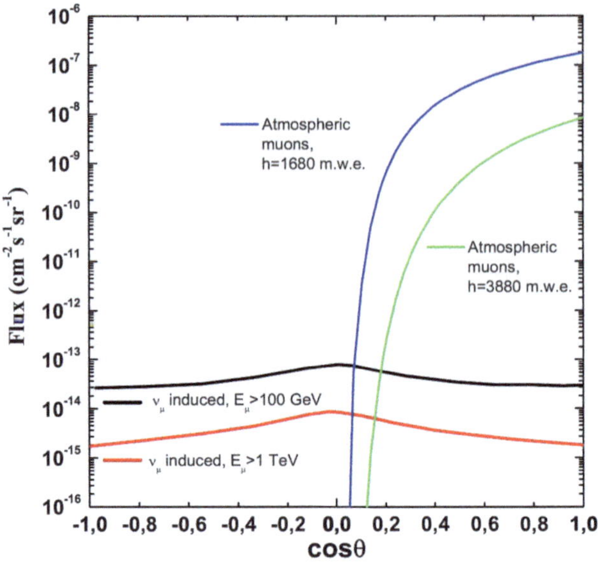

Fig. 10.4 Flux as a function of the cosine of the *zenith angle* of: (1) atmospheric muons (Sect. 11.3) for two different depths; (2) muons induced by CC interactions of atmospheric ν_μ (Sect. 11.7), for two different muon energy thresholds E_μ. Upgoing (downgoing) events have $\cos\theta < 0$ (> 0)

Atmospheric muons can be used for a real-time monitoring of the detector status and for detector calibration. However, they represent a major background source: downward-going particles wrongly reconstructed as upward-going and simultaneous muons produced by different CR primaries could mimic high-energy neutrino interactions.

10.4 Neutrino Detectors and Neutrino Telescopes

To detect a cosmic signal, huge detectors are necessary. In a large volume neutrino detector, one can distinguish between two main event classes: events with a long **track** due to a passing muon, and events with a **shower**, without the presence of a muon. Schematic views of ν_e, ν_μ and ν_τ charged current (CC) events and of a neutral current (NC) event are shown in Fig. 10.5. Neutrino and anti-neutrino reactions are not distinguishable; thus, no separation between particles and anti-particles can be made. Showers occur in all event categories shown in the figure. However, for CC ν_μ, often only the muon track is detected, as the path length of a muon in water exceeds that of a shower by more than three orders of magnitude for energies above 2 TeV. Therefore, such an event might very well be detected even if the interaction has taken place several km outside the instrumented volume, in the Earth's crust or in

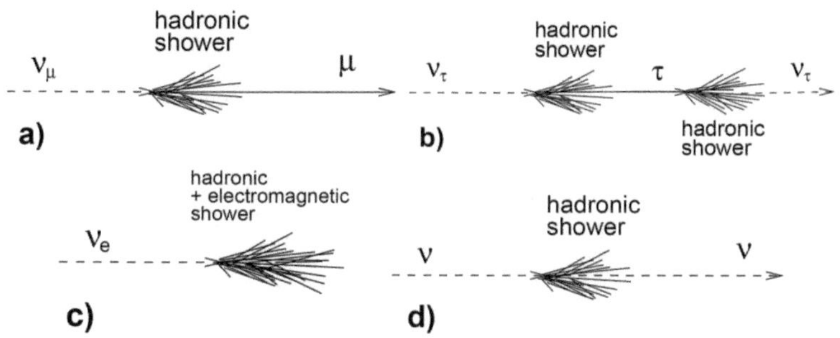

Fig. 10.5 Some event signature topologies for different neutrino flavors and interactions: (**a**) CC interaction of a ν_μ produces a muon and a *hadronic shower*; (**b**) CC interaction of a ν_τ produces a τ that decays into a ν_τ, tracing the double bang event signature. (**c**) CC interaction of ν_e produces both an EM and a *hadronic shower*; (**d**) a NC interaction produces a *hadronic shower*. Particles and anti-particles cannot be distinguished in large volume neutrino detectors

the surrounding transparent medium, provided that the muon traverses the detector. The properties of showers (total visible energy, a rough estimate of the neutrino direction) are obtained if the interaction occurs inside the instrumented volume.

Large volume detectors would identify a cosmic neutrino signal over the background using different methods:

1. For events with a track that can be accurately reconstructed, by the observation of an excess of events in a very small solid angle region $\Delta\Omega$ over the expected background.
2. For all neutrino interaction candidates, by the observation of an excess of events above a given observed energy. The expected cosmic signal is *harder* than the atmospheric neutrino spectrum. The spectrum of the cosmic signal is expected to be $\mathrm{d}\Phi_\nu/\mathrm{d}E \propto E^{-\alpha_\nu}$, with $\alpha_\nu \approx 2$ while $\alpha_\nu^A \approx 3.7$ in (10.6) for the background.

This latter method relies on the **calorimetric capabilities** of the detector. It is usually more efficient for showering events, due to the fact that most of the neutrino energy is released inside the instrumented volume. Large volume neutrino detectors can identify neutrino candidates from all directions, using a part of the detector as a *veto*, see Sect. 10.10.

The first method requires that neutrino detectors have good **tracking capabilities**: usually, only the track produced by a ν_μ CC interaction can be reconstructed with precision much better than 1°. In this case, the ν_μ detector behaves as a *"neutrino telescope"*. The sky visibility for neutrino telescopes is reduced to the bottom hemisphere to avoid contamination from the huge flux of atmospheric muons. The reason for looking for ν_μ's coming from "underneath", the ones that have crossed the Earth, stems from the need to avoid being swamped by the enormous background of atmospheric μ's. Neutrino telescopes, in contrast with every other instrument devoted to astronomy, are " downward looking".

A third method for identifying possible hadronic acceleration regions is:

3. The detection of a neutrino candidate in temporal/spatial coincidence with an external trigger, such as that resulting from a γ-ray burst observation, from a gravitational wave event, or from other transients observed by space- or ground-based observations (multimessenger strategy).

10.5 Reconstruction of Neutrino-Induced Tracks and Showers

Let us describe the different strategies for the determination of the neutrino direction and energy from track-like and shower-like events.

10.5.1 Muon Neutrino Detection

A muon crossing the detector gives a clean experimental signal, which allows for an accurate reconstruction of muon direction, closely correlated with the neutrino direction. Since neutrinos are not deflected by magnetic fields, it is possible to trace the muon back to the neutrino source. This is equivalent to traditional astronomy, in which photons point back to their sources. As an example, a muon with initial energy $E_\mu = 10\,\text{TeV}$ can travel $\sim$5 km in water keeping more than 1 TeV of residual energy. Figure 10.6 shows the path length of muons as a function of their energy. Muon energy losses and the muon range are the subject of *Extras # 5*.

The reconstruction ability for a muon track and the relation between neutrino and muon directions is essential for the concept of a neutrino telescope. The muon direction (and, optionally, muon energy) is determined by maximizing a likelihood that compares the time and position of fired PMTs with the expectation from the Cherenkov signal of a muon track. To do this, algorithms use functions that model the light propagation, giving the probability distribution function for a photon, radiated from a track with a given orientation, to reach a PMT at a given distance and orientation as a function of time. For this reason, photon scattering degrades the measurement of the muon direction.

Figure 10.7 shows the angular resolution estimated as the difference between the reconstructed and the true muon direction ($\overline{\theta}_{\mu_r\mu}$) as a function of the neutrino energy. The angular resolution estimated as the difference between the reconstructed muon and the true neutrino direction ($\overline{\theta}_{\mu_r\nu}$) is reported as well. The points are obtained with a pseudo-experiment with data produced by a Monte Carlo simulation of a detector in water similar to ANTARES (Sect. 10.9.2). The resolution $\overline{\theta}_{\mu_r\nu}$ for the neutrino is worse than $\sim$1° for $E_\nu \lesssim 0.5\,\text{TeV}$ because of kinematics. It decreases

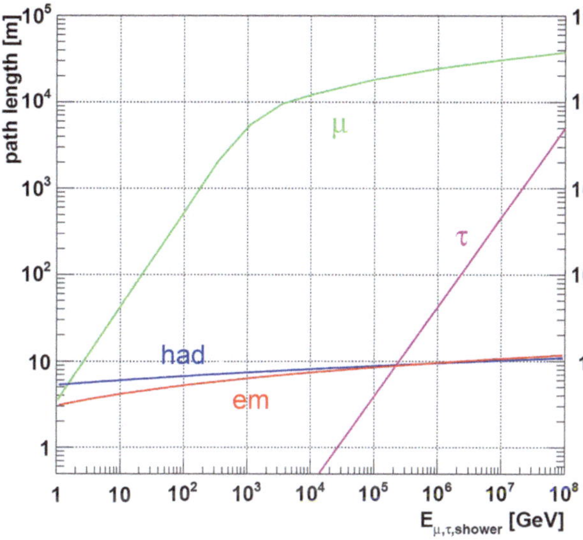

Fig. 10.6 Path length of particles produced by neutrino interactions in water: muons (μ), taus (τ), electromagnetic (em) and hadronic (had) showers, versus their respective energy. The shower lengths are calculated using a shower profile parameterization

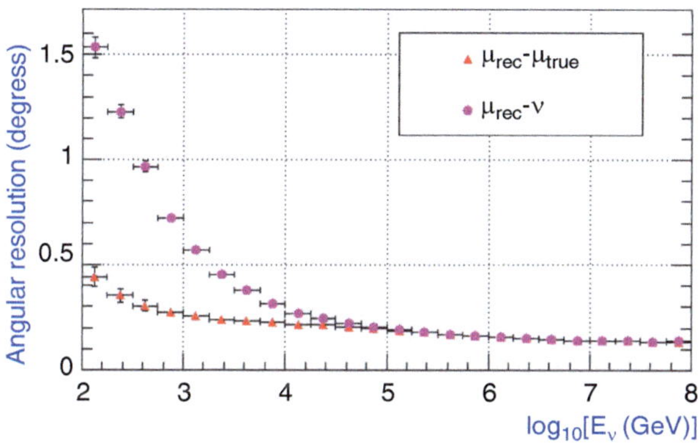

Fig. 10.7 Angular resolution (evaluated with a Monte Carlo simulation) as a function of event energy for the underwater ANTARES detector. The average differences between the true and reconstructed muon directions, as well as the difference with respect to the neutrino direction, is shown

as $\overline{\theta}_{\mu_r \nu} \propto E_\nu^{-1/2}$ at increasing energies, mainly depending on the reconstruction capabilities of the neutrino telescope and the propagation medium. The maximum attainable precision is on the order of $\sim 0.2°$. Multiple scattering affects the muon direction negligibly at these energies.

10.5.2 Showering Events

Electron Neutrinos A high-energy electron resulting from a charged current ν_e interaction radiates a photon via bremsstrahlung after a few tens of cm of water/ice (the radiation length in water is ~36 cm): this process leads to the development of an electromagnetic (EM) cascade. The longitudinal extension of an EM shower is on the order of a few meters and the lateral extension is negligible compared to the longitudinal. The shower length, defined as the distance within which 95% of the total shower energy has been deposited, slowly increases with the electron energy, Fig. 10.6. For instance, a 10 TeV electron induces a shower of ~8 m in length. Such a shower is short compared to the spacing of the PMTs in any existing or proposed high-energy neutrino detector. EM showers represent, to a good approximation, a point source of Cherenkov photons that are emitted almost isotropically along the shower axis. For this reason, pointing accuracy for showering events is greatly inferior, and energy determination greatly superior, to that achieved in the ν_μ channel. Additional photons are produced by the hadronic system X in (10.3) at the ν_e interaction vertex.

Neutral Current and Hadronic Showers The NC channel gives the same signature for all neutrino flavors. In this channel, a fraction of the interaction energy is always carried away unobserved by the outgoing neutrino, and therefore the error on the reconstructed energy of the primary neutrino increases accordingly. Even though EM and hadronic showers are different from each other in principle, the ν_e CC and the ν_x NC channels ($x = e, \mu$ or τ) are not distinguishable in large volume detectors. Hadronic cascades suffer event-to-event fluctuations that are much more important with respect to the EM ones. The dominant secondary particles in a hadronic shower are pions; kaons, protons or neutrons occur in variable fractions. Muons (from pion decay) can be present as well: as these usually leave the shower, they contribute significantly to the fluctuations. Monte Carlo simulations show that the longitudinal profile of hadronic showers is very similar to the EM one, Fig. 10.6.

Tau Neutrinos For ν_τ CC interactions, the produced τ-lepton travels some distance (depending on its energy) before decaying and producing a second shower. The Cherenkov light emitted by the charged particles in the showers can be detected if both the ν_τ interaction and the τ decay occur inside the instrumented volume of the detector. Below 1 PeV, the ν_τ CC channels (except for the case in which the τ produces a muon) also belong to the class of showering events, because the τ track cannot be resolved.

The τ lepton has a short lifetime, and within the energy range of interest, it travels (depending on the Lorentz factor $E_\tau/m_\tau c^2$) from a few meters up to a few kilometers before decaying (see Fig. 10.6). Most of the possible τ decay modes include the generation of a hadronic or an electromagnetic cascade. Thus, if the track of the τ is long enough to allow a separation of the primary interaction of the ν_τ from the decay of the tau (typically for τ energies above 1 PeV, see Fig. 10.6), the expected signatures for the ν_τ CC events are those of a shower, plus a track, followed by a second shower. This signature is called a *double bang event*, if the

Cherenkov light emitted by the charged particles in the first shower can be detected and separated from the light emitted by the particles produced in the τ decay.

10.6 Cosmic Neutrino Flux Estimates

The detection of neutrinos from a CR accelerator candidate would provide unambiguous ("smoking gun") evidence that hadrons are accelerated. Gamma-ray telescopes, at present, are providing essential information for neutrino astronomy, because they can tell where and/or when (in the case of transient sources or burst activities) to look for neutrinos. The specific characteristics (light curves and energy spectra) of each γ-ray source, either steady or transient like GRBs, can be used to make specific predictions for neutrino signals under the hypothesis of hadronic γ-ray production and to optimize their searches. In this section, we present some theoretical predictions concerning galactic and extragalactic neutrino source candidates, based on observations and constraints from γ-ray observations or from the CR flux. The status of the experimental results is presented in Sect. 10.10.

10.6.1 A Reference Neutrino Flux from a Galactic Source

If the hadronic mechanism is at work in a galactic source (for instance, in a SNR accelerating CRs), a flux of neutrinos comparable to that of γ-rays could be expected. It is useful to define a sort of *reference neutrino flux* from a galactic source.

The SNR RX J1713.7-3946 has been the subject of large debates about the nature of the process (leptonic or hadronic) that originates its γ-ray spectrum, Sect. 9.8. This source has been observed with high statistics by the HESS telescope up to $\sim$80 TeV, with a spectrum that can be reasonably well described by a power law with an exponential cut-off

$$E^2 \frac{\mathrm{d}\Phi_\gamma}{\mathrm{d}E} = \Phi^0_\gamma \exp[-(E/E_c)^{1/2}] \,, \tag{10.7}$$

where $\Phi^0_\gamma = 1.8 \times 10^{-11}\,\mathrm{TeV\,s^{-1}\,cm^{-2}}$ and the cut-off parameter $E_c = 3.7\,\mathrm{TeV}$, Fig. 10.8.

Based on the measured HESS γ-ray flux, or starting from a primary proton population with an energy spectrum presenting an exponential cut-off, different models exist that give predictions about the neutrino flux. Several assumptions are usually made in the modeling of the neutrino spectra:

- no significant contribution of nonhadronic processes to the measured γ-rays;
- no γ-ray absorption within the source, i.e., radiation and matter densities are sufficiently low for most of the γ-ray photons to escape;

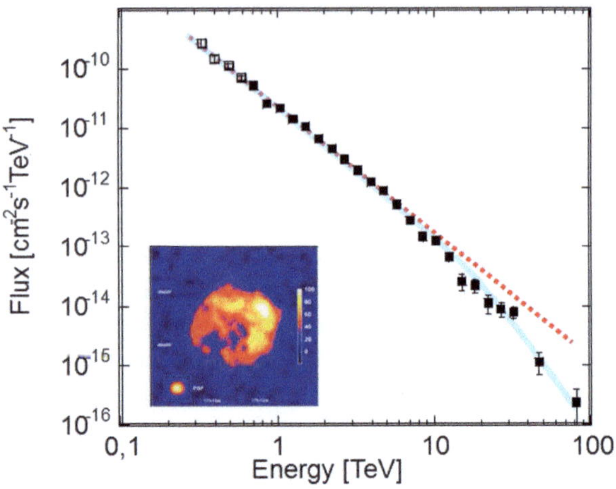

Fig. 10.8 HESS Measurement of the γ-ray flux from RX J1713.7-3946. It corresponds to the high-energy tail of Fig. 9.13. The *full line* represents the result of a fit using Eq. (10.7), while the *dashed line* represents the fit without the exponential cut-off. The fits are used to compute the ν_μ flux in the case of a hadronic mechanism

- $\pi^\pm$ decay before interacting (matter density is low) and also the $\mu^\pm$ decay without significant energy loss (magnetic field is low);
- the size of the emitting region within each source is large enough that oscillations (Sect. 12.8) will produce a fully mixed neutrino signal at the Earth, i.e., $\nu_e : \nu_\mu : \nu_\tau = 1 : 1 : 1$ (see Sect. 12.9).

Grossly, the expected number of ν_μ is equal to that of γ-rays, as in proton-proton interactions, the same number of π^0, π^+, π^- are produced. Two photons arise from the π^0 decay, and six neutrinos from the decay chains of the two charged pions. At a large distance from the source, two out of six arrive as ν_μ when neutrino oscillations are considered. However, the flux of high-energy neutrinos is about a factor of two lower with respect to the γ-rays of the same energy, because of the kinematics. In the decay of charged pions, a larger fraction of kinetic energy is transferred to the muon: in the pion rest frame, ~110 MeV to the muon and ~30 MeV to the neutrino (see Sect. 11.3.1). The results of the predictions agree on the fact that

$$\Phi_{\nu_\mu}(E) \simeq \frac{1}{2}\Phi_\gamma(E) \longrightarrow \frac{d\Phi_\nu}{dE_\nu} = \Phi_\nu^0 E_\nu^{-2} = 10^{-11} E_\nu^{-2}\,\mathrm{TeV\,s^{-1}\,cm^{-2}}\,. \tag{10.8}$$

Detailed computations produce energy-dependent normalization factors in (10.8) (Stegmann et al. 2007; Vissani and Aharonian 2012).

Predictions of the neutrino flux are available for many galactic sources. As reference neutrino flux for our following considerations, we will use Eq. (10.8) derived from RX J1713.7-3946. Higher neutrino fluxes could be expected in the case of hidden γ-ray sources. Lower neutrino fluxes are expected if one or more of the above conditions are not fulfilled.

10.6.2 Extragalactic Diffuse Neutrino Flux

The prediction of extragalactic high energy neutrinos is a direct consequence of CR observations (Becker 2008; Halzen 2006). As for the origin of UHE Cosmic Rays, Active Galactic Nuclei (AGN) or γ-ray bursts (GRBs) are the principal candidates as extragalactic neutrino sources.

Different relations between the observed UHECR flux and the possible flux of diffuse neutrinos exist. We consider here the simple approach to the problem found in Waxman and Bahcall (1998). The computation of the upper limit on the neutrino flux is based on the measured CR flux at 10^{19} eV and an assumed flat E^{-2} injection spectrum at sources. The authors derived the emissivity of UHECR in the Universe, 10^{44} erg/(Mpc^3 year) within the range 10^{19}–10^{21} eV, and with some assumptions about the interaction mechanism near sources where they estimated a maximum neutrino flux.

With the information provided in the previous chapters, we can reproduce this result. For a CR flux $\Phi_{CR}(E)$, Eq. (2.33a) gives the corresponding energy density $\rho(> E)$. The energy density, neglecting the constant factors, can be expressed in terms of

$$\rho(> E) \propto \int E\Phi_{CR}(E)\mathrm{d}E = E^2\Phi_{CR}(E) \,. \tag{10.9}$$

From Fig. 2.7, we obtain that the differential CR flux $\Phi_{CR}(E)$ decreases by $\sim$28 orders of magnitude when the minimum energy E increases from $\sim 10^9$ to 10^{19} eV. A more precise factor (1.4×10^{-28}) is obtained comparing Eq. (4.52) at an energy $E_1 = 10^{19}$ eV with Eq. (2.20a) at $E_0 = 10^9$ eV. Thus,

$$\frac{\rho(> E_1)}{\rho(> E_0)} = \left(\frac{E_1}{E_0}\right)^2 \cdot \frac{\Phi_{CR}(E_1)}{\Phi_{CR}(E_0)} \,. \tag{10.10}$$

By inserting the numerical values,

$$\rho_{19} \equiv \rho(> E_1) = 1\,\frac{\mathrm{eV}}{\mathrm{cm}^3} \cdot (10^{10})^2 \cdot (1.4\;10^{-28}) \simeq 1.4\;10^{-8}\,\frac{\mathrm{eV}}{\mathrm{cm}^3} \,. \tag{10.11}$$

This quantity corresponds to $\rho_{19} = 0.7 \times 10^{54}$ erg/Mpc^3. Taking into account that the CR are generated all over the cosmic evolution during a time comparable to the Hubble time ($t_H \sim 10^{10}$ y), then the emissivity of UHECR in the Universe corresponds to $(\rho_{19}/t_H) \simeq 0.7 \times 10^{44}$ erg/(y Mpc^3), in agreement with the Waxman and Bahcall (1998) derivation.

The next observation is that a fraction of these CRs would undergo photoproduction interactions (8.34) at their sources and a fraction of their energy loss $\epsilon < 1$ would be transferred to secondary γ-rays and neutrinos. According to (10.2), in photoproduction processes, the photon luminosity is about a factor three larger than that of neutrinos, thus to the latter is transferred about (1/4) of the CR energy. As the muon neutrinos (as usual, we refer to the sum of $\nu_\mu + \overline{\nu}_\mu$ as muon neutrinos) give a clear signature in a neutrino telescope, we tailor the calculation to this flavor, which corresponds to (2/3) of the emitted neutrinos in (8.37). We can define the maximum ν_μ flux $[E^2 \Phi_{\nu_\mu}(E)]_{\max}$ when $\epsilon = 1$ as

$$[E^2 \Phi_{\nu_\mu}(E)]_{\max} \simeq \left(\frac{2}{3}\right)\left(\frac{1}{4}\right)\left(\frac{c}{4\pi}\right)\rho_{19} = 0.6 \times 10^{-8} \frac{\text{GeV}}{\text{cm}^2\,\text{s sr}} . \tag{10.12}$$

The factor $(c/4\pi)$ converts a number density into a flux, as explained in Eq. (2.18a).

The maximal neutrino flux (10.12) must be corrected for two effects: (1) the redshift energy loss of neutrinos produced at cosmic time $t < t_H$ and the CR generation rate per unit of volume. These effects introduce a correction factor $\xi_z = (1+z)^3$, which corresponds to a factor $\xi_z \simeq 3$ when the cosmological evolution of the sources is considered (Waxman and Bahcall 1998). (2) Neutrino oscillations change the flavor of neutrinos during propagation, decreasing by a factor of two the number of cosmic muon neutrinos arriving on Earth, (Sect. 12.8). These two effects give a factor $\xi_z/2$, and the upper bound on the flux of neutrinos of a given flavor is

$$[E^2 \Phi_{\nu_\mu}(E)]_{\max} = 0.9 \times 10^{-8} \frac{\text{GeV}}{\text{cm}^2\,\text{s sr}} . \tag{10.13}$$

This upper limit is shown in Fig. 10.3 as a green line. The all-flavor upper limit (corresponding to the total $\nu_e + \nu_\mu + \nu_\tau$ flux) is a factor of three larger. The result (10.13) is lower than that reported in the original paper (Waxman and Bahcall 1998); the reason is that neutrino oscillation effects were not considered, and that the new measurements of the UHECRs from the PAO and TA experiments (Sect. 7.8) give a lower flux than that considered in 1998.

The diffuse flux is expected to exceed that of the atmospheric neutrino at energies above 100 TeV. This diffuse cosmic neutrino flux can be detected above the background using the calorimetric features of detectors, as mentioned in Sect. 10.3.

10.6.3 Neutrinos from GRBs

GRBs (Sect. 8.11) have been proposed as possible candidate sources for CRs above 10^{18} eV (Waxman and Bahcall 1997). In the fireball model, the observed electromagnetic radiation is explained by highly relativistic outflows of material, most likely collimated in jets pointing towards the Earth. Shock fronts emerge in the outflows in which electrons are accelerated. Within the framework of the fireball model, protons can also be shock-accelerated, yielding high-energy neutrinos that accompany the electromagnetic signal of the burst.

The neutrino fluence F_ν (fluence measures the flux integrated over time in units GeV/cm^2) can be obtained from theoretical models, starting from the GRB luminosity by a convolution of the accelerated proton distribution with the photon energy density, U_γ. The predictions are strongly constrained by the observed average burst fluence $F_\gamma \sim 3.6 \times 10^{-3}\,\text{GeV/cm}^2$. Assuming that the γ-rays are mostly produced by hadrons, from F_γ, the average neutrino fluence is obtained

$$F_\nu \sim 1.2 \times 10^{-3}\,\text{GeV/cm}^2 \tag{10.14}$$

using (10.2).

The maximum of this diffuse muon neutrino flux from GRBs can be obtained by the average fluence (10.14) with the GRB rate, r_{GRB} [s^{-1}], which is estimated to be 667 per year over 4π sr (Abbasi et al. 2011). Thus,

$$E^2[\Phi_{\nu_\mu}(E)]^{\text{GRB}}_{\text{max}} = \frac{F_{\nu_\mu} \times r_{\text{GRB}}}{4\pi} \simeq 1.5\frac{1.2\;10^{-3} \times 667}{4\pi(3.15\;10^7)} = 3\;10^{-9} \quad \frac{\text{GeV}}{\text{cm}^2\,\text{s}\,\text{sr}}\,, \tag{10.15}$$

where the factor 1.5 includes the effects of neutrino oscillations and cosmic evolution, as in Sect. 10.6.2.

The original computation of neutrino flux from GRBs (Waxman and Bahcall 1997) was based on the assumption of Fermi-accelerated protons in the relativistic ejecta of the burst interacting with the associated photon field to produce pions via Eq. (8.34). The subsequent decay of charged pions and muons leads to the emission of high-energy neutrinos. In the following, we work out the neutrino spectrum using simple approximations. According to the Band spectrum distribution (8.57), a doubly broken power-law spectrum for the neutrinos is obtained.

Because of the beamed emission of GRBs, both the accelerated protons and the photons are moving towards the observer collimated into a narrow cone of opening angle θ along the jet axis. The $p\gamma$ collisions are not head-on, i.e., $\cos\theta = -1$, as in (7.18), but are grazing collisions occurring at small angles, which can be approximated with that of the opening angle θ of the jet. The resonance energy for reaction (8.34) is given by (7.18):

$$(E^r_p E^r_\gamma) = \frac{m^2_\Delta - m^2_p}{2(1-\cos\theta)} \simeq \frac{0.64\,\text{GeV}^2}{\theta^2}\,. \tag{10.16}$$

Using relativistic arguments, in Sect. 5.8, it was shown that the relation $\theta \propto 1/\Gamma$ between the opening angle θ of the beamed emission in the laboratory frame and its Lorentz factor Γ holds. Using this relation, we can rewrite (10.16) as

$$E^r_p E^r_\gamma = 0.64\Gamma^2\,\text{GeV}^2\,. \tag{10.17}$$

Note that we again obtained the same Γ^2 factor derived in (8.25). This is a relativistic feature due to the change from the center-of-mass reference frame to the laboratory frame (the observer on our Earth). This Γ^2 factor has important consequences for the neutrino spectrum, because, from electromagnetic observations,

$\Gamma \sim \mathcal{O}(100)$. As the γ-ray spectrum (8.57) has a change of slope at ~1 MeV, inserting $E_\gamma^r = E_b = 1$ MeV, we obtain

$$E_p^r = \frac{0.6\Gamma^2 \,\mathrm{GeV}^2}{E_b} \simeq 6\,10^{15}\,\mathrm{eV}. \tag{10.18}$$

Thus, considering the relation (10.1) between the neutrino and the proton energy in photoproduction processes, we expect a break in the neutrino spectrum around

$$E_\nu^b \simeq 0.05 E_p^r \sim 10^{14}\,\mathrm{eV} \sim 100\,\mathrm{TeV}\,. \tag{10.19}$$

The neutrino spectrum continues to higher energies, and neutrinos in the PeV range are expected from the model.

The fluence $F_\nu(E)$ as a function of the neutrino energy can be derived assuming a Fermi-like spectrum for the proton energy distribution, $\mathrm{d}N_p/\mathrm{d}E_p \propto E_p^{-2}$. The photon energy distribution U_γ (in GeV/cm^3) is related to the observed γ-ray spectral index distribution (8.57) by

$$\begin{aligned} U_\gamma(E) = \frac{c}{4\pi} E^2 \frac{\mathrm{d}N_\gamma}{\mathrm{d}E} &\propto E^2 \cdot E^{-\alpha} \quad \text{for } E < E_b \\ &\propto E^2 \cdot E^{-\beta} \quad \text{for } E > E_b\,. \end{aligned} \tag{10.20}$$

For a GRB at a distance D, $F_{\nu_\mu}(E)$ can be obtained by the integration

$$F_{\nu_\mu}(E) \propto \frac{1}{4\pi D^2} \int \frac{\mathrm{d}N_p}{\mathrm{d}E_p} \cdot \sigma_{p\gamma} \cdot f_{p\pi} \cdot f_{\pi\nu} \cdot U_\gamma(E) \cdot \mathrm{d}E_p\,, \tag{10.21}$$

where, in addition to the already defined quantities, $\sigma_{p\gamma}$ represents the photoproduction cross-section as given in Fig. 7.7, $f_{p\pi}$ is the fraction of the proton energy transferred to pions, and $f_{\pi\nu}$ that transferred from pions to ν_μ. With the simple approximation that $(\mathrm{d}N_p/\mathrm{d}E_p)\sigma_{p\gamma} \sim 0$ outside the resonance region and $(\mathrm{d}N_p/\mathrm{d}E_p)\sigma_{p\gamma} \sim \sigma_0/E_p^r$ within a small range around the resonance energy E_p^r, the neutrino fluence is (considering only the energy dependence and including all the constants in the coefficient $a = \frac{f_{p\pi} f_{\pi\nu}\sigma_0}{4\pi D^2 E_p^r}$)

$$F_{\nu_\mu}(E) \propto a E^2 \cdot E^{-\alpha} \propto E^1 \quad \text{for } E < E_\nu^b\,, \tag{10.22a}$$

$$\propto a E^2 \cdot E^{-\beta} \propto E^0 \quad \text{for } E_\nu^b < E < E_\nu^s\,, \tag{10.22b}$$

$$\propto a E^2 \cdot E^{-\beta} \cdot E^{-1} \propto E^{-1} \quad \text{for } E > E_\nu^s\,. \tag{10.22c}$$

Pay attention to the last relation. At very high energies, the synchrotron radiation loss of secondary pions and muons prior to their decay also affects the neutrino spectrum. This energy loss becomes important when the synchrotron time scale is comparable to the pion's lifetime. It can be computed that this occurs at energies E_p^s about 100 times larger than that of (10.18). Thus, for $E > E_\nu^s \simeq 100\, E_\nu^s \simeq 10^{16}$ eV, an energy loss term proportional to E^{-1} must be included.

The expected diffuse ν_μ spectrum from GRBs is thus given by

$$E^2[\Phi_{\nu_\mu}(E)]^{\mathrm{GRB}} = \frac{F_{\nu_\mu}(E) \times r_{\mathrm{GRB}}}{4\pi} , \tag{10.23}$$

where the maximum given by (10.15) occurs in the energy interval between $E_\nu^b = 100\,\mathrm{TeV}$ and $E_\nu^s = 10\,\mathrm{PeV}$. The fluence increases linearly with energy when $E_\nu < E_\nu^b$, and decreases linearly with energy when $E_\nu > E_\nu^s$, as given by (10.22a)–(10.22c) and as shown in Fig. 10.3 with a dashed line. Consistent with the assumption, the diffuse contribution from GRBs is below the upper limit derived in Sect. 10.6.2. It should be noted that, from the experimental point of view, the contribution of GRBs to the diffuse flux is much easier to detect, because neutrinos can be observed in spatial and temporal coincidence with the electromagnetic observation of a burst.

As the normalization parameters in (10.22a)–(10.22c) are quite general and based on average quantities, the prediction derived following (Waxman and Bahcall 1997) does not depend on the specific GRB. This, of course, cannot be completely true, and theoretical models exist able to calculate individual neutrino fluxes for the observed bursts (Guetta et al. 2004). Recently, the development of dedicated Monte Carlo algorithms with a more detailed treatment of the particle physics involved in the calculation of the neutrino spectra reduces the resulting neutrino flux predictions compared to the above description (Adrián-Martínez et al. 2013). These particle physics processes include an additional high-energy component from K decays, the full energy-dependent photohadronic interaction cross-section, the energy dependence of the mean free path of protons, individual treatment of secondary particles (including energy losses), the use of the real average photon energy instead of the peak energy of the photon distribution, and detailed neutrino mixing. The combination of all these effects gives rise to a prediction concerning the neutrino yield that is about one order of magnitude below the result derived above. This, without new assumptions on the nature of GRBs in general, but making use in greater detail of known physics, together with the paradigm of the fireball model.

10.7 Why km^3-Scale Telescopes?

In this section, we use a simple calculation to work out why a detector having a volume of at least a cubic-kilometer is needed to detect cosmic neutrinos. In addition, we derive the minimum number of optical sensors in the instrumented volume required to track the events. In the following, we consider the reference neutrino flux for a galactic source given by Eq. (10.8). The event rate can be obtained by defining the *neutrino effective area*.

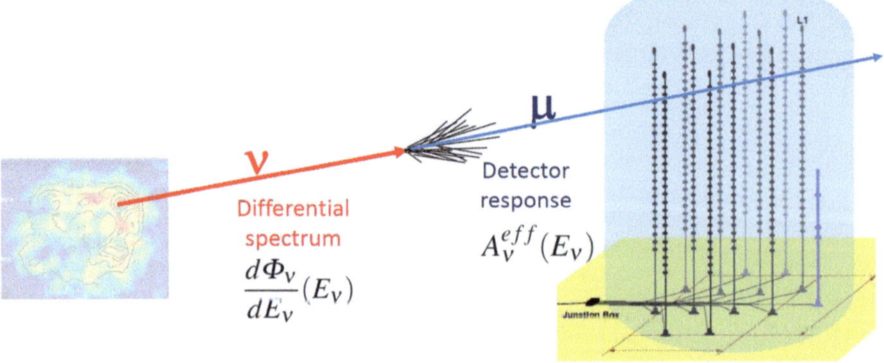

Fig. 10.9 The number of observed events in a large neutrino detector is given by the convolution of the differential neutrino flux and the effective neutrino area of the *detector*

The event rate during the observation time T can be expressed in terms of:

$$\frac{N_\nu}{T} = \int \mathrm{d}E_\nu \cdot \frac{\mathrm{d}\Phi_\nu}{\mathrm{d}E_\nu}(E_\nu) \cdot A_\nu^{\mathrm{eff}} \, , \tag{10.24}$$

where A_ν^{eff} (units: cm^2) is called the *effective area* of the neutrino detector. It corresponds to the quantity that, convoluted with the neutrino flux, gives the event rate, Fig. 10.9. A_ν^{eff} depends on the neutrino flavor and interaction type (if the neutrino interaction yields a muon or induces a cascade, the latter either through a CC or NC interaction); on the neutrino energy and incoming direction; on the status of the detector; and on the cuts that each particular analysis uses for the suppression of the background. Only detailed and dedicated Monte Carlo simulations can determine the neutrino effective area. In the following, we describe the ingredients necessary to construct, in a simplified analytic method, the effective area, $A_{\nu_\mu}^{\mathrm{eff}}(E_\nu)$, for the ν_μ charged current channel, assuming only a dependence on the neutrino energy E_ν.

To estimate $A_{\nu_\mu}^{\mathrm{eff}}(E_\nu)$ that embeds all the "complications" of the neutrino detection, we consider (Chiarusi and Spurio 2010) ν_μ interacting outside the instrumented volume of the telescope. To observe the event:

1. the ν_μ must survive the Earth's absorption and interact sufficiently near to the detector;
2. the daughter muon must reach the instrumented volume;
3. the muon must produce enough light to trigger the detector and to allow for the track reconstruction.

Thus, the effective neutrino area corresponds to

$$A_{\nu_\mu}^{\mathrm{eff}}(E_\nu) = A \cdot P_{\nu\mu}(E_\nu, E_{\mathrm{thr}}^{\mu}) \cdot \epsilon \cdot e^{-\sigma(E_\nu)\rho N_A Z(\theta)} \, . \tag{10.25}$$

Let us consider all the terms of Eq. (10.25).

- A [cm^2] is the geometrical projected detector surface.
- $P_{\nu\mu}(E_\nu, E^\mu_{\rm thr}) \equiv P_{\nu\mu}$ represents the probability that a neutrino with energy E_ν produces a muon arriving with a residual threshold energy $E^\mu_{\rm thr}$ at the detector. $P_{\nu\mu}$, in turn, can be expressed in terms of:

$$P_{\nu\mu} = N_A \int_0^{E_\nu} \frac{{\rm d}\sigma_\nu}{{\rm d}E_\mu}(E_\mu, E_\nu) \cdot R_{\rm eff}(E_\mu, E^\mu_{\rm thr}){\rm d}E_\mu \,, \tag{10.26}$$

where N_A is the Avogadro number, ${\rm d}\sigma_\nu/{\rm d}E_\mu$ is the differential neutrino cross-section for production of a muon of energy E_μ, and $R_{\rm eff}(E_\mu, E^\mu_{\rm thr})$ is the muon range for a muon of energy E_μ arriving at the detector with energy $E^\mu_{\rm thr}$. These quantities can be obtained with analytical approximations or with full Monte Carlo simulations. For our purposes, the following approximation is adequate (Gaisser et al. 1995):

$$P_{\nu\mu} \simeq 1.3\ 10^{-6} E_\nu^{2.2}, \quad \text{for } E_\nu = 10^{-3} - 1\,\text{TeV}\,, \tag{10.27a}$$

$$\simeq 1.3\ 10^{-6} E_\nu^{0.8}, \quad \text{for } E_\nu = 1 - 10^3\,\text{TeV}\,. \tag{10.27b}$$

The dependence of $P_{\nu\mu}$ in the two energy regimes reflects the energy dependence of the neutrino cross-section, $\sigma \propto E_\nu$, and the effective muon range, which depends linearly on the muon energy up to ~1 TeV. Above 1 TeV, muon radiative losses become dominant and the range is almost energy-independent. At higher energies, the cross-section is no longer linear as a function of energy and $P_{\nu\mu} \propto E_\nu^{0.4}$.
- The quantity ϵ corresponds to the fraction of muons with energy $E^\mu_{\rm thr}$ that are *detected*. In general, $\epsilon = \epsilon_t \cdot \epsilon_r \cdot \epsilon_c$, where ϵ_t represents the trigger efficiency (the muon must produce a sufficient number of photons to trigger the apparatus), ϵ_r the reconstruction efficiency and ϵ_c the probability of passing the analysis selection cuts. The quantity ϵ can be derived for each experiment only by detailed Monte Carlo simulations that take into account all the features of the detector.
- The term $e^{-\sigma(E_\nu)\rho N_A Z(\theta)}$, where $\sigma(E_\nu)$ is the total neutrino cross-section, (ρN_A) the target nucleon density and θ the neutrino direction with respect to the nadir, takes into account the absorption of neutrinos crossing along a path $Z(\theta)$ in the Earth. From the nadir $Z(0) = 6.4 \times 10^8$ cm, the absorption becomes not negligible for $\sigma > 10^{-34}$ cm^2, see Fig. 10.1, i.e., for $E_\nu > 100$ TeV.

To solve (10.24) analytically in our simplified model, we assume that muons arriving at the detector with residual energy > 1 TeV have constant probability ϵ to produce enough light to be triggered, reconstructed and to pass the analysis selection criteria. With these approximations, the event rate for the reference neutrino flux

$$\frac{N_{\nu_\mu}}{T} = \int_{1\text{ TeV}}^{10^3\text{ TeV}} \mathrm{d}E_\nu \cdot (\Phi_\nu^0 E_\nu^{-2}) \cdot A \cdot (P_\circ E_\nu^{0.8}) \cdot \epsilon = 0.5\ 10^{-16} \cdot A \cdot \epsilon \ \text{cm}^{-2}\,\text{s}^{-1}\,, \tag{10.28}$$

where $P_\circ = 1.3 \times 10^{-6}$ as in (10.27b). Assuming a cross-sectional area $A = 1\,\text{km}^2$ and $\epsilon \simeq 0.1$, the effective area at $E_\nu = 1$ TeV roughly corresponds to $A_{\nu_\mu}^{\text{eff}}(1\text{ TeV}) \simeq 10^3\,\text{cm}^2$ and the number of expected events is ~1.5 year. This expected rate is in agreement with more detailed computations.

According to (10.28), a detector of the km^3-size is required to detect a few neutrino interactions/year from a galactic source, based on neutrino flux models constrained by the TeV γ-ray observations. The effect of the possible exponential cut-off, as in (10.7), is to decrease the event rate by a factor of 5–10.

This expected signal is not background-free. This means that when considering the events from a point-like source arriving from an angular cone of semi-aperture ψ from the source direction, events resulting from atmospheric neutrinos have to be considered. This irreducible background is described by a differential flux $\frac{\mathrm{d}^2\Phi_\nu^{\text{Atm}}}{\mathrm{d}E_\nu \mathrm{d}\Omega}$ (Chap. 11). The rate of background events is given by

$$\frac{N_B}{T} = \int\int \mathrm{d}E_\nu \cdot \frac{\mathrm{d}^2\Phi_\nu^{\text{Atm}}}{\mathrm{d}E_\nu \mathrm{d}\Omega}(E_\nu, \theta) \cdot A_{\nu_\mu}^{\text{eff}}(E_\nu) \cdot \mathrm{d}\Omega \tag{10.29}$$

and it depends on the considered solid angle $\mathrm{d}\Omega$. It is easy to derive that, for small apertures, $\mathrm{d}\Omega \propto \psi^2$, and thus also that the background rate $N_B/T \propto \psi^2$. If the observational angle ψ is too large (i.e., >0.5°), the background could dominate over the signal. One possibility of increasing the signal-to-background ratio is to use the estimated energy of the events. As the signal behaves as $\mathrm{d}\Phi_\nu/\mathrm{d}E_\nu \propto E^{-2}$ and the background as $\mathrm{d}\Phi_\nu^{\text{Atm}}/\mathrm{d}E_\nu \propto E^{-3.7}$, the background can be reduced by selecting high-energy events.

10.7.1 The Neutrino Effective Area of Real Detectors

The energy-dependent effective area $A_\nu^{\text{eff}}(E_\nu)$ is obtained using Monte Carlo (MC) simulations for each experiment. $A_\nu^{\text{eff}}(E_\nu)$ is, in addition, usually different for different analyses, as it depends on the selection criteria. This quantity is also used to compare the sensitivity of different experiments for similar physics studies.

Figure 10.10 shows the average effective areas computed for the IceCube and ANTARES detectors (Sect. 10.9) for point-source searches of cosmic neutrinos. Since IceCube is located at the South Pole, the zenith angle θ and declination angle δ are simply related as $\theta = \delta + 90°$. At a different location, due to the

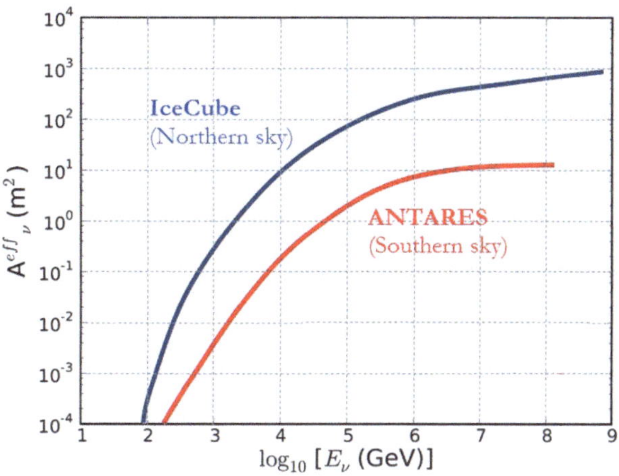

Fig. 10.10 Neutrino effective area as a function of the true simulated neutrino energy obtained for the events selected by the *IceCube* (Aartsen et al. 2013b) and *ANTARES* (Adrián-Martínez et al. 2012) detectors. A full Monte Carlo simulation is necessary to describe the triggering, tracking and selection efficiencies of the two detectors

Earth's motion, the declination in the detector frame of a given source in the sky changes with daytime. The effective area must thus be computed for each source as a function of the declination by averaging over the local coordinates (zenith and azimuth angles). IceCube is sensitive to sources optically visible in the Northern sky, while ANTARES, located in the Mediterranean Sea, is sensitive to those sources visible in the Southern sky.

Using the published IceCube values of the effective area provided in a tabular form (Aartsen et al. 2013b), it is straightforward to compute numerically the expected events from Eq. (10.24). In the case of IceCube for a source located in the Northern sky with the reference neutrino flux (Sect. 10.6.1), 13 events/year are expected for no cut-off, and 2.7 events/year for $E_c = 3.7$ TeV. Concerning the background of atmospheric neutrinos, IceCube detects 180 ν/days. Assuming a search cone of 0.5° centered on the source, the number of background events falling inside the cone is ~4 events/year. The background depends on the declination of the source, and it can be reduced with a cut on the reconstructed energy.

RX J1713.7-3946 and other galactic SNRs are located near the galactic center and visible in the Southern sky. The sensitivity of the IceCube telescope for this region is substantially null for neutrino energies below 100 TeV using the point-source search method described in Sect. 10.10.1. The galactic center region is studied in a better way with a detector in the Northern hemisphere, such as ANTARES. However, this detector has an effective area smaller than IceCube by a factor of ~40 and has marginal sensitivity to detect a source corresponding to the reference neutrino flux in an observation period of 10 years. A km^3 detector in the Northern hemisphere is necessary for the firm identification of galactic sources.

10.7.2 Number of Optical Sensors in a Neutrino Telescope

A cubic kilometer instrumented volume is necessary to detect cosmic neutrinos. How many PMTs are needed to measure a muon track accurately? This is one of the major impact factors on the cost of an experiment.

To solve the problem, we assume that a signal in at least 10 different PMTs is needed to reconstruct a muon track. The reconstruction algorithm has to determine 5 degrees of freedom (the coordinate of the impact point on the detector surface and two angles). The larger the number of PMTs, the better the quality of the track reconstruction. In the case of a muon, many photons arrive on the same PMT during the integration window of the electronics (which is on the order of 20–50 ns). The exact number depends on the PMT distance and orientation with respect to the track. Assuming that, on average, 10 photoelectrons (p.e.) are detected per PMT, the minimum number of p.e. to accurately reconstruct a muon track is $N_{p.e.} \sim 100$ p.e.

We assume that the telescope uses PMTs with 10" diameter (corresponding to detection area $A_{\rm pmt} \sim 0.05\,{\rm m}^2$) and quantum efficiency $\epsilon_{\rm pmt} \simeq 0.25$, see Fig. 10.2b. Similar PMTs have the advantage of fitting inside commercial pressure-resistant glass spheres (optical module, OM). This option has already been chosen by the IceCube and ANTARES collaborations. The overall efficiency of the PMT inside the OM is somewhat reduced, due to the presence of the glass, the glue between the glass and the PMT, and the mu-metal cage for magnetic shielding. We can assume that $\epsilon_{\rm om} \simeq 0.8 \cdot \epsilon_{\rm pmt} \simeq 0.2$.

Cherenkov photons emitted by the medium during muon propagation do not produce any signal if they are not intercepted by the photocathode of the PMT. Let us assume that only a photon propagating inside the *effective PMT volume* can induce a photoelectron (p.e.) with a probability $\epsilon_{\rm om} \simeq 0.2$ when arriving at the photocathode. The effective PMT volume $V_{\rm pmt}$ is defined as a cylinder of ice/water with base area $A_{\rm pmt}$ and height $\lambda_{\rm abs}$. The quantity $\lambda_{\rm abs}$ is the ice/water *absorption length*, introduced in the next section. For the following, we assume $\lambda_{\rm abs}= 50\,{\rm m}$ in the wavelength interval where the PMT quantum efficiency is not null, Fig. 10.2b. Thus, the effective PMT volume corresponds to $V_{\rm pmt} = A_{\rm pmt} \times \lambda_{\rm abs} \simeq 2.5\,{\rm m}^3$.

If $N_{\rm pmt}$ is the number of optical sensors inside the instrumented volume, the ratio R between the effective PMT volume of $N_{\rm pmt}$ and the instrumented volume ($1\,{\rm km}^3$) is:

$$R = \frac{V_{\rm pmt} \times N_{\rm pmt}}{10^9\,{\rm m}^3} = 2.5 \times 10^{-9} N_{\rm pmt}\,. \tag{10.30}$$

$N_{\rm pmt}$ is the unknown number to be determined.

The total number of Cherenkov photons emitted by a 1 km long muon track within the wavelength range of the PMTs' sensitivity (Sect. 10.2) is $N_C \simeq 3.5 \times 10^7$. Only the small fraction falling inside the effective volume of one PMT can be converted with efficiency $\epsilon_{\rm om}$ into photoelectrons and the total number of p.e.

produced by the muon track will be

$$N_{p.e.} = N_C \times R \times \epsilon_{\text{om}} \simeq (3.5 \times 10^7) \cdot (2.5 \times 10^{-9} N_{\text{pmt}}) \cdot \epsilon_{\text{om}}$$
$$= 1.8 \times 10^{-2} N_{\text{pmt}} . \qquad (10.31)$$

By requiring that $N_{p.e.} = 100$, from (10.31), the minimum number of optical sensors in a neutrino detector is

$$N_{\text{pmt}} \simeq 100/1.8 \times 10^{-2} \simeq 5000 . \qquad (10.32)$$

As we will show in Sect. 10.9.1, the IceCube collaboration has buried 5160 OMs in the ice. ANTARES (which is smaller than IceCube) uses 885 PMTs. The option planned by the KM3NeT collaboration is to use a larger number of smaller PMTs (3" diameter) inside each OM.

10.8 Water and Ice Properties

The effects of the medium (water or ice) on light propagation are absorption and scattering of photons. These affect the reconstruction capabilities of the telescope. Water is transparent only to a narrow range of wavelengths ($350\,\text{nm} \le \lambda \le 550\,\text{nm}$). The *absorption length* is the distance over which the light intensity has dropped to 1/e of the initial intensity I_0. Thus, the light intensity in a homogeneous medium reduces to a factor $I(x) = I_0 \exp(-x/\lambda_{\text{abs}})$ after traveling a distance x. For deep polar ice, the maximum value of $\lambda_{\text{abs}} \sim 100\,\text{m}$ is assumed in the blue-UV region, while it is about 70 m for clear ocean waters. Absorption reduces the amplitude of the Cherenkov wavefront, i.e., the total amount of light arriving on PMTs. Scattering changes the direction of the Cherenkov photons, and consequently delays their arrival time on the PMTs; this degrades the measurement of the direction of the incoming neutrino. *Direct photons* are those arriving on a PMT in the Cherenkov wavefront, without being scattered; the others are *indirect photons*.

Seawater has a smaller value of the absorption length with respect to deep ice, which is more transparent. The same instrumented volume of ice corresponds to a larger effective volume with respect to seawater. On the other hand, the effective scattering length for ice is smaller than water. This is a cause of a larger degradation of the angular resolution of detected neutrino-induced muons in ice with respect to water.

The South Pole ice has optical properties that vary significantly with depth and that need to be accurately modeled. Impurities trapped in the ice depend on the quality of the air present in the snows: Antarctic ice is laid down through a process of snowfall, hence trapping bubbles of air as it compacts itself. This happened over roughly the last 10^5 years. Because of variations in the long-term dust level in the atmosphere during this period, as well as occasional volcanic eruptions, impurity concentrations are depth-dependent. The ice is almost background-free from radioactivity.

The background in seawater has two main natural contributions: the decay of radioactive elements dissolved in water, and the luminescence produced by organisms, the so-called bioluminescence. The ^{40}K is by far the dominant of all radioactive isotopes present in natural seawater. The beta-decay of ^{40}K is above the threshold for Cherenkov light production. Bioluminescence is ubiquitous in oceans and results from two primary phenomena in deep sea: the steady glow of bacteria and flashes produced by animals. These can give rise to an optical background that occasionally can reach a level several orders of magnitude larger than that due to ^{40}K.

Optical properties of water depend on many factors. Environmental parameters such as water temperature and salinity are indicators of the aggregation state of H_2O molecules, which biases the diffusion of light. Water absorption and scattering also depend on the density and sizes of the floating particulate, which also affect the telescope response in terms of detector aging. Due to bio-fouling and sediments sticking to the optical modules, efficiency in photon detection can be reduced for long-term operations. In order to minimize the bias induced by external agents, the telescope sites must be far enough from continental shelf breaks and river estuaries, which can induce turbulent currents and spoil the purity of water. At the same time, the neutrino telescope should be close to scientific and logistic infrastructures on shore. With such requirements, the Mediterranean Sea offers optimal conditions on a worldwide scale.

10.9 Running and Planned Neutrino Detectors

The activities for the construction of a neutrino telescope started in the early 1970s. This was initially mainly a joint Russian-American effort that, after the Soviet invasion of Afghanistan, forced the teams to take separate paths (the DUMAND project for the Americans, BAIKAL for the Russian). The DUMAND project was canceled in 1995 (the indispensable submarine technology was not advanced enough at that time) and the efforts of the two communities turned toward an experiment in a lake with an iced surface and in the South Pole ice. In the sea, to minimize the noises induced by external agents, a telescope must be located far enough from continental shelf breaks and river estuaries. At the same time, the detector should be close to scientific and logistic infrastructures on shore. With such requirements, the Mediterranean Sea offers optimal conditions on a worldwide scale. For this reason, at the beginning of 1990s European groups began the exploration of the Mediterranean Sea as a possible site for an underwater neutrino telescope.

An exhaustive description of the history of neutrino telescope projects is in Spiering (2012).

10.9.1 Telescopes in the Antarctic Ice

An experiment at the South Pole, at the Amundsen-Scott station where the ice is about 2800 m deep, was pioneered by the Antarctic Muon And Neutrino Detector Array (AMANDA) collaboration. The researchers drilled holes in the ice using a hot water drill, and lowered strings of optical sensors before the water in the holes could freeze again. The first AMANDA string was deployed in 1993, at a depth of 800–1000 m. It was quickly found that at that depth, the ice had a very short scattering length, >50 cm. In 1995–1996, AMANDA deployed 4 strings at depths between 1500 and 2000 m. These detectors worked as expected, and AMANDA detected its first atmospheric neutrinos. This success led to AMANDA-II, which consisted of 19 strings holding 677 optical sensors. In 2005, after 9 years of operation, AMANDA officially became part of her successor project, the IceCube Neutrino Observatory (http://icecube.wisc.edu/).

IceCube is, at present, the only running km^3-scale neutrino observatory. Like its predecessor, it is located at the geographic South Pole, Fig. 10.11. The instrumented detector volume is a cubic kilometer of highly transparent Antarctic ice and it was built between 2005 and 2010. IceCube uses an array of 5160 Digital Optical

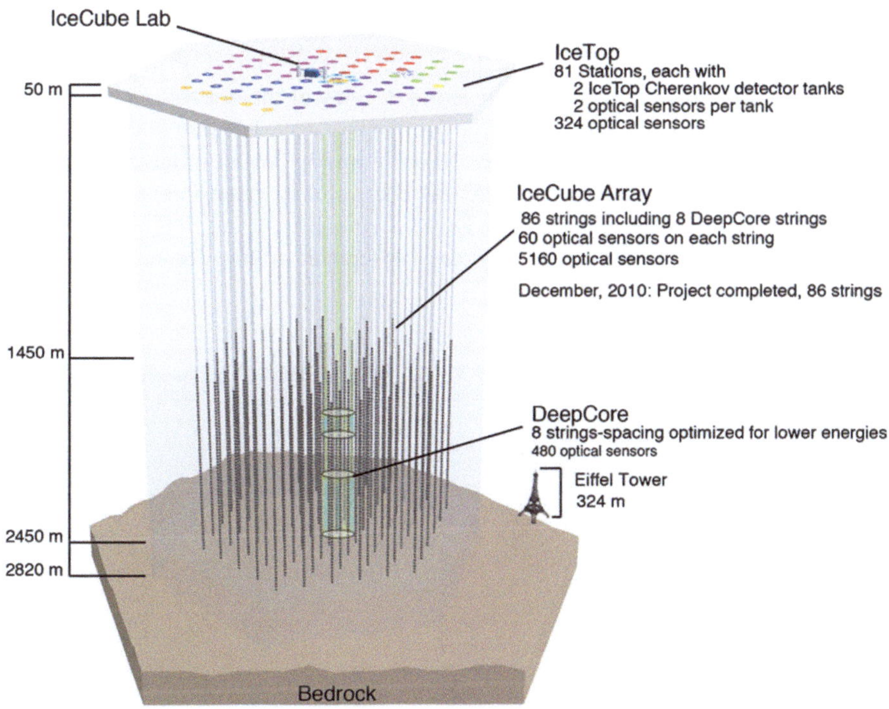

Fig. 10.11 Side view of the IceCube detector, consisting of 86 buried InIce strings. The *IceTop* surface array and the DeepCore are also shown. Credit: IceCube/NSF

Modules (DOMs) deployed on 86 strings at a depth of 1.5–2.5 km below the surface just above the bedrock in the clear, deep ice. Strings are arranged at the vertices of equilateral triangles that have sides of 125 m. The DOMs are spherical, pressure-resistant glass housings each containing a 25 cm diameter photomultiplier tube (PMT) plus electronics for waveform digitization, and vertically spaced 17 m from each other along each string. High quantum efficiency PMTs are used in a denser sub-array located in the center of the detector. This sub-array, called DeepCore, enhances the sensitivity to low energy neutrinos. Finally, a surface CR detector, called IceTop, completes the IceCube Observatory. Data acquisition with the complete configuration started in May 2011.

Because of the Antarctic weather, high altitude and remote location of the South Pole, logistics is a key issue. The construction season lasted from November through mid-February (during the Austral summer). Everything needed was flown to the Pole on ski-equipped transport planes. The main task in the construction of IceCube consisted in the drilling of holes for the strings (see Fig. 10.12). This was done with a 5 MW hot-water drill, melting a hole through the ice. Drilling a 2500 m deep, 60 cm diameter hole takes about 40 h.

IceCube-Gen2 The scientific results of IceCube (discussed in Sect. 10.10) made a good case for a new, larger observatory able to deliver statistically significant samples of very high-energy astrophysical neutrinos. This will enable detailed spectral studies, significant point source detections, and new discoveries.

The IceCube-Gen2 detector will have strings at distances of 250 m, and the construction will benefit from the successful designs of the hot water drill systems and the DOMs in the original IceCube project. In this way, the telescope will achieve a tenfold increase in volume to about 10 km^3, aiming for an order of magnitude increase in neutrino detection rates. A low-energy infill extension to the IceCube observatory, with 5 m spacing between DOMs in the same string, was also proposed. This sub-detector will feature a very large effective volume for neutrinos at an energy threshold of a few GeV.

Fig. 10.12 The *hot water hose* and *support cables* disappear down one of the many *boreholes* drilled into the *Antarctic ice* to construct the IceCube Neutrino Observatory. Credit: Jim Haugen, IceCube/NSF

10.9.2 Telescopes in the Mediterranean Sea

The ANTARES detector (http://antares.in2p3.fr) is, at present, the largest neutrino observatory in the Northern hemisphere (42 °N, 6 °E), offering a privileged view of the most interesting areas of the sky, like the galactic center, where many neutrino source candidates are expected. The ANTARES detector was completed in 2008, after several years of site exploration and detector R&D. The detector is located at a depth of 2475 m in the Mediterranean Sea, 40 km from the French town of Toulon. It comprises a three-dimensional array of 885 optical modules (OMs) looking 45° downward and distributed along 12 vertical detection lines, Fig. 10.13. An OM consists of a 10″ PMT housed in a pressure-resistant glass sphere together with its base, a special gel for optical coupling and a μ-metal cage for magnetic shielding, Fig. 10.2a. The OMs are grouped in 25 triplets (or storeys) on each line with a vertical spacing of 14.5 m between triplets. The total length of each line is 450 m; these are kept taut by a buoy located at the top of the line. The separation between the lines ranges from 60 to 75 m. Each line has been deployed by a ship and

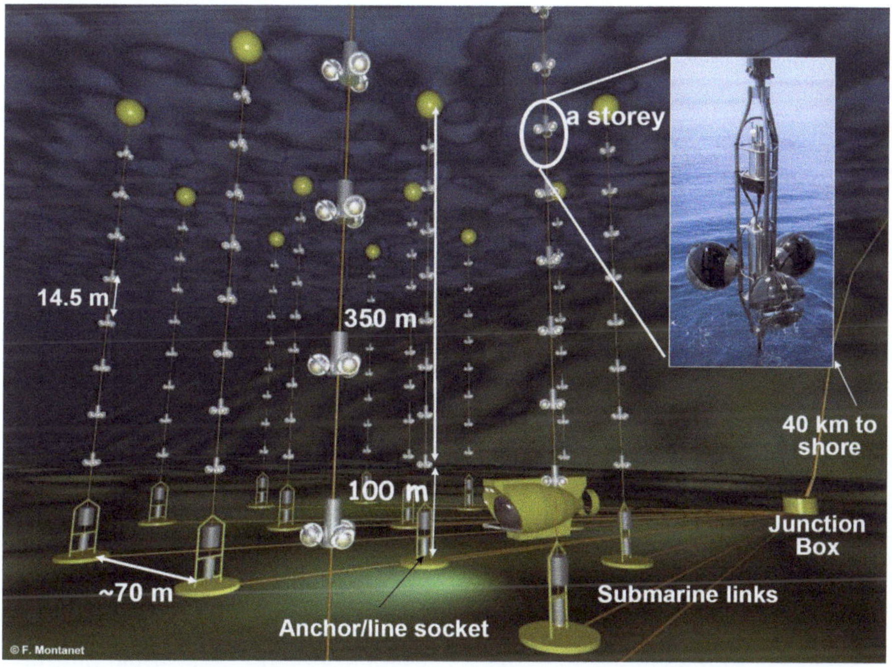

Fig. 10.13 Schematic view of the Astronomy Neutrino Telescope and Abyss environmental RESearch (ANTARES) detector. The *inset* shows the details of one of the PMT triplets mentioned in the text. Credit: F. Montanet and the ANTARES Collaboration

connected to a junction box by a remotely operated submarine vehicle. The junction box, in turn, is connected to shore via an electro-optical cable. Sea operations and sea technologies are of fundamental importance for the realization of underwater experiments.

The detector also includes several calibration systems. The lines slowly move due to the sea current (up to $\sim$15 m at the top of the line in cases of currents of 20 cm/s). A set of acoustic devices together with tiltmeters and compasses along the lines are used to reconstruct the shape of the lines and orientation of the storeys every 2 min. The acoustic system provides the position of each optical module with a precision better than 15 cm. The total ANTARES sky coverage is 3.5π sr, with an instantaneous overlap of 0.5π sr with that of the IceCube experiment. The galactic center is observed 67% of the time.

KM3NeT [Adrián-Martínez et al. (2016)] is a research infrastructure that will house the next generation neutrino telescope in the Mediterranean Sea. KM3NeT will consists of two different structures. The KM3NeT/ARCA telescope will be installed about 100 km off-shore Portopalo di Capo Passero on Sicily, Italy. ARCA will consist of two detector blocks of 115 vertical detection units (DUs) anchored at a depth of about 3500 m. The telescope will have an instrumented volume slightly larger than that of IceCube. A possible extension to several km^3 is also foreseen. The KM3NeT/ORCA detector will be located on the French site close to Toulon; the main scientific objectives are (similarly to the PINGU project) the determination of the neutrino mass hierarchy and the searches for dark matter. The facility will also house instrumentation for Earth and Sea sciences for long-term and on-line monitoring of the deep-sea environment.

Both KM3NeT structures will use the same DUs equipped with 18 optical modules, with each optical module comprising 31 small PMTs, as shown in Fig. 10.14. The technical implementation and solutions of ARCA and ORCA are almost identical, apart from the different spacing between DOMs.

10.9.3 A Telescope in Lake Baikal

Like a sort of platform, winter ice covers the surface of the Russian Lake Baikal (52 °N, 104 °E), which is the deepest lake in the world, reaching a depth of more than 1600 m. The ice layer can be used for assembly and deployment of instruments, instead of using ships and underwater remotely operating vehicles. The disadvantage of lake water is that the scattering length is much shorter than in seawater, with a consequently poorer determination of the neutrino direction. After research and development studies started in 1984, the NT200 was completed in 1998. It consisted of an array of 192 OMs deployed 3.6 km from shore at a depth of 1.1 km. NT200 has taken data for almost two decades.

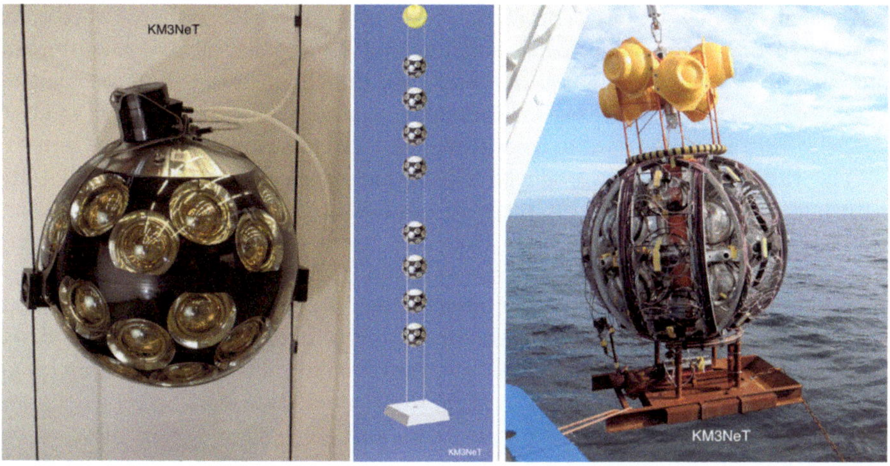

Fig. 10.14 The KM3NeT Digital Optical Modules (DOMs) and Detector Units (DUs). Left: a DOM consisting of a 17″ pressure-resistant glass sphere with 31 small (3″) PMTs. Middle: a sketch of a DU string with the breakout box and the fixation of the DOMs on the two parallel ropes. Right: Photo of a launch vehicle deployment containing a DU with 18 OMs. Credit: KM3NeT Collaboration Adrián-Martínez et al. (2016)

The second-generation is the Gigaton Volume Detector (Baikal-GVD): it is configured in "clusters", in which each cluster consists of eight strings, instrumented over a length of 520 m with 36 OMs each carrying a 10″ PMT with a high-sensitive photocathode. A first cluster was deployed in 2016, and in Spring 2017, a second cluster was added. Baikal-GVD will be built in two phases. Phase-1 will consist of eight clusters, each 120 m in diameter, with lateral distances of 300 m. The effective volume for neutrinos within the 10–100 TeV energy range inducing cascades will be $\sim$0.4 km^3. The sensitivity to ν_μ inducing tracks is negligible for $E_\nu < 1$ TeV, but rapidly raises in the multi-TeV range. Phase-1 is financed and planned to be completed in 2020/21. In a second phase, Baikal-GVD will be extended to 18 clusters, and thus surpass the cubic kilometer benchmark.

10.9.4 Ultra High Energy (UHE) Neutrino Detection

The Cherenkov technique described above is the viable solution for the investigation of astrophysical neutrino fluxes within the energy range $\lesssim$ 10 PeV. At higher energies, the expected fluxes are so low that km^3-scale detectors are also too small. For this reason, different and complementary techniques have been investigated, in particular, for searches of cosmogenic neutrinos in the EeV (or in the higher ZeV) energy scale, Sect. 10.11.4. These techniques (refer to Sapienza and Riccobene (2009) for a review) rely on the identification of radio or acoustic emissions from

UHE neutrino interactions. The induced cascades propagate in dense media for very large distances and produce coherent radiation. Radio waves have typical attenuation lengths of a few km in the ice and the attenuation length for acoustic waves in the sea is on the order of several km. Therefore, a sparse array of radio or acoustic sensors can be used to reconstruct the UHE neutrinos' interaction vertex.

The **radio emission** is produced by the fact that cascades induced by UHE neutrinos are very compact, displaced in a volume of a few dm^3, and longitudinal extension of a few tens of meters from the neutrino interaction vertex. During the EM cascade development, a negative charge is accumulated, due to Compton scattering of γ-rays and to positron annihilation. The charge asymmetry is about 20% at shower maximum. The relativistic motion of this negative charge in the medium produces coherent Cherenkov radiation within the radio frequency range 100 MHz–1 GHz. The *Askaryan radiation detection technique* exploits natural media such as the polar ice-cap, salt domes and the lunar regolith; these media provide huge target material for neutrino interaction, being, at the same time, optimal transparent radiators for electromagnetic radio frequency signals (Connolly and Vieregg 2017).

The **acoustic neutrino detection technique** is based on the fact that the energy deposition by a particle cascade leads to a local heating and a subsequent characteristic pressure pulse that propagates in the surrounding medium. The main advantage of using sound, as opposed to Cherenkov light, lies in the much longer attenuation length: several km for sound compared to several tens of meters for light in the respective frequency ranges of interest in seawater. As detection media for future detectors, water, ice, salt domes and permafrost have been discussed. Seawater and ice have been investigated most thoroughly by using existing arrays of acoustic receivers, mainly military arrays in various bodies of water, or by implementing dedicated acoustic arrays in Cherenkov neutrino telescopes (in IceCube at the South Pole, in the Lake Baikal experiment in Siberia and in ANTARES in the Mediterranean Sea). A summary review of the technique is provided in Lahmann (2016).

Another method for UHE neutrino detection uses **EAS arrays**. It is based on the reconstruction of quasi-horizontal atmospheric showers, initiated by CC or NC ν interaction close to the ground, or looking at up-going showers in the atmosphere initiated by the decay products of an emerging (Earth-skimming) τ lepton. This method has been described in Sect. 7.10.

10.10 Results from Neutrino Telescopes

We discuss here some recent results obtained in the study of neutrino fluxes presented in Sect. 10.6. Each analysis is based on appropriate cuts depending on the magnitude of the background and on the purity required to isolate an eventual signal. Neutrino candidates are selected using criteria that have been determined in a "blind" manner before performing the data analysis, i.e., using only simulations and/or a small subsample of data. The blinding policy limits the use of data during the optimization steps of analyses to avoid the selection procedure becoming inadvertently tuned towards a discovery. This procedure requires the simulation of a large number of *pseudo-experiments*, with the generation of both signal and background events. The background generally consists of atmospheric neutrinos and wrongly-reconstructed atmospheric muons.

10.10.1 Point-Like Sources

The ν_μ charged current interaction produces a long muon track that can be correlated with the parent neutrino direction, Sect. 10.5.1. No cosmic neutrino source has been identified so far. Only upper limits on neutrino flux from cosmic objects have been set. The number of signal events $N(s)$ from a given source in a given time T, Eq. (10.24), depends on the effective area $A^{\mathrm{eff}}_{\nu_\mu}$, Eq. (10.25), which, in turn, depends on the selection cuts of the analysis through the parameter ϵ. The same is true for the number of background events, $N(b)$, which is usually dominated by atmospheric neutrinos and can be calculated using Eq. (10.29). The selection cuts are optimized using the pseudo-experiments, in order to maximize the number $N(s+b)$ of signal events (which necessarily includes the irreducible background) over the number of background events. The procedure depends on the hypothesized energy spectrum from the source. Usually, an $\propto E^{-2}$ dependence is assumed, as expected from a Fermi acceleration mechanism. The number of signal and background events varies according to the angular aperture of the search cone, the cut imposed on the track quality parameters, and the quantities used as estimators of the energy of the event. This operation defines a set of optimal parameters, corresponding to an efficiency ϵ^*, which gives the best value $A^{*,\mathrm{eff}}_\nu$ for the considered analysis.

The selection criteria are chosen to optimize the so-called *sensitivity* or the *discovery potential*. These two quantities are generally used as figures of merit of the experiment, and well-defined statistical methods exist to perform such optimization using pseudo-experiments. The *discovery potential* is defined as the flux $\mathrm{d}\Phi_\nu/\mathrm{d}E_\nu$ in (10.24) needed to make a 5σ discovery in 50% of the pseudo-experiments. The *sensitivity* is defined as the average upper limit on the flux $\mathrm{d}\Phi_\nu/\mathrm{d}E_\nu$ using a detector with $A^{*,\mathrm{eff}}_\nu$ that would be obtained by an experiment with the expected background and no true signal, $N(s+b) = N(b)$.

Let us better define these quantities using the results of neutrino telescopes. Figure 10.15 shows, as a function of the sine of the declination δ, the upper limits on selected IceCube and ANTARES point source candidates and the sensitivity of ANTARES and IceCube experiments (full lines).

IceCube searched for neutrino candidates coming from the direction of 44 objects selected a priori, according to observations in γ-ray or astrophysical models of neutrino emission. The results in Fig. 10.15 refer to almost 3 years of live time, one each with a detector with 40, 59 and 79 strings, respectively. The strength of the analysis cuts has been varied as a function of the declination. From the

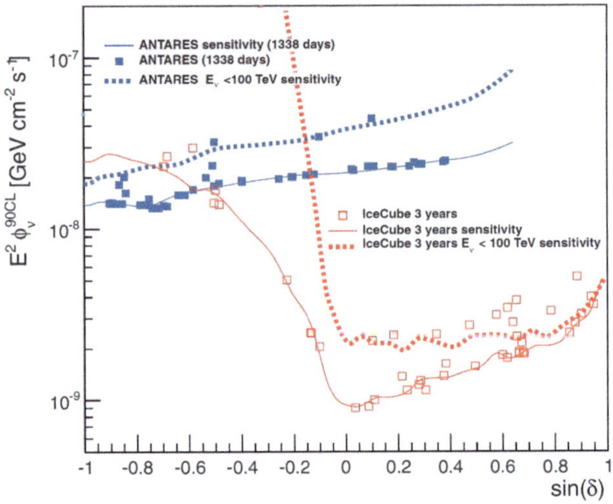

Fig. 10.15 Upper limits and sensitivities for an E^{-2} $\nu_\mu + \overline{\nu}_\mu$ spectrum from point sources as a function of sin(δ). The *open* (*full*) *squares* represent the 90% C.L. flux upper limits for 44 (50) objects considered by IceCube (ANTARES). The *full lines* are the sensitivity (at 90% C.L.) for a point-source with an E^{-2} spectrum for 3 years of IceCube (Aartsen et al. 2013b) and for 3.7 years of ANTARES (Adrián-Martínez et al. 2014). The *dashed lines* represent the same sensitivity, but for neutrino energies lower than 100 TeV. The IceCube sensitivity for sources in the Southern hemisphere is mostly due to events of higher energy

Northern hemisphere ($\sin\delta > 0$), most of the sensitivity is for neutrinos with $E_\nu > 100$ GeV. A different situation arises for sources located in the Southern hemisphere ($\sin\delta < 0$, which includes almost the whole galactic plane). Neutrino candidate events coming from negative declinations are observed as downward-going events in IceCube. They have been selected with strong topological cuts, with the veto capability of the surface array IceTop and with a cut on the reconstructed energy to reject the huge contamination of downward-going atmospheric muons. In fact, atmospheric muons are accompanied by extended air showers, which can produce early hits in the IceTop surface array. The corresponding sensitivity (red line) for sources located in the galactic center region is mostly for neutrinos with $E_\nu > 100$ TeV. When pseudo-experiment events with energy lower than 100 TeV are selected (red dashed line in figure), the IceCube sensitivity in the $\sin\delta < 0$ region is substantially null.

One neutrino observatory able to monitor the region of the sky with negative declinations with higher efficiency is the one located in the Mediterranean Sea. The sensitivity of the ANTARES neutrino telescope shown in Fig. 10.15 as a full line refers to 1339 days of live time with the full 12-string detector and refers to neutrinos with $E_\nu > 100$ GeV. ANTARES studied 50 selected sources, mainly in the galactic plane and assuming an E^{-2} energy spectrum.

The discovery potential corresponds to a curve almost parallel to the sensitivity and a factor of 3–5 higher.

In both the IceCube and ANTARES cases, after optimizing the cuts, the number of events in the search cone around each considered source was compared with the number of expected events. Three different possibilities can occur for source i at a given declination δ_i: the number of events in the data $\mathcal{N}_i$ is equal to, larger or smaller than $N_i(b)$, the number of expected background events. In the case of $\mathcal{N}_i = N_i(b)$, the flux upper limit (the point in the figure) exactly overlaps the sensitivity value of the detector at declination δ_i. The case of $\mathcal{N}_i < N_i(b)$ is simply interpreted as an under-fluctuation of the background. The upper limit would lie below the sensitivity value, but conventionally, in the plot, it matches the sensitivity as in the previous case. Finally, the number of data events could be larger than expected, $\mathcal{N}_i > N_i(b)$. This is attributed to an over-fluctuation of the background or to the presence of a real signal if $\mathcal{N}_i - N_i(b)$ exceeds a value pre-defined by the discovery potential.

For all sources studied by IceCube and ANTARES, the excesses are compatible with an over-fluctuation of the background. The corresponding upper limit for a source at declination δ_i lies above the corresponding sensitivity position. In Fig. 10.15, the positions of the 44 (50) upper limits obtained by IceCube (ANTARES) are indicated by the open (filled) markers. The larger the distance of the point from the sensitivity line, the larger the difference between $\mathcal{N}_i$ and $N_i(b)$. However, no source shows an excess of events incompatible with an over-fluctuation of the background.

In the absence of signals, different sources at the same declination can have under- or over-fluctuations; the upper limits can be above or below the sensitivity. For this reason, it is sometimes said that the sensitivity corresponds to the average upper limit that would be obtained by an ensemble of experiments with the expected background and no signal.

The IceCube result indicates that none of the considered sources located in the Northern hemisphere of the equatorial coordinate system generates a ν_μ flux larger than $E_\nu^2 \Phi_\nu \sim 0.5 \times 10^{-11}$ TeV/(cm^2 s). Note that this is below the value of our reference neutrino flux defined in Sect. 10.6.1. Concerning the limit from sources located in the Southern sky (the location of the galactic center is at $\delta \sim -29°$), IceCube has no sensitivity for $E_\nu < 100$ TeV. Galactic sources are better constrained by ANTARES. Also in this case, no significant statistical fluctuations have been observed. For the area of the sky that is always visible (that with declination $\delta < -48°$), the sensitivity is $E_\nu^2 \Phi_\nu \simeq 1.5 \times 10^{-11}$ TeV/(cm^2 s). The region with $\delta > 42°$ is never visible by a telescope in the Mediterranean sea. An experiment with a larger volume in the Northern hemisphere is needed to study galactic neutrino source candidates efficiently.

It is interesting to compare the discovery potential of neutrino telescopes with that of the space- and ground-based γ-ray experiments shown in Fig. 9.7. Neutrino astronomy is approaching the level of sensitivity of γ-ray astronomy experiments, in spite of the incredibly small neutrino cross-section and the consequent need for huge detectors.

10.10.2 Limits from GRBs and Unresolved Sources

The detection of neutrinos in spatial/temporal coincidence with GRBs would be an unambiguous proof of hadronic acceleration in cosmic sources, and could also serve to explain the origin of UHECRs. IceCube and ANTARES observe at least half the sky with a large duty cycle efficiency, and the requirement of temporal and spatial coincidence with a recorded GRB significantly reduces the number of expected background events. The observation of just one neutrino coincident within a few seconds with a γ-ray burst would be statistically significant. It is left as an exercise to show that the present limits on the neutrino mass do not significantly delay the neutrino arrival time with respect to that of a photon.

No neutrino events in IceCube were associated with one of the 200 GRBs observed when the detector was in operation with 40 and 59 lines. In Abbasi et al. (2012), the absence of neutrinos associated with GRBs is interpreted either as the fact that GRBs are not the only sources of CRs with energies exceeding 10^{18} eV or that the efficiency of neutrino production is much lower than predicted by the mechanism described in Sect. 10.6.3. Also, ANTARES reported no observation of neutrinos from about 300 GRBs in coincidence with electromagnetic observations. Due to the smaller size, the derived limit is about one order of magnitude less stringent than that of IceCube.

These null results motivated the more detailed Monte Carlo simulation of neutrino production in the GRB jet, also discussed in Sect. 10.6.3. The new evaluation reduces the neutrino yield by an order of magnitude below that foreseen by analytic computations. With this reduced prediction, the IceCube limit still does not exclude that a large fraction of CRs with energies $>10^{18}$ eV are produced by GRBs.

No steady extragalactic individual object is expected to produce a neutrino flux detectable as a point source in the current generation of neutrino telescopes. It is nevertheless possible that many sources, isotropically distributed throughout the Universe, could combine to make a detectable signal, Sect. 10.6.2. A cosmic diffuse flux is searched as an excess of high energy events over the expected atmospheric neutrino background above a certain value of the energy (Fig. 10.3). The difficult task is the estimate of the neutrino energy using the calorimetric properties of neutrino telescopes. Two different channels can be studied: the ν_μ-induced muons and the showering events.

The average muon energy loss per meter ($\mathrm{d}E/\mathrm{d}X$) is the observable correlated with the ν_μ energy. The energy loss $\mathrm{d}E/\mathrm{d}X$ is estimated from the observed collection of Cherenkov photoelectrons. The muon energy yields only a coarse proxy for the neutrino energy that, as already mentioned, is only partially transferred to the muon.

The interactions generating a shower provide little information about the neutrino's direction. For diffuse flux measurements, the poor directional resolution is not a major draw-back, and the resolution on the shower energy (about 30%) is better than that obtained in the muon channel. These diffuse studies led to the first evidence of cosmic neutrinos in the IceCube detector, as presented in the following section.

10.11 The First Evidences of Cosmic Neutrinos

The IceCube experiment has observed neutrinos of astrophysical origin in two different event samples. The first is characterized by neutrinos that interact inside a fiducial volume of the detector, with a contained interaction vertex, forming the *High Energy Starting Event (HESE)* sample. The second refers to upward-going muons induced by CC interaction of muon neutrinos crossing the Earth, the *passing muons*. These latter neutrinos arriving in IceCube originate from the Northern hemisphere.

10.11.1 The High-Energy Starting Events (HESE)

The first observation of an excess of high-energy astrophysical neutrinos over the expected background has been reported by IceCube using data collected from May 2010 to May 2013 and with 988 days live time (Aartsen et al. 2013a). This sample is continuously updated, and as of this writing (Spring 2018), results up to early 2016 are available for a total live time of 2078 days Aartsen et al. (2017).

The high-energy neutrino candidates have been selected with the requirement that the interaction vertex is contained within the instrumented ice volume, without any signal on the PMTs located on the top or sides of the detector. In such a way, the edges of IceCube are used as a veto for downgoing atmospheric muons. The muon rejection efficiency of the veto has been measured in data by using one region of IceCube to tag muons and then measuring their detection rate in a separate layer of PMTs equivalent to the veto.

The veto also provides a partial reduction of the downgoing atmospheric neutrino background. In fact, atmospheric neutrinos are produced by the same parent mesons that generate the shower muons. A high-energy atmospheric neutrino has a large probability of being accompanied by a downgoing atmospheric muon produced in the same cascade. To ensure a reliable trigger efficiency of the anticoincidence muon veto, an overall minimum number of 6000 photoelectrons (p.e.) has been required. From the number of p.e., the deposited energy E_{dep} in the detector is derived and, in turns, the true energy E_ν of the neutrino is estimated with the help of Monte Carlo simulation techniques. An event with 6000 p.e. corresponds to a deposited energy of $\sim$30 TeV. This minimum energy requirement provides rejection of all but one part in 10^5 of the cosmic ray muon background.

Figure 10.16 shows the distribution of the deposited energy (left plot) and of the sin(declination) (right plot) for HESE. In the 6-year sample, 82 events (data points with error bars) interacting within the IceCube fiducial volume and escaping the veto are present. Two events present problems and are removed from the sample, 22 are accompanied by a reconstructed muon track, and the remaining 58 have shower-like topology. Fifty-two of the events arrived from the Southern hemisphere, the region that contains most of the galactic plane.

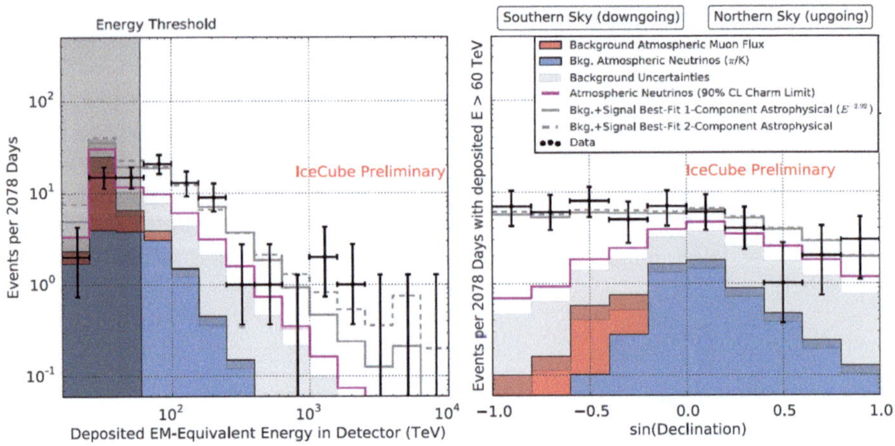

Fig. 10.16 Deposited energies E_{dep} (left panel) and arrival directions (right panel) of observed IceCube events (*crosses*), compared with predictions. The sample refers to 6 years of data. The *hashed region* shows uncertainties on the sum of all backgrounds, due to atmospheric muons and neutrinos. The contribution of an astrophysical $(\nu + \overline{\nu})$ flux for $E_{dep} > 60\,\text{TeV}$ is also indicated. See Aartsen et al. (2017) for details. Courtesy of the IceCube Collaboration

Referring to the distributions shown in Fig. 10.16, the expected background (filled histograms) corresponds to 25.2 ± 7.3 atmospheric muons escaping the veto and $15.6^{+11.4}_{-3.9}$ atmospheric neutrinos, including that from charmed meson decay (prompt neutrinos, Sect. 11.3.2). The atmospheric muon background would mostly appear to be low-energy track events in the Southern sky (region with negative declination). The events classified as tracks correlate with the parent ν_μ with an angular resolution of about 1°. The shower-like events correlate with the parent neutrino with an angular resolution of ∼15°.

The additional contribution in the data sample with respect to the background corresponds to a diffuse astrophysical signal, with topologies compatible with neutrino flavor ratio $\nu_e : \nu_\mu : \nu_\tau \sim 1 : 1 : 1$, as expected for a cosmic signal (Sect. 12.8). The best fit to the data above 60 TeV yields, for one neutrino flavor,

$$E_\nu^2 \Phi_\nu^D(E) = (2.46 \pm 0.3) \cdot 10^{-8} \left(\frac{E}{100\,\text{TeV}}\right)^{-0.92} \text{GeV cm}^{-2}\,\text{s}^{-1}\,\text{sr}^{-1} . \qquad (10.33)$$

The measurement highlights a significant excess with respect to the hypothesis that the data sample is due only to the atmospheric backgrounds. A fit of data without the astrophysical contribution is disfavored at >6 standard deviations. Most of the signal originates primarily from the Southern hemisphere, where neutrinos with $E_\nu \gg 100\,\text{TeV}$ are not absorbed by Earth.

The poor angular resolution (∼15°) of showering events prevents the possibility of accurate localization in the sky of the parent neutrino's direction. To identify any bright neutrino sources in the data, the usual maximum-likelihood clustering

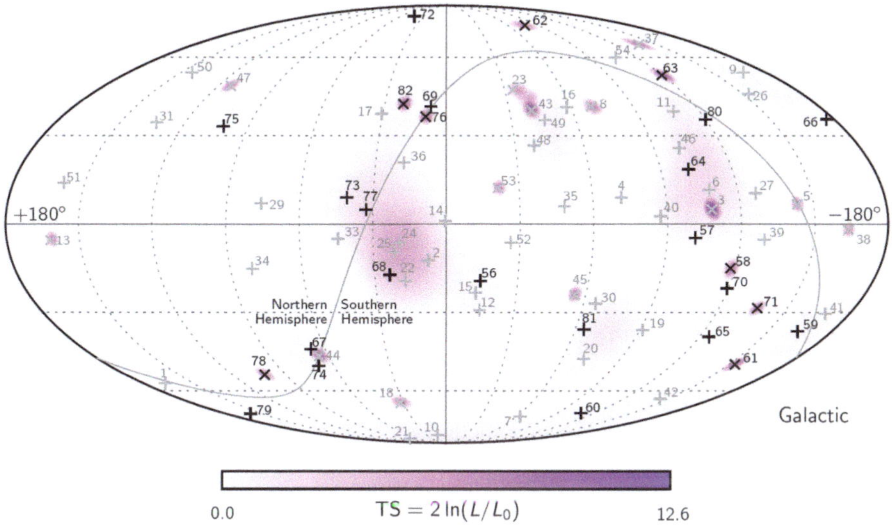

Fig. 10.17 Arrival directions of the IceCube HESE (6 year) in *galactic coordinates* (Aartsen et al. 2017). Shower-like events are marked with + and those containing muon tracks with ×. The *gray line* denotes the equatorial plane. The *color map* shows the test statistic for the point source clustering test at each location. A clustering of a fraction of the signal is present around the galactic center region, despite being statistically nonsignificant. Courtesy of the IceCube Collaboration

search has been used, as well as searches for directional correlations with TeV γ-ray sources. No hypothesis test has, at present, yielded statistically significant evidence of clustering or correlations. Given the high galactic latitudes of many of the highest energy events, the data seem to suggest an extragalactic origin with a possible Galactic contribution, Fig. 10.17.

10.11.2 The Passing Muons

The second IceCube sample that evidences a diffuse presence of cosmic neutrinos corresponds to CC upgoing muon neutrino events (Aartsen et al. 2016). The field of view for these events is restricted to the Northern hemisphere. This analysis has recently been extended with data collected up to 2017 (Aartsen et al. 2017). This last sample contains ~500,000 muon neutrino candidates with a negligible contribution of atmospheric muons.

For these events, the reconstructed energy E^i_{rec} of each individual neutrino i is a poor proxy of the true neutrino energy, E_ν. Thus, the reconstructed neutrino energy is used to produce a response matrix $P(E^i_{rec}; E_\nu)$, which must be inverted (as explained for a different situation in Sect. 4.10.1) to produce the posterior probability density function $P(E_\nu; E^i_{rec})$ (Aartsen et al. 2016). Finally, based on the

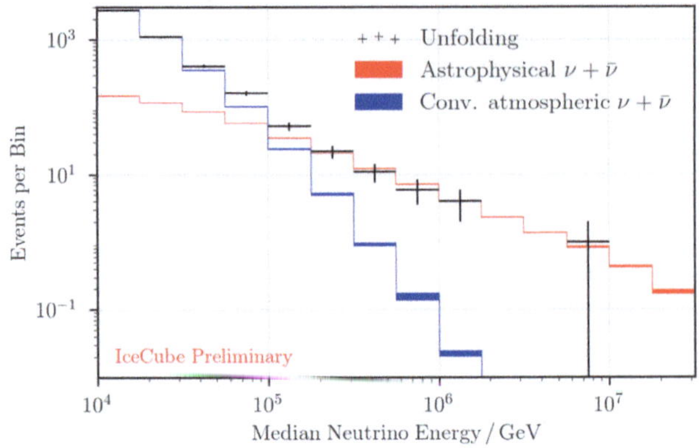

Fig. 10.18 Unfolded ν_μ energy spectrum for the passing muon sample in IceCube. The unfolded data (crosses) are compared to the best-fit fluxes for conventional atmospheric and astrophysical neutrinos. Courtesy of the IceCube Collaboration (Aartsen et al. 2017)

per-event probability density function $P(E_\nu;\, E^i_{rec})$, the median neutrino energy for each event is calculated. Figure 10.18 shows the distribution of the median neutrino energies for the 8-year sample. A clear excess above ∼100 TeV is visible and is not compatible with the atmospheric background expectation.

When the atmospheric neutrino background is removed, the best fit to the full data-set results in an astrophysical power-law flux for one neutrino flavor (Aartsen et al. 2017):

$$E^2_\nu \Phi^D_\nu(E) = (1.01^{+0.26}_{-0.23}) \cdot 10^{-8} \left(\frac{E}{100\,\text{TeV}} \right)^{-0.19} \text{GeV cm}^{-2}\,\text{s}^{-1}\,\text{sr}^{-1} . \quad (10.34)$$

This excludes a purely atmospheric origin of the observed events at 6.7 σ. The highest energy sample, with reconstructed energy above 200 TeV, corresponds to 35 events. No correlation with known γ-ray sources was found by analyzing the arrival directions of these 35 events.

10.11.3 Discussion of the Results and Perspectives for Neutrino Astrophysics

A Tension Between the Two IceCube Samples? As discussed in Sect. 10.6.2, for extragalactic neutrinos, an hard energy spectrum $\Phi_\nu(E) \propto E^{-\Gamma_\nu}$, with $\Gamma_\nu \sim 2$, is motivated by models of CR production at sources within the framework of the Fermi acceleration mechanism. The IceCube spectrum measured through passing muons, Eq. (10.34), has spectral index $\Gamma_\nu \sim 2.2$, close to the expected value. These events originate from the Northern sky, where the presence of the Galactic plane is marginal. Most of them could likely be of extragalactic origin.

On the other hand, the HESE presents a significantly softer spectrum, with $\Gamma_\nu \sim 2.9$, Eq. (10.33). The discrepancy between the best-fit for cosmic neutrinos from HESE and from the passing muon sample presents an approximately 3σ tension if a single unbroken power law is assumed (Aartsen et al. 2016). The possible origin of this discrepancy (the presence of two different extragalactic components; a Galactic plus an extragalactic component; more exotic combinations; a statistical fluctuation; etc.) is, at present, one intriguing research field in neutrino astrophysics.

The ANTARES Diffuse Flux Due to their relative proximity, the possibility of studying Galactic sources is particularly intriguing. The ANTARES detector in the Northern hemisphere can measure *upgoing events*, exploiting, with respect to IceCube, its complementing field-of-view, exposure, and lower energy threshold. In particular, it is also sensitive with these events to possible sources located in the Galactic plane. With a recent study, ANTARES reports a mild excess of high-energy events over the expected background in the searches for an all-flavor diffuse neutrino signal (Albert et al. 2018). The observed upward-going events in 9 years of data are due to a mixture of flavors, and are identified both in track-like (mainly ν_μ) and shower-like (mainly ν_e) events. The assumption of a single power-law cosmic neutrino spectrum yields a best-fit spectral index $\Gamma_\nu = 2.4^{+0.5}_{-0.4}$. The null cosmic flux assumption is rejected with a significance of 1.6σ.

The Next Generation Neutrino Telescope in the Mediterranean Sea The mission to identify galactic sources, to measure the neutrino energy spectrum and flavor distribution of events coming from the galactic region demands a neutrino telescope in the Northern sky. Such a detector will provide unique information about possible galactic accelerators. The flavor composition, in particular, will provide information on whether the flux is, as expected, coming from the decays of charged pions or from a different mechanism. Follow-up γ-ray, optical, and X-ray observations of the directions of individual high-energy neutrinos, which point to a sky region of angular size $<1°$, may also be able to identify neutrino sources, and cosmic ray accelerators, even from those objects whose neutrino luminosity is too low to allow identification from neutrino measurements alone. The KM3NeT Mediterranean neutrino telescope is actually in the construction phase. When completed, it will have an effective area much larger than that of the ANTARES telescope and will probably answer many questions about the nature of the unknown sources generating this astrophysical flux.

10.11.4 Cosmogenic Neutrinos

The prediction of a flux of *cosmogenic neutrinos* connected with the propagation of UHECRs and the GZK effect, Sect. 7.5.2, was discussed in Sect. 7.10 and qualitatively shown in Fig. 10.3. A measurement of the cosmogenic neutrino flux

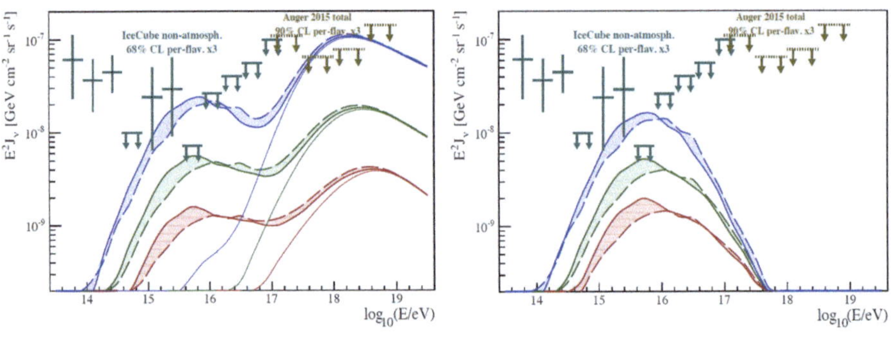

Fig. 10.19 Measurement and upper limits on the neutrino flux at energy >100 TeV from the IceCube neutrino telescope and the PAO air-shower array (Abreu et al. 2013). The theoretical model used in the left plot refers to neutrinos produced by UHECRs protons propagating in the CMB and in the EBL. The model used in the right plot uses a mixed composition of UHECRs in which protons are a minority component. The three different curves in each panel correspond to different assumptions about the cosmological evolution of UHECR sources. From bottom to top: no evolution (red), SFR (green) and AGN (blue). Refer to the text for more details. From Aloisio et al. (2017). Courtesy of Dr. R. Aloisio

is an important and complementary measurement to that of the charged UHECR spectrum and composition, and even the detection of a few events will considerably improve our knowledge of the features and origin of these extragalactic CRs (Allard 2012).

Figure 10.19 shows the measurement and limits of HE (>100 TeV) and UHE (in the EeV range) neutrinos and the expected fluxes according to the two scenarios discussed in Sect. 7.9.2: that in which protons dominate (the *dip model*), in the left panel, and that in which UHECRs are mainly nuclei, in the right panel. As explained, a flux of cosmogenic neutrinos is present only in the model dominated by UHE protons (left plot); the flux peaks at about 10^{18} eV and significantly declines at both lower and higher energy. The experimental points correspond to the above-mentioned IceCube HESE up to a few PeV and the upper limit from no-observations at higher energies (Aartsen et al. 2013a, 2017), and the PAO upper limits on EeV neutrino fluxes (Abreu et al. 2013), Sect. 7.10. The upper limits are usually represented with arrows pointing downwards. The three different fluxes for each scenario in Fig. 10.19 correspond to different assumptions about the cosmological evolution of UHECR sources, presented in Sect. 7.10. From bottom to top: no evolution (red); evolution as the star formation rate (SFR, green); evolution as active galactic nuclei (AGN, blue). The colored bands show the uncertainties due to the extragalactic background light (EBL) model considered. The thin solid lines are neutrino fluxes obtained taking into account only the CMB. The results start to disfavor some cosmological models.

10.12 Real-Time Alert and Multimessenger Follow-Up Programs

Cosmic neutrinos have been observed, but no steady sources of high-energy neutrinos have been identified so far. Thus, neutrinos produced during transient astrophysical events are intensively searched. Coincident detections of different messengers could provide a more complete picture of the inherent physical processes at the source. Neutrino telescopes are continuously in operation (almost 100% duty cycle) in both Earth hemispheres, covering, in combination, the full sky solid angle. At present, the IceCube and ANTARES collaborations run a number of alert and follow-up programs, which react to particular neutrino candidates identified in real-time. In case of a significant event, a brief message is automatically issued to the GCN Network (Sect. 8.11) that triggers radio, optical, X- and γ-ray telescopes. Conversely, neutrino candidates are searched for if an external alert is received; for example, neutrino telescopes have followed (with no positive results) all the detections of gravitational waves discussed in Chap. 13.

In addition to the GCN, another program that performs a real-time correlation analysis of the high-energy signals across all known astronomical messengers—photons, neutrinos, cosmic rays, and gravitational waves—is the Astrophysical Multimessenger Observatory Network (AMON) (Smith et al. 2013). AMON[1] is an effort to enhance the combined sensitivity of collaborating observatories to astrophysical transients by searching for coincidences in their sub-threshold data and enable rapid follow-up imaging or archival analysis of the putative astrophysical sources. Collaboration members of AMON are characterized as *triggering*, *follow-up*, or both. Triggering participants are generally observatories that monitor a large portion of the sky (such as neutrino telescopes, or Fermi-LAT) and feed a stream of sub-threshold events into the AMON system. These events are processed so as to search for temporal and spatial correlations, leading to secondary *AMON alerts*. Follow-up participants generally search for electromagnetic counterparts to the AMON alerts with narrower field-of-view telescopes. Among cosmic sources expected to emit in two or more such channels are AGNs, GRBs, supernovae, white dwarves, and neutron stars. Joint multimessenger searches are particularly important for sources dark in other messengers, namely those having little or no electromagnetic emission.

As an example, one of the most promising class of sources for searches for joint gravitational waves (GWs) and high-energy neutrinos (HENs) (Baret et al. 2011) is represented by gamma-ray bursts (GRBs), Sect. 8.11. Such potential sources include *choked GRBs*. The relativistic expanding fireball from the burst (i.e., the relativistic outflows or jets of plasma) is thought to produce internal shocks when the central engine releases a large amount of energy over a short time and small volume. Models predict that jets could be launched along the rotational axis of the

[1] http://www.amon.psu.edu/.

progenitor, powered by the gravitational energy. The relativistic outflows and shocks of magnetized plasma occurs for both classes of main cosmic progenitor candidates: stellar core-collapses that produce long GRBs and the merger of compact objects for short GRBs (Sect. 13.11). The γ-rays eventually emitted from its core will only be observable from the outside once the relativistic jet has broken out of the stellar envelope. If relativistic jets stall before reaching the surface of the star, observable neutrinos could still be emitted, while the envelope is transparent to GWs. This is a *choked GRB*. The result is a GW+HEN emission with little or no γ-ray counterparts. This mechanism has so far only been theoretically hypothesized; the eventual discovery of choked GRBs will have a fundamental impact on models of stellar evolution and on those for the generation of trans-iron elements.

References

M.G. Aartsen et al., Evidence for high-energy extraterrestrial neutrinos at the IceCube detector. Science **342**, 1242856–1 (2013a). Updated in arXiv:1405.5303

M.G. Aartsen et al., Search for time-independent neutrino emission from astrophysical sources with 3 years of IceCube data. Astrophys. J. **779**, 132 (2013b)

M.G. Aartsen et al., IceCube collaboration, Observation and characterization of a cosmic muon neutrino flux from the northern hemisphere using six years of IceCube data. Astrophys. J. **833**, 3 (2016)

M.G. Aartsen et al., The IceCube Collaboration, Contributions to ICRC 2017 Part II: properties of the atmospheric and astrophysical neutrino flux (2017). ArXiv:1710.01191. PoS(ICRC2017)981 and 1005

R. Abbasi et al., Limits on neutrino emission from gamma-ray bursts with the 40 string IceCube detector. Phys. Rev. Lett. **106**, 141101 (2011)

R. Abbasi et al., An absence of neutrinos associated with cosmic-ray acceleration in γ-ray bursts. Nature **484**, 351 (2012)

P. Abreu et al., Ultrahigh energy neutrinos at the Pierre Auger Observatory. Adv. High Energy Phys. **2013**, 708680 (2013). https://doi.org/10.1155/2013/708680

S. Adrián-Martínez et al., Search for cosmic neutrino point sources with four years of data from the ANTARES telescope. Astrophys. J. **760**, 53 (2012)

S. Adrián-Martínez et al. Search for muon neutrinos from gamma-ray bursts with the ANTARES neutrino telescope using 2008 to 2011 data. Astron. Astrophys. **559**, A9 (2013) . Also arXiv:1307.0304v1

S. Adrián-Martínez et al., Searches for point-like and extended neutrino sources close to the galactic center using the ANTARES neutrino telescope. Astrophys. J. Lett. **786**, L5 (2014)

S. Adrián-Martínez et al., Letter of intent for KM3NeT 2.0. J. Phys. G: Nucl. Part. Phys. **43**, 084001 (2016)

A. Albert et al., ANTARES Collaboration. All-flavor search for a diffuse flux of cosmic neutrinos with nine years of ANTARES data. Astrophys. J. Lett. **853**, L7 (2018)

D. Allard, Extragalactic propagation of ultrahigh energy cosmic-rays. Astropart. Phys. **39–40**, 33–43 (2012)

R. Aloisio, P. Blasi, I. De Mitri, S. Petrera, Selected topics in cosmic ray physics (2017). ArXiv:1707.06147

B. Baret et al., Bounding the time delay between high-energy neutrinos and gravitational-wave transients from GRBs. Astropart. Phys. **35**, 1 (2011)

J. Becker, High-energy neutrinos in the context of multimessenger astrophysics. Phys. Rep. **458**, 173–246 (2008)

S. Braibant, G. Giacomelli, M. Spurio, *Particle and Fundamental Interactions* (Springer, Berlin, 2011). ISBN: 978-9400724631

T. Chiarusi, M. Spurio, High-energy astrophysics with neutrino telescopes. Eur. Phys. J. **C65**, 649 (2010)

A.L. Connolly, A.G. Vieregg. Radio detection of high energy neutrinos, in *Neutrino Astronomy* (Word Scientific, Singapore, 2017). ISBN: 978-981-4759-40-3. ArXiv:1607.08232

T.K. Gaisser, F. Halzen, T. Stanev, Particle astrophysics, with high-energy neutrinos. Phys. Rep. **258**, 173–236 (1995). Erratum-ibid. **271**, 355–356 (1996)

D. Guetta et al., Neutrinos from individual gamma-ray bursts in the BATSE catalog. Astropart. Phys. **20**, 429–455 (2004)

F. Halzen, Astroparticle physics with high energy neutrinos: from AMANDA to IceCube. Eur. Phys. J. C **46**, 669–687 (2006)

R. Lahmann, Acoustic detection of neutrinos: review and future potential. Nucl. Part. Phys. Proc. **273–275**, 406–413 (2016)

P. Sapienza, G. Riccobene, High-energy neutrino astronomy. Riv. Nuovo Cimento **32**(12), 591 (2009)

M.W.E. Smith et al., The astrophysical multimessenger observatory network (AMON). Astropart. Phys. **45**, 56 (2013)

C. Spiering, Towards high-energy neutrino astronomy: a historical review. Eur. Phys. J. **H37**, 515–565 (2012). Also: arXiv:1207.4952v1

C. Stegmann, A. Kappes, J. Hinton, F. Aharonian, Potential neutrino signals in a northern hemisphere neutrino telescope from galactic gamma-ray sources. Astrophys. Space Sci. **309**, 429 (2007)

F. Vissani, F. Aharonian, Galactic sources of high-energy neutrinos: highlights. Nucl. Instrum. Meth. A **692**, 5–12 (2012)

E. Waxman, J. Bahcall, High energy neutrinos from cosmological gamma-ray burst fireballs. Phys. Rev. Lett. **78**, 2292–2295 (1997)

E. Waxman, J. Bahcall, High energy neutrinos from astrophysical sources: an upper bound. Phys. Rev. **D59**, 023002–1 (1998)

Chapter 11
Atmospheric Muons and Neutrinos

Abstract Muons are the most abundant charged particles arriving at sea level and the only ones able to penetrate deeply underground. The reason stems from their small energy loss, their relatively long lifetime, and their small interaction cross-section. The flux of muons with energy >1 GeV at sea level is on the order of 200 particles/(m^2 s). In this chapter, starting from the production of secondary nucleons and charged mesons by primary CRs interactions with atmospheric nuclei, we derive the energy spectra of atmospheric muons and atmospheric neutrinos. Atmospheric muons can penetrate up to ∼12 km of water. The knowledge of the underground muon flux is important for evaluating the background in searches for rare events in underground laboratories, as the proton decay predicted by Grand Unified Theories. The first generation of underground experiments immediately realized that atmospheric neutrinos represent the irreducible background. Because of the close relation between muon and neutrino production, the parameters characterizing the muon spectrum can provide important information on the atmospheric neutrino flux. These early searches for rare phenomena predicted by GUT theories failed, but these experiments discovered an unexpected phenomenon: the disappearance of atmospheric neutrino, explained by neutrino oscillations. The high-precision measurements of the oscillation parameters of atmospheric neutrinos represent the primary contribution of astroparticle experiments to particle physics, successively confirmed by accelerator experiments.

Muons are the most abundant charged particles arriving at sea level and the only ones able to penetrate deeply underground. The reason stems from their small energy loss (only ∼2 GeV across the whole atmosphere), their relatively long lifetime, and their fairly small interaction cross-section. The flux of muons with energy >1 GeV at sea level is on the order of 200 particles/(m^2 s). In this chapter, starting from the production of secondary nucleons (Sect. 11.1) and charged mesons (Sect. 11.2) by primary CRs interactions with atmospheric nuclei, we derive the energy spectra of atmospheric muons and atmospheric neutrinos, Sect. 11.3. The measurements of the muon flux and energy spectrum at sea level are presented in Sects. 11.4 and 11.5.

© Springer Nature Switzerland AG 2018

M. Spurio, *Probes of Multimessenger Astrophysics*, Astronomy and Astrophysics Library, https://doi.org/10.1007/978-3-319-96854-4_11

Atmospheric muons can penetrate up to ~12 km of water. As for high-energy cosmic neutrinos, for the study of low-energy astrophysical neutrinos, and in the search for rare events in the cosmic radiation presented in the following chapters, atmospheric muons represent the most dangerous background. The flux of underground muons as a function of depth, Sect. 11.6, is important for evaluating the background in searches for rare events. For instance, the simplest Grand Unified Theories (GUTs), the theories developed since the 1980s suggesting the unification of the electroweak interaction with the strong one, predicted proton lifetime values of $\tau_p \sim 10^{30}$ years and the existence of massive magnetic monopoles. This motivated the searches with kton- and 1000 m^2-scale detectors in underground laboratories. The first generation of underground experiments immediately realized that atmospheric neutrinos represent the irreducible background, Sect. 11.7. Because of the close relation between muon and neutrino production, the parameters characterizing the muon spectrum can provide important information on the atmospheric neutrino flux.

These early searches for rare phenomena predicted by GUT theories failed, but these experiments discovered an unexpected phenomenon: the disappearance of atmospheric ν_μ, explained by neutrino oscillations, Sect. 11.8. The high-precision measurements of the oscillation parameters of atmospheric ν_μ (Sect. 11.9) represent the primary contribution of astroparticle experiments to particle physics, successively confirmed by accelerator experiments, Sect. 11.10. The flux of atmospheric neutrinos at higher energies, up to 100 TeV, was measured by neutrino telescopes, Sect. 11.11.

11.1 Nucleons in the Atmosphere

Air showers (Chap. 4) are described by a set of coupled cascade equations with boundary conditions at the top of the atmosphere to match the primary spectrum. Using transport equations, analytic expressions of the cascade can be constructed. The solutions of these equations allow for computing the differential particle flux anywhere within the atmosphere. Concerning the muon component, some approximate analytic solutions are valid in the limit of high energies (Gaisser 1990, 2002; Lipari 1993). Numerical or Monte Carlo calculations are needed to account accurately for decay and energy loss processes, and for the energy dependence of the cross-sections.

We use here a simplified one-dimensional differential transport equation in the atmosphere following the approach of Gaisser (1990). Neutrons are stable during the transit time in the atmosphere, and the effect of their decay is completely negligible. An important parameter for describing the interactions and the subsequent propagation of the particles is the vertical atmospheric depth X_v (Sect. 4.2). The nucleon mean free path in the atmosphere is given in Eq. (3.2). Some aspects of the mathematical developments are similar to those used in Sect. 5.1 for the propagation in the interstellar matter of $\mathscr{M}$ and $\mathscr{L}$ nuclei.

Let the quantity $\mathcal{N}_N(E, X)dE$ represent the flux of nucleons (protons and neutrons) with energy in the interval E to $E + dE$ at the slant depth X in the atmosphere. The attenuation of nucleons traversing a layer dX of atmosphere is,

$$\frac{\partial \mathcal{N}_N(E, X)}{\partial X} = -\frac{\mathcal{N}_N(E, X)}{\lambda_{I_N}} + \int_E^\infty \frac{\mathcal{N}_N(E', X)}{\lambda_{I_N}} F_{NN}(E, E') \frac{dE'}{E} . \qquad (11.1)$$

Note that we assume as constant the cross-sections of nucleons on atmospheric nuclei, and correspondingly a nucleon mean free path $\lambda_{I_N} = \lambda_{I_p} = 85\ \mathrm{g\,cm^{-2}}$, as in (3.8a). The first term (with the − sign) in (11.1) is an attenuation term, indicating that the number of nucleons of a given energy E decreases as X increases. From baryon number conservation, it follows that the total number of nucleons is constant. The positive term takes into account the fact that an incident nucleon of energy E' can collide with an air nucleus and produce a nucleon with energy E. The quantity $F_{ac}(E_c, E_a)$ represents, in general, the probability that the particle c with energy E_c be produced by the particle a with energy $E_a > E_c$ during the process:

$$a + X \to c + Y . \qquad (11.2)$$

If a, c are hadrons, from the properties of hadronic interactions, namely from Feynman scaling (see *Extras # 3*), we have

$$F_{ac}(E_c, E_a) = F_{ac}(x^*) \quad \text{where} \quad x^* = \frac{E_c}{E_a} . \qquad (11.3)$$

This *scaling* condition represents the fact that in hadronic interactions, the probability that a 10 GeV secondary particle will be produced by a 100 GeV primary is the same as the probability of producing a 1 GeV secondary from a 10 GeV primary. Feynman scaling is violated in high-energy interactions.

Equation (11.1) has the boundary condition (2.20a) at the top of the atmosphere:

$$\mathcal{N}_N(E, 0) = \Phi(E) = K E^{-\alpha} . \qquad (11.4)$$

Note from this equation that the units of $\mathcal{N}_N$ are the same as those of $\Phi(E)$, namely ($\mathrm{cm^{-2}\,s^{-1}\,sr^{-1}\,GeV^{-1}}$). The same applies to the quantities $\mathcal{N}_\pi$, $\mathcal{N}_\mu$ defined in the following. Using Feynman scaling (11.3), Eq. (11.1) can be written in terms of the adimensional quantity x^* as

$$\frac{\partial \mathcal{N}_N(E, X)}{\partial X} = -\frac{\mathcal{N}_N(E, X)}{\lambda_{I_N}} + \frac{1}{\lambda_{I_N}} \int_0^1 \mathcal{N}_N\left(\frac{E}{x^*}, X\right) F_{NN}(x^*) \frac{dx^*}{x^{*2}} , \qquad (11.5)$$

where use has been made of the fact that $x^* = E/E'$, and thus $dx^* = \frac{E}{E'^2} dE' = x^{*2} \frac{dE'}{E}$, and finally $\frac{dE'}{E} = \frac{dx^*}{x^{*2}}$. We can solve the problem by assuming factorization as follows:

$$\mathscr{N}_N(E,X) = \Phi_N(E)\cdot H_N(X)\,. \tag{11.6}$$

Then, Eq. (11.5) becomes

$$\Phi_N\frac{dH_N}{dX} = -\frac{\Phi_N\cdot H_N}{\lambda_{I_N}} + \frac{H_N}{\lambda_{I_N}}\int_0^1 \Phi_N(E/x^*)\cdot F_{NN}(x^*)\cdot\frac{dx^*}{x^{*2}}\,. \tag{11.7}$$

Dividing the two sides by $\Phi_N\cdot H_N$

$$\frac{1}{H_N}\frac{dH_N}{dX} = -\left(\frac{1}{\lambda_{I_N}} - \frac{1}{\Phi_N\lambda_{I_N}}\int_0^1 \Phi_N(E/x^*)\cdot F_{NN}(x^*)\cdot\frac{dx^*}{x^{*2}}\right) \equiv -\left(\frac{1}{\Lambda_N}\right)\,. \tag{11.8}$$

The solution of Eq. (11.8) is

$$H_N(X) = H_N(0)\cdot\exp\left(-\frac{X}{\Lambda_N}\right)\,, \tag{11.9}$$

and the flux of nucleons of a given energy is exponentially attenuated during propagation in the atmosphere with attenuation length Λ_N. At a given depth X, the form of the energy spectrum $\Phi_N(E)$ at the top of the atmosphere is preserved. The quantity Λ_N has the dimension of an attenuation length, exactly as λ_{I_N}. The second term in (11.8) increases the effective mean free path, because of the regeneration of p, n with energy E from higher energy nucleons during propagation in the atmosphere. Λ_N depends on the energy spectrum $\Phi_N(E)$ and on the F_{NN} function. In general, the F_{ac} functions are parameterized by phenomenological models of hadronic interactions using accelerator data.

A deeper insight into (11.8) can be obtained assuming that the solution $\Phi_N(E)$ has the same dependence on energy as the boundary condition at $X = 0$, i.e.,

$$\Phi_N(E) = KE^{-\alpha}\,. \tag{11.10}$$

By replacing this trial solution in (11.8), we obtain

$$\begin{aligned}\frac{1}{\Lambda_N} &= \frac{1}{\lambda_{I_N}}\left(1 - \frac{1}{E^{-\alpha}}\int_0^1\left(\frac{E}{x^*}\right)^{-\alpha}\cdot F_{NN}(x^*)\cdot\frac{dx^*}{x^{*2}}\right)\\ &= \frac{1}{\lambda_{I_N}}\left(1 - \int_0^1 (x^*)^{\alpha-2}\cdot F_{NN}(x^*)\cdot dx^*\right) = \frac{1}{\lambda_{I_N}}(1 - Z_{NN})\,.\end{aligned} \tag{11.11}$$

Table 11.1 Atmospheric equivalent attenuation lengths for nucleons, pions and kaons from the definition (11.13)

Λ_N	Λ_π	Λ_K
120 g cm^{-2}	160 g cm^{-2}	180 g cm^{-2}

The quantities Z_{ac} depend on the hadronic interactions and are called *spectrum-weighted moments*. In general

$$Z_{ac} \equiv \int_0^1 (x^*)^{\alpha-2} \cdot F_{ac}(x^*) \cdot dx^* \,. \tag{11.12}$$

The quantity

$$\Lambda_a \equiv \frac{\lambda_{I_a}}{1 - Z_{aa}} \tag{11.13}$$

is the *equivalent attenuation length* of particle a when propagating in the atmosphere, under the effect of regeneration of higher energy particles of the same type. Table 11.1 shows the equivalent attenuation lengths for the main hadronic components of the atmospheric shower [from Gaisser 1990]. The values are about 20–30% higher than the corresponding values of λ_{I_a}.

Under the aforementioned simplifications (and in particular, under Feynman scaling), the elementary solution of the cascade equation preserves the power-law energy spectrum. The quantity Z_{NN} in (11.11) is a constant that produces an increase in the effective mean free path of nucleons in the atmosphere. The flux is attenuated as nucleons proceed in the atmosphere. The nucleons' flux in the atmosphere (11.6), using (11.9 and 11.10), is thus given by

$$\mathcal{N}_N(E, X) = H_N(0) \cdot e^{-X/\Lambda_N} \cdot KE^{-\alpha} \,. \tag{11.14}$$

With the boundary condition (11.4), $H_N(0) = 1$. Figure 11.1 shows the *vertical* flux of different secondary CR components as a function of the atmospheric depth X, and that of nucleons is indicated by $p+n$. We will show in the following that the flux of secondary particles in the atmosphere depends on their zenith angle θ. For this reason, the flux measured (or computed) in a small solid angle region centered on the vertical direction ($\cos\theta = 1$) is usually reported. This is called the *vertical flux*.

11.2 Secondary Mesons in the Atmosphere

All hadrons can be produced by a primary CR interaction with an air nucleus. The full shower development can be described today using Monte Carlo simulations, Sect. 4.5. A look into analytical solutions is useful for understanding the underlying processes. The two principal channels that produce atmospheric muons

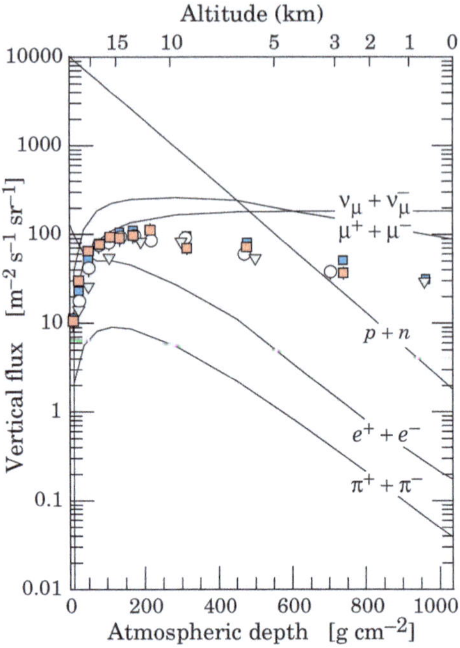

Fig. 11.1 Vertical fluxes as a function of the atmospheric depth X of different CR components with $E > 1$ GeV. The estimates derive from the nucleon flux (2.20a). The points show measurements of negative muons with $E_\mu > 1$ GeV. Figure from Beringer et al. (2012)

and neutrinos are the decays of charged pions and kaons. The development of $\pi^\pm$ (as well as of kaons) can be described mimicking Eq. (11.5), with the difference being that

- Charged pions can either interact or decay with characteristics lengths λ_{I_π} and d_π, respectively;
- Their mean free path $\lambda_{I_\pi} > \lambda_{I_N}$ in atmosphere, see (3.8b);
- At a given energy, pions can be produced either by interactions of nucleons or of more energetic pions;
- Pions are not present as primary radiation: this corresponds to the boundary condition $\mathcal{N}_\pi(E, 0) = 0$.

Including all these effects, the differential equation describing the pion propagation is, with obvious notation

$$\frac{\partial \mathcal{N}_\pi(E, X)}{\partial X} = -\left(\frac{1}{\lambda_{I_\pi}} + \frac{1}{d_\pi}\right)\mathcal{N}_\pi(E, X) \tag{11.15}$$

$$+ \int_0^1 \frac{\mathcal{N}_\pi(\frac{E}{x^*}, X)}{\lambda_{I_\pi}} \cdot F_{\pi\pi}(x^*) \cdot \frac{dx^*}{x^{*2}} + \int_0^1 \frac{\mathcal{N}_N(\frac{E}{x^*}, X)}{\lambda_{I_N}} \cdot F_{N\pi}(x^*) \cdot \frac{dx^*}{x^{*2}} .$$

This equation has no simple solution, because it does not factorize as Eq. (11.5) did. The reason is that the quantity d_π depends on E and on X. The pion decay length $d'_{\pi^\pm} = \Gamma c\tau_{\pi^\pm}$ was already introduced in Eq. (4.27) (units: cm). The quantity in units of (g cm^{-2}) is obtained by multiplying by the air density: $d_\pi = \rho(X_v) d'_{\pi^\pm}$.

Table 11.2 Lifetime τ_i and decay constants $\epsilon_i = mc^2 h_0/c\tau_0$ for secondary particles i produced by primary hadrons

Particles						
$i =$	$\mu^\pm$	$\pi^\pm$	π^0	$K^\pm$	$D^\pm$	D^0
τ_i (s)	2.19×10^{-6}	2.60×10^{-8}	8.4×10^{-17}	1.24×10^{-8}	1.04×10^{-12}	4.10×10^{-13}
ϵ_i (GeV)	1.0	115	3.5×10^{10}	850	4.3×10^7	9.2×10^7

The complication arises from the fact that the air density depends on the atmospheric depth X_v. From (4.8), we have $\rho(X) = X_v/h_0 \simeq \frac{X\cos\theta}{h_0}$ for zenith angles $\theta \lesssim 60°$. Thus, using the fact that $\Gamma = E/m_\pi c^2$, we have

$$\frac{1}{d_\pi} = \frac{1}{\Gamma c\tau_{\pi^\pm}\rho(X)} = \frac{m_\pi c^2 h_0}{Ec\tau_{\pi^\pm} X\cos\theta} = \frac{\epsilon_\pi}{EX\cos\theta} . \tag{11.16}$$

The quantity

$$\epsilon_\pi \equiv \frac{m_\pi c^2 h_0}{c\tau_{\pi^\pm}} \tag{11.17}$$

has the dimension of an energy and corresponds to the *characteristic pion decay constant*, see Table 11.2.

Explicitly inserting (11.16) into (11.15) and using (11.14) for the nucleon flux, we obtain

$$\begin{aligned}\frac{\partial \mathcal{N}_\pi(E,X)}{\partial X} &= -\left(\frac{1}{\lambda_{I_\pi}} + \frac{\epsilon_\pi}{EX\cos\theta}\right)\mathcal{N}_\pi(E,X) \\ &+ \int_0^1 \frac{\mathcal{N}_\pi\left(\frac{E}{x^*}, X\right)}{\lambda_{I_\pi}} \cdot F_{\pi\pi}(x^*) \cdot \frac{dx^*}{x^{*2}} + \frac{Z_{N\pi}}{\lambda_{I_N}}\mathcal{N}_N(E,0)e^{-X/\Lambda_N} .\end{aligned} \tag{11.18}$$

The term under the integral can be simplified assuming that the unknown $\mathcal{N}_\pi(E,X)$ can be factorized as a product of $E^{-\alpha}$ and a function of the depth. With this assumption, following (11.11)

$$\int_0^1 \frac{\mathcal{N}_\pi\left(\frac{E}{x^*}, X\right)}{\lambda_{I_\pi}} \cdot F_{\pi\pi}(x^*) \cdot \frac{dx^*}{x^{*2}} = \mathcal{N}_\pi(E,X) \cdot Z_{\pi\pi} ,$$

and thus we can define the term: $\frac{1}{\lambda_{I_\pi}}(1 - Z_{\pi\pi}) \equiv \frac{1}{\Lambda_\pi}$. Finally, we can write (11.18) as

$$\frac{\partial \mathcal{N}_\pi(E,X)}{\partial X} = -\mathcal{N}_\pi(E,X)\left(\frac{1}{\Lambda_\pi} + \frac{\epsilon_\pi}{EX\cos\theta}\right) + \frac{Z_{N\pi}}{\lambda_{I_N}}\mathcal{N}_N(E,0)e^{-X/\Lambda_N} . \tag{11.19}$$

As the variables X, E are strongly coupled, we can easily solve (11.19) only under the additional approximations of low energy ($E \ll \epsilon_\pi$) or high energy ($E \gg \epsilon_\pi$).

An equivalent equation holds for charged kaons; in this case, the subscript π must be replaced with K.

High Energy Limit in Pion Production In the high-energy (he) limit ($E \gg \epsilon_\pi$), we neglect, in (11.19), the energy-dependent decay term and

$$\frac{\partial \mathscr{N}_\pi^{he}(E,X)}{\partial X} = +\frac{Z_{N\pi}}{\lambda_{I_N}} \mathscr{N}_N(E,0) e^{-X/\Lambda_N} - \frac{\mathscr{N}_\pi^{he}(E,X)}{\Lambda_\pi} . \tag{11.20}$$

In this form, we recognize immediately the same equation (5.8) obtained for the $\mathscr{L}$ nuclei in the propagation of CRs in Sect. 5.1. $\mathscr{N}_\pi$ has the same boundary condition $\mathscr{N}_\pi(E,0) = 0$ as $N_{\mathscr{L}}(0)$. The equation $\mathscr{N}_N$ describing nucleons has the same exponential behavior as $N_{\mathscr{M}}$. Thus, after replacing

$$\begin{aligned} P_{\mathscr{M}\mathscr{L}} &\rightarrow Z_{N\pi} \\ \lambda_{I_{\mathscr{M}}} &\rightarrow \Lambda_N = \lambda_{I_N}/(1 - Z_{NN}) \\ \lambda_{I_{\mathscr{L}}} &\rightarrow \Lambda_\pi \\ N^0_{\mathscr{M}} &\rightarrow \mathscr{N}_N(E,0) = K E^{-\alpha} , \end{aligned} \tag{11.21}$$

we obtain the same solution as (5.11):

$$\mathscr{N}_\pi^{he}(E,X) = \left[\frac{Z_{N\pi}}{1 - Z_{NN}} \cdot \frac{\Lambda_\pi}{\Lambda_\pi - \Lambda_N} (e^{-X/\Lambda_\pi} - e^{-X/\Lambda_N}) \right] \cdot K E^{-\alpha} . \tag{11.22}$$

The moments Z_{ab} are defined in (11.12) and Λ_N, Λ_π in Table 11.1.

In conclusion, in the high-energy limit, the pion flux (11.22) can be factorized as a function $H_\pi^{he}(X)$—the quantity in the square bracket that depends only on the depth X—and a power-law energy function with the same spectral index α as that of primary nucleons:

$$\mathscr{N}_\pi^{he}(E,X) = H_\pi^{he}(X) \cdot K E^{-\alpha} . \tag{11.23}$$

Low Energy Limit in Pion Production In the low-energy (le) limit, when $E \cos\theta \ll \epsilon_\pi$, we can neglect the term λ_{I_π}. In the le limit, all pions are assumed to decay, and (11.19) becomes

$$\frac{\partial \mathscr{N}_\pi^{le}(E,X)}{\partial X} = -\mathscr{N}_\pi^{le}(E,X) \left(\frac{\epsilon_\pi}{E X \cos\theta} \right) + \frac{Z_{N\pi}}{\lambda_{I_N}} \mathscr{N}_N(E,0) e^{-X/\Lambda_N} . \tag{11.24}$$

It is straightforward to see that if we replace the term $\mathcal{N}_\pi^{le}$ on the right-hand side with

$$\mathcal{N}_\pi^{le}(E, X) = \frac{Z_{N\pi}}{\lambda_{I_N}} \mathcal{N}_N(E, 0) e^{-X/\Lambda_N} \cdot \frac{X E \cos\theta}{\epsilon_\pi} , \tag{11.25}$$

we obtain $\frac{\partial \mathcal{N}_\pi^{le}(E,X)}{\partial X} = 0$. This condition represents the fact that the number of decayed pions are regenerated by the production of new pions in the atmosphere by pions and nucleons. Equation (11.25) has a maximum at $X = \Lambda_N \simeq 120\,\mathrm{g\,cm^{-2}}$, which corresponds to an altitude of 15 km. The correctness of the hypothesis leading to (11.25) is confirmed by Monte Carlo simulations. Using (11.14), the pion flux $\mathcal{N}_\pi^{le}$ in the le limit can be factorized as

$$\mathcal{N}_\pi^{le}(E, X) = \left[\frac{Z_{N\pi}}{\lambda_{I_N}} \cdot e^{-X/\Lambda_N} \cdot X \right] \cdot \frac{E \cos\theta}{\epsilon_\pi} \cdot K E^{-\alpha} = H_\pi^{le}(X) \frac{\cos\theta}{\epsilon_\pi} \cdot K E^{-\alpha+1} . \tag{11.26}$$

The quantities inside the square brackets define $H_\pi^{le}(X)$.

In Fig. 11.1, particles with energy above 1 GeV are considered, and the pion contribution is dominated by the le solution. The main features of the pion component in the atmosphere are evident in the curve labeled $\pi^+ + \pi^-$. The maximum number of pions is at depth $\simeq 120\,\mathrm{g\,cm^{-2}}$. Using (11.26) and (11.14), the ratio between the pion and nucleon components as a function of X is

$$\frac{\mathcal{N}_\pi^{le}(E, X)}{\mathcal{N}_N(E, X)} = \frac{Z_{N\pi} X}{\lambda_{I_N}} \cdot \frac{E}{\epsilon_\pi} \cdot \cos\theta . \tag{11.27}$$

Numerically, the quantity $Z_{N\pi} \simeq 0.08$ (Gaisser 1990). For the energy threshold of 1 GeV considered in Fig. 11.1, the ratio is $(E/\epsilon_\pi) \sim 10^{-2}$. Using $\lambda_{I_N} = 85\,\mathrm{g\,cm^{-2}}$, $Z_{N\pi} = 0.08$, we obtain from (11.27), in the vertical direction, a ratio $\sim 2 \times 10^{-3}$ at the position $X = \Lambda_N = 120\,\mathrm{g\,cm^{-2}}$, where the number of pions in the shower has a maximum, and $\sim 10^{-2}$ at sea level. Compare this result with Fig. 11.1.

11.3 Muons and Neutrinos from Charged Meson Decays

Muons and neutrinos are produced by the decay of charged mesons (mainly pions and kaons). The muon flux as a function of X and E can be deduced from (11.19), folding with the kinematics for the decays:

$$\pi^+(K^+) \to \nu_\mu + \mu^+ \tag{11.28a}$$

$$\hookrightarrow \mu^+ \to \overline{\nu}_\mu + \nu_e + e^+$$

$$\pi^-(K^-) \to \overline{\nu}_\mu + \mu^- \\ \hookrightarrow \mu^- \to \nu_\mu + \overline{\nu}_e + e^- \,. \tag{11.28b}$$

We derive the spectrum of atmospheric muons using the above *le* and *he* limits obtained for the pion flux.

The number of muons and neutrinos (as well as the ratio between particles and antiparticles) is strictly correlated. On the other hand, the energy distribution of muons and neutrinos differs because of their different masses. The loss of pions due to the *decay* process in (11.15) is

$$\frac{d\mathcal{N}_\pi}{dX} = -\frac{\mathcal{N}_\pi}{d_\pi} \,, \tag{11.29}$$

with d_π defined in (11.16). The number of decaying charged pions in a layer dX of atmosphere is

$$d\mathcal{N}_\pi(E, X) = -\frac{\mathcal{N}_\pi(E, X) \cdot \epsilon_\pi}{E \cdot X \cdot \cos\theta} \cdot dX \,. \tag{11.30}$$

Equation (11.30) explains the meaning of the decay constant ϵ_π: if the particle energy is $E \gg \epsilon_\pi$, then the decay process is strongly suppressed with respect to the interaction. As 99.99% of pions decay into $\mu\nu$, the corresponding number of produced muons is

$$d\mathcal{N}_\mu(E_\mu, X) = -d\mathcal{N}_\pi(E, X) \,. \tag{11.31}$$

From this relation and (11.30), we obtain

$$\frac{\partial \mathcal{N}_\mu}{\partial X}(E_\mu, X) = \mathcal{N}_\pi(E, X) \cdot \frac{\epsilon_\pi}{E \cdot X \cdot \cos\theta} \,. \tag{11.32}$$

We use now the *le* and *he* limits of the previous section. In the low-energy limit, we use (11.26) and

$$\frac{\partial \mathcal{N}_\mu^{le}}{\partial X}(E_\mu, X) \simeq \left(H_\pi^{le} \frac{\cos\theta}{\epsilon_\pi} \cdot K E_\mu^{-\alpha+1} \right) \cdot \frac{\epsilon_\pi}{E_\mu \cdot X \cdot \cos\theta} = \frac{H_\pi^{le}}{X} \cdot K E_\mu^{-\alpha} \,. \tag{11.33}$$

In the case of the high-energy limit, we use (11.23):

$$\frac{\partial \mathcal{N}_\mu^{he}}{\partial X}(E_\mu, X) \simeq H_\pi^{he}(X) \cdot \frac{\epsilon_\pi}{E_\mu \cdot X \cdot \cos\theta} \cdot K E_\mu^{-\alpha} \,. \tag{11.34}$$

Note here that the dependence on energy is of the type $E_\mu^{-\alpha-1}$.

The general solution can be obtained as the superposition of the low- and high-energy solutions:

$$\frac{\partial \mathcal{N}_\mu}{\partial X}(E_\mu, X) = K E_\mu^{-\alpha}\left(\frac{A(X)}{1+\frac{B(X)E_\mu \cos\theta}{\epsilon_\pi}}\right), \tag{11.35}$$

with $A(X) = H_\pi^{le}(X)/X$ and $A(X)/B(X) = H_\pi^{he}(X)/X$.

The functional dependence on the atmospheric depth X is not easy to obtain analytically, including with the assumed simplifications regarding hadronic interactions. In general, the values assumed by $H_\pi(X)$ are obtained from full Monte Carlo simulations, although some analytical approximations exist [see for instance Gaisser 1990]. An equation similar to (11.35) holds for kaon decay.

The *differential muon intensity* $\Phi_\mu(E_\mu)$ is normally given at sea level ($X^{\text{sea}} = 1030\,\text{g cm}^{-2}$) and obtained by integration of (11.35) along the whole atmospheric depth. The units of $\Phi_\mu(E_\mu)$ are $\text{cm}^{-2}\,\text{s}^{-1}\,\text{sr}^{-1}\,\text{GeV}^{-1}$, as shown in (11.4). In Monte Carlo simulations, the muon energy loss in the atmosphere $-dE_\mu/dX$ is also taken into account. The so-called atmospheric muon flux at sea level mimics Eq. (11.35) and holds for zenith angles $\theta \lesssim 60°$, including the contributions from decays of pions, kaons, and charmed mesons:

$$\Phi_\mu(E) = K E^{-\alpha}\left(\frac{A_\pi}{1+\left(\frac{B_\pi E}{\epsilon_\pi}\right)\cos\theta} + \frac{A_K}{1+\left(\frac{B_K E}{\epsilon_K}\right)\cos\theta} + \frac{A_c}{1+\left(\frac{B_c E}{\epsilon_c}\right)\cos\theta}\right). \tag{11.36}$$

The coefficients A_i with $i = \pi, K, c$ depend on the ratio of muons produced by pions, kaons, and charmed hadrons. The A_i, B_i coefficients can be derived from Monte Carlo computations, numerical approximations or experimental data. Usually, the term resulting from the charmed mesons can be neglected, because $\epsilon_c \gg \epsilon_{\pi,K}$.

Different estimates of the parameters that enter into (11.36) for the conventional atmospheric muons have been published by several authors. For a review, see Lesparre et al. (2010). For the following, we will use the numbers reported in Beringer et al. (2012), which assumes (11.36) to be valid when muon decays are negligible ($E_\mu > 100/\cos\theta$ GeV) and when the curvature of the Earth can be neglected ($\theta \lesssim 60°$):

$$\begin{aligned} K A_\pi &= 0.14\,\text{cm}^{-2}\,\text{s}^{-1}\,\text{sr}^{-1}\,\text{GeV}^{-1} \\ A_K/A_\pi &= 0.054\,;\; B_\pi = B_K = 1.1\,, \end{aligned} \tag{11.37}$$

in addition to the already known values of $\alpha = 2.7$ and $\epsilon_\pi = 115$ GeV, $\epsilon_K = 850$ GeV, as given in Table 11.2. The contribution from the decay of charmed hadrons is discussed below.

11.3.1 The Conventional Atmospheric Neutrino Flux

As indicated by Eq. (11.28), the production mechanisms of atmospheric muons and neutrinos are strongly correlated; see Illanaet al. (2011) for a recent review. However, due to the two-body kinematics, the energy spectra of the μ's and ν_μ's from meson decays are different.[1]

Exercise Consider the pion decay in the center of mass (c.m.) system. Determine that ($m_\pi = 139.6$ MeV/c^2; $m_\mu = 105.7$ MeV/c^2): (a) The center-of-mass muon energy is given by $E^*_\mu = (m^2_\pi + m^2_\mu)/2m_\pi = 109.8$ MeV; (b) Find the relation for E^*_ν and verify that it corresponds to 29.8 MeV.

As shown in the exercise, muons carry a larger fraction of the meson energy than neutrinos. As a consequence, the energy spectrum of atmospheric ν_μ is given by an equation similar to (11.36), with different coefficients A_i, B_i in order to produce a distribution slightly shifted toward lower energies. In the laboratory system, the energies are boosted by the Lorentz factor $\Gamma = E_\pi/m_\pi c^2$. This gives the so-called *conventional atmospheric neutrino flux*.

Additional ν_μ are produced by the in-flight decay of muons, together with a ν_e and an electron/positron. Therefore, the ν_e flux also depends on the decay chain of charged mesons and muons. As the muon decay probability in the atmosphere decreases with increasing E_μ, the ν_e spectrum is reduced with respect to that of ν_μ at high energy, see Sect. 11.7.

11.3.2 The Prompt Component in the Muon and Neutrino Flux

At sufficiently high energies, another muon (and neutrino) production mechanism is possible. The so-called *prompt* (or *direct*) atmospheric muons are produced in the semileptonic decays [see Sect. 8.11 Braibant et al. 2012] of charmed mesons, like $D^\pm$, D^0, and baryons. As the lifetime of charmed particles is smaller than 10^{-12} s (*prompt decays*), prompt muons are produced before the parents lose energy in collisions and are, in general, highly energetic for kinematic reasons. Hence, for $E_\mu < \epsilon_c$, the spectrum for the prompt flux more closely follows the CR spectrum ($\Phi_\mathrm{prompt} \propto E^{-\alpha}$) and is about one power harder than the *conventional* flux at high

[1] In the following, when not explicitly stated, we use the symbol ν_μ or ν_e to indicate both neutrinos and antineutrinos of the given flavor.

energy. Since the production cross-section of charmed mesons in proton–nucleon interactions is rather small, D decays contribute significantly only at very high energies. The prompt flux (both of muons and neutrinos) has not yet been measured, but is expected to be important above $\sim$100 TeV. As for the conventional flux, predictions of the prompt one are dependent upon uncertainties in the normalization and spectral distribution of the primary CR flux. Additional sources of uncertainty for the prompt muons and neutrinos include charm production cross-sections and fragmentation functions, which have not been measured at these energies in accelerator experiments.

11.4 The Particle Flux at Sea Level

Muons are the dominant components of charged particles at sea level. As seen in the previous discussion, the bulk of the processes that yield muons are two-body decays with an associated ν_μ or $\overline{\nu}_\mu$ to satisfy conservation laws. Below the few GeV energy range, the probability of muon decay cannot be neglected. A 1 GeV muon has a Lorentz factor $\Gamma = E_\mu/m_\mu c^2 \sim 10$ and a mean decay length $d_\mu = \Gamma\tau_\mu c \sim 6$ km. Since low-energy pions are typically produced at altitudes of 15 km and decay relatively fast (for $\Gamma = 10$, the decay length is $d_\pi \sim 78$ m, which is almost the same value as λ_{I_π}), the daughter muons do not reach sea level, but rather decay themselves or are absorbed into the atmosphere.

The situation changes at higher energies. For 100 GeV pions ($d_\pi \sim 5.6$ km, corresponding to a column density of 160 g cm^{-2} measured from the production altitude), the interaction probability starts to dominate over that of decay. Pions of these energies will therefore produce further tertiary pions in subsequent interactions, which will eventually decay into muons, typically of lower energy. Therefore, the muon spectrum at high energies is always steeper compared to the parent pion spectrum.

Resuming the discussions of the previous sections, three different energy regions in the sea-level muon spectrum are distinguishable. The thresholds between different energy regimes are set by the values of the decay constants defined in Table 11.2:

- $E_\mu \leq \epsilon_\mu \sim 1$ GeV. Muon decay and muon energy loss are important and must be taken into account. Only full Monte Carlo simulations give accurate predictions. The energy spectrum is almost flat, gradually reproducing the energy dependence of the primary CR spectrum above 10 GeV.
- $\epsilon_\mu \leq E_\mu \leq \epsilon_{\pi,K}$. Above $\sim$100 GeV, the muon flux has the same power law as the parent mesons, and hence as the primary CRs. Below 100 GeV, the effect of the muon energy loss in the atmosphere is still important, particularly approaching the horizontal direction.
- $E_\mu \gg \epsilon_{\pi,K}$. The meson production spectrum has the same power-law dependence as the primary CRs, $\Phi_{\pi,K} \propto E^{-\alpha}$, but the rate of their decay steepens

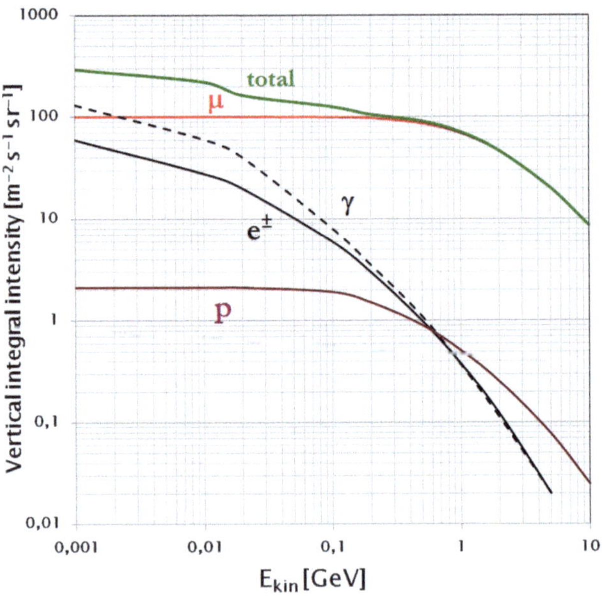

Fig. 11.2 Integral fluxes averaged over the 11-year solar cycles of μ, e, p and γ-rays arriving at geomagnetic latitudes $\sim$40° versus their kinetic energy. The lines include the contributions of both particles and antiparticles. From Cecchini and Spurio (2012)

by one power of E_μ since the pion and kaon decay length $d_{\pi,K} \propto 1/E$. The thickness of the atmosphere is not large enough for most pions to decay, because of the high Lorentz factor.

Plots of the integral flux of muons arriving at geomagnetic latitudes $\sim$40° versus their kinetic energy are shown in Fig. 11.2. The muon intensity depends on the incoming direction, due to the $\cos\theta$ dependence of (11.36). From the vertical direction and for $E_\mu > 1$ GeV, the intensity corresponds to $\sim$1 particle per cm^2 and per min: $I_v(E_\mu > 1\text{GeV}) \sim 70\,\text{m}^{-2}\,\text{s}^{-1}\,\text{sr}^{-1}$ (Grieder 2010). The mean energy of muons at ground level is about 3–4 GeV (Beringer et al. 2012).

The muon intensity from the horizontal directions at low energies is naturally reduced because of muon decays and absorption effects in the thicker atmosphere at large zenith angles. At high energy, the parent particles of muons travel relatively long distances in rare parts of the atmosphere. As a consequence, their decay probability is increased compared to the interaction probability.

Figure 11.3 gives a quantitative description of this effect. Muons below the few GeV/c momentum range fade fairly quickly with increasing zenith angle, with a dependence $\propto \cos^n\theta$, with $n \sim 2 \div 3$. The flux of muons within the 100 GeV/c range is relatively flat up to $\cos\theta \simeq 0.2$, and then quickly declines. At 1 TeV/c, the flux monotonically increases with the zenith angle, approaching the $1/\cos\theta$ dependence. The flux of TeV muons is particularly sensitive at large values of the

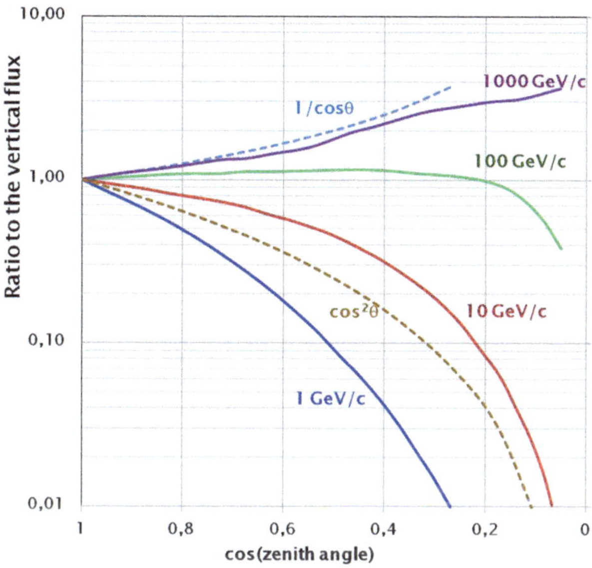

Fig. 11.3 Angular distribution of muons at ground level for different muon energies. The overall angular distribution of muons measured at sea level is $\propto \cos^2\theta$, which is characteristic of muons with $E_\mu \sim 3\,\text{GeV}$. At high energies, the flux approaches the $1/\cos\theta$ dependence, as predicted by (11.36). The estimate of the angular distribution is based on a Monte Carlo simulations and accounts for the curvature of the Earth's atmosphere

zenith angle. When approaching the horizontal direction, a small difference in $\cos\theta$ appreciably changes the thickness and the density profile of the atmosphere and the corresponding muon energy spectrum. For this reason, the measurement of almost horizontal muons is very difficult.

The **electromagnetic** component is made of electrons, positrons, and photons initiated by decays of neutral and charged mesons. At variance with the case of the upper atmosphere, where the decay of neutral pions is the dominant component, muon decay $\mu \rightarrow e\nu_e\nu_\mu$ is the dominant source of low-energy electrons at sea level. The integral vertical intensity of electrons plus positrons is very approximately 30, 6, and 0.2 $\text{m}^{-2}\,\text{s}^{-1}\,\text{sr}^{-1}$ above 10, 100, and 1000 MeV, respectively (Grieder 2010). The exact numbers depend sensitively on altitude, on the solar epoch of the measurement and on geomagnetic latitude. Figure 11.2 shows the integral fluxes of the electromagnetic component (lines with label $e^\pm$ for e^+, e^- and γ for the photons). Because we include here secondary particles produced by sub-GeV primary CRs, the fluxes are averaged over the 11-year solar cycles. The angular dependence is complex, because of the different altitude dependence of the muons decaying into electrons and the other different electron sources.

The **hadronic component** is made mainly of nucleons. The presence of other long-lived hadrons, such as the charged pions, is relatively small, $O(10^{-2})$ with respect to the nucleons, as derived in Sect. 11.2. The contributions of other hadrons can be neglected. Nucleons with momentum >1 GeV/c at ground level are degraded remnants of the primary cosmic radiation. The intensity is given approximately by (11.9) for $\theta \lesssim 60°$. At sea level, about 1/3 of the nucleons in the vertical direction are neutrons (neutrons are about ~10% at the top of the atmosphere). The integral intensity of vertical protons above 1 GeV/c at sea level is $\sim 0.6\,\mathrm{m^{-2}\,s^{-1}\,sr^{-1}}$ at the geomagnetic location of about 40° (see Fig. 11.2).

11.5 Measurements of Muons at Sea Level

The measurements of muons at ground level offer the advantage of a high stability, large collecting factor, and long exposure time due to relatively favorable experimental conditions. Sea level data offer the possibility of performing a robust check of the reliability of existing Monte Carlo codes.

Many experiments devoted to the measurement of the muon momentum spectra and intensity (vertical and inclined directions) have been carried out since the 1970s. Recently, new instruments, mainly spectrometers designed for balloon experiments or used primarily in CERN LEP and LHC experiments and used also for CR studies, have added new, valuable information.

The vertical muon intensity at sea level is a quantity that varies with the geomagnetic latitude, altitude, solar activity, and atmospheric conditions. The 11-year solar cycle modulates the CR flux up to energies of about 20 GeV. When comparing muon observations at such low energies, it is important to know the year when and location where the measurements were made (Cecchini and Spurio 2012).

The *geomagnetic field* tends to prevent low-energy CRs from penetrating through the magnetosphere down to the Earth's atmosphere, Sect. 2.9. Primary nuclei having rigidity lower than the cut-off (2.31) are deflected by the action of the geomagnetic field and do not produce muons. The geomagnetic effects are important for sea level muons up to about $E_\mu \sim 5$ GeV.

Different experimental methods have been used to measure the muon flux and energy spectrum. Muon telescopes are made of several planar detectors arranged horizontally parallel to each other. They are interlaid by one or more layers of absorbing material. In some experiments, the detector and absorber are in a rigid construction that could be rotated in zenith and azimuth, allowing for the selection of muons from a given direction of the celestial hemisphere. The quantity of material (in $\mathrm{g\,cm^{-2}}$) crossed by muons in such a telescope is approximately constant and it sets the muon energy threshold.

Multidirectional muon telescopes generally consist of at least two layers of segmented muon detectors. The coincidence of signals between two counters in the upper and bottom layers determines the arrival direction of muons. The quantity of material crossed by the particle in such detectors increases with increasing zenith angle, so the threshold energy for multidirectional muon telescopes depends on θ.

The muon energy spectrum has been extensively measured, mainly by iron magnet spectrometers. For these detectors, multiple scattering plays an important role in the momentum resolution, particularly at low energies. Recently, measurements have been performed using low-mass superconducting magnet spectrometers designed as a balloon-borne apparatus for cosmic ray studies, such as, for instance, in the BESS experiment (Sect. 3.3).

The atmospheric muon flux and energy spectrum were also measured using the precise muon spectrometer of the L3 detector, which was located at the LEP collider at CERN. This apparatus collected muons ~30 m below a stratified rock overburden, and with dimensions much larger than previous experiments (the volume of the 0.5 T magnetic field region was ~1000 m^3). Finally, information about the muon flux with momenta larger than a few TeV have been extracted from underground measurements, see Sect. 11.6. A comprehensive review of various types of particle detectors used for muon detection at sea level can be found in Dorman (2004).

The compilation of measurements of the momentum of vertical muons is presented in Fig. 11.4. The agreement between measurements is relatively good, and the largest contribution to the deviations is the systematic error due to incorrect knowledge of the acceptance, efficiency of the counters, and corrections for multiple scattering. Measurements of the muon momentum spectra for $p_\mu < 1$ TeV/c are particularly important for the comparison of nuclear cascade models with available data.

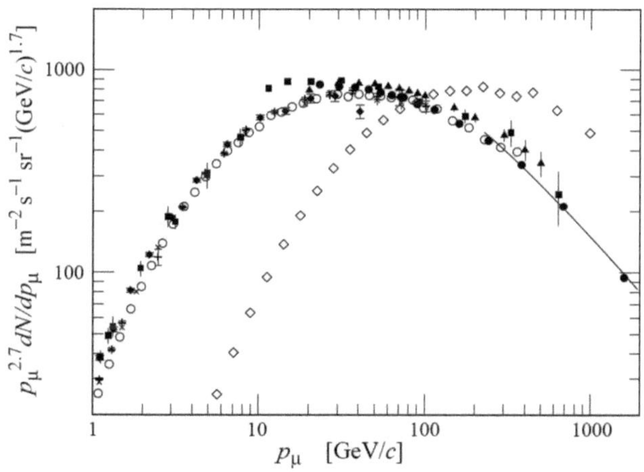

Fig. 11.4 Differential muon intensity $p_\mu^{2.7} \mathrm{d}N_\mu/\mathrm{d}E_\mu$ at sea level. The ordinate values have been multiplied by $p_\mu^{2.7}$ in order to compress the plot and to emphasize the differences. In this energy range, $p_\mu \simeq E_\mu$. All data points correspond to the vertical ($\theta = 0°$), with the exception of the set with the symbol (*open diamond*) corresponding to $\theta = 75°$. The line for $p_\mu > 200$ GeV/c shows the result from Eq. (11.36) for the vertical direction. From Beringer et al. (2012) and references therein

11.6 Underground Muons

Underground measurements offer the possibility of extending the energy range of the muon spectrum beyond 1 TeV. Deep underground detectors normally have large collecting areas and are not subject to the time restrictions of balloon experiments, hence they can measure the muon flux over long periods. The penetrating component of CRs underground depends on a complex convolution of different physics processes, such as muon production mechanisms and muon energy losses. Particularly important is the knowledge of the composition and the thickness of the material overburden above the detector.

11.6.1 The Depth–Intensity Relation

The muon spectrum at energies above a few TeV is deduced through underground measurements. The procedure used for this *indirect* measurement of the sea-level energy spectrum passes through the determination of the so-called *depth–intensity relation* $I^0_\mu(h)$. This quantity represents the muon intensity in the vertical direction as a function of the depth h. The number of muons reaching the depth h depends on their energy distribution at the surface. $I^0_\mu(h)$ is related with to muon intensity $\Phi_\mu(E) = \frac{d^2\varphi_\mu}{dE_\mu d\Omega}$ at the surface. The relation between the measured function $I^0_\mu(h)$ and the differential sea-level muon spectrum is expressed as:

$$I^0_\mu(h) = \int_0^\infty \frac{d^2\varphi_\mu}{dE_\mu d\Omega} \cdot P(E_\mu, h)\mathrm{d}E_\mu \ . \tag{11.38}$$

Here, $P(E_\mu, h)$ is the muon survival function. It represents the probability that muons of energy E_μ at the surface reach a given depth h and is usually determined via Monte Carlo calculations. Assuming (11.36), from (11.38), it is possible to unfold the sea level muon spectrum from the measured vertical muon intensity (Cecchini and Spurio 2012).

As underground detectors are at a fixed depth, in principle, only one point can be obtained. However, when measuring the muon intensity $I_\mu(h,\theta)$ at different zenith angle θ, the quantity of rock (or water) overburden changes. At high energy ($E_\mu > 1$ TeV) and for $\theta \lesssim 60°$, Eq. (11.36) provides a simple relationship between $I^0_\mu(h)$ and $I_\mu(h,\theta)$:

$$I_\mu(h,\theta) = I^0_\mu(h)/\cos\theta \ . \tag{11.39}$$

Using this expression, measurements of the muon intensity at different values of θ can be translated into an estimate of the vertical flux. Figure 11.5 shows the vertical

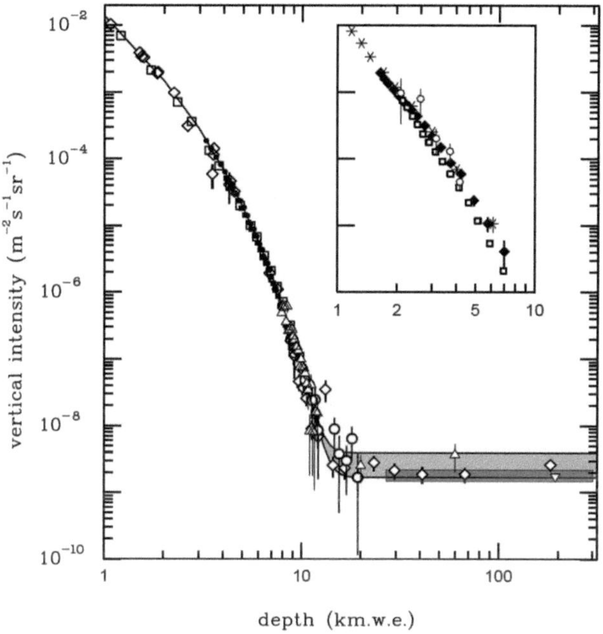

Fig. 11.5 Vertical muon intensity versus depth. The quantity on the x-axis is the depth, expressed in km of water equivalent: 1 km.w.e. = 10^5 g cm^{-2} of standard rock. The *shaded area* at large depths represents neutrino-induced muons of energy >2 GeV. The inset shows the *vertical* intensity curve measured under water and ice from neutrino telescopes (Chap. 10). From Beringer et al. (2012)

muon intensity as a function of depth as measured by underground experiments. The figure also shows the contribution from neutrino-induced muons for depths larger than ∼12 km.w.e. Assuming a threshold energy of 2 GeV, the flux of ν_μ induced upgoing events is $\sim 2 \times 10^{-13}$ cm^{-2} s^{-1} sr^{-1}.

11.6.2 Characteristics of Underground/Underwater Muons

Figure 11.6 shows a multiple muon event detected by the MACRO experiment, Sect. 11.9.3. Multiple events are closely packed bundles of parallel muons, usually of high energy, originating from the same primary CR. These muons are expected to arrive almost at the same time in the plane perpendicular to the shower axis. Multimuon events have been used to explore the properties of very high-energy hadronic interactions and to study the longitudinal development of showers. The multiplicity of produced secondary particles increases with the energy of the initiating particle. The muon multiplicity is an observable correlated with the mass of the primary CR: at a given total energy, heavier nuclei produce more muons than a primary proton.

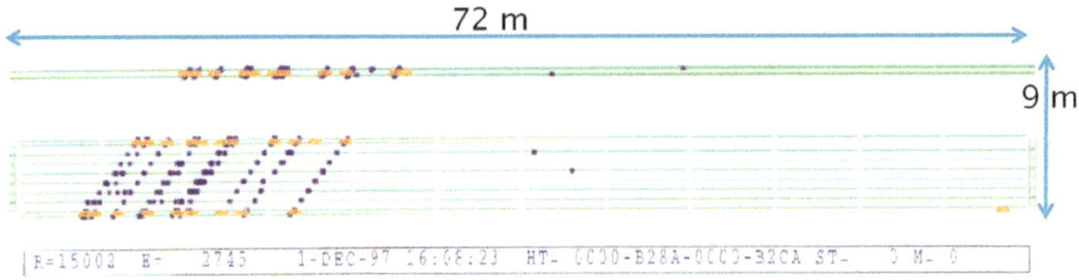

Fig. 11.6 A bundle of muons seen in the MACRO experiment, at a depth of ~3500 m.w.e. 10 different tracks are identified

The interaction vertex of the particles that initiate the air showers is typically at an atmospheric altitude of 15 km. Since secondary particles in hadronic cascades have small transverse momenta p_t (~300 MeV/c), high-energy muons are essentially collimated near the shower axis. Considering a primary nucleon producing mesons of energy $E_{\pi,K}$ with transverse momentum p_t at a height H_{prod}, the average separation of their daughter high-energy muons from the shower axis is given by

$$r \simeq \frac{p_t}{E_{\pi,K}} H_{prod} \,. \tag{11.40}$$

For primary energies around 10^{14} eV, lateral displacements of energetic muons (~1 TeV) of several meters are typically obtained underground. Displacements are almost exclusively caused by transferred transverse momenta in hadronic processes. Typical multiple scattering angles for muon energies around 100 GeV in thick layers of rock (50–100 m) are on the order of a few mrad.

The full characteristics of atmospheric muon bundles are particularly important for neutrino telescopes (Chap. 10) and other underground detectors. In fact, atmospheric muons usually represent the most abundant signal and can be used to calibrate the detectors and to check their expected response to charged particles. On the other side, atmospheric muons represent a dangerous background source. In neutrino telescopes, for instance, they can mimic high-energy neutrino interactions. The main features of muons reaching underground detectors can be reproduced with a full Monte Carlo simulation of atmospheric showers or with parametric formulae (Becherini et al. 2006). These parameterizations allow us to evaluate not only the total muon flux, but also the total number of muon bundles in deep detectors starting from the primary CR flux, CR composition, and interaction model that reproduces the MACRO data.

11.7 Early Experiments for Atmospheric Neutrinos

Starting from the 1980s, the search for proton decay, *Extras #7*, was the main reason for developing underground laboratories and large detectors. The simplest GUT model, SU(5), predicts a proton lifetime value of $\tau_p \sim 10^{30}$ years for the process

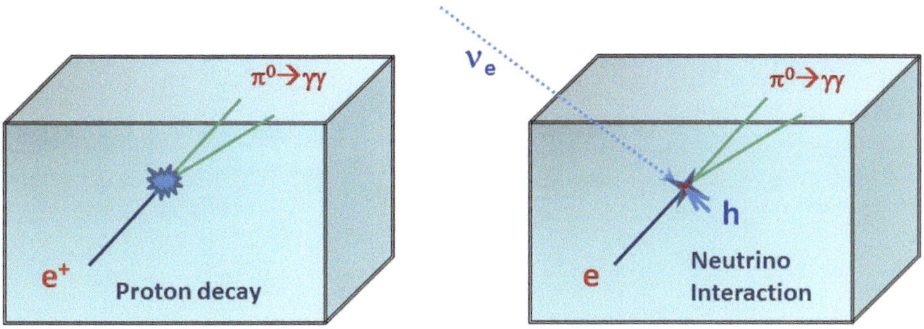

Fig. 11.7 Example of the expected signal (*left*) for the decay mode $p \to e^+\pi^0$ in a water Cherenkov detector or in a fine-grained calorimeter. Some interactions due to atmospheric neutrinos (*right*) are almost indistinguishable from the signal. The remaining hadronic system h might not emit Cherenkov radiation or might remain confined in the passive sectors of tracking experiments

$p \to e^+\pi^0$. From the experimental point of view, the prediction corresponds to many proton decay events in a kiloton-scale detector. The atmospheric neutrinos represented the irreducible background, Fig. 11.7. The experiments started to measure the fluxes of Φ_{ν_μ}, Φ_{ν_e} in the GeV range and their theoretical estimate became a fundamental aspect in astroparticle physics.

The detailed calculation of the atmospheric neutrino fluxes is a nontrivial problem. It requires a description of the primary CRs and a model of hadronic interactions to compute the multiplicity, energy, and angular distributions of the final-state particles. The measurements of the atmospheric muons described in the previous sections help to constrain the neutrino estimates. Usually, a Monte Carlo is used to follow the shower development taking into account all relevant processes, like the energy losses of charged particles, the competition between interaction and decay for unstable hadrons, and the weak decays of mesons and muons; see Gaisser and Honda (2002) for a review.

The uncertainties in the calculation of atmospheric neutrino fluxes differ between high and low energies. In most cases, neutrinos of a given energy E_ν are produced by primary CRs of energy at least a factor of 10 larger. For $E_\nu \sim 1$ GeV, the primary fluxes of CR components are relatively well-known. On the other hand, the low-energy CR flux is modulated by solar activity and affected by the geomagnetic field. As a consequence, the flux of low-energy neutrinos is higher for detectors located near the poles than for those near the equator. At higher energies ($E_\nu > 100$ GeV), solar activity and the rigidity cut-off do not affect the primary CRs, but larger uncertainties exist on the primary flux, in particular, about the chemical composition.

Independently from the details of the computation of $\Phi_{\nu_\mu}(E)$, $\Phi_{\nu_e}(E)$, one can obtain two fundamental and very robust properties:

- At energies below a few GeV, the flux of ν_μ is approximately twice as large as the flux of ν_e, i.e., $\Phi_{\nu_\mu} \simeq 2\Phi_{\nu_e}$.

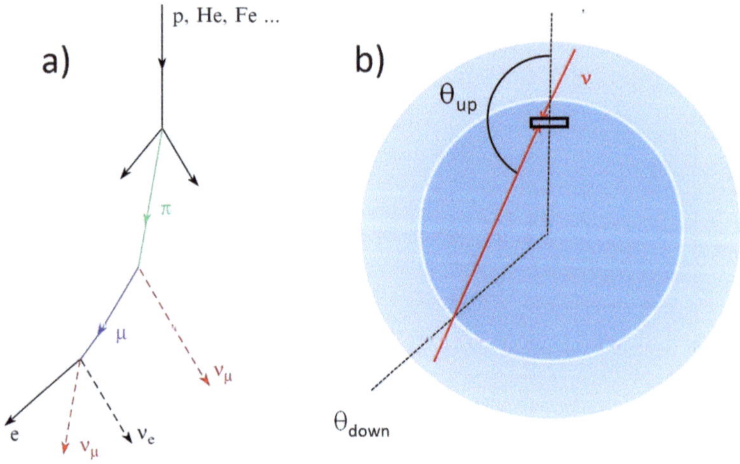

Fig. 11.8 (**a**) CR chain for atmospheric neutrino production. The neutrinos originate from an atmospheric layer of (10 ÷ 20) km thickness. (**b**) up/down symmetry for atmospheric neutrinos

- The fluxes of all neutrino species are up-down symmetric in the zenith angle θ, i.e., $\Phi_{\nu_\alpha}(E_\nu, \theta) = \Phi_{\nu_\alpha}(E_\nu, \pi - \theta)$.

These assumptions are valid if neutrinos do not change flavor during propagation from the production point to the detector. The condition $\Phi_{\nu_\mu}(E)/\Phi_{\nu_e}(E) \simeq 2$ is a simple consequence of (11.28): after the completion of the chain decay, for each π^+, there is one ν_μ, one $\overline{\nu}_\mu$, and one ν_e (and the charge conjugates for the π^-) that have approximately the same average energy, Fig. 11.8a. The ratio increases at high energy ($E_\nu \gtrsim 3$ GeV for vertical neutrinos) when, because of relativistic effects, the muon decay length becomes longer than the thickness of the atmosphere and muons reach ground level dissipating their energy by ionization without decaying. The prediction of the up-down symmetry is even more robust and is a consequence of the (quasi-exact) spherical symmetry of the Earth and the isotropy of the primary CR flux, see Fig. 11.8b.

A remarkably physical effect, originally predicted by Bruno Pontecorvo (1968) is that of *neutrino flavor oscillations*. This is a quantum-mechanical phenomenon connected with non-null neutrino masses. Both of the above predictions for the atmospheric neutrino fluxes offer a gold-plated method for studying flavor oscillations. As the Earth does not absorb neutrinos with energies below tens of TeV (see Sect. 10.7), the path lengths for upgoing and downgoing neutrinos are very different (by a factor of up to 10^3 in the vertical direction). In the presence of oscillations, one expects that the ν_μ, ν_e fluxes would be modified in different ways. Therefore, if oscillations occur, the up-down symmetry will be broken and the effect can be easily observed measuring a difference in the rates of upgoing and downgoing events of different flavors.

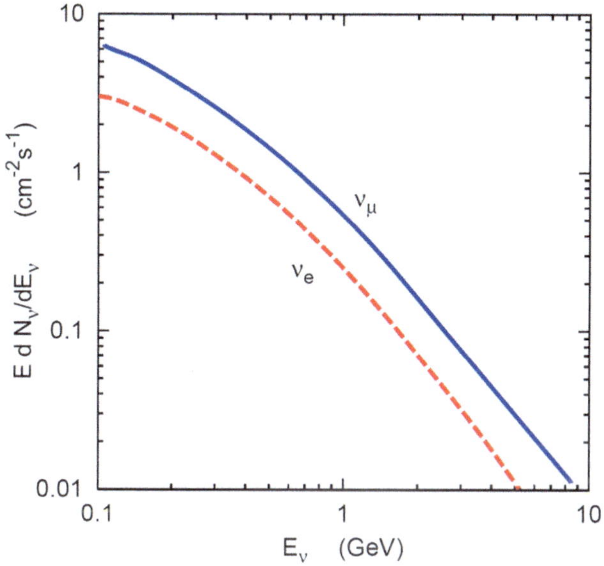

Fig. 11.9 Integral flux of atmospheric neutrinos ($= \nu + \overline{\nu}$) as a function of the threshold energy E_ν up to $\sim$10 GeV. The flux is integrated over the whole *solid* angle. The ν_μ flavor (*blue line*) is about a factor of two higher than the ν_e in the whole energy range. This computation is from the Bartol group (Barr et al. 2004)

Underground experiments of the kton scale, made ready in the 1980s for searches for proton decay, began to measure atmospheric ν_μ and ν_e charged current interactions. Two detection techniques were essentially utilized: tracking iron calorimeters with a segmentation on the order of 1 cm (such as Frejus, NUSEX and Soudan 2) and water Cherenkov detectors, such as Kamiokande and IMB. The flavor (electron or muon) of the charged lepton was identified through the different behaviors of these two particles in the detector.

The expected event rate of atmospheric neutrinos can be obtained from the flux computed via Monte Carlo by different groups Barr et al. (2004) and Honda et al. (2007), as in Fig. 11.9. With a threshold of about 1 GeV (relevant for background evaluation in proton decay searches), the Φ_{ν_μ} (>1 GeV) flux is $\sim$0.65 cm^{-2} s^{-1}, and that of Φ_{ν_e} (>1 GeV) a factor of two lower. The neutrino must interact inside the detector to produce a visible event. The event rate for $T = 1$ y and for the flavor $i = \mu, e$ is given by

$$N_{\nu_i} = \Phi_{\nu_i}(>1\,\text{GeV}) \cdot \sigma_\nu \cdot N_T \cdot T \ , \tag{11.41}$$

where $\sigma_\nu = \sigma_0 E_\nu$ is the neutrino cross-section. After averaging over ν and $\overline{\nu}$, we have $\sigma_0 = 0.5(0.667 + 0.334) \cdot 10^{-38} = 0.5 \cdot 10^{-38}$ cm^2/GeV [see Chapter 10 of Braibant et al. 2012]. $N_T = 6 \times 10^{32}$ nucleons/ton is the number of target nucleons in 1 ton of material, $T = 3.15 \cdot 10^7$ s is the number of seconds in 1 year. Thus, inserting the numerical values, we obtain

$$N_{\nu_\mu} = 66\,\text{kton/year} \quad ; \quad N_{\nu_e} = 34\,\text{kton/year} . \tag{11.42}$$

As a first estimate, the total number of contained events is ~100 events/(kton year).

The results of all the quoted experiments measuring atmospheric neutrinos were expressed in terms of the double ratio $R' = R_{obs}/R_{MC}$, where $R_{obs} = (N_{\nu_\mu}/N_{\nu_e})_{obs}$ is the ratio of observed contained μ and e events and $R_{MC} = (N_{\nu_\mu}/N_{\nu_e})_{MC}$ is the same ratio for Monte Carlo (MC) events. The R' double ratios from IMB and Kamiokande were smaller than expectations and pointed to a deficit of ν_μ, while the NUSEX and Frejus R' agreed with expectations. See Koshiba (1992) for a review.

The IMB and Kamiokande deficit of muon-like events was the first indication of atmospheric neutrino oscillations, discussed in the next section.

11.8 Oscillations of Atmospheric Neutrinos

The Standard Model of the microcosm includes three flavors of massless and left-handed neutrino, denoted as ν_e, ν_μ, ν_τ. The existence of three degenerate massless particles is surprising and cannot explain the difference amongst themselves and the reason for the separate L_e, L_μ, L_τ lepton number conservation.

If the neutrinos have non-null and nondegenerate masses, neutrino mixing and oscillations, as suggested by Pontecorvo, can be formally treated in the same manner as for the quark sector [Chapter 12 of Braibant et al. 2012]. The ν_e, ν_μ, ν_τ states are defined as *weak flavor eigenstates* that must be considered in particle decays (for example, in $\pi^+ \rightarrow \mu^+\nu_\mu$) and in particle interactions (e.g., $\nu_\mu n \rightarrow \mu^- p$). As far as the neutrino propagation is concerned, we shall consider the *mass eigenstates* denoted as ν_1, ν_2, ν_3. It is thus not correct to speak of the ν_μ or of the ν_e masses. According to quantum mechanics, the flavor eigenstates, $|\nu_f\rangle$ ($f = e, \mu, \tau$), are linear combinations of the mass eigenstates $|\nu_j\rangle$ ($j = 1, 2, 3$):

$$|\nu_f(t)\rangle = \sum_j U_{fj} |\nu_j(t)\rangle . \tag{11.43}$$

In vacuum, the mass eigenstates $|\nu_j\rangle$ propagate independently, that is,

$$|\nu_j(t)\rangle = e^{-E_j t} |\nu_j(0)\rangle . \tag{11.44}$$

For a given momentum, the eigenstates $|\nu_j\rangle$ propagate with different frequencies: in (11.44), the energies $E_j = \sqrt{p^2 + m_j}$ (in natural units with $c = 1$) of the mass eigenstates are slightly different for ν_1, ν_2, ν_3 for nondegenerate massive neutrinos.

We will return to the full mixing formula (11.43) in Sect. 12.6. The first indications of neutrino oscillations from atmospheric neutrinos were discussed in terms of the simple approximation of only two neutrino flavors, for example, the pair ν_μ, ν_τ. Each state is a linear combination of the two mass eigenstates, e.g., ν_2, ν_3. The flavor and mass eigenstates are related by a unitary transformation with one mixing angle ϑ:

$$\begin{pmatrix} \nu_\mu \\ \nu_\tau \end{pmatrix} = \begin{pmatrix} \cos\vartheta & \sin\vartheta \\ -\sin\vartheta & \cos\vartheta \end{pmatrix} \begin{pmatrix} \nu_2 \\ \nu_3 \end{pmatrix} . \tag{11.45}$$

Using simple algebra [see Sect. 12.6 of Braibant et al. 2012] it is possible to evaluate the probability $P(\nu_\mu \to \nu_\mu) \equiv |\langle \nu_\mu(t)|\nu_\mu(t)\rangle|$ that the ν_μ originated at $t = 0$ remains a ν_μ at a given time t. On the other hand, the probability that the ν_μ transforms itself into a ν_τ is $P(\nu_\mu \to \nu_\tau)$. These two probabilities are

$$\begin{cases} P(\nu_\mu \to \nu_\mu) = 1 - \sin^2 2\vartheta \cdot \left[\sin^2\left(\frac{E_3 - E_2}{2}\right) t\right] = 1 - \sin^2 2\vartheta \cdot \sin^2\left(\pi \frac{L}{L_{\rm osc}}\right) \\ P(\nu_\mu \to \nu_\tau) = 1 - P(\nu_\mu \to \nu_\mu) = \sin^2 2\vartheta \cdot \sin^2\left(\pi \frac{L}{L_{\rm osc}}\right) \end{cases} \tag{11.46}$$

where

$$L_{\rm osc} = \frac{4\pi p}{\Delta m^2} \simeq \frac{4\pi E}{\Delta m^2} = 2.48 \frac{E[\text{GeV}]}{\Delta m^2[\text{eV}]^2}[\text{km}] \tag{11.47}$$

is the *neutrino oscillation length*, Δm^2 is expressed in eV^2, and $L \simeq ct$ is the distance (in km) between the ν_μ production and the ν_μ (or, equivalently, the ν_τ) observation points; the neutrino energy E is expressed in GeV. The factor 2.48 results from this choice of units. Thus, in (11.46),

$$\left(\pi \frac{L}{L_{\rm osc}}\right) = \left(1.27 \frac{\Delta m^2[\text{eV}]^2 L[\text{km}]}{E[\text{GeV}]}\right) . \tag{11.48}$$

To observe the largest ν_μ disappearance (or ν_τ appearance), the argument of the sine function must be equal to $\pi/2$. For $E \simeq 1$ GeV, $\Delta m \simeq 0.05$ eV, the distance between the observer and the neutrino production point must be $L \simeq 10^3$ km. There are no theoretical estimates for ϑ.

11.9 Measurement of Atmospheric ν_μ Oscillations in Underground Experiments

In 1998, Super-Kamiokande (SK) (Fukuda et al. 1998), MACRO (Ambrosio et al. 1998), and Soudan 2 (Sanchez et al. 2003), at the same conference session in Japan, presented new results with definitive indications for atmospheric neutrino oscillations. In all experiments, the zenith angle distributions of the μ-like events (i.e., attributed to ν_μ charged current interactions) showed a clear deficit compared to the no-oscillation expectation. The number of e-like events (those induced by CC ν_e) was roughly in agreement with the prediction. None of the experimental techniques used by the detectors is able to measure the charge of the final-state leptons, and therefore neutrino and antineutrino induced events cannot be distinguished (Kajita 2012).

11.9.1 Event Topologies in Super-Kamiokande

The Super-Kamiokande (SK) detector consists of a stainless steel tank filled with 50,000 tons of ultrapure water, Fig. 11.10. The construction started in 1991, the observation began in 1996. The detector is divided into two regions; in the inner region, 11,200 photomultipliers (PMTs), 20 in. (50.8 cm) in diameter, detect faint flashes of light produced by the Cherenkov effect from charged particles passing through. The external region offers a shielded volume for atmospheric muons and is used for anticoincidence. SK is located 1000 m underground in the Kamioka

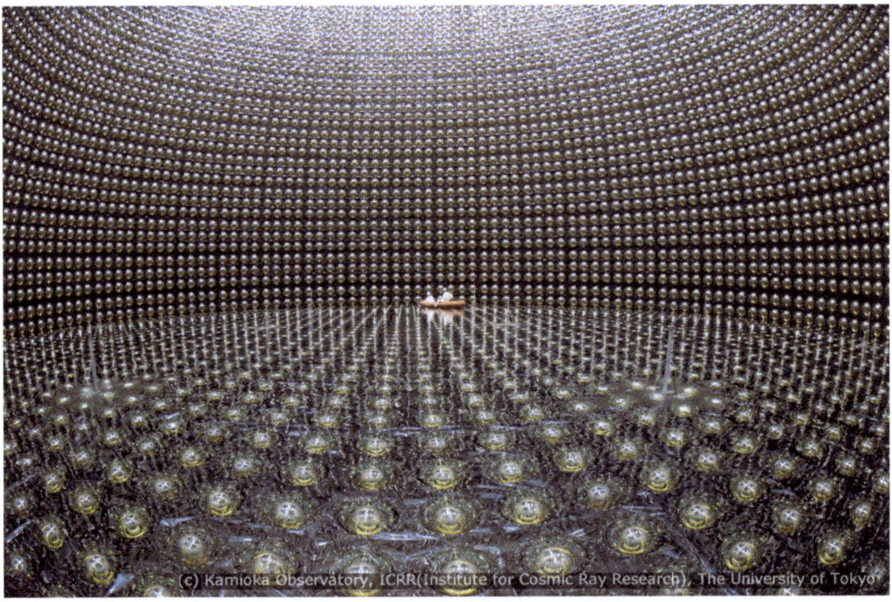

Fig. 11.10 The Super-Kamiokande detector during the construction phase. The water tank is 39 m in diameter and 42 m tall. In the inner detector, the 20-in. PMTs are placed at intervals of 70 cm, covering more than 40% of the *cylinder* surface. The picture was taken during the water filling phase. *Credit* Kamioka Observatory, ICRR (Institute for Cosmic Ray Research), The University of Tokyo (http://www-sk.icrr.u-tokyo.ac.jp/index-e.html)

mine in Japan. Here, the flux of atmospheric muons is reduced by a factor of $\sim 10^5$ with respect to the one at sea level. The experiment also measures solar neutrinos (Sect. 12.3). In 2001, an accident destroyed more than half of the PMTs, ending the SK-I phase (Apr. 1996–Jul. 2001). SK-II (Dec. 2002–Oct. 2005) came back into operation using about 5000 PMTs, with a slightly lower energy resolution. The full detector was restored for the SK-III phase, from Jul. 2006 to Aug. 2008. SK-IV started in Sep. 2008 and is still taking data.

The flavor assignment and the energy estimate of the incoming neutrino plays a crucial role in this experiment. The flavor identification of the neutrino occurs through pattern recognition of the event. The range of muons in water corresponds to about 5 m per GeV of energy. Muons lose energy continuously and emit Cherenkov light until they reach critical velocity, near the end of their range. This results in a "ring" of Cherenkov light at the surface of the detector. From the shape and "brightness" of the ring, it is possible to determine the neutrino interaction vertex and the muon's direction and energy. On the other hand, electrons in water undergo bremsstrahlung and originate an electromagnetic shower, so that their energy is dissipated in the ionization of several e^+, e^- particles. Each secondary particle generates a Cherenkov ring that overlaps in a single visible ring that is more "fuzzy" than that produced by muons, because low-energy electrons suffer considerable multiple Coulomb scattering in water. The difference between the sharp (μ-like) and fuzzy (e-like) rings allows us to determine the flavor of the charged lepton (see Fig. 11.11).

The neutrino energy estimate is done through the classification of the events into different *event topologies*: fully contained (FC) events, partially contained (PC) events, and upward-going muons. The FC events are required to have no activity

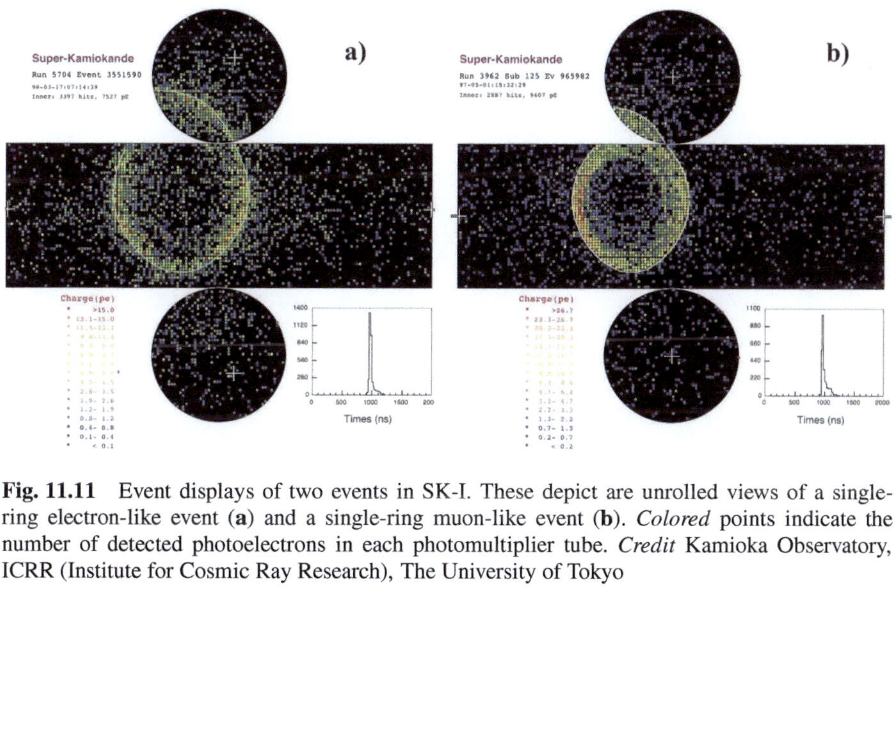

Fig. 11.11 Event displays of two events in SK-I. These depict are unrolled views of a single-ring electron-like event (**a**) and a single-ring muon-like event (**b**). *Colored* points indicate the number of detected photoelectrons in each photomultiplier tube. *Credit* Kamioka Observatory, ICRR (Institute for Cosmic Ray Research), The University of Tokyo

in the outer detector (veto). The neutrino interaction vertex for FC and PC events is required to be reconstructed within the *fiducial volume*, a region 2 m away from the detector walls with a *fiducial mass* of 22.5 kton. The FC events are classified into "sub-GeV" (visible energy, $E_{vis} < 1.33$ GeV) and "multi-GeV" ($E_{vis} > 1.33$ GeV). These events are further separated into subsamples based on the number of observed Cherenkov rings. Single-ring events have only one charged lepton, which radiates Cherenkov light in the final state, and particle identification (e- or μ-like) is particularly clean for single-ring FC events. All the PC events were assumed to be μ-like: in simulated events, the PC events comprise a 98% pure charged current ν_μ sample.

In addition to FC and PC events, higher energy neutrino events are observed as upward-going muons. Upward-going muons are classified into "upward-throughgoing muons" if they pass through the detector, or into "upward-stopping muons" if they enter into and stop inside the detector. The upward-throughgoing muons are further subdivided into "showering" and "nonshowering" based on whether their Cherenkov pattern is consistent with light emitted from an electromagnetic shower produced by a muon exceeding one TeV. The energy distributions of the parent neutrinos for different event topologies are shown in Fig. 11.12.

Figure 11.13 shows the zenith angle distributions of e-like and μ-like events from SK for 12 different event topologies (Takeuchi 2012). The value $\cos\theta = 1$ corresponds to the downward direction, while $\cos\theta = -1$ corresponds to the upward direction. The three panels in the first column show FC e-like events, while the corresponding events classified as μ-like are shown in the second. The third column shows the zenith distributions for two topologies of PC events. The fourth column shows event topologies measured only for upward-going events. These latter

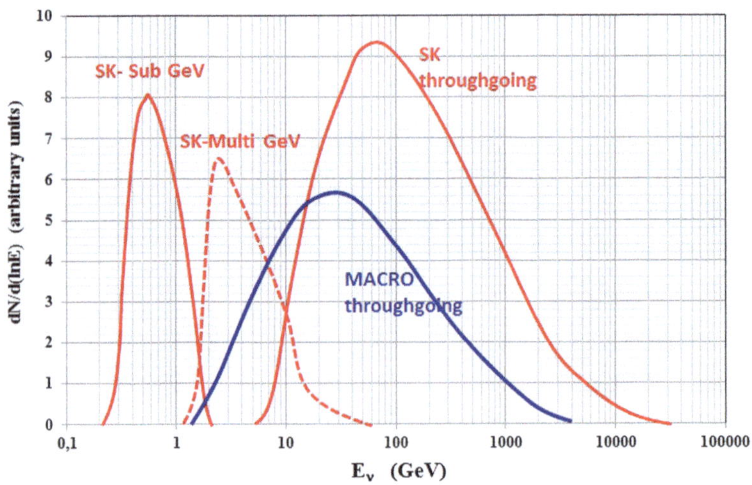

Fig. 11.12 Event rates as a function of neutrino energy for sub-GeV, multi-GeV, and through going muons in Super-Kamiokande and for up throughgoing muons in MACRO

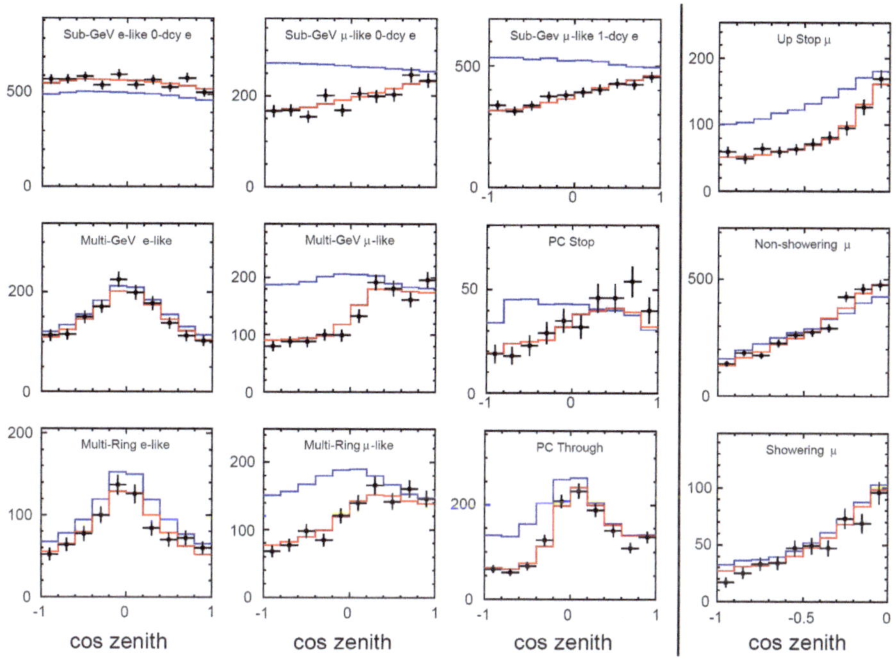

Fig. 11.13 Zenith angle distributions observed in SK-I+II+III (2806 days livetime). The panels in the first column show three topologies of *e*-like events, the remaining are μ-like. $\cos\theta < 0\,(>0)$ are vertical upward (*downward*)-going events. The *black* markers represent the data points; the *blue* histograms show the Monte Carlo prediction without neutrino oscillations. The *red* histograms show the best fit for $\nu_\mu \rightleftarrows \nu_\tau$ using the best-fit parameters ($\Delta m^2 = 2.11 \times 10^{-3}$ eV^2 and $\sin^2 2\vartheta = 1$). The last column shows upgoing events induced by higher energy ν_μ. *Credit* Kamioka Observatory, ICRR (Institute for Cosmic Ray Research), The University of Tokyo

topologies are due to μ-like events. The average energy of parent neutrinos increases from top to bottom, namely as: sub-GeV < multi-GeV ~ Multi-Ring < Up Stop ~ PC stop < PC through < non showering < showering. With the exclusion of sub-GeV events, the final-state leptons have good directional correlation with the parent neutrinos.

The results of SK are shown in Fig. 11.13, together with the corresponding Monte Carlo expectations (blue histograms) (Honda et al. 2007). Several aspects become quite clear from an inspection of the figure, in particular, the fact that the zenith angle distribution of the FC and PC μ-like events shows a strong deviation from the expectation. The most spectacular effect is the 50% reduction of the detected upgoing events ($\cos\theta < 0$) of the multi-GeV and of the Multi-Ring μ-like events. Deviations from expectations are almost nonvisible in the highest energy topologies in the last column. On the other hand, the zenith angle distributions of the *e*-like events are consistent with the expectation and exhibit the predicted up-down symmetry.

These characteristic features may be interpreted assuming $\nu_\mu \rightleftarrows \nu_\tau$ oscillations. The oscillation probability (11.46) depends on the neutrino energy and path length L. Upgoing ν_μ, having traveled $L \sim 10^4$ km, will have larger oscillation probabilities than downgoing ones, which have L ranging from a few to a few tens of km. Disappeared ν_μ should have oscillated into ν_τ neutrinos, because there is no indication of electron neutrino appearance (the e-like events from $\cos\theta < 0$ are consistent with those in the $\cos\theta > 0$ region).

As shown in Fig. 11.12, the atmospheric neutrinos corresponding to different topologies in Fig. 11.13 have energies over a wide range. The median energies vary from a fraction of GeV to more than 1 TeV, depending on the topology.

The neutrino path length L is very strongly correlated with the zenith angle $\cos\theta$ of the measured charged lepton. The outgoing charged lepton ℓ and the incoming neutrino ν directions are correlated for kinematic reasons, and the average angle $\langle\theta_{\ell\nu}\rangle$ between ℓ and ν shrinks with increasing energy as approximately E_ℓ^{-1}. For sub-GeV events, a value of $\langle\theta_{\ell\nu}\rangle \sim 60°$ was estimated by the SK collaboration; thus, the correlation between the charged lepton and neutrino directions is rather poor. For multi-GeV events, $\langle\theta_{l\nu}\rangle \sim 10°$, and the correlation is much more stringent.

Atmospheric neutrinos are produced at a typical height $h \sim 20$ km, with only a weak dependence on the energy, flavor, and zenith angle. Assuming that the neutrino's direction is given by the observed charged lepton, the neutrino path length L is therefore given, to a reasonable approximation, by Lipari (2001)

$$L \simeq -R_\oplus \cos\theta + \sqrt{(R_\oplus \cos\theta)^2 + 2R_\oplus h + h^2}\,. \tag{11.49}$$

This expression ranges from $L \sim h \sim 20$ km for vertically downgoing neutrinos ($\cos\theta = 1$) to $L \simeq 2R_\oplus + h \sim 13{,}000$ km for neutrinos crossing the Earth's diameter $2R_\oplus$ at $\cos\theta = -1$. For horizontal neutrinos, the path length is $L \sim \sqrt{2R_\oplus h} \sim 500\,\text{km}$.

For a first estimate of the oscillation parameters without the use of a full Monte Carlo simulation, we consider multi-GeV events that have traveled over a length $L_{\text{osc}} \sim 10^4$ km. The energy of the events can be assumed as the median energy $\overline{E}_\nu \sim 5$ GeV of the distribution. Thus, using (11.47), $\Delta m^2[\text{eV}]^2 \simeq 2.48\frac{E[\text{GeV}]}{L_{\text{osc}}[\text{ km}]} \simeq 10^{-3}$ eV^2. The E_ν of the parent neutrinos in multi-GeV events ranges from ~1 to ~100 GeV. For a fixed value of L (i.e., for neutrino directions in a small interval of $\cos\theta$), the argument of the function $\sin^2(1.27\Delta m^2 L/E)$ in (11.46) can assume values much larger than 1, and this produces fast oscillating probabilities. The observable quantity is simply the average $\langle\sin^2(1.27\Delta m^2 L/E)\rangle = 1/2$. Thus, the oscillation probability takes the simple form

$$P(\nu_\mu \to \nu_\mu) = 1 - \frac{\sin^2 2\vartheta}{2} \text{ for } E/L \gg \Delta m^2\,. \tag{11.50}$$

The observed 50% reduction for upgoing ν_μ suggests $\sin^2 2\vartheta \sim 1$.

In conclusion, SK data and predictions can be reconciled assuming the presence of $\nu_\mu \rightleftarrows \nu_\tau$ oscillations and choosing appropriate values of ϑ and Δm^2. A precise determination of the oscillation parameters can be obtained weighting the Monte Carlo predictions without the assumption of oscillations with the oscillation probability (11.46). The SK best fit within the framework of two neutrino oscillations to the data shown in Fig. 11.13 as red histograms gives us $\sin^2 2\vartheta = 1$ and $\Delta m^2 = 2.11 \times 10^{-3}$ eV^2.

11.9.2 The Iron Calorimeter Soudan 2 Experiment

The Soudan 2 was an iron-tracking calorimeter with 770 tons of fiducial mass that operated as a time projection chamber. The detector was active from 1989 to 2001 at a depth of 2070 m.w.e. in the Soudan Underground Mine State Park (USA). The active elements were 1 m long, 1.5 cm diameter drift tubes encased in a honeycomb matrix of 1.6 mm thick corrugated steel plates. Surrounding the tracking calorimeter on all sides but mounted on the cavern walls, there was a 1700 m^2 active shield array to veto the arrival of atmospheric muons.

Events were divided into two classes: fully contained events within the detector (FCE) and partially contained events, in which only the produced lepton exits the detector (PCE). The FCE data were further divided into topology classes to distinguish between e-like (if the highest energy secondary produces a shower) and μ-like (if the highest energy secondary is a nonscattering track) events, Fig. 11.14.

Differently from earlier experiments, probably because of the use of veto counters, Soudan 2 observed deviations from the no-oscillation hypothesis in an iron calorimeter (Sanchez et al. 2003). The e-like events behave as expected, while there is a deformation of the zenith angle distribution of the μ-like sample, Fig. 11.15. The event detection and reconstruction properties of Soudan 2 were different, and in many cases superior, to those of SK, but the exposure was much smaller. Also, the geographical locations and backgrounds of the two experiments were different.

11.9.3 Upward-Going Muons and MACRO

A different method for measuring the flux of atmospheric muon neutrinos (as usual, $\nu_\mu + \overline{\nu}_\mu$) is through the observation of upward-going muons. In these events, the neutrino interaction occurs in the rock around the detector. Because of the up-down symmetry of the atmospheric neutrino fluxes, one expects an equal number of upgoing and downgoing ν_μ-induced events. However, ν_μ-induced downgoing events cannot be distinguished from atmospheric muons (Fig. 10.4), and therefore only upgoing events can be used to study the neutrino flux, as in neutrino telescopes. In standard rock, a muon travels about 1.7 m for each GeV of energy. In

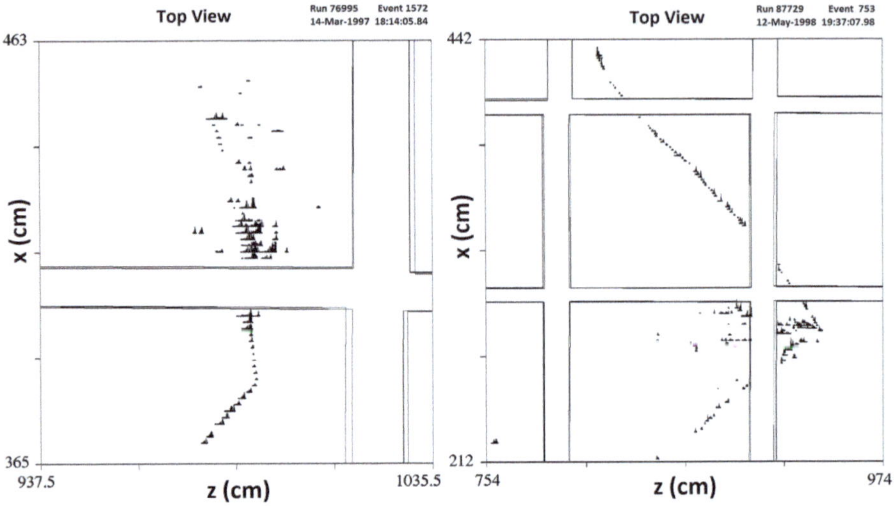

Fig. 11.14 Two neutrino interactions in the Soudan 2 detector. The event on the left is a $\nu_e + n \rightarrow p + e^-$ event. The electron travels about one radiation length before showering. The proton is easily recognizable by its heavy ionization (large symbols) and its lack of Coulomb scattering. The event on the *right* has a long noninteracting muon track, which shows typical Coulomb scattering, and a hadronic shower at the vertex

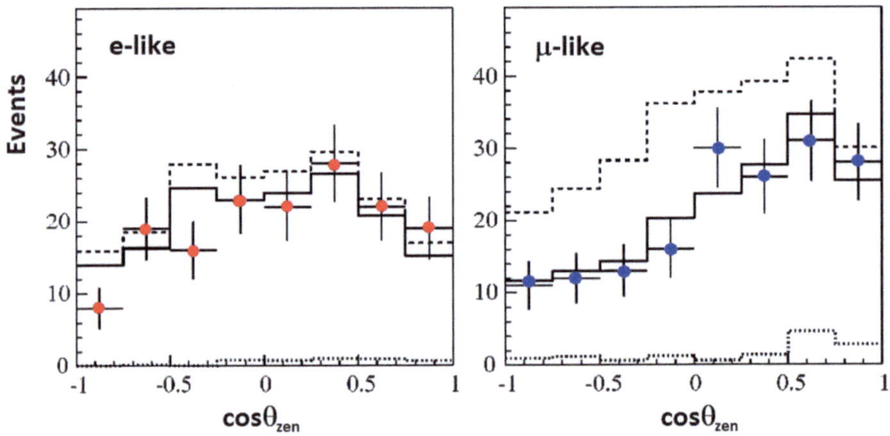

Fig. 11.15 Angular distributions for *e*-like (*left*) and μ-like events (*right*) in Soudan 2 as a function of the cosine of the zenith angle. The points with error bars are the data, the *dashed* histograms the predicted unoscillated neutrino distribution plus background; the *solid* histograms show the neutrino distribution weighted by the oscillation probability predicted by the best-fit parameters ($\Delta m^2 = 5.2 \times 10^{-3}$ eV^2 and $\sin^2 2\vartheta = 0.97$)

throughgoing events, the muons arrive and cross the detector, while in stopping events, the muons range out inside the detector volume. The average energy of observed events depends on the minimum track length required at the trigger level.

MACRO was a large area apparatus ($12 \times 9.3 \times 76.6$ m) in the Gran Sasso underground laboratory in Italy under a minimum rock overburden of 3150 m.w.e. It had a modular structure of six modules and collected data from 1994 to 2000. The bottom part contained layers of limited streamer tubes interleaved with passive material plus two layers of liquid scintillator detectors and one plane of nuclear track detectors. The top part of the apparatus was empty and had a "roof" with four horizontal planes of limited streamer tubes and another liquid scintillator layer. Vertically, the apparatus was surrounded by a liquid scintillator plane and six lateral planes of limited streamer tubes. The tracking was performed using the hits in streamer tubes; the time information provided by scintillation counters allowed for the determination of the direction by the time-of-flight (ToF) measurements. The MACRO detector measured three classes of atmospheric neutrino interaction, Fig. 11.16:

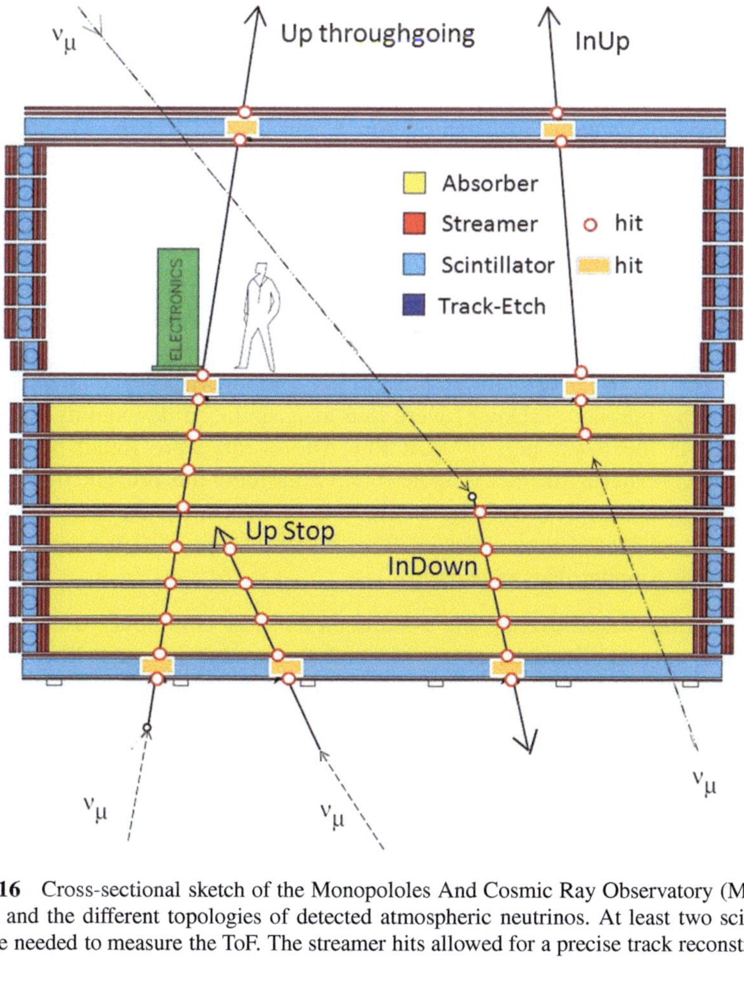

Fig. 11.16 Cross-sectional sketch of the Monopololes And Cosmic Ray Observatory (MACRO) detector and the different topologies of detected atmospheric neutrinos. At least two scintillator hits were needed to measure the ToF. The streamer hits allowed for a precise track reconstruction

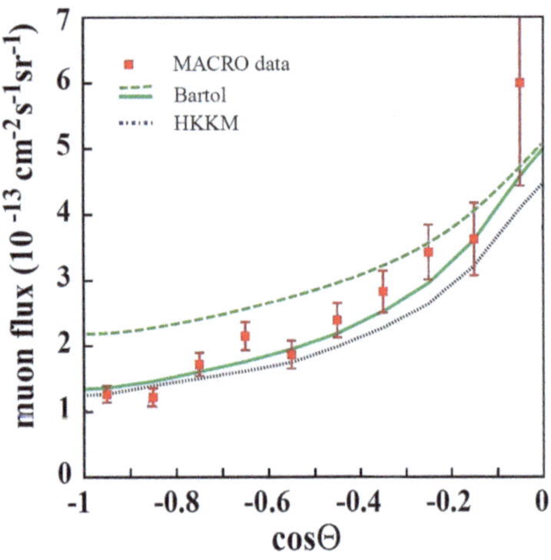

Fig. 11.17 Measured angular distribution of throughgoing muons in MACRO (*full squares*). The *lines* show the prediction assuming their best-fit values: $\Delta m^2 = 2.5 \times 10^{-3}$ eV^2 and $\sin^2 2\vartheta = 1$, using, as input, the Bartol (Barr et al. 2004) and HKKM (Honda et al. 2007) fluxes. The *green dashed line* represents the Bartol expectation without oscillations

- The "Up throughgoing muons" with $E_\mu > 1$ GeV. These were generated in the rock below the detector by ν_μ interactions with median energy of $\overline{E}_\nu = 50$ GeV. The measured angular distribution was found to deviate from the expectation without neutrino oscillations, Fig. 11.17. In particular, the vertical flux was 50% of the expected value.
- The "Internal Upgoing" (InUp) muons arose from ν_μ interactions with $\overline{E}_\nu \sim$4 GeV inside the lower apparatus. Since two scintillation counters were intercepted, the ToF method was applied to identify upward-going events. A 50% reduction was measured, without any distortion in the shape of the angular distribution. The "Upgoing Stopping muons" (Up Stop) are due to ν_μ interactions in the rock below the detector yielding upgoing muon tracks. The "semicontained downgoing muons" (InDown) were due to ν_μ-induced downgoing tracks with vertices in the lower part of MACRO. The events were found by means of topological criteria; the lack of at least two scintillator hits prevented a separation of the two subsamples. Without oscillations, an almost equal number of Up Stop and InDown events were expected. The measured ratio of (Up Stop + In down) was ~30% smaller than expected.

For upgoing muon events, the two methods most commonly used to reduce systematic uncertainties (the μ/e ratio and the up/down comparison) available for contained events cannot be applied. The uncertainty in the prediction of the absolute rate of the fluxes is on order of 15–20%. In this case, a robust method to disentangle the presence of $\nu_\mu \rightleftarrows \nu_\tau$ oscillations uses the shape of the zenith angle distribution of the measured ν_μ flux, which has a $1/\cos\theta$ dependence, as given in (11.36). In the presence of oscillations vertical neutrinos with $L \sim 10^4$ km undergo larger oscillations, and are more suppressed than horizontal ones. This distorts the expected shape of the up throughgoing muon flux, as shown in Fig. 11.17, with a best-fit point of $\sin^2 2\vartheta = 1$ and $\Delta m^2 = 2.5 \times 10^{-3}$ eV^2.

Also, the lower energies' topologies were consistent with the existence of $\nu_\mu \rightleftarrows \nu_\tau$ oscillations. In case of oscillations, a ~50% reduction in the flux of the up stopping events (Up Stop) and of the semicontained upgoing muons (InUp) is expected due to averaged oscillations (11.50). No reduction is instead expected for the semicontained downgoing (InDown) events, which come from neutrinos that have traveled ~20 km, in good agreement with the measured reductions.

11.10 Atmospheric ν_μ Oscillations and Accelerator Confirmations

The simplest explanation for the ν_μ disappearance observed by the three quoted experiments is the conversion of ν_μ into ν_τ, whose interactions are (almost) undetectable. Charged current ν_τ interactions occur at energies above the threshold for the formation of a τ lepton ($m_\tau \sim 1.7$ GeV). The τ flavor identification is not a simple task, as discussed in Sect. 10.4, as most events look like a neutral current interaction with a shower. The hypothesis of $\nu_\mu \rightleftarrows \nu_\tau$ oscillation is strengthened by the fact that the behavior of the atmospheric ν_e flavor is compatible with the no-oscillation hypothesis.

Starting from the expected neutrino flux and the disappearance probability $P(\nu_\mu \rightarrow \nu_\mu)$ given in (11.46), information on Δm^2 and on the mixing angle ϑ were obtained in each experiment by a global fit of the zenith angle distribution of all event topologies. Each set of ($\sin^2 2\vartheta$; Δm^2) parameters for a given topology gives a reduction of ν_μ events that depends on the zenith angle. The set of values giving the best reduced χ^2 when using all the event topologies represents the best fit of the experiment; statistical methods were then used to define the region on the parameter phase space that gives the 90% confidence level (c.l.). Other methods for obtaining the best parameters were also used, for instance, the fit of the number of events as a function of the observed L/E. The best-fit values for each experiment have already been quoted in the captions of Figs. 11.13, 11.15 and 11.17.

The remarkable result in 1998 of the study of atmospheric neutrino oscillations was that the high-significance result from SK had the same best-fit parameters obtained by MACRO. The two collaborations used different experimental techniques (water Cherenkov vs. tracking and scintillation) and measured neutrinos in different energy ranges. The simultaneous evidence for an unexpected process by two experiments is an important aspect in experimental physics.[2] Moreover, the successive Soudan 2 analysis was also compatible with the SK and MACRO results.

[2]In 2011, there was a huge emphasis on a high-statistical significance claim made by the OPERA collaboration that neutrinos travel with a speed exceeding that of light. Waiting for the confirmation from the MINOS experiment, a few months later, the collaboration discovered a problem in a hardware connection that invalidated the claim (Adam et al. 2014).

The discovery with atmospheric neutrinos of neutrino oscillations motivated a large effort to carry out independent checks using accelerator experiments. Before 1998, two *short baseline experiments* were carried out at CERN: CHORUS and NOMAD. These two detectors were exposed to a high-energy ν_μ beam, and had $L \sim 1$ km. Their goal was to search for neutrino oscillations with $\Delta m^2 \sim 1$ eV, by performing disappearance and appearance measurements. The results were null, because they had been designed to explore a range of Δm^2 values that Nature had not chosen.

Neutrino oscillations were then explored through accelerator-based long-baseline experiments, typically with $E_\nu \sim 1-10$ GeV and $L \sim$ several hundred km. With a fixed baseline distance and a narrower, well understood neutrino spectrum, the value of Δm^2 and also, with higher statistics, the mixing angle are potentially better constrained in accelerator experiments.

The KEK to Kamioka (K2K) long-baseline neutrino oscillation experiment in Japan was the first accelerator-based experiment with a neutrino path length covering hundreds of kilometers. K2K aimed at confirmation of the neutrino oscillation in ν_μ disappearance in the $\Delta m^2 > 2 \times 10^{-3}$ eV^2. The wideband muon neutrino beam had an average $L/E_\nu \sim 1.3$ GeV/250 km. The beam was produced from the KEK-PS accelerator and directed to the Super-Kamiokande detector from 1999 to 2004.

MINOS was the second long-baseline neutrino oscillation experiment, taking data from 2005 to 2012. Neutrinos were produced from the Fermilab Main Injector. MINOS comprised both a near and a far detector. The far detector was a 5.4 kton (total mass) iron-scintillator tracking calorimeter with a toroidal magnetic field, located underground in the Soudan mine. The baseline distance was 735 km. The near detector was also an iron-scintillator tracking calorimeter with a toroidal magnetic field, with a total mass of 0.98 kton. Both MINOS and K2K measured the neutrino-induced muon energy spectrum and confirmed a distortion consistent with that which is expected assuming neutrino oscillations. MINOS provided the most accurate measurement of the allowed region of the ($\sin^2 2\vartheta$; Δm^2) parameters, with best fit $\Delta m^2 = (2.32^{+0.12}_{-0.08}) \times 10^{-3}$ eV^2 and $\sin^2 2\vartheta > 0.90$ at 90% c.l. These values are very close to those found by SK and MACRO.

Although the atmospheric neutrino oscillations and accelerator long-baseline ν_μ disappearance data are fully consistent with $\nu_\mu \rightleftarrows \nu_\tau$ oscillations, the appearance of ν_τ remained to be confirmed in order to definitively rule out other exotic explanations (Beringer et al. 2012). For this purpose, an accelerator long-baseline experiment for studying the *appearance* of a τ lepton in a ν_μ beam was realized.

The OPERA experiment at Gran Sasso Laboratory started data taking in 2008 and stopped in 2012. The detector was a combination of nuclear emulsions and a magnetized spectrometer. The emulsion technique was used to identify short-lived τ leptons event-by-event in a ν_μ beam produced at CERN, with the baseline distance of 730 km. OPERA reported the observation of five ν_τ candidates, an excess of $\sim 5\sigma$ with respect to the nonoscillation hypothesis. The OPERA experiment has, in a definitive way, confirmed the oscillation scenario opened in 1998 by underground experiments.

Exercise Using the two flavor oscillation formula (11.46), determine the value of the muon neutrino energy which gives a 100% disappearance probability for neutrinos: (a) crossing the Earth's atmosphere ($L \sim 20$ km); (b) crossing the Earth's diameter ($L \sim 13{,}000$ km). Comment the results.

11.11 Atmospheric Neutrino Flux at High Energies

The leptonic or semileptonic decays of charged pions or kaons produce atmospheric ν_μ (ν_e) from a fraction of GeV up to about 100 (10) TeV (conventional neutrinos). Neutrinos from muon decay are important up to a few GeV. Oscillations modify the flux of different neutrino flavors within a relatively small energy range. Above $\sim$100 GeV, neutrino oscillations do not play any role on the scale of terrestrial path lengths, $L \sim 10^4$ km corresponding to $\cos\theta \sim -1$. Figure 11.18 shows the ν_μ disappearance probability for three different neutrino energies as a function of the zenith angle θ.

At sufficiently high energies, in addition to the conventional flux, another production mechanism is expected. The *prompt* atmospheric neutrino flux is originated by the semileptonic decays of charmed mesons and baryons with lifetimes of 10^{-12} s, before losing energy in collisions. Hence, the spectrum for the prompt flux more closely follows the CR spectrum and is harder by about one power than the conventional flux at high energy. The prompt flux could represent a dangerous background for cosmic neutrinos and has not yet been measured, but it is expected to be important above $\sim$100 TeV.

The ANTARES and IceCube neutrino telescopes presented in Chap. 10 were able to observe the disappearance of the atmospheric ν_μ with energies as low as a few tens of GeV. On the other hand, neutrino telescopes are the only devices large enough to detect a sizeable flux of atmospheric neutrinos above the TeV energy. Figure 11.19 shows the measurement of the ν_μ energy spectrum reported by the

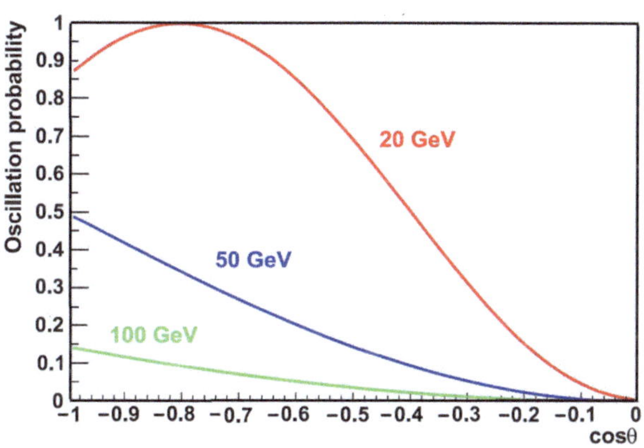

Fig. 11.18 $P(\nu_\mu \to \nu_\mu)$ disappearance probability as a function of the cosine of zenith angle θ for atmospheric ν_μ with $E_{\nu_\mu} =$ 20, 50 and 100 GeV. The curves were produced using Eq. (11.46) assuming $\Delta m^2 = 2.32 \times 10^{-3}$ eV^2 and $\sin^2 2\vartheta = 1$

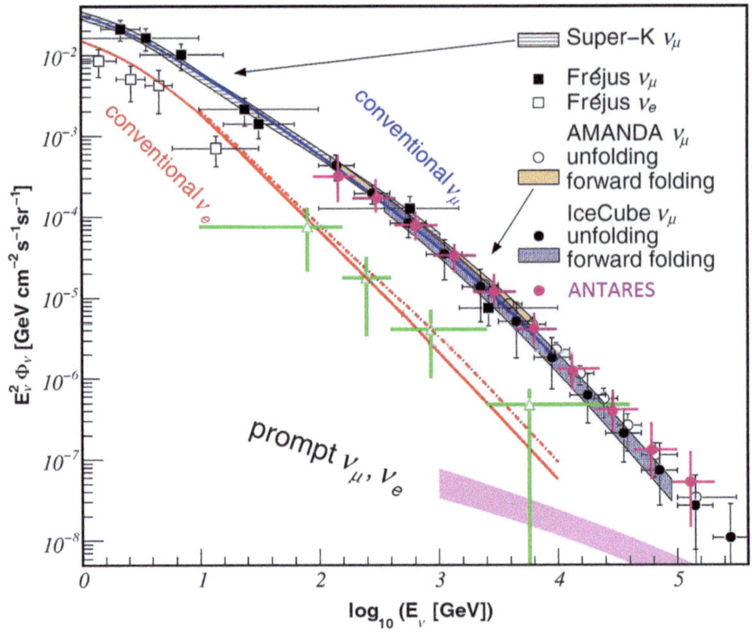

Fig. 11.19 Measured flux of atmospheric ν_μ and ν_e compared with expectations from the conventional flux of atmospheric neutrinos. The prompt component is, at present, too low to be measured. Courtesy of IceCube Collaboration

Frejus, AMANDA, IceCube and ANTARES experiments. IceCube (Aartsen et al. 2013) also recently measured the atmospheric ν_e component up to 1 TeV. The only existing measurement was between $\langle E_{\nu_e} \rangle \sim 0.4$–14 GeV by the Frejus underground experiment.

The data sets are consistent with current models of the atmospheric neutrino flux.

References

M.G. Aartsen et al., Measurement of the atmospheric ν_e flux in IceCube. Phys. Rev. Lett. **110**, 151105 (2013)

T. Adam et al., Measurement of the neutrino velocity with the OPERA detector in the CNGS beam (2014). Compare arXiv:1109.4897v1 with arXiv:1109.4897v4

M. Ambrosio et al., Measurement of the atmospheric neutrino-induced upgoing muon flux using MACRO. Phys. Lett. **434**, 451 (1998). (MACRO Collaboration)

G.D. Barr, T.K. Gaisser, P. Lipari, S. Robbins, T. Stanev, Three-dimensional calculation of atmospheric neutrinos. Phys. Rev. D **70**, 023006 (2004)

Y. Becherini, A. Margiotta, M. Sioli, M. Spurio, A Parameterisation of single and multiple muons in the deep water or ice. Astropart. Phys. **25**, 1 (2006)

J. Beringer et al., (Particle Data Group). The review of particle physics. Phys. Rev. **D86**, 010001 (2012)

S. Braibant, G. Giacomelli, M. Spurio, *Particles and Fundamental Interactions* (Springer, New York, 2012)

S. Cecchini, M. Spurio, Atmospheric muons: experimental aspects. Geosci. Instrum. Method. Data Syst. **1** 185–196 (2012). arXiv:1208.1171

L.I. Dorman, *Cosmic Rays in the Earth's Atmosphere and Underground* (Kluwer Academic Publisher, New York, 2004)

Y. Fukuda et al., Evidence for oscillation of atmospheric neutrinos. Phys. Rev. Lett. **81**, 1562 (1998). (SuperKamiokande Collaboration)

T.K. Gaisser, *Cosmic Rays and Particle Physics* (Cambridge University Press, Cambridge, 1990)

T.K. Gaisser, Semi-analytic approximations for production of atmospheric muons and neutrinos. Astropart. Phys. **16**, 285 (2002)

T.K. Gaisser, M. Honda, Flux of atmospheric neutrinos. Annu. Rev. Nucl. Part. Sci. **52**, 153–199 (2002)

P.K.F. Grieder, *Extensive Air Showers* (Springer, Berlin, 2010)

M. Honda, T. Kajita, K. Kasahara, S. Midorikawa, T. Sanuki, Calculation of atmospheric neutrino flux using the interaction model calibrated with atmospheric muon data. Phys. Rev. D **75**, 043006 (2007)

J.I. Illana, P. Lipari, M. Masip, D. Meloni, Atmospheric lepton fluxes at very high energy. Astropart. Phys. **34**, 663–673 (2011). arXiv:1010.5084

T. Kajita, Atmospheric neutrinos. Adv. High Ener. Phys. 504715 (2012). https://doi.org/10.1155/2012/504715

M. Koshiba, Observational neutrino astrophysics. Phys. Rep. **220**, 229–381 (1992)

N. Lesparre et al., Geophysical muon imaging: feasibility and limits. Geophys. J. Int. **183**, 1348 (2010)

P. Lipari, Lepton spectra in the earth's atmosphere. Astropart. Phys. **1**, 195 (1993)

P. Lipari, Introduction to neutrino physics. 1st CERN-CLAF School of High-energy Physics, Itacuruca, Brazil (2001). http://cds.cern.ch/record/677618/files/p115.pdf

B. Pontecorvo, Neutrino experiments and the problem of conservation of leptonic charge. Sov. Phys. **26**, 984–988 (1968). (English translation of a paper in Russian in 1967)

M. Sanchez et al., Measurement of the L/E distributions of atmospheric ν in Soudan 2 and their interpretation as neutrino oscillations. Phys. Rev. **D68**, 113004 (2003). (Soudan 2 Collaboration)

Y. Takeuchi, Results from Super-Kamiokande. Nucl. Phys. Proc. Supl. **79–84**, 229–232 (2012) (Super-Kamiokande Collaboration)

Chapter 12
Low-Energy Neutrino Physics and Astrophysics

Abstract Stellar evolution, the theory of how stars evolve, relies on observations of many stars of different masses, colors, ages, and chemical composition. The energy of stars is provided by nuclear fusion reactions in their core, and their evolution is strongly dependent upon their mass. The Sun, through the Standard Solar Model, is the only star for which the stellar evolution theory can be deeply tested through neutrinos emitted from various thermonuclear processes. The experimental study of solar neutrinos has made a fundamental contribution both to astroparticle and to elementary particle physics, offering an ideal test of solar models and providing, at the same time, fundamental indications concerning the physics of the neutrino sector. The solar neutrino experiments (with atmospheric neutrinos) have given compelling evidence for the existence of neutrino oscillations caused by nonzero neutrino masses and neutrino mixing. This has a huge impact on particle physics. It also has consequences on the prediction of the neutrino flavor composition from high-energy neutrino sources. Neutrinos do not only play a key role during the life of a star. When a massive star has exhausted its hydrogen, it evolves by producing energy through the fusion of heavier elements up to iron. Neutrinos produced during such reactions escape unimpeded from the stellar material and more and more intense nuclear burning is needed to replace the huge amount of energy carried away. Once the inner region of a star becomes primarily iron, further compression of the core no longer ignites nuclear fusion; the star collapses to form a compact object such as a neutron star or a black hole. A prominent prediction from theoretical models of the core-collapse of a massive star is that 99% of the gravitational binding energy of the resulting remnant is converted to neutrinos with energies of a few tens of MeV over a timescale of 10 s. Neutrinos were observed from the celebrated 1987A supernova in the Large Magellanic Cloud, the first event of multimessenger astrophysics.

Stellar evolution, the theory of how stars evolve, relies on observations of many stars of different masses, colors, ages, and chemical composition. Two of the principal successes of the stellar evolution theory are the prediction of the mass-luminosity relation in main sequence stars and the explanation of the Hertzsprung-Russell

© Springer Nature Switzerland AG 2018

M. Spurio, *Probes of Multimessenger Astrophysics*, Astronomy and Astrophysics Library, https://doi.org/10.1007/978-3-319-96854-4_12

diagram. This is a scatter graph of absolute magnitude or luminosity of stars versus temperature (color). The energy of stars is provided by nuclear fusion reactions in their core, Sect. 12.1, and their evolution is strongly dependent upon their mass.

The Sun, through the Standard Solar Model (SSM), is the only star for which the stellar evolution theory can be deeply tested through neutrinos emitted from various thermonuclear processes, Sect. 12.2. The experimental study of solar neutrinos (Sects. 12.3–12.5) has made a fundamental contribution both to astroparticle and to elementary particle physics, offering an ideal test of solar models and providing, at the same time, fundamental indications concerning the physics of the neutrino sector.

In conjunction with those for atmospheric neutrinos (Chap. 11), the solar neutrino experiments have given compelling evidence for the existence of neutrino oscillations, Sect. 12.6, transitions in flight between the three neutrino flavors, caused by nonzero neutrino masses and neutrino mixing. This has a huge impact on particle physics. It also has consequences on the prediction of the neutrino flavor composition from high-energy neutrino sources, Sect. 12.8.

Neutrinos do not only play a key role during the life of a star. When a massive star has exhausted its hydrogen, it evolves by producing energy through the fusion of heavier elements up to iron, Sect. 12.10. Neutrinos produced during such reactions escape unimpeded from the stellar material and more and more intense nuclear burning is needed to replace the huge amount of energy carried away. Once the inner region of a star becomes primarily iron, further compression of the core no longer ignites nuclear fusion; the star collapses to form a compact object such as a neutron star or a black hole, Sect. 12.11. A prominent prediction from theoretical models of the core-collapse of a massive star (Sect. 12.13) is that 99% of the gravitational binding energy of the resulting remnant is converted to neutrinos with energies of a few tens of MeV over a timescale of 10 s. Neutrinos were observed from the celebrated 1987A supernova (SN1987A) in the Large Magellanic Cloud, 50 kpc away from the Earth, Sect. 12.15. A few neutrino events in coincidence with the explosion were sufficient to confirm the baseline model of core-collapse and to provide a very wide range of constraints on neutrino physics and astrophysics. Today, more advanced and larger experiments are ready to witness the next galactic core-collapse supernova.

12.1 Stellar Evolution of Solar Mass Stars

The interplay between the gravitational, the electromagnetic and the nuclear forces determines the life evolution of stars. The force field that holds the nuclei in the star core is due to the gravitational pressure of matter present in the outer layers. A star is a system in equilibrium between pressure due to gravity and pressure due to radiation produced by fusion reactions in the core. The basic equations of stellar evolution and structure involve simple physics and can be found in Bahcall (1989).

The stars shine through nuclear fusion reactions. As light elements (starting from hydrogen) are burned, stars slowly evolve. Their central temperature is adjusted so that the average thermal energy of a nucleus is small compared to the Coulomb repulsion from potential fusion partners. Nuclear fusion reactions are possible because of the *tunneling effect* through a potential barrier. This effect was first discussed by Gamow in connection with the α-decay (see Chapter 14 of Braibant et al. 2011). For instance, in our Sun, the temperature is $T_\odot \simeq 1.5 \times 10^7$ K, corresponding to a proton kinetic energy of $E_p = kT_\odot \simeq 1.3$ keV. The large Coulomb repulsion, which corresponds to a few MeV, slows the nuclear reaction rates to long time scales, and the energy dependence of the fusion cross-section represents one of the main input factors for the theory of stellar evolution. Uncertainties about cross sections reflect into uncertainties about stellar parameters.

Approximately 80% of observed stars (called *main sequence stars*) lie along a path in the Hertzsprung-Russell diagram characterized by energy generation through proton burning. Main sequence stars fuse protons to produce energy contrasting the gravitational collapse via the proton–proton (pp) or the carbon-nitrogen-oxygen (CNO) cycles, two distinct reaction chains. The Sun provides a unique opportunity to test the theory of how stars evolve, as we can compare model predictions to solar properties that are precisely known.

The luminosity of a main sequence star is determined by the atomic opacity, the chemical composition, and the balance of gravitational and pressure forces. Stellar nucleosynthesis is responsible for the abundances of elements up to iron nuclei. The first reaction of the proton–proton cycle occurs through the weak interactions, that is,

$$p + p \rightarrow {}^2\mathrm{H} + e^+ + \nu_e \quad Q = 0.42\,\mathrm{MeV}. \tag{12.1}$$

The cross-section for this low energy process is extremely small, $\sigma \sim 10^{-55}\,\mathrm{cm}^2$. The reported Q value represents the sum of the e^+ kinetic energy and the neutrino energy. In the following, it also includes the total thermal energy released. ^{2}H can also be formed though the so-called pep reaction, which occurs at a very small rate (0.24%) with respect to (12.1)

$$p + e^- + p \rightarrow {}^2\mathrm{H} + \nu_e \quad Q = 1.442\,\mathrm{MeV}. \tag{12.2}$$

This reaction produces a monochromatic neutrino of energy equal to the Q-value.

The successive steps of the *pp* chain are shown in Fig. 12.1. The proton–proton cycle can be summarized as

$$4p \rightarrow {}^4\mathrm{He} + 2e^+ + 2\nu_e \quad Q = 26.73\,\mathrm{MeV}\,, \tag{12.3}$$

of which 2 MeV are due to the annihilation of the two positrons. In the chain, there are two important channels that produce ^{4}He with neutrinos as by-products. Neutrinos (each with average energy $\overline{E} \sim 0.3$ MeV) do not energetically contribute to

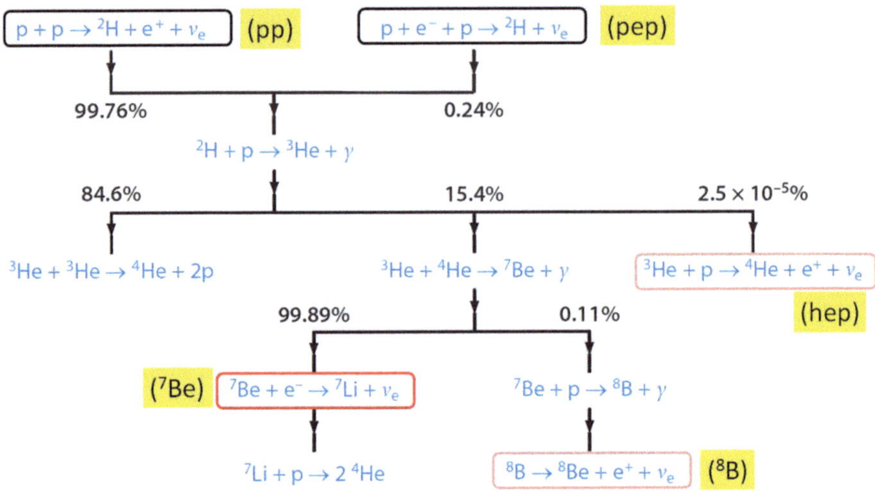

Fig. 12.1 The diagram shows the principal cycles comprising the proton–proton (pp) chain, and the associated neutrinos produced during the different branches. The pp processes occur in all stars of the main sequence. The probability of each branch reported in the figure is characteristic of our Sun during our epoch

stellar equilibrium, because they have a low interaction probability and immediately escape. It is interesting to note that the isotopes of ^{2}H, ^{3}H, $^7_3$Li $^7_4$Be, and $^8_4$B are collaborating spectators whose local number densities do not change when the chain is in equilibrium.

Due to many other physical processes, the energy produced by fusion propagates to the surface of the star, the photosphere. The diffusion time of photons from the stellar core to the photosphere (taking into account the $\gamma - p$ cross-section) is on the order of 10^5–10^6 years.

A star with roughly the solar mass, when hydrogen is exhausted, tends to contract and to increase its density; this happens because the radiation produced by the fusion reactions is no longer able to balance the gravitational pressure. During the contraction phase, gravitational energy is converted into kinetic energy of nuclei: the temperature increases and further fusion reactions may be ignited.

A critical point is the carbon formation. In a star composed mainly of ^{4}He nuclei, ^{8}Be is continuously formed. ^{8}Be has a mass that is slightly larger than twice the ^{4}He mass, that is, $^4\text{He} + {}^4\text{He} \rightarrow {}^8\text{Be}$; $Q = -0.09$ MeV. Once ^{8}Be is formed, it again splits into two ^{4}He nuclei. When the ^{4}He density is extremely high, a fusion reaction forming carbon nuclei in an excited state occurs with a resonant cross-section: $^4\text{He} + {}^8\text{Be} \rightarrow {}^{12}_{6}\text{C}^*$. The excited state C* immediately decays to the ground state. The carbon abundance in the Universe is relatively high, and it may also be present in stars that have not exhausted the proton cycle. In the presence of protons, the nucleus ^{12}C acts as a *catalyst* for another cycle, similar to the proton–proton

Table 12.1 The nuclear fusion reactions in the CNO cycle

Reaction	Q (MeV)
$p + {}^{12}\text{C} \rightarrow {}^{13}\text{N} + \gamma$	1.94
${}^{13}\text{N} \rightarrow {}^{13}\text{C} + e^{+} + \nu_e$	1.20
$p + {}^{13}\text{C} \rightarrow {}^{14}\text{N} + \gamma$	7.55
$p + {}^{14}\text{N} \rightarrow {}^{15}\text{O} + \gamma$	7.29
${}^{15}\text{O} \rightarrow {}^{15}\text{N} + e^{+} + \nu_e$	1.73
$p + {}^{15}\text{N} \rightarrow {}^{12}\text{C} + {}^{4}\text{He}$	4.96

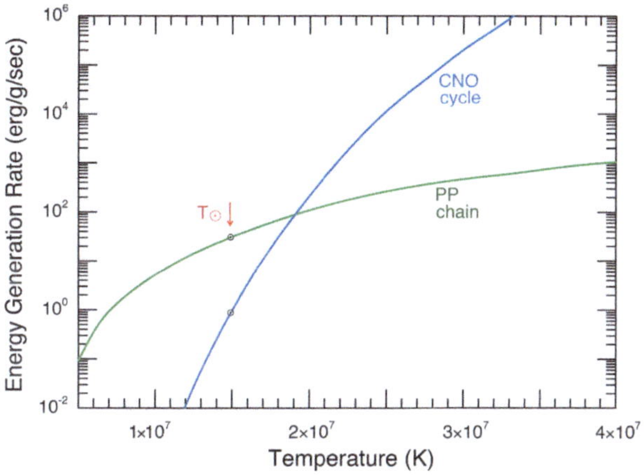

Fig. 12.2 The rate of energy generated by the *pp* chain and CNO cycle in stars as a function of the central temperature T. A density of 100 g/cm^3 is assumed with the solar metallicity. At the temperature range typically found in main sequence stars, the contribution due to the *pp* chain depends on T^4, whereas the CNO cycle depends on T^{17}

cycle, that produces energy transforming protons into helium nuclei: the CNO cycle (Table 12.1). At the end of the process, one has ${}^{12}\text{C} + 4p \rightarrow {}^{12}\text{C} + {}^{4}\text{He} + 2e^{+} + 3\gamma + 2\nu_e$ with a total energy released of about 26 MeV. The ${}^{12}_{6}\text{C}$ nucleus is strongly bound and is the starting point for the formation in massive stars of heavier nuclei through fusion, Sect. 12.10.

The temperature dependence of the reactions in the *pp* and CNO cycles is very different. The reaction rate raises quite significantly with temperature in both cases, but for the CNO cycle, this growth is much faster than that of the *pp* cycle. This means that for relatively low-mass stars ($\lesssim M_\odot$), the *pp* cycle largely dominates, while for very massive stars, the CNO cycle is believed to be the main source of energy. The only exception is for very *metal poor* stars, i.e., stars with a small content of elements heavier than helium. Figure 12.2 shows the energy production rate for a gram of material as a function of the temperature for the *pp* chain and the CNO cycle as produced by stellar models. At the Sun's temperature ($T_\odot \simeq 1.5 \times 10^7$ K), the energy predicted to be produced by the former is about 30 times larger. Even in the case of the Sun, the amount of energy produced by the reaction catalysed by the CNO is not precisely known. Only the measurement of solar neutrinos can offer a method for improving the knowledge of energy production in the stellar interior.

12.2 The Standard Solar Model and Neutrinos

The Sun has been converting hydrogen into helium for roughly 4.5×10^9 years. The values of the solar mass and the emitted power indicate that the process will continue for about as many years. The process shown in Fig. 12.1 produces energy, Eq. (12.3), and two neutrinos, which escape from the Sun, carrying away a fraction of the released energy. The kinetic energy of the other particles is the source of the thermal energy. The flux of solar neutrinos that reaches the Earth is then given by

$$\Phi_{\nu_e} \simeq \frac{1}{4\pi D_\odot^2} \frac{2L_\odot}{(Q - \langle E_\nu \rangle)} = 6 \times 10^{10} \text{cm}^{-2}\text{s}^{-1} , \tag{12.4}$$

where $L_\odot = 3.842 \times 10^{33}$ erg/s is the solar luminosity, $D_\odot = 1.495 \times 10^{13}$ cm is the Sun-Earth distance, and $\langle E_\nu \rangle \simeq 0.3$ MeV is the average energy of neutrinos produced in the fusion cycle. Main sequence stars burn their hydrogen into helium following the *pp* cycle and/or the CNO cycle. The same energy output can be produced by the two cycles that result in different energy distributions of neutrinos.

Exercise Estimate the Sun's lifetime assuming that the solar luminosity is provided: (a) by electromagnetic combustion reactions; (b) by the gravitational binding energy (the so-called Kelvin-Helmholtz contraction); (c) by the nuclear reactions (12.3).

In order to compute the relative importance of the different fusion cycles, and to obtain a precise prediction for the ν_e flux, it is necessary to compute in detail the structure of the Sun. This includes its density, temperature, composition profiles, and the rates of the different nuclear reactions, which also depend on the radial position inside the Sun. This is the task of the so-called Standard Solar Model (SSM), which uses well-known physics and some approximations, notably spherical symmetry, and the absence of rotation, to compute the structure of the Sun and predict the neutrino fluxes. The SSM calculations have been developed starting from the pioneering work of Bahcall (1989) and his collaborators about 50 years ago. The SSM calculations have been rather frequently updated to match new input data. For this reason, slightly different versions of the SSM exist today as a consequence of different interpretations of the best available physics and input data (Haxton 2009). The SSMs are based on the following assumptions (Antonelli et al. 2013; Bahcall 1989; Haxton et al. 2013):

- The Sun is in a state of hydrodynamic equilibrium, maintaining a local balance between the thermal pressure (outward) and the weight of the material above pressing downward (inward). To implement this condition, an equation of state is needed. Helioseismic information (see below) have provided important tests of the associated theory.
- The Sun produces its energy (12.3) by fusing protons into ^{4}He via the *pp* chain and with a small but yet not precisely known contribution from the CNO cycle. Nuclear cross-sections of different processes are taken from experiments or from theory.

- The mechanisms for energy transport from the core to the photosphere are radiation and convection. The inner region of the Sun (∼70% by radius) is radiative. In order to describe radiative transport, the *opacity* must be known as a function of temperature, density and composition. The opacity for the emitted γ-rays includes: Thomson scattering on electrons; interactions with fully ionized hydrogen and helium nuclei; complex processes such as bound-free scattering on metals. In the Sun's outer envelope, where the radiative gradient is larger, convection dominates the energy transport.
- The initial condition of the problem is a total mass of material and an *initial* (or *presolar*) composition. The latter is estimated as being equal to the present composition at the surface. The presolar composition is divided into hydrogen X_{ini}, helium Y_{ini}, and metals (=elements heavier than He) Z_{ini}, with $X_{\text{ini}} + Y_{\text{ini}} + Z_{\text{ini}} = 1$. Relative metal abundances can be determined from a combination of photospheric and meteoritic abundances and are generally assumed to have remained constant since the Sun formed. This observable is one of the key parameters of different versions of the SSM.
- The composition of the core evolves with time according to the rates of the nuclear reactions, and the structure slowly evolves according to these changes.
- Boundary conditions include the modern Sun's mass, age, radius $R_\odot$, and luminosity $L_\odot$.

The resulting solar models are dynamic and trace the evolution of the Sun to the present condition, thereby predicting contemporary solar properties such as the composition, temperature, pressure, sound-speed profiles, energy output, the relative weight of *pp* and CNO cycles and the neutrino fluxes. The luminosity of the Sun has increased by ∼40% over the solar lifetime. The interaction rates of different reactions shown in Fig. 12.1 depend on the quantity $\langle \sigma v \rangle$, where v is the relative velocity between two colliding nuclei and σ the cross-section. The $\langle \dots \rangle$ denotes an average over the Maxwell-Boltzmann velocity distribution in the solar plasma. The energy E_G for which the probability of a solar reaction reaches a maximum corresponds to the so-called *Gamow peak*. At energies higher than E_G, the number of particles able to induce the reaction becomes insignificant; at lower energies, the tunneling through the Coulomb barrier makes the reaction improbable, Fig. 12.3.

The most important observable for tests of the nuclear reactions occurring in the Sun's core are the neutrinos. The predicted spectrum of ν_e emitted by the Sun is shown in Fig. 12.4. Different experiments have different energy thresholds, and are sensitive to different regions of the neutrino spectrum. Referring to Fig. 12.1, the most abundant contribution comes from deuteron formation. The two reactions *pep* and ^{7}Be induced by electron capture, produce line sources of neutrinos of energy $E_\nu^{\max}$ that are only broadened by thermal effects.

The possibility of comparing model predictions to solar properties has inspired a great deal of laboratory work for the purpose of reducing uncertainties about atomic opacities and nuclear cross-sections (Broggini et al. 2010). The neutrino fluxes are a sensitive thermometer for the solar core, provided the associated nuclear physics is under control. Particularly important are input data from radiative opacities and from

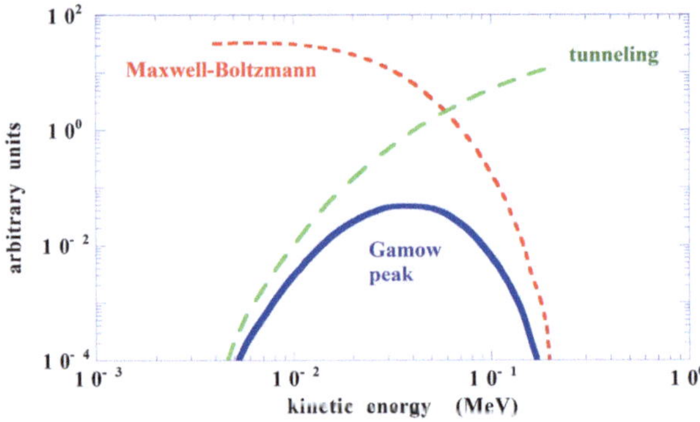

Fig. 12.3 The Gamow peak is the convolution of the Maxwell-Boltzmann distribution with the tunneling probability of nuclei through their Coulomb barrier. This plot shows (in arbitrary units) the Maxwell-Boltzmann distribution of protons at $T = 1.5 \times 10^7$ K (*dashed distribution*) and their tunneling probability through the Coulomb barrier (*large dashed line*). The product between the two curves (*full line*) gives the Gamow peak

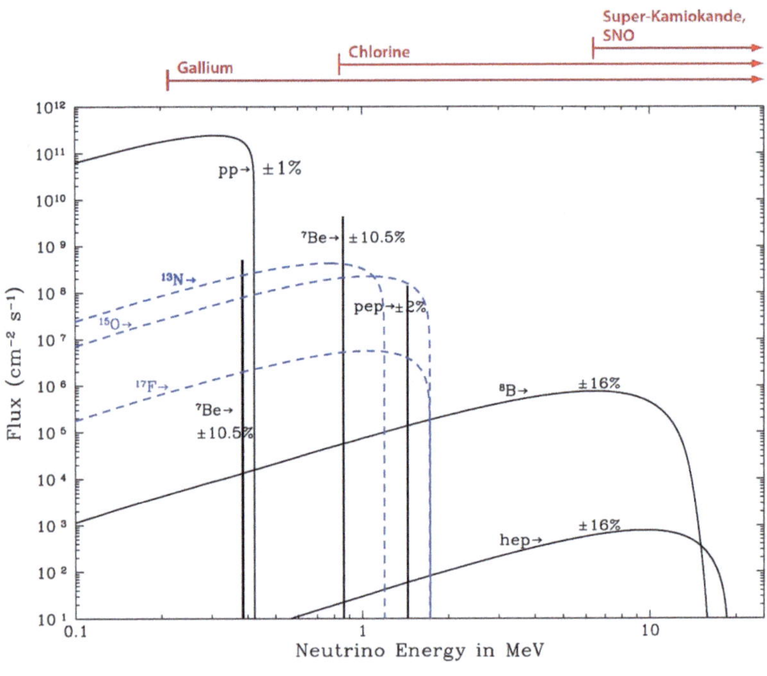

Fig. 12.4 The energy spectrum of solar neutrinos arriving on Earth along with the standard solar model uncertainties. Flux densities are shown for continuous sources: one must integrate over E to obtain fluxes in the indicated units. The abbreviations refer to reactions indicated in Fig. 12.1. The *solid lines* indicate the neutrinos coming from the most important reactions of the *pp* chain, the *dashed lines* indicate the neutrinos from the CNO cycle. The *arrows* at the *top* represent the thresholds of the indicated experiments. The numbers associated with the neutrino sources show theoretical errors of the fluxes. Adapted from the late John Bahcall's web site (http://www.sns.ias.edu/jnb/)

Table 12.2 Standard solar model (SSM) neutrino fluxes from the GS98 (Serenelli et al. 2011) and AGSS09 (Asplund et al. 2009), with associated uncertainties

Abbr.	Reaction	E_{ν}^{max} (MeV)	Φ_{ν_e} (GS98) (high Z/X) $cm^{-2}\ s^{-1}$	Φ_{ν_e} (AGSS09) (low Z/X) $cm^{-2}\ s^{-1}$
pp	$pp \rightarrow {}^2H\ e^+\ \nu_e$	0.42	$(5.98 \pm 0.6\%) \times 10^{10}$	$(6.03 \pm 0.6\%) \times 10^{10}$
pep	$pe^- p \rightarrow {}^2H\ \nu_e$	1.44	$(1.44 \pm 1.2\%) \times 10^8$	$(1.47 \pm 1.2\%) \times 10^8$
^7Be	$^7Be\ e^- \rightarrow {}^7Li\ \nu_e$	0.86 (90%)	$(5.0 \pm 7\%) \times 10^9$	$(4.6 \pm 7\%) \times 10^9$
		0.38 (10%)		
^8B	$^8B \rightarrow {}^8Be\ e^+\ \nu_e$	∼15	$(5.6 \pm 14\%) \times 10^6$	$(4.6 \pm 14\%) \times 10^6$
hep	$^3He\ p \rightarrow {}^4He\ e^+\ \nu_e$	18.77	$(8.0 \pm 30\%) \times 10^3$	$(8.3 \pm 30\%) \times 10^3$
^{13}N	$^{13}N \rightarrow {}^{13}C\ e^+\ \nu_e$	1.20	$(3.0 \pm 14\%) \times 10^8$	$(2.2 \pm 14\%) \times 10^8$
^{15}O	$^{15}O \rightarrow {}^{15}N\ e^+\ \nu_e$	1.73	$(2.2 \pm 15\%) \times 10^8$	$(1.6 \pm 15\%) \times 10^8$
^{17}F	$^{17}F \rightarrow {}^{17}O\ e^+\ \nu_e$	1.74	$(5.5 \pm 17\%) \times 10^8$	$(3.4 \pm 16\%) \times 10^8$

the determinations of solar abundances of heavy elements. Table 12.2 gives fluxes for two solar models that reflect the uncertainties about the metallicity in the Sun. There are significant differences between older composition models, with higher metallicity abundance (GS98) (Serenelli et al. 2011) and the newer ones, with lower heavy element abundances (AGSS09) (Asplund et al. 2009). The model labeled GS98 uses abundances derived from simple assumptions about the solar interior based on current analyses of the photosphere, yielding a ratio between metal and hydrogen $(Z/X) = 0.0229$. The model labeled AGSS09 takes abundances derived from updated techniques, a three-dimensional hydrodynamic model of the Sun's atmosphere and improved atomic physics, yielding $(Z/X) = 0.0178$.

The assumed percentage of metals in SSMs is an important parameter for evaluating the opacity. In the model with a low-metallicity ratio, the Sun's core is somewhat cooler (by ∼1%). A small variation of the central temperature induces large variations of high energy neutrinos, such as those from ^8B decay, which are reduced by ∼20% with respect to the model with high-metallicity (see Table 12.2). The uncertainties associated with the ν_e fluxes are linked with uncertainties assigned to approximately 20 model input parameters, which include the solar age, present day luminosity, opacities, the diffusion constant, the cross-sections for the *pp* chain and CNO cycle, and the abundances of different elements.

A unique tool for determining the structure of the solar interior is provided by the *helioseismology*, the study of the natural oscillations of the Sun. Measurements and analysis of Doppler shifts of photospheric absorption lines show that the Sun's surface oscillates with amplitudes of ∼30 m and velocities of $\sim 0.1\ m\ s^{-1}$, reflecting a variety of interior modes. The significant effort invested in helioseismological measurements and analysis has yielded a rather precise map of associated sound speed $c(r)$ over the outer 90% of the Sun by radius. This function depends on the same quantity used in the SSM to derive the neutrino fluxes, namely the Sun's quasi-static pressure, density, temperature, entropy, gravitational potential, and nuclear energy generation profiles, which are all functions of the radial coordinate r.

The comparisons of results of the SSM with helioseismic data pose a still unresolved problem, called the *solar abundance problem*. In short, all helioseismic predictions of SSMs with low *Z/X* ratio as the AGSS09 are in disagreement with helioseismic observations. On the other hand, the GS98, derived with simpler assumptions to model the solar interior and giving a higher *Z/X* ratio, is in better agreement with helioseismic data. The solar abundance problem thus represents the incompatibility between the model with the best description of the solar atmosphere and the helioseismologic description of the Sun's interior. A direct measurement of CNO neutrinos, providing experimental insight into the Sun's temperature and metallicity, would be pivotal to clarify the controversy.

12.3 Solar Neutrino Detection

The detection of low-energy electron neutrinos is extremely challenging. The electron neutrino interactions considered in the previous chapters occurred through the deep inelastic scattering process $\nu_e N \to e^- X$. Here, N is either a proton or a neutron, and X a hadronic system. This reaction occurs for neutrino energies much higher than the threshold for single pion production (∼140 MeV). Lower energy neutrinos (including the MeV-scale solar ν_e) interact only via quasi-elastic scattering

$$\nu_e + n \to p + e^- \tag{12.5}$$

or through elastic scattering (ES) on electrons, with a much smaller cross-section. The problem of (12.5) is that free neutrons do not exist in nature. Only neutrons bound in nuclei can be used in the reactions

$$\nu_e +^A_Z X \to^A_{Z+1} Y + e^- \, . \tag{12.6}$$

By changing the number of protons, ν_e capture transforms the nuclide into a new element. This requires an additional energy with respect to (12.5) due to the difference of the nuclear binding energies between the nucleus $^A X_Z$ and $^A Y_{Z+1}$. Reaction (12.6) can be used to detect solar neutrinos only if the very few Y atoms produced by the weak-interaction process can be separated from the huge number of X target atoms of the detector using *chemical extraction* techniques. Only a few elements X are thus suitable to be effectively used. The production rate of the Y atoms can be easily measured if the $^A Y_{Z+1}$ isotope is *radioactive* and, moreover, if the lifetime is neither too short nor too long. In this case, when extracted, the decay rates can be counted. The combination of these techniques give rises to the so-called *radiochemical* experiments.

The Chlorine Radiochemical Experiment Radiochemical detection of neutrinos using ^{37}Cl in (12.6) was suggested by Pontecorvo as early as 1946 and explored in more detail by Alvarez in 1949 (Lande 2009). It was Ray Davis (Nobel laureate in 2002) at Brookhaven who, in 1955, started to develop practical detectors, using about a 4000 l tank of perchloroethylene (C_2Cl_4) placed a few meters underground. The construction of an experiment with the scale necessary for solar neutrino detection (390,000 l of C_2Cl_4) began in the Homestake mine (South Dakota) in 1965, at a depth of 4000 m.w.e. The first results were announced in 1968, and the measurements continued until 2002, when the mine closed.

The "chlorine experiment" exploited the fortuitous properties of ^{37}Ar nuclei. These are produced with a threshold energy of 0.814 MeV in the reaction

$$\nu_e + {}^{37}_{17}\text{Cl} \rightarrow {}^{37}_{18}\text{Ar} + e^- . \tag{12.7}$$

The average solar neutrino reaction rate in the tank was 0.48 counts/day, above an estimated background of 0.09 counts/day. Argon is a noble gas that does not interact chemically, and it can be extracted with high efficiency (estimated as ∼95%) from large volumes of organic liquid. In addition, the ^{37}Ar isotope has a half-life of 35 days, long enough to allow them to build up their concentrations in the tank over a saturation time of about 2 months.

After extraction of ^{37}Ar nuclei from the tank, they decay via the capture of one orbital electron (usually from the K shell) returning to ^{37}Cl via the inverse reaction of (12.7). The newly formed ^{37}Cl, although still electrically neutral, is formed in an excited state with a missing electron in the inner shell. An outer shell electron will fill the empty inner level, thereby dropping to a lower state. During this process, the transient electron will emit an X-ray of 2.82 keV. Davis developed miniaturized gas proportional counters for counting such decays.

The chlorine experiment counted ∼25 Ar nuclei per year. Taking into account detector efficiencies and losses due to ^{37}Ar decaying in the tank, the counting rate can be converted to a capture rate that is the measurable quantity in the radiochemical experiments. This corresponds to the probability per unit time that a target nucleus captures a solar neutrino. The capture rate can be calculated as

$$\langle\sigma\Phi\rangle \equiv \int dE \frac{d\Phi_{\nu_e}}{dE}\sigma(E) \quad [\text{s}^{-1}] , \tag{12.8}$$

where $d\Phi_{\nu_e}/dE$ is the differential flux of solar neutrinos and $\sigma(E)$ is the cross-section for the reaction considered. It has become customary to use the unit 1 SNU $\equiv$ 1 Solar Neutrino Unit = 10^{-36} captures per second. The final result obtained by Davis was

$$\langle\sigma\Phi\rangle_{Cl} = 2.56 \pm 0.16_{\text{stat}} \pm 0.16_{\text{sys}} \text{ SNU}, \tag{12.9}$$

which is about a factor of three below the SSM best values: the GS98 predicts a rate of 8.00 ± 0.97 SNU. This result represents the beginning of the *solar neutrino problem*, a major discrepancy between measurements of the numbers of neutrinos reaching the Earth and theoretical predictions, lasting from the early results of the chlorine experiment to about 2002. The discrepancy between data end expectation about solar neutrinos had three possible interpretations: (1) experimental errors;

(2) an astrophysical overestimate of the solar neutrino production; (3) new physics in the neutrino sector. Over the years, the last interpretation of the measured deficit was found to be the correct one, as we show in the following.

As shown in Fig. 12.4, reaction (12.7) is mostly sensitive to the boron and beryllium neutrinos (above the threshold production of Ar), whose flux estimate originally had large uncertainties. These were connected with uncertainties about the central temperature of the Sun, on which the higher energy neutrino flux is primarily sensitive. For this reason, the solution to the solar neutrino problem was not immediately apparent and many explanations were proposed to change the SSM in order to produce a somewhat cooler core.

GALLEX/GNO, Sage Another possible element for reaction (12.6) is gallium through the reaction

$$\nu_e + {}^{71}_{31}\mathrm{Ga} \rightarrow {}^{71}_{32}\mathrm{Ge} + e^- . \tag{12.10}$$

The Soviet-American Gallium Experiment (SAGE) (Abdurashitov et al. 2009) and the Gallium Experiment (GALLEX) (Hampel et al. 1999) (successively: Gallium Neutrino Observatory (GNO) Altmann et al. 2005) began solar neutrino measurements in December 1989 and May 1991, respectively, exploiting the above reaction.

SAGE was located at the Baksan Neutrino Observatory in the Caucasus Mountains in Russia and used a target of 50 tons of Ga under the form of a molten metal at a temperature of 30 °C. It has reported results for 168 extractions through December 2007. GALLEX, which used 30 tons of Ga in the form of a $GaCl_3$ solution, ran between 1991 and 1997 at the Gran Sasso Laboratory (LNGS) in Italy. A number of improvements in Ge extraction procedures, electronics, counter efficiency calibrations, and radon event characterizations were incorporated into the follow-up experiment GNO, which continued through 2003.

As ^{71}Ge has a half-life of 11.4 days, a radiochemical experiment analogous to that done for chlorine is possible, although the chemistry of Ge recovery is considerably more complicated than that for Ar. GALLEX/GNO recovered Ge as $GeCl_4$ by bubbling nitrogen through the solution, and then scrubbing the gas. The Ge can be further concentrated and purified, converted into GeH_4, and then counted in miniaturized gas proportional counters similar to those used in the chlorine experiment. The separation of ^{71}Ge atoms in the liquid Ga metal of SAGE is different and more complex.

Reaction (12.10) has a low threshold (233 keV) and a strong cross-section for absorbing the low-energy *pp* neutrinos giving ^{71}Ge via a Gamow-Teller transition. In this transition (Chapter 8 of Braibant et al. 2011) the spin vectors of the initial and final nuclei change by one unit. Because of this lower threshold, reaction (12.10) also occurs for low energy ν_e from the deuteron formation (*pp* in Table 12.2). Assuming steady-state luminosity of the Sun and the standard weak interaction physics, 79 SNU is foreseen from this reaction. The remaining channels in Table 12.2 give additional 48 SNU, mainly from ^{7}Be and ^{8}B.

A unique aspect of the Ga experiments was the use, within a limited time interval, of an external ^{51}Cr artificial source with intensities of ~0.5 MCi. This radioactive isotope has a half-life of 27.7 days for electron capture that produces a ν_e. When inserted into a cavity of the detector, the external sources give an additional counting rate to the solar contribution, which can be evaluated from the source intensity. This additional event rate was correctly identified by the experiments. The procedure thus allowed an overall check of the experimental procedures: chemical extraction, counting, and analysis techniques.

After many years of operations, the weighted average of the SAGE (65.4 ± 5 SNU), GALLEX (73.1 ± 7 SNU) and GNO (62.9 ± 6 SNU) results is

$$\langle \sigma \Phi \rangle_{Ga} = 66.1 \pm 3.1\text{SNU}, \tag{12.11}$$

while the expected SSM (GS98) rate is 126.6 ± 4.2 SNU.

The Real-Time Experiments (Kamiokande and SK) A different detection strategy that confirmed the Cl and Ga deficit was used by the Kamiokande (Fukuda et al. 1996) experiment, and later on by Super-Kamiokande (SK), in Japan: neutrinos interacting via elastic scattering on electrons

$$ES: \quad \nu e^- \rightarrow \nu e^- \tag{12.12}$$

were detected in a large water tank. This reaction does not have an energy threshold, however, it is detectable above the natural radioactivity background only when the final state electron has a sufficiently high energy. This method can only reveal the highest energy neutrinos coming from the ^{8}B (Fig. 12.4).

The Kamiokande was an experiment led by M. Koshiba. It used a ~2.2 kt tank filled with purified water and viewed by 94820″ photomultipliers, providing ~20% surface coverage. The innermost 0.68 kt of the detector served as the fiducial volume for solar neutrino detection. The energy threshold varied from the initial 9 MeV to 7 MeV after subsequent detector improvements. The outer portion of the detector was instrumented with 123 PMTs to serve as a muon veto, and additional water was added to shield against γ-rays from the surrounding rock.

The electron scattering method is sensitive to all neutrino types; however, the cross-section for ν_e is approximately six times larger than for ν_μ or ν_τ. The reason is that reaction (12.12) occurs for all flavors through Z^0 exchange, Fig. 12.5a, while for ν_e also through $W^\pm$ exchange, Fig. 12.5b. The final state electron is emitted in a direction correlated with that of the incoming neutrino. Thus, by relating event directions with the position of the Sun, one can remove a large background uncorrelated with the solar position to reveal solar neutrino events in a forward cone, Fig. 12.6.

The first result of Kamiokande was based on a live time of 450 days through May 1988. The number of measured events was (46 ± 15)% with respect to the SSM prediction. The statistics increased with time until 1995, before it was succeeded by SK. Kamiokande was the first experiment to record solar neutrinos event by event, establish their solar origin through a correlation with the direction to the Sun, and to provide direct information on the ^{8}B energy spectrum.

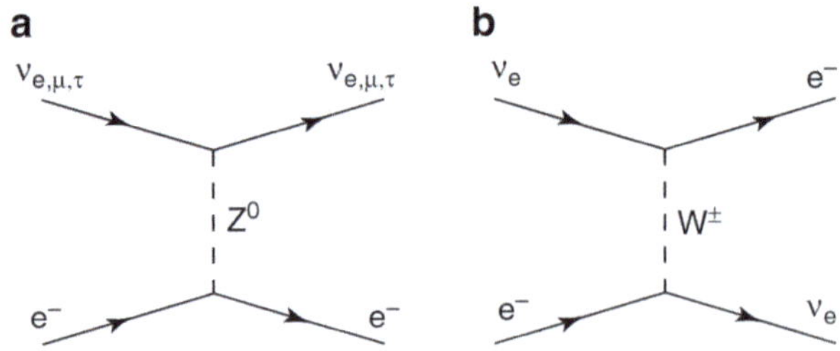

Fig. 12.5 Feynman diagrams for (**a**) the ν_e, ν_μ and ν_τ scattering on electrons through a neutral current weak interaction and (**b**) the charged current scattering of a ν_e on electron

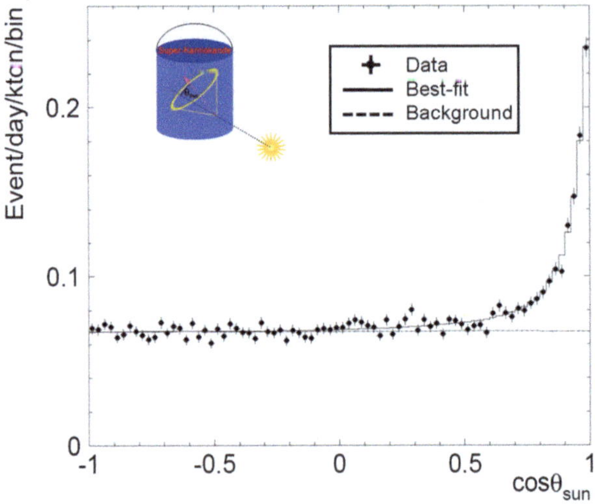

Fig. 12.6 Angular distribution of solar neutrino event candidates detected by SK-III within the energy range between 5.0 and 20.0 MeV. The area under the *dotted line* is the contribution from remaining background events. The area between the *solid* and *dotted line* indicates the elastic scattering peak. The definition of the angle θ_{sun} is in the inset. *Credit* Kamioka Observatory, ICRR (Institute for Cosmic Ray Research), The University of Tokyo

The most accurate measurement of solar ν_e through reaction (12.12) is due to SK-III, which ran from August of 2006 through August of 2008, with a lower total energy threshold of 5 MeV (Abe et al. 2011). The Kamiokande and SK results were expressed in terms of neutrino flux from the ^{8}B reaction. At present, the latest result from SK-III (Smy et al. 2013) is

$$\Phi_{\nu_e}(^8\text{B}) = (2.39 \pm 0.04_{\text{stat}} \pm 0.05_{\text{sys}}) \times 10^6\,\text{cm}^{-2}\text{s}^{-1}\,, \tag{12.13}$$

in agreement with the Kamiokande result of $(2.80 \pm 0.19 \pm 0.33) \times 10^6\,\text{cm}^{-2}\text{s}^{-1}$. The neutrino flux (12.13) is ∼50% smaller than expected from the SSM (compare with the predictions shown in Table 12.2).

All the above results indicate that there are "missing" neutrinos from the Sun, when data are compared to the SSM. Because neutrino oscillations were already observed in atmospheric neutrinos, they represent a natural explanation for the problem. However, none of the above experiments was able to conclusively prove that the lack of solar electron neutrinos was not connected with a combination of experimental problems, or to shortcomings of the theory. They were all ν_e *disappearance experiments*. Oscillations produce neutrinos of different flavors but conserve the total number. Neutrino *appearance experiments* should be able to observe neutrinos of a flavor different from ν_e. The ν_μ (or ν_τ) appearance through charged current (CC) interactions produces the corresponding charged lepton. Nevertheless, the muon (or tau) rest mass is much larger than the energy corresponding to solar neutrinos, and the CC reactions cannot occur. The problem was solved by the SNO experiment, which measured the fraction of $\nu_\mu + \nu_\tau$ in the neutrino flux from the Sun using their neutral current (NC) interactions.

12.4 The SNO Measurement of the Total Neutrino Flux

The Sudbury Neutrino Observatory (SNO) in Canada recorded data from 1999 until 2006. It was able to detect Cherenkov light emitted by charged particles crossing the detector. The detector was a 12-m diameter spherical acrylic vessel viewed by an array of 9500 20-cm PMTs, covering 56% of the spherical surface. It was filled with 1000 tons of heavy water (D_2O) contained in the inner volume, and surrounded by 1500 tons of normal water used for screening purposes. Heavy water is essential to the operation of CANadian Deuterium Uranium (CANDU) nuclear power reactors. D_2O is used as both a moderator and a heat transfer agent. The heavy water is extracted from the water of lake Ontario. The SNO experiment loaned heavy water from the Atomic Energy of Canada Limited. The reactions that occur in heavy water are:

- the elastic scattering (ES) on electron (12.12). As in the normal water, the ES is sensitive to neutrinos of any flavor, but the cross-section for ν_e is largely enhanced by $W^\pm$ exchange: $\sigma(\nu_e e \to \nu_e e) \simeq 6 \times \sigma(\nu_{\mu,\tau} e \to \nu_{\mu,\tau} e)$.
- the ν_e CC interaction on the proton of the deuteron $d = (pn)$:

$$CC: \quad \nu_e + d \to e^- + p + p, \tag{12.14}$$

  which only occurs for ν_e through a $W^\pm$ exchange. With the CC reaction, the flux of higher energy ν_e can be probed by detecting the produced electron. The deuterium breakup threshold is 1.44 MeV; the electrons carry off most of the energy, and thus provide significant information on the incident neutrino spectrum. The electron is detected through its Cherenkov emission in water.
- the deuterium dissociation through a Z^0 exchange:

$$NC: \qquad \nu_f + d \to \nu_f + p + n, \qquad \nu_f = \nu_e, \nu_\mu, \nu_\tau. \tag{12.15}$$

  A ~2.2 MeV photon is emitted as a result of the d dissociation in $p + n$.

Operations were carried out in three phases (Jelley et al. 2009). The first phase of SNO (SNO-I, from November 1999 through May 2001) operated with pure heavy water (Ahmad et al. 2002). No charged particles above the Cherenkov threshold are present in reaction (12.15). However, the neutron produced in the NC reaction can be captured on deuterium, releasing a 6.25 MeV γ-ray. This γ-ray can produce a signal

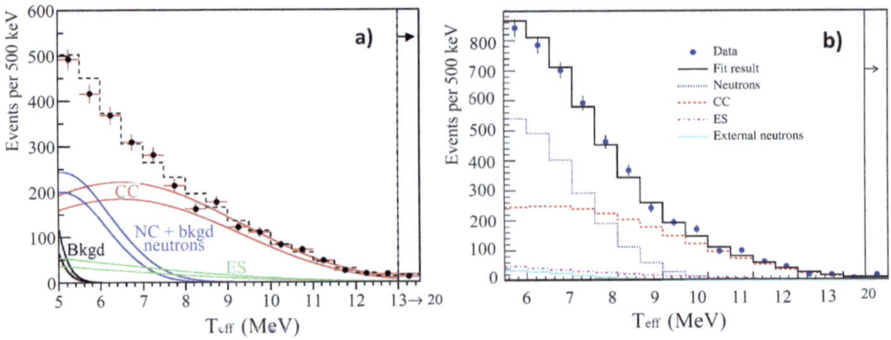

Fig. 12.7 (**a**) Kinetic energy spectrum (*brown* data points) for events with measured energy $T_{eff} \geq 5$ MeV occurring inside the fiducial region ($R \leq 550$ cm) in the SNO-I. Also shown are the Monte Carlo predictions for charged-current (CC = *red lines*), elastic scattering (ES = *green lines*), and neutral current (NC) and background (bkgd) neutron events (*purple lines*). The simulations are scaled to fit the results. The *dashed line* represents the summed components, and the bands show the statistical uncertainties from the signal-extraction fit. (**b**) The same, for the salt phase SNO-II which increased the neutron (*dotted line*) capture rate. Courtesy of Prof. Arthur B. McDonald and the SNO Collaboration

in the PMTs through a Compton scattering on electrons of the medium, yielding light for recoils above the Cherenkov threshold. The data contain events from all of the three above-mentioned reactions (CC, NC, ES), plus a small background. The result of an analysis aimed at detecting events above an effective kinetic energy of 5 MeV is shown in Fig. 12.7a. The CC and ES reactions could be resolved with the strong directional dependence of the ES reaction. The neutron capture efficiency was measured by deploying a ^{252}Cf source at various positions throughout the heavy water volume.

In the second SNO phase (SNO-II) (Aharmim et al. 2005), two tons of purified NaCl were dissolved in the water, in order to increase the neutron capture rate and energy release through the reaction $n + {}^{35}\text{Cl} \rightarrow {}^{35}\text{Cl}^* \rightarrow {}^{35}\text{Cl} + \gamma$. Data were accumulated from July 2001 through August 2003. Detector calibrations completed in SNO I were repeated and extended in SNO II. The analysis, aimed to unfold the solar ^{8}B spectrum shape from the data, was performed for a kinetic energy threshold of 5.5 MeV. The results of SNO-II are shown in Fig. 12.7b. In SNO-I and -II, the CC, ES, and NC rates were determined by a statistical analysis that decomposed the common signal, the Cherenkov light, into the three contributing components. The analysis made use of the angular correlations with respect to the Sun in ES events and of the energy differences in the CC-, ES-, and NC-associated light.

In a third phase (SNO-III, from November 2004 to November 2006), the separation of the NC and CC/ES signals was accomplished by direct counting of NC neutrons. The salt introduced in SNO-II was removed and an array of a specially designed proportional counters was deployed in the heavy water for neutron detection. The 40 strings of proportional counters were anchored to the inner surface of the acrylic vessel, forming a lattice on a 1-m grid. The counters were filled with a 85:15% mixture of ^{3}He and CF$_4$, and the NC detection occurred through the reaction $n + {}^3\text{He} \rightarrow p + {}^3\text{H}$; the final p^3H state has a total kinetic energy of 764 keV. The signals from each string were amplified and digitized. The neutron detection efficiency and the response of the arrays were determined with a variety of neutron calibration sources. Cherenkov light signals from CC, NC, and ES reactions were still recorded by the PMTs, though the rate of such NC events from reactions (12.15) was significantly suppressed due to neutron absorption in the ^{3}He proportional counters.

The SNO-I/II and SNO-III results are generally in good agreement, and have established the following, both separately and in combination:

- The total flux of active neutrinos $\nu_f = \nu_e + \nu_\mu + \nu_\tau$ from ^{8}B decay measured through NC interactions corresponds to

$$\Phi_{SNO}^{NC} = \Phi_{\nu_f}(^8\text{B}) = (5.25 \pm 0.16_{\text{stat}} \pm 0.13_{\text{sys}}) \times 10^6\,\text{cm}^{-2}\text{s}^{-1}, \qquad (12.16)$$

  in good agreement with SSM predictions, see Table 12.2.
- The flux of the ν_e flavor producing CC interactions is (SNO-II)

$$\Phi_{SNO}^{CC} = \Phi_{\nu_e}(^8\text{B}) = (1.68 \pm 0.06_{\text{stat}} \pm 0.09_{\text{sys}}) \times 10^6\,\text{cm}^{-2}\text{s}^{-1}. \qquad (12.17)$$

- The flux of the ES interactions is (SNO-II)

$$\Phi_{SNO}^{ES} = (2.35 \pm 0.22_{\text{stat}} \pm 0.15_{\text{sys}}) \times 10^6\,\text{cm}^{-2}\text{s}^{-1}\,, \qquad (12.18)$$

  with $\Phi^{ES} \simeq \Phi_{\nu_e} + (1/6)\Phi_{\nu_\nu+\nu_\tau}$ due to the relative weights on ES of different flavors.
- There is no statistically significant day-night effects (due to the passage of detected neutrinos through the Earth) or spectral distortions in the region of the ^{8}B neutrino spectrum above 5 MeV.

This result clearly indicates that $\Phi_{\nu_\mu+\nu_\tau} = \Phi_{\nu_f} - \Phi_{\nu_e}$ is nonzero, providing a definitive proof that 2/3 of the ^{8}B solar electron neutrinos, on their way to the Earth, changed flavor. On the other hand, the *total* number of solar neutrinos as measured through the NC is in agreement with the SSM expectation.

12.5 Oscillations and Solar Neutrinos

The solution to the solar neutrino problem in terms of new physics in the neutrino sector was clearly given by a nonsolar neutrino experiment in 2002. This year is very often denoted as the "annus mirabilis" of solar neutrino physics: in April, the first SNO result including NC showed that the total neutrino flux was compatible with the SSM; in October, the Nobel prize was awarded to R. Davis and M. Koshiba for the detection of cosmic neutrinos; in December, the first results of KamLAND offered the first clear terrestrial confirmation of the validity of the oscillation solution to the solar neutrino problem.

The **Kamiokande Liquid scintillator ANtineutrino Detector (KamLAND)** was a 1000-ton ultra-pure liquid scintillator detector located at the old Kamiokande site in Japan. The primary goal was a long-baseline (flux-weighted average distance of $\sim$180 km) neutrino oscillation studies using $\overline{\nu}_e$ emitted from a large number of nuclear reactors in the central region of Japan. Prior to the earthquake and tsunami of March 2011, Japan generated $\sim$30% of its electrical power from nuclear reactors (more than 60 GW). In commercial nuclear reactors, the energy is released in neutron-induced fissions with a nuclear fuel constituted by uranium enriched (to 2–5%) in the ^{235}U isotope. The most important processes are of the type $n + {}^{235}\text{U} \rightarrow X_1 + X_2 + 2n$ and result in the production of typically two neutrons (that can therefore sustain a chain reaction), two fragments X_1 and X_2 and thermal energy. The fragments of the fission are overly rich in neutrons (the ^{235}U has 92 protons and 143 neutrons) and to reach stability, they must undergo a succession of beta decays (on average, a total of 6), therefore emitting an average of 6 $\overline{\nu}_e$. From a knowledge of nuclear physics, it is possible to compute in detail the decay chains that are generated by the nuclear fissions, and from the knowledge of the reactor power, it is possible to estimate the flux and energy spectrum of the emitted $\overline{\nu}_e$. They have energy below 10 MeV, with an average value $E_{\overline{\nu}_e} \sim 3$ MeV.

As a $\overline{\nu}_e$ disappearance experiment, KamLAND studied the flux and the energy spectrum of positrons produced in the inverse beta-decay reaction

$$\overline{\nu}_e + p \rightarrow e^+ + n. \tag{12.19}$$

The prompt positron annihilation and the delayed coincidence of a 2.2 MeV γ-ray from neutron capture on a proton was used to identify the $\overline{\nu}_e$. With a prompt-energy analysis threshold of 2.6 MeV, this experiment had a sensitivity for neutrino oscillations down to $\Delta m^2 \sim 10^{-5}\,\text{eV}^2$. The result of this experiment is in relation to solar ν_e disappearance through the *CP* theorem (Chapter 6 of Braibant et al. 2011), which assumes that the properties of particles and antiparticles are strictly correlated, and in particular, that

$$P(\nu_\alpha \rightarrow \nu_\beta) = P(\overline{\nu}_\alpha \rightarrow \overline{\nu}_\beta), \tag{12.20}$$

where P represents the oscillation probability in vacuum. The search for the violating *CP* effects in the neutrino sector is a fascinating and very important topic in particle physics that will not be discussed in this book.

The first KamLAND result with a 162 ton· y exposure gave a ratio of observed to expected number events of $(0.611 \pm 0.085_{\text{stat}} \pm 0.041_{\text{sys}})$. With increased statistics, KamLAND observed not only the distortion of the $\overline{\nu}_e$ spectrum, but also, for the first time the oscillatory feature of the survival probability expected from neutrino oscillations, Fig. 12.8. The best fit for two-flavor oscillations using Eq. (11.46)

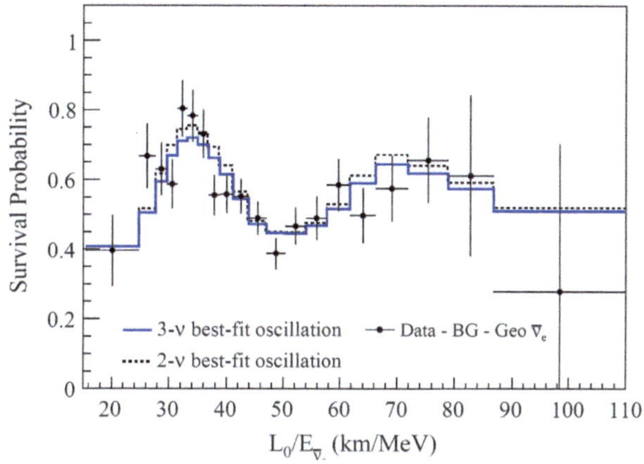

Fig. 12.8 The ratio of the measured $\overline{\nu}_e$ spectrum (after subtraction of the background) to the predicted one without oscillations as a function of L_0/E, where $L_0 = 180$ km. The histogram represents the survival probability based on the best-fit parameter values from the two- and three-flavor neutrino oscillation analyses. *Credit* Particle Data Group (Patrignani et al. 2016/2017)

gave us

$$\Delta m^2_{12} = (7.9 \pm 0.6) \times 10^{-5}\text{eV}^2; \quad \tan^2\theta_{12} = (0.40 \pm 0.10), \tag{12.21}$$

as shown by the dashed histogram in Fig. 12.8. Their results in terms of neutrino oscillation parameters (in this case, the $\overline{\nu}_e$ oscillations occur in vacuum) are in good agreement with the results obtained using solar neutrinos when matter effects are included, as discussed in the following.

Note that the above parameters are different from those derived from atmospheric neutrino experiments and accelerator long-baseline ν_μ disappearance, Sect. 11.10. This arises from the fact that three neutrino families are known; the mixing among different flavors depends on two different squared mass differences and on a 3×3 matrix with three mixing angles.

12.6 Oscillations Among Three Neutrino Families[1]

In the case of three flavors, the mixing between flavor and mass eigenstates is described by a 3×3 unitary matrix, similar to the Cabibbo-Kobahashi-Maskawa one describing mixing among quarks (Braibant et al. 2011). The unitary matrix for neutrino mixing is called the Pontecorvo-Maki-Nakagawa-Sakata (PMNS) matrix and it can be parameterized as ($f = e, \mu, \tau$; $j = 1, 2, 3$)

$$U_{fj} = \begin{pmatrix} c_{12}c_{13} & s_{12}c_{13} & s_{13}e^{-i\delta} \\ -s_{12}c_{23} - c_{12}s_{23}s_{13}e^{i\delta} & c_{12}c_{23} - s_{12}s_{23}s_{13}e^{i\delta} & s_{23}c_{13} \\ s_{12}c_{23} - c_{12}s_{23}s_{13}e^{i\delta} & -c_{12}s_{23} - s_{12}c_{23}s_{13}e^{i\delta_{13}} & +c_{23}c_{13} \end{pmatrix} \tag{12.22}$$

[1]This section can be skipped in the early reading steps.

Table 12.3 Summary of experiments studying neutrino oscillations

Source	Experiments	Neutrino type	$\overline{E}$ (GeV)	L (km)	Δm^2_{min} (eV^2)
Reactors	Chooz, Daya Bay, Reno	$\overline{\nu}_e$	10^{-3}	1	10^{-3}
Reactors	KamLAND	$\overline{\nu}_e$	10^{-3}	100	10^{-5}
Accelerator	Chorus, Nomad	$\nu_\mu, \overline{\nu}_\mu$	1	1	~ 1
Accelerator	K2K, MINOS, OPERA	$\nu_\mu, \overline{\nu}_\mu$	1	$300 \div 700$	10^{-3}
Atmospheric[a]	SK,Soudan	$\nu_{\mu,e}, \overline{\nu}_{\mu,e}$	1	$10 \div 10^4$	$10^{-1} \div 10^{-4}$
Atmospheric[b]	SK,MACRO	$\nu_\mu, \overline{\nu}_\mu$	10	$10^2 \div 10^4$	$10^{-1} \div 10^{-3}$
Sun	SK,SNO,GALLEX,...	ν_e	10^{-3}	10^8	10^{-11}

The columns report: the source of neutrinos; some of the most significant experiments; the neutrino flavor at the source; the mean path length from the source to the detector; the minimum value of Δm^2 that can be tested with the formalism of two-flavor oscillations
[a]The neutrino flavour is detected from the contained event topology
[b]Measured by ν_μ-induced upgoing muons

or, equivalently, as

$$U_{fj} = \begin{pmatrix} 1 & 0 & 0 \\ 0 & c_{23} & s_{23} \\ 0 & -s_{23} & c_{23} \end{pmatrix} \begin{pmatrix} c_{13} & 0 & s_{13}e^{-i\delta} \\ 0 & 1 & 0 \\ -s_{13}e^{i\delta} & 0 & c_{13} \end{pmatrix} \begin{pmatrix} c_{12} & s_{12} & 0 \\ -s_{12} & c_{12} & 0 \\ 0 & 0 & 1 \end{pmatrix} \tag{12.23}$$

(abbreviations, e.g., $s_{13} = \sin\theta_{13}$, are used). The phase angle δ would allow the *CP*-violation in the leptonic sector.

The limit of two-flavor oscillations was already discussed in Sect. 11.8, where the mixing angle was denoted as ϑ. Extending the discussion to three flavors, ϑ can indicate one of the above mixing angles $\theta_{12}, \theta_{23}, \theta_{13}$ in (12.22). The numerical values of the matrix components and the differences between mass squared have been measured with atmospheric and solar experiments, and using reactors and accelerators, as summarized in Table 12.3. The measured values of the mixing angles are reported in Sect. 12.8.

The more general formulas for the probability of three-flavor neutrino oscillations are rather complicated (Lipari 2001). They are obtained from Eq. (11.43) using the matrix U (12.22). The formulas can be greatly simplified if there is a *hierarchy* between the neutrino masses, for example[2]

$$m_3 \gg m_2 > m_1 > 0, \tag{12.24}$$

[2]This is usually called *normal ordering* or *Normal Hierarchy*. Another possible solution is the case with $0 < m_3 \ll m_1 < m_2$, which corresponds to an *inverted ordering* or *Inverted Hierarchy*. We do not consider these aspects of ν physics.

yielding

$$|\Delta m^2_{13}| \simeq |\Delta m^2_{23}| \gg |\Delta m^2_{12}| > 0. \tag{12.25}$$

This is exactly the situation depicted by measurements of atmospheric neutrinos on one hand (atm) and solar neutrinos and KamLAND on the other $(\odot)$, which demonstrated that $\Delta m^2_{\text{atm}} \gg \Delta m^2_{\odot}$. The first mass difference approximates $|\Delta m^2_{23}| \simeq |\Delta m^2_{13}|$ and the latter $|\Delta m^2_{12}|$, Sect. 12.8.

In this situation, there are basically two characteristic oscillation lengths given by Eq. (11.47), and that involving $|\Delta m^2_{12}|$ $(L_{12} \simeq E/\Delta m^2_{12})$ is longer. Then, there is a range of E and L values such that "short" fluctuations (i.e., those relating to $|\Delta m^2_{23}|$) are active, while the "long" oscillations have not yet developed. The probability of short oscillation $(\alpha \neq \beta)$ can be approximated by

$$P(\nu_\alpha \to \nu_\beta) = 4|U_{\alpha 3}|^2 |U_{\beta 3}|^2 \sin^2\left(\frac{\Delta m^2_{13}}{4E} L\right). \tag{12.26}$$

This formula is similar to that for two-flavor oscillations (11.46), and (for a given energy) the probability oscillates with a single frequency, related to the mass difference $|\Delta m^2_{13}| \simeq |\Delta m^2_{23}|$. For these "short" oscillations, the probability amplitudes only depend on the elements of the third column of the mixing matrix U (12.22). Explicitly, one has

$$P(\nu_e \to \nu_\mu) = 4|U_{e3}|^2 |U_{\mu 3}|^2 \sin^2\left(\frac{\Delta m^2_{13}}{4E} L\right)$$

$$= s^2_{23} \sin^2 2\theta_{13} \sin^2\left(\frac{\Delta m^2_{13}}{4E} L\right) \tag{12.27a}$$

$$P(\nu_e \to \nu_\tau) = 4|U_{e3}|^2 |U_{\tau 3}|^2 \sin^2\left(\frac{\Delta m^2_{13}}{4E} L\right)$$

$$= c^2_{23} \sin^2 2\theta_{13} \sin^2\left(\frac{\Delta m^2_{13}}{4E} L\right) \tag{12.27b}$$

$$P(\nu_\mu \to \nu_\tau) = 4|U_{\mu 3}|^2 |U_{\tau 3}|^2 \sin^2\left(\frac{\Delta m^2_{13}}{4E} L\right)$$

$$= c^4_{13} \sin^2 2\theta_{23} \sin^2\left(\frac{\Delta m^2_{13}}{4E} L\right) \tag{12.27c}$$

Because in atmospheric neutrinos only (12.27c) gives observable results and the electron neutrinos are practically not affected by oscillations, $\sin^2 2\theta_{13}$ should be very small. The small θ_{13} mixing angle was recently measured by dedicated experiments, Sect. 12.6.2.

The survival probability for diagonal transitions ($\alpha = \beta$) also takes the simple form (for instance, for ν_e)

$$P(\nu_e \to \nu_e) = 1 - [1 - |U_{e3}|^2]|U_{e3}|^2 \sin^2\left(\frac{\Delta m^2_{13}}{4E}L\right)$$

$$= 1 - \sin^2 2\theta_{13} \sin^2\left(\frac{\Delta m^2_{13}}{4E}L\right). \tag{12.28}$$

This coincides with the ν_e survival probability in the two-flavors case (11.46) with the substitutions: $\Delta m^2 \to \Delta m^2_{13}$ and $\vartheta \to \theta_{13}$.

12.6.1 Three-Flavor Oscillation and KamLAND

When the condition $x = \frac{\Delta m^2_{13}}{4E}L \gg 1$ holds, the "short" oscillations formulas (12.27a)–(12.27c) and (12.28) are averaged out and "long" oscillations are active. This corresponds to the other extreme case, the detection of $\overline{\nu}_e$ from KamLAND and solar neutrinos. If the argument x of the $\sin^2(x)$ function is rapidly oscillating, only its average value is important. Thus, long-type oscillations are observed, where the electron neutrino survival probability is given by

$$P(\nu_e \to \nu_e) \simeq c^4_{13} P + s^4_{13} \tag{12.29}$$

with

$$P = 1 - \sin^2 2\theta_{12} \sin^2\left(\frac{\Delta m^2_{12}}{4E}L\right). \tag{12.30}$$

Because of the small value of θ_{13} in (12.29), $c^4_{13} \simeq 1, s^4_{13} \simeq 0$ and the electron neutrino oscillations can be described by a formula that looks like that used for two-flavor oscillations. The fact that the θ_{13} mixing angle is small can be appreciated in Fig. 12.8 by comparing the fit using (12.29) (the full blue line) with that obtained in the limit of two-flavor oscillations assuming $\theta_{13} = 0$ (dashed line).

The KamLAND observations are thus simply explained by vacuum oscillations. To interpret the results of solar neutrino experiments, we must consider (1) the numerical value of θ_{13} (Sect. 12.6.2), recently measured; (2) the propagation of neutrinos in matter from the core of the Sun to the surface and in vacuum, on their way to the Earth, Sect. 12.7.

12.6.2 Measurements of θ_{13}

Reactor $\overline{\nu}_e$ disappearance experiments with $L \sim 1$ km, $E \sim 3$ MeV are sensitive to $\sim E/L \sim 3 \times 10^{-3}$ eV^2 $\sim |\Delta m^2_{\text{atm}}|$. At this baseline distance, the reactor $\overline{\nu}_e$ oscillations driven by $\Delta m^2_{\odot}$ are negligible. Therefore, as can be seen from Eq. (12.28), the quantity $\sin^2 2\theta_{13}$ can be directly measured from ν_e disappearance.

A reactor neutrino oscillation experiment in the 1990s at the Chooz nuclear power station in France was the first experiment of this kind. The detector was located in an underground laboratory with 300 m.w.e. rock overburden, about 1 km from the reactor. It consisted of a central 5-ton target filled with 0.09% gadolinium loaded liquid scintillator, surrounded by an intermediate 17-ton and outer 90-ton regions filled with undoped liquid scintillator. Reactor $\overline{\nu}_e$'s were detected via the reaction (12.19). Gadolinium-doping was chosen to maximize the neutron capture efficiency. The Chooz experiment found no evidence for $\overline{\nu}_e$ disappearance and the 90% c.l. upper limit on the θ_{13} mixing angle was $\sin^2 2\theta_{13} < 0.15$.

The mixing angle θ_{13} can also be measured using the $\nu_\mu \rightarrow \nu_e$ appearance method through Eq. (12.27a) using a conventional neutrino beam at an accelerator (see details on Chapter 8 of Braibant et al. 2011). The K2K was the first long-baseline experiment to search for the ν_e appearance signal in a ν_μ beam. Also, MINOS searched for a nonnull value of this mixing angle.

Only in 2011, experimental indications of $\nu_\mu \rightarrow \nu_e$ oscillations and a nonzero θ_{13} were reported by the T2K experiment. The baseline distance is 295 km between the J-PARC in Tokai, Japan and Super-Kamiokande. The T2K experiment is the first off-axis long-baseline neutrino oscillation experiment (Chapter 12 of Braibant et al. 2012). A narrow-band ν_μ beam produced by 30 GeV protons from the J-PARC Main Ring is directed 2.5° off-axis to SK. In this configuration, the ν_μ beam is tuned to the energy corresponding to the first oscillation maximum. Before the earthquake of March 2011, T2K observed six candidate ν_e events having all the characteristics of being due to $\nu_\mu \rightarrow \nu_e$ oscillations, while the expectation for $\theta_{13} = 0$ was 1.5 events. In 2013, T2K announced the observation of 28 ν_e appearance events with 4.6 predicted background events for $\theta_{13} = 0$. This result means that $\theta_{13} = 0$ is excluded with a significance of 7.5σ, and $\sin^2 2\theta_{13} = 0.140^{+0.038}_{-0.032}$.

In 2012, three reactor neutrino experiments (Daya Bay, Double Chooz and RENO), much improved with respect to the old Chooz experiment, reported their first results on reactor $\overline{\nu}_e$ disappearance. Daya Bay and RENO measured reactor $\overline{\nu}_e$s with near and far detectors. The first results of Double Chooz was obtained with only a far detector. The $\overline{\nu}_e$ detectors of all the three experiments have similar structures; an antineutrino detector consisting of three layers and an optically independent outer veto detector. The innermost layer of the antineutrino detector is filled with Gd-doped liquid scintillator; it is surrounded by a *γ–catcher* layer of liquid scintillator and an outermost layer filled with mineral oil. The outer veto detector is filled with purified water (Daya Bay and RENO) or liquid scintillator (Double Chooz).

The Daya Bay experiment measured $\overline{\nu}_e$s from the Daya Bay nuclear power complex (six 2.9,GW reactors) in China with six functionally identical detectors deployed in three underground halls, two near (470 and 576 m of flux-weighted baselines) and one far away (1648 m). With only 55 days of live time, Daya Bay reported evidence at a level of 5.2σ for nonzero θ_{13}. Data taking is still in progress, and the latest result reported in Patrignani et al. (2016/2017) is $\sin^2 2\theta_{13} = 0.0841 \pm 0.0027_{stat} \pm 0.0019_{sys}$. The RENO experiment measured $\overline{\nu}_e$s from six 2.8,GW reactors at Yonggwang Nuclear Power Plant in Korea with two identical detectors located at 294 and 1383 m from the reactor array center. The last RENO measurements give $\sin^2 2\theta_{13} = 0.086 \pm 0.006_{stat} \pm 0.005_{sys}$ from 1500 live days of data taking. Finally, Double Chooz measured $\overline{\nu}_e$s from two 4.25,GW reactors with a far detector at 1050 m from the two reactor cores, giving $\sin^2 2\theta_{13} = 0.090^{+0.032}_{-0.029}$ with 470 live days of running.

12.7 The Neutrino Flux from the Sun

12.7.1 *Matter Effect in the Sun*

The presence of matter drastically changes the pattern of neutrino oscillations: neutrinos interact with protons, neutrons and electrons. In particular, the presence of electrons significantly affects the propagation of ν_e due to the charged current processes in Fig. 12.5b. This induces a situation analogous to the electromagnetic process that leads to the refractive index of light in a medium. This means that neutrinos in matter have a different effective mass than neutrinos in vacuum, and since neutrino oscillations depend upon the squared mass difference of the neutrinos, neutrino oscillations are different in matter than they are in vacuum.

For a pedagogical and beautiful description of quantum-mechanical physics of neutrino oscillation in vacuum and matter, we refer to Lipari (2001). For the latest results, refer to the *Neutrino mass, Mixing, and Oscillations* section of http://pdg.lbl.gov/. For a presentation of a simple two-neutrino flavors scheme, refer to *Extras # 6*. Here, a resonant condition for the neutrino energy is derived, usually referred to as the *resonance energy*:

$$E_{res} = \frac{\Delta m^2_{12} \cos 2\theta_{12}}{2\sqrt{2} N_e G_F} , \tag{12.31}$$

which depends on the Fermi constant G_F, neutrino physics properties (Δm^2_{12} and $\cos 2\theta_{12}$) and solar properties (the electron number density in the Sun, N_e). In the Sun, N_e changes considerably along the neutrino path. In the core, the matter density is about $\rho \sim 150\,\text{g/cm}^3$ and it decreases monotonically towards a small value at the surface.

As a general result, the electron neutrino survival probability is a function of the neutrino energy E and depends on solar and neutrino physical properties:

$$P_{ee}(E) \equiv P_{(\nu_e \to \nu_e)}(E) = f(N_e; \Delta m^2_{12}, \theta_{12}, \theta_{13}) . \tag{12.32}$$

Using the best-fit values obtained by KamLAND (12.21) in (12.31) with the density of the Sun's core, we obtain the minimum energy for which the resonance condition occurs, which is on the order of $E_{res} \lesssim 1\,\text{MeV}$. This means that below ~1 MeV (corresponding to the lower-energy solar neutrinos, those of the pp reaction with $E < 0.42\,\text{MeV}$), oscillations occur as in vacuum. Matter effect strongly affects neutrinos arising from the ^{8}B transitions observed by the SK and SNO experiments, with $E \gtrsim 5\,\text{MeV}$.

Different measurements of the ratio between the data and the SSM prediction by solar neutrino experiments tightly constrained the P_{ee} values in the high-energy (matter-dominated) region, in the low-energy (vacuum-dominated) region and in the transition region, between 1 and 3 MeV. The Borexino experiment played a particularly important role.

12.7.2 The Borexino Experiment at Gran Sasso Lab

The Borexino experiment (Calaprice et al. 2012) has a layout similar to SNO, but it measures the $\nu_e e$ elastic scattering (ES) by using a liquid scintillator as the active target. It is located at the Laboratori Nazionali del Gran Sasso, Italy, and started its data-taking in 2007. One fundamental Borexino characteristic consists in the reduction of the radioactivity of the scintillator itself. This has been achieved using several radio-purification techniques developed by the physicists, chemists and engineers of the collaboration. This allows them to test the SSM using neutrinos of lower energy than those of the ^{8}B branch. As evident in Table 12.2, the uncertainty on the ^{7}Be flux is smaller than that on ^{8}B. As usual, the measured ES rate would depend on all the neutrino flavors, despite being dominated by the ν_e flavor.

The detector consists of 278 tons of a high-purity liquid scintillator. The scintillation light yield is a measure of the energy imparted to the electron of the ES, but has no sensitivity to direction. The scintillation photons are detected with an array of 2200 PMTs mounted on the inside surface of the stainless-steel sphere. With a light yield of 500 detected photons per MeV, the energy resolution is approximately 5% at 1 MeV, and the position resolution is ~10–15 cm. The 0.862 MeV ^{7}Be neutrinos produce a recoil electron spectrum with a distinctive cut-off edge at 665 keV.

The Borexino results are summarized in Table 12.4. In particular, in 2014, Borexino reported the first direct measurement of pp neutrinos (Borexino Collaboration 2014), although not yet with sufficient precision as would be needed to probe the solar luminosity constraint in a strong way. Borexino was the first experiment cable of making spectrally resolved measurements of solar neutrino interactions below 3 MeV. The measured interaction rates for the different channels are shown in the second column. The ratio between the data and the SSM predictions (using the AGSS09 as reported in Table 12.2) is shown in the third column.

Table 12.4 Summary of the interaction rates (column 2, with statistical and systematic errors) of the different neutrino species measured by Borexino and the ratios with respect to SSM (column 3)

ν React.	Interaction rate counts (day 100 ton)$^{-1}$	$(\frac{Data}{SSM})$	$\Phi_{\nu_e}(E)$ (10^8 cm^{-2} s^{-1})	$(\frac{Data}{SSM})/P_{ee}$ ratio
pp	$144 \pm 13 \pm 10$	0.64 ± 0.12	$(6.6 \pm 0.7) \times 10^2$	1.10 ± 0.22
^{7}Be	$46.0 \pm 1.5 \pm 1.6$	0.51 ± 0.07	48.4 ± 2.4	0.97 ± 0.09
pep	$3.1 \pm 0.6 \pm 0.3$	0.62 ± 0.17	1.6 ± 0.3	1.1 ± 0.2
^{8}B	$0.22 \pm 0.04 \pm 0.01$	0.31 ± 0.15	0.05 ± 0.01	0.91 ± 0.23
CNO	< 7.9	–	< 7.7	< 1.5

After the correction for neutrino oscillations, the derived neutrino flux and its ratio with respect to SSM are reported in columns 4 and 5

12.7.3 Summary of Solar Experimental Results

To summarize: solar neutrinos have been measured by different experiments in different energy ranges. In particular, Borexino (Table 12.4) measured the mono-energetic neutrino fluxes from ^{7}Be and *pep*, and the continuous ^{8}B spectrum with two different thresholds at 3 and 5 MeV, as well as providing the first direct measurement of the continuous *pp* spectrum. Additional contributions from the ν_e measurement arise from Gallex/GNO, Homestake, Sage, SNO and SuperKamiokande. The results indicate that the reduction with respect to the SSM predictions ranges between 1/3 and 1/2, depending on the energy. These discrepancies are all corrected when neutrino oscillations that include matter effect in the Sun are considered.

Using the oscillation probability derived from the best-value parameters from a global fit of all available data (Patrignani et al. 2016/2017), one can compare the experimental results to the SSM expectations corrected for oscillations. Once the energy-dependent oscillation probability $P_{ee}(E)$ is considered, the observed ν_e flux on Earth at a given energy, $Data(E)$, is related to the flux originated in the center of the Sun, $\Phi_{\nu_e}(E)$, through the relation

$$Data(E) = P_{ee}(E) \cdot \Phi_{\nu_e}(E). \tag{12.33}$$

The ν_e fluxes in the core of the Sun from different reactions after the correction for oscillation effects are reported in column four of Table 12.4. Finally, the ratio between the measured neutrino flux corrected for the neutrino oscillation effects and the SSM prediction is shown in the last column.

Figure 12.9 shows the values of different $Data/SSM$ ratio as measured from the experiments described in Sects. 12.3, 12.4, and 12.7.2 as a function of energy. The gray band corresponds to the expected P_{ee} from three-flavor neutrino oscillations, including the matter effect in the Sun. For the computation, the oscillation parameters reported in Table 12.5 are used. The band is due to the uncertainties on the parameters.

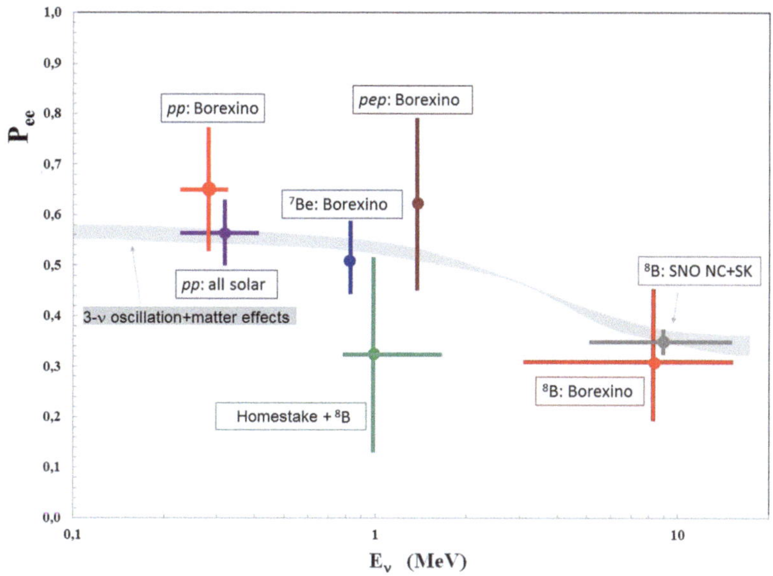

Fig. 12.9 Survival probability as a function of the energy of ν_e produced by the different nuclear reactions in the Sun. The gray band represents the ratio between the expected ν_e flux on Earth and the flux prediction at the Sun from the SSM when neutrino oscillations in the Sun matter are considered. Points with error bars represent measurements from different experiments. The horizontal uncertainty on data points shows the neutrino energy range used in the measurement. Refer to Borexino Collaboration (2014) for those related to Borexino

12.8 Neutrino Oscillation Parameters

All existing compelling data on neutrino oscillations can be described assuming three-flavor neutrino mixing. In this case, there are only two independent neutrino mass squared differences. The numbering of massive neutrinos ν_j is arbitrary. It proves convenient, from the point of view of relating the mixing angles θ_{12}, θ_{23} and θ_{13} to observables, to identify $|\Delta m^2_{21}|$ with the smaller of the two neutrino mass squared differences. As it follows from the data, this mixing angle is responsible for the solar ν_e and for the reactor $\overline{\nu}_e$ oscillations observed by KamLAND. Thus, the larger neutrino mass square difference $|\Delta m^2_{31}|$ or $|\Delta m^2_{32}|$, can be associated with the experimentally observed oscillations of the atmospheric neutrinos and accelerator long-baseline experiments (Sect. 11.9).

As $\sin\theta_{13}$ is relatively small, it makes it possible to identify the angles θ_{12} and θ_{23} with the neutrino mixing angles associated with the solar ν_e and the dominant atmospheric $\nu_\mu, \overline{\nu}_\mu$ oscillations, respectively. For this reason, the angles $\theta_{12} = \theta_\odot$ and $\theta_{23} = \theta_{\text{atm}}$ are often called *solar* and *atmospheric* neutrino mixing angles, while $\Delta m^2_{21} = \Delta^2_m\odot$ and $\Delta m^2_{23} \simeq \Delta m^2_{13} = \Delta m^2_{\text{atm}}$ are often referred to as the *solar* and *atmospheric* neutrino mass squared differences.

Table 12.5 The best-fit values of the 3-neutrino oscillation parameters, derived from a global fit of the current neutrino oscillation results

Parameter	Best-fit value ($\pm 1\sigma$)
$\Delta m^2_{\odot} \equiv \Delta m^2_{21}$	$(7.58^{+0.22}_{-0.26}) \times 10^{-5}\,\mathrm{eV}^2$
$\Delta m^2_{atm} \equiv \Delta m^2_{31} \simeq \Delta m^2_{23}$	$(2.35^{+0.12}_{-0.09}) \times 10^{-3}\,\mathrm{eV}^2$
$\sin^2 \theta_{12}$	$0.306^{+0.018}_{-0.015}$
$\sin^2 \theta_{23}$	$0.42^{+0.08}_{-0.03}$
$\sin^2 \theta_{13}$	0.0251 ± 0.0034

From Beringer et al. (2012)

As a summary, we present in Table 12.5 the best fit values of the PMNS matrix mixing parameters and mass squared differences derived from a global fit of the current neutrino oscillation data (Beringer et al. 2012; Fogli et al. 2012). Refer to Patrignani et al. (2016/2017) for a more detailed review. The possibility of *CP*-violation in the leptonic sector depends on the phase parameter δ, which is presently unknown.

12.9 Effects of Neutrino Mixing on Cosmic Neutrinos

Neutrino oscillations have effects on the flavor composition of astrophysical neutrinos arriving on Earth. High-energy neutrinos (Chap. 10) in galactic or extragalactic systems follow mainly from the production and decay of unstable hadrons, mainly charged pions. These hadrons may be produced when the accelerated protons in these environments interact with the ambient photon field in $p\gamma$ and/or protons pp interactions, Sect. 8.4. The $\nu_\tau, \overline{\nu}_\tau$ may also be produced by the decay of heavy charmed mesons. However, the high energy threshold and low cross-section for such reactions imply that the ratio of charmed meson to pion production is $\sim 10^{-4}$ and the fraction of ν_τ/ν_μ another factor of ten smaller. Thus, in $p\gamma$ and pp collisions, one typically obtains the following ratio of intrinsic high-energy cosmic neutrinos in proximity to sources:

$$\Phi^0(\nu_e) : \Phi^0(\nu_\mu) : \Phi^0(\nu_\tau) = 1 : 2 :< 10^{-5} \,. \simeq 1 : 2 : 0 \tag{12.34}$$

Vacuum neutrino mixing modifies the observed ratios as described below, following the elegant method presented in Athar et al. (2000). As usual, we count both neutrinos and antineutrinos in the symbol for neutrinos. In most situations, the matter effect plays no role, because high-energy cosmic neutrinos originate in regions of sufficiently low densities around the sources. The presence of relatively dense objects between the distant high-energy neutrino sources and neutrino telescopes is unlikely. A different scenario arises when neutrinos are produced in dense environments, as in the core-collapse supernova, Sect. 12.13. In addition, when distant sources are involved, the change in the flavor composition of the high-energy cosmic neutrinos due to vacuum mixing is essentially energy-independent over the entire energy range relevant for observations. This occurs because, in the oscillation formulas (12.27a–12.27c), the energy E appears to be an argument for an oscillating function $[\sin^2(\Delta m^2 L/E)]$ whose effects are averaged out for large L.

The flavor oscillation probability in vacuum (12.26) and in the limit $L \to \infty$ (after averaging the rapid oscillating $\sin^2$ function) can be written as

$$P_{\alpha\beta} \equiv P(\nu_\alpha \to \nu_\beta) = \sum_j |U_{\alpha j}|^2 |U_{\beta j}|^2 . \tag{12.35}$$

We can represent the oscillation probability as a symmetric matrix $\mathscr{P}$:

$$\mathscr{P} = \begin{pmatrix} P_{ee} & P_{e\mu} & P_{e\tau} \\ P_{e\mu} & P_{\mu\mu} & P_{\mu\tau} \\ P_{e\tau} & P_{\mu\tau} & P_{\tau\tau} \end{pmatrix} \equiv AA^T, \text{ with} \quad A = \begin{pmatrix} |U_{e1}|^2 & |U_{e2}|^2 & |U_{e3}|^2 \\ |U_{\mu1}|^2 & |U_{\mu2}|^2 & |U_{\mu3}|^2 \\ |U_{\tau1}|^2 & |U_{\tau2}|^2 & |U_{\tau3}|^2 \end{pmatrix} . \tag{12.36}$$

Now, the neutrino flux from a far cosmic source can be expressed as a product of $\mathscr{P}$ and the intrinsic flux $\Phi^0(\nu_\alpha)$, $\alpha = e, \mu, \tau$:

$$\begin{pmatrix} \Phi(\nu_e) \\ \Phi(\nu_\mu) \\ \Phi(\nu_\tau) \end{pmatrix} = \mathscr{P} \begin{pmatrix} \Phi^0(\nu_e) \\ \Phi^0(\nu_\mu) \\ \Phi^0(\nu_\tau) \end{pmatrix} = AA^T \begin{pmatrix} \Phi^0(\nu_e) \\ \Phi^0(\nu_\mu) \\ \Phi^0(\nu_\tau) \end{pmatrix} . \tag{12.37}$$

We assume the standard ratio of the intrinsic cosmic neutrino flux (12.34), so that we obtain

$$A^T \begin{pmatrix} \Phi^0(\nu_e) \\ \Phi^0(\nu_\mu) \\ \Phi^0(\nu_\tau) \end{pmatrix} = \begin{pmatrix} |U_{e1}|^2 & |U_{\mu1}|^2 & |U_{\tau1}|^2 \\ |U_{e2}|^2 & |U_{\mu2}|^2 & |U_{\tau2}|^2 \\ |U_{e3}|^2 & |U_{\mu3}|^2 & |U_{\tau3}|^2 \end{pmatrix} \begin{pmatrix} 1 \\ 2 \\ 0 \end{pmatrix} \Phi^0(\nu_e) \tag{12.38}$$

$$= \begin{pmatrix} 1 \\ 1 \\ 1 \end{pmatrix} \Phi^0(\nu_e) + \begin{pmatrix} |U_{\mu1}|^2 - |U_{\tau1}|^2 \\ |U_{\mu2}|^2 - |U_{\tau2}|^2 \\ |U_{\mu3}|^2 - |U_{\tau3}|^2 \end{pmatrix} \Phi^0(\nu_e) \tag{12.39}$$

where we have used the unitarity condition, i.e., $\sum_j |U_{\alpha j}|^2 = 1$. When $|U_{e3}|^2 \ll 1$, there is $s_{13} \simeq 0$, $c_{13} \simeq 1$ and $|U_{\mu j}| \simeq |U_{\tau j}|$. In fact, using the values reported in Table 12.5, it can be easily obtained that the numerical values of matrix elements in (12.22) are $|U_{\mu1}| \simeq |U_{\tau1}| \simeq 0.4$, $|U_{\mu2}| \simeq |U_{\tau2}| \simeq 0.6$, and $|U_{\mu3}| \simeq |U_{\tau3}| \simeq 0.7$. For this reason, the second term in Eq. (12.39) is negligible. Hence, with the constraints of the solar and atmospheric neutrino and the reactor data, we obtain, from (12.39),

$$\begin{pmatrix} \Phi(\nu_e) \\ \Phi(\nu_\mu) \\ \Phi(\nu_\tau) \end{pmatrix} = A \begin{pmatrix} 1 \\ 1 \\ 1 \end{pmatrix} \Phi^0(\nu_e) \simeq \begin{pmatrix} 1 \\ 1 \\ 1 \end{pmatrix} \Phi^0(\nu_e) , \tag{12.40}$$

where we have used the unitarity condition again. Therefore, we conclude that the ratio of the cosmic high-energy neutrino fluxes at far distances from us is (1:1:1). This only depends on the fact that the PMNS matrix elements in the second and third row are almost identical and on the flavor flux ratio (1:2:0) at the source.

The message of this excursus in the neutrino oscillation formalism is clear. The observation of solar neutrinos contributed to the understanding of the physics properties of these eluding particles. The neutrino oscillation effects must be considered when astrophysical properties are studied, as in the case of energy production in the Sun. Propagation effects also change the flavor composition ratio for high-energy neutrinos arriving on Earth from astrophysical sources.

12.10 Formation of Heavy Elements in Massive Stars

Nucleosynthesis from nuclear fusion proceeds until the formation of nuclei with $A \leq 60$. Nuclei with $A \sim 60$ (the *iron peak*) have the highest binding energy, Fig. 3.8. The luminosity of a main sequence star of mass M changes very little until it begins to move off the main sequence when the helium core has a mass of about 10% of M. The central temperature T is $T \propto M$ and if $M \gtrsim 1.3\,M_\odot$, the CNO cycle dominates, see Fig. 12.2. In yet more massive stars, post-main sequence evolution proceeds by successive core and shell burning to produce nuclei with higher and higher binding energies.

For the most massive stars ($M > 8M_\odot$), the sequence continues with carbon and oxygen burning to produce silicon, which can eventually be burned to create elements around the iron peak. It is therefore expected that in the final stages of evolution, a very massive star will take up an "onion-skin" structure. The central core of iron peak elements is surrounded by successive shells of silicon, carbon and oxygen, helium and hydrogen, Fig. 12.10.

As the nuclear reactions proceed through the sequence of carbon, neon, oxygen and silicon burning, the temperature in the core increases, and consequently the time-scale for nuclear burning decreases. In particular, after helium burning, the time-scales are drastically reduced. The reason is due to the enormous neutrino luminosity that exceeds the optical luminosity of the star. As $T \gtrsim 10^9$ K, thermal populations of e^+e^- pairs are created, which, in turn, can annihilate into $\nu + \overline{\nu}$. Neutrinos escape unimpeded from the stellar material and nuclear burning is needed to replace the huge amount of energy carried away. The time-scales for the silicon burning last only for a period of roughly tens of days.

Once the star's inner region is made primarily of Fe, further compression of the core no longer ignites nuclear fusion; the star is unable to thermodynamically support its outer envelope made of concentric shells, Fig. 12.10. The rest of the star, without the support of the radiation, collapses, compressing the star's nucleus, producing a *core-collapse supernova*.

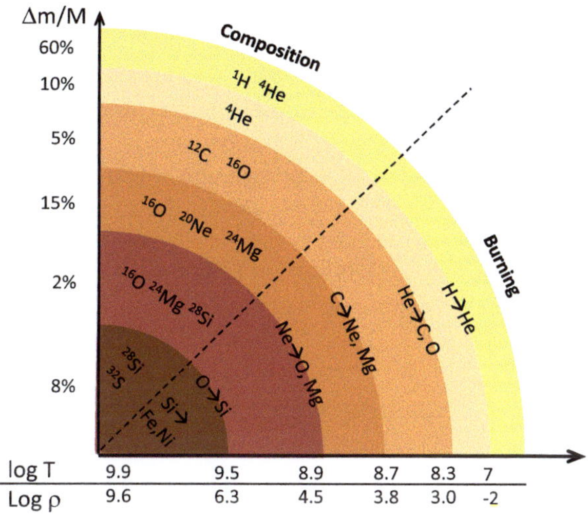

Fig. 12.10 The onion-like structure in the final stage of a massive star (25 $M_\odot$). The outermost envelope is composed of hydrogen and helium, and progressively heavier nuclei (up to iron) are layered, due to successive fusion reactions. Typical values of the mass, density ρ (in g/cm^3) and temperature T (in K) of the different *shells* are indicated along the axes (Kippenhahn and Weigert 1990)

12.11 Stellae Novae

At least as early as 185 AD, Chinese astronomers observed and recorded "guest stars". These objects suddenly appeared in the sky, were visible for a certain amount of time, and then faded away. The most brilliant guest star (or supernova in modern language) that ever appeared was recorded on May, 1st 1006, by Chinese and Arab astronomers. The remnant of this explosion was first identified at radio wavelengths on the basis of historical evidence, and are now observed with different instruments at different wavelengths up to TeV γ-rays (Acero et al. 2010). The most famous old supernova is that of 1054, also recorded by Chinese astronomers and not reported anywhere in Europe, despite being in the field of view. The appearance of a guest star was in contradiction to Aristotle's view of the heavens mentioned in Chap. 8. The remnant of the 1054 supernova (SN) is the Crab nebula, Sect. 9.5. At the center of the Crab there is a pulsar, which emits electromagnetic radiation. The remnants of the 185 and 1006 SN have no pulsars in the center.

Another historical SN occurred in 1572 and was recorded by Tycho Brahe, Sect. 6.3.2. Kepler, in October 1604, saw another *Stella Nova* (as the title of his book published in Prague in 1606), less bright than that of Tycho but remaining visible for a whole year. Another supernova, Cassiopeia A or Cas A, exploded in our galaxy between 1650 and 1680. Its remnant is a very strong radio source and can be observed at different wavelengths (Sect. 9.7), but it was not reported by contemporary observers. The advent of modern instruments and photographic plates has allowed the observation of supernovae in other galaxies since 1885. See Bethe (1990) and Marschall (1988) for an introduction to SN.

A systematic study of supernovae was started around 1930 by Zwicky and Baade with newly developed Schmidt telescopes, which allowed for photographing a large area of the sky. Since then, astronomers have discovered between 10 and 30 SN each year, and for most of them, they measured the spectrum and the *light curve*, i.e., the optical luminosity as a function of time. SN are designated by the year of discovery and a capital letter.

Connected to the possibility of measuring the spectra of the discovered SNe, the astronomers found that there are at least two topologies. If the spectrum contains hydrogen spectral lines, the supernova is classified as a *core-collapse* SN (or Type II). The remnants usually have a neutron star or a black hole at their center. Supernovae whose spectra do not contain hydrogen lines are classified as *thermonuclear supernovae* or Type I supernovae. The inherent physical processes of the two SN topologies are completely different.

12.12 Accreting White Dwarf: Type I Supernovae

Type I supernovae play a fundamental role in cosmology. In fact, about 80% of Type I SNe (designated as Type Ia) have a characteristic light curve. The SNe of this subset are used as *standard candles* for determining the absolute magnitudes of galaxies, and hence their distance. There now appears to be agreement in theoretical models that Type Ia SN are due to the thermonuclear disruption of white dwarfs (WD, Sect. 6.7.1). A WD consists mainly of carbon and oxygen; if it accretes material from a companion star, carbon (or possibly helium) is ignited under highly degenerate conditions. Within a few seconds, a substantial fraction of the matter in the WD undergoes nuclear fusion, releasing enough energy ($1-2\times10^{51}$ erg) to unbind the star in a supernova explosion (Hoeflich et al. 1993).

The end product of this nuclear burn is Fe, which is very abundant in the optical spectrum of Type I SNe. The total optical energy observed can be calculated from the assumption that most of the light is generated by the successive β^- decays of ^{56}Co $\rightarrow ^{56}$Ni $\rightarrow ^{56}$Fe and that essentially all the mass of the star burns to these end products. Because the WD mass at the Chandrasekhar limit is fixed, the model explains why the light curves of all Type Ia SNe are similar. The absence of hydrogen lines is because the hydrogen accreted from the companion is quickly converted into helium, before the supernova explosion. Hence, some experimental aspects do not completely fit the model and the mechanism of the burn is still unclear. In summary, the mechanism of Type I supernovae is less well understood than core-collapse (or Type II SNe).

12.13 Core-Collapse Supernovae (Type II)

A completely different physics mechanism is involved in Type II supernovae. Since the 1960s, it has been generally accepted that the core of a massive star collapses at the end of its lifetime to something like a neutron star. After collapse, this incompressible core would bounce back the in-falling material and would start a shock, which would then propagate into the mantle and propel most of the star's mass to the interstellar medium. As for Type I SN, the total energy release observable with astronomical tools is of the order of 10^{51} erg. However, this quantity represents in Type II SNe only $\sim$1% of the total estimated energy balance. About 99% of the energy must be emitted in a form not easily detectable: neutrinos. This fundamental function in the supernova mechanism stands in great in contrast to the usual marginal position of neutrinos due to their weak interaction processes.

The kinetic energy of the matter emitted by the SN explosion can be estimated from the mass of the pre-collapsing star and the observed velocity of the ejecta ($v \sim 10^{-2}c$). This typically gives $\sim 10^{51}$ ergs for a $M = 10\,M_\odot$ star. The bolometric observations of electromagnetic radiation add a negligible contribution (about 1%) to this energy balance. The final stage is a neutron star ($M_{NS} = 1.4 M_\odot$, $R_{NS} \simeq$ 10 km, see Sect. 6.7.2) and the explosion destroys the remains of the star. Thus, the work U done by gravity in compressing the core represents a total energy of

$$U = |\frac{3GM^2}{5R} - \frac{3GM_{NS}^2}{5R_{NS}}| \simeq \frac{3GM_{NS}^2}{5R_{NS}} \simeq 3 \times 10^{53}\text{erg}\,, \tag{12.41}$$

where G is the gravitational constant and R is the radius of the pre-collapsing star. This quantity is almost independent of the initial mass M, as $R \gg R_{NS}$. According to energy conservation, and based on the interaction theory among the constituents of the collapsing star, about 99% of the gravitational energy must be converted into kinetic energy of massless particles different from the photons: the neutrinos.

The theoretical relationship between neutrinos and supernovae was experimentally confirmed in February 1987, when at least two underground detectors recorded a burst of neutrinos (Sect. 12.15) and a spectacular supernova was later observed by astronomers worldwide. That observation was a breakthrough in the long history of supernovae.

12.13.1 Computer Simulations of Type II Supernovae

Analytic models and more and more sophisticated 1-D, 2-D and three-dimensional computer simulations of Type II SN have been developed to reproduce the observations. Simulating gravitational collapse is a very active area of numerical astrophysics that needs an equation of state, both at densities below normal nuclear density $\rho_N \sim 10^{14}\,\text{g/cm}^3$ and above that density; a detailed knowledge of three-dimensional hydrodynamics; neutrino transport; realistic nuclear physics; magnetic fields; and rotation. For instance, concerning neutrinos: the weak interaction processes must be accurately modeled, namely charged current capture and emission by nuclei; neutral current scattering by nuclei and nucleons; scattering by electrons; production of neutrino pairs from electron pairs; and its inverse reaction, neutrino pair annihilation into e^+e^-. The output of simulations provides information on visible quantities, such as light curves, energies of the emitted material, and neutrino yields; some simulations also predict the energy and spectral characteristics of gravitational waves.

However, it is still beyond the capabilities of computers to completely simulate a core-collapse supernova: simulations typically have great difficulty in producing the explosion with the observed light-curves and energetics. The role of neutrinos is fundamental: as mentioned, in a typical supernova, simulations suggest that 99% of the gravitational energy must be converted into neutrino kinetic energy. The beginning of the collapse is characterized by a short, bright neutronization (or breakout burst) with a time duration $\Delta t \simeq 1 - 10\,\text{ms}$, in which 10% of neutrinos (mainly ν_e) are emitted. The cooling phase lasts of the order of 10 s, and the core looses most of its gravitational potential energy by production pairs of neutrino and antineutrinos of all flavors. The results of these studies, still not definitive, are summarized in the next subsection. For a detailed description, see Bethe (1990) and Janka et al. (2007).

Finally, it must be mentioned that gravitational waves emitted in the burst might extract a very small fraction of the total available mass-energy, between about 10^{-5} and 10^{-7}. As the burst duration of $\Delta t = 1 - 10\,\text{ms}$, we expect typical frequencies of gravitational waves (see Chap. 13) on the order of $\nu_{gw} = 2/\Delta t \simeq 200 - 2000$ Hz. The detection possibility of core-collapse supernova through gravitational waves is studied in Sect. 13.11

12.13.2 Description for a Type II Supernovae

Near the end of its life, a massive star consists of concentric shells that are the relics of its previous burning phases, Fig. 12.10. Iron is the final stage of nuclear fusion, as the synthesis of any heavier element does not release energy. The dynamics of the collapse is very sensitive to the equation of state of the system, and in particular, to the number of leptons per baryon, Y_e. In the early stage of the collapse, Y_e decreases through electron capture on protons bound inside Fe nuclei

$$p + e^- \rightarrow n + \nu_e \,. \tag{12.42}$$

This reaction is energetically favorable when the electrons have energies of a few MeV at the densities involved in the star center. Reaction (12.42) reduces the electron pressure, and produces nuclei in the core that are more neutron-rich. Consequently, some of the nuclei undergo β^--decay, producing $\overline{\nu}_e$. Iron-group nuclei can also suffer partial photodisintegration to α particles. The above processes reduce the core energy and its lepton density. The electron degeneracy pressure can no longer stabilize the core and the star collapses (*core-collapse*).

An important change in the physics of the collapse occurs when the central density reaches $\rho_t \approx 10^{12}\,\text{g/cm}^3$. Under these conditions, neutrinos produced by (12.42) are essentially trapped in the core, Fig. 12.11a. We can estimate the mean free path of ν_e of energy E_ν assuming that the scattering process is due to NC interactions on protons. The cross-section of $\nu_e p \rightarrow \nu_e p$ corresponds (Fig. 12.12) to $\sigma^{NC}(E) \simeq 2 \times 10^{-42}(E_\nu/10\text{ MeV})^2\,\text{cm}^2$. The interaction length (Sect. 3.2.3)

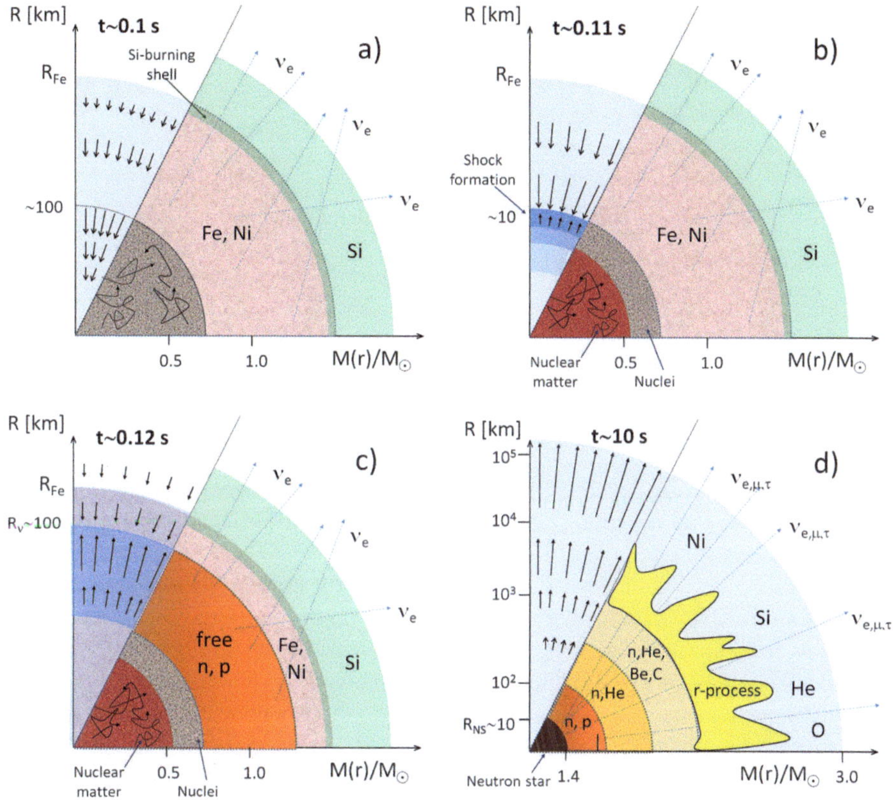

Fig. 12.11 Schematic representation of the evolutionary stages of the core-collapse supernova. At $t = 0$, the in-fall dynamics starts, because the gravitational pressure is no longer supported by radiation pressure. Plot (**a**) shows the beginning of the neutrino trapping phase, $t \simeq 0.1$ s; (**b**) the bounce and shock formation, $t \simeq 0.11$ s; (**c**) the shock propagation and the neutronization ν_e burst, $t \simeq 0.12$ s; (**d**) the neutrino-driven wind during the neutrino-cooling phase of the proto-neutron star, $t \simeq 10$ s. Each *panel* displays the dynamical conditions on the *left*, with *arrows* representing velocity vectors. The nuclear composition, as well as the nuclear and weak processes, are indicated in the *lower half* of each *panel*. The *horizontal axis* gives mass information in units of the solar mass $M_\odot$. The *vertical axis* (not in scale) shows corresponding radii. The proto-neutron star has densities above that of nuclear matter ρ_N

of neutrinos is given by $\lambda_\nu = 1/(\sigma^{NC} \cdot N_A \cdot \rho)$ [cm] where N_A is the Avogadro number and ρ the matter density. At a density ρ_t, the ν_e mean free path is given by

$$\lambda_\nu(E_\nu) = \frac{1}{\sigma^{NC} \cdot N_A \cdot \rho_t} = \frac{1}{2 \times 10^{-42} \left(\frac{E_\nu}{10\,\mathrm{MeV}}\right)^2 \times 6 \times 10^{23} \times 10^{12}}$$
$$\simeq 10\,\mathrm{km}\left(\frac{E_\nu}{10\,\mathrm{MeV}}\right)^{-2}. \tag{12.43}$$

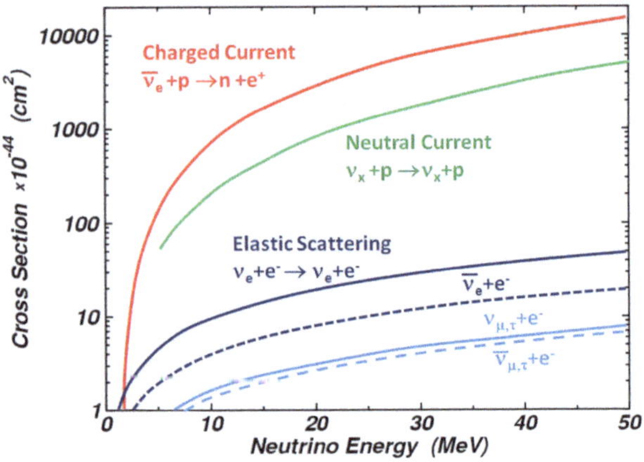

Fig. 12.12 Cross-section for different processes at energies of interest for SN neutrinos

This quantity is smaller than the radius of a sphere with density ρ_t and one solar mass; remember that the final state will be a 10 km neutron star with $M = 1.4 M_\odot$. This situation is called the *neutrino trapping phase*.

After this phase, the collapse proceeds essentially homologously, until nuclear densities ($\rho_N \approx 10^{14}$ g/cm^3) are reached in the core. Since nuclear matter has a much lower compressibility, the core decelerates and the in-falling matter bounces in response to the increased nuclear matter pressure. This drives a shock wave into the outer core, i.e., the region of the iron core, which lies outside of the homologous core and, in the meantime, has continued to fall inwards at supersonic speed. The core bounce and the formation of a shock wave trigger the supernova explosion, Fig. 12.11b. The exact mechanism of the explosion and the crucial ingredients are still uncertain and controversial. A large fraction of the neutrinos produced in the neutronization phase by electron captures (12.42) leave the star quickly in the so-called *neutrino burst* at shock break-out. Energy is carried away with ν_es and the shock is weakened so much that it finally stalls and turns into an accretion shock at a radius between 100 and 200 km, because the matter downstream of the shock has velocities toward the center and continues falling inward, Fig. 12.11c. After the core bounce, a compact remnant forms at the center of the collapsing star, rapidly growing by the accretion of in-falling stellar material. This nascent remnant will evolve into a neutron star or collapse into a black hole, depending on whether the progenitor star had a mass below or above $\sim$25 $M_\odot$. The newly born neutron star is initially still proton-rich and contains a large number of degenerate electrons and trapped neutrinos. The region with matter density high enough to trap neutrinos is called the *neutrinosphere*.

The modeling of stellar collapse and subsequent explosion using computer simulations started at the end of the 1960s. It was immediately recognized that the prompt shock following collapse alone is not sufficient to explain the SN explosion, as the shock uses its energy in the outer core mostly by the dissociation of heavy nuclei into nucleons. The shock can be revived by absorption of the neutrinos emitted by the hot star that is formed at the center, the proto-neutron star. This *delayed shock* starts about half a second after the collapse.

In the delayed shock explosion scenario, the stalled shock wave can be revived by the neutrinos streaming off the neutrinosphere. Near the center, the temperature is very high, $kT \sim 10\,\text{MeV}$ or more. Under these conditions, electron pairs (which are already plentiful at lower temperature) transform rapidly into neutrino pairs of each flavor ($f = e, \mu, \tau$)

$$e^+ + e^- \rightarrow \overline{\nu}_f + \nu_f\ . \tag{12.44}$$

The rate of energy transfer by this process from electron to neutrino pairs has been calculated and found to be proportional to T^9. Therefore, during the in-fall, neutrino pair production is negligible. During the shock phase, it is only important in the inner part of the core, typically the inner solar mass or less. The production of neutrino pairs stops when the neutrino density has become high enough so that the inverse process to (12.44) reaches equilibrium. These neutrinos carry most of the energy set free in the gravitational collapse of the stellar core and deposit some of their energy in the layers between the nascent neutron star and the stalled shock front mainly through CC interactions on free nucleons

$$\nu_e + n \rightarrow e^- + p; \quad \overline{\nu}_e + p \rightarrow e^+ + n \tag{12.45}$$

and NC interaction of ν_f. These processes increase the thermal energy of the stellar medium and the pressure behind the shock. At high matter density, ν_f continue to be trapped. But as the heated layers begin to expand, the pressure increases and drives the shock outwards again. Thus, the density at its front decreases, ultimately becoming smaller than the trapping density. At this point, neutrinos will be released: those close behind the shock escape immediately. Neutrinos from deep inside are free, as they are able to make their way by diffusion. This delayed shock explosion scenario requires that a few percent of the radiated neutrino energy be converted to thermal energy of nucleons, leptons, and photons. This corresponds to the kinetic energy observed in the SN ejecta. The further cooling of the hot interior of the proto-neutron star then proceeds by neutrino-pair production and diffusive loss of neutrinos of all three lepton flavors, as depicted in Fig. 12.11d. After several tens of seconds, the compact remnant becomes transparent to neutrinos and the neutrino luminosity drops significantly.

12.13.3 Supernovae Producing Long GRBs

Some long-duration GRBs (Sect. 8.11) have been correlated with SN explosions, sometimes denoted as hypernovae (HNe). HNe have exceptional brightness, originated by high Ni production; they are now considered to be core-collapse with unusually high ejecta velocities, and therefore high kinetic energies. In the HNe case, rapid stellar rotation is thought to be relevant.

GRBs are understood to have originated in ultrarelativistic, collimated jets. Some emission lines observed in many HNe suggest strong global asymmetry. Such events are interpreted as signatures of black hole-forming stellar collapses (*collapsars*). Matter around rapidly spinning black holes sets free energy in neutrinos, electromagnetic flux, and mass outflow with an efficiency of up to roughly 40% of the rest-mass energy of accreted material. A similar energy flux could be originated by fast rotating ($P \simeq 1$ ms) neutron stars with a dynamo-generated magnetic field (*magnetars*). The jet and stellar explosion could be powered either by the rotational energy of the magnetar or by the gravitational and rotational energy of the collapsar.

Such requirements favor particular stars (such as the so-called Wolf-Rayet stars) as progenitors, with special initial conditions such as high angular momentum. In the present-day Universe, HNe and GRB are rare (with a frequency of 10^{-3} with respect to normal supernovae). Core-collapse events forming black holes and GRBs could be very common in the early Universe.

12.14 Neutrino Signal from a Core-Collapse SN

Despite enormous recent progress, the physics of core-collapse is not completely understood. The only experimental verification of the theoretical scenario depicted in the previous section occurred in 1987 (Sect. 12.15). Worldwide capabilities for supernova neutrino detection have increased by orders of magnitude since then. The next multimessenger observation of a nearby core-collapse supernova will provide a great deal of information for both physics and astrophysics.

12.14.1 Supernova Rate and Location

The standard method for estimating the SN rate in our Galaxy is to scale optical observations from external galaxies. Another classic approach is to extrapolate the five historical SNe of the past millennium to the entire Galaxy. A third estimate can be made using measurements of γ-rays emitted by ^{26}Al, which traces explosions of massive stars. All these observations give about 1–3 core-collapse SNe per century in our Galaxy and its satellites.

To estimate the typical distance for a galactic SN, it is considered that core-collapses mark the final evolution of massive stars, and thus they must occur in regions of active star formation, i.e., in the galactic spiral arms, Sect. 2.7. Active star formation regions in other galaxies or in our Galaxy show significant presence of pulsars, SN remnants, molecular and ionized hydrogen, and hot and massive stars (OB-stars). All of these observables are consistent with a deficit of SNe in the inner Galaxy and a maximum at 3.0–5.5 kpc from the galactic center. Detailed models indicate that the characteristic distance of a possible SN event in our Galaxy is about 10 kpc from Earth.

12.14.2 *The Neutrino Signal*

At the beginning of a collapse, one expects a short, bright neutronization or breakout burst dominated by ν_e from electron capture (12.42), with a time duration $\Delta t \sim \mathcal{O}(10\,\text{ms})$. This burst is followed by an accretion phase, $\Delta t \sim \mathcal{O}(100\,\text{ms})$, with $\nu_e, \overline{\nu}_e$ produced by reactions (12.45). Finally, during the cooling, $\Delta t \sim \mathcal{O}(10\,\text{s})$, the core loses most of its gravitational potential energy through production of $\nu_f \overline{\nu}_f$ pairs. About 90% of the energy is released during this phase and, as neutrinos escape, the temperatures gradually decrease. An overall feature of the neutrino flux is that the luminosity is roughly equally divided among the three flavors. The neutrinos have an expected energy distribution whose average energies follow the hierarchy $\langle E_{\nu_e}\rangle < \langle E_{\overline{\nu}_e}\rangle < \langle E_{\nu_\mu,\nu_\tau}\rangle$. This ordering reflects the strength of interaction with matter: ν_e have more interactions than $\overline{\nu}_e$ because of the excess of neutrons in the core; in turn, $\overline{\nu}_e$ have more interactions than ν_μ, ν_τ, which are restricted to neutral currents. These expected average energies are largely model-dependent, ranging between ~12 and 20 MeV.

Figure 12.13 shows an example of a flux prediction (Gava et al. 2009). There may be significant variations in the expected flux from supernova to supernova due to differences in the mass and composition of the progenitor, and possibly asymmetries, rotational effects, or magnetic field effects.

Matter oscillation effects modify the spectra as the neutrinos traverse dense matter, so the neutrino spectra arriving on Earth will depend both on supernova matter profiles and on neutrino oscillation parameters. Matter propagation affects neutrinos and antineutrinos differently (in particular, $\nu_e, \overline{\nu}_e$). Also, the mass hierarchy (Sect. 12.5) plays a role: in the Normal Hierarchy, there are two light and one heavy neutrino mass states; in the Inverted Hierarchy, there are two heavy and one light neutrino mass states. The observation of neutrinos from a galactic SN would provide fundamental information on the neutrino physics (Raffelt 1999), improving the constraints obtained with SN1987A.

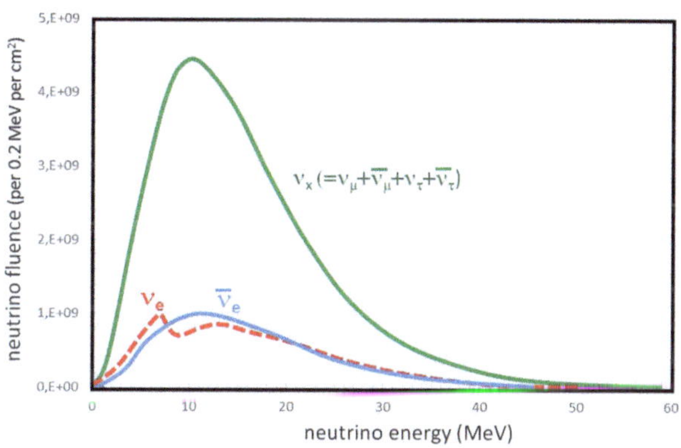

Fig. 12.13 Example of supernova neutrino fluence (the time integrated neutrino flux) for the different flavor components. The prediction includes collective effects, which are responsible for the structure observed in the ν_e *dashed line*

Supernova neutrinos have energies (10–20 MeV) intermediate between those of solar (1 MeV) and atmospheric neutrinos (>100 MeV), Fig. 1.6. In addition, the signal contains roughly equal amounts of all neutrino flavors. However, ν_μ, ν_τ are below the threshold for CC interactions. Thus, most of the signal produced by a supernova is provided by $\overline{\nu}_e + p$ inverse beta decay reaction (12.19), as was the case with SN1987A. Interaction on electrons and NC interactions of all neutrino flavors can add only a small contribution to the signal, as evident from the cross-sections shown in Fig. 12.12.

Although the cross-section is relatively small compared to that for interaction with nucleons, neutrino-electron elastic scattering could be important because of its directionality. As supernova neutrinos are more energetic than solar neutrinos, the electron is scattered closer to the direction of the incoming neutrino; ES events can be thus used to point in the direction of the supernova, although this represents a difficult experimental task.

Neutrinos also interact with nucleons in nuclei via CC and NC processes, although cross-sections are typically somewhat smaller for bound than for free nucleons. The kinematic threshold depends on the binding energies of the initial and final nucleus. Neutrino interactions on nuclei in the tens-of-MeV range are not completely theoretically understood; see Scholberg (2012) for a review.

Relatively cheap detector materials such as water and hydrocarbon-based scintillators have many free protons. The neutrino energy threshold for reaction (12.19) is 1.8 MeV and the positron's energy loss is usually observed. The neutron may also be captured on free protons, after thermalization and capture time of $\sim$200 μs, producing MeV-scale γ-rays, as discussed for the KamLAND detector.

12.14.3 Detection of Supernova Neutrinos

Starting from the 1980s, different experiments have been or are sensitive to a supernova neutrino burst in our Galaxy. As derived below, typical event yields for current detectors are a few hundred events per kiloton of detector material for a core-collapse SN 10 kpc away from Earth. The expected number of events from a supernova should scale simply with distance to the supernova D as $1/D^2$. An ideal detector would measure the flavor, energy, time, and direction of the neutrinos on an event-by-event basis and no background. A real detector would settle for imperfectly reconstructed events and inferred statistical information. To a good approximation for most technologies, event rates scale linearly with detector mass.

The detection techniques use the fact that a high fraction of neutrinos from a SN burst exceeds 5–7 MeV, the typical energy threshold of real-time experiments used for solar neutrino studies. Liquid scintillator detectors (such as Borexino, KamLAND, LVD, MACRO, Baksan) are composed of hydrocarbons, which have the approximate chemical formula C_nH_{2n}. The energy loss of charged particles is proportional to the light emitted from de-excitation of molecular energy levels. The interaction vertices may be reconstructed using the timing information of the photons. Because of the presence of free protons in scintillators, inverse beta decay (12.19) is largely dominant for a supernova burst signal. Elastic scattering on e^- will contribute a few percent to the total supernova burst event rate.

Water is an inexpensive medium with an abundance of free protons, and detectors with large volumes are possible (Kamiokande, IMB, SK, future detectors). As for scintillators, interaction rates in water are dominated by reaction (12.19). Secondary charged particles are detected via their Cherenkov light emission. The neutrino energy is estimated through particle energy loss, which is proportional to the number of detected photons.

Backgrounds for SN neutrino detection vary by detector type and location. Ambient radioactivity, from the environment or detector materials, produces irreducible signals whose decay products rarely have energies >10 MeV. Nevertheless, radioactivity can be troublesome for measurements of the low-energy end of the signal, possibly at late times of SN burst, and for low-threshold detectors. Cosmic ray related backgrounds could be suppressed by constructing detectors deep underground. Nuclear fragments produced by spallation or capture processes of surviving atmospheric muons can still be produced in the detector or in the surrounding materials. Muon spallation events can produce fake bursts over timescales of tens of seconds and represent a potential background. However, for current underground detectors, background rates should be very low for the duration of a galactic SN burst that typically lasts $\mathcal{O}(10\,\text{s})$.

Expected event rates can be computed by folding a given supernova neutrino flux with the NC and CC interactions of different neutrino flavors on the different targets (p, e^-, nuclei) in the detector, and taking into account the detector response. In the following, we present a first-order estimate of the expected number of signal events in a typical 1 kiloton detector using only $\overline{\nu}_e$ and the dominant inverse beta decay reaction (12.19).

The energy released by a core-collapse SN is $E_{SN} = 3 \times 10^{53}$ erg, Eq. (12.41). As average neutrino energy, we assume $\langle E_\nu \rangle \sim 15$ MeV $\simeq 2.5 \times 10^{-5}$erg. The total number of neutrinos of all flavors, N_{ν_f}, at the source in 10 s interval is

$$N_{\nu f} = \frac{E_{SN}}{\langle E_\nu \rangle} = \frac{3 \times 10^{53}}{2.5 \times 10^{-5}} = 1.2 \times 10^{58} \quad \rightarrow \quad N_{\overline{\nu}_e} = \frac{N_{\nu f}}{6} = 2 \times 10^{57} . \tag{12.46}$$

$N_{\overline{\nu}_e}$ represents the number of $\overline{\nu}_e$, producing most of the signal in a water Cherenkov detector. This is derived from N_{ν_f} assuming equipartition of neutrino flavors during the cooling phase, Fig. 12.13.

The neutrino fluence $F_{\overline{\nu}_e}$ on Earth depends on the distance D of the SN. For a typical distance $D = 10\,\text{kpc} = 3 \times 10^{22}\,\text{cm}$

$$F_{\overline{\nu}_e} = \frac{N_{\overline{\nu}_e}}{4\pi D^2} = 1.7 \times 10^{11}\,\text{cm}^{-2} . \tag{12.47}$$

Note that the average neutrino flux (the fluence divided by the SN time of 10 s) is about a factor of 3 smaller than the flux of solar neutrinos, Eq. (12.4). SN neutrinos are, however, much easier to detect, due to their higher energy.

The expected number of events in a detector depends on the medium, on the ν cross-section $\sigma(E_\nu)$ and, critically, on the detection efficiency $\varepsilon(E_\nu)$. Water, for instance, contains two free protons per molecule. The number of free protons in a kt of water is thus $N_p = (2/18) \times 10^9\,\text{cm}^3 \times N_A\,\text{cm}^{-3} = 7 \times 10^{31}$. The cross-section for the inverse beta decay (12.19) at 15 MeV corresponds to $\sigma_0 = \sigma(15\,\text{MeV}) \simeq 2 \times 10^{-41}\text{cm}^2$ (see Fig. 12.12). The expected number of interactions with a positron in the final state in one kt of water is thus given by (using average values):

$$N_{e^+} = F_{\overline{\nu}_e} \times \sigma_0 \times N_p \times \overline{\varepsilon} = 1.7 \times 10^{11} \times 2 \times 10^{-41} \times 7 \times 10^{31} \times \overline{\varepsilon} \simeq 230\,\overline{\varepsilon} \text{ events.} \tag{12.48}$$

Indicatively, the number of expected events for some of the quoted experiments (some of them operated in the past) for a SN event 10 kpc away are (with the fiducial detector mass in parenthesis): 7000 events in SK (32 kt); 300 events in LVD, MACRO, KamLAND ($\sim$1 kt); 100 events in Borexino (0.3 kt); 50 events in Baksan (0.33 kt). Some future proposed experiments can have hundreds of kt of fiducial mass (Scholberg 2012), possibly opening the field of extra-galactic SN observations.

A prompt alert from a SN neutrino signal would give astronomers valuable time to catch the electromagnetic signal from the supernova and to study the environment immediately surrounding the progenitor star during the initial stages of the event. Advance warning could enable observation of ultraviolet and soft X-ray flashes, which are predicted at very early times. Particularly intriguing is the possibility of having an early warning from the detection of a gravitational wave signal. There could also be entirely unexpected effects at early times. A galactic supernova is rare enough that it will be critical to save all available information.

The SN1987A neutrino events described in the next section were recorded approximately 2.5 h before the inferred time of the supernova's first light. In reality (due to lack of on-line monitoring at that time), the experimentalists found the neutrino signal in their data only by a search triggered from the optical discovery. The situation will be different for the next nearby supernova event. The SuperNova Early Warning System (SNEWS) (Antonioli et al. 2004) is an international network of detectors that aims to provide an early alert to astronomers of a supernova's occurrence.

12.15 The SN1987A

SN1987A was the first supernova since 1604 visible with the naked eye. The progenitor was Sanduleak-69202, a main-sequence star of mass $M = 16 - 22 M_\odot$. It was located in the Large Magellanic Cloud (LMC), a small galaxy satellite of our Milky Way at a distance of about $50 \pm 5\,\text{kpc}$ from the Earth. The gamma line emission successively observed from SN1987A gave confirmation that heavy elements up to iron, cobalt, and nickel were synthesized by the exploded star, in agreement with predictions of the core-collapse supernova model. SN1987A was also the first (and unique, up to May 2018) supernova from which neutrinos were observed, in fact, the first extraterrestrial identified neutrino source other than the Sun.

Two water Cherenkov detectors, Kamiokande-II and the Irvine Michigan Brookhaven experiment (IMB), observed 12 and 8 neutrino interaction events, respectively, over a 13 s interval. This time interval is consistent with the estimated duration of a core-collapse. The IMB detector was a 5 kt underground detector located at a depth of 1570 m.w.e. It was bigger than Kamiokande (2.2 kt, energy threshold of 7–8 MeV) but used smaller PMTs (8-inch), with an overall detection threshold of about 20 MeV. For this reason, IMB was not suited for solar neutrinos. The signals of the two experiments were almost simultaneous, although in 1987, the technology of time dissemination through GPS was not available. The relative time precision between the signals in the two experiments was about 1 min (according to the Kamiokande synchronization procedure). Nearly all the SN1987A events were consistent with $\overline{\nu}_e$ interactions (Bethe 1990; Koshiba 1992).

Two smaller scintillator detectors, Baksan and LSD, also reported observations. Baksan reported five counts, but the first came 25 s after IMB. The LSD report was controversial, because the events were recorded several hours early (Bethe 1990).

Although the Kamiokande and IMB experiments collected a small sample of neutrino events from the SN1987A, they were sufficient to give an exact time for the start of the explosion to which the light curve can be normalized and to confirm the baseline model of core-collapse. In particular, referring to Fig. 12.14:

- the time distribution of the observed events is in reasonable agreement with theoretical predictions of a $\sim$10 s burst;
- their energy distribution gives a measure of the temperature of the neutrinosphere $T \sim 4.2\,\text{MeV}$. The average energies of detected neutrinos is $\sim$15 MeV;
- the number of observed events is in agreement with the expected 3×10^{53} erg luminosity of a core-collapse burst.

This last point is the crucial one. By scaling the number of expected events for a 10 kpc SN, Eq. (12.48), to the LMC distance, the number of expected events for a 100% efficient detector is $N_{e^+}^{LMC} = (10/50)^2 N_{e^+} = 9$ events/kt. Kamiokande had a mass of 2.2 kt, and about 20 events were thus expected. The experiment had a threshold of about 7–8 MeV for the emitted positron; thus, the detection efficiency $\overline{\varepsilon}$ averaged over the neutrino energy spectrum is large (on the order of $\sim$0.5). This is in agreement with the 11 observed events (one event was estimated as being due to background), assuming that 99% of the SN energy is carried away by neutrinos. IMB had a mass about $\times 2$ larger, but a higher detection energy threshold and average detection efficiency about $\times 1/4$ that of Kamiokande.

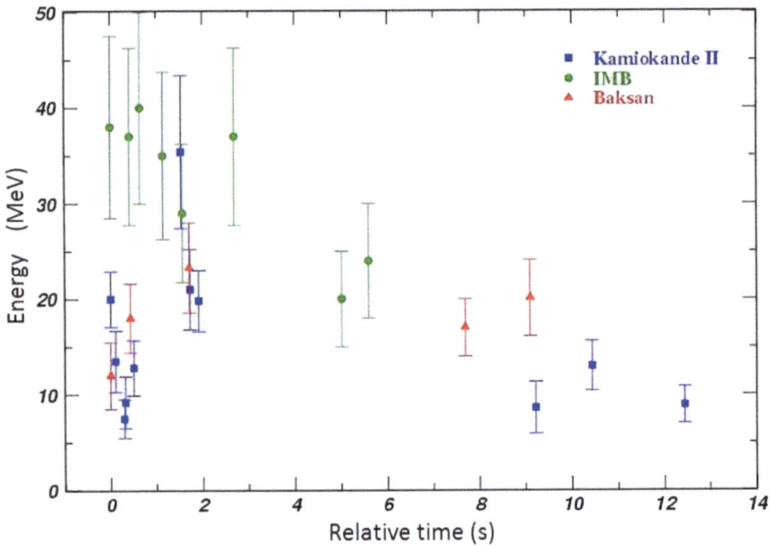

Fig. 12.14 Relative time and energy of SN1987A neutrino events observed by Kamiokande, IMB and Baksan. The time of the first event was arbitrarily set as $t = 0$

Exercise SN1987A also allowed for investigations into particle physics that were hitherto inaccessible to laboratories. Using information contained in Fig. 12.14, estimate: (a) an upper limit on the neutrino mass magneton and a lower limit on its lifetime; (b) a limit on the difference of neutrino speed w.r.t. c, $\frac{|v_\nu - c|}{c}$.

12.16 Stellar Nucleosynthesis and the Origin of Trans-Fe Elements

One of the most important interconnections between nuclear physics and astrophysics is that needed to explain the origin and the abundance of elements in the Periodic Table 12.15. The abundance of chemical elements in the Universe is dominated by hydrogen and helium, which were produced in the Big Bang. The remaining elements, making up only about 2% of the Universe, have been produced as the result of stellar activities. Nuclear fusion in stars synthetize elements with mass number A up to 56. ^{56}Fe is one of the highest binding energies of all of the isotopes, and is the last element that releases energy by nuclear fusion, exothermically. Elements of higher mass number become progressively rarer, because they increasingly absorb energy in being produced. The abundance of elements in the Solar System is thought to be similar to that in the Universe, as discussed in Sect. 3.6

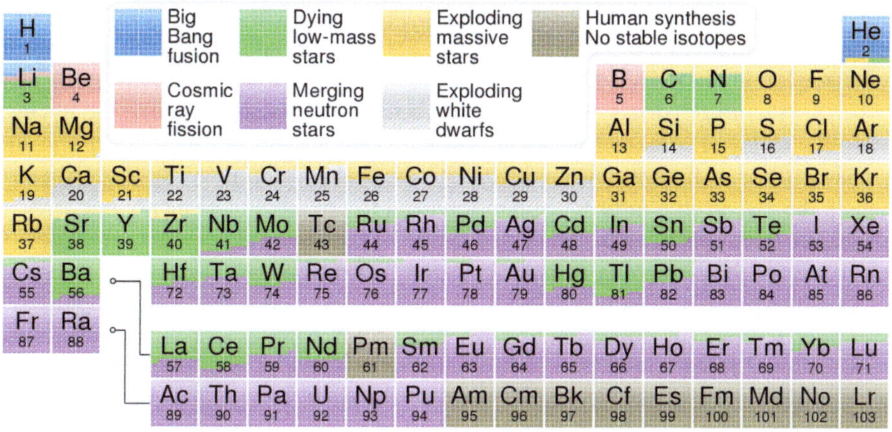

Fig. 12.15 A version of the periodic table indicating the main origin of elements found on Earth. The elements with $Z > 94$ are mainly of human synthesis. From https://commons.wikimedia.org/w/index.php?curid=31761437

Supernova nucleosynthesis is the theory of the releasing in the Universe of elements up to iron ($Z = 26$) and nickel ($Z = 28$) in supernova explosions, first advanced by F. Hoyle in 1954. During the supernova explosion, the outer shells of the star are ejected. Core-collapse supernovae are the main contributors of the heavy elements ($A \geq 12$) in galaxies, and the elements that have been produced during the various stellar burning stages are mixed into the interstellar medium. For a review of stellar nucleosynthesis, see Woosley et al. (2002) and Thielemann et al. (2017).

Referring to Fig. 12.15, the different elements are released in the Universe by different processes. Two different exploding stellar scenarios occur. The first involves a *white dwarf* star, which undergoes a nuclear-based explosion after it reaches its Chandrasekhar limit after absorbing matter from a neighboring star. The second cause is when a massive star, usually a supergiant, reaches ^{56}Ni and ^{56}Fe in its nuclear fusion processes.

Elements heavier that iron are produced by neutron capture in neutron-rich astrophysical environments, followed by β decay, $n \to pe^-\overline{\nu}_e$, of some neutrons in the forming nuclei. The so-called *s-process* is believed to occur mostly in asymptotic giant branch stars. During the late stages of the life of main-sequence stars, before the formation of a white dwarf, free neutrons are produced. Example of reactions producing free neutrons are ${}^{13}_{6}\mathrm{C} + {}^{4}_{2}\mathrm{He} \to {}^{16}_{8}\mathrm{O} + n$, or ${}^{22}_{10}\mathrm{Ne} + {}^{4}_{2}\mathrm{He} \to {}^{25}_{12}\mathrm{Mn} + n$. Heavy nuclei X (mainly iron and nickel, the starting material), left by a supernova during a previous generation of stars, capture neutrons and produce heavier elements via *slow* neutron capture

$$n + {}^{A}_{Z}\mathrm{X} \to {}^{A+1}_{Z}\mathrm{X} + \gamma ; \quad {}^{A+1}_{Z}\mathrm{X} \to {}^{A+1}_{Z+1}\mathrm{Y} + e^- + \nu_e \, . \tag{12.49}$$

In this way, relatively heavy elements are produced. This slow neutron capture process occurs when the characteristic time for neutron capture τ_n for the formation of the instable nucleus ${}^{A+1}_{Z}$X is $\tau_n \gg \tau_\beta$. The newly formed nucleus has enough time to undergo β^- decay with lifetime τ_β and tends to stabilize along the nuclear stability valley. The extent to which the s-process moves up the elements in the periodic table to higher mass numbers is essentially determined by the degree to which the star is able to produce neutrons and to the amount of heavy nuclei X in the star's initial abundance. The s-process is believed to occur over time scales of thousands of years, passing decades between successive neutron captures.

Another process by which heavier nuclei are produced is via proton capture (or *p-process*). Coulomb repulsion ensures that proton capture is a much rarer event than neutron capture, and the process only takes place at very high temperatures ($>2\times10^9$ K) in very massive stars ($\sim 25 M_\odot$) during supernova explosions.

The passage of the huge flux of neutrinos through a star experiencing core-collapse would cause interesting transmutation of the elements, even in the relatively cool outer regions. However, because of the small ν cross-section, production of new elements in this *ν-process* is restricted to rare species made from abundant target elements.

The last and probably more important process for the formation of stable elements up to uranium is the so-called *r-process*, which involves neutron capture, as in reaction (12.49). However, here the neutron capture time is much smaller than the nucleus decay time, $\tau_n \ll \tau_\beta$, due to the high neutron density. The newly formed nucleus does not decay immediately; after subsequent captures, the isotopes move away from the stability valley (${}^{A}_{Z}\mathrm{X} \rightarrow {}^{A+1}_{Z}\ \mathrm{X} \rightarrow {}^{A+2}_{Z}\ \mathrm{X} \rightarrow \cdots$). The number of neutrons increases, and thus the instability favoring β^- decay. Since the probability of neutron capture is large and the reactions occur at a high rate, these are called *r-processes* (*r* =rapid). The neutron captures must be rapid: the newly formed nucleus does not have time to undergo β-decay before another neutron arrives to be captured. Thus, necessarily, the r-process occurs in astrophysical locations where there is a high density of free neutrons. Which are those neutron-rich astrophysical regions is a matter of ongoing research.

There are large uncertainties at the site where r-processes occur, as they are believed to occur over time scales of seconds in explosive environments. Until the observation of GW170817 (Sect. 13.7), the environment around core-collapse SN was the most plausible candidate, as illustrated in Fig. 12.11d, with alternative models involving a neutrino-powered wind of a young neutron star, or a very asymmetric explosion and jetlike outflows, such as the explosion producing a gamma-ray burst. As discussed in Sect. 13.8, GW170817 showed that the most suited ambient for r-processes is probably the neutron-rich matter thrown off from a binary neutron star merger (the so-called *kilonova*).

References

J.N. Abdurashitov et al., Measurement of the solar neutrino capture rate with gallium metal. III. Results for the 2002–2007 data-taking period. Phys. Rev. C **80**, 015807 (2009)

K. Abe et al., Solar neutrino results in Super-Kamiokande-III. Phys. Rev. D **83**, 052010 (2011)

F. Acero et al., HESS Collab., First detection of VHE gamma-rays from SN1006 by H.E.S.S. Astron. Astrophys. **516**, A62 (2010)

B. Aharmim et al., Electron energy spectra, fluxes, and day-night asymmetries of 8B solar neutrinos from the 391-day salt phase SNO data set. Phys. Rev. C **72**, 055502 (2005)

Q.R. Ahmad et al., Direct evidence for neutrino flavor transformation from neutral-current interactions in the Sudbury Neutrino Observatory. Phys. Rev. Lett. **89**, 011301 (2002)

M. Altmann et al., Complete results for five years of GNO solar neutrino observations. Phys. Lett. B **616**, 174 (2005)

P. Antonioli et al., SNEWS: the SuperNova Early Warning System. New J. Phys. **6**, 114 (2004)

V. Antonelli, L. Miramonti, C. Peña Garay, A. Serenelli, Solar neutrinos. Adv. High Energy Phys. **2013**, 34 pp (2013); Article ID 351926. https://doi.org/10.1155/2013/351926

M. Asplund, N. Grevesse, A.J. Sauval, P. Scott, The chemical composition of the Sun. Annu. Rev. Astron. Astrophys. **47**, 481 (2009)

H. Athar, M. Jezabek, O. Yasuda, Effects of neutrino mixing on high-energy cosmic neutrino flux. Phys. Rev. **D62**, 103007 (2000)

J.N. Bahcall, *Neutrino Astrophysics* (Cambridge University Press, Cambridge, 1989). ISBN: 978-0521379755. The Solar Standard Model is also described on the website: http://www.sns.ias.edu/jnb/

J. Beringer et al., Particle data group, The review of particle physics. Section: 13. Neutrino mass, mixing, and oscillations. Phys. Rev. **D86**, 010001 (2012)

H.A. Bethe, Supernova mechanisms. Rev. Mod. Phys. **62**, 801 (1990)

Borexino Collaboration, Neutrinos from the primary proton-proton fusion process in the Sun. Nature **512**, 383 (2014)

S. Braibant, G. Giacomelli, M. Spurio, *Particle and Fundamental Interactions* (Springer, Berlin, 2011). ISBN: 978-9400724631

S. Braibant, G. Giacomelli, M. Spurio, *Particles and Fundamental Interactions: Supplements, Problems and Solutions* (Springer, Dordrecht, 2012)

C. Broggini, D. Bemmerer, A. Guglielmetti, R. Menegazzo, LUNA: nuclear astrophysics deep underground. Annu. Rev. Nucl. Part. Sci. **60**, 53–73 (2010)

F. Calaprice, C. Galbiati, A. Wright, A. Ianni, Results from the Borexino solar neutrino experiment. Annu. Rev. Nucl. Part. Sci. **62**, 315–336 (2012)

G.L. Fogli et al., Global analysis of neutrino masses, mixings and phases: entering the era of leptonic CP violation searches. Phys. Rev. **D86**, 013012 (2012)

Y. Fukuda et al., Solar neutrino data covering solar cycle 22. Phys. Rev. Lett. **77**, 1683 (1996)

J. Gava, J. Kneller, C. Volpe, G.C. McLaughlin, A dynamical collective calculation of supernova neutrino signals. Phys. Rev. Lett. **103**, 071101 (2009)

W. Hampel et al., GALLEX solar neutrino observations: results for GALLEX IV. Phys. Lett. B **447**, 127 (1999)

W. Haxton, The scientific life of John Bahcall. Annu. Rev. Nucl. Part. Sci. **59**, 1–20 (2009)

W.C. Haxton, R.G. Hamish Robertson, A.M. Serenelli, Solar neutrinos: status and prospects. Annu. Rev. Astron. Astrophys. **51**, 21–61 (2013)

A. Hoeflich, E. Mueller, P. Hoeflich, Light curves of type IA supernova models with different explosion mechanisms. Astron. Astrophys. **270**, 223–248 (1993)

H.Th. Janka et al., Theory of core-collapse Supernovae. Phys. Rep. **442**, 38 (2007)

N. Jelley, A.B. McDonald, R.G.H. Robertson, The sudbury neutrino observatory. Annu. Rev. Nucl. Part. Sci. **59**, 431–465 (2009)

R. Kippenhahn, A. Weigert, *Stellar Structure and Evolution* (Springer, Berlin, 1990)

M. Koshiba, Observational neutrino astrophysics. Phys. Rep. **220**, 229–381 (1992)

K. Lande, The life of Raymond Davis, Jr. and the beginning of neutrino astronomy. Annu. Rev. Nucl. Part. Sci. **59**, 21–39 (2009)

P. Lipari, Introduction to neutrino physics, in *1st CERN—CLAF School of High-energy Physics*, Itacuruca, Brazil (2001). http://cds.cern.ch/record/677618/files/p115.pdf

L.A. Marschall, *The Supernova Story* (Princeton Science Library, Princeton, 1988). ISBN: 978-0691036335

C. Patrignani et al. (Particle data group), Chin. Phys. C **40**, 100001 (2016/2017)

G. Raffelt, Particle physics from stars. Annu. Rev. Nucl. Part. Sci. **49**, 163 (1999)

K. Scholberg, Supernova neutrino detection. Annu. Rev. Nucl. Part. Sci. **62**, 81 (2012)

A.M. Serenelli, W.C. Haxton, C. Pena-Garay, Solar models with accretion. I. Application to the solar abundance problem. Astrophys. J. **743**, 24 (2011)

M.B. Smy et al., SK Collab., Super-Kamiokande's solar ν. Nucl. Phys. B Proc. Suppl. **49**, 235–236 (2013)

F.-K. Thielemann et al., Neutron star mergers and nucleosynthesis of heavy elements. Annu. Rev. Nucl. Part. Sci. **67**, 253 (2017)

S.E. Woosley, A. Heger, T.A. Weaver, The evolution and explosion of massive stars. Rev. Mod. Phys. **74**, 1015 (2002)

Chapter 13
Basics on the Observations of Gravitational Waves

Abstract On February 11, 2016, the LIGO collaboration announced the discovery of gravitational radiation due to the merger of two black holes. As the event was observed on September 14, 2015, its official designation is GW150914. This discovery represents a major scientific breakthrough for physics, astrophysics, and cosmology. Probably even more important, on October 16, 2017, the LIGO/Virgo collaboration announced, together with a large number of other experiments, the first coincident observation of GWs and electromagnetic radiation. These observations are connected with a collision of two neutron stars 40 Mpc away from Earth, producing almost simultaneously both gravitational radiation (GW170817) and a short gamma ray burst (GRB170817A). The electromagnetic observations in the following days revealed signatures of recently synthesized material, including gold and platinum, solving a decades-long mystery concerning where about half of all elements heavier than iron are produced. The purpose of this chapter is to explain key features of the observed gravitational radiation in terms of introductory physics. Gravitational waves carry a form of radiant energy that the current generation of laser interferometers was finally able to detect. We use data on figures reported in the discovery papers to make estimates of the astrophysical parameters. Simple arguments based on Newtonian gravity, dimensional analysis and analogies with electromagnetic waves are employed. Key parameters obtained in this way (masses of merging objects, distances, emitted energy) are compared with the parameters reported in the discovery papers, in which they were extracted by fitting data to templates generated by numerical relativity. In the near future, networks of interferometers will help researchers to determine the locations of sources in the sky and trigger "traditional" astronomical observations and neutrino telescopes for the study of high-energy processes in the Universe. Combining observations in this way is the basis of multimessenger astrophysics.

On February 11, 2016, the LIGO collaboration announced the discovery of gravitational radiation due to the merger of two black holes. The binary system was located 400 Mpc away from Earth (Abbott et al. 2016). As the event was observed on September 14, 2015, its official designation is now GW150914. This discovery

© Springer Nature Switzerland AG 2018

M. Spurio, *Probes of Multimessenger Astrophysics*, Astronomy and Astrophysics Library, https://doi.org/10.1007/978-3-319-96854-4_13

represents a major scientific breakthrough for physics, astrophysics, and cosmology. The 2017 Nobel Prize in Physics was awarded to R. Weiss, B. Barish and K. Thorne (all three members of the LIGO/Virgo Collaboration) for their decisive contributions to the LIGO detector and the observation of gravitational waves (GWs).

Probably even more important, on October 16, 2017, the LIGO/Virgo collaboration announced (Abbott et al. 2017a), together with a large number of other experiments (Abbott et al. 2017e), the first coincident observation of GWs and electromagnetic radiation. These observations are connected with a collision of two neutron stars 40 Mpc away from Earth, producing almost simultaneously both gravitational radiation (GW170817) and a short gamma ray burst (GRB170817A). The electromagnetic observations in the following days revealed signatures of recently synthesized material, including gold and platinum, solving a decades-long mystery concerning where about half of all elements heavier than iron are produced.

The purpose of this chapter is to explain key features of the observed gravitational radiation in terms of introductory physics. In Einstein's general theory of relativity, gravity is an effect due to the curvature of space-time. This curvature is caused by the presence of masses. As massive objects move around, the curvature changes to reflect the variated positions of those objects. The changes in this curvature propagate outwards at the speed of light in a wave-like manner, Sect. 13.1. These propagating phenomena are known as gravitational waves; they carry a form of radiant energy similar to electromagnetic radiation, Sect. 13.2.

To a good approximation, in the case of orbiting bodies, the masses follow simple Keplerian orbits before merging. Such an orbit represents a changing quadrupole moment that induces GWs, Sect. 13.3. The energy carried away by the emitted radiation induces an *inspiral*, or decrease in orbit. The current generation of laser interferometers, presented in Sect. 13.4, was finally able to detect the signals produced by the coalescence of two black holes (the first of them, GW150914, is described in Sect. 13.5), and two neutron stars (GW170817, Sect. 13.7). We use data on figures reported in the discovery papers to make estimates of the astrophysical parameters. Simple arguments based on Newtonian gravity, dimensional analysis and analogies with electromagnetic waves are employed. Key parameters obtained in this way (masses of merging objects, distances, emitted energy) are compared with the parameters reported in the discovery papers (Abbott et al. 2016, 2017a,e), in which they were extracted by fitting data to templates generated by numerical relativity. Similar pedagogical efforts have been carried out in LIGO Scientific and VIRGO Collaborations (2017) and Mathur et al. (2017).

The network of Earth-based laser interferometers (and in future, space-based interferometer such as eLISA) has opened a new observational window onto the Universe. The discovery of the production sites of trans-iron elements, discussed in Sect. 13.8, is an example. Networks of interferometers will help researchers to determine the locations of sources in the sky and trigger "traditional" astronomical observations and neutrino telescopes for the study of high-energy processes in the Universe, Sect. 13.10. Combining observations in this way is the basis of multimessenger astrophysics; the physics of the gravitational stellar collapse would greatly benefit from a joint GW-electromagnetic radiation observation, as shown in

Sect. 13.11. Some of the new insights in physics, astrophysics and cosmology that can be studied in the near and far future in conjunction with the detection of GWs are mentioned in the final section of this chapter and in Sect. 1.8.

13.1 From Einstein Equation to Gravitational Waves

13.1.1 A Long Story Short

Gravitational waves were firstly proposed in 1907 by the French physicist Henri Poincaré ("ondes gravifiques") as emanating from massive bodies and propagating at the speed of light. The mathematical framework necessary for their description is that of the theory of general relativity, published afterwards in 1915. Einstein himself, based on various approximations, derived three types of propagating solutions from the field equations, designed as longitudinal-longitudinal, transverse-longitudinal, and transverse-transverse oscillations. However, the nature of Einstein's approximations led many (including Einstein himself) to doubt the result. In 1922, Arthur Eddington showed that two types of wave were artifacts resulting from the choice of coordinate system (a sort of "gauge effect"), and could be made to propagate at any speed by choosing appropriate coordinates. The famous Eddington's joking sentence that GWs "propagate at the speed of thought" appears today in the title of a strongly recommended monography (Kennefick 2007) on the subject. For the historical path toward a theoretical understanding of GWs, see also the recent (Cervantes-Cota et al. 2016; Cheng et al. 2017).

In 1936, Einstein and Nathan Rosen submitted a paper to the Physical Review Letter with the title *Are there any gravitational waves?* The original version of the manuscript does not exist today, but Einstein's epistolary documents show that the answer to the title was "they do not exist". The editor sent the manuscript to be reviewed by an anonymous referee (in the usual peer review process), who questioned the conclusion of the paper (today, we know that the anonymous referee was Howard P. Robertson). Einstein angrily withdrew the manuscript, asserting that he would never publish in the Physical Review again.[1] By some fortuitous circumstance, Leopold Infeld (at that time, an assistant of Einstein) met Robertson at a conference, the latter subsequently convincing Infeld that the conclusion in his presentation (that contained in the Einstein-Rosen paper) was incorrect. Ultimately, Infeld similarly convinced Einstein that the criticism was correct; the paper was rewritten with the same title, the opposite conclusion and published elsewhere.

The question whether the waves carry energy (and are thus "physical" objects) or are instead a "gauge" effect remained controversial up to the end of the 1950s. Finally, Pirani showed that gravitational waves would exert tidal forces on

[1]The LIGO Collaboration has published the GW's discovery paper (Abbott et al. 2016) on PRL!

intervening matter, producing a strain in the material with a quadrupole oscillation pattern. This stimulated experimental searches for gravitational radiation, which started in the 1960s with the work of Weber. He began to speculate as to the way in which GWs might be detected, also motivated by incorrect predictions concerning the possibility of waves with amplitude (or *strain*, a dimensionless quantity defined in the following sections) on the order 10^{-17} at frequencies near 1 kHz. At the University of Maryland, Weber built an aluminium bar 2 m in length and 0.5 m in diameter, with resonant mode of oscillation of $\sim$ 1.6 kHz. The bar was fitted with piezo-electric transducers to convert its motion into an electrical signal. In 1971, with the coincident use of two similar detectors (the second was in Illinois), Weber claimed detection of GWs from the direction of the galactic center. This led to the construction of many other bar detectors of comparable or better sensitivity, which never confirmed his claims.

Improved theoretical models and calculations that appeared in the 1970s showed that gravitational wave strains were likely to be of the order on 10^{-21} or less and could encompass a wide range of frequencies. The correctness of such theoretical results remained a matter of controversy into the 1980s. The question would ultimately be solved by the observation of the Hulse-Taylor binary pulsar system: the rate of decrease of orbital period is 76.5 ms per year, in accord with the predicted energy loss due to gravitational radiation. Thus, with respect to resonant bars, a more sensitive and wider-band detection technique was necessary. Such a technique became available with the development of laser interferometers. After the prototype demonstrations at Caltech, Glasgow, and Garching, funding agencies in the USA and Europe committed to the construction of large, kilometer-scale laser interferometers: LIGO (USA, 4 km), Virgo (France and Italy, 3 km) and GEO (UK and Germany, 600 m). The length of their arms today allows for a strain sensitivity on the order of 10^{-22} over a 100 Hz bandwidth, a development that finally led to the discovery in 2015.

13.1.2 Summary of the Mathematical Background

In General Relativity, space-time is considered as a four-dimensional manifold, and gravity is a manifestation of the manifold's curvature. We recall here some fundamental concepts from general relativity, remanding to more specialized texts for a detailed description (Maggiore 2007). The content of this subsection can be skipped if you are not familiar with general relativity, the only relevant equation for the following being Eq. (13.15).

The differential line element ds at space-time point $\mathbf{x}$ has the form:

$$ds^2 = g_{\mu\nu}(\mathbf{x}) dx^\mu dx^\nu \,, \tag{13.1}$$

where $g_{\mu\nu}$ is the symmetric metric tensor, and repeated indices imply summation. For example, for a flat Cartesian coordinate metric $[\mu = (ct, x, y, z)]$

$$g_{\mu\nu}^{Cart}(\mathbf{x}) = \begin{pmatrix} -1 & 0 & 0 & 0 \\ 0 & 1 & 0 & 0 \\ 0 & 0 & 1 & 0 \\ 0 & 0 & 0 & 1 \end{pmatrix} . \tag{13.2}$$

If the space is not *flat*, the form of the metric tensor is much more complicated.

Starting from the observed equivalence of gravitational and inertial mass, which was elevated to the status of a fundamental physical principle, Einstein interpreted gravity as the physical manifestation of curvature in the geometry of space-time. The mathematical way adopted in general relativity is to quantify the curvature of a metric via the covariant equation of motion for a test particle. Thus, space-time curvature is associated with matter and energy:

$$G_{\mu\nu} \equiv R_{\mu\nu} - \frac{1}{2} R\, g_{\mu\nu} = \frac{8\pi G}{c^4} T_{\mu\nu} \, . \tag{13.3}$$

On the left-hand side, $G_{\mu\nu}$ is the Einstein tensor, which is formed from the Ricci curvature tensor $R_{\mu\nu}$ and the space-time metric $g_{\mu\nu}$; the matrix $G_{\mu\nu}$ is symmetric, and $R = g^{\mu\nu} R_{\mu\nu}$ is called the curvature scalar. On the right-hand side, $T_{\mu\nu}$ is the stress-energy tensor of matter fields, and G is Newton's gravitational constant. Equation (13.3) derived by Einstein, quantifies how energy density leads to curvature and, in turn, how curvature influences energy density. Though simple in appearance, the Einstein equation is a nonlinear function of the metric and its first and second derivatives; this very compact geometrical statement disguises ten coupled, nonlinear partial differential equations.

In order to give a very simple mechanical analogy of (13.3), consider the potential energy connected with the spatial deformation of a spring:

$$kx = \nabla U \, . \tag{13.4}$$

Here, x takes the place of the metric tensor and U that of the stress-energy tensor. Thus, the equivalent of the spring's constant k in (13.3) is

$$k \longrightarrow \quad \frac{c^4}{8\pi G} = 5.6 \times 10^{45} \text{ kg m s}^{-2} \, . \tag{13.5}$$

This is equivalent to saying that the energy required to distort space is analogous to that required to induce an elastic deformation of rigid materials, but to a much greater degree, because space is extremely stiff.

Generation of GWs is implicit in the Einstein equations. In fact, if we consider a small and flat region far from a non-static source (for instance, two massive objects orbiting each other), the gravitational field should vary with time. This can be thought of as an effect that perturbs the flat Cartesian metric by only a small amount, $h_{\mu\nu}$:

$$g_{\mu\nu}(\mathbf{x}) = g^{Cart}_{\mu\nu} + h_{\mu\nu}(\mathbf{x}) \,. \tag{13.6}$$

Under these assumptions, the left side of the Einstein equation (13.3) can be greatly simplified by keeping only first order terms in $h_{\mu\nu}$ and applying a gauge condition analogous to that applied on the electromagnetic potential. The choice of a particular gauge (gauge fixing) denotes the mathematical procedure for coping with redundant degrees of freedom in field variables.[2] In vacuum ($T_{\mu\nu} = 0$), one obtains the homogeneous wave equation

$$\left(-\frac{1}{c^2}\frac{\partial^2}{\partial t^2} + \nabla^2\right)h_{\mu\nu}(\mathbf{x}) \equiv \Box h_{\mu\nu}(\mathbf{x}) = 0 \tag{13.7}$$

that has familiar space- and time-dependence solutions, for example, for a fixed wave vector $\mathbf{k}$:

$$h_{\mu\nu}(\mathbf{x}) = h^0_{\mu\nu} e^{[i(\mathbf{k}\cdot\mathbf{x}-\omega t)]} \,, \tag{13.8}$$

but describes a tensor perturbation. The constant $h^0_{\mu\nu}$ is a symmetric 4×4 matrix and $\omega = kc$. A particular useful solution for the GW in vacuum is obtained by choosing the z-axis along the direction of the wave vector $\mathbf{k}$; this condition is known as the *transverse-traceless (TT) gauge* and leads to the relatively simple form

$$h_{\mu\nu}(\mathbf{x}) = \begin{pmatrix} 0 & 0 & 0 & 0 \\ 0 & h_+ & h_\times & 0 \\ 0 & h_\times & -h_+ & 0 \\ 0 & 0 & 0 & 0 \end{pmatrix} e^{[i(\mathbf{k}\cdot\mathbf{x}-\omega t)]} \,, \tag{13.9}$$

where h_+ and $h_\times$ are constant amplitudes. For illustration, Fig. 13.1 depicts the nature of these two polarizations as gravitational waves propagating along the z-axis impinge upon a ring of "free" test masses in a plane perpendicular to the wave direction $\mathbf{k}$.

Equation (13.9) can be used to explain the effect of a GW impinging on free-fall test masses of a detector on Earth. We now need to determine the relation of GWs to their source. This is defined by the inhomogeneous Einstein equation (13.3). Under the assumptions of a weak field in a nearly flat space-time, Cartesian coordinates

[2] In the electromagnetic theory, the Lorenz gauge condition (or Lorenz gauge) is a partial gauge fixing of the four-vector potential. The condition is that $\partial_\mu A^\mu = 0$. In ordinary vector notation and SI units, the gauge condition is written as $\nabla\cdot\mathbf{A} + \frac{1}{c^2}\frac{\partial\varphi}{\partial t} = 0$. This does not completely determine the gauge: one can still make a gauge transformation $A^\mu \to A^\mu + \partial^\mu f$, where f is a scalar function satisfying $\partial_\mu\partial^\mu f = 0$.

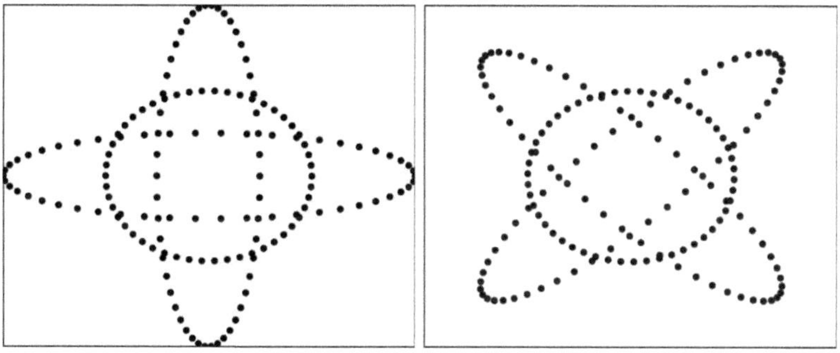

Fig. 13.1 In the weak field, the gravitational waves have two independent polarizations called h_+ and $h_\times$. The effect on the separations of test masses displaced in a circular ring in the (x, y)-plane, perpendicular to the direction of the wave, is shown on the left for the h_+ wave and on the right for $h_\times$. The ring continuously gets deformed into one of the ellipses and back during the first half of a gravitational wave period and gets deformed into the other ellipse and back during the next half

and the transverse-traceless gauge, one has an inhomogeneous wave equation:

$$\Box h_{\mu\nu}(\mathbf{x}) = \frac{16\pi G}{c^4} T_{\mu\nu} \ . \tag{13.10}$$

This source equation is analogous to the wave equation originating from a relativistic electrodynamic field:

$$\Box A^{\mu}(\mathbf{x}) = -\mu_0 J^{\mu} \ , \tag{13.11}$$

where $A^\mu = (\phi/c, \mathbf{A})$ is the four-vector with the scalar and vector potential functions and $J^\mu = (c\rho, \mathbf{J})$ with the electric scalar charge and current density. In the case of electrodynamics, the Green function formalism is applied to derive the solution; for instance, the vector potential is written as an integral over a source volume:

$$\mathbf{A}^{\mu}(t, \mathbf{x}) = \frac{\mu_0}{4\pi} \int d^3x' \frac{[\mathbf{J}(t', \mathbf{x}')]_{ret}}{|\mathbf{x} - \mathbf{x}'|} \ , \tag{13.12}$$

where $[\ldots]_{ret}$ indicates evaluation at the *retarded time* $t' \equiv t - |\mathbf{x} - \mathbf{x}'|/c$. Similarly, the solution for the waves (13.10) produced by variations of the mass configuration can be written as

$$h_{\mu\nu}(t, \mathbf{x}) = \frac{4G}{c^4} \int d^3x' \frac{[T_{\mu\nu}(t', \mathbf{x}')]_{ret}}{|\mathbf{x} - \mathbf{x}'|} \ . \tag{13.13}$$

In the following, we are interested in some particular solutions, namely that originated by a source with scale dimension R that varies harmonically over time with characteristic frequency ν_s, wavelength $\lambda = c/\nu_s$ and with the energy tensor dominated by the rest mass of the rotating objects. This includes the systems with two massive objects (two black holes, or two neutron stars, or a black hole and a neutron star) orbiting around one another. In addition, we assume that:

1. $\lambda \gg R$, i.e., the *long-wavelength approximation*, and
2. $r \gg R$, where r is the distance of the observer from the source (the *distant-source approximation*).

Under these approximations, the connection (13.13) between the tensor h and source reduces to

$$h_{\mu\nu}(t, \mathbf{x}) \simeq \frac{4G}{rc^4} \int d^3x' \; T_{\mu\nu}(t - r/c, \mathbf{x}') \; . \qquad (13.14)$$

This relation further simplifies if we assume that the energy density of the source is dominated by its rest-mass density ρ_m (non-relativistic internal velocities), obtaining a relation for the spatial coordinates:

$$h_{ij} \simeq \frac{4G}{rc^4} \frac{d^2 Q_{ij}}{dt^2} \; , \qquad (13.15)$$

where Q_{ij} is a 3×3 tensor of the mass quadrupole moment:

$$Q_{ij} = \int d^3x \left(x_i x_j - \frac{1}{3} r^2 \delta_{ij} \right) \rho_m(\mathbf{x}) \; . \qquad (13.16)$$

Here, δ_{ij} is the Kronecker-delta matrix (diagonal elements =1, off-diagonal elements =0). Although h_{ij} is a tensor quantity, in the following, we indicate with h the order-of-magnitude of its elements, i.e., the *effect* of the GW.

A tensorial object similar to (13.16) appears in advanced courses of electromagnetism in the multipole expansions of charge distributions. It is simple to introduce it if you are familiar with the moment of inertia tensor, I, introduced in mechanics (see, for instance, Feynman et al. 1964 and Kittel et al. 1965). For a system of n particles with masses m_α and positions $(x_\alpha, y_\alpha, z_\alpha)$, the elements of I are

$$I_{xx} = \sum_{\alpha=1}^{n} m_\alpha (y_\alpha^2 + z_\alpha^2) \; ; \quad I_{xy} = - \sum_{\alpha=1}^{n} m_\alpha x_\alpha y_\alpha \; , \qquad (13.17)$$

and the other diagonal and off-diagonal components can be written down by analogy. The quadrupole tensor is similar: the off-diagonal components have the form $Q_{ij} = -I_{ij}$ and the diagonal components $Q_{xx} = -I_{xx} + (1/3)\overline{I}$, and similarly for Q_{yy} and Q_{zz}. Here, $\overline{I} = I_{xx} + I_{yy} + I_{zz}$.

13.2 Energy Carried by a Gravitational Wave

The effect of accelerated charges is to produce an electromagnetic wave with oscillating electric and magnetic fields propagating at the light speed. The effect of accelerated matter is to produce a GW propagating at light speed that distorts the local metric. This means that the distance L between two free masses can be stretched or shrunk by a quantity ΔL such that $h = \Delta L/L$. This quantity ΔL oscillates with the frequency of the GW.

To the lowest order, gravitational radiation is a quadrupolar phenomenon. In electromagnetism, radiation induced by electric dipole and magnetic dipole processes is supported, while "monopole" radiation is prohibited by electric charge conservation. "Monopole" gravitational radiation is prohibited by energy conservation; dipole radiation is related to the source's center of mass; momentum conservation ensures that a closed system's center of mass cannot accelerate and, correspondingly, there is no dipole contribution to GWs. Note that, as for electrodynamics, gravitational radiation intensity is not spherically symmetric (isotropic) about the source.

The problem as to how small h is, which oscillation frequencies are typical and which methods could be used to experimentally observe ΔL are the subjects of the following section. Here, we concentrate on the problem of energy carried out by a GW. As we mentioned before, a long discussion took place in the community about the energy flux implicit in GWs. The computation is not easy, and we report only the salient results. The evaluation of the GW energy flux is easier if considered in a spatial volume encompassing many wavelengths, but small in dimension compared to the characteristic radius of curvature of the space. Under this assumption, the GW energy flux corresponds to

$$\mathscr{F} = \frac{1}{32\pi}|\dot{h}|^2 \frac{c^3}{G} . \tag{13.18}$$

The SI unit of the $\mathscr{F}$ vector is the Watt per square meter (W/m^2). It has the same units as the electromagnetic Poynting vector, $\mathbf{S} = \frac{1}{\mu_0}\mathbf{E} \times \mathbf{B}$. The Poynting vector represents the directional energy flux (the energy transfer per unit area per unit time). We do not derive (13.18) (see, for instance, Maggiore 2007); however, it is easy to verify that the quantity c^3/G has dimensions of [Energy Time/Area]; the quantity $|\dot{h}|$ [Time^{-1}] takes the place of the derivative of the electromagnetic potential, i.e., the electric and magnetic fields, and thus $(c^3|\dot{h}|^2/G)$ with dimensions of [Energy /(Area Time)] has the role of $\mathbf{S}$. Finally, the numerical term $1/32\pi$ is the result of heavy computation.

As a general result (Hartle 2003; Maggiore 2007; Saulson 1994), the total luminosity (in Watt) of GWs in the radiation zone, $\mathscr{L}$, depends on the third time derivative of the mass quadrupole moment averaged over several cycles:

$$\mathscr{L} = \frac{1}{5}\frac{G}{c^5} \sum_{i,j=1}^{3} \frac{d^3 Q_{ij}}{dt^3} \frac{d^3 Q_{ij}}{dt^3} . \tag{13.19}$$

In the following sections, we specify the above general formulas to the case of two-body systems. With some approximations, we can produce simple and reasonably accurate predictions for the frequency, duration, and strength of gravitational radiation from these systems. Before turning to this, it is useful to consider some comparisons between gravitational radiation and electromagnetic radiation:

- In most astrophysical cases, the electromagnetic radiation emitted is an incoherent superposition of light from sources much larger than the radiation wavelengths; in contrast, gravitational radiation that is likely to be detectable (below a few kHz) comes from systems with sizes R smaller (or, in some cases, comparable) to the emitted wavelength λ. Hence, the signal reflects the coherent motion of extremely massive objects.
- Solutions to Maxwell's equations for a localized oscillating source of dimension R at a distance r in a homogeneous material (e.g., vacuum or air) are $\mathbf{E}$ and $\mathbf{B}$ fields that decay as $\frac{1}{r}$ when $r \gg R$. These are the *radiating fields*, and the condition $r \gg R$ defines the *far field*. Similarly, the quantity describing the gravitational field (the strain, Eq. (13.15)) decreases as $\frac{1}{r}$.
- Detectors of electromagnetic radiation are sensitive to the flux intensity (i.e., to the Poynting vector, $\mathbf{S}$) that decreases as $\frac{1}{r^2}$. This, because work must be done on electric charges (for example, in an antenna). Contrastingly, gravitational interferometer detectors register waves coherently by following the *phase* of the wave and not just measuring its intensity. The phase of the wave is contained in the strain h that decreases as $\frac{1}{r}$.
- The number N of astrophysical sources within a given distance D from the Earth increases as the volume of a sphere, $N \propto D^3$. Consider a telescope for the electromagnetic radiation with sensitivity that allows for the detection of sources with intensity $\mathscr{L}$; if the sensitivity of the telescope is improved by a factor of (e.q.) three, then the horizon of the telescope increases as a factor $\sqrt{3}$, because $\mathscr{L} \propto D^{-2}$. Thus, following the improvement, the number of observable sources increases as $N \propto 3^{3/2} \simeq 5$. If the sensitivity of an interferometer for GWs is increased by a factor of three, the number of detectable sources increases by a factor of $3^3 = 27$.
- The frequencies of detectable GWs are below the few kHz range, and thus graviton energies $h\nu_{gw}$ are very small, making detection of individual quanta extremely challenging (if not almost impossible in the near future).
- Gravitational radiation suffers a very small absorption when passing through ordinary matter. As a result, GWs can carry to us information about violent processes occurring in very dense environments. In the context of detection, in comparison, even neutrinos have large scattering cross-sections with matter.
- It is almost impossible with current technology to produce and detect manmade GWs. In a classic example from Saulson (1994), if we consider a dumbbell consisting of two 1-ton compact masses with their centers separated by 2 m and spinning at 1 kHz (this is the limit for its stability), the strain h to an observer 300 km away (in the far field) is $h \sim 10^{-38}$.

13.3 The Two-Body System

The quadrupole moment (13.16) of a system of two point-like masses m_1 and m_2 in a binary orbit can be calculated. Here, a simple Newtonian approach is used that holds for low velocities. When the velocities become relativistic, the Newtonian framework used to derive relations between quantities no longer applies. One example is the case of Kepler's third law,

$$\omega_s^2 = \frac{GM}{R^3} \quad \text{where } M = m_1 + m_2 \,, \tag{13.20}$$

connecting angular velocity ω_s^2 with the orbit size R. For small R and high velocities (such as the later stages of the inspiral, as discussed in the following), further *post-Newtonian* approximations are necessary. The post-Newtonian approximations are expansions terms of v/c and are used to find an approximate solution to the Einstein field equations for the metric tensor in the case of weak fields.

We assume that the two-body system lies in the (x, y)-plane shown in Fig. 13.2; the quadrupole moment Q_{ij} is computed using the Cartesian coordinate system $\mathbf{x} = (x_1, x_2, x_3) = (x, y, z)$, whose origin is the center-of-mass; r_i is the distance of the mass $i = 1, 2$ from the origin. Thus, according to the definition (13.17):

$$Q_{ij} = \sum_{\alpha=1,2} m_\alpha \begin{pmatrix} \frac{2}{3}x_\alpha^2 - \frac{1}{3}y_\alpha^2 & x_\alpha y_\alpha & 0 \\ x_\alpha y_\alpha & \frac{2}{3}y_\alpha^2 - \frac{1}{3}x_\alpha^2 & 0 \\ 0 & 0 & -\frac{1}{3}r_\alpha^2 \end{pmatrix} . \tag{13.21}$$

In the simple case of a circular orbit at separation $R = r_1 + r_2$, frequency ν_s, and angular velocity $\omega_s = 2\pi\nu_s$, it is easy to derive, with the help of Fig. 13.2,

$$Q_{ij}^\alpha = \frac{m_\alpha r_\alpha^2}{2} J_{ij} \,, \tag{13.22}$$

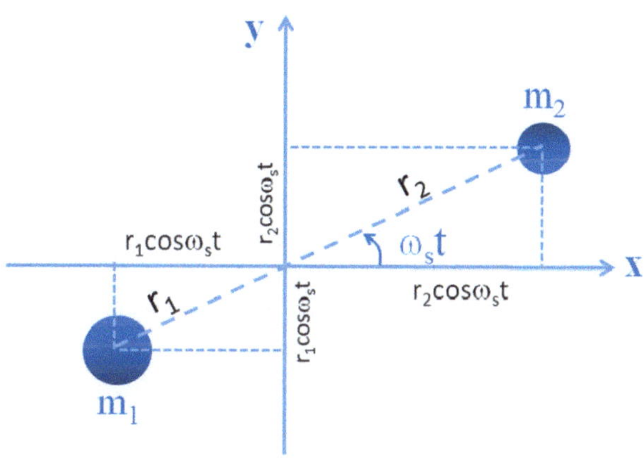

Fig. 13.2 A two-body system, m_1 and m_2 orbiting in the (x, y)-plane around their center of mass

where the elements J_{ij} of the 3 × 3 matrix are:

$$\begin{pmatrix} \cos(2\omega_s t) + \frac{1}{3} & \sin(2\omega_s t) & 0 \\ \sin(2\omega_s t) & \frac{1}{3} - \cos(2\omega_s t) & 0 \\ 0 & 0 & -\frac{2}{3} \end{pmatrix} . \tag{13.23}$$

By summing up the contribution of the two masses, we obtain:

$$Q_{ij} = \sum_{\alpha=1,2} Q^{\alpha}_{ij} = \frac{1}{2}\mu R^2 J_{ij} , \tag{13.24}$$

where

$$\mu \equiv \frac{m_1 m_2}{m_1 + m_2} \tag{13.25}$$

is the *reduced mass* of the system.

It is clear from Eq. (13.15) that the intensity h_{ij} of the GW depends on the relative orientation of the observer with respect to the (x, y)-plane of the source. However, as given in (13.9), in the direction perpendicular to the wave vector **k**, there are only two degrees of freedom, expressed by the h_+ and $h_\times$ constant amplitudes. To give a first-order estimate of the GW effect, let us assume that $J_{ij} = \cos(2\omega_s t)$ in (13.24), i.e., $Q_{ij} \sim Q$, and that

$$h_\circ \sim h_+ \sim h_\times . \tag{13.26}$$

Thus,

$$\frac{d^2 Q}{dt^2} = \frac{1}{2}\mu R^2 \cdot (4\omega_s^2) \cdot \cos(2\omega_s t) . \tag{13.27}$$

The time-dependent wave amplitude is derived from (13.15):

$$h(t) \simeq \frac{4G}{rc^4} \cdot (2\mu R^2 \omega_s^2) \cdot \cos(2\omega_s t) = h_\circ \cos(\omega_{gw} t) , \tag{13.28}$$

where

$$\omega_{gw} = 2\omega_s . \tag{13.29}$$

Notice that because the quadrupole moment is symmetric under rotations of an angle π about the orbital axis, **the radiation has a frequency, ν_{gw}, twice that of the orbital frequency of the source, ν_s**. Now, by using Kepler's third law (13.20), we can remove the angular velocity from (13.28) and obtain

$$h_\circ \simeq \frac{4G}{rc^4} \cdot (2\mu R^2)\frac{GM}{R^3} , \tag{13.30}$$

or, equivalently,

$$h_\circ = 2\left(\frac{2GM}{c^2 r}\right)\left(\frac{2G\mu}{c^2 R}\right) = 2\frac{\mathcal{R}_{S_1} \cdot \mathcal{R}_{S_2}}{r \cdot R} . \tag{13.31}$$

This is a relevant result: the strain $h_\circ$ derived from the quadrupole formula can be written into a manifestly dimensionless form by recognizing that the mass times $2G/c^2$ corresponds to the Schwarzschild radius $\mathcal{R}_{S_i}$ of the object, Eq. (6.84). Notice, at the denominator, the distance R is an *internal* parameter of the system, while r is the distance of the source from the observer. If the binary system consists of two neutron stars ($m_1 \simeq m_2 \simeq 1.4 M_\odot$), then the Schwarzschild radius is ~4 km for both objects. If we consider two close-by neutron stars approaching their merging when $R \simeq 100$ km and at a distance of 40 Mpc from the Earth, we obtain

$$h_\circ = 2\left(\frac{(4000 \text{ m})^2}{10^5 \times 1.2 \times 10^{24} \text{ m}^2}\right) \simeq 3 \times 10^{-22} . \tag{13.32}$$

Let us summarize the salient results in terms of observable quantities. As a GW passes an observer, that observer will find spacetime distorted by the effects of strain. Distances L between objects increase and decrease rhythmically as the wave passes, with a maximum amplitude ΔL such that

$$\frac{\Delta L}{L} \simeq h_\circ , \tag{13.33}$$

with the pattern shown in Fig. 13.1 and at a frequency corresponding to that of the wave. To get a feeling for this, Eq. (13.33) means that the distance of the Earth from the Sun is changed by the distance of one atom during the passage of such a GW. The frequency of the wave depends on the relative distance R of the merging objects (in the Newtonian regime, according to Kepler's third law (13.20)). The frequency interval 10–1000 Hz is particularly relevant. Thus, the quantity (13.32) represents the order-of-magnitude of a detector's sensibility to detect GW signals.

Exercise Verify that for the above conditions, the wavelength $\lambda_{gw} \gg R$, as required by the long-wavelength approximation.

Let us now compute the total luminosity (13.19) of the source. The third derivative of (13.24) yields the matrix

$$\dddot{Q}_{ij} = \frac{1}{2}\mu R^2 (2\omega_s)^2 \begin{pmatrix} \cos(2\omega_s t) & \sin(2\omega_s t) & 0 \\ \sin(2\omega_s t) & -\cos(2\omega_s t) & 0 \\ 0 & 0 & 0 \end{pmatrix} . \tag{13.34}$$

The double summation in (13.19) yields a scalar (the sum of the product of the first line by the first column + second line times second column), explicitly

$$\sum_{i,j=1}^{3} \frac{d^3 Q_{ij}}{dt^3} \frac{d^3 Q_{ij}}{dt^3} = [\cos^2(2\omega_s t) + \sin^2(2\omega_s t)] + [\sin^2(2\omega_s t) + \cos^2(2\omega_s t)] = 2\,. \tag{13.35}$$

Thus, the scalar quantity of Eq. (13.19) becomes

$$\mathscr{L} = \frac{1}{5}\frac{G}{c^5} \cdot \left(\frac{1}{2}\mu R^2\right)^2 \cdot (2\omega_s)^6 \cdot 2 = \frac{32}{5}\frac{G}{c^5} \cdot [\mu R^2 \omega_s^3]^2\,. \tag{13.36}$$

In a similar way, the energy flux (13.18) of a sinusoidal wave of angular frequency ω_{gw} and amplitude $h_\circ$ as obtained by using (13.28) is

$$\mathscr{F} = \frac{1}{32\pi}\frac{c^3}{G} h_\circ^2 \omega_{gw}^2\,, \tag{13.37}$$

which for $\omega_{gw} = 400\,\mathrm{s}^{-1}$ and $h_\circ = 3\;10^{-22}$, corresponds to $\mathscr{F} = 7\;10^{-5}\,\mathrm{W\,m^{-2}} = 0.07\,\mathrm{erg\,cm^{-2}\,s^{-1}}$. For comparison (Sect. 8.10), typical fluxes of sources measured by Fermi-LAT in the γ-ray band for steady sources are on the order of $10^{-11}\,\mathrm{erg\,cm^{-2}\,s^{-1}}$. Hence, during the time interval $\Delta t \sim 1/\nu_{gw}$ when the waves of a coalescing binary neutron star system 40 Mpc away pass the Earth, the energy flux is on the order of 10^{10} that for a steady source of γ-rays. However, as shown below, detecting the passage of this energy flux is a formidable experimental challenge.

13.4 Ground-Based Laser Interferometers

To enable sensitivity to a wide range of astrophysical GW sources, ground-based interferometers must thus be designed to achieve strain down to $\sim 10^{-22}$, or better, possibly over the widest frequency range within 10–5000 Hz.[3]

Ground-based interferometers are arranged in the Michelson configuration (L-shaped). They consist of a laser, a beam splitter, a series of mirrors and photodetectors that record the interference pattern, see Fig. 13.4. The laser beam passes through a *beam splitter* that splits a single beam into two identical beams, one of which is at 90°. Each beam then travels down an arm of the interferometer. At the end of each arm, a mirror acting as a *test mass* reflects each beam back to the

[3]As the standard range of human audible frequencies is from 20 to 20,000 Hz, the signal of the passage of a GW can be transduced to a sound audible by human ears. There are different examples on the educational resources webpages of the experiments, https://www.ligo.caltech.edu/. However, remember that this is just a didactic and sociological trick, and that GWs are not detected by acoustic devices.

Fig. 13.3 Left: Aerial view of the LIGO gravitational wave detector in Livingston, Louisiana. (Credit: Caltech/MIT/LIGO Lab). A similar detector exists in the Washington State (LIGO Hanford). Right: Aerial view of the Virgo gravitational wave interferometer in Italy (Credit: EGO/Virgo)

beam splitter, where the two beams merge back into a single beam. In 'merging', the light waves from the two beams interfere with each other before reaching a *photodetector*. GW interferometers are set up so that the interference is destructive at the photodetector. Any change in light *intensity* due to a different interference pattern indicates that something (noise or signal) happened to change the distance L travelled by one or both laser beams. Moreover, the interference pattern can be used to calculate precisely $\Delta L/L$, i.e., the signal strain (13.33). This point is of fundamental importance: **the interferometer is sensitive to the phase of the quantity h(t) (the strain, such as that given in Eq. (13.28)), and not to the GW energy flux, Eq. (13.18). The former decreases as 1/r, the latter as $1/r^2$.**

A GW observatory cannot operate alone. A coincident detection with two interferometers suppresses the noise background and ameliorates the possibility of source localization. These objectives become even more improved when interferometers are connected in a network, as in the present configuration of GW observatories.

LIGO consists of two widely separated (about 3000 km) identical detector sites in the USA working as a single observatory: one in southeastern Washington State and the other in rural Livingston, Louisiana, Fig. 13.3 left. The LIGO Scientific Collaboration (LSC) includes scientists from both LIGO laboratories and collaborating institutions. LSC members work in strict contact with the GEO 600 detector in Germany. Virgo is a 3 km interferometer located close to Pisa, Italy, funded by the European Gravitational Observatory (EGO), a collaboration between the Italian INFN and the French CNRS, Fig. 13.3 right. While the LSC and the Virgo Collaboration are separate organizations, they cooperate closely; they are referred to as LVC, and they collectively sign the research papers.

Initial LIGO (iLIGO) took data between 2001 and 2010, almost contemporaneously with initial Virgo, without detecting GWs. The redesign, construction, preparation and installation of the Advanced LIGO (aLIGO) took 7 years, from 2008 to 2015, and those for the Advanced Virgo from 2010 to 2017. The improvements had the objective of making the observatories ten times more sensitive with respect

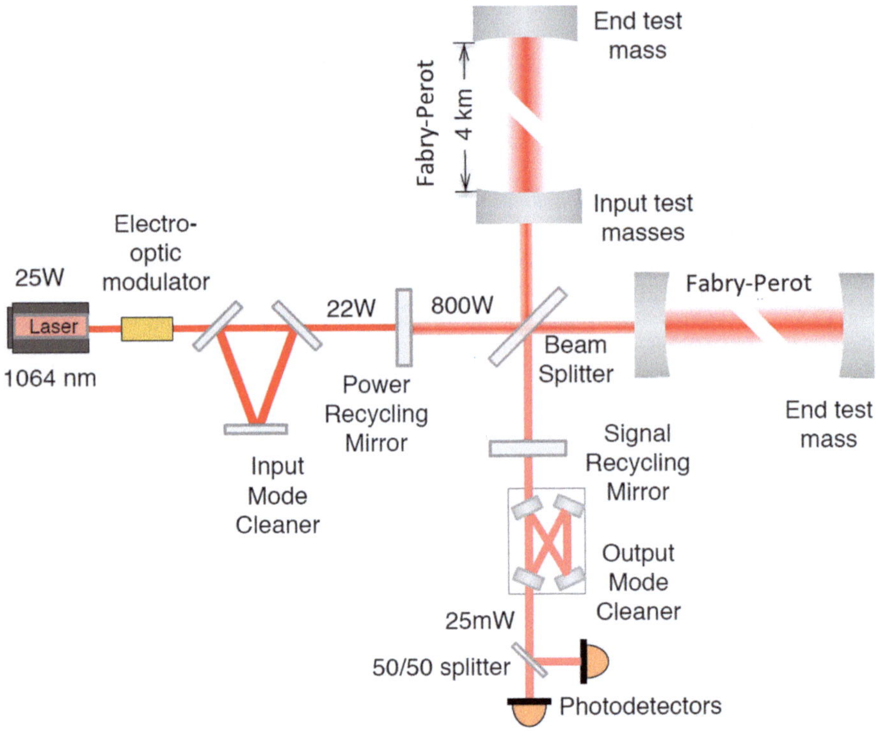

Fig. 13.4 Layout of an aLIGO detector. See text for details. Credit: The LIGO Collaboration. Adapted from Martynov et al. (2016)

to the initial phase, allowing them to increase the volume of the observable universe by a factor of 10^3. In September 2015, aLIGO began the era of GW astronomy with its first observation run (O1) and detections, collecting data until January 2016. The interferometers were not yet operating at final design sensitivity during O1. The second observing run (O2) of aLIGO started on November 30, 2016. aVirgo joined the O2 run on August 1, 2017. Both ended O2 operations on August 25, 2017.

In the following, some details of the current design of the interferometers are described, referring, in particular, to Fig. 13.4. The most impressive technology resides in their laser, seismic isolation systems necessary to remove unwanted vibrations, vacuum systems, optics components to preserve laser light and power, and computing infrastructure to handle collected data in real time. Some quantities (such as the number of reflections, laser power, etc.) changed slightly from run O1, O2 and the final design. We tailor the description to the aLIGO setup; the Virgo interferometer works similarly.

13.4.1 The Advanced LIGO Interferometers

The Optics System of GW interferometers consists of lasers, a series of mirrors, and photodetectors. If LIGO's interferometers were basic Michelson's, even with arms 4 km long, they would still not be long enough to be sensitive to GWs. One of the fundamental tools is the *Fabry-Perot cavities*. A Fabry-Perot cavity is created by adding mirrors near the *beam splitter* that continually reflect parts of each laser beam back and forth within the long arms. In aLIGO, this occurs about 270 times before the laser beams are merged together again, making LIGO's interferometer arms effectively 1080 km long.

A second design factor important for improving the interferometer's resolution is the laser power. The more photons that merge at the beam splitter, the sharper the resulting interference pattern becomes. To reach the sensitivity necessary for a discovery ($h \sim 10^{-22}$), the laser must reach a much higher power (see the discussion in Sect. 13.4.2). For this reason, an additional device, the *power recycling mirrors*, is placed between the laser source and the beam splitter to boost the power of the LIGO laser: in the O1 run, this power was increased by a factor of $\sim$ 40. Similarly to the beam splitter itself, the power recycling mirror is only partly reflective and the light from the laser first passes through the mirror to reach the beam splitter. The instrument is accurately aligned in such a way that the largest fraction of the reflected laser light from the arms follows a path back to the recycling mirrors, rather than to the photodetector. These 'recycled' photons add to the ones just entering. As a further difference with simple Michelson interferometers, aLIGO has *signal recycling* mirrors that, like power recycling, enhance the output signal.

Before entering the power recycling mirror, the *input mode cleaner* is a suspended, triangular Fabry-Perot cavity needed to clean up the spatial profile of the laser beam, clean polarization, and help stabilize the laser frequency. Similarly, before the photodetector, an *output mode cleaner* is present at the antisymmetric port, to reject unwanted spatial and frequency components of the light, before the signal is detected.

The Laser The heart of LIGO is its Nd:YAG laser, with wavelength $\lambda = 1064$ nm. The maximum power is $\sim$200 W by design, but only 22 W were used in the O1 run. It takes different steps to amplify its power and refine its wavelength to the level necessary for the experiment. The first step is a laser diode generating an 808 nm near-infrared beam of $\sim$4 W (about 800 times more powerful than standard laser pointers). Then, the 4 W beam enters a device consisting of a small boat-shaped crystal, and it bounces around inside this crystal and stimulates the emission of a 2 W beam with a wavelength of 1064 nm, in the infrared part of the spectrum. Another amplifying device boosts the beam from 2 to 35 W. Finally, a High Powered Oscillator performs further amplification and refinement, and generates the final beam.

Mirrors The suspended primary mirrors act as the *test masses*, and must be of the highest quality available, both in material and shape. LIGO's mirrors weigh 40 kg each and are made of very pure fused silica glass. The mirrors were polished so precisely that the difference between the theoretical design and the actual polished surface was measured in atoms. They reflect most of the laser light and absorb just one in 3×10^6 hitting photons, avoiding mirror heating. The heating could alter the mirrors' shapes enough that they degrade the quality of the laser light. The mirrors also refocus the laser, keeping the beam traveling coherently throughout its multiple reflections before arriving at the photodetector.

Seismic Isolation Laser interferometers are extremely sensitive to all vibrations near (such as trucks driving on nearby roads) and far (earthquakes, nearby and far away). The suspended primary mirrors must be as free as possible, i.e., decoupled from any man made or earthly vibrations. For this reason, *active* and *passive* damping systems are used to eliminate vibrations. The active damping consists of a system of sensors designed to feel different frequencies of ground movements. These sensors work side-by-side and send their feedback to a computer that generates a net counter-motion to cancel all of the vibrations simultaneously. The passive damping system holds all test masses (its mirrors) perfectly still through a 4-stage pendulum called a *quad*. At the end of the quad, LIGO's mirrors are suspended by 0.4 mm thick fused-silica (glass) fibers. The configuration absorbs any movement not completely canceled out by the active system.

Vacuum The laser beam travels in one of the largest and purest sustained vacuums on Earth (10^{-6} Pa). The presence of dust in the path of the laser, or worse, on the surface of a mirror, can cause some of the light to scatter (i.e., be reflected in some random direction away from its path). The presence of air produces an index of refraction that could affect the apparent distance between the mirrors. In addition, molecules of air hitting the mirrors due to Brownian motion can cause them to move, masking the signal strain. Many techniques can be used to remove all the air and other molecules from vacuum tubes; for instance, in this case, the tubes were heated (between 150 C and 170 C) for 30 days to drive out residual gas molecules, and turbo-pumps sucked out the bulk of the air contained in the tubes. Finally, ion pumps operating continuously maintained the vacuum by extracting individual remaining gas molecules. It took about 40 days to remove $\sim 10^4\ \mathrm{m}^3$ of air and other residual gases from each of the vacuum tubes, before starting the physics runs.

Computation and Data Collection Computers are required both to run the LIGO instruments and to process the data that they collect. When it is in *observing* mode, an interferometer generates TB of data every day that must be transferred to a network of supercomputers for storage and archiving. Because much of the astrophysical information are extracted from the phase of the GW, different kinds of data analysis method are employed than those normally used in astronomy. These are based on matched filtering and searches over large parameter spaces of potential signals. This style of data analysis requires the input of pre-calculated template

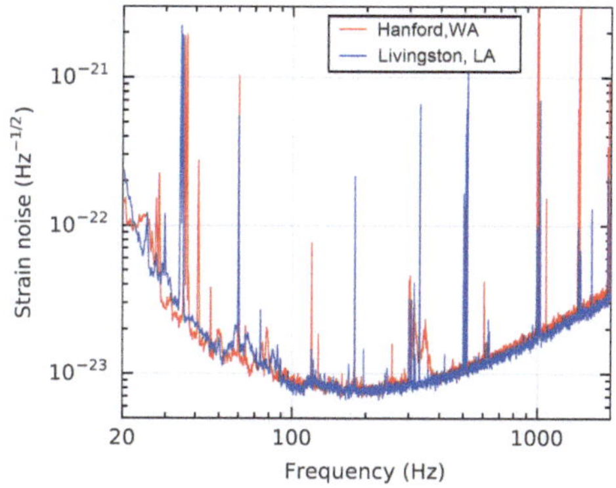

Fig. 13.5 The aLIGO instrument noise at Hanford, WA, and Livingston, LA, at the time of GW150914. On the y axis there is an amplitude spectral density, expressed in terms of equivalent gravitational-wave strain amplitude. The sensitivity is limited by photon shot noise at frequencies above 150 Hz, and by a superposition of other noise sources at lower frequencies. Narrow-band features include calibration lines (33–38, 330, and 1080 Hz), vibrational modes of suspension fibers (500 Hz and harmonics), and 60 Hz electric power grid harmonics. Credit: The LIGO Collaboration. Adapted from Abbott et al. (2016)

signals, which means that GW detection depends more strongly than most other branches of astronomy on theoretical input modeled on a computer.

13.4.2 Sensitivity of Ground-based Interferometers[4]

A genuine GW signal must be extracted from the large background due to noise sources. These noise sources can be divided into two categories: **displacement noises**, such as thermal noise, ground vibrations and gravity gradient noises. These dominate at the frequencies below ~100 Hz. The **sensing noises**, such as shot noise and quantum effects, are associated with the conversion of a small displacement into a readout signal and dominates in the frequency region above 100 Hz. The sensitivity of the detectors at different frequencies is represented by plots like that in Fig. 13.5.

The thermal noise, collective modes of motion of components of the apparatus, represents a generalization of Brownian motion, which arises from a coupling of a macroscopic element to its environment. Interferometers perform measurements at frequencies far from the resonant frequencies (pendulum suspensions in the few Hz range; internal vibrations of the mirrors at several kH), where the amplitude of thermal vibrations is largest. Thermal effects also produce other disturbances. Some of the mirrors (such as the beam splitters) are partly transmissive, and they absorb a small amount of light power during transmission. This absorption raises the

[4]This section can be skipped in the early reading steps.

temperature of the mirror and changes its index of refraction. The effect degrades the optical properties of the system, and effectively limits the amount of laser power that can be used in the detector.

Among sensing noises, the shot noise occurs because the photons used for interferometry are quantized: light arriving at the beamsplitter in bunches on N photons will be subject to Poisson statistics with uncertainty decreasing (for large N) as $\sqrt{N}$. Thus, shot noise is minimized by maximizing the photon arrival rate, or equivalently, the laser power. On the other hand, as the laser power is increased, the accuracy of position-sensing improves, with a final limit due to the Heisenberg uncertainty principle: the momentum transferred to the mirror by the measurement leads to a disturbance that can mask a gravitational wave (the *quantum effects* noise).

It is instructive from a didactic point of view to estimate the sensitivity limits on the strain h due to the aforementioned shot noise. Let consider a laser bunch of N photons of wavelength λ and wave number $k = 2\pi/\lambda$. According to Poisson statistics, the uncertainty about the distance Δx arising from a wave packet formalism is

$$\Delta x \cdot k = \frac{1}{\sqrt{N}}$$

or, equivalently,

$$\Delta x \simeq \frac{\lambda}{2\pi\sqrt{N}} \,. \tag{13.38}$$

To measure a GW with frequency ν_{gw}, one has to make at least $2\nu_{gw}$ measurements per second, so one can accumulate photons for a time Δt_{gw}, such as

$$\Delta t_{gw} \simeq \frac{1}{2\nu_{gw}} \,. \tag{13.39}$$

If we use a laser with power P (Watt) with photons of energy $E_\gamma = hc/\lambda = 2\pi\hbar c/\lambda$ (we will use the reduced Planck constant $\hbar$ to avoid confusion with the strain h), the number of photons N in a bunch of length Δt_{gw} is

$$N = \frac{P \cdot \Delta t_{gw}}{E_\gamma} = \frac{P \cdot \Delta t_{gw} \cdot \lambda}{2\pi\hbar c} = \frac{P \cdot \lambda}{4\pi \cdot \hbar c \cdot \nu_{gw}} \,. \tag{13.40}$$

The strain h from a GW induces a variation ΔL on the test masses that depends on the interferometer length L and on the number of reflections of the laser light n_{rif} in the Fabry-Perot cavities (see Sect. 13.4):

$$\Delta L = h \cdot n_{rif} \cdot L \quad \longrightarrow \quad h = \frac{\Delta L}{n_{rif} L} \,, \tag{13.41}$$

equivalent to Eq. (13.33) when the reflections are taken into account. The sensitivity of the interferometer to the strain h above the shot noise corresponds to the condition $\Delta L \gtrsim \Delta x$, where Δx is given by Eq. (13.38). Thus, in order to overcome this noise, the value of h in the interferometer must be

$$h \gtrsim \frac{\lambda}{2\pi \cdot n_{rif} \cdot L \cdot \sqrt{N}} \tag{13.42}$$

or, using (13.40),

$$h \gtrsim \frac{1}{n_{rif} \cdot L} \sqrt{\frac{\hbar c \cdot \lambda}{\pi} \cdot \frac{\nu_{gw}}{P}} \,. \tag{13.43}$$

To insert numerical values into (13.43), we refer to parameters in the O1 run, reported in Fig. 13.4. First, we rearrange the expression assuming $n_{rif} = 270$, $P = 22\,\mathrm{W}$ and a GW frequency $\nu_{gw} = 100\,\mathrm{Hz}$:

$$h \gtrsim \frac{100^{1/2}}{270 \times 22^{1/2}} \sqrt{\frac{\hbar c \cdot \lambda}{\pi L^2}} \cdot \left(\frac{270}{n_{rif}}\right) \cdot \left(\frac{22\,\mathrm{W}}{P}\right)^{1/2} \cdot \left(\frac{\nu_{gw}}{100\,\mathrm{Hz}}\right)^{1/2} , \tag{13.44}$$

and, after inserting numerical constants ($c = 3 \times 10^8$ m/s, $\hbar = 1.05 \times 10^{34}$ J s) and parameter values for LIGO ($\lambda = 1064$ nm, $L = 4 \times 10^3$ m), we obtain

$$h \gtrsim 4 \times 10^{-22} \cdot \left(\frac{22\,\mathrm{W}}{P}\right)^{1/2} \cdot \left(\frac{\nu_{gw}}{100\,\mathrm{Hz}}\right)^{1/2} . \tag{13.45}$$

The recycling of the laser light in aLIGO increases the power of the laser with a gain factor of ~ 40, thus reducing the sensibility to a further $40^{-1/2}$, i.e.,

$$h \gtrsim \frac{4 \times 10^{-22}}{\sqrt{40}} = 6 \times 10^{-23} \cdot \left(\frac{\nu_{gw}}{100\,\mathrm{Hz}}\right)^{1/2} . \tag{13.46}$$

Compare with the value in Fig. 13.5 at the frequency of 100 Hz.

13.5 GW150914

"On September 14, 2015 at 09:50:45 UTC the two detectors of the Laser Interferometer Gravitational-Wave Observatory simultaneously observed a transient gravitational-wave signal. The signal sweeps upwards in frequency from 35 to 250 Hz with a peak gravitational-wave strain of 1.0×10^{-21}*. It matches the waveform predicted by general relativity for the inspiral and merger of a pair of black holes and the ringdown of the resulting single black hole. The signal was*

observed with a matched-filter signal-to-noise ratio of 24 and a false alarm rate estimated to be less than 1 event per 203,000 years, equivalent to a significance greater than 5.1 σ. The source lies at a luminosity distance of 410^{+160}_{-180} Mpc *corresponding to a redshift* $z = 0.09^{+0.03}_{-0.04}$. *In the source frame, the initial black hole masses are* $36^{+5}_{-4}M_\odot$ *and* $29^{+4}_{-4}M_\odot$, *and the final black hole mass is* $62^{+4}_{-4}M_\odot$, *with* $3.0^{+0.5}_{-0.5}M_\odot c^2$ *radiated in gravitational waves. All uncertainties define 90% credible intervals. These observations demonstrate the existence of binary stellar-mass black hole systems. This is the first direct detection of gravitational waves and the first observation of a binary black hole merger.*"

The text reproduced above is the abstract of one of most important papers in the history of science (Abbott et al. 2016), opening the field of astrophysics with gravitational waves.

The theoretical work started in the 1970s led to the understanding of GWs produced by the merging of two black holes (BHs) through the so-called "quasinormal" emission. Mathematically, the solutions of the Einstein equations foresaw complex frequencies, with the real part representing the actual frequency of the oscillation and the imaginary part representing a damping. In the 1990s higher-order post-Newtonian calculations preceded extensive analytical studies. These improvements, together with the significant contribution of *numerical relativity*, have enabled modeling of binary BH mergers and accurate predictions of their gravitational waveforms.

Binary BH mergers take place in three stages, as evident in Fig. 13.6 and sketched in the top part of Fig. 13.7. Initially, they circle their common center of mass in essentially circular orbits: because they lose orbital energy in the form of gravitational radiation, they spiral inward (*inspiral*). In the second stage (*merging*), the two objects coalesce to form a single BH. In the third stage (*ringdown*), the merged object relaxes into its equilibrium state, a Kerr BH. The LIGO/Virgo collaboration in the search for a GW signal in the data stream makes use of a formalism that defines many templates of *matched-filter signal-to-noise ratio* combining results from the post-Newtonian approach with results from perturbation theory and numerical relativity. In particular, GW emission from binary systems with $h \gtrsim 10^{-22}$, individual masses from 1 to 99 $M_\odot$, and dimensionless spins (see Sect. 13.5.5) up to $\chi = 0.99$ were searched for. For GW150914, approximately 250,000 template waveforms have been used to cover the parameter space.

We shall try to derive, through inspection of the detector data in Fig. 13.6 and the physics of GWs produced by binary systems described in the previous sections, the main results reported in the abstract.

13.5.1 Inspiral Stage

The initial inspiral phase occurs when the BHs non-relativistically rotate their common center of mass in circular orbits, as in Fig. 13.2. Thus, Newtonian

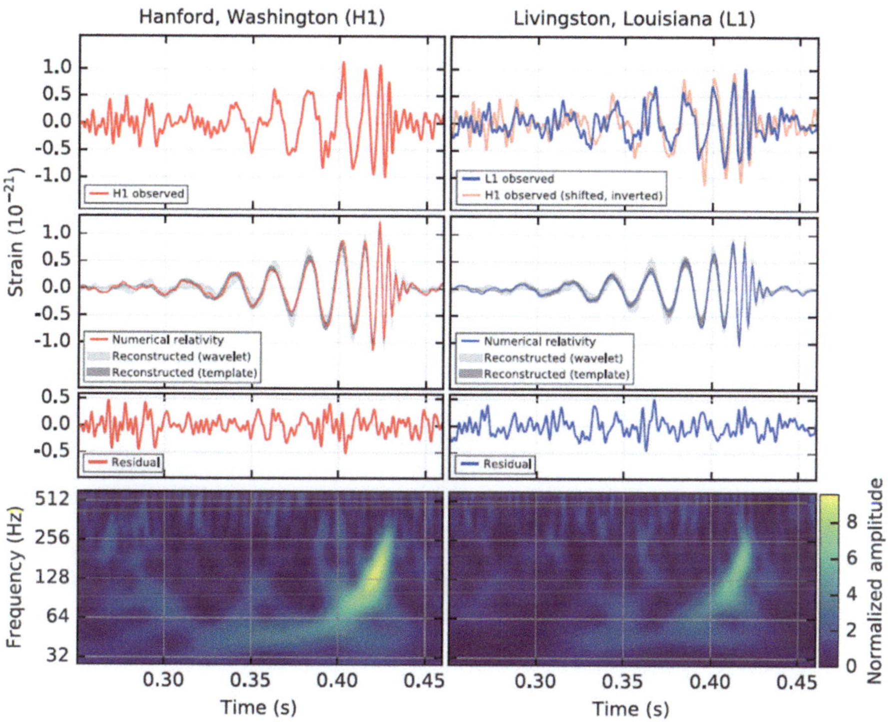

Fig. 13.6 Summary of LIGO data (Fig. 1 of Abbott et al. 2016) for GW150914. The top left (right) panel shows the strain h observed by the Hanford (Livingston) detector as a function of time. Spectral noise features in the detectors have been filtered. The second row shows a fit to the data using sine-Gaussian wavelets (light gray) and a different waveform reconstruction (dark gray). Also shown in color are the signals obtained from numerical relativity using the best-fit parameters to the data. The third row shows the residuals obtained by subtracting the numerical relativity curve from the filtered data in the first row. The fourth row gives a time-frequency representation of the data and shows the signal frequency increasing over time (the *chirp effect)*

mechanics apply and the angular frequency ω_s is related to the separation of the two black holes, R, via Kepler's third law (13.20).

Let consider now the *orbital energy* and its variation with time. The total energy is the sum of the kinetic, K, and potential, U, energies. In the gravitationally bound system of Fig. 13.2, we have

$$E_{tot} = K + U = \frac{1}{2}\mu\omega_s^2 R^2 - \frac{Gm_1m_2}{R} = -\frac{GM\mu}{2R} = -\frac{Gm_1m_2}{2R} . \tag{13.47}$$

This is a well-known equation called the *virial theorem*; in our case, it gives the total energy of the system as a function of the BH separation.

Classically, there is no gravitational radiation, and the circular orbit will persist forever. In general relativity, the orbiting BHs will emit gravitational radiation

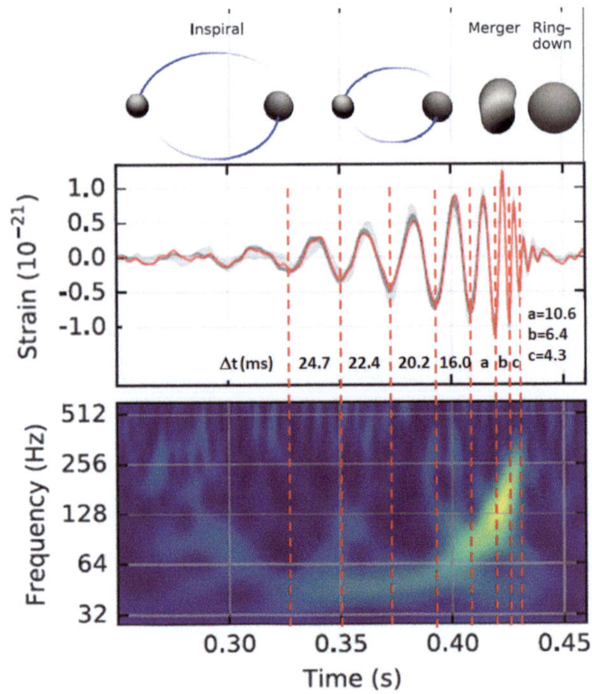

Fig. 13.7 Detail of Fig. 13.6 with the strain h observed by the Hanford detector. The vertical dashed lines are used to compute the time interval Δt (ms) between two successive minima; the values are reported in Table 13.1 and used to obtain ν_{gw} and $\dot{\nu}_{gw}$, and thus the chirp mass. On top, a sketch of the three stages of the event

thereby losing energy and spiraling towards each other.[5] At large distances between objects (or, equivalently, low ω_s), it is easy to see from (13.28) that $h_\circ$ is small and not measurable in a detector. As the BHs lose orbital energy in the form of gravitational radiation, they spiral inward. If the radius of the orbit decreases, the total energy (13.47) also decreases at a rate

$$\frac{dE_{tot}}{dt} = \frac{GM\mu}{2R^2}\frac{dR}{dt} = \frac{GM\mu}{2R}\frac{\dot{R}}{R} \tag{13.48}$$

that must be numerically equal to the power emitted as gravitational radiation, Eq. (13.36). According to Kepler's third law, the angular velocity also changes, by increasing in time, as obtained by differentiation of (13.20):

$$\frac{\dot{\omega}_s}{\omega_s} = -\frac{3}{2}\frac{\dot{R}}{R}. \tag{13.49}$$

If we want to know the mass of the system that produces the wave, we must correlate it to the observables in Fig. 13.6, namely: the measured strain h, the

[5]Circular orbits are used for simplicity, but careful analysis shows that even if the orbits were initially elliptical, emission of GWs would quickly produce circular orbits.

frequency of the wave ν_{gw}, and its derivative, $\dot{\nu}_{gw}$. Thus,

$$\frac{dE_{gw}}{dt} = -\frac{dE_{tot}}{dt} = -\frac{GM\mu}{2R}\frac{\dot{R}}{R} = \frac{GM\mu}{3R}\frac{\dot{\omega}_s}{\omega_s} \,. \tag{13.50}$$

The left-hand side of this equation can be replaced with the total luminosity $\mathscr{L}$ of the gravitational wave obtained in (13.36):

$$\frac{32}{5}\frac{G}{c^5}\mu^2 R^4 \omega_s^6 = \frac{GM\mu}{3R}\frac{\dot{\omega}_s}{\omega_s} \,, \tag{13.51}$$

which numerically depends on the masses, the radius R, the frequency and its time derivative. We can make $\dot{\omega}_s$ explicit in (13.51):

$$\dot{\omega}_s = \frac{96}{5}\frac{\mu\omega_s^7 (GM/\omega_s^2)^{5/3}}{Mc^5} \,, \tag{13.52}$$

where we removed the R^5 term using Kepler's third law. This equation can be rewritten as

$$\dot{\omega}_s^3 = \left(\frac{96}{5}\right)^3 \frac{G^5}{c^{15}}\mu^3 M^2 \omega_s^{11} = \left(\frac{96}{5}\right)^3 \frac{\omega_s^{11}}{c^{15}} \cdot (G\mathscr{M})^5 \,, \tag{13.53}$$

where the so-called *chirp mass* $\mathscr{M}$ is defined as

$$\mathscr{M} \equiv (\mu^3 M^2)^{1/5} = \frac{(m_1 m_2)^{3/5}}{(m_1 + m_2)^{1/5}} \,. \tag{13.54}$$

The value of the chirp mass is a crucial scale in the inspiral process, and it can be derived by inverting (13.53):

$$\mathscr{M} = \frac{c^3}{G}\left[\left(\frac{5}{96}\right)^3 \omega_s^{-11}\,\dot{\omega}_s^3\right]^{1/5} \,. \tag{13.55}$$

In order to obtain the chirp mass from data, it helps to rewrite Eq. (13.55) in terms of the frequency ν_{gw} of the *observed* radiation. Remembering that $\omega_{gw} = 2\omega_s$, Eq. (13.29), thus $\pi\nu_{gw} = \omega_s$. Making this substitution in Eq. (13.55), we obtain

$$\mathscr{M} = \frac{c^3}{G}\left(\frac{5}{96}\pi^{-8/3}\,\nu_{gw}^{-11/3}\,\dot{\nu}_{gw}\right)^{3/5} \,, \tag{13.56}$$

which corresponds to the only equation present in the discovery paper (Abbott et al. 2016). Equation (13.56) shows that as the BHs spiral inward, the frequency of the GW increases rapidly. This is the famous *chirp* effect, visible in the bottom panel of Fig. 13.6.

Table 13.1 The first column reports the value of the Δt between successive minima as obtained from Fig. 13.7; then, the frequency $\nu_{gw} = 1/\Delta t$ (second column) and the frequency change rate of the gravitational wave, $\dot{\nu}_{gw} = \Delta\nu_{gw}/\Delta t$ (third column)

Δt (ms)	ν_{gw} (Hz)	$\dot{\nu}_{gw}$ (Hz s^{-1})	$\nu_{gw}^{11}\ \dot{\nu}_{gw}^{3}$	$\mathcal{M}$ (kg)	$\mathcal{M}/M_{\odot}$	R (km)
24.7	40	–	–	–	–	630
22.4	45	186	4.6E−12	6.0E+31	30	590
20.2	50	241	3.2E−12	5.6E+31	28	550
16.0	63	812	9.4E−12	7.0E+31	35	470
10.6	94	3004	5.1E−12	6.2E+31	31	360
6.4	156	9673	6.7E−13	4.1E+31	21	255
4.3	233	17746	5.2E−14	2.5E+31	12	200

In the following columns: $\nu_{gw}^{11}\ \dot{\nu}_{gw}^{3}$, the chirp mass $\mathcal{M}$, and its ratio with the solar mass, $M_{\odot}$. Finally, in the last column, the distance R between the two objects evaluated with Kepler's third law (remember: $\omega_s = \pi\nu_{gw}$)

We can compute the chirp mass $\mathcal{M}$ by extracting the values of time Δt between successive minima in the strain h from Fig. 13.7, also reported in the first column of Table 13.1. Then, $\nu_{gw} = 1/\Delta t$ and $\dot{\nu}_{gw} = \Delta\nu_{gw}/\Delta t$ are reported in the second and third columns of the table. According to (13.56), the product $\nu_{gw}^{11}\ \dot{\nu}_{gw}^{3}$ (fourth column) must be constant and connected to the value of the chirp mass $\mathcal{M}$ at different phases (fifth column). Thus, the characteristic mass scale $\mathcal{M} \simeq 30M_{\odot}$ of the radiating system is obtained by direct inspection of the time-frequency behavior of data, in agreement with the value reported in Abbott et al. (2016). The last column contains the distance R between BHs during the different cycles reported in the figure. It can be noticed that R is incredibly small with respect to normal length scales for stars.

The chirp mass is a quantity that depends on the two BH masses, but by itself, it does not reveal their individual values. For identical objects ($m_1 = m_2$, likely condition for two NSs but not for BHs), then the total mass $M = m_1 + m_2$ corresponds to $M = 4^{3/5}\mathcal{M} \simeq 2.3\mathcal{M}$. More generally, the total mass of the pair has to be greater than $4^{3/5}\mathcal{M}$. In fact, if

$$m_1 = \alpha M \quad ; \quad m_2 = (1-\alpha)M \,, \tag{13.57}$$

then, from the definition (13.54)

$$M = \frac{\mathcal{M}}{[(\alpha(1-\alpha)]^{3/5}} \,. \tag{13.58}$$

The denominator is maximum for $\alpha = 1/2$, and thus M is minimum for a system with equal masses. If the two BHs in GW150914 are equal, then the minimum total mass of the system is $M = 2.3\mathcal{M} \simeq 69\ M_{\odot}$.

When the two BHs approach, the values of $\mathscr{M}$ in the last two rows of Table 13.1 significantly deviate from previous values: the validity of the Newtonian approach no longer holds, and spin effects also start to be significant. The observables in the second stage can be used to derive the values of the two individual masses.

Exercise Estimate the speed of the masses in Table 13.1).

13.5.2 *Coalescence Stage: Individual Masses*

In the second stage of the recorded signal of GW150914, both the frequency and the strain increase, and the BHs coalesce to form a single BH.

The gravitational radiation emitted during the inspiral stage can be described through the simple Newtonian approach; as the distance between objects decreases and angular velocity increases, the radiation luminosity increases, see Eq. (13.36). Thus, the computation of observables during the merger is less simple than in the inspiral stage. The merger presents a formidable problem that has only recently begun to be faced with numerical relativity.

A rational choice for the beginning of the coalescence is the moment when the separation of the two BHs is equal to the sum of their Schwarzschild radii. This can be expressed, using (6.84), as

$$\mathscr{R} = \frac{2G}{c^2}(m_1 + m_2)\,. \tag{13.59}$$

For $M = m_1 + m_2 \simeq 70M_\odot$, the corresponding Schwarzschild radius is $\mathscr{R} \simeq$ 200 km. This agrees with the minimum observable distance reported in Table 13.1. At this value of $\mathscr{R}$ and $M = m_1 + m_2$, from Kepler's third law (13.20), an angular velocity corresponds of

$$\omega_{Schw} = \frac{1}{\sqrt{8}}\frac{c^3}{GM}\,. \tag{13.60}$$

From inspection of the bottom panel of Fig. 13.7, a signal is visible up to (roughly) the half of the bin between 256 and 512 Hz. This corresponds (because of the non-linear scale) to a maximum visible frequency of the gravitational wave of

$$\nu_{gw}^{max} \simeq 330 \text{ Hz}\,. \tag{13.61}$$

By inverting (13.60) and using the maximum observable frequency to estimate $\omega_{Schw} \simeq \pi\nu_{gw}^{max}$ (always keep in mind the factor of two between the frequency of the wave and that of the system), we obtain

$$M = \frac{1}{\pi\sqrt{8}}\frac{c^3}{G\nu_{gw}^{max}} = 1.38 \times 10^{32}\text{ kg} \simeq 70M_\odot\,, \tag{13.62}$$

a value close to the minimum M. Thus, from inspection of the data at the detector on Earth (and for this reason, we add now a superscript to the values), we have obtained $\mathscr{M}^{det} = 30M_\odot$ and $M^{det} = 70M_\odot$. Those two values can be used to determine the individual masses of the BHs. Using (13.58), we derive a value of $\alpha \simeq 0.6$, and thus

$$m_1^{det} = \alpha M^{det} = 42M_\odot \quad ; \quad m_2^{det} = (1-\alpha)M^{det} = 28M_\odot \,. \tag{13.63}$$

After the correction for cosmological effects (next subsection), these values are compatible, within errors, with that obtained from the LIGO/Virgo Collaboration and reported in the abstract of the paper.

13.5.3 Luminosity Distance and Cosmological Effects

An estimate of the distance of the system can be obtained through the relation between the intrinsic and observed luminosity, as explained in Sect. 7.1. The luminosity distance D_L has been defined in Eq. (7.3) through the ratio between the effective luminosity of the object, $\mathscr{L}$, and its energy flux, $\mathscr{F}$. Neglecting, as a first approximation for GW150914, cosmological corrections (to be verified a posteriori), and using $\mathscr{F}$ from (13.37) and $\mathscr{L}$ from (13.36), we obtain

$$D_L^2 \frac{1}{2\pi}\frac{c^3}{G} h_\circ \omega_s^2 = \frac{32}{5}\frac{G}{c^5}\mu^2 R^4 \omega_s^6 \tag{13.64}$$

(always remembering that $\omega_{gw} = 2\omega_s$), and thus

$$D_L = \frac{8}{\sqrt{5}}\frac{G}{c^4}\frac{1}{h_\circ}(\mu R^2 \omega_s^4) \,. \tag{13.65}$$

Let us now insert the value determined in our computation for this event; the reduced mass corresponds to $\mu = 17M_\odot$. The values of angular velocity and distance at different Δt are reported in Table 13.1, and the strain $h_\circ$ in Fig. 13.7. We insert into Eq. (13.65) the values corresponding to $\Delta t = 16.0$ ms: $h_\circ \simeq 0.8 \times 10^{-21}$, $\omega_s = \pi \nu_{gw} = \pi\, 63\,\text{Hz} \simeq 200\,\text{s}^{-1}$, $R = 4.7 \times 10^5$ m. We obtain

$$D_L \simeq 1.1 \times 10^{25}\ \text{m} = 0.4\ \text{Gpc} \,, \tag{13.66}$$

value in agreement with the luminosity distance of 410 Mpc reported in the paper (notice the large error on this estimate).

The redshift of an object cannot be directly measured using GWs. If the source producing the GW is identified through a different measurement (as part of a multimessenger program, as we will see for the case discussed in Sect. 13.7), the redshift measured with different instruments can be used. Otherwise (as in the case of GW150914), the z can be determined assuming standard cosmology, e.g., from

Fig. 7.1. From inspection of this figure, from $D_L \simeq 400\,\text{Mpc}$, the corresponding redshift is $z \simeq 0.1$. From such (relatively) small redshift value, the relation (13.64) is affected by a correction smaller than the uncertainties on the measured quantities.

The quantity that can be measured with a relatively small uncertainty is the chirp mass, and this value can be corrected for the redshift, as shown below.

Like the electromagnetic radiation, GWs are stretched by the expansion of the Universe. This increases the wavelength (at *redshift* z), decreases the frequency of the waves detected ("det") on Earth compared to their values when emitted at the source ("s") and time intervals are "redshifted" at the location of the observer as

$$\Delta t^{det} = (1+z)\Delta t^{s} \ . \tag{13.67a}$$

Thus, redshift has the following effects on observables:

$$\nu^{det} = \frac{\nu^{s}}{1+z} \ , \tag{13.67b}$$

$$\dot{\nu}^{det} = \frac{\Delta \nu^{det}}{\Delta t^{det}} = \frac{\Delta \nu^{s}}{\Delta t^{s}} \frac{1}{(1+z)^2} = \dot{\nu}^{s} \frac{1}{(1+z)^2} \ . \tag{13.67c}$$

The effect on the chirp mass *at the source frame* can be derived using Eq. (13.56), which correspond to the *detected* value:

$$\begin{aligned}
\mathscr{M}^{det} &\propto (\nu_{gw}^{det})^{-11/3}\,(\dot{\nu}_{gw}^{det})^{3/5} \\
&= \frac{(\nu^{s})^{-11/5}}{(1+z)^{-11/5}} \frac{(\dot{\nu}^{s})^{3/5}}{(1+z)^{6/5}} = (1+z)(\nu^{s})^{-11/5}(\dot{\nu}^{s})^{3/5} \\
\mathscr{M}^{det} &= (1+z)\mathscr{M}^{s} \ .
\end{aligned} \tag{13.68}$$

Consequently, the individual masses of the involved objects as measured on Earth are scaled up by a similar factor as the chirp mass:

$$m_1^{det} = (1+z)m_1 \quad ; \quad m_2^{det} = (1+z)m_2 \ , \tag{13.69}$$

as can be easily verified from the definition of chirp mass, Eq. (13.54). The direct inspection of the detector data yields mass values from the red-shifted waves, and thus the values we derived in (13.63) must be scaled down by $(1+z)$ to obtain the values at the source frame (those reported in the abstract of the paper).

In conclusion, from the derived redshift of $z \simeq 0.1$, the masses at the source frame are about 10% smaller than that derived in (13.63) at the detector frame.

13.5.4 Total Emitted Energy

Another impressive consequence of the observation of GW150914 is the surprising amount of energy emitted in the form of gravitational radiation in binary BH mergers.

We can evaluate the total gravitational energy radiated starting from the value of the total energy of the orbiting BHs given by (13.47). We assume an initial, very large distance of the black holes, $R \to \infty$, and a final separation given by the sum of their Schwarzschild radii, Eq. (13.59). From this, we have

$$\Delta E = E^f_{tot} - E^i_{tot} = -\frac{Gm_1m_2}{2\mathscr{R}} = -\frac{Gm_1m_2c^2}{4GM} = \frac{\mu c^2}{4} \simeq 4M_\odot c^2 \qquad (13.70)$$

or 5×10^{47} J as the estimate of the total amount of gravitational wave energy radiated, in agreement with the value of $3M_\odot$ c^2 determined in Abbott et al. (2016). Equation (13.70) also shows that, for a fixed total mass $M = m_1 + m_2$, the radiated energy depends on the reduced mass μ of the system, and thus it is maximum when the merging BH masses are equal.

This enormous amount of energy is emitted, according to the waveform in Fig. 13.7, in a tenth of a second. During its 10^{10} year of lifetime, a star like the Sun is expected to convert less than 1% of its mass into light and radiation. Thus, the energy emitted by the two BHs during ∼0.1 s as GWs is ∼ 300 times as much energy as the electromagnetic radiation emitted by the Sun during its history.

13.5.5 Ringdown Stage: Spin of the BHs

The above Newtonian approximations ignore polarization of the gravitational radiation and the intrinsic angular momentum (spin) of the BHs. Their spin leads to additional velocity-dependent interactions during inspiral. Incorporation of these effects and other refinements is not straightforward in terms of an elementary presentation.

For an object with mass m and spin $\mathbf{S}$, the dimensionless spin parameter is defined as

$$\chi = \frac{c}{G}\frac{|\mathbf{S}|}{m^2}\,. \qquad (13.71)$$

The spin modifies the radius of the event horizon with respect to the Schwarzschild radius: for an object with $\chi = 1$, the event horizon corresponds to Gm/c^2, half of the value of $\mathscr{R}$ for a non-spinning BH. Thus, for two $\chi > 0$ rotating BHs, the system is more compact than for $\chi = 0$ objects. The spins of the initial BHs can be inferred using templates modelled on the inspiral data. From this, the LIGO/Virgo Collaboration determined that the spin of the primary BH (the more massive one) is constrained to have $\chi < 0.7$, while the spin of the secondary is only weakly constrained.

The effects introduced by the BH spin are more important in the third and final stage, called ringdown. During this stage, the merged object relaxes into its equilibrium state, a Kerr black hole. The ringdown process can still be analytically treated with general relativity formulas. As mentioned, during the ringdown phase, the strain h in Fig. 13.7 looks like the transients of a damped harmonic oscillator (the "quasinormal" mode). The damping rate and ringing frequency of the quasinormal mode depend only on the mass and spin of the quiescent Kerr BH that forms after the merging.

The final spin of the black hole was estimated to be $\chi = 0.67^{+0.05}_{-0.07}$. Thus, the spins of the initial BHs, determined using the inspiral data, and the spin of the final merged object, determined using a numerical analysis of the ringdown, agree with each other. Although still plagued by significant uncertainty, this result represents the first experimental test of general relativity in the hitherto inaccessible *strong field regime*, and it constitutes another significant outcome of the LIGO/Virgo discovery.

13.5.6 Source Localization in the Sky

Gravitational wave interferometers are linearly-polarized quadrupolar detectors and do not have good directional sensitivity. As a result, two antennas are necessary in order to obtain minimum directional information on the source position using the relative arrival time of the signal. The two LIGO antennas have a separation baseline of $L \sim 3 \times 10^6$ m; thus, the gravitational wave at 200 Hz (the frequency at which the signal has maximum strain) has wavelength $\lambda = 1.5 \times 10^6$ m, and thus the detector has a resolution of

$$\Delta\theta \simeq \frac{\lambda}{L} = 0.5 \text{ rad} \sim 28^\circ \,. \tag{13.72}$$

The uncertainty on the source position corresponds to about $\Delta\theta^2 \sim 800\,\text{deg}^2$. The 90% credible region mentioned in Abbott et al. (2016) corresponds to approximately $600\,\text{deg}^2$. The localization improves significantly using three detectors. By measuring the time differences in signal arrival times at various detectors in a network (triangulation technique), the $\Delta\theta^2$ reduces by an order of magnitude or more.

13.6 Astrophysics of Stellar Black Holes After GW150914

GW150914 is not the only binary BH merger observed by LIGO/Virgo. As of this writing, another five merging events (one of them is a candidate with a lower-significance) have been reported. The key parameters of the observed six events

Table 13.2 Summary table of binary BHs merger detected in O1 and O2 runs by the LIGO/Virgo Collaboration

Name	m_1	m_2	$\mathscr{M}$	M_{fin}	ΔE	χ	D_L
			$M_\odot$		$M_\odot c^2$		Mpc
GW150914	36	29	30	62	3	0.67	410
LVT151012	23	13	15.1	35	1.5	0.66	1000
GW151226	14.2	7.5	8.9	20.8	1.0	0.74	440
GW170104	31.2	19.4	21.1	48.7	2.0	0.64	880
GW170608	12	7	7.9	18	0.85	0.69	340
GW170814	30.5	25.3	24.1	53.2	2.7	0.7	540

The event labelled LVT is a lower-significance event. m_1 and m_2 are the initial masses, and M_{fin}, χ the BH mass and dimensionless spin after merging. ΔE is the total emitted energy as GWs (in unit of solar masses), and D_L is the luminosity distance

are reported in Table 13.2. Until 2016, there had only been a couple dozens of stellar BHs indirectly detected via electromagnetic radiation, mainly X-rays. The largest of them was $\sim 20 M_\odot$; the more likely mass was 5–10 $M_\odot$. The common characteristic of almost all BHs reported using GWs is that the masses are larger than expected from previous observations and theoretical astrophysical models (biased by observations). Figure 13.8 shows the distribution of the masses of stellar remnants measured in many different ways. Each observation through the merger of binary systems corresponds to three objects: the individual two before merging, and the final state. The last event announced by LIGO/Virgo, GW170608, is the lightest black hole binary system, the only one matching the pre-discovery prejudice about BH masses. The simple distribution of masses (initial and final) in the table probably requires some revisions of astrophysical models of stellar evolution.

13.7 GW170817, GRB170817A and AT 2017gfo: One Event

If sufficiently close to the Earth, the merger of two neutron stars (NSs) is predicted to produce three observable phenomena: a gravitational wave (GW) signal; a short burst of γ-rays (GRB) and, possibly, neutrinos; and a transient optical-near-infrared source. Such a transient (also called a "kilonova") would be powered by the synthesis of large amounts of very heavy elements such as gold and platinum via rapid neutron capture (the so-called astrophysical *r-process*, Sect. 12.16).

On August 17, 2017, 12:41:04 universal time (UT), the LIGO-Virgo detector network observed a GW signal from the inspiral of two low-mass compact objects consistent with a binary NS merger (GW170817). Independently, a γ-ray burst (GRB170817A) was observed less than 2 s later by the Gamma-ray Burst Monitor on board the Fermi satellite, and by the INTEGRAL satellite. This joint GW/GRB detection was followed by the most extensive worldwide observational campaign ever performed up to that time, with the use of space- and ground-based telescopes,

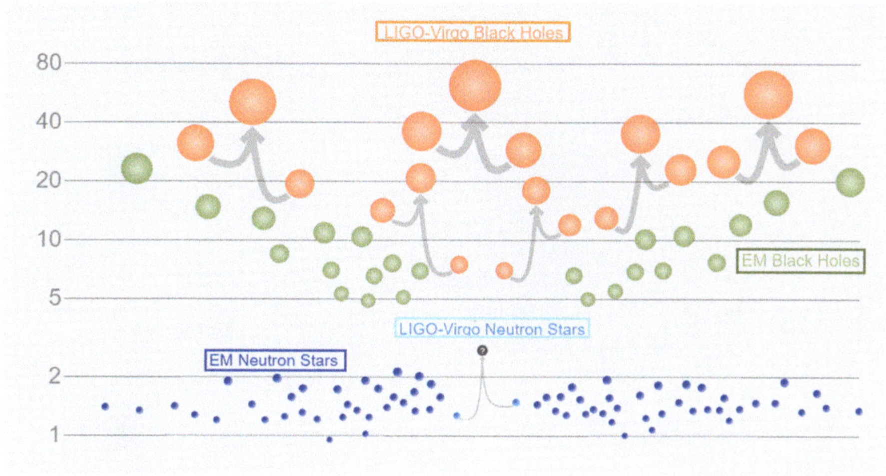

Fig. 13.8 Distribution of stellar BH and NS masses, in units of $M_\odot$. The masses for BHs detected through electromagnetic observations (Sect. 6.8) are in green; the BHs measured by gravitational-wave observations are in orange. Neutron stars measured with electromagnetic observations (Sect. 6.7.3) are in blue; the masses of the NSs that merged in GW170817 are in the center. Credit: LIGO-Virgo/Frank Elavsky/Northwestern

to scan the sky region where the event was detected, see Fig. 1.4. The localization in the sky of the GW, GRB, and optical signals is presented in Fig. 13.9. Also, underwater/ice neutrino telescopes searched for a neutrino counterpart to the signal. Less than 12 h later (when the Sun no longer illuminated the signal region), a new point-like optical source was reported by different optical telescopes. The source was located in the galaxy NGC 4993 at a distance of 40 Mpc from Earth, consistent with the luminosity distance of the GW signal. Its official designation in the International Astronomical Union (IAU) is AT 2017gfo. The source was intensively studied in the subsequent weeks with every type of traditional astronomical instrument from radio to X-rays. The interest and effort have been global: a large number of papers on different observations was published in the same issue of The Astrophysical Journal Letters (Vol. 848, n. 2) on October 20, 2017. This includes one paper describing the multimessenger observations (Abbott et al. 2017e), which was coauthored by almost 4000 physicists and astronomers from more than 900 institutions, using 70 observatories on all continents and in space.

13.7.1 GW170817

Binary NS systems produce GWs with luminosity (in the Newtonian approach) given by Eq. (13.36). As the orbit of a binary NS system gets smaller, the GW luminosity increases, accelerating the inspiral. This process has long been predicted

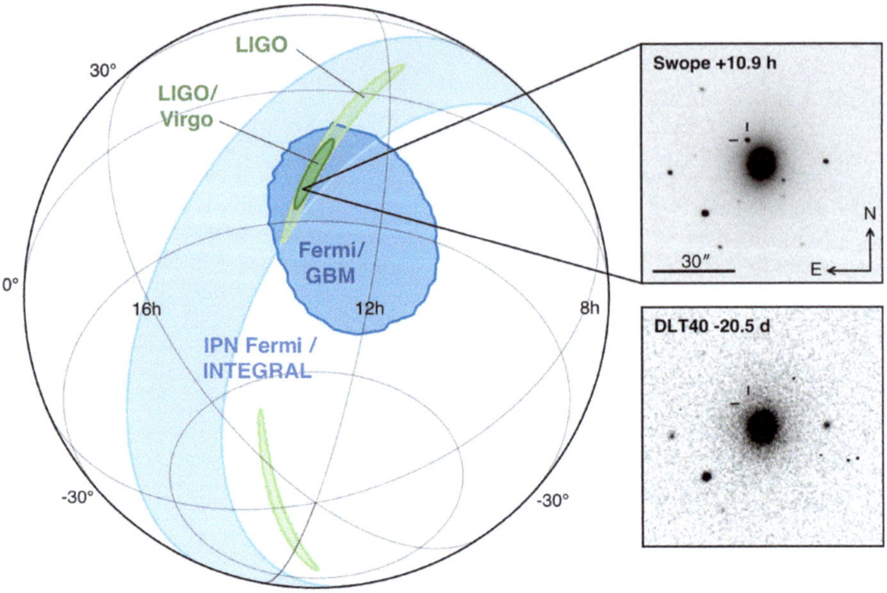

Fig. 13.9 Localization of the GW, GRB, and optical signals. On the left, the orthographic projection of the 90% credible regions from LIGO alone (190 deg^2, light green); the initial LIGO-Virgo localization (31 deg^2, dark green); the result from the triangulation from the time delay between Fermi and INTEGRAL (light blue); and Fermi-GBM (dark blue). The inset shows the location of the apparent host galaxy NGC 4993 in the Swope optical discovery image 11 h after the merger (top right). Below, the pre-discovery image from 20.5 days prior to the merger from another telescope, the DLT40 (bottom right). The reticle marks the position of the transient in both images. From Abbott et al. (2017e)

to produce a GW signal observable by ground-based detectors in the final minutes[6] before the massive objects collide. Models of the population of compact binaries have predicted a number of possible observations for the network of advanced GW detectors, ranging from $\mathcal{O}(0.1)$ to $\mathcal{O}(100)$ every year.

The first indirect observation of a binary NS system releasing energy in the form of gravitational radiation came in 1974, with the discovery of the first system with two rotating NSs by Hulse and Taylor. They found that this binary NS system was losing energy at a rate equal to that foreseen by the emission of gravitational waves.

Exercise (The Hulse and Taylor Pulsar) PSR B1913+16 is a pulsar that, together with another NS, is in orbit around a common center of mass, thus forming a binary star system. It is also known as the *Hulse-Taylor binary system* after its discoverers. The period of the orbital motion is $T = 7.7517\,\text{h}$, and the period decay with a rate of $\dot{T} = (-3.2 \pm 0.6) \times 10^{-12}\,\text{s s}^{-1}$.

[6]Note this: *minutes*. This means that GW interferometers could, under favorable circumstances, pre-alert satellites and earth-based observatories!

(1) Compute the energy emitted by the system, assuming $m_1 = m_2 = 1.4 M_\odot$ and a circular orbit. (2) Estimate the decay rate of the period, $\dot{T}$, assuming emission of GWs.
The above estimate needs to be revised to allow for the non-negligible eccentricity of the orbit, $\epsilon = 0.617$. This yields an additional multiplicative factor on $\mathscr{L}$ given by $f(\epsilon) = (1 + 7/24\epsilon^2 + 37/96\epsilon^4)(1 - \epsilon^2)^{-7/2}$ (see Schutz 2009). The factor $f(\epsilon = 0.617) = 12$ explains why the orbits of binary systems are circular before merging. The luminosity $\mathscr{L}$ depends on the angular velocity of the system to a high power, and the system rearranges its orbit to a circular one to minimize the energy loss in terms of gravitational radiation.

Toward the end of the O2 data run of aLIGO and aVirgo, a binary NS signal, GW170817, was identified by matched filtering of the data. They were based on post-Newtonian waveform models. The signal was observed for about ~100 s in the sensitive frequency band of GW interferometers (at frequency >24 Hz). During the few minutes needed by the matched filters to pick-up the signal from the data stream, a γ-ray burst (GRB) was observed and reported by satellites. The GRB occurred 1.7 s after the coalescence time, derived by the GW signal. The combination of data from GW detectors allowed for a sky position localization to an area of 28 deg^2 within a few hours, enabling the electromagnetic follow-up campaign that identified an optical counterpart in the galaxy NGC 4993.

The time evolution of the frequency of the GW emitted by a binary NS system before merging is determined primarily by the chirp mass, Eq. (13.54). We can estimate $\mathscr{M}$, according to Eq. (13.55), extracting numerical values from the time-frequency representation of the signal shown in the bottom panel of Fig. 13.10. Table 13.3 reports, for different Δt from the time of coalescence, the derived values of chirp mass $\mathscr{M}$ and radius R of the system.

As the orbital separation R approaches the size of the bodies, the gravitational wave is increasingly influenced by relativistic effects related to the mass ratio $q = m_2/m_1$, where $m_1 \geq m_2$, as well as spin-orbit and spin-spin couplings. One can note that, in the last rows of Table 13.3, the derived values of the chirp mass differ from the values at earlier times. This means that the details of the objects' internal structure become important. For neutron stars, the tidal field of the companion induces a mass-quadrupole moment and accelerates the coalescence. Tidal effects have not been considered in the above discussion of GW150914, because there is no internal structure in a BH (it has no-hairs.)

As for GW150914, the properties of the GW source have been inferred by matching the data with predicted waveforms. The results of the LIGO/Virgo collaboration, reported in Table 13.4 and discussed below, include dynamical effects from tidal interactions, point-mass spin-spin interactions, and couplings between the orbital angular momentum and the orbit-aligned dimensionless spin components of the stars, χ.

Chirp Mass Our simple Newtonian approach gives, in Table 13.3, a value of $\mathscr{M} \sim 1.1 M_\odot$ (a part the last two rows). In the detailed analysis of Abbott et al. (2017a), the chirp mass is the best-determined quantity. The value obtained from the GW phase, $\mathscr{M}^{det} = 1.1977 M_\odot$, corresponds to the detector frame, and it is related to

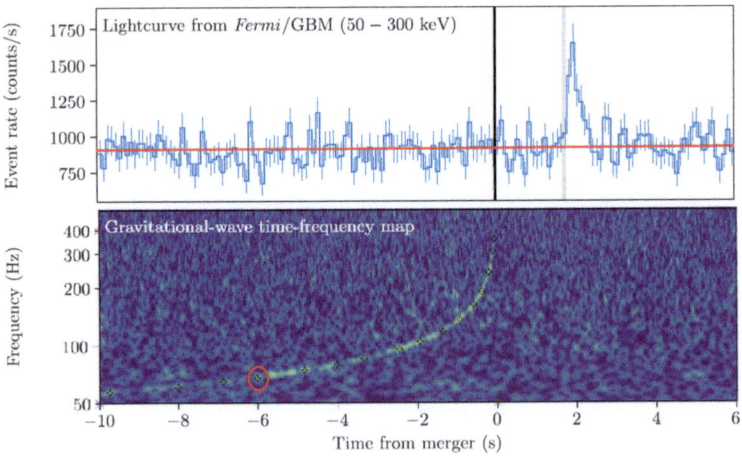

Fig. 13.10 Part of the joint, multimessenger detection of GW170817 and GRB170817A close to the merger time. Top: the summed Fermi-GBM light curve within the 50–300 keV energy range. Bottom: the time-frequency map of GW170817. All times here are in reference to the GW170817 coalescence time T_0. The markers on the bottom panel (one of which is highlighted by a red circle) are used in the present analysis to infer the frequency-time values reported in Table 13.3. Adapted from Abbott et al. (2017b)

Table 13.3 Distance Δt from time of the coalescence and frequency ν_{gw}, as obtained from the GW time-frequency map in Fig. 13.10; from the third column: the frequency change rate, $\dot{\nu}_{gw}$, evaluated as $\Delta\nu_{gw}/\Delta t_{gw}$; the value of the chirp mass, $\mathscr{M}$, as obtained from Eq. (13.56); the distance R between the two NSs evaluated with the Kepler's third law (remember: $\omega_s = \pi\nu_{gw}$)

Δt (s)	ν_{gw} (Hz)	$\dot{\nu}_{gw}$ (Hz s^{-1})	$\mathscr{M}$ (kg)	$\mathscr{M}/M_\odot$	R (km)
−9.74	57.1	–	–	–	166
−6.87	64.8	2.7	2.1E+30	1.0	153
−4.83	74.3	4.7	2.2E+30	1.1	140
−3.33	85.7	7.6	2.1E+30	1.1	127
−2.45	95.7	11.4	2.1E+30	1.1	118
−1.93	104.7	17.2	2.2E+30	1.1	111
−1.37	118.2	23.8	2.1E+30	1.0	102
−0.94	136.3	42.8	2.1E+30	1.1	93
−0.59	163.1	75.1	2.0E+30	1.0	83
−0.21	239.7	201.1	1.6E+30	0.8	64
−0.06	359.9	810.0	1.5E+30	0.7	49

value assumed at the rest-frame of the source by its redshift z, as given in (13.68). A redshift of $z = 0.008$ is derived from the luminosity distance and the cosmology parameters, which is consistent with the known distance of galaxy NGC 4993. The values of masses reported in Table 13.4 are corrected for this redshift value.

Luminosity Distance According to the discussion in Sect. 13.5.3, the luminosity distance D_L can be obtained from the masses of the system and the strain h. In the case of GW170817, $h \sim 10^{-22}$ and D_L is obtained with a 20–30% uncertainty. Refer to Eq. (13.32), which uses the values derived from this NS system.

Table 13.4 Source properties for GW170817

	$\lvert\chi_{NS}\rvert < 0.05$	$\lvert\chi_{NS}\rvert < 0.89$
Chirp mass $\mathscr{M}$	$1.188^{+0.004}_{-0.002}\ M_\odot$	$1.188^{+0.004}_{-0.002}\ M_\odot$
Luminosity distance D_L	40^{+8}_{-14} Mpc	40^{+8}_{-14} Mpc
Mass ratio $q = m_2/m_1$	0.7–1.0	0.4–1.0
Total mass $M = m_1 + m_2$	$2.74^{+0.04}_{-0.01} M_\odot$	$2.82^{+0.47}_{-0.09} M_\odot$
Primary mass m_1	1.36–1.60 $M_\odot$	1.36–2.26 $M_\odot$
Secondary mass m_2	1.17–1.36 $M_\odot$	0.86–1.36 $M_\odot$
Viewing angle Θ	$\leq 55°$	$\leq 56°$
Using NGC 4993 location	$\leq 28°$	$\leq 28°$
Tidal deformability $\Lambda(1.4 M_\odot)$	≤ 800	≤ 1400
Radiated energy E_{rad}	$> 0.025\ M_\odot c^2$	$> 0.025\ M_\odot c^2$

The central values encompass the 90% credible intervals for different assumptions of the waveform model to bound systematic uncertainty. The masses are quoted within the frame of the source, accounting for uncertainty in the source redshift. Adapted from Abbott et al. (2017e)

Individual Masses: Mass Ratio and Total Mass While $\mathscr{M}$ is well-constrained, the estimates of the component masses are affected by the degeneracy between mass ratio q and the aligned spin components of the two NSs. These latter values are very poorly constrained from the data, also combined with external information about the total angular momentum, **J**, of the system. In fact, **J** corresponds to the sum of the orbital angular momentum of the two rotating masses and the individual spins of the NSs. Due to low masses of NSs, the NS spins have little impact on the total angular momentum. While the dimensionless spin parameter (13.71) assumes values $\chi < 1$ for black holes, realistic NS equations of state typically imply $\chi < 0.7$. Thus, in Table 13.4, two different assumptions (or "priors") have been considered: a high-spin value ($\lvert\chi_{NS}\rvert \leq 0.89$) and a low-spin value ($\lvert\chi_{NS}\rvert \leq 0.05$). The mass ratio, $q = m_2/m_1$, changes according to these two priors. The central values of the total mass, M, of the system are very close in the two cases and always compatible with the presence of two equal objects with masses close to $1.4 M_\odot$.

Inclination Angle The total angular momentum, **J**, is (almost) perpendicular to the plane of the orbit. The luminosity distance is correlated with the inclination angle

$$\cos\theta_{JN} = \frac{\mathbf{J}\cdot\hat{\mathbf{N}}}{J}\,, \tag{13.73}$$

where $\hat{\mathbf{N}}$ is the unit vector from the source towards the Earth. Data are consistent with an antialigned source: $\cos\theta_{JN} \leq -0.54°$. The relevant quantity is the viewing angle

$$\Theta \equiv min(\theta_{JN};\, 180° - \theta_{JN})\,, \tag{13.74}$$

which corresponds, in this case, to $\Theta \leq 56^{\circ}$. However, since D_L can be determined using the multimessenger association with the galaxy NGC 4993, Eq. (13.73) can be further constrained to $\cos\theta_{JN} \leq -0.88^{\circ}$, and thus $\Theta \leq 28^{\circ}$.

Tidal Effects and Energy Emitted in GW Tidal effects, and in particular, the *tidal polarizability parameters*, are important because they contain information on the nuclear equation of state (EOS) for NSs.

Tides are well-known effects in the study of a planet's motions. As early as the 1910s, Augustus E. Love introduced two dimensionless parameters (k_1, k_2) to characterize the rigidity of a planetary body and the susceptibility of its shape to change in response to a tidal potential. In particular, k_2 encodes information about the body's internal structure and is defined as the ratio between the tidally-induced quadrupole moment Q_{ij} and the companion's perturbing tidal gradient (the external field). The tidal deformability (or polarizability) is

$$\Lambda = \frac{2}{3}k_2\left(\frac{c^2}{G}\frac{R}{m}\right)^5 \tag{13.75}$$

(we do not give any derivation of this; see Abbott et al. (2017a) and referred papers). Both R (the stellar radius) and k_2 are fixed for a given stellar mass m by the EOS. For neutron-star matter (according to the discussion in Abbott et al. 2017a), $k_2 \simeq 0.05$–0.15, while black holes have $k_2 = 0$. Tidal effects increasingly affect the phase of the GW and become significant above $\nu_{gw} \simeq 600$ Hz, so they are potentially observable by ground-based interferometers. Unfortunately, interferometers in the O2 run were not sufficiently sensitive above 400 Hz.

Gravitational wave observations alone are able to set a lower limit on the compactness of the NS system and provide information on the EOS through an estimate of the deformability (13.75). The values of Λ for GW170817 reported in the table disfavor EOS that predict less compact stars; objects more compact than neutron stars, such as quark stars, black holes, or more exotic objects, are not excluded. The energy emitted, E_{rad}, depends critically on the EOS. For this reason, only a lower bound on the energy emitted before the onset of strong tidal effects at $\nu_{gw} \sim 600$ Hz is derived, which is consistent with that obtained from numerical simulations.

Final State After the Collision One interesting subject (not discussed in the discovery paper or in Table 13.4) is the fate of the system after the collision (Abbott et al. 2017c). After such a merger, a compact remnant is left over whose nature depends primarily on the masses of the inspiralling objects and on the EOS of nuclear matter. This could be either a BH or a NS, with the latter being either long-lived or too massive for stability, implying delayed collapse to a BH (Fig. 13.11). Depending on the mass of the intermediate state (hypermassive NS or supramassive NS), short (< 1 s) or intermediate-duration (< 500 s) GW emission is expected. No signal was found in this case, so no particular mechanism for the formation of the final state is defined.

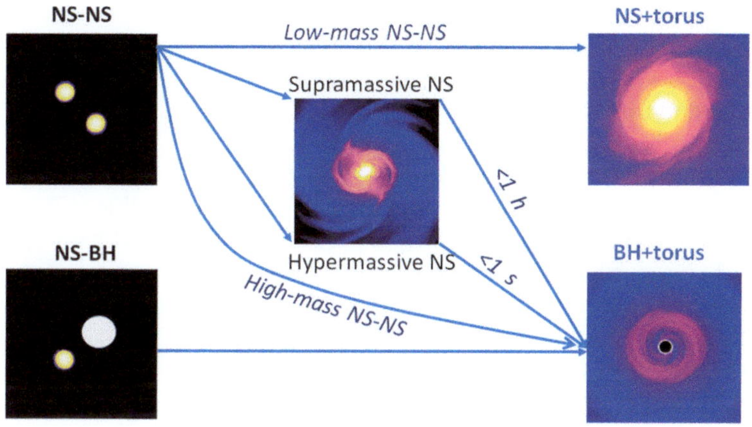

Fig. 13.11 Possible final state of a binary NS-NS or NS-BH system

This particular observation was thus not able to identify whether a NS or a small BH is formed in the final state (this is the reason for the question mark in correspondence to this event in Fig. 13.8). However, models shows that post-merger emission from a similar event may be detectable when advanced detectors reach design sensitivity or with next-generation detectors.

13.7.2 GRB170817A

Gamma-ray bursts (GRBs, Sect. 8.11) are extremely intense and relatively short bursts of gamma radiation observed by dedicated satellite experiments, coordinated in the Gamma-ray Coordinates Network (GCN).

In a GRB, after the initial flash of γ-rays, a longer-lived "afterglow" is usually emitted at longer wavelengths (X-ray, ultraviolet, optical, infrared, microwave and radio). Since the observation of first afterglow from the Beppo-SAX satellite in 1997, we have known that GRBs are of extragalactic origin and that they are the brightest electromagnetic events known to occur in the Universe.

As discussed in Sect. 8.12, GRBs are classified as *short* or *long* depending on the duration Δt of their prompt γ-ray emission. This division is based on the observed bimodal distribution of Δt and on differences in the γ-ray spectra, Fig. 8.12. This empirical division was accompanied by hypotheses that the two classes have different progenitors. While long GRBs have been firmly connected to the collapse of massive stars through the detection of associated Type Ibc core-collapse supernovae, prior to GRB170817A, the connection between short GRBs and mergers of binary NSs (or NS-BH binaries) has only been supported by weak indirect observational evidence. A GRB (either short or long) consists of a *prompt emission*, followed by several components, such as an *extended emission*,

X-ray flares, and a plateau emission, which are usually referred to as the *afterglow emission*. The prompt emission is attributed to internal energy dissipation inside the relativistic jet, whereas the afterglows are caused by forward shocks propagating in the surrounding ambient material.

GRB170817A was autonomously detected in-orbit by the Fermi-GBM flight software +1.74 s after the GW coalescence (see Fig. 13.10). A similar coincidence was observed by an instrument on board the INTEGRAL satellite.[7] The onset of γ-ray emission from a binary NS merger progenitor is predicted by models to be within a few seconds after the merger, given the expected formation time of the central engine and the jet propagation delays that are on the order of the GRB duration Δt. The unambiguous joint detection of GW and electromagnetic radiation from the same event (the occurrence of an observation by chance has an estimated probability of 5×10^{-8}) confirms that binary NS mergers are progenitors of (at least some) short GRBs.

The prompt γ-ray emission from GRB 170817A had an observed energy of $E_{iso} \sim 4 \times 10^{46}$ erg, as recorded by Fermi-GBM. The subscript "iso" means that the computation assumes that the gammas are isotropically emitted by the source. This is at least three orders of magnitude below typical observed short GRB energies. As discussed below (Sect. 13.10), a plausible explanation is the presence of a beamed emission, with the Earth off-axis with respect to the jet.

13.8 The Kilonova: Electromagnetic Follow-up of AT 2017gfo

A *kilonova* is a transient event observable through traditional astronomical methods occurring when two NSs (or a NS-BH system) merge into each other. The term kilonova (an alternative to *macronova* or *r-process supernova*) was introduced in Metzger et al. (2010) to characterize the peak brightness of the isotropic emission that reaches 10^3 times that of a classical nova. As the authors says in the abstract: "*Because of the rapid evolution and low luminosity of NS merger transients, electromagnetic counterpart searches triggered by GW detections will require close collaboration between the GW and astronomical communities. ... Because the emission produced by NS merger ejecta is powered by the formation of rare r-process elements, current optical transient surveys can directly constrain the unknown origin of the heaviest elements in the Universe*" (Metzger et al. 2010). This was exactly the situation that occurred on August 17, 2017; the details of the spectral identification and the physical properties of the bright kilonova associated with the GW170817 and GRB170817A are in Pian et al. (2017).

[7]The GCN circulars for GRB170817A/GW170817 follow-up are available at the GCN website: https://gcn.gsfc.nasa.gov/other/G288732.gcn3.

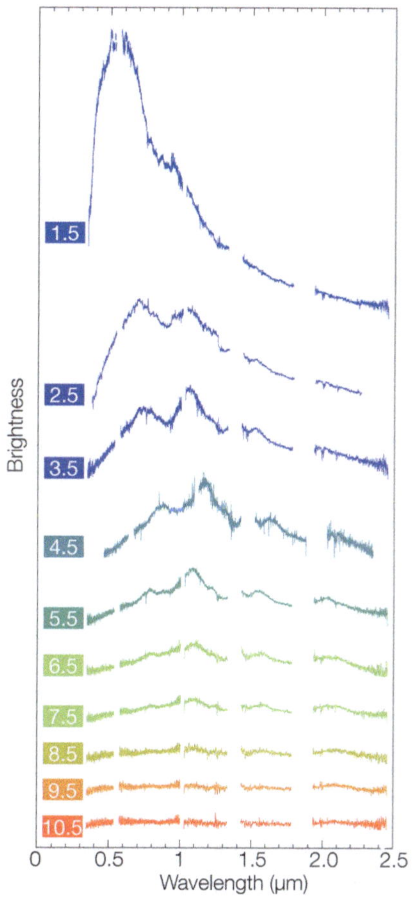

Fig. 13.12 Composition of spectra from the near-ultraviolet to the near-infrared taken using the X-shooter instrument on ESO's Very Large Telescope over a period of ~11 days after the NS merging. Credit: ESO/E. Pian et al./S. Smartt & ePESSTO

Following the joint GW/GRB detection, an extensive observational campaign across the electromagnetic spectrum was launched, leading to the discovery less than 11 h after the merger of a bright optical transient, now with the IAU identification of AT 2017gfo in the galaxy NGC 4993 (Abbott et al. 2017e). Subsequent observations targeted the object and its environment. Early ultraviolet observations revealed a blue transient that faded within 48 h. Optical and infrared observations showed a redward evolution over ~10 days.

These observations support the hypothesis that, after the merger of two NSs, a kilonova powered by the radioactive decay of r-process nuclei synthesized in the ejecta was produced. The information are derived from the series of spectra presented in Fig. 13.12 over a period of ~11 days after the NS merging from ground-based observatories covering the wavelength range from the ultraviolet to the near-infrared. This multi-wavelength campaign shows that observations are consistent with the presence of an optically thick ejecta at early stages, with a speed of $\sim 0.2c$. As the ejecta expands, broad absorption-like lines appear on the spectral continuum, indicating new atomic species synthesized by nucleosynthesis. A fraction of the newly formed nuclei is radioactive; their presence is revealed by the fact that, while decaying, they heat the ejecta. Consequently, the ejected material radiates thermally. All the atomic species present in the ejecta have various degrees of excitation and ionization; the absorption from the continuum causes the formation of lines. The models that aim to reproduce these lines assume a total explosion energy, a density profile and an abundance distribution of the ejecta. The spectral characteristics and their time evolution thus result in a good match with the expectations for kilonovae, suggesting that the merger ejected 0.03–0.05 $M_\odot$ of material, including high-opacity lanthanides (Pian et al. 2017).

The same conclusion that a minimum of 0.05 $M_\odot$ was produced under the form of heavy elements is independently derived by another analysis (Drout et al. 2017). Typical solar abundance (by mass fraction) for the r-process elements with mass number A$>$ 100 is $\sim 10^{-7}$ (see Fig. 3.7). To explain this value, in our Galaxy, the r-processes need to produce heavy elements at a rate of $\sim 3\times 10^{-7} M_\odot$ y^{-1} (Metzger et al. 2010).

If neutron star mergers dominate r-process production over other mechanisms (see Sect. 12.16), and thus if we assume that all the galactic heavy elements are produced by NS merger events, this production rate requires an event like GW170817 in our Galaxy every 20,000–80,000 years. This corresponds to a volume density of such events equal to $(1-4)\times 10^{-7}$ Mpc^{-3} yr^{-1}. At their design sensitivity, the network of laser interferometers will be able to detect binary NS mergers out to $\sim$200 Mpc, leading to a possible detection rate of 3–12 such events per year, or less than one event per year as nearby as GW170817. If this estimate is correct, in the following few years, we will have an answer about the long-lasting problem of the origin of the heavy elements. On the other hand, if the observed rate of GW170817-like events were to be larger, some refinements regarding the theoretical models would be necessary. If the GW interferometers end up observing fewer events, other r-process mechanisms will probably have to be considered.

13.9 Perspectives for Observational Cosmology After GW170817

GW170817 represents the first event for which both gravitational and electromagnetic waves from a single astrophysical source have been observed, thereby also opening new perspectives in fields other from astrophysics, as discussed in Abbott et al. (2017b). For instance, the combined observation was used to constrain the difference between the speed of light, c, and the speed of gravity, v_{gw}, by improving the previous estimate by about 14 orders of magnitude. In addition, the observation of GW170817 allowed for investigation of the equivalence principle and Lorentz invariance.

Exercise Using the information in Fig. 13.10 and the measured luminosity distance $D_L = 44$ Mpc of the source, show that the difference $v_{gw} - c$ is constrained to stay within $-3 \times 10^{-15}c$ and $+7 \times 10^{-16}c$.

In cosmology, GWs provide a novel approach to measuring the expansion of the Universe: the distance estimate using GWs is completely independent of the cosmic distance ladder derived from electromagnetic observations. The coalescence of a binary NS system, in fact, represents a "standard siren", which is the gravitational analogue of the electromagnetic "standard candle", Sect. 7.1: their intrinsic luminosity distance D_L can be inferred directly from observations (masses of compact objects and other parameters of the system). These information can be used to determine the Hubble constant H_0. In the case of GW170817, the analysis of the waveform yielded $D_L = 44$ Mpc, assuming that the sky position of GW170817 was exactly coincident with its optical counterpart. The associated uncertainty on H_0 corresponds to $\sim$ 15%, resulting from a combination of instrumental noise in the detectors and the poor determination of the inclination of the orbital plane of the binary neutron star system with respect to the Earth. To estimate H_0, the luminosity distance to NGC4993 was combined with the galaxy's radial velocity, a quantity affected (after correction for the *peculiar velocity* due to local irregularities and "clumpiness") by the Hubble expansion.

The obtained value of $70\,\mathrm{km\,s^{-1}\,Mpc^{-1}}$ (Abbott et al. 2017d) can be compared with the two state-of-the-art analyses that solely use electromagnetic data: the analysis of cosmic microwave background radiation from the Planck satellite and the SHoES analysis combining the Cepheid variable and type Ia supernovae data from the relatively nearby universe. The Planck and SHoES results are not in agreement with each other at the 95.4% probability level. Due to the large amount of uncertainty on this particular GW measurement, the derived H_0 is consistent with both the Planck and SHoES values. However, this marks an important milestone in the fundamental problem of measuring the expansion rate of the Universe, and future GW observations will be able to make increasingly precise measurements of this quantity.

13.10 GW170817: The Axis Jet, the Afterglow and Neutrinos

X-ray and radio emission were discovered at the AT 2017gfo position about 9 and 16 days after the merger, respectively. Both the X-ray and radio emission likely arise from a physical process that is distinct from the one that generates the UV/optical/near-infrared emission discussed in Sect. 13.8.

The most plausible model for the delayed X-ray and radio afterglow emission, consistent with the kilonova description of the NS merger, is the presence of an off-axis jet, that is, one pointing away from Earth. The details have still not been completely determined. The delayed X-ray and radio production are consistent with different scenarios: with the presence of a simple uniform jet observed on Earth; or

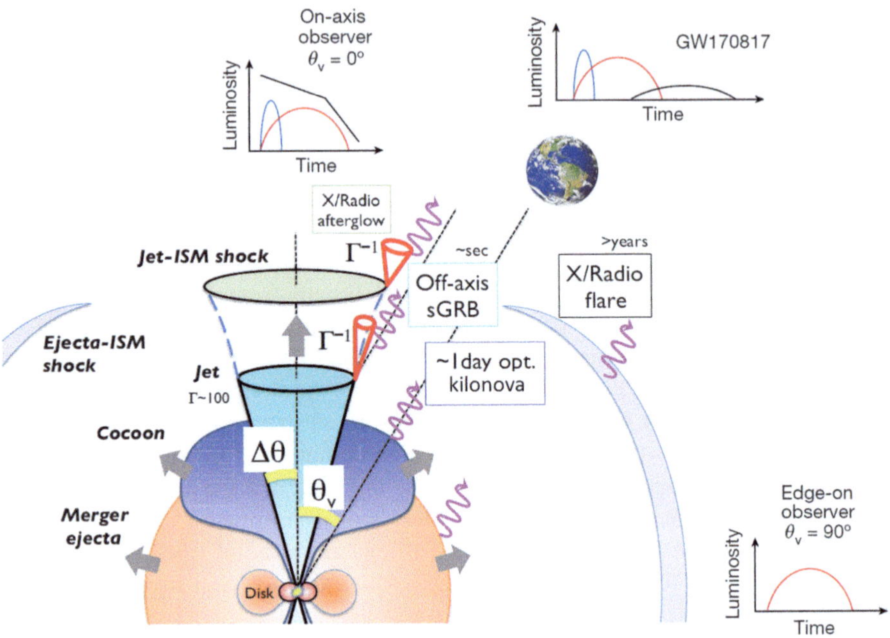

Fig. 13.13 Sketch of the geometry of GW170817 and production of electromagnetic transients. See text for details. Adapted from Metzger et al. (2010), Troja et al. (2017) and Ioka and Nakamura (2017)

with the presence of a more complex, structured jet in which the energy decreases with the angular distance from the axis; or with the presence of a *cocoon* accelerated quasi-isotropically at mildly relativistic velocities by the jet. In all cases, the Earth's location is a relatively large angle θ_v with respect to the jet axis, with a value in agreement with the GW observation given by Eq. (13.74).

Referring to Fig. 13.13, the collimated jet (black solid cone) emits synchrotron radiation visible at radio, X-ray and optical wavelengths. This afterglow emission (black line in the luminosity vs. time plots at the top of the figure) outshines all other components if the jet is seen on-axis. However, to an off-axis observer, the afterglow emission appears as a low-luminosity component delayed by several days or weeks (luminosity in the top-right plot). The jet opening angle, θ_v, is related to the Lorentz Γ factor of the particles in the jet as $\theta_v = 1/\Gamma$, see Eq. (5.64). As the jet slows, the opening angle broadens. Following the NS merger, a fast-moving *merger ejecta*, with speed of $\sim 0.2\,c$ and neutron-rich (orange shells), emits an isotropic kilonova peaking in the infrared (red lines in the luminosity-time plots). Edge-on observations ($\theta_v \sim 90°$) detect only this component. A larger mass neutron-free wind (*cocoon*) along the polar axis (blue arrows) produces an emission (blue lines in the luminosity-time plots) peaking at optical wavelengths. This emission, although isotropic, is not visible to edge-on observers, because it is only visible within a range of angles and otherwise shielded by the high-opacity ejecta.

In a GRB, neutrinos and γ-rays are expected to be produced by the central engine's activity, which results in fluctuations of the relativistic outflow, creating internal shocks in the ejecta. These internal shocks accelerate electrons and protons in the outflow through the Fermi acceleration process, Sect. 6.1.3. Shock-accelerated electrons radiate their energy through synchrotron or inverse-Compton radiation, producing γ-rays. Shock-accelerated protons interact with ambient photons and γ-rays ($p\gamma$ process), as well as with other, non-relativistic protons (pp process), producing charged pions and kaons. Secondary pions and kaons decay into high-energy neutrinos through

$$\pi^{\pm},\ k^{\pm} \rightarrow \mu^{\pm} + \nu_{\mu}(\overline{\nu}_{\mu})\ . \tag{13.76}$$

Since internal shocks in the relativistic outflow result in both γ-ray and high-energy neutrinos, the latter are expected to be produced at the same time as the GRB emission. In addition, since efficient production of neutrinos requires high target density of radiation and/or particles, typical neutrino production is likely to take place close to the central engine.

The radiation observed in the afterglow phase is mainly produced by synchrotron emission of shock-accelerated electrons. The energy distribution of protons is expected to be similar to that of electrons. Therefore, the softer emission spectrum during the afterglows indicates lower proton energies and lower neutrino production probability. However, because of the longer time for Fermi acceleration, some models predict that GRBs can accelerate protons to energies up to 10^{11} GeV. This corresponds to the maximum energy of observed charged cosmic rays. Consequently, a few ultra-high-energy neutrinos of energies $\sim 10^8$–10^9 GeV might be emitted during the afterglow phase.

In the case of the off-axis scenario of GW170817, the active neutrino telescopes (ANTARES, IceCube) and the Pierre Auger large air shower array searched for high-energy neutrino emission in time windows of ± 500 s from the coalescence time (Albert et al. 2017). The method used by the Auger array to detect neutrinos is described in Sect. 7.11. The most promising neutrino-production mechanism seems to be related to the extended γ-ray emission phase during the afterglow: the (relatively) low Lorentz factor of the expanding material results in high meson production efficiency. The models for the neutrino flux associated with the prompt GRB emission seem to be less favorable for neutrino production. Finally, a search extended to 14 days after the merger was also performed to account for neutrinos produced at the end of a (possible) acceleration of protons up to the highest energies. In all cases, no neutrino candidates have been found from this merger event.

13.11 Bursts of GWs from Stellar Gravitational Collapses

Neutron stars and stellar black holes are formed from the core collapse of an accreting white dwarf (Sect. 12.12) or the gravitational collapse of a highly massive star (Sect. 12.13). There is increasing evidence that some gravitational collapses (hypernovae and collapsars) also produce most of the observed long gamma-ray bursts, Sect. 12.13.3. Many pulsars present large measured speeds relative to their neighbors: this observation suggests that some supernovae do exhibit substantial non-spherical motion, perhaps because of dynamical instabilities in rapidly rotating, massive progenitor stars. If the collapse occurs non-spherically (a spherically symmetric explosion or implosion does not have a quadrupole moment), gravitational waves are produced.

However, core collapse supernovae produce bursts whose time evolution is theoretically poorly known. Although computer simulations are available (Sect. 12.13.1), predictions of strains h of the produced gravitational waves remain subject to large uncertainties. Thus, although algorithms for searching for bursts of GWs have been developed, they are necessarily less sensitive than matched-filter approaches, in which known phase evolution can be exploited.

To make a rough estimate of the amplitude h (following Sathyaprakash and Schutz 2009), we can start from the energy flux carried by the GW. If the GW burst can be approximated with a constant increase of the strain h from 0 to $h_\circ$, and then a similar decrease to 0, in a time window $\Delta t \sim 1 - 10\,\text{ms}$, then

$$\dot{h} \simeq 2h_\circ/\Delta t \simeq h_\circ \nu_{gw} , \tag{13.77}$$

where we used the fact that the characteristic frequency of a GW signal produced in a time interval is $\nu_{gw} \simeq 2/\Delta t$. Finally, using the relation for the luminosity distance (7.3), assuming that $\mathscr{L} = E/\Delta t$, and from the energy flux Eq. (13.18), we obtain (neglecting the numerical factor)

$$h_\circ \simeq \frac{(G/c^3)^{1/2}}{D_L \nu_{gw}} \sqrt{\frac{E}{\Delta T}} . \tag{13.78}$$

Here, E is the total energy radiated as gravitational waves. According to simulations (see Sect. 12.13.1), for a massive star of ten solar masses, $E \sim (10^{-7} - 10^{-5}) \times 10 M_\odot$.

Using representative values for a supernova burst in the central region of our Galaxy, at $D_L = 10\,\text{kpc}$, lasting for 1 ms, emitting the (conservative) energy equivalent of $10^{-6} M_\odot$ at a frequency of 1 kHz, the strain amplitude (13.78) would be

$$h_\circ \simeq 10^{-20} \left(\frac{E}{10^{-6} M_\odot}\right)^{1/2} \left(\frac{1\,\text{ms}}{\Delta t}\right)^{1/2} \left(\frac{1\,\text{kHz}}{\nu_{gw}}\right) \left(\frac{10\,\text{kpc}}{D_L}\right) . \tag{13.79}$$

This amplitude is large enough for current ground-based detectors to observe a galactic supernova with a reasonably high confidence. As we discussed in Sect. 12.14, the event rate within 10 kpc is expected to be far too small to make an early detection likely. Supernovae of Type II are believed to occur at a rate of three per century in a galaxy similar to the Milky Way. The Virgo supercluster has diameter of $\sim$ 30 Mpc and contains about 2500 large galaxies (see Sect. 7.2), thus we might expect an event rate of $\sim$50 per year. Hypernova events are considerably rarer. Note that until one reaches the Andromeda galaxy ($\sim$800 kpc), there is relatively little additional stellar mass beyond the edge of the Milky Way: nearby dwarf galaxies (such as the Large and Small Magellanic Clouds) contribute only a few percent additional mass. At the distance of Andromeda, the strain (13.79) decreases by about two orders of magnitude with respect to a Galactic event.

Due to the large uncertainties about modeling stain amplitudes, it seems to be difficult to effect an easy detection of gravitational waves from SN bursts over a short timescale. Maybe the multimessenger astrophysics approach can work in the opposite direction in this case: if a supernova is observed optically in a galaxy sufficiently close to us, the gravitational wave imprint can be extracted off-line from the data. This requires the localization in space of the event and, with a greater degree of difficulty, a temporal localization of the event. In this case, we can learn about the burst mechanism by analyzing the gravitational wave strain $h(t)$.

The LIGO and Virgo detectors stopped data acquisition at the end of August 2017 so that their sensitivities might be further improved. A new observing run, O3, is expected to begin early in 2019. We are now prepared for new discoveries.

References

B.P. Abbott et al., Observation of gravitational waves from a binary black hole merger. Phys. Rev. Lett. **116**, 061102 (2016)

B.P. Abbott et al., GW170817: observation of gravitational waves from a binary neutron star inspiral. Phys. Rev. Lett. **119**, 161101 (2017a)

B.P. Abbott et al., Gravitational waves and gamma-rays from a binary neutron star merger: GW170817 and GRB 170817A. Astrophys. J. Lett. **848**, L13 (2017b)

B.P. Abbott et al., Search for post-merger gravitational waves from the remnant of the binary neutron star merger GW170817. Astrophys. J. **851**, L16 (2017c). arXiv:1710.09320

B.P. Abbott et al., A gravitational-wave standard siren measurement of the Hubble constant. Nature **551**, 85 (2017d). arXiv:1710.05835

B.P. Abbott et al., Multi-messenger observations of a binary neutron star merger. Astrophys. J. Lett. **848**, L12 (2017e)

A. Albert et al., Search for high-energy neutrinos from binary neutron star merger GW170817 with ANTARES, IceCube, and the Pierre Auger observatory. Astrophys. J. **850**, L35 (2017). arXiv:1710.05839

J.L. Cervantes-Cota, S. Galindo-Uribarri, G.F. Smoot, A brief history of gravitational waves. Hist. Philos. Phys. **2**, 22 (2016). arXiv:1609.09400

C. Cheng et al., A brief history of gravitational wave research. Chin. J. Phys. **55**, 142–169 (2017)

M.R. Drout et al., Light curves of the neutron star merger GW170817/SSS17a: implications for r-process nucleosynthesis. Science (2017). https://doi.org/10.1126/science.aaq0049. arXiv:1710.05443

R.P. Feynman, R.B. Leighton, M. Sands, *The Feynman Lectures on Physics*, vol. 2, Chap. 31 (Addison-Wesley, New York, 1964)

J.B. Hartle, *Gravity - An Introduction to Einstein's General Relativity* (Addison Wesley, San Francisco, 2003)

K. Ioka, T. Nakamura, Can an off-axis gamma-ray burst jet in GW170817 explain all the electromagnetic counterparts? (2017). arXiv:1710.05905

D. Kennefick, Traveling at the speed of thought, in *Einstein and the Quest for Gravitational Waves* (Princeton University Press, Princeton, 2007). ISBN: 9780691117270

C. Kittel, W.D. Knight, M.A. Ruderman, *Mechanics (Berkeley Physics Course)*, vol. 1, Chap. 8 (McGraw, New York, 1965)

LIGO Scientific and VIRGO Collaborations, The basic physics of the binary black hole merger GW150914. Ann. Phys. **529**(1-2), 1600209 (2017). https://doi.org/10.1002/andp.201600209

M. Maggiore, *Gravitational Waves: Theory and Experiments* (Oxford University Press, Oxford, 2007). ISBN-13: 978-0198570745

D.V. Martynov et al., The sensitivity of the advanced LIGO detectors at the beginning of gravitational wave astronomy. Phys. Rev. D **93**, 112004 (2016). arXiv:1604.00439

H. Mathur, K. Brown, A. Lowenstein, An analysis of the LIGO discovery based on introductory physics. Am. J. Phys. **85**, 676 (2017)

B.D. Metzger et al., Electromagnetic counterparts of compact object mergers powered by the radioactive decay of r-process nuclei. Mon. Not. R. Astron. Soc. **406**, 2650 (2010). arXiv:1001.5029

E. Pian et al., Spectroscopic identification of r-process nucleosynthesis in a double neutron-star merger. Nature **551**, 67-70 (2017). https://doi.org/10.1038/nature24298. arXiv:1710.05858

F. Pirani, On the physical significance of the Riemann tensor. Gen. Relativ. Gravit. **41**, 1215–1232 (2009); republication of the 1956 paper

B.S. Sathyaprakash, B.F. Schutz, Physics, astrophysics and cosmology with gravitational waves. Living Rev. Relativ. **12**, 2 (2009). arXiv:0903.0338

P.R. Saulson, *Fundamentals of Interferometric Gravitational Wave Detectors* (World Scientific, Singapore, 1994)

B. Schutz, *A First Course in General Relativity* (Cambridge University Press, Cambridge, 2009). ISBN-13: 978-0-511-53995-4

E. Troja et al., The X-ray counterpart to the gravitational-wave event GW170817. Nature **551**, 71–74 (2017). https://doi.org/10.1038/nature24290

J. Weber, Detection and generation of gravitational waves. Phys. Rev. **117**, 306–313 (1960)

Chapter 14
Microcosm and Macrocosm

Abstract Particles and their fundamental interactions, astrophysics, and cosmology have become closely related fields. The submicroscopic phenomena allow us to better understand the cosmic evolution, and vice versa. The theory of the electromagnetic, weak, and strong interactions, which mediate the dynamics of the known subatomic particles, is called the Standard Model (SM) of particle physics. The structure of the SM suggests the existence of a Grand Unified Theory (GUT) at a very-high energy scale. Are all interactions (eventually excluding gravitation) really unified at high energies? Which symmetry governs this unification? Most likely, no answer can be provided by accelerator experiments, while some experimental tests of GUT predictions, such as the searches for baryon number nonconservation and the existence of relic particles from the Big Bang, are performed in underground laboratories. The subject within which particle physics, astrophysics, and cosmology are more strictly correlated is that connected with dark matter. The combination of many observations, including galactic rotation curves, gravitational lensing, the cosmic microwave background, and primordial light element abundances, cannot be explained without new, non-SM objects, which may annihilate or decay to ordinary particles detectable far from their source or be scattered by ordinary matter. Although there are other motivations for physics beyond the Standard Model, astrophysics and cosmology give direct evidence for new physics, thus making the search for signatures of dark matter particles an especially compelling area of research. Many dedicated experimental searches (also described in this chapter) have been developed. No conclusive results have been obtained so far from these experiments, nor for signals of physics beyond the Standard Model at accelerators. The next decade will probably be decisive concerning the solution of this joint astroparticle physics-cosmology problem.

Particles and their fundamental interactions, astrophysics, and cosmology have become closely related fields. The submicroscopic phenomena allow us to better understand the cosmic evolution, Sect. 14.1, and vice versa. When the Universe was very small, it could be considered as a hot gas of highly energetic particles. As it expanded (in four dimensions), it cooled down (the average energy of its

© Springer Nature Switzerland AG 2018

M. Spurio, *Probes of Multimessenger Astrophysics*, Astronomy
and Astrophysics Library, https://doi.org/10.1007/978-3-319-96854-4_14

constituents decreased) and passed through several phase transitions in which the nature of the particles involved in the "gas of the Universe" changed.

The theory of the electromagnetic, weak, and strong interactions, which mediate the dynamics of the known subatomic particles, is called the *Standard Model (SM) of particle physics*. The predictions of the SM have been verified with great precision, particularly at LEP and LHC experiments. The SM explains the weak and electromagnetic unification at the electroweak energy scale, $\mathcal{O}(100\,\text{GeV})$. The gauge structure of the SM suggests the existence of a Grand Unified Theory (GUT) of strong and electroweak interactions at an energy scale of roughly 10^{16} GeV, Sect. 14.2. Are all interactions (eventually excluding gravitation) really unified at high energies? Which symmetry governs this unification? Most likely, no answer can be provided by accelerator experiments, while some experimental tests of GUT predictions, such as the searches for baryon number nonconservation and the existence of relic particles from the Big Bang, are performed in underground laboratories.

The subject within which particle physics, astrophysics, and cosmology are more strictly correlated is that connected with dark matter (DM), Sects. 14.3 and 14.4. The combination of many observations, including galactic rotation curves, gravitational lensing, the cosmic microwave background (CMB), and primordial light element abundances, cannot be explained without new, non-SM objects, which may annihilate or decay to ordinary particles detectable far from their source or be scattered by ordinary matter. Although there are other motivations for physics beyond the Standard Model, astrophysics and cosmology give direct evidence for new physics, thus making the search for signatures of DM particles an especially compelling area of research.

Supersymmetric (SUSY) theories, Sect. 14.5, provide perhaps the most promising candidates for solving the DM enigma, based on a common paradigm for particle physics and cosmology. Although speculative, supersymmetric DM is very well motivated and based on a simple physical principle. This "coincidence" between cosmological observations and new physics at the electroweak scale is highly suggestive and has stimulated a large experimental effort over the last 20 years on the search for a weakly interacting, stable, and massive particle (the WIMP).

The two processes that can underline the presence of DM in the Universe are annihilation of WIMP pairs and their elastic scattering with ordinary matter, Sect. 14.6. Based on the WIMP-matter cross-section, Sect. 14.7, dedicated experimental searches have been developed in the so-called *direct detection methods* Sect. 14.8, and *indirect detection methods*, Sect. 14.9. No conclusive results have been obtained so far from these experiments, nor for signals of physics beyond the Standard Model at accelerators. The next decade will probably be decisive in regard to the solution of this joint astroparticle physics-cosmology problem, Sect. 14.10.

14.1 The Standard Model of the Microcosm: The Big Bang

The fundamental scenario that describes the Universe as a system evolving from a highly compressed state is the so-called *Big Bang model*. The expansion of the Universe originated around 10^{10} years ago from a primordial event, the "Big Bang", from a space-time singularity of very high density and temperature. Starting from the Big Bang, the Universe can be considered as a gas of particles rapidly expanding. Three fundamental ingredients are needed to build a cosmological model: (1) the equations relating the geometry of the Universe with its matter and energy content. These are provided by the Einstein field equations. (2) The metrics, describing the symmetries of the problem. The properties of statistical homogeneity and isotropy of the Universe greatly simplify the mathematical analysis. (3) The equation of state, specifying the physical properties of the matter and energy content. Particle physics enters into this last aspect. The theory of fundamental interactions (the Standard Model of particle physics) and its possible extensions have been applied to describe the evolution of the Universe immediately following the Big Bang.

Einstein initially introduced a *cosmological constant* term Λ in his Eq. (13.3) by replacing the curvature scalar term $R/2 \longrightarrow (R/2 - \Lambda)$. This was done with the intention to obtain a stationary solution. The term Λ represents a "vacuum energy" associated with space-time itself, rather than its matter content, that is, a source of a gravitational field even in the absence of matter. Einstein removed the Λ term when the expansion of the Universe was discovered.

Assuming that the matter content can be treated as a perfect fluid, the Einstein field equations simplify to the Friedmann-Lemaître equations, which closely connect the evolution and the curvature of the Universe with its matter density. The matter density is defined in terms of the *critical density of the Universe*, ρ_c:

$$\rho_c \equiv \frac{3H^2}{8\pi G_N} = 1.88 \times 10^{-29} h^2 \text{ [g cm}^{-3}] = 1.05 \times 10^{-5} h^2 \text{ [GeV cm}^{-3}] , \tag{14.1}$$

where the scaled Hubble parameter, $h \sim 0.7$, is defined in terms of the Hubble constant $H \equiv 100\, h$ km s^{-1} Mpc^{-1}, Sect. 7.1. The complete state of the homogeneous Universe can be described by giving the current values of all the density parameters $\Omega_i \equiv \rho_i/\rho_c$ for the various matter species, and reintroducing a term Ω_Λ for the cosmological constant. Thus, the Friedmann equation describing the time evolution of a scalar quantity representing the size of the Universe, $R = R(t)$, can be written as:

$$\sum_i \Omega_i + \Omega_\Lambda - 1 = \frac{k}{R^2 H^2} . \tag{14.2}$$

Its ultimate evolution is determined by the constant k (called the *curvature constant*) that, in turn, depends on $\Omega = \sum_i \Omega_i + \Omega_\Lambda$. For $k = +1$ ($\Omega > 1$), the Universe will recollapse in a finite time, whereas for $k = 0, -1$ ($\Omega = 1$ or < 1, respectively), the

Universe will expand indefinitely. Typically, contributions to Ω arise from baryons (Ω_b), photons (Ω_γ), neutrinos (Ω_ν), and cold dark matter (Ω_c).

The knowledge of these parameters also allows us to track the history of the Universe back in time, at least until an epoch when interactions allow interchanges between the densities of the different species. This interchange is believed to have happened until the neutrino-decoupling epoch, shortly before Big Bang Nucleosynthesis. To look further back into the history of the Universe is only a matter of guessing at present.

The goal of observational cosmology is to make use of astrophysical information to derive cosmological parameters, Sect. 7.1. Starting from 1998, two teams of astrophysicist began to study the distance-redshift relation by observing Type Ia SNe, Sect. 12.12. Although not perfect standard candles, their luminosity at maximum brightness is used as an efficient distance indicator, see Fig. 7.1. The two teams found that very far galaxies are moving away from us more slowly than expected from the Hubble law. Saul Perlmutter, Brian P. Schmidt and Adam G. Riess were awarded the Nobel Prize in 2011 following this discovery.

In the Big Bang model, the attraction due to gravity slows down the motion of bodies that are moving away from each other. Therefore, young "objects" should move faster away from us than older ones. The results from Type Ia SNe indicate the opposite, namely, that the Universe is now expanding faster than in the past. This is an outstanding result, which has forced the reintroduction of something similar to the " vacuum energy" term in the Einstein equations. The origins of this vacuum energy contribution are not understood in the Standard Models of the microcosm and macrocosm.

One of the major achievements of experimental cosmology is the series of precision measurements of the cosmic microwave background (CMB). The CMB radiation is well explained as radiation left over from an early stage in the development of the Universe. The first nuclei were formed about 3 min after the Big Bang, through a process called *nucleosynthesis*. It was then that hydrogen and helium (and traces of heavier Li, Be and B) nuclei formed. The key parameter necessary to calculate the effects of Big Bang nucleosynthesis is the ratio $n_{\gamma/p}$ of photons to baryon.

After nucleosynthesis, the energy of particles and radiation was so high that neutral atoms could not be formed. Charged particles were in a plasma state. As the Universe expanded, both the plasma and the radiation cooled. When the energy of the photons in the radiation field was not sufficient to ionize atoms, a transition phase occurred and nuclei and electrons combined to form neutral atoms (the so-called *recombination epoch*). These atoms could no longer absorb the thermal radiation, and consequently the Universe became transparent. Photons started to travel freely through space, rather than constantly being scattered by electrons and protons (the *photon decoupling* transition phase). These photons have been propagating ever since, becoming less and less energetic as the expansion of the Universe caused their wavelength to increase.

In 1992, the COBE satellite measured the black body spectrum (at a temperature of 2.725 K) of the CMB. The measurements confirmed that any deviation from a

Planck curve was smaller than 1/10,000, as expected (Mather and Smoot, Nobel laureates in 2006). The deviations from a perfect black body spectrum were first measured with experiments placed on balloons (BOOMERanG, MAXIMA).

Recently, the WMAP and Planck satellite experiments measured the temperature differences in the CMB radiation with high precision. When the temperature of the Universe was ~3000 K, electrons and protons combined to form neutral hydrogen (the so-called *recombination* process). Before this epoch, free electrons acted as glue between photons and baryons through Thomson and Coulomb scattering, so the cosmological plasma was a tightly coupled photon-lepton-baryon fluid. The "initial conditions", the energy contents of the Universe before recombination, can be studied using basic fluid mechanics equations, which express the temperature T as a function of two angular variables, $T = T(\theta, \phi)$ (Hu and Dodelson 2002). The precise observation of the CMB radiation in different directions (θ, ϕ) in the sky provides a map that allows us to determine temperature anisotropies. These are used to estimate the Universe's geometry, content, and evolution; to test the Big Bang model and the cosmic inflation theory.

The Planck satellite, in 2013 (Ade et al. 2014) released the most accurate results to date on the CMB spectrum, with a measurement of the temperature power spectrum corresponding to a precision of the angular size smaller than 0.1°. In combination with other data (see Sects. 22: Big-Bang cosmology, 23. Inflation, 24: Big-Bang nucleosynthesis, 25: Cosmological parameters of Patrignani et al. 2016/2017) Planck and WMAP observations are consistent with spatial flatness, corresponding to $k = 0$ in Eq. (14.2), or, equivalently, to $\Omega = 1$. They also provide a precise measurement of the age of the Universe, which is about 13.7 billion years old (with a 1 % error margin). This experimental scenario is well-reproduced by inflation models of the Universe's evolution, which automatically generate negligible spatial curvature with $k = 0$.

The combination of cosmological observations of CMB data gives a baryon density $\Omega_b \sim 5\,\%$ of the critical density. This value is consistent with that coming from Big Bang Nucleosynthesis. The observed baryon-to-photon ratio is equal to $\eta = n_B/n_\gamma \simeq 10^{-9}$–$10^{-10}$. The photons in the Universe are mainly the photons of the CMB radiation, with average energy of $\sim 10^{-4}$ eV. Thus, $\Omega_\gamma \lesssim 10^{-4}\Omega_b$. The density parameter in neutrinos is predicted to be $\Omega_\nu h^2 = \sum m_\nu/93\,\text{eV}$. Different upper limits on the sum of the mass of *active neutrinos* exist. Active neutrinos are those interacting through standard weak interactions. These limits give $\sum m_\nu \ll 1$ eV, and thus $\Omega_\nu \ll 1$. Active neutrinos contribute negligibly to Ω.

From these observations, the best scenario for the Universe's composition is the so-called Λ with Cold Dark Matter (ΛCDM) model. The Universe's evolution depends mainly on the cosmological constant and nonrelativistic (cold) dark matter, with a density term denoted Ω_c. The approximate values of some of the key parameters are (see Sect. 25: Cosmological parameters of Patrignani et al. (2016/2017) for more details and a different combination of experimental data)

$$\Omega_b \sim 0.05; \quad \Omega_c \sim 0.25; \quad \Omega_\Lambda \sim 0.70\,, \tag{14.3}$$

and a scaled Hubble constant $h \sim 0.70$. The spatial geometry is very close to being flat (and is usually assumed to be precisely flat). The nature of the dark energy remains a mystery.

14.2 The Standard Model of Particle Physics and Beyond

The *"Standard Model"* (SM) of the microcosm is a gauge theory in which the fundamental fermions are leptons and quarks. The SM provides an excellent description of the phenomena of the microcosm (at least until $\sqrt{s} \simeq$ few TeV), with the triumph of the discovery of the Higgs boson.

There are many reasons, however, to believe that the SM is incomplete and represents a valid theory at relatively low energies only, Sect. 1.1. For these reasons, models that contain the SM in the low energy limit were sought. Of particular interest are the models based on complete symmetries, such as those attempting a true unification between leptons and quarks in terms of a single symmetry group (*Grand Unified Theories, GUTs*) and those attempting unification between fermions and bosons, such as the *supersymmetry*. Finally, some models are even trying to include gravity (*supergravity, SUGRA*).

In the Standard Model structure, quarks and leptons are placed in separate multiplets. In the first family, there are two quarks [the (u, d)] and two leptons $[(\nu_e, e)]$. Baryon number conservation forbids proton decay. However, there is no known gauge symmetry that generates baryon number conservation. Therefore, the validity of baryon number conservation must be considered as an experimental question. On the other hand, Grand Unified Theories place quarks and leptons in the same multiplets; we may think that quarks and leptons are, at the low energies of our laboratories, different manifestations of a single particle. At very high energies, therefore, quark $\leftrightarrow$ lepton transitions are possible.

Starting from the 1980s, the search for proton decay was the main reason for developing large detectors and underground laboratories (Perkins 1984). As shown in *Extras # 7*, no proton decay events have been observed so far. Despite its beauty, the simplest GUT model is rejected.

Exercise The Kamiokande detector had a fiducial volume of 1000 tons of water. Calculate the number of protons in the detector's fiducial volume. Assuming a proton lifetime of 10^{32} years, evaluate how many protons would decay in 1000 tons of water each year.

Theories beyond the SM of fundamental interactions have been applied to describe the evolution of the Universe after the Big Bang. This model predicts that the huge initial temperature decreased as the Universe expanded (see Fig. 1.1) with an evolution of the interactions among particles. Collisions at LEP have reproduced situations that were typical some 10^{-10}–10^{-9} s after the Big Bang, while the collisions studied at the LHC ($\sqrt{s}$ = 14 TeV) correspond to typical situations of about 10^{-12}–10^{-11} s after the Big Bang. For subnuclear physicists, the first moments of the Universe represent the equivalent of a limitless energy accelerator!

Researches related to supersymmetric particles refer to the TeV scale, which is probably accessible in the LHC and/or in astroparticle experiments, as discussed in this chapter. Energies associated with the GUT theories cannot be reached with accelerators on Earth. GUTs foresee that electroweak and strong interactions were unified from $\sim 10^{-44}$ to $\sim 10^{-35}$ s after the Big Bang, in a highly symmetric Universe. As the temperature decreased, some phase transitions took place. A situation similar to that occurring for a magnetic substance happened: at a high temperature, there is no preferred direction; when the temperature decreases below the Curie point, the material loses its rotational symmetry. Magnetic domains appear: this corresponds to a more ordered phase, but with a lower degree of symmetry. Many important events probably occurred in the Universe's evolution at $t = 10^{-35}$ s, corresponding to a temperature of 10^{15} GeV ($\simeq 10^{28}$ K).

GUT theories require the existence of supermassive *magnetic monopoles* (MMs) (Giacomelli 1984). They could be created as point-like topological defects at the time of the Grand Unification symmetry, breaking into subgroups at 10^{15} GeV. These MM should have a mass m_M equal to the mass of the massive X, Y bosons, divided by the unified coupling constant α at 10^{15} GeV, $m_M \simeq m_X/\alpha \sim 10^{15}/0.03 \sim 3 \times 10^{16}$ GeV/c^2.

Many different experimental searches for magnetic monopoles have been performed (Patrizii and Spurio 2015). The largest apparatus constructed to detect GUT MMs was the MACRO experiment (Sect. 11.9). No MM candidate was observed (see *Extras # 7*) and MMs are excluded as a significant component of *dark matter* in the Universe.

14.3 Gravitational Evidence of Dark Matter

Most of the information about our Galaxy, as well as the rest of the Universe, mainly comes from the electromagnetic emission: not only in the optical band, but also in the radio, infrared, X-ray and γ-ray bands. The existence of matter in the Universe that does not emit electromagnetic radiation, the dark matter, was indirectly highlighted through its gravitational interaction with ordinary matter, which does emit electromagnetic radiation.

Deviations from trajectories expected from Newton's law of gravitation have proved very effective in deepening our understanding of the Universe. Observed anomalies were regarded in the past as an indication of the existence of unseen ("dark") objects, such as the anomalous motion of the planet Uranus, which led to the discovery of Neptune. In other situations, they induced deep revisions of the theory, as in the case of the attempt to explain the anomalies in the motion of Mercury as being due to the existence of a new planet. This interpretation failed, and the final solution had to wait for the advent of Einstein's theory of general relativity.

The modern problem of dark matter is conceptually very similar to the old problem of unseen planets. We observe in large astrophysical systems, at scales ranging from galactic to cosmological, some "anomalies" that can only be explained either by assuming the existence of a large amount of unseen dark matter or by assuming a deviation from the known laws of gravitation and general relativity.

The first hints of the presence of dark matter (in the modern sense) was inferred in 1933 by Zwicky from measurements of the velocity dispersion of the galaxies in the Coma cluster. He derived a mass-to-light ratio of around 400 solar masses per unit of solar luminosity. This ratio exceeds that observed in the solar neighborhood by two orders of magnitude. Today, the mass of a galaxy cluster can be determined via several methods, including application of the virial theorem to the observed distribution of radial velocities, by gravitational lensing, and by studying the profile of X-ray emission that traces the distribution of hot emitting gas in rich clusters.

The most robust evidence for dark matter emerges from the analysis of the revolution speeds of stars and gas clouds in the galactic halo as a function of the distance from the center of the galaxy (Trimble 1987). Spiral galaxies, like our own Galaxy, contain $\sim 10^{11}$ stars, arranged in the form of a central nucleus and a flattened rotating disk. The revolution speed of a mass m star around the center of the galaxy is determined by the condition of stable orbits, resulting from the equality of the gravitational and centrifugal force, that is,

$$\frac{G_N m M_r}{r^2} = \frac{m v^2}{r} , \tag{14.4}$$

from which

$$v(r) = \sqrt{\frac{G_N M_r}{r}} . \tag{14.5}$$

M_r is the total mass of stars and interstellar material within the distance r from the center of the galaxy. Most of the stars of a spiral galaxy are located in the central spherical bulge with radius r_s. If $\overline{\rho}$ is the average density of stars in the bulge, we have $M_r = \overline{\rho} \cdot \frac{4}{3}\pi r^3$ for $r < r_s$; it follows that

$$v(r) = \sqrt{\frac{4}{3}\pi G_N \overline{\rho}} \cdot r \propto r \quad \text{for } r < r_s . \tag{14.6a}$$

If all the galaxy mass is assumed to lie inside the bulge, then $M_r \simeq$ constant for $r > r_s$, and

$$v(r) \propto 1/\sqrt{r} \quad \text{for } r > r_s . \tag{14.6b}$$

Using the neutral hydrogen 21-cm emission line, the circular velocities of clouds of neutral hydrogen can be measured as a function of r. In almost all cases, after a rise near $r = 0$, the velocities remain constant as far as can be measured, as shown in Fig. 14.1 for the spiral galaxy NGC3198 (Begeman 1989). The luminous disk extends no further than about 5 kpc from the center of the galaxy. Observations

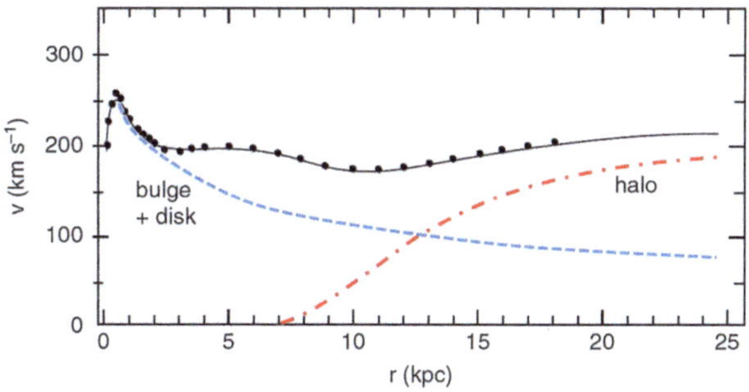

Fig. 14.1 The *points* indicate the measured rotation speeds of stars in the spiral galaxy NGC3198, versus their distance r from the galactic center. The *dashed line* represents the expected contribution based on the visible matter in the galactic nucleus (*bulge*) and in the galactic disk; the *dotted-dashed line* includes the contribution of a halo of dark matter. The *solid line* is a fit to experimental data assuming visible matter in the bulge and disk and dark matter in the halo

of tracers other than neutral hydrogen give similar results, leading to an expected revolution speed of the stars as shown by the dashed line. The figure also shows the contribution of a possible dark matter halo. The experimental data obtained using the Doppler effect are indicated with black dots. From the figure, it is evident that at high r, the measured velocities $v \simeq 200\,\text{km s}^{-1}$ are larger than those obtained using (14.5), also including the additional contribution of stars in the disk. There must be a contribution from a halo of dark matter. Similar results are obtained by measuring the speed of stars in other spiral galaxies and in elliptical galaxies.

The evidence for dark matter in the Universe from the dynamics of objects is compelling at all observed astrophysical scales, from galaxies to galaxy clusters.

14.4 Dark Matter

The ΛCDM model, whose predictions are given in Eq. (14.3), is largely accepted by cosmologists as the best description of the present data. Leaving aside for the moment the problem of dark energy, in the following, we describe one of the major issues of astroparticle physics: *what is dark matter made of*?

Several categorization schemes have been defined in the past to organize the DM candidates and to suggest possible searches. The first is the *baryonic* versus *nonbaryonic* distinction. Although Big Bang Nucleosynthesis rules out the possibility that baryonic objects contribute to the nonobserved DM, searches for baryonic DM candidates have been performed. The baryonic fraction of DM could be due to astrophysical bodies in their terminal phase (*remnants*) as white dwarfs, neutron stars and black holes. They could also be smaller objects, with insufficient mass to become stars (Jupiter-like planets). The search in galactic halos (via effects like *gravitational lensing*) indicates that such objects with $m \lesssim 0.1 M_{\odot}$ exist, but that their total mass is much smaller than that of the dark matter halo, confirming that dark baryonic objects contribute negligibly to Ω.

Among the nonbaryonic candidates, an important distinction is made between "hot" and "cold" objects. A dark-matter candidate is called "hot" if it was relativistic at the time when galaxies started to form. It is called "cold" if its motion was nonrelativistic at that time. This classification has important consequences for structure formation, and studies of galaxy formation have provided clues as to whether dark matter is hot or cold. In fact, N-body computer simulations of structure formation in a universe dominated by hot dark matter do not reproduce the observed structure. Relativistic particles smooth out (moving from high to low density regions) fluctuations, preventing the formation of larger structures from smaller ones. The N-body simulations require that the mass of thermal relics be above $\sim$1 keV.

The nonbaryonic cold dark-matter candidates are elementary massive particles that have not yet been discovered. There is no shortage of candidates (see Feng (2010) for a review), and the leading nonbaryonic cold-dark-matter candidates are sterile neutrinos, axions, Kaluza-Klein states, superheavy particles and Weakly Interacting Massive Particles (WIMPs).

A possible DM candidate is a **"sterile" neutrino**, a hypothetical particle that does not interact via weak interactions. The term "sterile" neutrino usually refers to neutrinos with right-handed chirality, which may be added to the Standard Model, to distinguish them from the known "active" neutrinos that couple with the Z^0 and $W^{\pm}$ bosons. The existence of right-handed neutrinos is theoretically well-motivated, as all other known fermions have been observed with left and right chirality, and they can explain the small value of active neutrino masses in a natural way. In some models, the introduction of a sterile neutrino also solves the matter/antimatter asymmetry in the baryogenesis. The number of sterile neutrino types is unknown and their mass could have any value between 1 eV and 10^{15} GeV. Searches for sterile neutrinos are in progress, and most experimental techniques rely on the mixing between active and sterile neutrinos that would produce particular oscillation patterns in active neutrinos.

The hypothesis that **axions** are the main DM component would also represent a possible solution to the so-called strong-CP problem. The strong CP problem (for the definition of the parity, P, and charge conjugation, C, operators, see Chapter 6 of Braibant et al. 2011) is the puzzling question as to why quantum chromodynamics (QCD) does not seem to break the CP-symmetry. There are, in fact, natural terms in the QCD Lagrangian that are able to break the CP-symmetry in the strong interactions. However, there is no experimentally known violation of the CP-symmetry in strong interactions. One of the solutions to this "fine tuning" problem involves the existence of a new scalar particle, the axion. See the Section on Axions in Patrignani et al. (2016/2017) for further details.

The **Kaluza–Klein states** rely on the concept of extra dimensions. This idea received great attention after the attempt by Kaluza, in 1921, to unify electromagnetism with gravity. Although our world appears to consist of 3+1 (three space and one time) dimensions, it is possible that other dimensions exist and appear at higher energy scales. Motivations for the study of theories with extra dimensions come from string theory, searching for a consistent theory of quantum gravity and a unified description of all interactions. It appears that such theories may require the presence of six or even more extra-dimensions. A general feature of extra-dimensional theories is that, upon "compactification" of the extra dimensions, a set of modes, called Kaluza-Klein (KK) states, appears. From our point of view in the four-dimensional world, these KK states appear as a series of states with definite masses. Each of these new states has the same quantum numbers, such as charge, color, etc.

Some **superheavy particles**, such as the GUT magnetic monopoles, have already been excluded as a major component of DM. In the past, a common motivation for superheavy dark matter candidates has come from the claim of a large excess of UHECRs in the AGASA data above the GZK cutoff, Sect. 7.5.2. As discussed, this originated top-down models in which the UHECRs are the decay product of superheavy particles. These models are ruled out by the Auger Observatory and Telescope Array results, Sect. 7.11.

The **WIMPs** represent the most investigated class of cold-dark-matter candidates. WIMPs are stable particles that arise in extensions of the Standard Model of particle physics. They are colorless (no strong interactions) and electrically neutral (no electromagnetic interactions). They interact with ordinary matter with the coupling characteristic of weak interactions, in addition to gravity. A well-motivated WIMP candidate is the Lightest Supersymmetric Particle (LSP). Predicted WIMP masses are typically within the range from 10 GeV/c^2 to few TeV/c^2.

Due to the importance of this DM candidate, we introduce in the next section how supersymmetric extensions of the Standard Model justify one of the most promising WIMP candidates, the so-called *neutralino*.

14.5 Supersymmetry (SUSY)

Super SYmmetry (SUSY) is a proposed extension of spacetime symmetry that relates bosons and fermions. Each particle from one group is associated with a particle from the other, called its superpartner (or *sparticle*), whose spin differs by a half-integer. If supersymmetric transformations were to exist, they would imply that bosons and fermions are different manifestations of a unified state. A supersymmetric operation changes the spin of particles by 1/2, leaving the electric and color charges unchanged. Supersymmetry has a cultural interest in itself; it also addresses some of the difficulties of Grand Unified Theories. Without supersymmetry, it is indeed difficult to understand why the known fundamental particles are so light with respect to the Grand Unification scale at $\sim 10^{15}$ GeV.

An example of this last type of problem is the *gauge hierarchy problem*, connected to the reason why the Higgs boson mass m_h is so small. From the known fundamental constants, one combination could be obtained with dimensions of mass, called the Planck mass

$$M_{Pl} = \sqrt{\frac{\hbar c}{G_N}} \simeq 1.2 \times 10^{19} \mathrm{GeV/c^2} \,. \tag{14.7}$$

We therefore expect parameters connected with a mass to have values of either 0, if enforced by a symmetry, or on the order of M_{Pl}. In the SM, electroweak symmetry is broken, and the Higgs boson mass is nonzero. The gauge hierarchy problem is the question of why $m_h \simeq 126\,\mathrm{GeV/c^2} \ll M_{Pl}$. The problem could possibly be solved by considering higher order corrections to Feynman diagrams. If an energy scale Λ exists at which the SM is no longer a valid description of nature, the gauge hierarchy problem may be eliminated if $\Lambda < 1$ TeV, implying new physics at the weak scale $m_{\rm weak} \sim 10$ GeV–1 TeV. Supersymmetric models offer a natural solution, and with the right energy scale for this problem.

Due to the fact that in the Standard Model there is no connection between fundamental bosons and fermions, the sparticles must be new objects. Since no sparticle has been observed up to now, a new quantum number, *R-parity*, was introduced in order to provide the supersymmetric particles with some properties that make them (currently) inaccessible. *R-parity* is equal to $R = (-1)^{3(B-L)+2S}$, where B is the baryon number, L the lepton number, and S the spin. This means that $R = 1$ for ordinary particles and $R = -1$ for their superpartners. Supersymmetric particles are expected to be heavier than any known particle. We consider SUSY models in which the lightest R-odd particle will be absolutely stable. The models predict the existence of a stable supersymmetric particle with a minimum mass (the "Lightest Supersymmetric Particle," LSP).

14.5.1 Minimal Standard Supersymmetric Model (MSSM)

Among SUSY models, the Minimal Standard Supersymmetric Model (MSSM) represents the simplest one. The MSSM is minimal in the sense that it contains the smallest possible object content necessary to give rise to all the particles of the Standard Model. The MSSM requires at least two complex doublets of Higgs bosons to generate the masses of quarks and charged leptons. Supersymmetric neutral states should mix themselves, similarly to neutrinos. The four neutral supersymmetric fermions (the sparticles of the photon, of the Z^0 and of the two neutral Higgs bosons) are not mass eigenstates. These states mix into four Majorana fermionic mass eigenstates, called *neutralinos*. A Majorana particle is equivalent to its own antiparticle. These neutralinos are indicated with $\tilde{\chi}_1^0$, $\tilde{\chi}_2^0$, $\tilde{\chi}_3^0$, $\tilde{\chi}_4^0$, in order of increasing mass. In the following, we will refer to $\tilde{\chi}_1^0$, i.e., the lightest of the four neutralinos, as the *neutralino*.

Neutralinos are expected to be extremely nonrelativistic in the present epoch. The neutralino interactions relevant for the purposes of dark matter detection are self-annihilation and the elastic scattering off nucleons. Calculations in the MSSM use the same Feynman rules as the Standard Model, adding the contribution of diagrams in which particles are replaced by their supersymmetric partners. Since, in the MSSM model, *R-parity* is a conserved quantum number, all vertices include supersymmetric partner pairs. This implies that SUSY partners are produced in pairs from normal particles, and that there is always a SUSY particle in the decay products of a SUSY particle. It is usually assumed that the LSP is the neutralino, $\tilde{\chi}_1^0$, which is neutral both in terms of electric and color charges. This is the reason that the neutralino became an important DM candidate. As the Standard Model, supersymmetric theories cannot predict the values of the particle masses.

Although relatively simple in many respects, the MSSM has a huge number of free parameters. Most of these parameters represent masses and mixing angles, as in the case of the Standard Model. To allow for a practical phenomenological study of the MSSM, the number of parameters that are considered must be reduced. This is done with theoretically well-motivated assumptions, which reduce the free parameters from more than 100 to a more tractable quantity. In many cases, experimental results can be interpreted in terms of some widely considered supersymmetric scenarios, such as mSUGRA (often called the constrained MSSM) or a phenomenologically simplified MSSM (called the phenomenological, or, pMSSM). In all models, however, the coupling of neutralinos with ordinary matter is a free and unknown parameter.

14.5.2 Cosmological Constraints and WIMP

The stable LSP[1] could have a significant cosmological abundance today. The basic idea is simple. If it exists, such a particle must have been in thermal equilibrium in the early Universe, when the temperature T exceeded the mass of the particle, $kT > m_\chi$. The equilibrium abundance was maintained by annihilation of the sparticle with its anti-sparticle $\overline{\chi}$ into lighter particles ℓ ($\chi\overline{\chi} \rightarrow \ell\overline{\ell}$), and vice versa ($\ell\overline{\ell} \rightarrow \chi\overline{\chi}$). If the LSP is a Majorana object, then $\chi = \overline{\chi}$. When, at a given time t^*, the Universe cooled to a temperature such that $kT < m_\chi c^2$, the rate Γ for the annihilation reaction falls below the Universe expansion rate $H(t^*)$ when expressed in units of [s^{-1}]. The rate $\Gamma = \sigma_\chi \cdot v \cdot n_\chi$ depends on the annihilation cross-section σ_χ, the relative velocity v between the two WIMPs and their number density n_χ. At the given time t^*, the interactions that maintained thermal equilibrium came to an

[1]The LSP is also denoted as χ. However, some of the following related discussions can also be extended to other non-SUSY WIMP candidates.

end and a relic cosmological abundance of WIMPs froze in. This condition can be expressed as

$$H(t^*) = \Gamma = \langle \sigma_\chi v \rangle n_\chi \quad [\mathrm{cm}^2][\mathrm{cm/s}][\mathrm{cm}^{-3}] \,, \tag{14.8}$$

where $\langle \sigma_\chi v \rangle$ represents the convolution of the $\chi\overline{\chi}$ annihilation cross-section times the relative velocity v over their thermal distribution spectrum. When condition (14.8) was reached, the χ's ceased to annihilate, fell out of equilibrium, and their total number in the Universe no longer changed significantly. Freezing out happened at a temperature $kT \sim m_\chi/20$, almost independently of the properties of the WIMP. Assuming this kinetic energy, the corresponding velocity is $v_{dec} \sim 0.3c$. This means that WIMPs were already nonrelativistic when they decoupled from the thermal plasma. Because of Eq. (14.8), the abundance today is inversely proportional to the WIMP annihilation cross-section σ_χ.

The present density of a generic χ particle can be derived by applying Eq. (14.8) using a Maxwell–Boltzmann distribution of the particle velocities. One can find the standard calculation, for instance, in Jungman et al. (1996). Following this computation, if such a stable particle of mass m_χ exists, its *relic abundance* (that is, the present mass density in units of the critical density ρ_c, Eq. 14.1), is given by

$$\Omega_\chi h^2 \equiv \frac{m_\chi n_\chi}{\rho_c} = \frac{3 \times 10^{-27} \mathrm{cm}^3 \mathrm{s}^{-1}}{\langle \sigma_\chi v \rangle} \,. \tag{14.9}$$

The above constraint on Ω_χ is only derived from cosmological conditions.

To reproduce the observed DM density of our Universe, Eq. (14.3), the condition $\Omega_\chi \equiv \Omega_c$ is required, and thus $\Omega_\chi h^2 \sim 0.1$. To obtain an order of magnitude estimate of the annihilation cross-section in (14.9), we assume $\langle \sigma_\chi v \rangle \sim \sigma_\chi v_{\mathrm{dec}}$, where $v_{\mathrm{dec}} \sim c/3 = 10^{10}$ cm/s is the average velocity of the WIMPs at the time of decoupling. Thus, the annihilation cross-section of our hypothetical χ particle should be on the order of

$$\sigma_\chi \sim \frac{3 \times 10^{-27} \mathrm{cm}^3 \mathrm{s}^{-1}}{0.1 \cdot v_{dec}} \simeq 3 \times 10^{-36} \mathrm{cm}^2 \,. \tag{14.10}$$

From the particle physics side, let us assume that a new, not yet detected, neutral particle with weak-scale interaction exists. The reasonable mass for this particle is that of the electro-weak symmetry breaking, i.e., $m_{EW} \sim 100\,\mathrm{GeV/c}^2$. The annihilation cross-section for such a particle has the same behavior as that of the e^+e^- annihilation in two fermions $f\overline{f}$ (for instance, the process $e^+e^- \to \mu^+\mu^-$, Chapter 9 of Braibant et al. 2011):

$$\sigma(e^+e^- \to f^+f^-) \simeq \frac{\alpha_{EM}^2 (\hbar c)^2}{s} \simeq \frac{100}{s[\mathrm{GeV}^2]} [\mathrm{nb}] \,, \tag{14.11}$$

where $\alpha_{EM} \sim 1/137$ is the electromagnetic coupling constant and s is the square of the center-of-mass energy. One may worry about the assumptions in (14.11): we are estimating the annihilation cross-section assuming *charged* particles and using the *electromagnetic* coupling constant α_{EM}! However, at energies above the Z^0 pole (∼90 GeV), the electromagnetic and weak coupling constants are unified, $\alpha_W \sim \alpha_{EM}$. The symmetry below that energy is broken, while above the annihilation proceeds through a Z^0 exchange with almost the same probability as an electromagnetic process. At the energy $\sqrt{s} = m_\chi$, our hypothetical neutral, heavy and weakly interacting particles annihilate, behaving almost like e^+e^- pairs, and (14.11) yields

$$\sigma(\chi\overline{\chi} \to \ell^+\ell^-) \simeq \frac{\alpha_W^2(\hbar c)^2}{s} \simeq \frac{100[\text{nb}]}{100^2} \simeq 10^{-35}[\text{cm}^2] \,. \tag{14.12}$$

The cross-section of such a particle is remarkably close to the value required to account for the dark matter in the Universe, Eq. (14.10). There is no a priori reason for a particle with a weak-scale interaction to have anything to do with closure density, a cosmological requirement that produces the condition (14.10). This striking coincidence suggests that if there is a stable particle associated with new physics at the electroweak scale, that WIMP would likely represent the main component of dark matter. *Se non è vero, è ben trovato.*

14.6 Interactions of WIMPs with Ordinary Matter

WIMPs must have some unknown, small but finite coupling to ordinary matter. This requirement follows from the fact that, otherwise, WIMPs would not have annihilated in the early Universe and would be unacceptably overabundant today. They can annihilate yielding *normal* particles into the final state, which are accessible to so-called *indirect experiments*, Sect. 14.9. By crossing symmetry, the amplitude for WIMPs annihilation into quarks is related to the amplitude for elastic scattering of WIMPs from quarks. Although unknown in strength, the WIMP-matter coupling has motivated different experimental strategies to search for such relic objects. These strategies rely on the small, but nonzero, coupling of WIMPs to nuclei in a detector that would provide a finite (albeit small) event rate in the so-called *direct experiments*, Sect. 14.7.

The key ingredients for an estimate of the signal rate in detection experiments are the density and the velocity distributions of WIMPs in the solar neighborhood, the WIMP-nucleon scattering cross-section, and the annihilation cross-section into different "normal" particle pairs.

There are numerous arguments that lead to the confidence that our Galaxy, like most other spiral galaxies, is embedded in a DM halo that exceeds the luminous component by about a factor of ten. The crucial quantities for experimentalists are the DM density in the region of our solar system, or the *local density* ρ_0, and the distribution of DM particle velocities. These quantities are determined by observing the rotation curves of the Galaxy with some difficulty due to the location of the Sun within it. N-body simulations suggest the existence of a universal DM density profile, with the same shape for all masses, epochs and input power spectra (Navarro et al. 1996).

The velocity is usually assumed to have a Maxwellian distribution, corresponding to an isothermal and spherical model of the DM halo. In the following, we will use its average value, $\overline{v} = \langle v^2 \rangle^{1/2}$. As canonical values for the local density and average speed (see Sect. 27. Dark Matter of Patrignani et al. (2016/2017)) we shall use:

$$\rho_0 = (0.39 \pm 0.03)\,\mathrm{GeV\,cm^{-3}} \quad \text{and} \quad \overline{v} = 220\,\mathrm{km/s}\,, \tag{14.13}$$

although there is considerable uncertainty and model-dependence in these numbers.

14.6.1 WIMPs Annihilation

The annihilation cross-section σ_χ is the relevant quantity related to the relic abundance of cosmological DM. In the assumption of a weakly interacting massive particle, the cross-section should correspond to that given in (14.10). This assumption could not hold in different DM scenarios (KK, axions, …).

WIMPs can annihilate into numerous final states. The most studied ones are those referring to the neutralino, the LSP. For a more general discussion and extension to other DM candidates, see Cirelli et al. (2011). The dominant annihilation processes are those at the lowest order in perturbation theory, with two vertices (the so-called "tree" level). All these processes are characterized by two-body final states: fermion-antifermion pairs ($f\overline{f}$), W^+W^-, Z^0Z^0, two Higgs bosons, one ordinary gauge boson and a Higgs boson, see Bertone et al. (2005); Jungman et al. (1996). Several Feynman diagrams contribute to each process, so the computation of the total annihilation cross-section is a difficult task. All the terms computed at the tree level (as well as those computed at higher orders) contain unknown supersymmetric parameters.

The annihilation of neutralinos to a fermion-antifermion pair, Fig. 14.2, has several important features. The neutralino mass is expected to be on the order of, or greater than, $10\,\mathrm{GeV/c^2}$. Thus, the annihilation channel into light fermions will always be accessible. For many interesting neutralino masses, other channels will be forbidden or suppressed, so that $f\overline{f}$ final states are often the only open channels. One important feature of this channel is the helicity constraint. Neutralinos are Majorana fermions, and in the limit of zero relative velocity, they are in a relative s-wave (i.e., null relative orbital angular momentum). Consequently, by Fermi statistics, they must have spins oppositely directed, and the total angular momentum is null. Therefore, the two fermions $f\overline{f}$ in the final state must have spins oppositely directed as well. This configuration introduces a helicity factor in the probability decay into $f\overline{f}$ that is proportional to the mass of the fermion m_f. The situation is

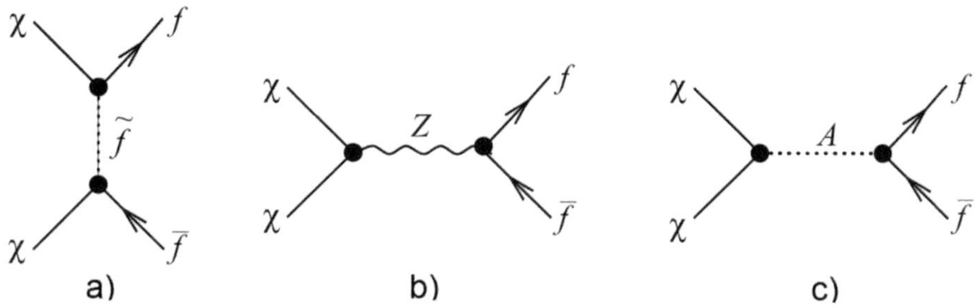

Fig. 14.2 Tree level diagrams for neutralino annihilation into fermion pairs, ($\chi\chi \rightarrow f\overline{f}$). (**a**) Annihilation through the t-channel, with the exchange of a sfermion, (**b**) s-channel exchange of a Z^0 boson, and (**c**) of a pseudoscalar Higgs boson

analogous to the decay of a spinless particle as the charged pion, as presented in § 8.10 of Braibant et al. (2011). Thus, neutralinos prevalently decay into the highest mass accessible fermions: annihilation into light quarks (i.e., u, d, s, and c) and leptons (e and μ) is negligible in comparison with annihilation into heavy quarks (i.e., b, and t) and into the τ lepton. Direct decay into massless particles is forbidden (and that into neutrino thus completely negligible). If the neutralino is heavy enough to annihilate into a top quark ($m_\chi > m_t = 173\,\text{GeV/c}^2$), then annihilation occurs essentially entirely into $t\overline{t}$ pairs in most models usually considered.

14.6.2 WIMPs Elastic Scattering

The elastic scattering of a WIMP with a nucleus in a detector can be seen as the interaction of the WIMP with a nucleus as a whole, causing it to recoil. The energy of the recoil nucleus can be measured, if large enough. The WIMP-nucleus elastic scattering cross-section is the quantity studied in direct experiments. This cross-section also determines the rate at which particles from the Galactic halo accrete onto the Sun (or other massive objects) and contributes to the signal yield in the indirect detection experiments.

The cross-section for WIMP-proton or WIMP-neutron elastic scattering (in the following, always denoted as $\sigma_0 \equiv \sigma_{\chi p} \sim \sigma_{\chi n}$) depends on the WIMP-quark interaction strength. The interaction of WIMPs with quarks and gluons (=*partons*) of the nucleon is quantified in the cross-section using Feynman diagrams, such as those shown in Fig. 14.3. For supersymmetric models, the effective interactions of neutralinos at the microscopic level depend on the masses of the exchanged particles and on other important SUSY parameters. Under general conditions, the elastic-scattering cross-section σ_0 is related to the WIMP annihilation cross-section σ_χ and

$$\sigma_0 \ll \sigma_\chi \simeq 10^{-36}\text{cm}^2 = 1\text{pb}\,. \tag{14.14}$$

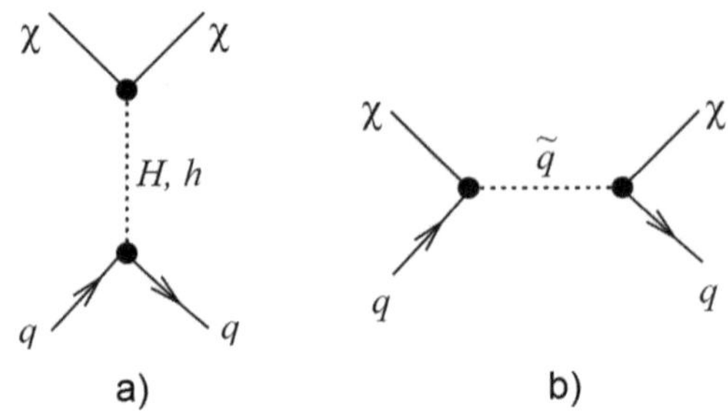

Fig. 14.3 Tree level Feynman diagrams for neutralino-quark scalar (spin-independent) elastic scattering. (**a**) t-channel exchange of a Higgs boson, (**b**) s-channel exchange of a squark

Even in simplified SUSY versions, there are typically many possible values allowed. In MSSM, in a wide scan of the model parameters, $\sigma_0 \sim 10^{-3}$–$10^{-13}\sigma_\chi$. In DM models different from SUSY, the relation between annihilation and interaction cross-sections could be more complicated. Experimental results are thus used to infer limits on σ_0, sometimes under particular conditions.

Important simplifications in the estimates of σ_0 occur when the nonrelativistic limit holds. This is exactly the situation for local WIMPs, with the average velocity given in (14.13). Only two general cases need to be considered: the spin-spin (or *spin-dependent*) interaction and the scalar (or *spin-independent*) interaction. In the case of the spin-spin interaction, the WIMP couples to the spin of the nucleus; in the case of the scalar interaction, the WIMP couples to the mass of the nucleus.

Spin-Independent Interactions In the nonrelativistic limit, the spin-independent (SI) term of the WIMP cross-section on a proton or a neutron (we assume here $m_p = m_n$) can be parameterized as

$$\sigma_0^{SI} = \frac{4m_\chi^2 m_p^2}{\pi(m_\chi + m_p)^2} f_{p,n}^2 = \frac{4m_{r_p}^2}{\pi} f_{p,n}^2 , \tag{14.15}$$

where $m_{r_p} \equiv (m_p m_\chi)/(m_p + m_\chi)$ is the reduced mass of the WIMP-proton system ($m_{r_p} \simeq m_p$ for $m_\chi \gtrsim 10\,\text{GeV/c}^2$). The quantity $f_{p,n}$ represents the WIMP spin-independent coupling to protons or neutrons, as derived from Feynman diagrams on the partons constituents. In general, as in SUSY models, this quantity depends on the coupling of the WIMP with quarks and gluons. Equation (14.15) is called *spin-independent*, SI, or *scalar* cross-section.

Usually, the interaction of a WIMP with a nucleus A is experimentally studied. Thus, the cross-sections $\sigma_{\chi A}$ must be related to σ_0, taking into account the distribution of quarks in the nucleon and the distribution of nucleons in the nucleus. To relate $\sigma_{\chi A}$ to σ_0, we need to consider the de Broglie wavelength of the scattering system. For reasonable values of the WIMP mass ($m_\chi = 10\,\text{GeV/c}^2$ to a few TeV/c^2) the average momentum transferred to a nucleus of mass $m_A \sim Am_p$ is $p = m_{r_A}\overline{v}$, where m_{r_A} is the reduced mass of the WIMP-nucleus system and $\overline{v}$ their relative velocity given in (14.13).

Assuming nuclei within the range $A = 10 - 100$, the reduced mass is $m_{r_A} \sim m_\chi$ when $m_\chi \ll m_A$, and $m_{r_A} \sim m_A$ when $m_\chi \gg m_A$. Thus, the transferred momentum p ranges between 10 and 50 MeV/c. Hence, elastic scattering occurs in the extreme nonrelativistic limit. The de Broglie wavelength corresponding to a momentum transfer of $p = 10$ MeV/c is

$$\lambda = \frac{h}{p} = \frac{hc}{pc} = \frac{197\,\text{MeV fm}}{10\,\text{MeV}} = 20\,\text{fm}\,, \tag{14.16}$$

which is larger than the radius of the nucleus $r_A \sim 1.25 A^{1/3}$ fm (see Chapter 14 of Braibant et al. 2011). The extrapolation of the scattering amplitudes (14.15) to nuclei with Z protons and $A - Z$ neutrons will then add up coherently as

$$\sigma^{SI}_{\chi A} = \frac{4m^2_{r_A}}{\pi}[Zf_p + (A-Z)f_n]^2 \simeq \left(\frac{m_{r_A}}{m_{r_p}}\right)^2 A^2 \sigma_0^{SI}\,. \tag{14.17}$$

The last equality holds if $f_p - f_n$. The main feature of the SI scattering is the increase with the squared mass number A of the target nuclei. Current experiments using heavy atoms as targets are typically dominated by spin-independent scattering.

Spin-Dependent Interactions Axial-vector interactions result from couplings of WIMP semi-integer spin to the spin content of a nucleon. The equation similar to (14.15) that holds for *spin-dependent*, SD, interactions on protons and neutrons is

$$\sigma^{SD}_{\chi A} = \frac{32m^2_r}{\pi} G_F^2 \frac{J+1}{J}[a_p\langle S_p\rangle + a_n\langle S_n\rangle]^2\,. \tag{14.18}$$

G_F is the Fermi constant and a_p, a_n are the effective WIMP-couplings to proton, $\langle S_p\rangle$, and neutron, $\langle S_n\rangle$, expectation values of the spin operators. In general, the SD cross-section is proportional to $J(J+1)$, where J is the total angular momentum of the nucleus. No relevant gain of the spin-dependent cross-section is obtained using heavy target nuclei.

Nonelastic Interactions of WIMPs include inelastic scattering, such as the interaction with orbital electrons in the target or the interactions with the target nuclei yielding an excited nuclear state. They are usually not considered, because they give signatures that have to compete with huge backgrounds of natural radioactivity.

14.7 Direct Detection of Dark Matter: Event Rates

The idea that WIMPs can be detected by elastic scattering off nuclei in a terrestrial detector, Fig. 14.4a, dates back to 1985 (Goodman and Witten 1985). It was immediately extended to include the fact that the Earth's motion around the Sun would produce an annual modulation in the expected signals (Drukier et al. 1986).

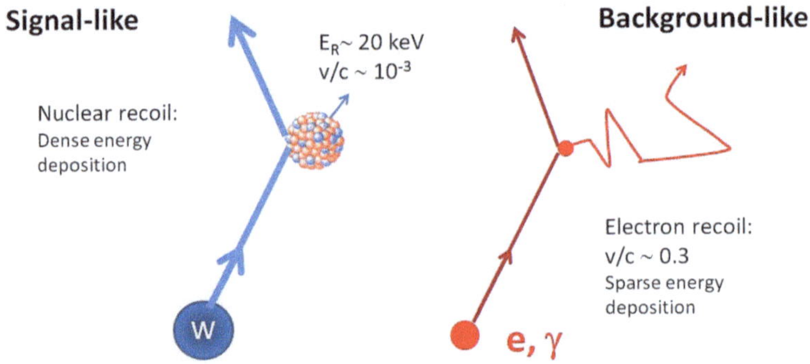

Fig. 14.4 Sketch of the elastic WIMP scattering off a nucleus. The nucleus recoil (**a**) produces excitation/ionization of the medium, different from that produced by the background interaction (**b**) of electrons/photons on the electrons of the medium

When considering a nucleus of mass $m_A \simeq Am_p$, using nonrelativistic kinematic arguments, the energy that is transferred to the recoiling nucleus is

$$E_R = \frac{p^2}{2m_A} = \frac{m_{r_A}^2 v^2}{m_A}(1 - \cos\theta), \tag{14.19}$$

where p is the transferred momentum, v the WIMP velocity and θ the scattering angle. Numerically, assuming $m_\chi = 100\,\text{GeV/c}^2$, $v \simeq \overline{v} \simeq 10^{-3}c$ and a nucleus with $A \sim 100$ such that $m_A \sim m_\chi$ and $m_{r_A} = m_\chi/2$, we obtain, at maximum $(\cos\theta = -1)$,

$$\langle E_R \rangle = \frac{1}{2} m_\chi \overline{v}^2 \sim 50\,\text{keV}\,. \tag{14.20}$$

Specialized detectors able to measure recoils of considerably lower energy, as low as a few keV, and to distinguish them from background, may make such direct detection possible. The event rate in a detector depends on many input factors: (1) the nature of the interacting particle, related to the unknown scattering cross section $\sigma_{\chi A}^{SD,SI}$; (2) the nuclear form factors of the detection material; (3) the astrophysical density distribution of WIMPs and their velocity distribution $f(v)$ in the Galactic halo; (iv) the response of the detector as a function of the nucleus recoil energy.

If the halo of our Galaxy consists of WIMPs with the density and average velocity given in (14.13), their flux on Earth is given by:

$$\Phi_\chi = \frac{\rho_0 \cdot \overline{v}}{m_\chi} \simeq 7 \times 10^4 \left(\frac{100}{m_\chi[\text{GeV/c}^2]} \right) \text{cm}^{-2}\text{s}^{-1}\,. \tag{14.21}$$

Let us compute an order-of-magnitude of the event rate, which can be simply expressed as the product of (14.21) by the cross-section on nuclei A of a homogeneous detector of mass M_T. The corresponding number of target nuclei is given by $N_T = M_T N_A/(Am_p)$, where N_A is Avogadro's number. The event rates are usually expressed for 1 kg detector, such as $N_T = 6 \times 10^{26}/A\ \mathrm{kg}^{-1}$. Assuming a spin-independent interaction, the WIMP-nucleus cross-section is given by (14.17). Thus, a first-order estimate of the event rate is

$$R = N_T \cdot \frac{\rho_0 \cdot \overline{v}}{m_\chi} \cdot \sigma^{SI}_{\chi A} = \frac{6 \times 10^{26}}{A} \cdot \left(\frac{7 \times 10^4}{m_\chi [100\,\mathrm{GeV/c^2}]}\right) \cdot \left(\frac{m_{r_A}}{m_{r_p}}\right)^2 A^2 \sigma_0 \quad \mathrm{kg}^{-1}\mathrm{s}^{-1}\,. \tag{14.22}$$

Assuming a WIMP having mass $m_\chi = 100\,\mathrm{GeV/c^2}$, cross-section $\sigma_0 = 10^{-9}\ pb = 10^{-45}\,\mathrm{cm}^2$, 1 day (=86,400 s), and using the definition of reduced mass ($m_\chi \gg m_p$) we obtain

$$R = 3.6 \times 10^{-9} \left(\frac{A^{3/2} m_\chi}{Am_p + m_\chi}\right)^2 \left(\frac{\sigma_0}{10^{-45}\mathrm{cm}^2}\right) \quad \mathrm{kg}^{-1}\mathrm{d}^{-1}\,. \tag{14.23}$$

For materials used in typical detectors, the factor depending on A in brackets is about 4.0×10^5 for Xe ($A = 131$), 1.2×10^5 for Ge ($A = 73$), and 0.3×10^5 for Ar ($A = 40$). Thus, from (14.23) the corresponding event rates are on the order of $\sim 10^{-4}$–10^{-3}/kg/day for argon and xenon, respectively.

Direct-detection experiments measure the number of signals equivalent to a given nuclear recoil E_R per day per kilogram of detector material as a function of deposited energy. It should be noted that many background sources can simulate events with a deposited energy equivalent to E_R, see Fig. 14.4b. For a real experiment, the detailed distribution of velocities $f(v)$, the nuclear form factor $F^2(E_R)$ and the detector efficiency must be considered, significantly decreasing the number of signal events. The nuclear form factor term is about unit for light nuclei, while it suppresses high energy recoils for heavy nuclei, see Fig. 14.5a, reducing the number of observable events. The nuclear form factor depends on nuclear physics properties only. The integral event rate as a function of the nuclear recoil computed for some peculiar nuclei (expressed as events/kg/year) is shown in Fig. 14.5b. For a realistic 100 kg Xe detector (such as XENON100), the number of expected events above a detection threshold of 5 (15) keV is 18 (8) events per year, assuming a cross-section $\sigma_0 = 10^{-45}\,\mathrm{cm}^2$ and a WIMP mass of $m_\chi = 100\,\mathrm{GeV/c^2}$.

In the above example, the experimental observable (the event rate) depends on m_χ and σ_0, according to (14.23). For this reason, the results of experimental searches are usually expressed as a contour in the plane of parameters of cross-section vs. WIMP mass. A closed contour expresses the range of parameters that produce the observed signal compatible with errors (in the case of a positive claim); a curve delimits the region of parameter values that are excluded, as shown in Sect. 14.8.4.

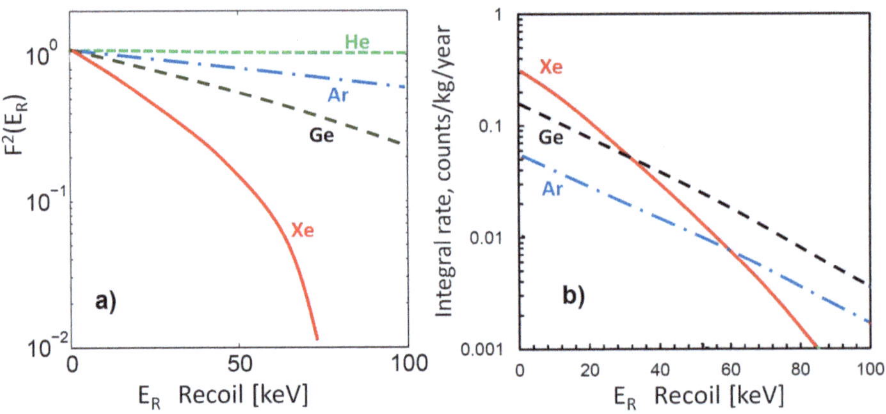

Fig. 14.5 (**a**) Nuclear form factors $F^2(E_R)$ for four different nuclei as a function of the nucleus recoil energy E_R. The more extended is the nuclear distribution, the stronger is the fall-off as a function of the recoil energy. (**b**) Integral energy spectrum of a spin-independent elastic scattering WIMP-nucleus for four different nuclei, assuming perfect energy resolution of the detector. The heavier nuclei show a higher interaction rate at low recoil energies. The assumptions are: $m_\chi = 100\,\mathrm{GeV/c^2}$, $\sigma_0 = 10^{-45}\,\mathrm{cm^2}$ and the astrophysical conditions given by (14.13)

14.8 Direct Searches for WIMPs

Direct detection experiments in astroparticle physics play a complementary role with respect to collider experiments. A new, massive, and long-lived particle eventually discovered at LHC cannot automatically be associated with a DM candidate. Accelerator detectors cannot record the cosmological abundance of an observed WIMP. On the other hand, the detection of dark matter particles in astroparticle physics experiments will not be sufficient to identify the nature of these particles conclusively.

To observe WIMPs, detectors with a low energy threshold, an ultra-low background noise and a large target mass are mandatory (Saab 2012). There is currently ongoing development of larger and more sophisticated detectors sensitive to WIMP-nucleus interactions yielding ionization and/or excitation. See Gaitskell (2004) for a review of early experimental techniques.

In a detector, the kinetic energy of a nucleus after a WIMP elastic scattering is converted into a measurable signal: depending on experimental techniques, the signal corresponds to (1) *ionization*, (2) *scintillation light*, (3) *vibration quanta (phonons)*. The main experimental problem is to distinguish the genuine nuclear recoil induced by a WIMP from the huge background due to environmental radioactivity. All detectors have a threshold energy E_R^{th} above which they are sensitive, typically within the 5–40 keV range. The simultaneous detection of two observables strengthens the discrimination against background events.

Highly granular detectors and/or with good timing and position resolution are used to distinguish the WIMP localized energy depositions. Some experimental techniques are sensitive to only a fraction of the recoil energy of the nucleus. The measured energy thus has to be converted into the true recoil energy through an energy-dependent factor called the *quenching factor*. The measured energy is often labeled in units of keVee (keV electron equivalent), while the nuclear recoil energy is often labeled with units of keV, keVr, or keVnr.

In the energy range of interest for WIMP detection (deposited energy below 100 keV), the main contributions to background are electromagnetic interactions originated from α, β and γ-decays of environmental radioactivity interacting with electrons of the medium, Fig. 14.4b. Radioactive isotopes are present in the material surrounding the detectors, in airborne contaminants that can be deposited on the surfaces, or within the detectors themselves. Neutron interactions, such as genuine WIMP signals, scatter off nuclei. Sometimes, neutron sources are used to simulate the signal. Particularly dangerous are neutrons from natural radioactivity (the so-called radiogenic neutrons, with energies below 10 MeV), or induced by α capture of a nucleus, or produced by secondary CRs. Regardless of the experimental technique, all the DM detectors are located at a deep underground site to reduce the flux of muon-induced energetic neutrons, Fig. 1.8.

Appropriate shielding with passive and/or active materials and sometimes the use of veto detectors around the experiment can significantly suppress the background event rate. Lead, copper, and other high-Z materials are necessary to absorb electrons and γ-rays, whereas water, polyethylene, and other hydrogen-rich materials are well-suited for moderating the neutron background. Direct detection experiments have made tremendous progress in the last three decades. Figure 14.6

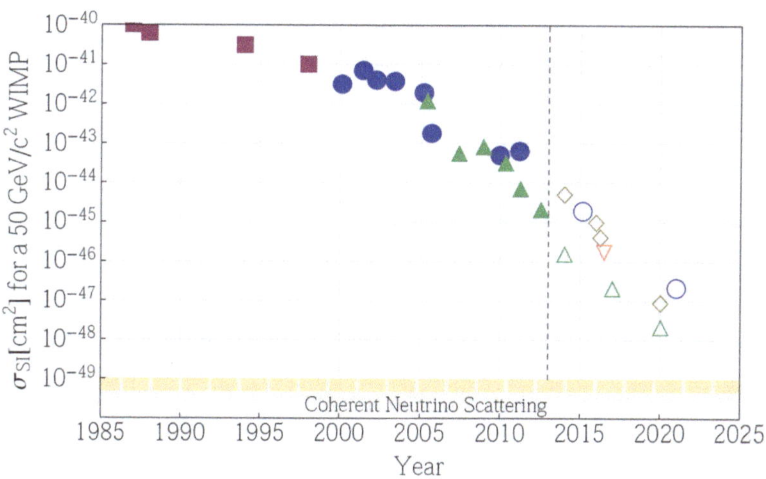

Fig. 14.6 History and projected evolution with time of SI WIMP-nucleon cross-section limits for a 50 GeV WIMP. The symbols used to denote the different technologies are: cryogenic solid state (*blue circles*), crystal detectors (*purple squares*), liquid argon (*brown diamonds*), liquid xenon (*green triangles*), and threshold detectors (*orange inverted triangle*). Below the *yellow dashed line*, WIMP sensitivity is limited by coherent neutrino-nucleus scattering. From Bauer et al. (2014)

shows the sensitivity reached on the SI cross-section as a function of year. This rapid progress has been driven by remarkable innovations in detector technologies that have provided extraordinary active rejection of backgrounds.

Solar (Sect. 12.2) and atmospheric neutrinos (Sect. 11.7) represent the ultimate, irreducible background. Nuclear recoils induced by coherent neutrino-nucleus scattering cannot be distinguished from a genuine WIMP-induced signal. This background could dominate the measured event rate if the WIMP-nucleon cross-section is below 10^{-49} cm^2 (Baudis 2012), see Fig. 14.6.

14.8.1 Solid-State Cryogenic Detectors

Cryogenic detectors operate at sub-Kelvin temperatures and allow us to perform a calorimetric energy measurement down to very low energies ($E_R < 10\,\text{keV}$), with unsurpassed energy resolution and the ability to differentiate nuclear from electron recoils on an event-by-event basis.

The operational principles rely on the fact that the heat capacity of a dielectric crystal depends on temperature as $\sim T^3$. A small energy deposition, such as that induced by a nuclear recoil, significantly changes the temperature of the absorber, yielding collective excitations (phonons) of the lattice of the detector crystal. The phonon (or thermal) response of the detector determines the total recoil energy of an interaction. The signals induced by a nuclear recoil and by ionization from an electron (the two cases in Fig. 14.4) are different. For this reason, if a second detector response is available in addition to the thermal response, such as those dependent on scintillation or ionization, the background rejection is highly improved.

The leading cryogenic detectors are the CDMS, CRESST and EDELWEISS experiments, all operating underground. All experiments have yielded negative results, and the corresponding upper limits are reported in Fig. 14.8. Some early positive claims have been removed with the increased statistics. The CoGeNT experiment uses the ionization signal from high-purity, low-radioactivity germanium as detection mechanism. Like other germanium ionization detectors, it can reach sub-keV energy thresholds and low backgrounds, but it lacks the ability to distinguish nuclear recoils from the signals produced by ionization. A possible excess that had originally been observed using data from a 18.5 kg$\times$days exposure has been understood and rejected.

14.8.2 Scintillating Crystals

A second technique uses scintillator crystals encased in a low-radioactivity enclosure and read out with photomultiplier tubes (PMTs). Electron and nuclear recoils induce signals in the PMTs with different pulse shapes, but normally, this effect is too weak to exploit on an event-by-event basis. For this reason, some experiments are looking for a time-dependent modulation of a WIMP signal in their data.

The *seasonal (or annual) modulation* effect is a discrimination method based on the expected annual variation in the WIMP event rates. The WIMPs are assumed stationary within the galactic frame. As the Earth moves around the Sun, their flux, if any, should be maximum in June (when the revolution velocity of the Earth adds to the velocity of the solar system in the Galaxy) and minimum in December (when the two velocities have opposite directions), with an expected amplitude variation of a few percent. To produce an acceptable statistics, this method requires an experiment with large exposure and long data acquisition periods (many years).

The DAMA/LIBRA (and the former DAMA/NaI) is the first experiment using this detection strategy. The apparatus is made of 25 highly radio-pure NaI(Tl) crystals, each with a mass of 9.7 kg, arranged in a 5×5 grid. Each detector is encapsulated in a pure copper housing with quartz light guides coupled to a PMT at two opposing faces of the crystal. The detector is placed in the center of a multilayer Cu/Pb/Cd-foils/polyethylene/paraffin structure for shielding at the Gran Sasso laboratory.

The DAMA/LIBRA result uses a total exposure of 1.17 ton$\times$years covering a period of 13 annual cycles, with an event rate of $\sim$1 count/kg/keV/day above an energy threshold of $\sim$2 keV. The experiment has detected, with high statistical significance, an annual modulation of the event rate that is consistent with the phase, amplitude, and spectrum expected from a nonrotating WIMP halo. Although the WIMP-nucleon cross-section calculated from this data is compatible with some MSSM, it is incompatible with the current upper-limits from other experiments (Fig. 14.8). A number of possible explanations for this discrepancy has been proposed: effects due to non-MSSM dark matter candidates; variations in the dark matter halo model; unaccounted physical effects in the detectors; unaccounted background sources; issues with the data and/or its interpretation. At present, none of the proposed explanations seems to offer a satisfactory solution to the incompatibility of the various experimental results.

Future projects will also look for this seasonal effect using the same NaI(Tl) or different crystals, such as the SABRE and COSINE projects; the DM-ICE at South Pole, the KIMS at the Yangyang laboratory in South Korea and the ANAIS at the Canfranc laboratory (all underground experiments) are currently collecting data. Some of them have now reached the needed maturity to test the DAMA result.

14.8.3 Noble Liquid Detectors

Noble elements in liquid state such as argon ($A = 40$) and xenon ($A = 131$) offer excellent media for building nonsegmented, homogeneous, compact, and self-shielding detectors. Liquid xenon (LXe) and liquid argon (LAr) are good scintillators and ionizers in response to the passage of radiation. They can operate in single readout mode as a scintillation-only detector (similar to the scintillator crystals). Detectors can also be arranged in such a way as to allow a strong rejection of electron recoils on an event-by-event basis. The features of LXe and LAr, together with the relative facility of scaling-up to large masses, have contributed to make noble gases/liquids powerful targets for WIMP searches. An interaction in the liquid produces ionization and excitation of the target atoms. Using pulse-shape discrimination of the signal in the PMTs, the nuclear-recoil induced by a WIMP can be separated from a background electron-recoil. Examples of these single-phase experiments are XMASS and miniCLEAN.

Two-phase time projection chambers (TPC) have also been developed for dark matter detection using xenon (see Aprile and Doke 2010) and argon, and several large detectors are in the process of commissioning or construction. All experiments have a common design/operation principle, which is shown schematically in Fig. 14.7. A low-radioactivity vessel is partially filled with liquid xenon (or argon), with the rest of the vessel containing xenon (argon) gas. Electric fields of $\sim$1 keV/cm and $\sim$10 keV/cm are established across the liquid and gas volumes, respectively by applying a voltage bias to the electrodes (shown as dashed lines in the figure). An interaction in the LXe produces excitation and ionization processes. Photomultiplier tube arrays are present at the top (in the gas volume) and bottom (in the liquid). A first signal (called S1) after the interaction is due to the de-excitation processes, yielding prompt photons near the interaction vertex in the LXe. The associated electrons produced by ionization drift under the electric field; when they reach the liquid surface, they are extracted into the gas phase by the higher electric field. As electrons accelerate through the gas, their interactions produce a second scintillation signal (S2). The right side of Fig. 14.7 shows the S1 and S2 signals for two events: an electron recoil caused by a background event (here, a γ-ray interaction) and a nuclear recoil caused by a neutron interaction, simulating a genuine WIMP process. The ratio of the S2 and S1 signals is used as a discriminator between these types of event.

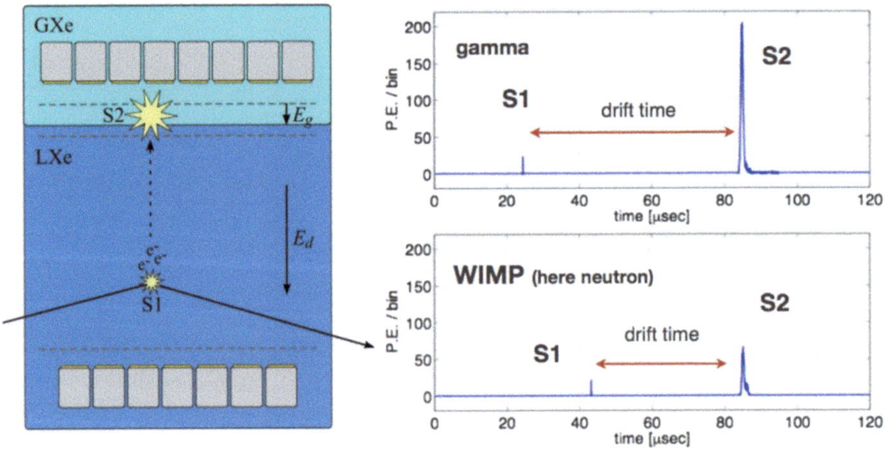

Fig. 14.7 Schematic of a xenon two-phase time-projection chamber, showing the recorded signals for a γ-ray interaction and a WIMP (actually a neutron used for calibration). The ratio S2/S1 is used to discriminate between electron and nuclear recoils. Courtesy Prof. T. Saab

Neutron background cannot be removed using this procedure. The reconstruction of the interaction vertex inside a fiducial volume of the detector offers an additional tool to suppress the neutron contamination that piles up near the detector surface. The TPC configuration allows for the measurement of the vertex position. The pattern of hits in the PMTs determines the $(x; y)$ coordinates; the time difference between the S1 and S2 pulses measures the electron drift time, and thus determines the z position (that along the electric field). The low energy threshold of the experiment is determined by the smaller S1 signal, and is typically set at a few photons per event.

14.8.4 Present Experimental Results and the Future

WIMPs could have scalar, σ_0^{SI}, and/or spin-dependent, σ_0^{SD}, interactions with nuclei, and experimental results are discussed in terms of either interaction type. Figure 14.8 illustrates the current best limits on WIMP spin-independent scattering cross-sections as a function of the WIMP mass. The results of experiments using different target nuclei are normalized to scattering on a single nucleon, σ_0, using Eq. (14.17). Null results from different experiments yield upper limits plotted as lines. The parameter space above a line is excluded. The enclosed areas labeled DAMA/LIBRA and CDMS-Si are regions of interest from possible signal events. The region of space parameters colored in orange will be subject to the background due to neutrino interactions. The region not excluded by searches of typical SUSY models at LHC is that colored in yellow. As the field is in constant evolution, refer to Patrignani et al. (2016/2017) and updates for newer results.

The upper limits as a function of m_χ are a consequence of the above discussion about the event rate: the sensitivity is maximum for WIMP masses near the mass of the recoiling nucleus (50–100 GeV/c^2). At lower WIMP mass, the sensitivity drops because of effects connected with the energy threshold of detectors. At higher masses, the sensitivity worsens because the WIMP flux decreases as $\propto 1/m_\chi$.

The limits from direct detection are competitive with the limits obtained at accelerator experiments. At present, the absence of any signal of physics beyond the SM at the LHC, as well as the discovery of a SM-like Higgs boson with the relatively high mass ~126 GeV/c^2, constrains many well-motivated WIMP models towards large WIMP masses and low σ_0.

Several ton-scale direct detection experiments are starting taking science data, using the same (or improved) detection techniques illustrated above (Baudis 2012). Those experiments can reach a discovery if the WIMP-nucleon cross-section is larger than 10^{-46} cm^2 within the mass range $20 < m_\chi < 1000$ GeV/c^2. Sensitivities down to $\sigma_0 \sim 10^{-49}$ cm^2 are needed to probe nearly all of the MSSM parameter space at WIMP masses above 10 GeV/c^2. Below this limit, the irreducible neutrino-induced background would probably cover any DM signature. Such sensitivities will be reached with multi-ton mass detectors with superb background discrimination capabilities.

Alternative solutions to huge massive experiments are detectors capable of measuring the direction of the recoiling nucleus. This represents a signature that would unequivocally confirm the Galactic origin of a signal. Because of the correlation between the direction of the incoming WIMP and that of the recoiling nucleus, signal events should point in the direction of the WIMP wind. The relative directions of the laboratory frame and the WIMP wind has a 24-h modulation, due to the Earth's rotation around its axis. With an ideal detector able to reconstruct the tracks of individual nuclei, a challenging discrimination technique using this 24-h modulation can be developed. Background events are not expected to exhibit anisotropies. Because nuclear recoils have a small energy, probably only gaseous detectors can precisely measure the very limited nuclear range. Several directional detectors are presently in the research and development phase. See also Sect. Dark Matter of Patrignani et al. (2016/2017) and the subsequent updates for details on future projects.

14.9 Indirect Searches for WIMPs

WIMPs annihilate into standard model particles with the cross-section given in (14.10) to explain cosmological observations. This small value no longer affects the overall WIMP abundance after decoupling. However, WIMPs annihilation continues, and may be large enough to be observed indirectly if the end products include photons, neutrinos, electrons, protons, deuterium, and their corresponding antiparticles. As a consequence of this long particle list, there are many indirect detection methods being pursued. The relative sensitivities of indirect detection methods are highly dependent on which WIMP candidate is being considered. The difficulties in determining backgrounds and systematic uncertainties also vary greatly from one method to another. The most exploited methods refer to the detection of neutrinos from massive objects, the study of γ-rays, and the searches for antimatter in cosmic rays.

14.9.1 Neutrinos from WIMP Annihilation in Massive Objects

Massive objects (such as our Sun) act as amplifiers for DM annihilations by capturing DM particles as they lose energy through elastic scattering with nuclei. Once gravitationally captured, DM particles settle into the core, where their densities and annihilation rates are greatly enhanced. Most of their annihilation products are immediately absorbed, and only neutrinos escape these dense objects. The centers of massive objects represent the natural places to look for a possible neutrino excess from DM annihilation using neutrino telescopes. In the following, we specialize the discussion to the case of the Sun ($\odot$). The extension to other massive objects (the Earth's core, the Galaxy's center, nearby dwarf spheroidal galaxies) exists in the literature.

The DM capture rate $C^{\odot}$ [s^{-1}] can be written as:

$$C^{\odot} \simeq \Phi_{\chi} \cdot \left(\frac{M_{\odot}}{m_p}\right) \cdot \sigma_0 \,, \tag{14.24}$$

where Φ_{χ} is the local DM flux (14.21), the ratio $(M_{\odot}/m_p) \simeq 10^{57}$ is an estimate of the number of target nucleons in the Sun and σ_0 represents, as usual, the WIMP-nucleon cross-section. As discussed, this quantity could be spin-dependent, spin-independent or both. The current bounds from direct searches are $\sigma_0^{SD} \lesssim 10^{-39}\,\mathrm{cm}^2$ and $\sigma_0^{SI} \lesssim 10^{-45}\,\mathrm{cm}^2$ (see Fig. 14.8). Thus, inserting the relevant numbers in (14.24) and arbitrarily assuming a reference cross-section of $10^{-42}\,\mathrm{cm}^2$, intermediate between the above two SI and SD limits, we obtain:

$$C^{\odot} \simeq \frac{3 \times 10^{20}}{[\mathrm{s}]} \cdot \left(\frac{\rho_0}{0.3\,\mathrm{GeV\,cm^{-3}}}\right) \cdot \left(\frac{\overline{v}}{220\,\mathrm{km\,s^{-1}}}\right) \cdot \left(\frac{100\,\mathrm{GeV}}{m_{\chi}}\right) \cdot \left(\frac{\sigma_0}{10^{-42}\,\mathrm{cm}^2}\right) . \tag{14.25}$$

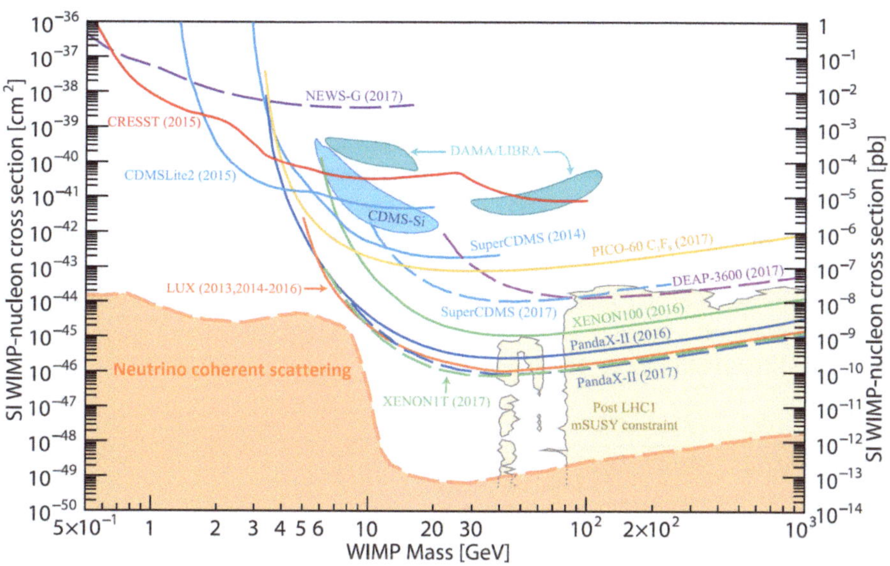

Fig. 14.8 A compilation of WIMP-nucleon SI cross-section limits (*solid curves*) as a function of m_{χ}. The regions of a possible signal arising from DAMA/LIBRA and CDMS-Si are represented by enclosed areas. An approximate band where coherent scattering of solar neutrinos, atmospheric neutrinos and diffuse supernova neutrinos with nuclei will begin to limit the sensitivity of direct detection experiments to WIMPs is also reported. From Sect. 27 of Patrignani et al. (2016/2017), which we refer to for the references to experiments

The number of DM particles present in the Sun depends on time: they are accumulating at a constant rate (14.25), but some of them undergo annihilation processes that depend on the annihilation cross-section σ_χ. Under reasonable assumptions, it is possible to show (follow the details of this exercise in Bertone et al. 2005 and Profumo 2014) that an equilibrium solution is reached. At equilibrium, the annihilation rate $\Gamma_{\rm ann}$ of WIMP particles in the Sun is

$$\Gamma_{\rm ann} = \frac{C^{\odot}}{2} [{\rm s}^{-1}] \, . \tag{14.26}$$

The rate of energy release through this hypothetical process (not included in the Solar Standard Model) corresponds, from (14.25), to $\Gamma_{\rm ann} \cdot 100\,{\rm GeV} = 10^{19}$ erg/s (for the assumed cross-section of $10^{-42}\,{\rm cm}^2$), which is negligible with respect to the solar luminosity $L_\odot \simeq 3.8 \times 10^{33}$ erg/s. The existence of DM does not perturb the energy balance due to thermonuclear reactions in the main sequence stars.

Neutrinos (of all flavors) could be produced in the Sun core by decays or interactions of standard model particles produced in the WIMPs scattering with matter and/or by their annihilation into $\chi\overline{\chi} \rightarrow \ell\overline{\ell}\ ;\ q\overline{q}\ ;\ W^+W^-\ ;\ Z^0Z^0\ ;\ \gamma\gamma;\ hh$. Here, ℓ represents a charged lepton, q a quark, $W^{\pm}$, Z^0 the gauge bosons, γ the photon and h the Higgs scalar boson.

Exercise Discuss the reason why neutrinos could be directly produced by $\chi\overline{\chi} \rightarrow \nu_f\overline{\nu}_f$ annihilations only if the WIMP is not the neutralino or another Majorana particle.

The corresponding neutrino energy spectrum dN_ν/dE_ν (GeV^{-1}) depends on the assumed DM candidate. These neutrinos typically have energies much larger than those of thermonuclear origin (MeV or fraction). There are no astrophysical processes in the main sequence stars that produce GeV (or higher energies) neutrinos. The spectrum dN_ν/dE_ν depends on the capture rate $C^{\odot}$, mass m_χ and, for a given candidate, the main decay channel and annihilation cross-section. Details can be found in Baratella et al. (2014). From the energy spectrum dN_ν/dE_ν at the Sun, the differential flux on Earth can be derived:

$$\frac{d\Phi_\nu}{dE_\nu} = \frac{\Gamma_{ann}}{4\pi d^2}\frac{dN_\nu}{dE_\nu} = \frac{C^{\odot}}{8\pi d^2}\frac{dN_\nu}{dE_\nu} \quad [{\rm GeV}^{-1}{\rm cm}^{-2}{\rm s}^{-1}] \, , \tag{14.27}$$

where d is the Sun-Earth distance.

The flux prediction (14.27) depends strongly on the assumed preferred WIMP annihilation channel. Conventionally, two reference channels are used: the *hard channel* assumes that all WIMP annihilations produce a W^+W^- pair (sometimes, the $\tau^+\tau^-$ leptons). This assumption gives rise to a harder neutrino energy spectrum, with the largest fraction of high-energy neutrinos. The so-called *soft channel*, on the other hand, assumes 100 % production of $b\overline{b}$ pairs, giving rise to a softer neutrino energy spectrum. In SUSY models, they are exactly the preferred final states for the lightest neutralinos. The interpretation of the neutrino flux from the Sun in terms of WIMP parameters depends on the assumed WIMP decay channel (if hard, soft or intermediate). The effective area of neutrino telescopes increases with E_ν (Fig. 10.10): the hypothesis of a preferred hard spectrum for WIMP annihilations produces a larger event rate.

The number of events expected in neutrino telescopes (typically within the energy range $E_\nu \gtrsim 100\,\text{GeV}$) such as IceCube, ANTARES or the planned KM3NeT is given by inserting the predicted flux (14.27) into Eq. (10.24). A precise measurement of the neutrino direction (for events from the position of the Sun) can be achieved only for ν_μ, thus only the ν_μ fraction in the flux is considered. As usual, atmospheric neutrinos represent the irreducible background and a statistically significant signal excess over a flat background is searched for. At present, no excess of neutrino events from the direction of the Sun or other massive objects is reported.

For a given WIMP mass, m_χ, a null result can be translated on a 90 % C.L. upper limit on the WIMP-nucleon cross-section, assuming a spin-independent, σ_0^{SI}, or spin-dependent, σ_0^{SD}, interaction. The limits are different depending on the assumption of a hard or soft WIMP decay channel. In general, the spin-independent limit obtained with this method is not competitive with that of direct experiments. In fact, the WIMP-nucleus cross-section enters in the capture rate (14.24). The Sun is made of light elements (H and He) and there is no gain due to the A^2 behavior of the SI cross-section.

In Fig. 14.8, in which limits from direct experiments are reported assuming a SI interaction, the actual best result from neutrino telescopes (Aartsen et al. 2013) would produce an upper limit (under the hypothesis of a hard channel) of $\sigma_0^{SI} \sim 10^{-42}$–$10^{-43}$ cm^2 within the energy range between 100 GeV and 1 TeV. The hypothesis of WIMP annihilations in a soft channel gives a limit two orders of magnitude worse.

On the other hand, indirect measurements using neutrinos provide the best result if the WIMP-nucleon interaction occurs through the spin-dependent coupling. The 90 % C.L. upper limits obtained by the ANTARES and IceCube neutrino telescopes under the SD assumption are shown in Fig. 14.9. Direct detection experiments (such as COUPP, Simple, PICASSO) give less restrictive upper limits. Refer to Sect. Dark Matter of Patrignani et al. (2016/2017) for further information.

14.9.2 Gamma-Rays from WIMPs

Gamma rays eventually produced in DM annihilations or decays are not deflected by magnetic fields. If produced in transparent regions, they can travel to us from anywhere in the Galaxy and local Universe, effectively indicating the direction of their source. However, disentangling a possible DM signal from astrophysical backgrounds is not straightforward (Porter et al. 2011). Spectral information is the only method for distinguishing between a DM signal and that of astrophysical origin. It is, in fact, likely that DM annihilation or decay produces a γ-ray spectrum

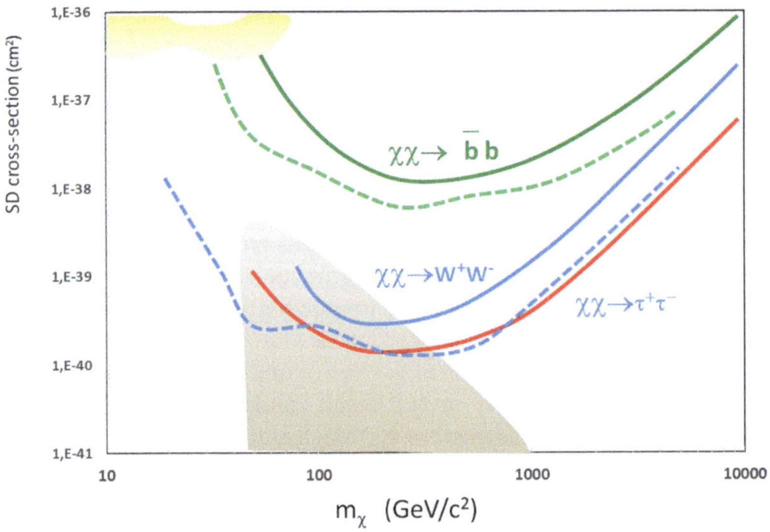

Fig. 14.9 Upper limits (at 90 % C.L.) on WIMP-nucleon cross-sections versus the WIMP mass m_χ under the assumption of a spin-dependent coupling and for the three self-annihilation channels. The limits are obtained by the ANTARES (*full lines*) (Adrían-Martínez et al. 2013) and IceCube (*dashed lines*) (Aartsen et al. 2013) neutrino telescopes searching for an excess of high-energy ν_μ from the direction of the Sun. The assumption of WIMPs annihilation into the *hard* channels $\tau^+\tau^-$ (*red line*) or W^+W^- (*blue line*) provides the most stringent limits. The *soft* $b\bar{b}$ channel (*green lines*) is less restrictive. As in Fig. 14.8, the *bottom* colored region represents the space of parameters allowed by MSSM and the *yellow* region at the *top* left the DAMA allowed region

falling with increasing energy at a slower rate than that produced by a typical astrophysical source and with a bump (or edge) near the WIMP mass.

As shown in Sect. 8.9, the production and propagation of CR protons, nuclei, and electrons induce a diffuse nonthermal emission of photons from radio to γ-rays. For a while, the GeV excess seen by EGRET (the anomalous signal in the diffuse Galactic emission observed at γ-ray energies $\gtrsim$1 GeV) received attention as a possible DM signature. The subsequent result of Fermi-LAT suggests that the EGRET GeV excess was likely the result of instrumental errors.

The cleanest and most convincing DM signal that could be measured would be annihilation into final states that include photons, such as $\gamma\gamma$, γZ^0 or γh. These processes can provide a mono-energetic feature giving a distinctive line in the γ-ray spectrum that is potentially distinguishable from otherwise challenging astrophysical backgrounds. However, DM models predict branching fractions into such states that are typically $\sim 10^{-4} - 10^{-1}$ compared to the total annihilation or decay rate, placing them below the flux sensitivity of any existing instrument.

A search for γ-ray lines from the direction of the galactic center was done by EGRET within the range of 0.1–10 GeV, and from the Fermi-LAT within the energy range of 30–200 GeV. Another search for a γ-ray flux originating from DM candidates was made by Fermi-LAT from the directions of nearby dwarf spheroidal galaxies. They represent the most attractive candidate objects for DM searches, because the observations of these structures show very high mass-to-light ratios. All the above searches gave null results.

14.9.3 The Positron Excess: A WIMP Signature?

DM can annihilate into charged particles in the final state, which add to the cosmic radiation from astrophysical origin. This additional contribution to cosmic rays is, at best, subdominant to the observed CR energy density, much smaller than an $\mathcal{O}(1\,\%)$ effect. Typical dark matter models are, however, democratic in producing as much matter as antimatter in the annihilation or decay final products. From the experimental point of view, the search for stable antiparticles (namely, $\overline{p}$ and e^+) in the cosmic radiation is a promising method for searching for a possible DM signal. Antimatter is not abundant as primary radiation.

The antiproton flux (see Fig. 3.11) does not shown any unexpected feature with respect to the hypothesis of pure secondary production of $\overline{p}$, Sect. 3.9. On the contrary, one of the most intriguing recent results is the excess of positrons in the CR spectrum, shown in Fig. 3.13, found by HEAT, PAMELA, Fermi-LAT, and finally, by AMS-02 experiments, Sect. 3.10. Particularly important is the measurement of the positron fraction within the energy range 5–350 GeV presented in 2013 by the AMS-02 collaboration. After this result, obtained with a large statistical sample, the likelihood that the excess is the result of experimental artifacts is negligible.

Most DM annihilation or decay models can naturally reproduce the observed rise of the positron fraction, and this has been a widely conjectured explanation for the e^+ excess, as shown by the green curve in Fig. 3.13. However, models must also explain several unexpected characteristics of these data. In particular, the positron excess over that expected from CR propagation (the black full line in Fig. 3.13) requires, if due to DM annihilation, a cross-section $\sigma_\chi \sim 10^2 \div 10^3$ times larger than given by Eq. (14.10). In addition, the nonobservation of a similar antiproton excess requires the hypothesis that hadron production is suppressed: the DM must be "leptophilic".

It is plausible that both the positron excess and the $e^\pm$ spectrum (Fig. 3.12) can be explained by modifying the assumption usually made in models of CR propagation, as for instance in GALPROP, Sect. 5.4. This computer code (like similar ones) has proven very successful in describing a wide range of CR data in different sections of this book: antiprotons, Fig. 3.11, or other heavier antinuclei; stable secondary nuclei, Fig. 5.4; radioactive nuclei; diffuse γ-rays, Fig. 8.6. The only exception seems to correspond to the case of electrons and positrons, in which a smooth spatial distribution of sources could not be completely adequate. For the $e^\pm$ component, details of the discrete source distributions in the local Galaxy are probably important. In particular, the presence of a nearby pulsar seems to explain the positron excess (the red line in Fig. 3.13). However, no unique cosmic object has been identified that is able to explain the measured data. We can argue about what the requirements are regarding the age and distance of a pulsar that could contribute to the observed positron anomaly.

The rate of energy loss of an electron (or positron) in a region with a high magnetic field is given by Eq. (8.9c). Typical values of the magnetic field in the region surrounding a pulsar is about $B \sim 100\ \mu$G; see Fig. 9.11 for the case of the Crab. For 100 GeV electrons, the dissipated power from (8.9c) is $|dE/dt| = 0.4 \times 10^{-20} \cdot (10^2\,\text{GeV})^2 \cdot (100\,\mu\text{G})^2 \sim 10^{-12}\,\text{GeV/s}$. Thus, the characteristic time τ_{loss} for which a $E = 100$ GeV electron loses its energy is

$$\tau_{\text{loss}} \simeq \frac{E}{|dE/dt|} \longrightarrow \tau_{\text{loss}} \simeq \frac{100\,\text{GeV}}{10^{-12}\,\text{GeV/s}} = 10^{14}\,\text{s} \sim 3\text{My}\,. \tag{14.28}$$

Thus, the pulsar age T_{psr} must be shorter than this characteristic time, $T_{\text{psr}} < \tau_{\text{loss}}$.

Cosmic electrons/positrons propagate in the interstellar matter in a manner analogous to CR protons, following a diffusion process, Sect. 5.3.2. Using the diffusion coefficient D derived in Eq. (5.32) and the pulsar age T_{psr}, we can derive the characteristic length scale L. In this case, this corresponds to the distance L_{psr} from which electrons/positrons can arrive, i.e., the pulsar distance from the Earth:

$$L_{\text{psr}} = \sqrt{D \cdot T_{\text{psr}}} < \sqrt{D \cdot \tau_{\text{loss}}} = (3 \times 10^{27} \times 10^{14})^{1/2} \sim 10^{21}\text{cm} = 0.3\text{kpc}. \tag{14.29}$$

The above condition gives a sufficient estimate for our subsequent considerations, although it can be improved considering the energy-dependence of the diffusion coefficient $D = D(E)$. A possible candidate pulsar originating the positron excess is younger than about a few My and closer than a kpc from us. Note that this result is in good agreement with that obtained in Sect. 5.8.2, Eq. (5.76).

Astrophysical models ensure that enough power can be injected under the form of accelerated $e^{\pm}$ pairs. Some authors find a remarkable agreement between the data and the hypothesis that the positrons excess is originated from a single nearby pulsar, such as Geminga (a pulsar 0.25 kpc away and $\sim 3 \times 10^5$ year old) or Monogem (also known as B0656+14, a $\sim 10^5$-year old pulsar at 0.3 kpc), with the spectral index $\Gamma_e \sim 2$. The possibility that the positron excess is due to multiple known pulsars is also considered viable.

In years to come, AMS-02 will increase the energy range over which positrons and electrons are measured, as well as the total number of collected events. If the positron excess is originated from a single astrophysical source, such as a nearby pulsar, it would probably give an anisotropy in the arrival direction of e^+, e^- of the highest energy. If the origin is due to a few nearby pulsars, the corresponding spectrum should have characteristic structures that can be observed with a large statistics data sample (Yin 2013). On the other hand, a sharp cut-off in the positron fraction would probably be the signature of a DM origin of the positron excess.

14.10 What's Next?

The decades-long quest to discover the Higgs boson was completed at the CERN LHC in 2012. In the SM scenario, the Higgs represented the last missing piece. Also, after the 2013 Nobel Prize award for this discovery, the particle physics community is still feeling unsatisfied. There are deficiencies and open questions related to experimental data that the SM cannot explain (the neutrino mass problem), and to experimental data that can be explained, but only for seemingly unnatural choices of parameters (the strong CP problem, Sect. 14.4, and the gauge hierarchy problem, Sect. 14.5). SUSY theories seem to offer a natural solution for the latter type of problems.

Supersymmetry offers some leading dark matter candidates and provides guidance for dark matter searches. For this reason, the hunt for signals beyond the SM at the LHC has an impressive priority now. In parallel, direct searches for WIMPs in underground laboratories are covering the parameter phase space of Fig. 14.8 in the region of $m_\chi \sim 10$–$100\,\text{GeV/c}^2$ and cross-section $\sigma_0 \sim 10^{-46} \div 10^{-47}\,\text{cm}^2$. A third method relies on the fact that WIMPs can annihilate and their annihilation products can be detected. These secondary particles include neutrinos, γ-rays, positrons, antiprotons, and antinuclei. The searches for unexpected excess of those particles are complementary to direct detection and might be able to explore higher masses and different coupling scenarios. *Smoking gun* signals for indirect detection are GeV neutrinos coming from the center of the massive objects, such as the Sun or Earth center, and mono-energetic γ-rays from WIMP annihilation in space.

Beyond the present phase of LHC operation improvements, the priority in the world strategy for particle physics is the exploitation of the full potential of the LHC, including the high-luminosity upgrade of the machine and detectors. This will allow for collecting ten times more data than in the initial design, by around 2030. The next step for direct matter experiments would probably be multi-ton experiments, or detectors able to point out the direction of the nuclear recoil. Those experiments would cover the parameter phase-space until the irreducible background induced by neutrinos is reached.

In conjunction with the LHC results, direct and indirect searches can confirm (or exclude) a discovery over a large region of the parameters allowed by SUSY models. What will occur if both, accelerator, and astroparticle experiments for WIMP searches, give null evidence of supersymmetric partners? The lack of SUSY signals at the LHC and the increasingly stronger limits from direct and indirect searches may have weakened the argument for WIMPs being embedded in a larger theory that addresses the hierarchy problem. However, the evidence for DM is so strong that a plethora of new models of particle DM has been generated that must be tested by the new generation experiments (Feng 2010; Patrignani et al. 2016/2017).

References

M.G. Aartsen et al., IceCube Collaboration, Search for dark matter annihilations in the Sun with the 79-string IceCube detector. Phys. Rev. Lett. **110**, 131302 (2013)

P.A.R. Ade et al., Planck Collaboration (2014). arXiv:1303.5062v1

S. Adrían-Martínez et al., ANTARES Collaboration, First results on dark matter annihilation in the Sun using the ANTARES neutrino telescope. J. Cosmol. Astropart. Phys. **11**, 032 (2013). arXiv:1302.6516

E. Aprile, T. Doke, Liquid xenon detectors for particle physics and astrophysics. Rev. Mod. Phys. **82**, 2053–2097 (2010)

P. Baratella et al., PPPC 4 DMν: a poor particle physicist cookbook for neutrinos from DM annihilations in the sun (2014). arXiv:1312.6408

L. Baudis, Direct dark matter detection: the next decade. Dark Univ. **1**, 94–108 (2012)

D. Bauer et al., Snowmass CF1 summary: WIMP dark matter direct detection (2014). arXiv:1310.8327v2

K.G. Begeman, H I rotation curves of spiral galaxies. I - NGC 3198. Astron. Astrophys. **223**, 47–60 (1989)

G. Bertone, D. Hooper, J. Silk, Particle dark matter: evidence, candidates and constraints. Phys. Rep. **405**, 279–390 (2005)

S. Braibant, G. Giacomelli, M. Spurio, *Particle and Fundamental Interactions* (Springer, Berlin, 2011). ISBN: 978-9400724631

M. Cirelli et al., PPPC 4 DM ID: a poor particle physicist cookbook for dark matter indirect detection. J. Cosmol. Astropart. Phys. **11**(03), 051 (2011)

A. Drukier, K. Freese, D. Spergel, Detecting cold dark matter candidates. Phys. Rev. **D33**, 3495–3508 (1986)

J.L. Feng, Dark matter candidates from particle physics and methods of detection. Annu. Rev. Astron. Astrophys. **48**, 495–545 (2010)

R.J. Gaitskell, Direct detection of dark matter. Annu. Rev. Nucl. Part. Sci. **54**, 315–359 (2004)

G. Giacomelli, Magnetic monopoles. La Rivista del Nuovo Cimento **7**(12), 1 (1984)

M.W. Goodman, E. Witten, Detectability of certain dark matter candidates. Phys. Rev. **D31**, 3059 (1985)

W. Hu, S. Dodelson, Cosmic microwave background anisotropies. Ann. Rev. Astron. Astrophys. **40**, 171–216 (2002)

G. Jungman, M. Kamionkowski, K. Griest, Supersymmetric dark matter. Phys. Rep. **267**, 195–373 (1996)

J.F. Navarro, C.S. Frenk, S.D. White, The structure of cold dark matter halos. Astrophys. J. **462**, 563 (1996)

C. Patrignani et al. (Particle Data Group), Chin. Phys. C **40**, 100001 (2016/2017)

L. Patrizii, M. Spurio, Status of searches for magnetic monopoles. Annu. Rev. Nucl. Part. Sci. **65**, 279–302 (2015)

D.H. Perkins, Proton decay experiments. Annu. Rev. Nucl. Part. Sci. **34**, 1–50 (1984)

T.A. Porter, R.P. Johnson, P.W. Graham, Dark matter searches with astroparticle data. Annu. Rev. Astron. Astrophys. **49**, 155–194 (2011)

S. Profumo. TASI 2012 Lectures on astrophysical probes of dark matter (2014). arXiv:1301.0952

T. Saab, An introduction to dark matter direct detection searches and techniques (2012). arXiv:1203.2566

V. Trimble, Existence and nature of dark matter in the universe. Annu. Rev. Astron. Astrophys. **25**, 425–472 (1987)

P.-F. Yin et al., Pulsar interpretation for the AMS-02 result. Phys. Rev. D **88**, 023001 (2013)

Chapter 15
Conclusions

The study of the Universe with probes different from the electromagnetic radiation has only recently reached its maturity. The joint effort to understand high-energy astrophysics phenomena using cosmic rays, γ-rays, neutrinos, gravitational waves, in addition to electromagnetic radiation, is called *multimessenger astrophysics*.

Conventionally, the birth date of multimessenger astrophysics can be set on February 23, 1987, with the detection of neutrinos from a supernova explosion in the Large Magellanic Cloud. Since then, advanced detectors have been constructed and carried into space, located on the ground, installed under-ground/-water/-ice, in order to increase the observational possibilities for particles of cosmic origin and to open the window for astrophysical studies using gravitational waves. The outstanding results obtained with the observation of the merging of two neutron stars on August 17, 2017, is the demonstration of capabilities of multimessenger astrophysics.

The beginning of the twenty-first century has been characterized by fundamental experiments, relevant discoveries and fast improvements in astroparticle physics. In this book, I tried to cover the status of the impressive amount of information collected in the last ~20–30 years. In some cases, the present generation of experiments represents a sort of excellence that will be very difficult to overcome in the near future. In others, the experimental activity is in adolescence: it will mature with planned experiments. In general, any improvement in the performances of the present generation of detectors would be extremely expensive, and the community must adequately ponder the priorities.

The charged *cosmic radiation* was discovered more than 100 years ago (Chap. 2) and an extended study has been carried out since then. Experiments carrying spectrometers in space (Chap. 3) are measuring the flux of CR nuclei, electrons and antimatter with high sensibility, and are enabling a detailed modeling of Galactic sources. The observed excess of the positron flux up to 350 GeV with respect to the standard expectation is particularly intriguing for future observations. AMS 02 is

© Springer Nature Switzerland AG 2018

M. Spurio, *Probes of Multimessenger Astrophysics*, Astronomy and Astrophysics Library, https://doi.org/10.1007/978-3-319-96854-4_15

still running, and it is improbable that its sensitivity can be surpassed in the near future. However, projects extending the energy range of AMS-02 are under study.

Information arising from direct CR observations, in conjunction with the indirect measurements presented in Chap. 4, have been used to produce a sort of *standard CR model*. This paradigm (Chaps. 5 and 6 for the propagation and acceleration, respectively) assumes that supernova remnants provide strong shocks able to accelerate CRs with a E^{-2} spectrum up to the knee region ($\sim 3 \times 10^{15}$ eV). However, the ever increasing quality of experimental data reveals aspects that require improvements in the quality of the theoretical models, which, in turn, require more detailed observations. This is the case, for instance, with the energy spectra of individual chemical elements at energies above 100 TeV, which need an improvement in the experimental results.

In the high-energy extreme of the CR flux, the PAO and TA (Chap. 7) observed the suppression of the CR flux at energies above a few 10^{19} eV; their studies required an extended operation time. The upgrade of the present generation of experiments, for instance, with additional counters to disentangle the muonic and electromagnetic components, promises improvements in the estimate of the mass composition of UHECR. Probably only a different technique (such as the observation from space satellites of CR-induced cascades in the Earth's atmosphere) could significantly improve the state-of-the-art provided by these two experiments.

Gamma-ray astronomy is another young discipline. The observation of MeV-GeV γ-rays in space needed detection methods borrowed from accelerator experiments. After EGRET on the CGRO satellite, the Fermi-LAT (Chap. 8) reached such maturity that it will be difficult to be surpassed in the near future. In any case, after the end of LAT operations, the presence in space of (at least) one satellite with similar sensitivity is mandatory for multimessenger observations. TeV γ-ray observations, after the results in the last decade obtained by HESS, MAGIC and VERITAS (Chap. 9), are waiting for (probably) the definitive experiment, the CTA, the construction phase of which has just started.

Neutrino observations (Chap. 10) are complementary with GeV-TeV γ-ray astrophysics. The neutrino production is strictly related to the hadronic mechanism of γ-ray production. Neutrinos are much more penetrating than photons; this represents the disadvantage: huge detectors are required to catch very few neutrino interactions. On the other hand, neutrinos allow for the study of cosmic regions that are completely opaque to the electromagnetic radiation. The first observation of a cosmic neutrino flux from IceCube is extremely encouraging, strongly motivating the construction of multi-km^3 detectors in the Northern hemisphere and the upgrade of the Antarctic telescope.

Neutrino interactions produce rare events that must be studied underground to shield experiments from the penetrating radiation of secondary cosmic rays (Chap. 11). Underground experiments allowed the study of the energy source of our Sun and thus of the nuclear reactions occurring in main-sequence stars; always underground experiments detected low-energy neutrinos from the 1987 Supernova (Chap. 12). The present generation of advanced detectors is really waiting for another Galactic supernova event.

A real breakthrough in our knowledge of the high-energy Universe occurred with the first observations of gravitational waves (GWs), Chap. 13. As demonstrated by the observational campaign after the detection of GW170817, the discovery opportunities opened by the conjunction of GW, γ-rays, lower energy photons and (possibly) neutrino information are of fundamental importance. The community should be grateful to the LIGO Scientific Collaboration and the Virgo Collaboration for their decades-long efforts, for the setting up of an impressive follow-up observation program, and for sharing invaluable scientific information for the benefit of the emerging multimessenger astronomy.

Underground laboratories also host experiments for the direct searches of the elusive *dark matter (DM)*, Chap. 14. The presence of DM is required by the observation of motion of large-scale structures in the Universe (from stars in galaxies to the motion of galaxies in large galaxy clusters). Really, the study of DM coalesces the interests and efforts of particle physicists, cosmologists and astrophysicists. The next decade will probably experience some final answer concerning the existence or not of the candidates (probably) preferred by particle physicists, namely those connected with SuperSymmetric models.

The goals of future astroparticle experiments include not only astrophysics, but also studies (not covered in this book) more related to particle physics, general physics and cosmology. This includes, for instance: the measurement of the neutrino mass hierarchy through experiments using atmospheric neutrinos crossing the Earth; the search in the cosmic radiation for particles not included in the Standard Model; the measurement of particle's (protons, photons, neutrinos) cross- sections at energies unattainable in Earth-bound accelerators; the search for baryon number violation; the study of general relativity in the *strong field regime*; an alternative distance scale-ladder using GWs; the measurement of the extragalactic background light using the attenuation of γ-rays; the understanding of the cosmic history of star formation; the search for hints on the origin of the matter-antimatter asymmetry of the Universe; the exploration of the fundamental nature of spacetime, ...

The first half part of the twentieth century saw a strict interconnection of particle and astroparticle physics. The advent of accelerators decoupled the two fields in the second part of the century. From the 1950s, the study of the microcosm had an impressive growth, forced by the increasing energy of accelerators from the MeV to the TeV scale. To go far beyond the energy scale (10 TeV) reached by the LHC, efforts probably at the limit of human (financial) possibilities are required. The return to the use of cosmic accelerators will probably be a necessity. From the particle physics point of view, the possibility of using cosmic beams to improve our understanding of Nature will depend upon either the detailed understanding of cosmic acceleration and on the development of methods for controlling systematic errors introduced by our lack of understanding of these processes. Thus, the combined information arising from gravitational waves, from the measurements of γ-rays with high-resolution instruments, from high-statistics measurements of charged CRs and from neutrino telescopes is mandatory. For this purpose, multimessenger observations are not just an advantage, but also a necessity.

Index

© Springer Nature Switzerland AG 2018

M. Spurio, *Probes of Multimessenger Astrophysics*, Astronomy and Astrophysics Library, https://doi.org/10.1007/978-3-319-96854-4

MIX
Papier aus verantwortungsvollen Quellen
Paper from responsible sources
FSC® C105338

If you have any concerns about our products,
you can contact us on
ProductSafety@springernature.com

In case Publisher is established outside the EU,
the EU authorized representative is:
Springer Nature Customer Service Center GmbH
Europaplatz 3, 69115 Heidelberg, Germany

Printed by Libri Plureos GmbH
in Hamburg, Germany